90 0755989 0

The BIOLOGY of VIBRIOS

The BIOLOGY of VIBRIOS

EDITED BY

Fabiano L. Thompson, Brian Austin, and Jean Swings

ASM
PRESS

WASHINGTON, D.C.

UNIVERSITY OF PLYMOUTH

9007559890

Copyright © 2006   ASM Press
American Society for Microbiology
1752 N St., N.W.
Washington, DC 20036-2904

Library of Congress Cataloging-in-Publication Data

The biology of vibrios / edited by F. L. Thompson, B. Austin, and J. Swings.
     p. ; cm.
   Includes bibliographical references and index.
   ISBN 10: 1-55581-365-8 (alk. paper)
   ISBN-13: 978-1-55581-365-9 (alk. paper)
   1. Vibrio.   2. Vibrio infections.   I. Thompson, F. L. (Fabiano L.)   II. Austin, B. (Brian), 1951– .
   III. Swings, J. G.   IV. American Society for Microbiology.
   [DNLM: 1. Vibrio. 2. Vibrio—pathogenicity. 3. Vibrio Infections—etiology.   QW 141 B615 2006]

QR82.S6B56 2006
579.3'25—dc22

                                                                                            2005032511

All rights reserved
Printed in the United States of America

10 9 8 7 6 5 4 3 2 1

Address editorial correspondence to ASM Press, 1752 N St., N.W., Washington, DC 20036-2904, U.S.A.

Send orders to: ASM Press, P.O. Box 605, Herndon, VA 20172, U.S.A.
Phone: 800-546-2416; 703-661-1593
Fax: 703-661-1501
E-mail: books@asmusa.org
Online: http://estore.asm.org

Cover figure: Colonial morphology of vibrios in different media (courtesy of Gomez-Gil and Roque [see Chapter 2]).

To Huai-Shu Xu

CONTENTS

CONTRIBUTORS

Luis A. Actis
Department of Microbiology, Miami University, Oxford, Ohio 45056

Brian Austin
School of Life Sciences, John Muir Building, Heriot-Watt University, Riccarton, Edinburgh, EH14 4AS, Scotland, United Kingdom

Douglas H. Bartlett
Marine Biology Research Division (0202), Scripps Institution of Oceanography, University of California, San Diego, La Jolla, California 92093-0202

Yan Boucher
Department of Chemistry and Biomolecular Sciences, Macquarie University, Sydney, NSW 2109, Australia

J. Grant Burgess
School of Marine Science and Technology, Armstrong Building, University of Newcastle, Newcastle Upon Tyne NE1 7RU, United Kingdom

Nancy Busico-Salcedo
College of Veterinary Medicine, University of Southern Mindanao, Kacacan, 9407 Cotabato, Philippines

Ana Coelho
Department of Genetics, Instituto de Biologia, Universidade Federal do Rio de Janeiro, Rio de Janeiro, CEP 21944-970, Brazil

R. R. Colwell
Center for Bioinformatics and Computational Biology (CBCB), Agriculture/Life Sciences Surge Bldg. #296, Room 3103, University of Maryland, College Park, Maryland 20740

Jorge H. Crosa
Department of Molecular Microbiology and Immunology, School of Medicine, Oregon Health and Science University, Portland, Oregon 97201-3098

Shah M. Faruque
Molecular Genetics Laboratory, International Centre for Diarrhoeal Disease Research, Bangladesh, Mohakhali, Dhaka-1212, Bangladesh

Dirk Gevers
Laboratory of Microbiology, Ghent University, K. L. Ledeganckstraat 35, B-9000 Ghent, Belgium

Bruno Gomez-Gil
CIAD, A.C. Mazatlán Unit for Aquaculture and Environmental Management, A.P. 711, Mazatlán, Sin. 82000, México

Takeshi Honda
Department of Bacterial Infections, Research Institute for Microbial Diseases, Osaka University, 3-1 Yamadaoka, Suita, Osaka 565-0871, Japan

Tetsuya Iida
Department of Bacterial Infections, Research Institute for Microbial Diseases, Osaka University, 3-1 Yamadaoka, Suita, Osaka 565-0871, Japan

Staffan Kjelleberg
School of Biotechnology and Biomolecular Sciences, Centre for Marine Biofouling and Bio-Innovation, University of New South Wales, Sydney 2052, Australia

Karl E. Klose
South Texas Center for Emerging Infectious Diseases and Department of Biology, The University of Texas at San Antonio, San Antonio, Texas 78249

Omry Koren
Department of Molecular Microbiology and
Biotechnology, George S. Wise Faculty of Life
Sciences, Tel Aviv University, Ramat Aviv 69978,
Israel

Ken Kurokawa
Laboratory of Comparative Genomics, Graduate
School of Information Science, Nara Institute of
Science and Technology, 8916-5 Takayamacho,
Ikoma 630-0192, Japan

Frédérique Le Roux
Laboratoire de Génétique et Pathologie, Ifremer,
BP 133, Ronce les bains, 17390 La Tremblade,
France

Didier Mazel
Unité postulante "Plasticité du Génome Bactérien,"—
CNRS URA 2171, Dept. Structure et Dynamique
des Génomes, Institut Pasteur, 75724 Paris, France

Linda L. McCarter
Microbiology Department, The University of Iowa,
Iowa City, Iowa 52242

Diane McDougald
School of Biotechnology and Biomolecular Sciences,
Centre for Marine Biofouling and Bio-Innovation,
University of New South Wales, Sydney 2052,
Australia

G. Balakrish Nair
Laboratory Sciences Division, International Centre
for Diarrhoeal Disease Research, Bangladesh,
Mohakhali, Dhaka-1212, Bangladesh

Mitsuaki Nishibuchi
Center for Southeast Asian Studies, Kyoto
University, Yoshida, Sakyo-ku, Kyoto 606-8501,
Japan

James D. Oliver
Department of Biology, University of North
Carolina at Charlotte, Charlotte, North Carolina
28223

Leigh Owens
Microbiology and Immunology, School of
Veterinary and Biomedical Sciences, James Cook
University, Townsville 4811, Australia

Kwan-Sam Park
Department of Bacterial Infections, Research
Institute for Microbial Diseases, Osaka University,
3-1 Yamadaoka, Suita, Osaka 565-0871, Japan

Martin F. Polz
Department of Civil and Environmental
Engineering, Massachusetts Institute
of Technology, 77 Massachusetts Avenue,
Cambridge, Massachusetts 02139

Michael G. Prouty
South Texas Center for Emerging Infectious Diseases
and Department of Biology, The University of Texas
at San Antonio, San Antonio, Texas 78249

Irma Nelly G. Rivera
Department of Microbiology, Biomedical
Science Institute, University of São Paulo,
São Paulo, CEP 05508-900, Brazil

Ana Roque
Instituto de Recerca i Tecnologia Agroalimentaries,
Centre d'Aquicultura, AP200 Sant Carles de la
Rapita 43540, Spain

Eugene Rosenberg
Department of Molecular Microbiology and
Biotechnology, George S. Wise Faculty of Life
Sciences, Tel Aviv University, Ramat Aviv 69978,
Israel

Dean A. Rowe-Magnus
Department of Microbiology, Sunnybrook &
Women's College Health Sciences Centre,
Toronto, Ontario M4N 3N5, Canada

Tomoo Sawabe
Laboratory of Microbiology, Graduate
School of Fisheries Sciences, Hokkaido
University, 3-1-1 Minato-cho,
Hakodate 041-8611, Japan

Eric V. Stabb
University of Georgia, Department of
Microbiology, 828 Biological Sciences,
Athens Georgia 30602

Hatch W. Stokes
Department of Chemistry and Biomolecular
Sciences, Macquarie University, Sydney, NSW
2109, Australia

Jean Swings
Laboratory of Microbiology and BCCM/LMG
Bacteria Collection, Ghent University, K. L.
Ledeganckstraat 35, B-9000 Ghent, Belgium

Fabiano L. Thompson
Microbial Resources Division and Brazilian
Collection of Environmental, and Industrial
Micro-organisms (CBMAI), CPQBA, UNICAMP,
Alexandre Caselatto 999, CEP 13140000,
Paulínia, Brazil

Janelle R. Thompson
Department of Civil and Environmental
Engineering, Massachusetts Institute of Technology,
77 Massachusetts Avenue, Cambridge,
Massachusetts 02139

Marcello E. Tolmasky
Department of Biology, College of Natural Sciences
and Mathematics, California State University—
Fullerton, Fullerton, California 92834-6850

Hidetoshi Urakawa
Center for Advanced Marine Research, Ocean
Research Institute, The University of Tokyo,
1-15-1 Minamidai, Nakano, Tokyo 164-8639, Japan

Yves Van de Peer
BioInformatics & Evolutionary Genomics,
Ghent University/VIB Technologiepark 927,
B-9052 Ghent, Belgium

Ana Carolina P. Vicente
Department of Genetics, Instituto Oswaldo Cruz,
Rio de Janeiro, CEP 21045-900, Brazil

Michelle D. Vieira
Department of Genetics, Instituto de Biologia,
Universidade Federal do Rio de Janeiro,
Rio de Janeiro, CEP 21944-970, Brazil

Mohammed Zouine
Unité Postulante "Plasticité du Génome
Bactérien,"—CNRS URA 2171, Dept. Structure et
Dynamique des Génomes, Institut Pasteur,
75724 Paris, France

PREFACE

Two decades has passed since the last dedicated text-book on the vibrios, i.e. *Vibrios in the Environment*, which was edited by R. R. Colwell. Since then, there have been tremendous developments in the knowledge of the vibrios, including improvements in the taxonomy, ecology, and pathogenicity of the group. Indeed, vibrios are the best studied of all aquatic bacteria. Many new species have been described, and exciting concepts have been proposed. Improved detection, characterization, and identification tools have been developed to enable the rapid screening of strains. Molecular biology analyses and, more recently, whole genome sequencing of several vibrios, have shed much light on the biology of these microbes in their natural habitats and opened up new avenues for basic and applied research. It is therefore timely to compile a volume containing data about the current status of research and understanding of the vibrios. For this, we are grateful to the co-operation of the numerous authors, all of whom have produced manuscripts within a tight time frame. ASM Press was especially helpful during all stages of the book, from the nurturing of the original idea to the professional editing of the text, to the production of the finished volume. The result is a book that is primarily targeted at bacterial taxonomists, microbial ecologists, genome researchers, health management workers, and postgraduate and senior undergraduate students.

We are grateful to the following publishers, who have given permission to use copyrighted material: ASM Press, Blackwell Publishing, Boxwood Press, Elsevier Science BV, and the *Proceedings of the National Academy of Sciences, USA*. Numerous scientists have provided original photographs, and for these we acknowledge J. Bina, M. N. Guentzel, J. Mekalanos, J. Oakey, J. Reidl and F. Yildiz.

We hope the book will be a fitting tribute to those who have worked assiduously to improve the understanding of this fascinating group of vibrios.

F. L. Thompson, B. Austin, and J. Swings
August 2005

I. INTRODUCTION

The Biology of Vibrios
Edited by F. L. Thompson et al.
© 2006 ASM Press, Washington, D.C.

Chapter 1

A Global and Historical Perspective of the Genus *Vibrio*

R. R. COLWELL

Vibrios have played a significant role in human history. Outbreaks of cholera, caused by *Vibrio cholerae*, can be traced back in time to early recorded descriptions of enteric infections. Indeed, the path of human history has been influenced significantly by this organism (Wendt, 1885; Pollitzer, 1959). First described by Pacini (1854) while he was a medical student in Italy and at a time when the germ theory of disease was in dispute, *V. cholerae* was subsequently identified and described in greater detail by Robert Koch (1883, 1884), to whom credit for the discovery of the causative agent of cholera traditionally has been given. However, Pacini was rescued from obscurity and provided recognition for his pioneering work; his stained microscope slides remain on display at the University of Verona, Italy. Volumes have been published on cholera, its origins, pathology, and epidemiology, rendering *V. cholerae* one of the most studied of the bacterial species (Wachsmuth et al., 1994).

The germ theory of disease was developed in the 19th century, based on the British queen's physician John Snow's tracing an 1849 cholera outbreak to a single contaminated well in the Broad Street area of central London; it remains a canonical example of epidemiology. Snow's demonstration that the contaminated communal hand-operated pump was supplied by a particular water company remains equally powerful today. Snow's book was an important milestone in public health, correctly identifying the fecal-oral route to human infection and offering powerful arguments for the germ theory (Snow, 1855). Many advances in the prevention and treatment of infectious diseases during the latter half of the 19th century and the first half of the 20th century follow directly from the acceptance of Snow's point of view. Yet, in 2002, >120,000 cases of cholera were reported, and >3,700 resulted in death (World Health Organization, 1992).

The vibrios have also received the attention of marine microbiologists who observed that the readily cultured bacterial populations in near-shore waters and those associated with fish and shellfish were predominantly *Vibrio* spp. For example, the "gut group" vibrios were described by Liston (1954, 1957), working at the Marine Laboratory in Aberdeen, Scotland. Fish diseases caused by vibrios have been reviewed extensively by many investigators (Austin, in press) and, among the many fish pathogens, *Vibrio anguillarum* has been recognized historically as a major pathogen of marine animals.

In the 1950s, *Vibrio parahaemolyticus* was first isolated and described by Japanese medical scientists, and the major epidemics caused by this *Vibrio* species have been extensively documented (Takeda, 1988). *V. parahaemolyticus* was subsequently shown to have an annual cycle of abundance in near-shore marine waters and estuaries, particularly in association with zooplankton (Kaneko and Colwell, 1973, 1975, 1978). The biology, ecology, and pathogenicity of *V. parahaemolyticus* have been extensively reviewed and are addressed in this book. A third significant human pathogenic species of the genus, *Vibrio vulnificus*, has stimulated extensive research, and the literature is rich with descriptions of its ecology, pathogenicity, and biology (Oliver, 1995). Shared characteristics of these vibrios include requirement for salt for growth (either that sufficient in prepared bacteriological media or requiring amendment to concentrations of 1 to 3% [wt/vol] NaCl), chitin digestion, and general morphological features, i.e., curved rods. Vibrios are fermentative in metabolism and, most recently, have been found to carry two chromosomes (Heidelberg et al., 2000). Many *Vibrio* spp. are bioluminescent, including *V. cholerae* and *V. fischeri*, the genes for bioluminescence having been characterized and detected in *Vibrio* spp. by employing gene probes and

Rita R. Colwell • Center for Bioinformatics and Computational Biology (CBCB), Biomolecular Sciences Bldg. #296, Room 3103, University of Maryland, College Park, MD 20742.

genomic sequencing (see Palmer and Colwell, 1991; Heidelberg et al., 2000). Over the past 20 years, many nonpathogenic species of *Vibrio* have been described, including *V. diazotrophicus* (Guerinot et al., 1982), a nitrogen-fixing species, and species associated with marine mammals, e.g., *V. carchariae* (Grimes et al., 1989). Clearly, a wider role of vibrios in the environment, notably in nutrient cycling, has begun to be appreciated.

The complete genome sequences of *V. cholerae*, *V. parahaemolyticus*, and *V. vulnificus* have been determined, providing a rich set of data illuminating the metabolic versatility of these species. Recently, a *Vibrio* sp. isolated from a hydrothermal vent has been sequenced (unpublished data). Interestingly, extensive sequence similarity among genes of the three previously sequenced *Vibrio* spp. and the vent vibrio has been observed (Fig. 1), suggesting a historical origin in the deep sea. Thus, comparative genomics of the vibrios, based on complete genomic sequences, is now possible and highly useful in establishing both a phylogenomic taxonomy and a biogeography for the genus (manuscript in preparation).

Vibrios are clearly very important inhabitants of the riverine, estuarine, and marine aquatic environments. For this reason, by taking the perspective of a global microbial ecology of the vibrios, a deeper understanding of microbial ecological systems can be gained. *V. cholerae* provides a useful example and is therefore discussed here in the context of a general pattern of environmental pathogens and their close

linkage with climate, weather systems, seasonality, and physical and chemical parameters. Through such an analysis, the vibrios can be more fully appreciated in their many activities and functions (Colwell, in press; unpublished data).

Today, the study of infectious disease, whether of humans, animals, or plants, draws insight from a series of contexts, each nesting like one concentric circle within the last, from nanoscience to genomics and from mathematics, ecology, geography, and social science to climatology. The connections between cholera—an ancient and extensively studied waterborne disease—and the environment provide a valuable paradigm for this perspective. Fully dimensional understanding of an infectious disease, whether cholera, hantavirus, or malaria, reaches from countries to continents and beyond and connects medicine to many viewpoints across science and engineering, and even to daily life. A global context indisputably frames all human health issues in the 21st century. This context is formed of several realities: the worldwide movement of people and goods, the new recognition that earth processes operate on a global scale, and a dynamic international scientific enterprise.

Science and engineering have always flourished across national borders, but the current global scale of research is unprecedented. As research grows increasingly interdisciplinary, more scientific questions surmount national borders. A study of the vibrios suitably must address the concentric circles surrounding the diseases caused by vibrios, as well as their

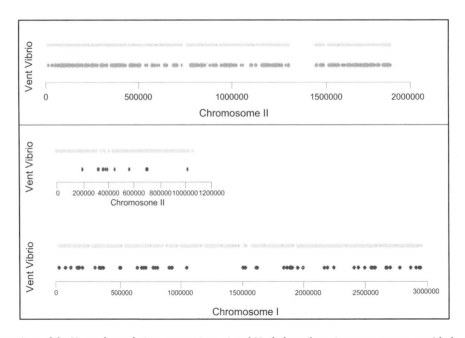

Figure 1. Comparison of the *V. parahaemolyticus* genome (upper) and *V. cholerae* (lower) genome sequences with the genome sequence of a recently isolated *Vibrio* sp. from a hydrothermal vent in the East Pacific Rise.

many functions and capabilities—notably, the international setting and the philosophical construct of biocomplexity (Colwell, 2002; R. R. Colwell, Editorial, *EcoHealth* **1**:6–7, 2004). Therefore, it can be instructive to compare selected cases of infectious diseases, whether of humans, animals, or plants, in their ecological and climatological environments, providing context for the case of cholera and the geographical distribution of vibrios. *V. cholerae* serves as a paradigm for this perspective and a model, perhaps, of a larger perspective for understanding the more general role of vibrios in nature. The concentric circles of many disciplines provide insights from mathematics, ecology, oceanography, and the space and social sciences. World Health Organization (WHO) data provide a useful starting point (Fig. 2). Infectious diseases cause about one-quarter of deaths worldwide (not including cancer and cardiovascular and respiratory diseases, many of which have recently been shown to be caused by infections). Broken down into the six leading infectious killers, diarrheal diseases, not long ago number one, are currently third overall—but still rank second for children under age 5 (Fig. 3). The major cause of death for children 4 years old and under is infectious disease, which causes almost two-thirds, or 63%, of these deaths (Fig. 4), and outbreaks of cholera substantially exceed those of any other disease. Thus, *V. cholerae* ranks near the top of the list of human pathogens, and these data constitute some of the largest concentric circles that frame today's global context for environment and health. International travel has skyrocketed in the past half century, with more than 500 million inter-

national arrivals per year by 2000, and continuing to climb. Thus, the ubiquity of selected pathogens in the environment and the ease of transmissibility by the migration of people and goods justify consideration of the vibrios, commonly present in riverine, estuarine, and coastal systems, at a global level.

The international arena is one context, and another is conceptual—the framework termed biocomplexity, which denotes the study of complex interactions in biological systems, including humans, and their physical environments (Colwell, 2002; R. R. Colwell, Editorial, *EcoHealth* **1**:6–7, 2004). Ecosystems do not respond linearly to environmental change, nor do the microorganisms that live in them. It is important to underscore the point that understanding demands observing at multiple scales, from the nano to the global. Complexity principles emerge at each level: the cell, organism, community, and ecosystems. With the perspective of biocomplexity, disciplinary worlds, formerly discrete, intersect to form fuller, more nuanced viewpoints and integrate across disciplines and scales, a perspective that roots epidemiology firmly in ecology. As signals from climate models are recognized and incorporated into health measures, new opportunities arise for proactive—rather than reactive—approaches to public health, providing the basis for a new kind of medicine, a predictive, hence a preemptive, medicine.

Indeed, ecology has immediate lessons for epidemiology. One useful model is the mosquito that lays its eggs in North American carnivorous plants, the pitcher plants, which are similar to plants that harbor mosquitoes in Southeast Asia. Although the

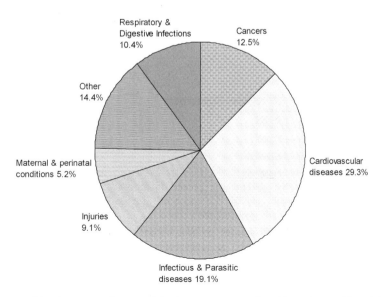

Figure 2. Leading causes of death (57.02 million) worldwide in 2002. Reproduced from the World Health Organization, with permission.

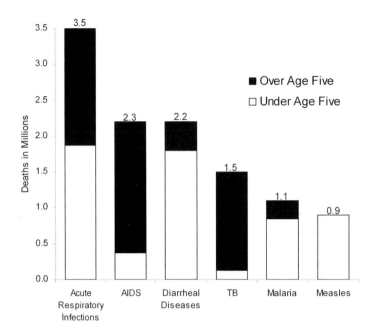

Figure 3. Leading infectious killers according to millions of deaths worldwide for all ages in 1999. Reproduced from the World Health Organization, with permission.

pitcher plant mosquito is not a disease vector, adaptation of this formerly tropical insect to the climatic gradient of North America carries insights for the spread of disease. The mosquito, like many organisms, uses day length to regulate seasonal development (Mathias et al., 2005), and its populations have adapted to the climate of North America, from Florida to Canada. Disease-carrying invaders, like the Asian tiger mosquito, similarly adapt to cold and to different day lengths. As spring arrived earlier and the growing season lengthened over the latter half of the last century, the pitcher plant mosquito adapted to shorter photoperiods, especially in the northern United States. In this example of a documented genetic shift due to warming, as latitude increases, the mosquito's genetic shift to shorter photoperiods has occurred over as short a time interval as 5 years, essentially "evolution at breakneck speed" (Mathias et al., 2005).

A pathogen similarly intertwined with climate began to cause human disease outbreaks in 1993 in the Four Corners area of the United States, when young and otherwise healthy people began dying from an unknown affliction. At the time, in fact, bioterrorism was suspected. The agent proved to be a New World hantavirus, unknown until the outbreak. The mortality rate of those infected with the virus was 70% in the first few weeks. At the time of the outbreak, the question was whether the new virus was a

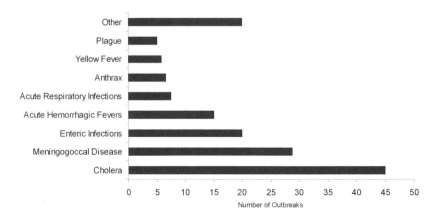

Figure 4. Reported outbreaks of known infectious diseases, 1998 to 1999. Reproduced from the World Health Organization, with permission.

mutant or in the environment naturally. The carrier turned out to be a rodent, the deer mouse. The virus was detected in mouse tissue that had been archived years before (Yates et al., 2002), and Native American legends corroborated a history of outbreaks. Most important, a link between climate and outbreak of disease was demonstrated. Mild and wet winters associated with the El Niño/Southern Oscillation had provided an increased supply of food for the rodents, whose populations increased dramatically in 1993. Eventually, a time lag between rodent population increase and increase in human infections was determined, and the time lag occurred between peaking rodent populations and increase in disease incidence in the rodents. Densities of hanta-infected mice were correlated with total deer mouse densities 1 year earlier.

The key predictor of disease was not the increase in numbers of rodents, but the increase in infected rodents. This led to development of a trophic cascade hypothesis, explaining changing levels of human risk for zoonotic diseases associated with climate variability with the cascade of disease through the trophic levels set off by El Niño (Yates et al., 2002). A predictive model, indicating areas of highest risk for hantavirus, has been developed, and the Canyon del Muerto has been proposed as a historical site for the hantavirus, its presence occurring there years before human awareness (Fig. 5).

Figure 5. Map of New World hantaviruses discovered since 1993. Reproduced from Yates et al. (2002), with permission of the publisher.

Campylobacter, emerging as a leading cause of gastroenteritis in some countries more than a quarter century ago, was addressed by public health interventions focused on food-borne transmission, but without decreased disease incidence. In fact, an annual rise over a 10-year period was recorded, and the disease displays a striking and consistent seasonal pattern. A significant correlation between increased temperature and the seasonal peak of *Campylobacter* infections in England and Wales has been reported (Louis et al., 2005).

Tularemia, another disease with very complex environmental links, occurs throughout the Northern Hemisphere and caused a million human cases over the past 70 years. A plague-like disease, it occurs in >100 wild mammals, as well as in birds, insects, and humans. Like cholera, waterborne outbreaks of tularemia have resulted from contaminated drinking water, muskrats, and hares as reservoirs in nature.

These diseases and their environmental interactions and reservoirs provide a backdrop for cholera, the understanding of which has evolved from a linear reductionist model of a waterborne bacterium and a human host to a much more complex picture. The current characterization of the ecology of cholera includes global weather patterns, aquatic reservoirs, phages, zooplankton, collective behavior of surface-attached cells, an adaptable genome, and the deep sea, together with the bacterium and its host, emerging from the perspective of biocomplexity.

Clearly, endemic cholera cannot be understood with an ecology built on a 19th century model of disease. Prevention and treatment of cholera mandate appreciation of the subtlety and variety of interactions between *V. cholerae* and the aquatic environment and depend on an inherently interdisciplinary and international research strategy, informed by an appreciation of complexity at every scale of length and time: in fact, a systems approach.

Ecological studies of vibrios began in the early 1960s; in 1977, *V. cholerae* was cultured in water samples from the upper and mid-Chesapeake Bay in the United States in late fall and early spring (Joseph et al., 1982). Similar estuaries around the world have also proved to be reservoirs for *V. cholerae*. An interesting observation at that time was that *V. cholerae* could not be isolated from Chesapeake Bay when the water temperatures dropped in the winter. Direct fluorescent assays demonstrated that the vibrios were, in fact, present throughout the year, but during the winter months they were found to be in a dormant or "viable but not culturable" or "not yet culturable" state (Roszak and Colwell, 1987). Since those early studies, many gram-negative bacteria are now known to enter this dormant state when faced with adversity, mainly extreme conditions of temperature, salinity, or nutrient. A 3-year study in the late 1980s showed that the toxigenic 01 serotype of *V. cholerae* is a year-round inhabitant of the Bay of Bengal (Tamplin et al., 1990), entering the viable but not culturable state and re-emerging in culturable form with the arrival of the monsoon season. Thus, the premise that the characteristics of *V. cholerae* derived solely from an adaptation to the human host was not supported by the data, and critical environmental variables were subsequently identified. Temperature and salinity (Louis et al., 2005) and the association of vibrios with shellfish and plankton were found to be critical factors (Colwell and Huq, 1994). Together, temperature and salinity accounted for about 80% of the seasonal variability in the Chesapeake Bay population (Louis et al., 2003). Similar relationships between *V. parahaemolyticus* and the Chesapeake blue crab had been shown earlier (Kaneko and Colwell, 1978). Thus, by the mid-1980s, the ecology of the cholera bacteria had begun to be considered in a new context.

Zooplankton in estuarine and riverine systems were concluded to play a significant role in the *V. cholerae* and cholera cycle, similar to that of a vector in diseases such as malaria or tularemia (Huq et al., 1983). Interestingly, the two epidemic variants of *V. cholerae*, O1 and O139, were found not to compete equally well for space on the copepod surface. *V. cholerae* O1 appears to outcompete *V. cholerae* O139 (T. Rawlings et al., unpublished data), yet another example of the idiosyncrasy of the vibrio and its environmental capabilities, as well as further evidence of its autochthonous aquatic nature.

Adhesive ability is an important attribute of *V. cholerae*, whether in the environment or in the human gut. Capabilities of *V. cholerae* useful in environmental settings include its ability to secrete a powerful chitinase (Dastidar and Narayanaswami, 1968), which assists its growth on chitin surfaces and mucinase. This enables the bacterium to penetrate mucus barriers, both on plankton (Islam et al., 1999) and on that which covers the gastrointestinal epithelium (Zhu and Mekalanos, 2003). In fact, Pruzzo et al. (2003) reported that serum from mussel (*Mytilus*) hemolymph increases attachment in the intestinal epithelial cells. The upshot is that when *V. cholerae* is ingested with seafood, cholera acquires "bridging molecules" that make cholera highly adhesive in the intestine. As Pruzzo et al. explain, "virulence and infectivity depend on both bacterial properties and environmental factors." The conclusion is that *V. cholerae* originally evolved commensally with marine animals, such as copepods, both as their gut flora and as a surface for growth, nutrition, and other mutual benefits.

Environmental variables influencing cholera outbreaks in Bakerganj, Bangladesh, have been shown to include not only water temperature and salinity, but also water depth, rainfall, conductivity, and copepod counts (Alam et al., in press; Huq et al., 2005). The lag periods between increases or decreases in units of these factors, notably temperature and salinity, and occurrence of cholera were strongly correlated with plankton population blooms (Huq et al., 2005) (Fig. 6).

By using remotely sensed temperature data to help estimate phytoplankton and zooplankton blooms in the Bay of Bengal, Bangladesh, a global warning system for cholera was constructed (Lobitz et al., 2000). That is, when plankton blooms "crash," vibrios are released into the water columns in abundance, and a correlation between sea surface temperature and cholera outbreaks in Bangladesh, with a 6-week time lag, has been established (Huq et al., 2005; Lobitz et al., 2000; Colwell, 1996). The incidence of cholera outbreaks related to sea surface measurements for the period 1992 to 1995 is shown in Fig. 7. The global weather pattern, El Niño, has been found to bring new cholera outbreaks with each warm water cycle (Koelle et al., 2005).

Continuing studies of monsoons in the Bay of Bengal and El Niño in South America provide even more useful data to achieve the goal of a model sufficiently powerful to predict outbreaks locally, allowing public health interventions to be directed to where they are most needed (Gil et al., 2004).

In the Southern Hemisphere off the coast of Peru, some new insights on cholera have already been obtained. Cholera surfaced in Peru in 1991, after a century of absence in Latin America. It has recurred there ever since, following a seasonal pattern, with the greatest number of cases in summer (January to March) in Lima and in other major cities along the coast. Interestingly, both El Niño events and cholera outbreaks have increased since the 1970s, a pattern characteristic for both Peruvian coastal areas and the Bay of Bengal. Sea surface temperature and height, as well as plankton blooms, can be remotely sensed and used to forecast outbreaks (Colwell, in press). The climate–cholera link documented for the 1997 to 1998 El Niño year confirms that an early warning system for cholera risk can be established. Furthermore, even more straightforward and simple measures can be taken. Reducing the number of zooplankton and particulates in drinking water and, therefore, reducing exposure to the attached bacteria by simple filtration, using four to eight folds of sari cloth, yielded about a 50% reduction in cholera cases in Bangladesh villages where water treatment was unavailable. This was an unexpected application of ba-

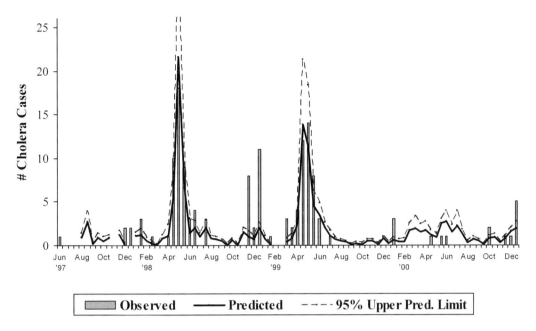

Figure 6. Observed number of cholera cases versus number of cases predicted by the Poisson regression model and upper 95% prediction limit when using water temperature (lag, 6 weeks), *ctx* gene probe count (lag, 0 weeks), conductivity (lag, 0 weeks), and rainfall (lag, 8 weeks) for Lake 2 in Bakerganj, Bangladesh. Reproduced from Huq et al., 2005, with permission of the publisher.

sic research findings to improving human health (Colwell et al., 2003). Results of field trials in Bangladesh showed that this simple procedure reduced the incidence of cholera by more than a factor of 2. Thus, the vibrios, exemplified by *V. cholerae*, play a significant and global role in human health and well-being, in the context of biocomplexity.

In summary, infectious diseases have distinctive biographies, and each has a complex relationship with the environment, amply demonstrated by the example of *V. cholerae*. In a world of ever more rapid change, the ecological patterns of microorganisms expand across scales, and explanations must draw upon interdisciplinary research in the biological, physical,

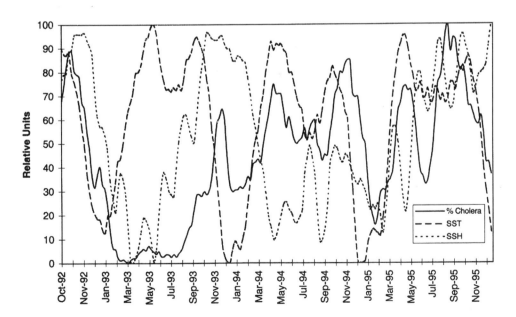

Figure 7. Cholera outbreaks and increases in sea surface temperature (SST) and sea surface height (SSH). Reproduced from Lobitz et al., 2000, with permission of the publisher.

and social sciences. For the first time, the complexity of these patterns can be integrated if new paradigms are explored in order to reach new dimensions.

REFERENCES

Alam, M., M. Sultana, G. B. Nair, R. B. Sack, D. A. Sack, A. K. Siddique, A. Huq, and R. R. Colwell. 2006. Toxigenic *Vibrio cholerae* in natural environmental setting of Bangladesh. *Appl. Environ. Microbiol.* 72:2849–2855.

Alam, M., A. Sadique, Nur-A-Hasan, N. A. Bhuiyan, S. Ahsan, G. B. Nair, D. A. Sack, A. K. Siddique, A. Huq, R. Bradley Sack, and R. R. Colwell. 2006. Effect of transport at ambient temperature on detection and isolation of *Vibrio cholerae* from environmental samples. *Appl. Environ. Microbiol.* 72:2185–2190.

Austin, B. Bacterial pathogens of marine fish. *In* S. Belkin and R. R. Colwell (ed.), *Oceans and Health: Pathogens in the Marine Environment*, in press. Kluwer Academic/Plenum Publishers, New York, N.Y.

Colwell, R. R. 1996. Global climate and infectious disease: the cholera paradigm. *Science* 274:2025–2031.

Colwell, R. R. 2002. Biocomplexity, p. 145–146. *In* H. A. Mooney and J. G. Canadell (ed.), *The Earth System: Biological and Ecological Dimensions of Global Environmental Change*, vol. 2. *Encyclopedia of Global Environmental Change*. John Wiley & Sons, Ltd., Chichester, England.

Colwell, R. R. Global microbial ecology of *Vibrio cholerae*. *In* S. Belkin and R. R. Colwell (ed.), *Oceans and Health: Pathogens in the Marine Environment*, in press. Kluwer Academic/Plenum Publishers, New York, N.Y.

Colwell, R. R., and A. Huq. 1994. Vibrios in the environment: viable but nonculturable *Vibrio cholerae*, p. 117–133. *In* I. K. Wachsmuth, O. Olsvik, and P. A. Blake (ed.), Vibrio cholerae *and Cholera: Molecular to Global Perspectives*. ASM Press, Washington, D.C.

Colwell, R. R., A. Huq, M. S. Islam, K. M. A. Aziz, M. Yunus, N. H. Khan, A. Mahmud, R. B. Sack, G. B. Nair, J. Chakraborti, D. A. Sack, and E. Russek-Cohen. 2003. Reduction of cholera in Bangladeshi villages by simple filtration. *Proc. Natl. Acad. Sci. USA* 100:1051–1055.

Dastidar, S. G., and A. Narayanaswami. 1968. The occurrence of chitinase in vibrios. *Indian J. Med. Res.* 56:654–658.

Gil, A. I., V. R. Louis, I. N. G. Rivera, E. Lipp, A. Huq, C. F. Lanata, D. N. Taylor, E. Russek-Cohen, N. Choopun, R. Bradley Sack, and R. R. Colwell. 2004. Occurrence and distribution of *Vibrio cholerae* in the coastal environment of Peru. *Environ. Microbiol.* 6:699–706.

Grimes, D. J., J. Burgess, J. Crunkleton, P. Brayton, and R. R. Colwell. 1989. Potential invasive factors associated with *Vibrio carchariae*, an opportunistic pathogen for sharks. *J. Fish Dis.* 12:69–72.

Guerinot, M. L., P. A. West, J. V. Lee, and R. R. Colwell. 1982. *Vibrio diazotrophicus*, sp. nov., a marine N_2-fixing bacterium. *Int. J. Syst. Bacteriol.* 32:350–357.

Heidelberg, J. F., J. A. Eisen, W. C. Nelson, R. A. Clayton, M. L. Gwinn, R. J. Dodson, D. H. Haft, E. K. Hickey, J. D. Peterson, L. Umayam, S. R. Gill, K. E. Nelson, T. D. Read, H. Tettelin, D. Richardson, M. D. Ermolaeva, J. Vamathevan, S. Bass, H. Qin, I. Dragoi, P. Sellers, L. McDonald, T. Utterback, R. D. Fleishmann, W. C. Nierman, O. White, S. L. Salzberg, H. O. Smith, R. R. Colwell, J. J. Mekalanos, J. C. Venter, and C. M. Fraser. 2000. DNA sequence of both chromosomes of the cholera pathogen *Vibrio cholerae*. *Nature* 406:477–483.

Huq, A., E. B. Small, P. A. West, M. I. Huq, R. Rahman, and R. R. Colwell. 1983. Ecological relationships between *Vibrio cholerae* and planktonic crustacean copepods. *Appl. Environ. Microbiol.* 45:275–283.

Huq, A., R. B. Sack, A. Nizam, I. M. Longini, G. B. Nair, A. Ali, J. G. Morris, Jr., M. N. H. Khan, A. K. Siddique, M. Yunus, M. J. Albert, D. A. Sack, and R. R. Colwell. 2005. Critical factors influencing the occurrence of *Vibrio cholerae* in the environment of Bangladesh. *Appl. Environ. Microbiol.* 71:4645–4654.

Islam, M. S., Z. Rahim, M. J. Alam, S. Begum, S. M. Moniruzzaman, A. Umeda, K. Amako, M. J. Albert, R. B. Sack, A. Huq, and R. R. Colwell. 1999. Association of *Vibrio cholerae* O1 with the cyanobacterium, *Anabaena* sp., elucidated by polymerase chain reaction and transmission electron microscopy. *Trans. R. Soc. Trop. Med. Hyg.* 93:36–40.

Joseph, S., J. Kaper, and R. Colwell. 1982. *Vibrio parahaemolyticus* and related halophilic vibrios. *Crit. Rev. Press* 10:77–124.

Kaneko, T., and R. R. Colwell. 1973. Ecology of *Vibrio parahaemolyticus* in Chesapeake Bay. *J. Bacteriol.* 113:24–32.

Kaneko, T., and R. R. Colwell. 1975. Incidence of *Vibrio parahaemolyticus* in Chesapeake Bay. *Appl. Microbiol.* 30:251–257.

Kaneko, T., and R. R. Colwell. 1978. The annual cycle of *Vibrio parahaemolyticus* in Chesapeake Bay. *Microb. Ecol.* 4:135–155.

Koch, R. 1883. Der zweite Bericht der Deutschen Cholera-Commission. *Dtsch. Med. Wochenschr.* 9:743–744.

Koch, R. 1884. An address on cholera and its bacillus. *Br. Med. J.* 2:403–407; 453–459.

Koelle, K., X. Rodo, M. Pascual, M. Yunus, and G. Mostafa. 2005. Refractory periods and climate forcing in cholera dynamics. *Nature* 436:696–700.

Liston, J. 1954. A group of luminous and nonluminous bacteria from the intestine of flatfish. *J. Gen. Microbiol.* 12:i.

Liston, J. 1957. The occurrence and distribution of bacterial types on flat fish. *J. Gen. Microbiol.* 16:205–216.

Lobitz, B., L. Beck, A. Huq, B. Wood, G. Fuchs, A. S. G. Faruque, and R. Colwell. 2000. Climate and infectious disease: use of remote sensing for detection of *Vibrio cholerae* by indirect measurement. *Proc. Natl. Acad. Sci. USA* 97:1438–1443.

Louis, V. R., E. Russek-Cohen, N. Choopun, I. N. G. Rivera, B. Gangle, S. C. Jiang, A. Rubin, J. A. Patz, A. Huq, and R. R. Colwell. 2003. Predictability of *Vibrio cholerae* in Chesapeake Bay. *Appl. Environ. Microbiol.* 69:2773–2785.

Louis, V. R., I. A. Gillespie, S. J. O'Brien, E. Russek-Cohen, A. D. Pearson, and R. R. Colwell. 2005. Temperature-driven *Campylobacter* seasonality in England and Wales. *Appl. Environ. Microbiol.* 71:85–92.

Mathias, D., L. Jacky, W. E. Bradshaw, and C. M. Holzapfel. 2005. Geographic and developmental variation in expression of the circadian rhythm gene, timeless, in the pitcher-plant mosquito, *Wyeomyia smithii*. *J. Insect Physiol.* 51:661–667.

Oliver, J. D. 1995. The viable but non-culturable state in the human pathogen *Vibrio vulnificus*. *FEMS Microbiol. Lett.* 133:203–208.

Pacini, F. 1854. Osservazioni microscopiche e deduzioni patologiche sul colera asiatico. *Gaz. Med. Italiana* 6:405–412.

Palmer, L., and R. R. Colwell. 1991. Detection of luciferase gene sequence in nonluminescent *Vibrio cholerae* by colony hybridization and polymerase chain reaction. *Appl. Environ. Microbiol.* 57:1286–1293.

Pollitzer, R. 1959. *Cholera*, p. 11–50. World Health Organization. Geneva, Switzerland.

Pruzzo, C., R. Tarsi, M. del Mar Lleò, C. Signoretto, M. Zampini, L. Pane, R. R. Colwell, and P. Canepari. 2003. Persistence of adhesive properties in *Vibrio cholerae* after long-term exposure to sea water. *Environ. Microbiol.* 5:850–858.

Roszak, D. B., and R. R. Colwell. 1987. Survival strategies of bacteria in the natural environment. *Microbiol. Rev.* **51:**365–379.

Snow, J. 1855. *On the Mode of Communication of Cholera*, p. 1–38. John Churchill, London, England.

Takeda, Y. (ed.). 1988. Vibrio cholerae *and Cholera*. KTK Scientific Publishers, Tokyo, Japan.

Tamplin, M., A. Gauzens, A. Huq, D. Sack, and R. Colwell. 1990. Attachment of *Vibrio cholerae* serogroup O1 to zooplankton and phytoplankton of Bangladesh waters. *Appl. Environ. Microbiol.* **56:**1977–1980.

Wachsmuth, I. K., P. A. Blake, and O. Olsvik. 1994. Vibrio cholerae *and Cholera: Molecular to Global Perspectives*. ASM Press, Washington, D.C.

Wendt, E. C. (ed.). 1885. *A Treatise on Asiatic Cholera*. William Wood and Co., New York, N.Y.

World Health Organization. 1992. *Global Health Situation and Projections, Estimates*. World Health Organization, Geneva, Switzerland.

Yates, T., L., J. N. Mills, C. A. Parmenter, T. G. Ksiazek, R. R. Parmenter, J. R. Vande Castle, C. H. Calisher, S. T. Nichol, K. D. Abbott, J. C. Young, M. L. Morrison, B. J. Beaty, J. L. Dunnum, R. J. Baker, J. Salazar-Bravo, and C. J. Peters. 2002. The ecology and evolutionary history of an emergent disease: hantavirus pulmonary syndrome. *BioScience* **52:**989–998.

Zhu, J., and J. J. Mekalanos. 2003. Quorum sensing-dependent biofilms enhance colonization in *Vibrio cholerae. Dev. Cell.* **5:**647–656.

II. ISOLATION, ENUMERATION, AND PRESERVATION

The Biology of Vibrios
Edited by F. L. Thompson et al.
© 2006 ASM Press, Washington, D.C.

Chapter 2

Isolation, Enumeration, and Preservation of the *Vibrionaceae*

Bruno Gomez-Gil and Ana Roque

ISOLATION

Vibrios are fairly easy to isolate from both clinical and environmental material, though some species may require growth factors and/or vitamins. Worth special mention is the need for NaCl, although some species can grow with minimum NaCl concentrations, i.e., *Vibrio cholerae, Vibrio mimicus, Vibrio hispanicus,* and some strains of *Vibrio fluvialis, Vibrio furnissii,* and *Vibrio metschnikovii* (Alsina and Blanch, 1994; Gomez-Gil et al., 2004b). Optimal Na$^+$ concentration for many marine bacteria is between 70 and 300 mM (Reichelt and Baumann, 1974). The optimal requirement of NaCl for vibrios is ~2.0 to 2.5% (wt/vol), although some might require other ions, especially magnesium and potassium (Donovan and van Netten, 1995).

Media can select for vibrios when appropriate selective agents are incorporated. The main agents employed are bile salts, teepol, tellurite, and polymyxin B and E (colistin). Alkaline pH and high (or sometimes low) concentrations of NaCl have also been used to select for certain *Vibrio* species, based on the ability of most vibrios to grow at pH values above 8.0 and at 3.0% or higher concentrations of NaCl. The usefulness of high-pH media has been questioned as a selective factor for vibrios (Gilmour et al., 1976).

Bile salts (oxgall, sodium cholate, and taurocholate) have often been used in the formulation of media because they can inhibit the growth of gram-positive and some gram-negative bacteria (Monsur, 1963), but not members of the *Enterobacteriaceae.* Tellurite (potassium tellurite) inhibits gram-negative and gram-positive bacteria, but some staphylococci, streptococci, enterococci, corynebacteria, and lactic acid bacteria are resistant (Donovan and van Netten, 1995; Holt and Krieg, 1994). To avoid the conversion of tellurite to tellurate, it has been recommended

to add freshly prepared solutions of potassium tellurite and to use the medium within 24 h (Chatterjee et al., 1977). Teepol is an anionic detergent made of sodium and potassium salts and alkyl sulfates that inhibit the growth of gram-positive bacteria and has been used as a replacement for bile salts; it can also encourage the development of larger colonies (Jameson and Emberley, 1958). Polymyxin B and polymyxin E (colistin) are bacteriocidal agents (fatty acyl decapeptide antibiotics) against some gram-negative bacteria; they interfere with the structure and function of the outer and cytoplasmic membranes (Donovan and van Netten, 1995; Holt and Krieg, 1994). Both polymyxins differ only in a single amino acid, and their modes of action and microbiological activities are identical (Sogaard, 1982). One or a combination of these agents is often present in most of the selective media commonly used for the selective isolation of vibrios.

A scheme for the isolation of pathogenic vibrios often includes enrichment prior to plating on a selective agar, especially when vibrios are at low levels and competing with other bacteria that can outgrow them in general media. As a confirmatory step, biochemical or molecular identification is required, since most isolation procedures are at best presumptive.

Enrichment Media

Several enrichment media have been used for the selection of one group or species of *Vibrio*. Special effort has been devoted to enhance the presence of potential pathogenic species in contaminated food or in clinical samples.

For enrichment of *V. cholerae, V. parahaemolyticus,* and *V. vulnificus,* the most widely used medium (Farmer et al., 2003), and the one recommended by the U.S. Food and Drug Administration (FDA) (DePaola and Kaysner, 2004), is alkaline peptone water

Bruno Gomez-Gil • CIAD, A.C. Mazatlán Unit for Aquaculture and Environmental Management, A.P. 711, Mazatlán, Sin. 82000 México.   **Ana Roque** • Instituto de Recerca i Tecnologia Agroalimentaries, Centre d'Aquicultura, AP200 Sant Carles de la Rapita 43540, Spain.

(APW), a medium in use since 1887 (Donovan and van Netten, 1995). Another enrichment medium is taurocholate tellurite peptone water (Monsur, 1963), which has been recommended as a replacement for selecting *V. cholerae* from clinical samples. It is a simple medium to prepare and inhibits the growth of competing intestinal bacteria owing to the presence of bile salts and tellurite in peptonated water. Other, less frequently used media for the enrichment of *V. cholerae* are derivates of taurocholate tellurite peptone water, either by adding only one of the inhibitory agents or by changing their concentrations (Pal et al., 1967; Spira, 1984).

Enrichment media predominantly intended for isolating *V. parahaemolyticus* include glucose-salt-teepol broth (GSTB) (Sakazaki, 1965), salt-teepol buffer (Chun et al., 1974), salt colistin broth (SCB) (Nakanishi and Murase, 1973), and salt-polymyxin broth (SPB) (Hara-Kudo et al., 2001). SCB and SPB (at 250 U/ml of polymyxin B) proved to be better choices for detecting *V. parahaemolyticus* than GSTB or tryptic soy agar (TSA) from raw fish or environmental samples (Nakanishi and Murase, 1973). SCB is not useful for other pathogenic vibrios, and GSTB does not appear to improve APW results and is substantially worse for enrichment of other pathogenic vibrios (Spira, 1984).

Peptone salt cellobiose or peptone salt cellobiose colistin (PNCC) are two enrichment broths devised for the selective growth of *V. vulnificus* (Hsu et al., 1998). Peptone salt cellobiose enhanced the growth of this bacterium as compared to APW, while PNCC permitted the growth of *V. vulnificus* present at low levels and can inhibit non-target bacteria. Starch gelatin polymyxin B broth is another broth formulated for enrichment of *V. vulnificus* from food samples (Kitaura et al., 1983). *V. vulnificus* serotype E (biotype 2) is an important eel pathogen, and a broth (saline eel serum broth) has been developed for the selective isolation of virulent strains (Sanjuan and Amaro, 2004).

Vibrio fluvialis enrichment medium (Nishibuchi et al., 1983) is a modification of APW with the introduction of novobiocin as a selective agent and the concentration of NaCl raised to 4%. This medium proved to be better than APW when samples were collected at low salinities (0.6%), like those of sewage (Nishibuchi et al., 1983).

Alkaline peptone water

Alkaline peptone water (APW) is the preferred enrichment medium for vibrios, especially devised for *V. cholerae* but also used for *V. parahaemolyticus* and other species. The high pH of the medium (pH close to 9) and NaCl concentration inhibit many other bacteria and favor vibrios. Peptone concentration can range between 1 and 2%, with the latter being more appropriate for marine species. NaCl can be omitted to favor the growth of *V. cholerae* and *V. mimicus*. Electrolyte supplements (NaCl, $MgCl_2$, KCl) may be included to stimulate growth of *V. cholerae* and other pathogenic vibrios (Furniss et al., 1978). Enrichment can be accomplished in 6 to 8 h at 35 to 36°C and then again at 16 to 24 h (DePaola and Kaysner, 2004; Farmer et al., 2003).

APW is peptone, 10.0 g; NaCl, 10.0 g; distilled water, 1,000 ml. pH has to be adjusted to 8.5 ± 0.1 with 1 N NaOH. Autoclave at 121°C for 10 min.

Media Used for the Cultivation of Vibrios

The most commonly used medium for the cultivation of vibrios from the environment is marine agar, originally devised by Zobell and named marine agar 2216E (Zobell, 1941). Marine agar is a nonselective medium on which almost all vibrios will grow, but many other environmental bacteria will grow on it as well. Diverse media have been developed for the selective isolation and/or differentiation of a specific group or species of vibrios, focusing primarily on human pathogenic vibrios (Table 1). Thiosulfate citrate bile salt sucrose agar (TCBS) is the most widely used medium to isolate vibrios from a variety of sources, whether clinical, environmental, or contaminated food.

Selective or differential media have been devised principally for human pathogenic vibrios, but almost none are capable of permitting the growth of only one species; even fewer can permit the growth of only pathogenic strains. The most widely used medium for *V. cholerae* and many other vibrios is TCBS (see below). Taurocholate tellurite gelatin agar (Monsur, 1961; O'Brien and Colwell, 1985) is an inexpensive medium that produces transparent gray colonies (2 to 3 mm in diameter) of *V. cholerae* with a surrounding halo; the colonies are difficult to distinguish from those of *V. parahaemolyticus*. To overcome this problem, a modified taurocholate tellurite gelatin agar (O'Brien and Colwell, 1985) was developed with the addition of β-galactosidase (4-methyllumbelliferyl-β-Δ-galactosidase); *V. cholerae* produces a very strong reaction in 24 h or less, showing brilliant-blue fluorescence. Sucrose teepol tellurite (Chatterjee et al., 1977) is a modification of TCBS whereby the bile salts are replaced with teepol. Bile salts are responsible for variations in the quality of TCBS (Donovan and van Netten, 1995), a problem commonly associated with this medium (Chatterjee et al., 1977;

Table 1. Media used for growth of medically important vibrios (adapted from Donovan and van Netten, 1995)

Medium	pH	Selective agent(s)	Reference(s)
Selective media of general use			
TCBS	8.6	Thiosulfate, ferric citrate, ox bile	Nakanishi, 1963; Kobayashi et al., 1963
Bromothymol blue teepol (BTBT)	7.8	Teepol, NaCl	Sakazaki, 1972
Modified taurocholate tellurite gelatin agar (mTTGA)	8.5	Taurocholate, tellurite, pH	O'Brien and Colwell, 1985
Media devised for *V. cholerae*			
Taurocholate tellurite gelatin agar (TTGA)	8.4	Taurocholate, tellurite, pH	Monsur, 1961; O'Brien and Colwell, 1985
Modified vibrio agar		Sodium deoxycholate, ox bile, sodium lauryl sulfate	Tamura et al., 1971
Sucrose teepol tellurite (STT)	8.0	Teepol, tellurite, no NaCl	Chatterjee et al., 1977
SV medium	8.5	Ox bile, ferric citrate	Salles et al., 1976
Cellobiose polymyxin B colistin	7.6	Polymyxin B, colistin	Massad and Oliver, 1987
Polymyxin mannose tellurite (PMT)	8.4	Polymyxin B, tellurite, sodium dodecyl sulfate	Shimada et al., 1990
Media devised for *V. parahaemolyticus*			
Wagatsuma agar		None	Miyamoto et al., 1969
MT agar	8.0	NaCl (7%)	Vanderzant and Nickelson, 1972
Tryptone soya agar + triphenyltetrazolium chloride (TSAT)	7.1	Bile salts, NaCl (3%)	Kourany, 1983
Chromogenic agar medium (CHROMagar Vibrio, CV agar)	9.0	Chromogenic mix	Hara-Kudo et al., 2001
Vibrio parahaemolyticus sucrose agar (VPSA)	6.8	Bile salts, pH	Entis and Boleszczuk, 1983
Media devised for *V. vulnificus*			
Vibrio vulnificus agar (VV agar)	8.6	Tellurite, crystal violet, ox gall, pH	Brayton et al., 1983
Sodium dodecyl sulfate polymyxin B sucrose medium (SPS)		Polymyxin B, sodium dodecyl sulfate	Kitaura et al., 1983
Cellobiose polymyxin B colistin agar (CPC)	7.6	Polymyxin B, polymyxin E (colistin), 40°C	Massad and Oliver, 1987
Vibrio vulnificus enumeration medium (VVE)	8.5	Sodium cholate, sodium taurocholate, ox gall, potassium tellurite	Miceli et al., 1993
Cellobiose-colistin (CC) agar	7.6	Polymyxin E (colistin), 40°C	Hoi et al., 1998
Vibrio vulnificus medium (VVM)	8.5	Polymyxin B, polymyxin E (colistin)	Cerda-Cuellar et al., 2000

Nicholls et al., 1976; Spira, 1984; West et al., 1982). Sucrose teepol tellurite does not contain NaCl, thus favoring *V. cholerae*, which permitted the growth of more El Tor colonies than TCBS (Chatterjee et al., 1977). Cellobiose polymyxin B colistin (CPC) (Massad and Oliver, 1987) was developed for differential isolation of *V. cholerae* and *V. vulnificus* owing to the presence of the polymyxins (B and E), high incubation temperature (40°C), and cellobiose as a differential trait between both species. Polymyxin mannose tellurite (Shimada et al., 1990) permits the differentiation between *V. cholerae* serogroups O1 and non-O1 based on the ability of the O1 type to ferment mannose. However, this character is variable, since many non-O1 serogroups are also capable of fermenting mannose, especially the pathogenic O139. Other media have also been developed with variable results (Adzhieva, 2000; De et al., 1977; Ozsan

and Mercangoz, 1980; Salles et al., 1976; Tamura et al., 1971).

Bromothymol blue teepol (BTBT) agar (Sakazaki, 1972) is a TCBS-based medium for *V. parahaemolyticus* in which bile salts are replaced by teepol. This medium does not allow a clear observation of the Kanagawa phenomenon (hemolysis halo) of pathogenic strains; therefore, a modified BTBT medium was devised that permitted a better production of hemolysins by pathogenic strains, thus allowing isolation of *V. parahaemolyticus* and identification of the Kanagawa phenomenon in a single plate (Honda et al., 1982). Wagatsuma agar (Miyamoto et al., 1969) is the standard medium for detection of hemolysis of *V. parahaemolyticus* pathogenic strains. The hemolysis halo is detected in this medium only if it is prepared with defibrinated human or rabbit blood, and with a NaCl concentration of no less than 5%.

Chromogenic agar (CHROMagar Vibrio) (Hara-Kudo et al., 2001) is a differential medium that clearly distinguishes colonies of *V. parahaemolyticus* (violet color, Color Plate 1h) from *V. vulnificus* and *V. cholerae*, which are pale blue, and from other vibrios which grow as milky white or colorless colonies. *V. parahaemolyticus* sucrose agar (DePaola and Kaysner, 2004; Entis and Boleszczuk, 1983) is the preferred medium for the enumeration of this species in contaminated food samples through a hydrophobic grid membrane filtration procedure, as recommended by the FDA. *V. parahaemolyticus* and *V. vulnificus* colonies on *V. parahaemolyticus* sucrose agar are green to blue.

Other media used less frequently for isolation or differentiation of *V. parahaemolyticus* are MT agar (Vanderzant and Nickelson, 1972), mannitol salt agar (Carruthers and Kabat, 1976), and TSA plus triphenyltetrazolium chloride (Kourany, 1983).

TCBS proved to be somewhat inhibitory to *V. vulnificus* (West et al., 1982); therefore, *V. vulnificus* agar (Brayton et al., 1983) was developed by introducing some modifications to TCBS. Sucrose was replaced with salicin as the primary source of carbohydrates, because *V. vulnificus*, unlike other vibrios, is capable of fermenting it. Inhibiting agents are tellurite, crystal violet, ox gall, and a high pH. *V. vulnificus* colonies are large, light gray, and transluscent with a dark gray or black center. Some colonies of *V. parahaemolyticus* and of *V. fluvialis* also might have similar characteristics (Brayton et al., 1983), making the use of this medium less effective as a primary differentiation medium. *V. vulnificus* enumeration (VVE) (Miceli et al., 1993) medium has almost the same inhibitory agents as *V. vulnificus* agar; the differentiating agent is 5-bromo-4-chloro-3-indolyl-β-D-galactopyranoside, which permits β-galactosidase-positive colonies to grow with a blue-green color. The level of specificity of VVE is modest, with a mean of 82% reduction of unwanted bacteria; colonial morphology is not sufficient to distinguish *V. vulnificus* from other species (Miceli et al., 1993). VVE is part of a procedure that involves other media and tests that can successfully be used for enumeration of this species in contaminated oysters.

Several media have been formulated on the basis of the ability of *V. vulnificus* to utilize cellobiose and the resistance of vibrios to polymyxins, e.g., CPC agar (Massad and Oliver, 1987), modified CPC agar (Tamplin et al., 1991), cellobiose colistin agar (Hoi et al., 1998), PNCC (Hsu et al., 1998), *V. vulnificus* medium (VVM) (Cerda-Cuellar et al., 2000), and modified VVM (mVVM) (Cerda-Cuellar et al., 2001). Fermentation of cellobiose is also a characteristic common in strains of *Vibrio aestuarianus*, *Vibrio algino-*

lyticus, *Vibrio anguillarum*, *Vibrio campbellii*, *Vibrio harveyi*, and *Vibrio navarrensis* (Alsina and Blanch, 1994); therefore, a further characterization step is required for the correct isolation of *V. vulnificus*. Plating efficiencies varied greatly between media, those media with high concentrations of polymyxins being the most inhibitory for environmental *V. vulnificus* strains (Hoi et al., 1998). Modified CPC and cellobiose colistin media are recommended by the FDA for isolation of *V. vulnificus* from food samples (DePaola and Kaysner, 2004). *V. vulnificus* colonies on both media are 1 to 2 mm in diameter, yellow, round, flat, and opaque on these media.

Other media for *V. vulnificus* are sodium dodecyl sulfate polymyxin B sucrose medium (Kitaura et al., 1983) and direct plating medium (VVE medium) (Miceli et al., 1993). *V. vulnificus* seems to be the species for which more media have been specifically designed. The use, advantages, and limitations of 11 of these media are reviewed by Harwood et al. (2004).

Media for isolation of aquatic animal pathogens include *V. harveyi* agar (Harris et al., 1996), *V. anguillarum* medium (VAM) (Alsina et al., 1994), and a differential medium for *Vibrio proteolyticus* (Muniesa-Perez et al., 1996). *V. harveyi* agar employs the ability of *V. harveyi* to utilize cellobiose and ornithine to differentiate it from other vibrios; *V. harveyi* colonies on this medium appear as small, light-green with dark centers and a yellow halo. It was proven recently that ornithine decarboxylase is not a reliable characteristic to differentiate *V. harveyi* from its close species *V. campbellii* (Gomez-Gil et al., 2004a) and thus, this medium should be reevaluated taking these findings in consideration. VAM selects *V. anguillarum* and related species due to the high salinity of the medium and the resistance to ampicillin, while it differentiates *V. anguillarum* from this related species due to sorbitol fermentation. Colonies of *V. anguillarum* appear yellow, round, flat, and with a yellow halo after 48 or 72 h of incubation. Some strains of *V. fluvialis*, *V. metschnikovii*, and *V. harveyi* form similar colonies, making this medium a presumptive one (Alsina et al., 1994). *V. proteolyticus* medium is also a presumptive medium that needs a specific probe to identify *V. proteolyticus* strains. Similarly to VVM, this medium is based on sorbitol fermentation; thus, it cannot differentiate between the same species, as is the case with VAM.

Marine agar

Marine agar generally allows the growth of very healthy colonies after 1 to 2 days, although some strains may require up to a week (Color Plate 1a and d). A simplified Zobell medium (Carlucci and Pramer, 1957; Jannasch and Jones, 1959) has also been used by many laboratories with acceptable results, although some precipitation may occur after autoclaving (see below). Marine agar is not recommended for further analysis of bacteria (purification, identification, antibiotic susceptibility) because the high concentration of ions may alter results.

Simplified Zobell medium is bactopeptone or polipeptone, 5.0 g; yeast extract, 1.0 g; ferric chloride (FeCl$_2$, 1.2 g/liter, H$_2$O), 1.0 ml; or FePO$_4$, 0.01 g; distilled water, 250 ml; aged filtered seawater, 750 ml; bacteriological agar, 15.0 g. pH has to be adjusted to 7.5 ± 0.1 before adding the agar. Autoclave at 121°C for 20 min and stir before pouring to minimize or evenly distribute precipitations. Do not add the agar if you want a broth. For aged seawater, filter clean seawater through 0.45 μm and keep in a dark container for more than a month.

TCBS agar

TCBS (Oxoid, Difco, Merck) is an ideal medium for the selective isolation and purification of vibrios. Although this medium was originally designed for the isolation of *V. cholerae* and *V. parahaemolyticus* (Kobayashi et al., 1963; Nakanishi, 1963), most vibrios grow to healthy large colonies with many different colonial morphologies (Color Plate 1b). Gram-positive and coliform bacteria are strongly inhibited owing to the presence of bile salts. Vibrios that are able to use sucrose will form yellow colonies (Color Plate 1f), while sucrose-negative strains will form green colonies (Color Plate 1g). Care should be taken, since, in older cultures (more than 48 h), refrigerated plates, or heavily grown plates, the color of the colonies may change; therefore, colonial color has to be registered only in recent and well-isolated colonies. Since TCBS is composed of many ingredients not easily acquired, we recommend obtaining it from commercial sources. Marked variations have been registered between brands (Nicholls et al., 1976). It is not necessary to add extra salt (NaCl), since TCBS has an adequate amount for the growth of the majority of vibrios.

Until now, TCBS was the best selective medium for isolation of vibrios, though some strains of *Staphylococcus, Flavobacterium, Pseudoalteromonas, Streptococcus, Aeromonas,* and *Shewanella* may present slight growth on it (Nicholls et al., 1976); but usually, these colonies can be observed as very small and poorly developed (Color Plate 1c). Some species do not grow on TCBS, e.g., *Vibrio penaeicida,* or grow very poorly, e.g., *Vibrio cincinnatiensis, V. metschnikovii,* and *Grimontia hollisae* (reclassified from *Vibrio hollisae*).

Tryptone soya medium

TSA and TSB (for the agar and broth, respectively) are perhaps the most useful media for after-sample analysis of marine bacteria, provided an adequate concentration of salt (NaCl) is added. It is necessary to obtain a final concentration of NaCl between 1.5 and 2.5%; TSA and TSB already have 0.5% NaCl. Vibrios grow as big creamy colonies (Color Plate 1e and i) after just 24 h at temperatures between 15 and 30°C, depending on the strain under analysis.

Luminescent agar

Many vibrios are luminescent; an appropriate medium that contains glycerol stimulates this characteristic, although some vibrios are capable of showing luminescence in general media (marine agar or TSA) or even in complex media (TCBS). Luminescence can be lost after prolonged incubation or after several passes in the lab. A simple medium contains glycerol (3.0 ml), peptone (5.0 g), yeast extract (3.0 g), agar (15.0 g), aged seawater (750 ml), and distilled water (250 ml) adjusted to pH 7.8 (Schneider and Rheinheimer, 1988). Complex media have also been described that can permit luminescence in vibrios and other bacterial groups (Baumann and Schubert, 1983; Nealson and Hastings, 1979; West and Colwell, 1984). Luminescence should be observed in complete darkness and after a 5-min adaptation period.

Isolation of Vibrios from Clinical Samples

Isolation of vibrios from clinical samples is best accomplished if the stools are plated immediately after collection, preferably within the first 24 h of the onset of symptoms (diarrhea) and before any antimicrobial treatment. Rectal swabs or stool specimens can be plated directly onto TCBS agar. If it is not possible to plate the sample immediately, it can be transported for a short period of time in a closed container, as vibrios are susceptible to desiccation. For longer periods, inoculation in APW or in Cary and Blair transport medium is recommended. Blood agar is also a suitable medium for stool samples so that hemolysis of some vibrios can be observed. Other enteric plating media (e.g., MacConkey) should not be employed, as vibrios might grow poorly in these media. There are no special procedures for collecting vibrios from extraintestinal specimens such as blood, wound, or tissue.

Isolation of Vibrios from Contaminated Food Products

The most important food products that can be contaminated with pathogenic vibrios are raw or undercooked shellfish, especially during months with high water temperatures. *V. parahaemolyticus* is the leading cause of bacterial diarrhea associated with the consumption of these products, but other species are also responsible. *V. cholerae* and *V. vulnificus* are also very important species that have to be monitored in food products, as they can be fatal. Nonfermentative gram-negative bacteria, other vibrios, different enterobacteria, and some gram-positive bacteria are the principal groups that can also be detected in foods and sampling sites (Donovan and van Netten, 1995) and should, therefore, be suppressed with the aid of selective media.

Processing of shellfish samples can be done by following the methodologies of the FDA's *Bacteriological Analytical Manual,* which has an updated online version (http://www.cfsan.fda.gov/~ebam/bam-toc.html).

Samples should be cooled immediately after collection (at 7 to 10°C), but direct contact with ice should be avoided, as vibrios are susceptible to extreme temperatures. Samples could be cut or pooled and blended at high speed for up to 2 min under sterile conditions in a buffer (phosphate-buffered saline or APW). Serial dilutions can be made in phosphate-buffered saline or APW and plated onto TCBS. An overnight enrichment in APW at 35 ± 2°C is preferable; after that, a loopful of the surface pellicle could be streaked onto TCBS to obtain separate colonies. Typical colonies of *V. cholerae* are large (2 to 3 mm), smooth, yellow, and slightly flattened with opaque centers and translucent borders. *V. parahaemolyticus* colonies are large (2 to 3 mm), round, opaque, and green or bluish (Color Plate 1g). *V. vulnificus* colonies are mostly green, although some could be yellow.

Isolation of Vibrios from Environmental Samples

Several commercial media may be used for the isolation of vibrios from marine environments, but Zobell 2216E agar (Zobell, 1941), commercial name Bacto Marine Agar 2216 (MA; Difco 0979), is considered the best medium for primary isolation and quantification of marine heterotrophic bacteria. Some species, e.g., the *Vibrio halioticoli* group and *Vibrio agarivorans*, require supplementation with sodium alginate (0.5%) (Sawabe et al., 1995). TCBS agar is also an excellent medium for the selective isolation of vibrios from environmental sources, but differences in media formulation between manufacturers might be reflected in the enumeration of vibrios.

If the bacterial density of the sample is high, serial dilutions in sterile 2.5% (wt/vol) NaCl are easy, reliable, and economical. To appreciate differences in colony morphology, especially in TCBS agar, it is crucial that colonies grow apart, preferably <100 per plate. Colonies in TCBS that seem similar at 24 h might change after 48 h. For purification and after-sample culture, TSA or TSB supplemented with NaCl to achieve a final concentration of 1.5 to 2.5% is more than adequate. Other general media can be used, provided NaCl is added.

Psychrophilic vibrios, i.e., *V. logei, V. wodanis,* and *V. salmonicida,* will grow poorly at temperatures higher than 20°C; therefore, it is recommended that these strains be grown in Luria-Bertani agar (Difco) supplemented with 1 to 3% (wt/vol) NaCl at 15°C.

Isolation of Vibrios from Aquatic Organisms

Isolation of vibrios from aquatic animals can be a valuable diagnostic tool because vibriosis is a common disease that affects many marine and estuarine animals under culture. Sampling can be done in the same media as those used for environmental samples (marine agar and TCBS), but other media can also be employed, e.g., brain heart infusion agar or TSA. Supplementation with 1 to 3% (wt/vol) NaCl or a seawater base is strongly recommended, especially for primary isolation (Whitman, 2004).

Isolation of vibrios from fish

To isolate vibrios from fish, live or recently dead specimens must be used. If the specimen is alive, it must be euthanized, such as with an overdose of MS222, before sampling. When sampling fish for veterinary reasons, observable external lesions, abnormal internal organs, and anterior kidney are the most common targets. If fish are very small, whole organisms can be used.

Access to the kidney is gained through the abdominal area, as it allows examination of the internal organs. First, the mucus from the fish must be wiped off, and the skin surface must be disinfected with 70% (vol/vol) ethanol. Using a pair of sterile scissors, make three incisions to open the fish. The first incision goes from the anus toward the head, ending before the operculum. The second incision goes from the anus up to the middle lateral line and then should run parallel to the first incision and all the way to the operculum. The last incision joins both cuts at the anterior end. Once these incisions are made, the sidewall is lifted to expose the internal organs. The organs should then be removed, including the swimming bladder, leaving the kidney exposed. To collect the sample, the easiest way is to use a sterile swab and smear it onto the agar plates. Plates should be incubated for 24 to 48 h at 20 to 30°C, depending on the geographical location.

Isolation of vibrios from crustaceans

The main tissue to analyze for the presence of vibrios is the hemolymph, but the hepatopancreas can also provide valuable information. Hemolymph should be collected from live organisms with the aid of a sterile insulin syringe. The bases (coxae) of the pereiopods should be thoroughly cleaned and disinfected with 70% (vol/vol) alcohol, and the needle is inserted a couple of millimeters to reach the ventral sinus. Sufficient hemolymph can be obtained by this method (e.g., 20 to 30 μl from a 5-g shrimp) to plate on TCBS agar. It is best to determine the number of CFU per milliliter of hemolymph, as is not uncommon for healthy crustaceans to carry limited numbers of vibrios in the hemolymph (Gomez-Gil et al., 1998; Scott and Thune, 1986). Vibrios in the hepatopancreas of crustaceans can reach several thousands per gram in healthy organisms; therefore, quantification of bacteria is advisable to detect unusually high numbers. Sampling of the hepatopancreas should be done carefully to avoid contamination with intestinal bacteria, and serial dilutions can be done in sterile saline solution. Incubation should be done as described for the isolation of vibrios from fish.

Sampling of exoskeletal lesions is of little diagnostic value, since it will produce many species and strains that are not necessarily implicated in infection processes but are merely natural inhabitants of the surrounding environment.

Isolation of vibrios from mollusks

After collection, samples must be rinsed to remove mud from the shells. Samples can be transported in a moist chamber at approximately 8°C for several hours before significant changes in the composition of the bacterial flora occur. Bivalves should not be re-immersed in water after collection, since they will continue filtering and thus cause alterations in their bacterial flora. If the bivalves are to be examined for public health purposes, it must be kept in mind that, apart from rinsing off the mud, the organisms examined must have their shells intact; furthermore, there are legal constraints on the time and temperature of storage prior to examination (http://www.cefas.co.uk).

Aseptic opening, briefly, involves inserting a sterile blade (of appropriate size and strength) between the two valves and cutting off the adductor muscle (Whitman, 2004). Care must be taken when severing the adductor muscle to avoid trauma to the internal organs and cross-contamination. Isolation can be made from hemolymph or internal organs (pooled or an individual organ). If the collection is made from organs, the organs are generally homogenized in appropriate diluents (sterile saline solution or sterile peptone water) and then inoculated on the agar plates. Plates should be incubated as previously described.

ENUMERATION

It is often difficult to assess the precise number of bacterial groups in a sample from quantitative data. Commonly, such information is obtained from the use of selective procedures.

Traditional Enumeration Techniques

Plate count

The plate count technique assumes that each viable bacterium present in the sample will produce an individual colony when inoculated in an appropriate culture medium.

This technique can be used to enumerate potential vibrios by inoculating a selective medium with the sample. A 10-fold serial dilution is made of the sample, and 100 μl from each dilution tube is inoculated in duplicate in the selected medium plates. Plates are incubated for 24 h at 25 to 30°C, and the number of colonies is counted. The dilution sample producing between 30 and 300 colonies is the one to be counted, and the final number of CFU is estimated from the arithmetic mean counts of the replicate plates, taking into account the dilution factor multiplied by 10 in order to express the result in CFU/ml (Gerhardt et al., 1994). Serial dilutions can be done in sterile saline solution.

When a selective medium is used to enumerate vibrios or a particular species of *Vibrio*, molecular, immunochemical, or biochemical confirmation of the presumptive identity should be performed (Miceli et al., 1993). Biochemical confirmation using commercial systems such as API 20E or Biolog GN2 systems is commonly performed (Vandenberghe et al., 1999, 2003; Wright et al., 1993), although some authors find these systems inaccurate (Colodner et al., 2004). Immunological confirmation of identity has been performed for some species, such as *V. vulnificus*, using either polyclonal antibodies (Nishibuchi and Seidler, 1985) or monoclonal antibodies (Tamplin et al., 1991). Molecular techniques include the use of PCR analysis (Hara-Kudo et al., 2003; Pfeffer et al., 2003) or DNA probes for colony hybridization (Pfeffer et al., 2003).

Flow cytometry

Flow cytometry has been successfully applied to the enumeration of bacteria. Measurements of each particle or cell are made separately, and results represent the cumulative individual cytometric features. Some flow cytometers are able to physically separate cell subsets on the basis of their cytometric characteristics (Alvarez-Barrientos et al., 2000). The scattered light and fluorescence emission of each particle are collected by detectors and sent to a computer, where the distribution of the population with respect to the different parameters is represented. To obtain additional information, samples can be stained using different fluorochromes. Therefore, this technique has the potential to enumerate vibrios or even a particular *Vibrio* species present in a sample. However, to our knowledge, there are not any publications available on this subject. A single study demonstrated the use of flow cytometry to recognize *V. cholerae* O1 from a mixture of microorganisms with high sensitivity and specificity in a few hours (Alvarado-Aleman et al., 1994).

Molecular Enumeration Techniques

FISH

The application of culture-independent techniques, such as direct extraction of nucleic acids from samples (water, tissue, and sediments), followed by fluorescence in situ hybridization (FISH) of filter-fixed cells with oligonucleotide probes targeting 16S rRNA and subsequent visualization by epifluorescence microscopy, provides the means to identify and quantify marine bacteria, including vibrios (Thompson et al., 2004). This approach contributed to overcoming the problem of the difference in magnitude between direct cell counts and CFU counts on general culture medium (Azam, 2001). FISH was used to identify and quantify the abundance of vibrios in samples of pelagic bacteria from the North Sea. The probes used were cluster and genus specific. Samples were collected all year round; this fraction (gamma proteobacteria) was relatively constant (6 to 9%), and vibrios never surpassed 1% (Eilers et al., 2000).

Changes in natural bacterial and viral assemblages were studied in seawater mesocosms manipulated with the addition of inorganic (nitrate plus phosphate) and inorganic plus organic (glucose) nutrients. Based on FISH, the major bacterial response was identified as an increase in the population of gamma proteobacteria. A specific FISH probe, designed from a band sequence affiliated with *Vibrio splendidus*, linked a large-celled bacterial morphotype to the denaturing gradient-gel bands dominating in glucose-amended mesocosms (Oevreas et al., 2003).

Colony hybridization

Colony hybridization technique is not exclusively molecular; rather, it is a combination of plate count and confirmation of the identity of the colony through DNA hybridization. Most hybridization uses species-specific probes based on variable regions of the 16S rRNA; however, not all probes are sufficiently specific (Thompson et al., 2004).

This technique has been commonly used for confirmation of enumeration of *V. vulnificus*, targeting a portion of the cytolysin gene labeled with alkaline phosphatase covalently linked to the DNA (Wright et al., 1993); a hemolysin probe has also been used (Pfeffer et al., 2003). The two latter works focused on the ecology of this bacterium in different study areas, and both probes worked well.

Colony hybridization has also been used to confirm the presence and number of total and pathogenic *V. parahaemolyticus* (DePaola et al., 2000, 2003), targeting the genes *tl* (total *V. parahaemolyticus*) with a probe labeled either with alkaline phosphatase or with digoxigenin, *tdh* and *trh* (potentially pathogenic *V. parahaemolyticus*), with digoxigenin-labeled probes. The identity of isolates hybridizing with either *tdh* or *trh* probes was further confirmed by screening them for the production of Tdh by the Kanagawa phenomenon or by testing them for the somatic (O) serotype (DePaola et al., 2003).

Colony hybridization was developed to isolate and identify *G. hollisae* from seafood samples. The suggested protocol is to carry out an initial *toxR*-specific reaction, followed by isolation on differential medium and hybridization with both *htpG*- and *toxR*-specific probes (Vuddhakul, 2000).

Real-time PCR

In real-time PCR, reactions are characterized by the point in time during cycling when amplification of a target is first detected rather than the amount of target accumulated after a fixed number of cycles. The real-time PCR system is based on the detection and quantification of a fluorescent reporter. The signal increases in direct proportion to the amount of PCR product in the reaction. The higher the starting copy number of nucleic acid target, the sooner a significant increase in fluorescence will be observed (http://dorakmt.tripod.com/genetics/realtime.html). Normally, fixed fluorescence threshold is set above the baseline, and the parameter C_T (threshold cycle) is defined as the cycle number at which fluorescence emission exceeds the fixed threshold. Normally, a standard curve is generated by plotting the log of the initial target copy number for a set of standards versus the C_T values obtained during real-time PCR. Cycle thresholds of unknowns are then compared with the standard curve to determine the initial target number in a given reaction. There are three main fluorescence-monitoring systems.

Hydrolysis probes (TaqMan probes) seem to be the most commonly used when real-time PCR is applied to vibrios such as *V. cholerae* (Lyon, 2001) and *V. vulnificus* (Campbell and Wright, 2003). A TaqMan assay for quantification of *V. vulnificus*, targeting the *toxR* gene, can detect as few as 10 microbes per ml, in both seawater and oyster samples (Takahashi et al., 2005). Another protocol for *V. vulnificus* uses SYBR green (DNA-binding dye) for the detection of a gene specific to *V. vulnificus* (Panicker et al., 2004). Real-time PCR has also been used to enumerate *V. parahaemolyticus* from oysters by comparing different matrices to overcome the facts that mantle fluid seems to be a PCR inhibitor and that it is difficult to work with internal organs and tissues (Kauffman et al., 2004). To study the ecology of vibrios in Barnegat Bay, N.J., quantitative PCR with primers specific for the genus *Vibrio* was combined with separation and quantification of amplicons by constant denaturant capillary electrophoresis. These populations were then used for identification through cloning and sequencing of the 16S rRNA genes (Thompson et al., 2004). Use of this technique is still not widespread, but it is becoming more common owing to the advantages it presents. However, some investigators argue that this method is not accurate enough for situations where there is a need to enumerate live cells, since it may be detecting nonviable DNA. Alternative methodologies include real-time RT-PCR for detecting RNA, usually transcripts of genes specific for the bacterium under study (Fischer et al., 2002).

PRESERVATION

Temporal or Short-Term Preservation

Vibrios can be preserved in agar slants for several weeks, or up to 5 months under ideal conditions. Inoculation of a fresh strain in almost any general medium with the addition of NaCl is sufficient. Good results have been obtained with TSA plus 2.0% (wt/vol) NaCl slants covered with sterile mineral oil after the strain is fully grown (usually in less than 24 h); minimal media are also recommended because the metabolic rate of the organisms is reduced and the transfers can be prolonged. The slants can be inoculated on the surface and/or stabbed deep in the agar (properly called slants and stabs, respectively). Storage of slants is preferred inside a refrigerator at 5 to 8°C or at room temperature, but dehydration, light, and temperature variations affect the cultures negatively (West and Colwell, 1984). Serious disadvantages of this subculturing method are the risk of contamination, genetic variation, loss of culture, storage

space, and mislabeling (Gherna, 1994; Schneider and Rheinheimer, 1988).

Long-Term Preservation

Freeze-drying

Most vibrios (except *V. ezurae*, *V. gallicus*, *V. pectenicida*, *V. penaeicida*, *V. salmonicida*, and *V. tapetis*) tolerate the freeze-drying process very well. Coincidentally, these species are also difficult to grow on any culture media. Ampoules containing freeze-dried cultures prepared nearly 30 years ago have yielded viable and healthy colonies on TSA. Normally, these ampoules are filled with 0.01 g of bacterial culture previously suspended in 0.5 ml of cryoprotectant mix (horse serum–D-glucose–nutrient broth–MilliQ water, 3:0.3:0.3:1).

Cryopreservation

Vibrio strains may be maintained viably in cryopreservation at −70 to −80°C for years. The cryopreservation methodology described by Gherna (1994) has been routinely used in our laboratory for more than 10 years without any noticeable reduction in viability. A fresh bacterial culture is mixed with a cryprotectant, usually 15% (vol/vol) glycerol, to make a dense suspension. A cryovial with glass beads is filled with the suspension, thoroughly shaken, and left standing for some minutes; the suspension is removed from the vial with a Pasteur pipette, leaving only the soaked beads. The vial can be immediately put into the freezer.

To recover the strain, one or two beads are removed from the vial and placed in TSB plus 2.0% (wt/vol) NaCl or streaked directly onto TSA plus 2.0% (wt/vol) NaCl (Color Plate 1i). Sometimes, some difficult-to-grow strains can be recovered in marine broth or agar; it is advised to employ both, since some strains prefer solid rather than liquid media. Often, vibrios recovered directly from cryopreservation require at least two passes to fully recuperate and longer incubation periods.

Culture Collections

Many culture collections have vibrios deposited; perhaps the most complete is the Collection of the Laboratory for Microbiology of the University of Ghent, Belgium, part of the Belgian Co-ordinated Collections of Micro-organisms, where all the type strains and hundreds of isolates are preserved either freeze-dried or cryopreserved at −80°C (http://www.belspo.be/bccm/index.htm). The Collection of Aquatic Important Microorganisms also houses the majority of type strains of the *Vibrionaceae* and more than a thousand strains, mainly from cultured aquatic organisms and aquaculture systems (http://www.ciad.mx/caim). The Spanish Collection of Type Cultures (Colección Española de Cultivos Tipo; http://www.cect.org) also has a considerable number of type strains and environmental isolates.

REFERENCES

Adzhieva, A. A. 2000. A dried nutrient medium for the isolation of *Vibrio cholerae. Zh. Mikrobiol. Epidemiol. Immunobiol.* (Mar-Apr):29–31.

Alsina, M., and A. R. Blanch. 1994. A set of keys for biochemical identification of environmental *Vibrio* species. *J. Appl. Bacteriol.* 76:79–85.

Alsina, M., J. Martinez-Picado, J. Jofre, and A. R. Blanch. 1994. A medium for presumptive identification of *Vibrio anguillarum. Appl. Environ. Microbiol.* 60:1681–1683.

Alvarado-Aleman, F. J., J. Kumate-Rodríguez, J. Sepulveda-Amor, and C. Wong-Arámbula. 1994. Identification of *Vibrio cholera* O1 by flow cytometry. *Rev. Latinoamer. Microbiol.* 36:283–293.

Alvarez-Barrientos, A., J. Arroyo, R. Canton, C. Nombela, and M. Sanchez-Perez. 2000. Applications of flow cytometry to clinical microbiology. *Clin. Microbiol. Rev.* 13:167–195.

Azam, F. 2001. Introduction, history, and overview: the methods to our madness. *Methods Microbiol.* 30:1–12.

Baumann, P., and R. H. W. Schubert. 1983. *Vibrionaceae,* p. 516–550. *In* N. R. Krieg and J. G. Holt (ed.), *Bergey's Manual of Systematic Bacteriology.* Lippincott Williams & Wilkins, Baltimore, Md.

Brayton, P. R., P. A. West, E. Russek, and R. R. Colwell. 1983. New selective plating medium for isolation of *Vibrio vulnificus* biogroup 1. *J. Clin. Microbiol.* 17:1039–1044.

Campbell, M. S., and A. C. Wright. 2003. Real-time PCR analysis of *Vibrio vulnificus* from oysters. *Appl. Environ. Microbiol.* 69:7137–7144.

Carlucci, A. F., and D. Pramer. 1957. Factors influencing the plate method for determining abundance of bacteria in sea water. *Proc. Soc. Exp. Biol. Med.* 96:392–394.

Carruthers, M. M., and W. J. Kabat. 1976. Isolation of *Vibrio parahaemolyticus* from fecal specimens on mannitol salt agar. *J. Clin. Microbiol.* 4:175–179.

Cerda-Cuellar, M., J. Jofre, and A. R. Blanch. 2000. A selective medium and a specific probe for detection of *Vibrio vulnificus. Appl. Environ. Microbiol.* 66:855–859.

Cerda-Cuellar, M., L. Permin, J. L. Larsen, and A. R. Blanch. 2001. Comparison of selective media for the detection of *Vibrio vulnificus* in environmental samples. *J. Appl. Microbiol.* 91:322–327.

Chatterjee, B. D., P. K. De, and T. Sen. 1977. Sucrose teepol tellurite agar: a new selective indicator medium for isolation of *Vibrio* species. *J. Infect. Dis.* 135:654–658.

Chun, D., J. K. Chung, and S. Y. Seol. 1974. Enrichment of *Vibrio parahaemolyticus* in a simple medium. *Appl. Microbiol.* 27:1124–1126.

Colodner, R., R. Raz, I. Meir, T. Lazarovich, L. Lerner, J. Kopelowitz, Y. Keness, W. Sakran, S. Ken-Dror, and N. Bisharat. 2004. Identification of the emerging pathogen *Vibrio vulnificus* biotype 3 by commercially available phenotypic methods. *J. Clin. Microbiol.* 42:4137–4140.

De, S. P., D. Sen, P. C. De, A. Ghosh, and S. C. Pal. 1977. A simple selective medium for isolation of vibrios with particular reference to *Vibrio parahaemolyticus. Indian J. Med. Res.* 66:398–399.

DePaola, A., and C. A. Kaysner. 2004. *Vibrio.* [Online.] *In Bacteriological Analytical Manual,* Food and Drug Administration, Rockville, Md. http://www.cfsan.fda.gov/~ebam/bam-9.html.

DePaola, A., C. A. Kaysner, J. Bowers, and D. W. Cook. 2000. Environmental investigations of *Vibrio parahaemolyticus* in oysters after outbreaks in Washington, Texas, and New York (1997 and 1998). *Appl. Environ. Microbiol.* 66:4649–4654.

DePaola, A., J. L. Nordstrom, J. C. Bowers, J. G. Wells, and D. W. Cook. 2003. Seasonal abundance of total and pathogenic *Vibrio parahaemolyticus* in Alabama oysters. *Appl. Environ. Microbiol.* 69:1521–1526.

Donovan, T. J., and P. van Netten. 1995. Culture media for the isolation and enumeration of pathogenic *Vibrio* species in foods and environmental samples. *Int. J. Food Microbiol.* 26:77–91.

Eilers, H., J. Pernthaler, F. O. Glockner, and R. Amann. 2000. Culturability and in situ abundance of pelagic bacteria from the North Sea. *Appl. Environ. Microbiol.* 66:3044–3051.

Entis, P., and P. Boleszczuk. 1983. Overnight enumeration of *Vibrio parahaemolyticus* in seafood by hydrophobic grid membrane filtration. *J. Food Prot.* 46:783–786.

Farmer, J. J., J. M. Janda, and K. Birkhead. 2003. *Vibrio,* p. 706–718. *In* P. R. Murray (ed.), *Manual of Clinical Microbiology,* 8th ed. ASM Press, Washington, D.C.

Fischer, M., D. Hervio, S. C. R. Loaec, and M. Pommepuy. 2002. Detection of cytotoxin-hemolysin MRNA in non-culturable populations of environmental and clinical *Vibrio vulnificus* strains in artificial seawater. *Appl. Environ. Microbiol.* 68:5641–5646.

Furniss, A. L., J. V. Lee, and T. J. Donovan. 1978. *The Vibrios.* Monograph series. Her Majesty's Stationery Office, London, England.

Gerhardt, P., R. G. E. Murray, W. A. Wood, and N. R. Krieg (ed.). 1994. *Methods for General and Molecular Bacteriology.* ASM Press, Washington, D.C.

Gherna, L. R. 1994. Culture preservation, p. 278–292. *In* P. Gerhardt, R. G. E. Murray, W. A. Wood, and N. R. Krieg (ed.), *Methods for General and Molecular Bacteriology.* ASM Press, Washington, D.C.

Gilmour, A., M. C. Allan, and M. F. McCallum. 1976. The unsuitability of high pH media for the selection of marine *Vibrio* species. *Aquaculture* 7:81–87.

Gomez-Gil, B., S. Soto-Rodriguez, A. Garcia-Gasca, A. Roque, R. Vazquez-Juarez, F. L. Thompson, and J. Swings. 2004a. Molecular characterization of *V. harveyi* related isolates associated with diseased aquatic organisms. *Microbiology* 150:1769–1777.

Gomez-Gil, B., F. L. Thompson, C. C. Thompson, A. Garcia-Gasca, A. Roque, and J. Swings. 2004b. *Vibrio hispanicus* sp. nov., isolated from *Artemia* sp. and sea water in Spain. *Int. J. Syst. Evol. Microbiol.* 54:261–265.

Gomez-Gil, B., L. Tron-Mayen, A. Roque, J. F. Turnbull, V. Inglis, and A. L. Guerra-Flores. 1998. Species of *Vibrio* isolated from hepatopancreas, haemolymph and digestive tract of a population of healthy juvenile *Penaeus vannamei. Aquaculture* 163:1–9.

Hara-Kudo, Y., Y. Kasuga, A. Kiuchi, T. Horisaka, T. Kawasumi, and S. Kumagai. 2003. Increased sensitivity in PCR detection of tdh-positive *Vibrio parahaemolyticus* in seafood with purified template DNA. *J. Food Prot.* 66:1675–1680.

Hara-Kudo, Y., T. Nishina, H. Nakagawa, H. Konuma, J. Hasegawa, and S. Kumagai. 2001. Improved method for detection of *Vibrio parahaemolyticus* in seafood. *Appl. Environ. Microbiol.* 67:5819–5823.

Harris, L., L. Owens, and S. Smith. 1996. A selective and differential medium for *Vibrio harveyi*. *Appl. Environ. Microbiol.* **62:** 3548–3550.

Harwood, V. J., J. P. Ghandi, and A. C. Wright. 2004. Methods for isolation and confirmation of *Vibrio vulnificus* from oysters and environmental sources: a review. *J. Microbiol. Methods* **62:**3548–3550.

Hoi, L., I. Dalsgaard, and A. Dalsgaard. 1998. Improved isolation of *Vibrio vulnificus* from seawater and sediment with cellobiose-colistin agar. *Appl. Environ. Microbiol.* **64:**1721–1724.

Holt, J. G., and N. R. Krieg. 1994. Enrichment and isolation, p. 179–215. *In* P. Gerhardt, R. G. E. Murray, W. A. Wood, and N. R. Krieg (ed.), *Methods for General and Molecular Bacteriology*. ASM Press, Washington, D.C.

Honda, T., C. Sornchai, Y. Takeda, and T. Miwatani. 1982. Immunological detection of the Kanagawa phenomenon of *Vibrio parahaemolyticus* on modified selective media. *J. Clin. Microbiol.* **16:**734–736.

Hsu, W. Y., C. I. Wei, and M. L. Tamplin. 1998. Enhanced broth media for selective growth of *Vibrio vulnificus*. *Appl. Environ. Microbiol.* **64:**2701–2704.

Jameson, J. E., and N. W. Emberley. 1958. Teepol in substitution for bile salts. *J. Gen. Microbiol.* **18:**1–238.

Jannasch, H. W., and G. E. Jones. 1959. Bacterial populations in sea water as determined by different methods of enumeration. *Limnol. Oceanog.* **4:**128–139.

Kauffman, G. E., G. M. Blackstone, M. C. Vickery, A. K. Bej, J. Bowers, M. D. Bowen, and R. F. Meyer. 2004. Real-time PCR quantification of *Vibrio parahaemolyticus* in oysters using an alternative matrix. *J. Food Prot.* **67:**2424–2429.

Kitaura, T., S. Doke, I. Azuma, M. Imaida, K. Miyano, K. Harada, and E. Yabuuchi. 1983. Halo production by sulfatase activity of *V. vulnificus* and *V. cholerae* O1 on a new selective sodium dodecyl sulfate-containing agar medium: a screening marker in environmental surveillance. *FEMS Microbiol. Lett.* **17:**205–209. [Online.] http://www.sciencedirect.com/.

Kobayashi, T., S. Enomoto, R. Sakazaki, and S. Kuwahara. 1963. A new selective isolation medium for the *Vibrio* group; on a modified Nakanishi's medium (TCBS agar medium). *Nippon Saikingaku Zasshi* **18:**387–392.

Kourany, M. 1983. Medium for isolation and differentiation of *Vibrio parahaemolyticus* and *Vibrio alginolyticus*. *Appl. Environ. Microbiol.* **45:**310–312.

Massad, G., and J. D. Oliver. 1987. New selective and differential medium for *Vibrio cholerae* and *Vibrio vulnificus*. *Appl. Environ. Microbiol.* **53:**2262–2264.

Miceli, G. A., W. D. Watkins, and S. R. Rippey. 1993. Direct plating procedure for enumerating *Vibrio vulnificus* in oysters (*Crassostrea virginica*). *Appl. Environ. Microbiol.* **59:**3519–3524.

Miyamoto, Y., T. Kato, Y. Obara, S. Akiyama, K. Takizawa, and S. Yamai. 1969. In vitro hemolytic characteristic of *Vibrio parahaemolyticus*: its close correlation with human pathogenicity. *J. Bacteriol.* **100:**1147–1149.

Monsur, K. A. 1961. A highly selective gelatin-taurocholate-tellurite medium for the isolation of *Vibrio cholerae*. *Trans. R. Soc. Trop. Med. Hyg.* **55:**440–442.

Monsur, K. A. 1963. Bacteriological diagnosis of cholera under field conditions. *WHO Bull.* **28:**387–389.

Muniesa-Perez, M., J. Jofre, and A. R. Blanch. 1996. Identification of *Vibrio proteolyticus* with a differential medium and a specific probe. *Appl. Environ. Microbiol.* **62:**2673–2675.

Nakanishi, H., and M. Murase. 1973. Enumeration of *Vibrio parahaemolyticus* in raw fish meat, p. 117–121. *In International Symposium on Vibrio parahaemolyticus*. Tokyo, Japan.

Nakanishi, Y. 1963. An isolation agar medium for cholerae and enteropathogenic halophilic vibrios. *Modern Media* **9:**246.

Nealson, K. H., and J. W. Hastings. 1979. Bacterial bioluminescence: its control and ecological significance. *Microbiol. Rev.* **43:**496–518.

Nicholls, K. M., J. V. Lee, and T. J. Donovan. 1976. An evaluation of commercial thiosulphate citrate bile salt sucrose agar (TCBS). *J. Appl. Bacteriol.* **41:**265–269.

Nishibuchi, M., N. C. Roberts, H. B. Bradford, Jr., and R. J. Seidler. 1983. Broth medium for enrichment of *Vibrio fluvialis* from the environment. *Appl. Environ. Microbiol.* **46:**425–429.

Nishibuchi, M., and R. J. Seidler. 1985. Rapid microimmunodiffusion method with species-specific antiserum raised to purified antigen for identification of *Vibrio vulnificus*. *J. Clin. Microbiol.* **21:**102–107.

O'Brien, M., and R. Colwell. 1985. Modified taurocholate-tellurite-gelatin agar for improved differentiation of *Vibrio* species. *J. Clin. Microbiol.* **22:**1011–1013.

Oevreas, L., D. Bourne, R. A. Sandaa, E. O. Casamayor, S. Benlloch, V. Goddard, G. Smerdon, M. Heldal, and T. F. Thingstad. 2003. Response of bacterial and viral communities to nutrient manipulations in seawater mesocosms. *Aquat. Microb. Ecol.* **31:**109–121.

Ozsan, K., and F. Mercangoz. 1980. New media for the isolation of *Vibrio cholerae*. *Zentbl. Bakteriol A* **247:**71–73.

Pal, S. C., G. V. Murty, C. G. Pandit, D. K. Murty, and J. B. Shrivastav. 1967. A comparative study of enrichment media in the bacteriological diagnosis of cholera. *Indian J. Med. Res.* **55:** 318–324.

Panicker, G., M. L. Myers, and A. K. Bej. 2004. Rapid detection of *Vibrio vulnificus* in shellfish and Gulf of Mexico water by real-time PCR. *Appl. Environ. Microbiol.* **70:**498–507.

Pfeffer, C. S., M. F. Hite, and J. D. Oliver. 2003. Ecology of *Vibrio vulnificus* in estuarine waters of eastern North Carolina. *Appl. Environ. Microbiol.* **69:**3526–3531.

Reichelt, J. L., and P. Baumann. 1974. Effect of sodium chloride on growth of heterotrophic marine bacteria. *Arch. Microbiol.* **97:**329–345.

Sakazaki, R. 1965. Vibrio parahaemolyticus, *Isolation and Identification*. Nihon Eiyo Kagaku Co. Press, Tokyo, Japan.

Sakazaki, R. 1972. *BTB Teepol Agar in Media for Bacteriological Examination*. Kindai Igaku Co., Tokyo, Japan.

Salles, C. A., S. Voros, E. C. Marbell, and L. Amenuvor. 1976. Colony morphology of *Vibrio cholerae* on SV medium. *J. Appl. Bacteriol.* **40:**213–216.

Sanjuan, E., and C. Amaro. 2004. Protocol for specific isolation of virulent strains of *Vibrio vulnificus* serovar E (biotype 2) from environmental samples. *Appl. Environ. Microbiol.* **70:**7024–7032.

Sawabe, T., Y. Oda, Y. Shiomi, and Y. Ezura. 1995. Alginate degradation by bacteria isolated from the gut of sea urchins and abalones. *Microb. Ecol.* **30:**192–202.

Schneider, J., and G. Rheinheimer. 1988. Isolation methods, p. 73–94. *In* B. Austin (ed.), *Methods in Aquatic Bacteriology*. John Wiley & Sons, Chichester, England.

Scott, J., and R. L. Thune. 1986. Bacterial flora of hemolymph from red swamp crawfish, *Procambarus clarkii* (Girard), from commercial ponds. *Aquaculture* **58:**161–165.

Shimada, T., R. Sakazaki, S. Fujimura, K. Niwano, M. Mishina, and K. Takizawa. 1990. A new selective, differential agar medium for isolation of *Vibrio cholerae* O1: PMT (polymyxin-mannose-tellurite) agar. *Jpn. J. Med. Sci. Biol.* **43:**37–41.

Sogaard, H. 1982. The pharmacodynamics of polymyxin antibiotics with special reference to drug resistance liability. *J. Vet. Pharmacol. Ther.* **5:**219–231.

Spira, W. M. 1984. Tactics for detecting pathogenic vibrios in the environment, p. 251–268. *In* R. R. Colwell (ed.), *Vibrios in the Environment*. John Wiley & Sons, New York, N.Y.

Takahashi, H., Y. Hara-Kudo, J. Miyasaka, S. Kumagai, and H. Konuma. 2005. Development of a quantitative real-time polymerase chain reaction targeted to the *toxR* for detection of *Vibrio vulnificus*. *J. Microbiol. Methods* **61**:77–85.

Tamplin, M. L., A. L. Martin, A. D. Ruple, D. W. Cook, and C. W. Kaspar. 1991. Enzyme immunoassay for identification of *Vibrio vulnificus* in seawater, sediment, and oysters. *Appl. Environ. Microbiol.* **57**:1235–1240.

Tamura, K., S. Shimada, and L. M. Prescott. 1971. Vibrio agar: a new plating medium for isolation of *Vibrio cholerae*. *Jpn. J. Med. Sci. Biol.* **24**:125–127.

Thompson, F. L., T. Iida, and J. Swings. 2004. Biodiversity of vibrios. *Microbiol. Mol. Biol. Rev.* **68**:403–431.

Vandenberghe, J., F. L. Thompson, B. Gomez-Gil, and J. Swings. 2003. Phenotypic diversity amongst *Vibrio* isolates from marine aquaculture systems. *Aquaculture* **219**:9–20.

Vandenberghe, J., L. Verdonck, A. R. Robles, G. Rivera, A. Bolland, M. Balladares, B. Gomez-Gil, J. Calderon, P. Sorgeloos, and J. Swings. 1999. Vibrios associated with *Litopenaeus vannamei* larvae, postlarvae, broodstock, and hatchery probionts. *Appl. Environ. Microbiol.* **65**:2592–2597.

Vanderzant, C., and R. Nickelson. 1972. Procedure for isolation and enumeration of *Vibrio parahaemolyticus*. *Appl. Microbiol.* **23**:26–33.

Vuddhakul, V., T. Nakai, C. Matsumoto, T. Oh, T. Nishino, C. H. Chen, M. Nishibuchi, and J. Okuda. 2000. Analysis of *gyrB* and *toxR* gene sequences of *Vibrio hollisae* and development of *gyrB*- and *toxR*-targeted PCR methods for isolation of *V. hollisae* from the environment and its identification. *Appl. Environ. Microbiol.* **66**:3506–3514.

West, P. A., and R. R. Colwell. 1984. Identification and classification of *Vibrionaceae*—an overview, p. 285–363. *In* R. R. Colwell (ed.), *Vibrios in the Environment*. John Wiley & Sons, New York, N.Y.

West, P. A., E. Russek, P. R. Brayton, and R. R. Colwell. 1982. Statistical evaluation of a quality control method for isolation of pathogenic *Vibrio* species on selected thiosulfate-citrate-bile salts-sucrose agars. *J. Clin. Microbiol.* **16**:1110–1116.

Whitman, K. A. 2004. *Finfish and Shellfish Bacteriology Manual. Techniques and Procedures.* Iowa State Press, Blackwell Publishing Company, Ames, Iowa.

Wright, A. C., G. A. Miceli, W. L. Landry, J. B. Christy, W. D. Watkins, and J. G. Morris, Jr. 1993. Rapid identification of *Vibrio vulnificus* on nonselective media with an alkaline phosphatase-labeled oligonucleotide probe. *Appl. Environ. Microbiol.* **59**:541–546.

Zobell, C. E. 1941. Studies on marine bacteria. 1. The cultural requirements of heterotrophic aerobes. *J. Mar. Res.* **4**:42–75.

III. CLASSIFICATION AND PHYLOGENY

The Biology of Vibrios
Edited by F. L. Thompson et al.
© 2006 ASM Press, Washington, D.C.

Chapter 3

Taxonomy of the Vibrios

Fabiano L. Thompson and Jean Swings

INTRODUCTION

Prokaryotic taxonomy deals with the classification (i.e., taxa description), identification (i.e., isolate allocation) and nomenclature of microorganisms (Vandamme et al., 1996). This science has provided different fields of microbiology with a sound framework that is stable, predictable, informative, and objective. The beginning of the taxonomy of vibrios can be traced back to the very beginning of prokaryotic taxonomy itself. Vibrios were one of the first groups to be recognized in nature and were further taxonomically described (Pacini, 1854). In 1883, Robert Koch isolated *Vibrio cholerae* strains for the first time during cholera outbreaks in India and Egypt and had enough evidence to suggest this microbe as the causative agent of cholera, one of the most serious diseases at that time. Initially, the taxonomy of vibrios was based on very few morphological features, including flagellation, morphology and curvature of the cells, and cultural aspects. This approach led to the description of many new, poorly characterized species. Of the 34 *Vibrio* species listed in the 7th edition of *Bergey's Manual of Determinative Bacteriology* (Breed et al., 1957) under the family *Spirillaceae*, only *V. cholerae* (*V. comma*) and *V. metschnikovii* were retained in the Approved List of Bacterial Names (Skerman et al., 1980). All the other species were later reclassified into other genera, e.g., *Campylobacter* (*C. fetus*, *C. jejuni*, *C. sputorum*), *Comamonas* (*C. terrigena*), *Pseudomonas* (*P. fluorescens*), or were no longer accepted as validly described species. The genus *Photobacterium* comprised only *P. phosphoreum* and was allocated into the genus *Bacterium* of the family *Bacteriaceae* (Breed et al., 1957).

The proposal for polyphasic taxonomy ~35 years ago (Colwell, 1970) and the advent of new technologies—particularly DNA-DNA hybridization; DNA nucleotide composition; measurements of amino acid sequence differences by microcomplement fixation; and screening of phenotypic features, including various carbohydrates, lipids, proteins, amino acids, acids, and alcohols as sources of carbon and/or energy, enzyme activity, salt tolerance, luminescence, growth at different temperatures and salinities, antibiograms, and morphological features—have provided a firm basis for the current taxonomy of *Vibrio*. Extensive screenings of fresh isolates following this approach resulted in the description of several new species, e.g., *V. campbellii* (Baumann et al., 1971), *V. nereis* (Baumann et al., 1971), *V. fluvialis* (Lee et al., 1981), and *V. parahaemolyticus* (Fujino et al., 1974). The DNA-DNA hybridization studies of Baumann and coworkers underpinned the taxonomy of vibrios (Baumann and Schubert, 1984; Baumann et al., 1984; Reichelt et al., 1976). They found a core group of related vibrios, i.e., the *V. harveyi* group consisting of *V. harveyi*, *V. campbellii*, *V. natriegens*, *V. alginolyticus* and *V. parahaemolyticus*. *V. harveyi* and *V. campbellii* were found to have 61 to 74% DNA-DNA similarity, whereas *V. parahaemolyticus* and *V. alginolyticus* had 61 to 67% similarity. Sequencing of molecular chronometers, e.g., 5S and 16S RNA, in the late 1980s revolutionized prokaryotic taxonomy. Today, polyphasic taxonomy operates on the 16S rRNA backbone. Nevertheless, there is a growing need for alternative phylogenetic and identification markers for every prokaryotic group, including vibrios. Additional phylogenetic markers within the ~50 to 100 genes in the bacterial core genome are now being studied in order to complement the phylogenetic picture obtained with 16S rRNA (Charlebois and Doolittle, 2004; Harris et al., 2003; Zeigler, 2003). Ideally, bacterial taxonomy would be performed on the whole core genome sequences, but this seems not yet feasible. The application of multilocus sequence

Fabiano L. Thompson • Microbial Resources Division and Brazilian Collection of Environmental and Industrial Micro-organisms (CBMAI), CPQBA, UNICAMP, Alexandre Caselatto 999, CEP 13140000, Paulínia, Brazil. **Jean Swings** • Laboratory of Microbiology and BCCM/LMG Bacteria Collection, Ghent University, K.L. Ledeganckstraat 35, Ghent 9000, Belgium.

analyses (MLSA) to vibrios is a further step toward this end. We discuss here the historical underpinnings of vibrio taxonomy and the traditional phenotypic basis of the taxonomy of this group. Subsequently, we review the recent improvements in the taxonomy of vibrios that are mainly due to the application of genomic methodologies, including amplified fragment length polymorphism (AFLP), repetitive extragenic palindromic PCR (rep-PCR), DNA-DNA hybridization, and MLSA. We conclude the chapter by discussing the development and application of an electronic prokaryotic taxonomy with vibrios as a prototype.

CURRENT STATUS OF THE TAXONOMY OF VIBRIOS

According to the outline of *Bergey's Manual*, the family *Vibrionaceae* comprises eight genera: *Vibrio* (65 species), *Allomonas* (1 species), *Catenococcus* (1 species), *Enterovibrio* (2 species), *Grimontia* (1 species), *Listonella* (2 species), *Photobacterium* (8 species), and *Salinivibrio* (1 species) (Table 1, Fig. 1). The novel species *Enterovibrio coralii* and *Photobacterium rosenbergii* (Thompson et al., 2005a), *Photobacterium aplysiae* (Seo et al., 2005a), *Photobacterium frigidiphilum* (Seo et al., 2005b), *Photobacterium indicum* (Xie and Yokota, 2004), and *Photobacterium lipolyticum* (Yoon et al., 2005) have been recently proposed. *Catenococcus thiocyclus* TG5-3 (AF139723) (Sorokin, 1992) shares 97.9% 16S rRNA similarity with *V. natriegens* LMG 10935^T (LMG Bacteria Collection, Ghent University, Belgium), while the *Listonella* species are grouped within *Vibrio*, clearly putting in doubt the taxonomic validity of these genera. *Allomonas enterica* (Kalina et al., 1984) and *V. fluvialis* have nearly identical 16S rDNA sequences and phenotypic features (Farmer, 1986). The subcommittee on the taxonomy of *Vibrionaceae* has concluded that *A. enterica* is indeed a synonym of *V. fluvialis* (Farmer, 1986).

There has been a tremendous improvement in the taxonomy of vibrios in the last 3 years, particularly due to the description of new species through the application of genomic techniques. We have described several new *Vibrio* species, mainly in the phylogenetic neighborhood of *V. diazotrophicus* (Gomez-Gil et al., 2004a), *Grimontia hollisae* (Thompson et al., 2002b), *V. harveyi* (Gomez-Gil et al., 2003a), *Vibrio halioticoli* (Hayashi et al., 2003; Sawabe et al., 2004a,b), *Vibrio pelagius* (Thompson et al., 2003d), *Vibrio splendidus* (Faury et al., 2004; Gomez-Gil et al., 2003b; Thompson et al., 2003b,c), and *Vibrio tubiashii* (Ben-Haim et al., 2003; Thompson et al., 2003a).

Newly described species constitute commonly found bacterial strains in the marine environment.

Thompson et al. (2004) proposed, on the basis of concatenated 16S rRNA, *recA*, and *rpoA* gene sequences and phenotypic data, that the family *Vibrionaceae* be split into four new families: *Enterovibrionaceae* (comprising the genera *Enterovibrio* and *Grimontia*), *Photobacteriaceae* (comprising the genus *Photobacterium*) and *Salinivibrionaceae* (comprising the genus *Salinivibrio*) (Table 1). This proposal was intended to facilitate further studies on vibrios. The creation of these three new families resulted in a more compact family, *Vibrionaceae*. Allocation of strains into different families is obtained by 16S rRNA sequences and phenotypic analysis, while allocation of strains into species is currently best achieved by using fluorescent AFLP, rep-PCR, or MLSA gene sequences. Phenotypically, *Salinivibrionaceae* strains are readily distinguishable from all the other vibrios in that they grow at 45°C and in media with 20% (wt/vol) NaCl. *Enterovibrionaceae* strains have the fatty acids 16:0 ω9c and 18:0 ω9c, but *Photobacteriaceae* strains do not. *Photobacteriaceae* strains accumulate poly(3-hydroxybutyrate) and show arginine dihydrolase activity, but *Salinivibrionaceae* strains do not.

The newly proposed family *Vibrionaceae* comprises only the genus *Vibrio*. Different types of phenotypic and molecular data show that *Vibrio* is highly heterogeneous (Thompson et al., 2001a, 2005). We consider only *V. cholerae* and *V. mimicus* as the bona fide members of the genus *Vibrio*. *Vibrio fischeri*, *V. fluvialis*, *V. halioticoli*, *V. harveyi*, *V. splendidus*, and *V. tubiashii* should be split further into several genera. In most of these cases, e.g., within the *V. harveyi* group, there are no conspicuous phenotypic features to warrant the creation of genera on this basis. An exception is the *V. fischeri* group. These organisms are yellow-pigmented and have multiple polar flagella, but are generally unable to grow at 35 to 40°C (Bang et al., 1978; Farmer, 1992). Amino acid sequence divergence analyses of glutamine synthetase and superoxide dismutase also indicated that *V. fischeri* and *Vibrio logei* are apart from the genus *Vibrio* (Baumann et al., 1980, 1983).

PHENETIC BASIS OF THE TAXONOMY OF VIBRIOS

Vibrios are gram negative, usually motile rods, mesophilic, chemo-organotrophic, and have a facultatively fermentative metabolism (Farmer, 1992). Overall, cells are 1 μm in width and 2 to 3 μm in length and motile by at least one polar flagellum. Vib-

Table 1. List of species, place and date of isolation[a]

Species name	Type strain	Place and date of isolation	Source
Enterovibrio coralii	LMG 22228[T]	Magnetic Island (Australia), 2002	Water extract of bleached coral (*Merulina ampliata*)
E. norvegicus	LMG 19839[T]	AARS, Austevoll (Norway), 1997	Gut of turbot larvae (*Scophthalmus maximus*)
Grimontia hollisae	LMG 17719[T]	Maryland (U.S.A.)	Human feces
Photobacterium angustum	LMG 8455[T]	Hawaii (U.S.A.)	Sea water
P. aplysiae	KCTC 12383[T]	Mogiyeo Island, South Sea (Korea), 2003	*Aplysia kurodai* eggs
P. damselae subsp. *damselae*	LMG 7892[T]	U.S.A.	Diseased damsel fish (*Chromis punctipinnis*)
P. eurosenbergii	LMG 22223[T]	Magnetic Island (Australia), 2003	Tissue extract of bleached coral (*Pachyseris speciosa*)
P. frigidiphilum	KCTC 12384[T]	Edison Seamount (Papua New Guinea)	Deep-sea sediment
P. iliopiscarius	LMG 19543[T]	Norway	Gut of fish
P. indicum	ATCC 19614[T]		
P. leiognathi	LMG 4228[T]	Thailand	Leiognathidae fish (family *Leiognathidae*)
P. lipolyticum	KCTC 10560BP[T]	Kaehwa-do (Korea)	Intertidal sediment at the Yellow Sea
P. phosphoreum	LMG 4233[T]	Hawaii (U.S.A.)	Sea water
P. profundum	LMG 19446[T]	Ryukyu Trench (Japan)	Sediment
Salinivibrio cosicola subsp. *costicola*	LMG 11651[T]	Australia	Bacon-curing brine
Vibrio aerogenes	LMG 19650[T]	Nanwan bay (Taiwan)	Sediment of sea-grass bed
V. aestuarianus	LMG 7909[T]	Oregon (U.S.A.)	Oyster
V. agarivorans	LMG 21449[T]	Valencia (Spain)	Sea water
V. alginolyticus	LMG 4409[T]	Japan	Spoiled horse mackerel (*Trachurus trachurus*)
V. anguillarum	LMG 4437[T]	Norway	Diseased cod (*Gadus morhua*)
V. brasiliensis	LMG 20546[T]	LCMM, Florianópolis (Brazil), 1999	Bivalve larvae (*Nodipecten nodosus*)
V. calviensis	LMG 21294[T]	Bay of Calvi (France)	Sea water of the Mediterranean sea
V. campbellii	LMG 11216[T]	Hawaii (U.S.A.)	Sea water
V. chagasii	LMG 21353[T]	AARS, Austevoll (Norway), 1997	Gut of turbot larvae (*Scophthalmus maximus*)
V. cholerae	LMG 21698[T]	Asia	Clinical
V. cincinnatiensis	LMG 7891[T]	Ohio (U.S.A.)	Human blood and cerebrospinal fluid
V. coralliilyticus	LMG 20984[T]	Indian Ocean (near Zanzíbar), 1999	Diseased *Pocillopora damicornis*
V. crassostreae	LMG 22240[T]	IFREMER, La Tremblade (France)	Hemolymph of diseased reared oysters (*Crassostera gigas*)
V. cyclitrophicus	LMG 21359[T]	Washington (U.S.A.)	Creosote-contaminated sediment
V. diabolicus	LMG 19805[T]	East Pacific Rise, 1991	Dorsal integument of polychaete (*Alvinella pompejana*)
V. diazotrophicus	LMG 7893[T]	Nova Scotia (Canada)	Sea urchin (*Strongylocentrotus*)
V. ezurae	LMG 19970[T]	Kanagawa (Japan), 1999	Gut of abalone (*Haliotis diversicolor supertexta*)
V. fischeri	LMG 4414[T]	Massachusetts (U.S.A.), 1933	Dead squid
V. fluvialis	LMG 7894[T]	Bangladesh	Human feces
V. fortis	LMG 21557[T]	Ecuador, 1996	*Litopenaeus vannamei* larvae
V. furnissii	LMG 7910[T]	Japan	Human feces
V. gallicus	LMG 21330[T]	Brest (France), 2001	French abalone *Haliotis tuberculata*
V. gazogenes	LMG 19540[T]	Massachusetts (U.S.A.)	Mud from salt marsh
V. halioticoli	LMG 18542[T]	Kumaishi (Japan), 1991	Gut of abalone (*Haliotis discus hanai*)
V. harveyi	LMG 4044[T]	Massachusetts (U.S.A.), 1935	Dead amphipod (*Talorchestia* sp.)
V. hepatarius	LMG 20362[T]	CENAIM (Ecuador), 2000	Digestive gland of white shrimp (*Litopenueus vannamei*)
V. hispanicus	LMG 13240[T]	Barcelona (Spain), 1990	Culture water
V. ichthyoenteri	LMG 19664[T]	Hiroshima (Japan)	Gut of diseased Japanase flounder (*Paralichtys olivaceus*)
V. kanaloaei	LMG 20539[T]	IFREMER (France), 1998	Diseased oyster larvae (*Ostrea edulis*)
V. lentus	LMG 21034[T]	Valencia (Spain)	Oysters in the Mediterranean coast
V. logei	LMG l9806[T]	U.S.A.	Gut of arctic scallop
V. mediterranei	LMG 11258[T]	Valencia (Spain)	Coastal sea water
V. metschnikovii	LMG 11664[T]	Asia	Diseased fowl
V. mimicus	LMG 7896[T]	North Carolina (U.S.A.)	Infected human ear
V. mytili	LMG 19157[T]	Valencia (Spain)	Bivalve (*Mytilus edulis*)
V. natriegens	LMG 10935[T]	Sapeto Island (U.S.A.)	Salt marsh mud

Continued on following page

Table 1. *Continued*

Species name	Type strain	Place and date of isolation	Source
V. navarrensis	LMG 15976[T]	Villa Franca, Navarra (Spain), 1982	Sewage
V. neonatus	LMG 19972[T]	Kanagawa (Japan), 1999	Gut of abalone (*Haliotis discus discus*)
V. neptunius	LMG 20536[T]	LCMM, Florianópolis (Brazil), 1998	Bivalve larvae (*Nodipecten nodosus*)
V. nereis	LMG 3895[T]	Hawaii (U.S.A.)	Sea water
V. nigripulchritudo	LMG 3896[T]	Hawaii (U.S.A.)	Sea water
V. ordalii	LMG 13544[T]	Washington (U.S.A.), 1973	Diseased coho salmon (*Oncorhynchus rhoddurus*)
V. orientalis	LMG 7897[T]	Yellow Sea (China)	Sea water
V. pacinii	LMG 19999[T]	Dahua (China), 1996	Healthy shrimp larvae (*Penaeus chinensis*)
V. parahaemolyticus	LMG 2850[T]	Japan	Diseased human
V. pectenicida	LMG 19642[T]	Brittany (France), 1991	Diseased bivalve larvae (*Pecten maximus*)
V. pelagius	LMG 3897[T]	Hawaii (U.S.A.)	Sea water
V. penaeicida	LMG 19663[T]	Kagoshima (Japan)	Diseased kuruma prawn (*Penaeus japonicus*)
V. pomeroyi	LMG 20537[T]	LCMM, Florianópolis (Brazil), 1998	Healthy bivalve larvae (*Nodipecten nodosus*)
V. ponticus	CECT 5869[T]	Mediterranean coast (Spain), 1986	Gilthead sea bream (*Sparus aurata*), mussels and seawatcr
V. proteolyticus	LMG 3772[T]	U.S.A.	Intestine of isopod (*Limnoria tipunctala*)
V. rotiferianus	LMG 21460[T]	ARC, Ghent (Belgium), 1999	Rotifer in recirculation system (*Brachionus plicatilis*)
V. ruber	LMG 2l676[T]	Keelung (Taiwan)	Sea water
V. rumoiensis	LMG 20038[T]	Japan	Drain pool of a fish processing plant
V. salmonicida	LMG 14010[T]	Norway	Diseased Atlantic salmon (*Salmo salar*)
V. scophthalmi	LMG 19158[T]	Spain	Turbot juvenile (*Scophthalmus maximus*)
V. splendidus	LMG 19031[T]	North Sea	Marine fish
V. superstes	LMG 21323[T]	Australian coast	Gut of abalone (*Haliotis laevigata* and *H. rubra*)
V. tapetis	LMG 19706[T]	Landeda (France)	Clam (*Tapes philippinarum*)
V. tasmanienis	LMG 20012[T]	MPL (Tasmania)	Atlantic salmon (*Salmo salar*)
V. tubiashii	LMG 10936[T]	Milford, Conn. (U.S.A.)	Hard clam (*Mercenaria mercenaria*)
V. vulnificus	LMG 13545[T]	U.S.A.	Human wound infection
V. wodanis	LMG 21011[T]	Norway, 1988	Salmon suffering of winter ulcer (*Salmo salar*)
V. xuii	LMG 21346[T]	Dahua (China), 1995	Shrimp culture water

[a] Abbreviations: AARS, Austevoll Aquaculture Research Station, Austevoll, Norway; ARC, Artemia Reference Center, Ghent, Belgiun; LCMM, Laboratory for Culture of Marine and Molluscs, Florianópolis, Brazil; CENAIM, National Center for Marine and Aquaculture Research, Guayaquil, Ecuador; IFREMER, French Institute for Exploration of the Sea, Brittany, France; MPL, Mount Pleasant Laboratories in Tasmania. Modified from Thompson et al., 2004.

rios display a wide variation in colony morphology (round to irregular) and color (beige to red) on tryptic soy agar. Colony variation is also a common feature within vibrios (Austin et al., 1996; Hickman et al., 1982). Vibrios are generally able to grow on marine agar and on the selective medium, e.g., TCBS, and are mostly oxidase positive. Most vibrios do not grow at 4°C and in media with 0 or 12% (wt/vol) NaCl. Overall, vibrios require 2 to 3 % (wt/vol) NaCl for optimum growth. Most vibrios utilize D-glucose, dextrin, glycogen, N-acetyl-D-glucosamine, D-fructose, maltose, D-trehalose, methyl pyruvate, L-asparagine aconitate, L-proline, or inosine as the sole carbon source. Most vibrios reduce nitrate, produce acetoin, and are susceptible to the vibriostatic agent 0/129. Most vibrios do not utilize N-acetyl-D-galactosamine, L-erythritol, *m*-inositol, xylitol, α-hydroxybutyric acid, D-saccharic acid, D,L-carnitine, and phenyl ethylamine as sole carbon source. The most abundant fatty acids are 16:1ω7*c* and/or 15 iso 2-OH, 16:0, 18:1ω7*c*, 14:0, 12:0 and 16:0 iso corresponding

to >70% of all fatty acids in most species. Other fatty acids occur in trace amounts. Most vibrios show leucine arylamidase, acid and alkaline phosphatase activity, but not urease, tryptophane deaminase, α-mannosidase, α-fucosidase, and β-glucuronidase. Obviously, there are exceptions in all these phenotypic traits, as can be seen in, for example, Baumann et al. (1984), Farmer (1992), Farmer and Hickman-Brenner (1992), and Thompson et al. (2004).

Fatty acid methyl ester and Biolog analyses were used frequently for the screening of large collections of isolates and strain allocation into genera in the 1990s. However, because several species have indistinguishable fatty acid methyl ester and Biolog profiles, these techniques are of limited value for species identification. We compared the phenotypic identification of vibrios by Biolog (Vandenberghe et al., 2003) versus AFLP identification (Thompson et al., 2001a). Different *Vibrio* species appear within the same Biolog group. For example, strains misidentified as *V. harveyi* by Biolog were later correctly identified

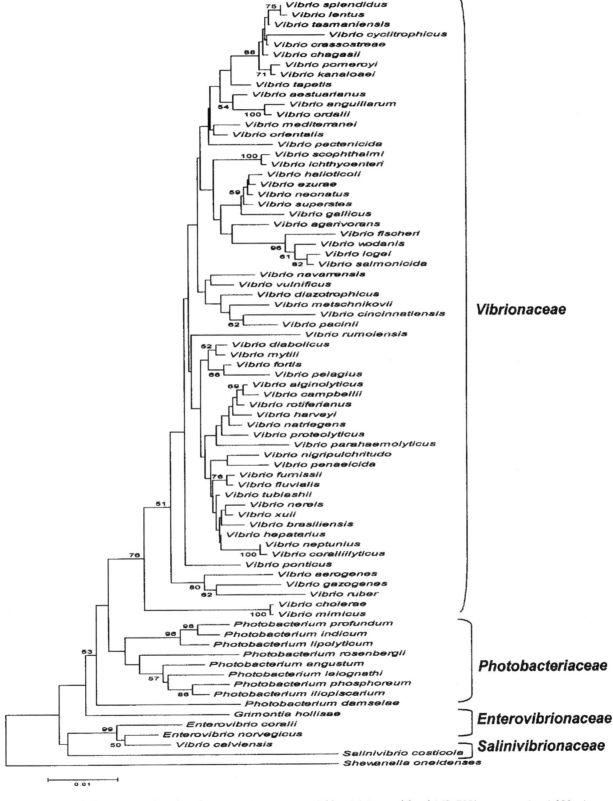

Figure 1. Phylogenetic tree based on the Kimura 2-parameter neighbor-joining model and 16S rRNA sequences (ca. 1,300 nt) of the type strains. Bootstrap values were obtained after 500 repetitions.

as *V. campbellii* or classified as the new species, *V. rotiferianus*, by AFLP and DNA-DNA hybridizations. Indeed, *V. campbellii*, *V. harveyi*, and *V. rotiferianus* have nearly indistinguishable phenotypes. Several strains phenotypically misidentified as *V. harveyi* turned out to be *V. campbellii* (Gomez-Gil et al., 2004b). Strains misidentified by Biolog as *V. campbellii* turned out to be *V. chagasii*, whereas strains believed to be *V. splendidus* were classified into *V. kanaloaei*. On the other hand, many strains identified by AFLP as, e.g., *V. cincinnatiensis*, *V. splendidus*, and *V. tubiashii* correspond to multiple Biolog groups. The phenotypic features of vibrios had a preponderant role in former classification schemes (Baumann et al., 1984; Farmer and Hickman-Brenner, 1992), but as new species are described, it becomes clear that there is huge variability of phenotypic features within and among species of vibrios, making it difficult to develop classifications on this basis.

GENOMIC BASIS OF THE TAXONOMY OF VIBRIOS

Vibrios belong to the *Gammaproteobacteria* and are neighbors of the families *Enterobacteriaceae*, *Pasteurellaceae*, *Aeromonadaceae*, and *Alteromonadaceae* according to 16S rRNA gene sequence analysis (see http://www.bergeys.org). The current taxonomy of vibrios is based mainly on genomic data. The advantage of this approach is the establishment of a highly informative, reproducible, and stable classification frame. The application of various techniques, including AFLP, rep-PCR, DNA-DNA hybridization, and 16S rRNA sequencing, indicates the occurrence of several species groups within the current family *Vibrionaceae*. In other bacterial genera, e.g., *Aeromonas*, *Burkholderia*, and *Xanthomonas*, such groups have been deemed homology groups or genomic species (Coenye et al., 1999; Huys et al., 1996; Rademaker et al., 2000). On the basis of molecular evidence gathered in the last few years, we opted to consider each of the members of the species groups as separate bona fide species.

DNA-DNA Hybridization and %G+C of DNA

The definition and circumscription of bacterial species in vibrios follows the polyphasic approach in which DNA-DNA reassociation and 16S rRNA data have a pivotal role (Stackebrandt et al., 2002). Both types of data had a tremendous, positive impact in the development of bacterial taxonomy and placed it on firm scientific ground. These techniques measure intra- and interspecific genomic relatedness of strains.

DNA-DNA hybridizations indirectly measure the similarity of two single-stranded DNA molecules on the basis of the physical-chemical properties of these molecules (Gillis et al., 1970); results obtained reflect the whole genome similarity among different bacterial strains as calculated in silico (Zeigler, 2003). Several researchers consider this technique a "black box" and point out several drawbacks, the most important being that DNA-DNA data are not cumulative (Stackebrandt, 2003; C. C. Thompson et al., 2004).

The taxonomy of vibrios has been shaped by massive DNA-DNA hybridization experiments (e.g., Baumann and Baumann, 1977; Baumann et al., 1984; Reichelt et al., 1976; C. C. Thompson et al., 2004). Different species groups are recognized on the basis of DNA-DNA similarity and 16S rRNA (Table 2). Overall, species within each of these groups also have highly related phenotypes (Thompson et al., 2003a,b). For example, within the *V. harveyi* group, *V. campbellii*, *V. harveyi*, and *V. rotiferianus* have nearly indistinguishable phenotypes (Gomez-Gil et al., 2003a, 2004b). The utilization of α-cyclodextrin, *cys*-aconitic acid, citric acid, glucose-6-phosphate, hydroxy-L-proline, *N*-acetyl-D-galactosamine, quinic acid, and sucrose was thought to be useful to discriminate *V. campbellii* and *V. harveyi* (Baumann et al., 1984), but Gomez-Gil et al. (2004a,b) have shown that strains of both species cannot be differentiated on this basis.

AFLP

The amplified fragment length polymorphism (AFLP) technique consists of three main steps: (i) digestion of total genomic DNA with two restriction enzymes and subsequent ligation of the restriction halfsite-specific adaptors to all restriction fragments; (ii) selective amplification of these fragments with two PCR primers that have corresponding adaptor and restriction site sequences as their target sites, and (iii) electrophoretic separation of the PCR products on polyacrylamide gels with selective detection of fragments that contain the fluorescently labeled primer and computer-assisted numerical analysis of the band patterns (Huys and Swings, 1999; Vos et al., 1995). AFLP indexes variation in the whole genome and thus is considered to give useful information on the short- and long-term evolution of bacterial strains (Lan and Reeves, 2002).

AFLP has been used to study various vibrios, including *V. alginolyticus* (Vandenberghe et al., 1999), *V. cholerae* (Jiang et al., 2000a,b; Lan and Reeves, 2002), *V. harveyi* (Pedersen et al., 1998; Gomez-Gil et al., 2004b), *V. vulnificus* (Arias et al., 1997a,b), *V. wodanis* (Benediktsdóttir et al., 2000) and *Photo-*

Table 2. DNA-DNA similarity and mol% G+C of the main species groups of vibrios

Species group	DNA-DNA similarity (%)	mol% G+C of DNA
V. anguillarum		
V. anguillarum and *V. ordalii*	58–69	43–44
V. gazogenes		
V. aerogenes, V. gazogenes, and *V. ruber*	52–56	46–47
V. cholerae		
V. cholerae and *V. mimicus*	73–79	47–49
V. fischeri		
V. fischeri, V. logei, V. salmonicida, and *V. wodanis*	60	39–42
V. fluvialis		
V. fluvialis and *V. furnissii*	60	49–51
V. halioticoli		
V. halioticoli, V. ezurae, V. gallicus, V. neonatus, and *V. superstes*	16–70	44–49
V. harveyi		
V. harveyi, V. alginolyticus, V. campbellii, V. natriegens, V. parahaemolyticus, and *V. rotiferianus*	29–69	44–48
V. nigripulchritudo		
V. nigripulchritudo and *V. penaeicida*	18	46–47
V. pelagius		
V. pelagius and *V. fortis*	55–66	45–46
V. splendidus		
V. splendidus, V. chagasii, V. crassostreae, V. cyclitrophicus, V. kanaloaei, V. lentus, V. pomeroyi, and *V. tasmaniensis*	36–65	44–45
V. tubiashii		
V. tubiashii, V. brasiliensis, V. coralliilyticus, V. nereis, V. neptunius, and *V. xuii*	16–66	44–47
P. phosphoreum		
P. angustum, P. leiognathi, and *P. phosphoreum*	36–61	39–44
P. profundum		
P. profundum and *P. frigidiphilum*	52	41–44

bacterium damselae (Thyssen et al., 2000). Thompson et al. (2001a) analyzed 506 strains from all species of the vibrios by AFLP using the enzyme combination HindIII/TaqI (Fig. 2). The AFLP band patterns consisted of 102 ± 24 bands (50 to 536 bp in size), and the mean reproducibility of these patterns was 91 ± 3%. Sixty-nine groups (A1 to A69) were defined using a cutoff of 45% similarity. This study confirmed that, in most cases, the species were clearly different from one another. Each correctly identified species showed a characteristic genome pattern and fell in separate clusters. *Vibrio trachuri* and *Vibrio shilonii* had AFLP patterns highly related to *V. harveyi* and *Vibrio mediterranei*, respectively, indicating that they were synonyms. Indeed, subsequent studies confirmed this hypothesis (Thompson et al., 2001b, 2002a). The study of Thompson et al. (2001a) culminated in the description of many new vibrio species because numerous isolates (*n* = 236) had genomes unrelated (<45% AFLP pattern similarity) to any of the known type strains.

We have recently proposed AFLP as an alternative to the classification and identification of vibrios (C. C. Thompson et al., 2004). We can indeed predict DNA-DNA similarity values from the AFLP similarities. AFLP band pairwise similarities of ~70% correspond to DNA-DNA similarities of ~80 to 100%. In fact, this very pattern is recognizable in many other bacterial genera, including *Aeromonas* (Huys et al., 1996), *Agrobacterium* (Mougel et al., 2002), *Burkholderia* (Coenye et al., 1999), and *Xanthomonas* (Rademaker et al., 2000). In our hands, AFLP is the most reliable genomic fingerprinting identification and classification tool for vibrios to date. AFLP data can also be accumulated in local laboratory databases. Because AFLP indexes variation in the whole genome, including regions of unknown function (e.g., pseudogenes and mobile elements), it cannot be used as a phylogenetic marker.

MULTILOCUS SEQUENCE ANALYSES

In 1998, Maiden and co-workers (Maiden et al., 1998) proposed the use of multilocus sequence typing (MLST) for studying the population biology and epidemiology of *Neisseria meningitidis*. MLST is a development of multilocus enzyme electrophoresis, indexing variation directly in housekeeping gene sequences (Feil and Spratt, 2001). MLST is a suitable

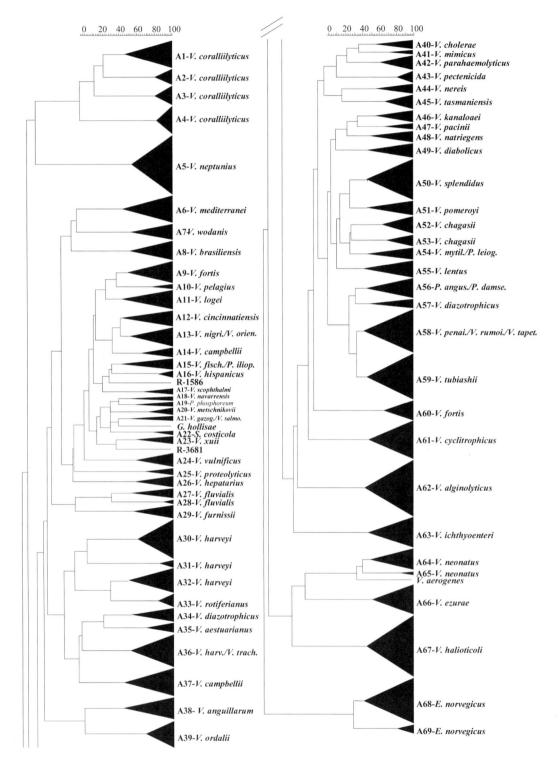

Figure 2. Dendrogram of the fluorescent AFLP patterns of 506 *Vibrionaceae* strains. A band-based (Dice) cluster analysis (Ward) was used. The threshold for cluster delineation was 45%. Adapted from Thompson et al. (2001a).

tool for microevolution studies on bacteria, in that it detects synonymous and nonsynonymous changes as well as recombinational events between strains (Feil, 2004). Data are reproducible, accurate, and portable, allowing population biology inferences (Jolley et al., 2005) and the creation of Internet databases for on-line strain identification (Urwin and Maiden, 2003).

The phylogenetic picture obtained by the "gold standard" 16S rRNA has been confirmed but also refined by additional chronometers. For example, the phylogenetic tree based on the concatenated 16S rRNA, *rpoA*, *recA*, and *pyrH* gene sequences (3,324 nucleotides [nt]) corroborates the creation of the three new families of vibrios (Fig. 3). The different species groups, e.g., *V. harveyi* and *V. splendidus*, seen in the 16S rRNA tree are also evident within the family *Vibrionaceae*. Notably, the *V. fischeri* species group is positioned at the outskirts of the family *Vibrionaceae*.

So far, most MLST schemes have been applied to only single and/or pairs of related species, e.g., *Bacillus cereus* group (Priest et al., 2004), *Burkholderia mallei–B. pseudomallei* (Godoy, 2003), *N. meningitidis* (Maiden et al., 1998), *V. cholerae* (Farfán et al., 2002; Garg et al., 2003), *V. vulnificus* (http://pubmlst.org/vvulnificus/), and *V. parahaemolyticus* (Chowdhury et al., 2004). The sequencing of six genes (*asd*, *cadA*, *idh*, *lap*, *mdh*, and *epd*) on a subset of 31 *V. cholerae* serogroup O139 strains revealed four distinct groups of strains (Farfán et al., 2002). No signs of recombination were detected among them. Garg et al. (2003) carried out a more comprehensive study on 96 *V. cholerae* O139 strains isolated in India between 1992 and 2000. They analyzed the sequences of *dnaE*, *lap*, *recA*, *pgm*, *gyrB*, *cat*, *chi*, *rstR*, and *gmd* and concluded that conspecific homologous recombination may have occurred in *gmd*, *recA*, and *lap*. Garg et al. (2003) think that such events lead to the cohesion of the species. MLST of four loci, i.e., *gyrB*, *recA*, *dnaE*, and *gnd*, applied to 81 isolates of *V. parahaemolyticus* revealed that pandemic strains are clonal (Chowdhury et al., 2004). In the *V. vulnificus* MLST scheme (http://pubmlst.org/vvulnificus/), the sequences of 10 loci, i.e., *pyrC*, *pntA*, *dtd*, *glp*, *tnA*, *purM*, *lysA*, *metG*, *gyrB*, and *mdh*, were examined in a collection of 159 isolates. These genes had 29 to 46 alleles and a sequence similarity of >96%, except for *pyrC*, *tnA*, and *lysA*, in which the sequence similarity was >94%. Overall, the gene sequence similarity was >98% for the different loci examined in all these MLST studies, except for *recA*, *asd*, and *epd*, which were around 97%, >96%, and >93%, respectively. Given the in-depth sampling of strains within these three human pathogenic vibrios, we may tentatively state that strains of the same species will have >94% *gyrB*, *dnaE*, *gnd*, *lap*, *pgm*, *cat*, *chi*, *rstR*, *gmd*, *idh*, *mdh*, *glp*, *metG*, *purM*, *dtdS*, *lysA*, *pntA*, *pyrC*, *tnaA*, and *dtd*. The definition of bacterial species by gene sequence similarity has been encouraged by an ad hoc committee (Stackebrandt et al., 2002).

Our recent MLSA on the four families of vibrios has clearly shown that species may be defined on the basis of gene sequences (Thompson et al., 2005b). We analyzed a collection of well-documented strains of vibrios. They all fulfill the criterion that strains of the same species have >60% AFLP band pattern similarity, >70% DNA-DNA similarity, and >97% 16S rRNA sequence similarity. We sequenced fragments of the *rpoA* (928 to 931 nt), *recA* (613 to 797 nt), and *pyrH* (443 to 549 nt) from a collection of 192 *Vibrionaceae* strains. For *atpA* (1,266 to 1,542 nt) and *obg* (1,001 to 1,188 nt) we analyzed ~100 strains. We determined the intraspecies variation of the different loci by including several strains per species. All loci had mean synonymous substitutions per synonymous site/mean nonsynonymous substitutions per nonsynonymous site ratio (*ds/dn*) >21, indicating that these are indeed housekeeping genes and are under neutral selective pressure (Table 3). The high correlation between the pairwise similarities of *rpoA*, *atpA*, *pyrH*, *recA*, and the 16S rRNA (Fig. 4) and the agreement with former polyphasic taxonomic studies suggest that these genes may be used as alternative phylogenetic and identification markers. The five currently known genera of vibrios can all be differentiated on the basis of these loci. According to *rpoA*, *recA*, and *pyrH* sequences, the genus *Vibrio* was heterogeneous and polyphyletic, with *V. fischeri*, *V. logei*, and *V. wodanis* grouping closer to *Photobacterium*. *V. halioticoli*-, *V. harveyi*-, *V. splendidus*-, and *V. tubiashii*-related species formed distinct groups within the genus *Vibrio*. Overall, the genetic loci examined were more discriminatory among species than the 16S rRNA (Fig. 4). *rpoA* was the least discriminatory gene, in contrast to *recA* and *pyrH*. In many cases, e.g., within *V. splendidus* and *V. tubiashii* group *rpoA*, gene sequences were less discriminatory than *recA*. In these cases, the combination of several loci will yield the most robust identification. Each species clearly formed separated clusters with at least 99% *rpoA*, 94% *recA*, 96% *pyrH*, 98% *atpA*, and 96% *obg* sequence similarity. We can conclude that these similarity values will be the thresholds for species circumscription in vibrios. This type of data offer a readily accessible, reliable and practical alternative to the "gold standards" DNA-DNA hybridization and 16S rRNA applied in the current taxonomy of vibrios. An evolving website devoted to the electronic taxonomy of *Vibrionaceae* based on the MLSA data

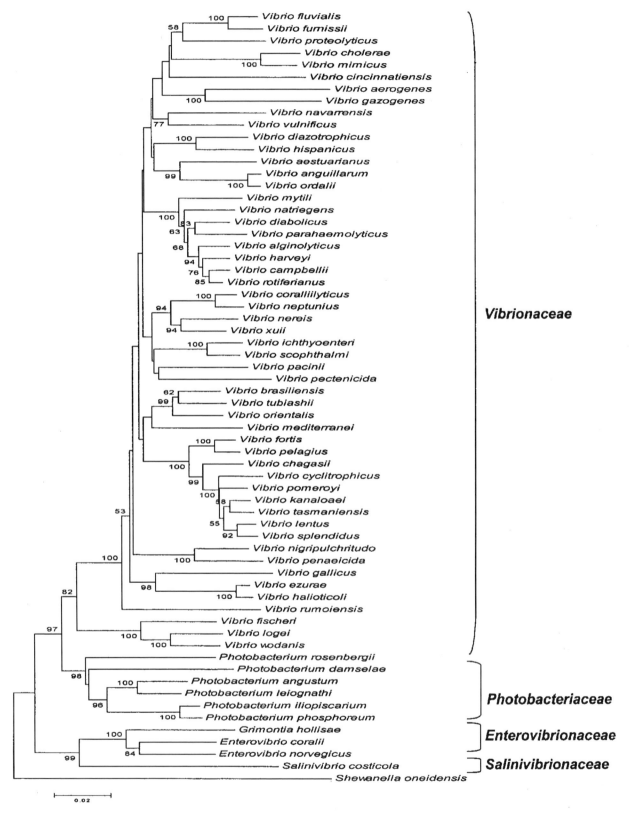

Figure 3. Phylogenetic tree based on the Kimura 2-parameter neighbor-joining model. Bootstrap values were obtained after 500 repetitions. Concatenated sequences (3,284 nt) of the genes 16S rRNA (1,300 nt), *recA* (613 nt), *rpoA* (928 nt), and *pyrH* (443 nt). The type strains of 64 species were included in the tree.

Table 3. Genes used in MLSA of vibrios and some important features

Genes	% G+C	ds/dn	Similarity (%)		Recombination[a]
			Species	Genus	
rpoA	46.0 ± 0.6	29	99	81	+
recA	46.6 ± 1.9	29	94	70	+
pyrH	48.7 ± 0.9	39	96	73	+
atpA	48.2 ± 1.3	25	98	75	+
obg	47.6 ± 1.8	23	96	67	+
16S rRNA	53.7 ± 0.0		97	89	+

[a] Recombination was detected using different techniques. +, evidence of recombination by at least one of the following methods: homoplasy ratio, maximum chi^2, Sawyer's test, and Splits tree decomposition.

is under development (http://lmg.ugent.be/bnserver/ MLSA/Vibrionaceae/).

PERSPECTIVES

Bacterial taxonomy has been frequently deemed an arbitrary science in that the concept, definition, and circumscription of bacterial species, genera, and higher taxa follow a highly pragmatic approach that generates an artificial framework (Smith et al., 2000; Staley, 1997; Ward, 1998). In a recent philosophical paper, Bapteste et al. (2004) argued that current taxonomies based on a few genes, e.g., 16S rRNA of the core bacterial genome, do not reflect the natural history of bacterial isolates because they look only at 5 to 10% of total information content in the genome. These genes are present in all bacteria and are thought to be refractory to horizontal gene transfer (HGT). Species trees constructed on the basis of this reduced gene set reveal a central trend in genome evolution that is in agreement with the Tree of Life using 16S rRNA, but do not show a full evolutionary picture (Koonin, 2003). According to Bapteste et al. (2004), such a reductionist framework can be mis-leading with respect to the biology of the bacterial isolates because some related genomes may share more genes due to HGT than to vertical inheritance. These authors propose an alternative system called the Synthesis of Life that embraces both HGT and vertical inheritance.

HGT played a probable central role in the evolution of cells, particularly before the so-called Darwinian threshold, at which evolution was communal rather than generating organismal lineages (Woese, 2004). Genes involved in the central metabolism and in pathogenesis and antibiotic resistance are good examples of candidate loci where HGT can operate, while various genes involved in the DNA metabolism, e.g., replication, transcription, and translation apparatuses, can be considered quite refractory to HGT (Woese, 2004). Genes less prompted to HGT tend to be fixed in the genome and generate the organismal lineages that are detectable in phylogenetic trees (Woese et al., 2000). But foreign DNA acquired via HGT may launch an organism into a new life style and improve its fitness in a previously unexplored niche that may result in speciation (Lawrence, 1999). According to Lawrence (2002), HGT is the main driving force in bacterial speciation. Splits tree decomposition analyses of *rpoA*, *recA*, *atpA*, *obg*, and *pyrH* of all species of vibrios showed a star-like shape. Networks became evident when the trees based on the five different genes were pruned, indicating recombination among conspecific strains of certain species (Fig. 5). Although splits decomposition analysis suggests that recombination within species may occur in different loci, including *rpoA*, *recA*, *atpA*, *obg*, and *pyrH*, there is no evidence that recombination will hamper species identification.

Cohan (2004) viewed bacterial species as clusters of strains with identical gene sequences in several loci. These clusters are deemed to be adapted to different ecological niches and are thus named ecotypes. The ecotype genetic diversity is periodically purged by natural selection. Although the incorporation of ecological information in bacterial taxonomy is unusual practice currently, this approach is welcomed by many microbiologists (Rossello-Mora and Amann, 2001; DeLong, 2004; Ward, 1998). Cohan argues that mutation is the main driving force in the evolution of bacterial species. Recombination, on the other hand, may act as a cohesive force to keep conspecific strains tightly related, with the spread of successful genetic loci throughout the bacterial population (Lawrence, 2002). Importantly, the ecotype concept implies that the currently named bacterial species harbor, in fact, many clusters that equate species. The application of this principle within the *Vibrionaceae* may yield a multiplication of the species but a more

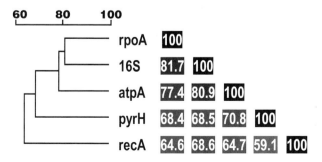

Figure 4. Congruence among *rpoA*, 16S rRNA, *atpA*, *pyrH*, and *recA* loci. Pearson correlation values are shown.

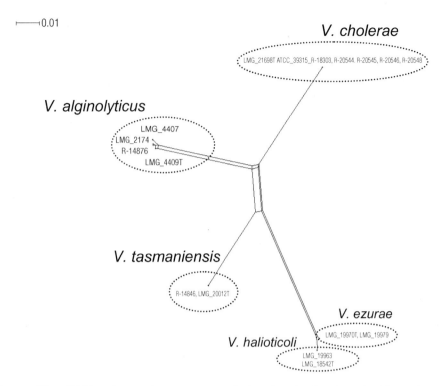

Figure 5. Splits tree (Fit = 100) based on *rpoA* gene sequences of selected strains that showed evidence of recombination by Sawyer's, maximum chi^2, or homoplasy tests. *V. alginolyticus* LMG 4409^T, *V. cholerae* R-20544 and R-18303, and *V. tasmaniensis* LMG 20012^T recombined with one another according to Sawyer's test (SSCF; $P < 0.01$).

objective way to detect sequence clusters (i.e., species) worldwide.

With the flood of whole-genome sequences in the last few years, it becomes clear that the traditional polyphasic taxonomy has to move toward a genomic taxonomy (Gevers et al., 2005). This new heretical electronic taxonomy will be a more accessible, reproducible, and transparent system (Dawyndt, 2004). New types and new names will be entered into the taxonomic landscape automatically; taxonomists will just have to find the right settings to define their new species. Of course, the right settings are already emerging from MLSA studies (http://lmg.ugent.be/bnserver/MLSA/Vibrionaceae/). The use of core genome genes, i.e., ca. 50 to 100 conserved genes across the prokaryotic world, will improve both phylogeny and taxonomy. Ideally, all the core genome genes could be used, but that might prove to be redundant, since various core genes may well be under similar selective pressure and so will reveal (reassuring) similar pictures. It has already been shown in various bacterial groups that the use of 1 to 3 genes surpasses the discriminatory power of both DNA-DNA hybridization and 16S rRNA for identification and classification purposes (Naser et al., 2005a,b). Nevertheless, the advantage of using, e.g., 7 to 10 genes, as in

the case of MLST schemes, is that taxonomy on this basis becomes much more robust. The taxonomy of vibrios is being rapidly improved owing to the application of modern molecular techniques. New data gathered by these techniques will lead to a revision and improvement of the taxonomy of vibrios in the next few years.

Acknowledgments. F.L.T. acknowledges a young researcher grant (2004/00814-9) from FAPESP, Brazil. J.S. acknowledges grants from the Fund for Scientific Research (FWO), Belgium.

REFERENCES

Arias, C. R., L. Verdonck, J. Swings, R. Aznar, and E. Garay. 1997a. Intraspecific differentiation of *Vibrio vulnificus* biotypes by amplified fragment length polymorphism and ribotyping. *Appl. Environ. Microbiol.* **63:**2600–2606.

Arias, C. R., L. Verdonck, J. Swings, R. Aznar, and E. Garay. 1997b. A polyphasic approach to study the intraspecific diversity amongst *Vibrio vulnificus* isolates. *Syst. Appl. Microbiol.* **20:**622–633.

Austin, B., D. A. Austin, V. M. Falconer, K. Pedersen, J. L. Larsen, J. Swings, and L. Verdonck. 1996. Dissociation of *Vibrio anguillarum* and *V. ordalii* cultures into two or three discrete colony types. *Bull. Eur. Fish. Pathol.* **16:**101–103.

Bang, S. S., P. Baumann, and K. H. Nealson. 1978. Phenotypic characterization of *Photobacterium logei* (sp. nov.), a species related to *P. fischeri. Curr. Microbiol.* **1:**285–288.

Bapteste E., Y. Boucher, J. Leigh, and W. F. Doolittle. 2004. Phylogenetic reconstruction and lateral gene transfer. *Trends Microbiol.* **12:**406–411.

Baumann, P., L. Baumann, and M. Mandel. 1971. Taxonomy of marine bacteria: the genus *Beneckea. J. Bacteriol.* **107:**268–294.

Baumann, P., and L. Baumann. 1977. Biology of the marine enterobacteria: genera *Beneckea* and *Photobacterium. Annu. Rev. Microbiol.* **31:**39–61.

Baumann, P., L. Baumann, S. S. Bang, and M. J. Woolkalis. 1980. Reevaluation of the taxonomy of *Vibrio, Beneckea,* and *Photobacterium:* abolition of the genus *Beneckea. Curr. Microbiol.* **4:**127–132.

Baumann, P., L. Baumann, M. J. Woolkalis, and S. S. Bang. 1983. Evolutionary relationships in *Vibrio* and *Photobacterium:* a basis for a natural classification. *Annu. Rev. Microbiol.* **37:**369–398.

Baumann, P., and R. H. W. Schubert. 1984. *Vibrionaceae,* p. 516–550. *In* N. R. Krieg and J. G. Holt (ed.), *Bergey's Manual of Systematic Bacteriology.* Lippincott Williams & Wilkins, Baltimore, Md.

Baumann, P., A. L. Furniss, and J. L. Lee. 1984. Genus I: *Vibrio* (Pacini, 1854), p. 518–538. *In* N. R. Krieg and J. G. Holt (ed.), *Bergey's Manual of Systematic Bacteriology.* Lippincott Williams & Wilkins, Baltimore, Md.

Benediktsdóttir, E., L. Verdonck, C. Sproer, S. Helgason, and J. Swings. 2000. Characterization of *Vibrio viscosus* and *Vibrio wodanis* isolated at different geographical locations: a proposal for reclassification of *Vibrio viscosus* as *Moritella viscosa* comb. nov. *Int. J. Syst. Evol. Microbiol.* **50:**479–488.

Ben-Haim, Y., F. L. Thompson, C. C. Thompson, M. C. Cnockaert, B. Hoste, J. Swings, and E. Rosenberg. 2003. *Vibrio coralliilyticus* sp. nov., a temperature-dependent pathogen of the coral *Pocillopora damicornis. Int. J. Syst. Evol. Microbiol.* **53:**309–315.

Breed, R. S., E. G. D. Murray, and N. R. Smith. 1957. *Bergey's Manual of Determinative Bacteriology,* 7th ed., p. 229–249. Lippincott Williams & Wilkins, Baltimore, Md.

Charlebois, R. L., and W. F. Doolittle. 2004. Computing prokaryotic gene ubiquity: rescuing the core from extinction. *Genome Res.* **14:**2469–2477.

Chowdhury, N. R., O. C. Stine, J. G. Morris, and G. B. Nair. 2004. Assessment of evolution of pandemic *Vibrio parahaemolyticus* by multilocus sequence typing. *J. Clin. Microbiol.* **42:**1280–1282.

Coenye, T., L. M. Schouls, J. R. W. Govan, K. Kersters, and P. Vandamme. 1999. Identification of *Burkholderia* species and genomovars from cystic fibrosis patients by AFLP fingerprinting. *Int. J. Syst. Bacteriol.* **49:**1657–1666.

Cohan, F. M. 2004. Concepts of bacterial biodiversity for the age of genomics, p.175–194. *In* C. M. Fraser, T. Read, and K. E. Nelson (ed.), *Microbial Genomes,* Humana Press, Totowa, N.J.

Colwell, R. R. 1970. Polyphasic taxonomy of the genus *Vibrio:* numerical taxonomy of *Vibrio cholerae, Vibrio parahaemolyticus,* and related *Vibrio* species. *J. Bacteriol.* **104:**410–433.

Dawyndt, P. 2004. Knowledge accumulation of microbial data aiming at a dynamic taxonomic framework. Ph.D. thesis, Ghent University, Ghent, Belgium.

DeLong E. F. 2004. Microbial population genomics and ecology: the road ahead. *Environ. Microbiol.* **6:**875–878.

Farfán, M., D. Minana-Galbis, M. C. Fuste, and J. G. Loren. 2002. Allelic diversity and population structure in *Vibrio cholerae* O139 Bengal based on nucleotide sequence analysis. *J. Bacteriol.* **184:**1304–1313.

Farmer, J. J. 1986. International Committee on Systematic Bacteriology. Subcommittee on the Taxonomy of *Vibrionaceae.* Minutes of the meetings. *Int. J. Syst. Bacteriol.* **39:**210–212.

Farmer, J. J. 1992. The family *Vibrionaceae,* p. 2938–2951. *In* A. Balows, H. G. Trüper, M. Dworkin, W. Harder, and K. H.

Schleifer (ed.), *The Prokaryotes. A Handbook on the Biology of Bacteria: Ecophysiology, Isolation, Identification, and Applications,* 2nd ed. Springer-Verlag. Berlin, Germany.

Farmer, J. J., III, and F. W. Hickman-Brenner. 1992. The genera *Vibrio* and *Photobacterium,* p. 2952–3011. *In* A. Balows, H. G. Trüper, M. Dworkin, W. Harder, and K. H. Schleifer (ed.), *The Prokaryotes. A Handbook on the Biology of Bacteria: Ecophysiology, Isolation, Identification, and Applications,* 2nd ed. Springer-Verlag. Berlin, Germany.

Faury, N., D. Saulnier, F. L. Thompson, M. Gay, J. Swings, and F. Le Roux. 2004. *Vibrio crassostreae* sp. nov., isolated from the hemolymph of oysters (*Crassostrea gigas*). *Int. J. Syst. Evol. Microbiol.* **54:**2137–2140.

Feil, E. J., and B. G. Spratt. 2001. Recombination and the population structures of bacterial pathogens. *Annu. Rev. Microbiol.* **55:**561–590.

Feil, E. J. 2004. Small change: keeping pace with microevolution. *Nat. Rev. Microbiol.* **12:**483–495.

Fujino, T., R. Sakazaki, and K. Tamura. 1974. Designation of the type strains of *Vibrio parahaemolyticus* and description of 200 strains of the species. *Int. J. Syst. Bacteriol.* **24:**447–449.

Garg, P., A. Aydanian, D. Smith, J. G. Morris, G. B. Nair, and O. C. Stine. 2003. Molecular epidemiology of O139 *Vibrio cholerae:* mutation, lateral gene transfer, and founder flush. *Emerg. Infect. Dis.* **9:**810–814.

Gevers, D., F. M. Cohan, J. G. Lawrence, B. G. Spratt, T. Coenye, E. J. Feil, E. Stackebrandt, Y. Van de Peer, P. Vandamme, F. L. Thompson, and J. Swings. 2005. Opinion: Re-evaluating prokaryotic species. *Nat. Microbiol. Rev.* **3:**733–739.

Gillis, M., J. De Ley, and M. De Cleene. 1970. The determination of molecular weight of bacterial genome DNA from renaturation rates. *Eur. J. Biochem.* **12:**143–153.

Godoy, D., G. Randle, A. J. Simpson, D. M. Aanensen, T. L. Pitt, R. Kinoshita, and B. G. Spratt. 2003. Multilocus sequence typing and evolutionary relationships among the causative agents of melioidosis and glanders, *Burkholderia pseudomallei* and *Burkholderia mallei. J. Clin. Microbiol.* **41:**2068–2079. (Erratum, **41:**4913.)

Gomez-Gil, B., F. L. Thompson, C. C. Thompson, and J. Swings. 2003a. *Vibrio rotiferianus* sp. nov., isolated from cultures of the rotifer *Brachionus plicatilis. Int. J. Syst. Evol. Microbiol.* **53:**239–243.

Gomez-Gil, B., F. L. Thompson, C. C. Thompson, and J. Swings. 2003b. *Vibrio pacinii* sp nov., from cultured aquatic organisms. *Int. J. Syst. Evol. Microbiol.* **53:**1569–1573.

Gomez-Gil, B., F. L. Thompson, C. C. Thompson, A. Garcia-Gasca, A. Roque, and J. Swings. 2004a. *Vibrio hispanicus* sp. nov., isolated from *Artemia* sp and sea water in Spain. *Int. J. Syst. Evol. Microbiol.* **54:**261–265.

Gomez-Gil, B., S. Soto-Rodríguez, A. García-Gasca, A. Roque, R. Vazquez-Juarez, F. L. Thompson, and J. Swings. 2004b. Molecular identification of *Vibrio harveyi*-related isolates associated with diseased aquatic organisms. *Microbiology* **150:**1769–1777.

Harris, J. K., S. T. Kelley, G. B. Spiegelman, and N. R. Pace. 2003. The genetic core of the universal ancestor. *Genome Res.* **13:**407–412.

Hayashi, K., J. Moriwaki, T. Sawabe, F. L. Thompson, J. Swings, N. Gudkovs, R. Christen, and Y. Ezura. 2003. *Vibrio superstes* sp nov., isolated from the gut of Australian abalones *Haliotis laevigata* and *Haliotis rubra. Int. J. Syst. Evol. Microbiol.* **53:**1813–1817.

Hickman, F. W., J. J. Farmer, III, D. G. Hollis, G. R. Fanning, A. G. Steigerwalt, R. E. Weaver, and D. J. Brenner. 1982. Identification of *Vibrio hollisae* sp. nov. from patients with diarrhea. *J. Clin. Microbiol.* **15:**395–401.

Huys, G., R. Coopman, P. Janssen, and K. Kersters. 1996. High-resolution genotypic analysis of the genus *Aeromonas* by AFLP fingerprinting. *Int. J. Syst. Bacteriol.* **46:**572–580.

Huys, G., and J. Swings. 1999. Evaluation of a fluorescent amplified fragment length polymorphism methodology for the genotypic discrimination of *Aeromonas* taxa. *FEMS Microbiol. Lett.* **177:**83–92.

Jiang, S. C., V. Louis, N. Choopun, A. Sharma, A. Huq, and R. R. Colwell. 2000a. Genetic diversity of *Vibrio cholerae* in Chesapeake Bay determined by amplified fragment length polymorphism fingerprinting. *Appl. Environ. Microbiol.* **66:**140–147.

Jiang, S. C., M. Matte, G. Matte, A. Huq, and R. R. Colwell. 2000b. Genetic diversity of clinical and environmental isolates of *Vibrio cholerae* determined by amplified fragment length polymorphism fingerprinting. *Appl. Environ. Microbiol.* **66:**148–153.

Jolley, K. A., D. J. Wilson, P. Kriz, G. McVean, and M. C. J. Maiden. 2005. The influence of mutation, recombination, population history, and selection on patterns of genetic diversity in *Neisseria meningitides. Mol. Biol. Evol.* **22:**562–569.

Kalina, G. P., A. S. Antonov, T. P. Turova, and T. I. Grafova. 1984. *Allomonas enterica* gen. nov., sp. nov., deoxyribonucleic-acid homology between *Allomonas* and some other members of the *Vibrionaceae. Int. J. Syst. Bacteriol.* **34:**150–154.

Koonin, E. V. 2003. Comparative genomics, minimal gene-sets and the last universal common ancestor. *Nat. Rev. Microbiol.* **1:**127–136.

Lan, R., and P. R. Reeves. 2002. Pandemic spread of cholera: genetic diversity and relationships within the seventh pandemic clone of *Vibrio cholerae* determined by amplified fragment length polymorphism. *J. Clin. Microbiol.* **40:**172–181.

Lawrence, J. G. 1999. Gene transfer, speciation, and the evolution of bacterial genomes. *Curr. Opin. Microbiol.* **2:**519–523.

Lawrence, J. G. 2002. Gene transfer in bacteria: speciation without species? *Theor. Popul. Biol.* **61:**449–460.

Lee, J. V., P. Shread, A. L. Furniss, and T. N. Bryant. 1981. Taxonomy and description of *Vibrio fluvialis* sp. nov. (synonym group F vibrios, group EF6). *J. Appl. Bacteriol.* **50:**73–94.

Maiden, M. C., J. A. Bygraves, E. Feil, G. Morelli, J. E. Russell, R. Urwin, Q. Zhang, J. Zhou, K. Zurth, D. A. Caugant, I. M. Feavers, M. Achtman, and B. G. Spratt. 1998. Multilocus sequence typing: a portable approach to the identification of clones within populations of pathogenic microorganisms. *Proc. Natl. Acad. Sci. USA* **95:**3140–3145.

Mougel, C., J. Thioulouse, G. Perriere, and X. Nesme. 2002. A mathematical method for determining genome divergence and species delineation using AFLP. *Int. J. Syst. Evol. Microbiol.* **52:**573–586.

Naser, S., F. L. Thompson, B. Hoste, D. Gevers, K. Vandemeulebroecke, I. Cleenwerck, C. C. Thompson, M. Vancanneyt, and J. Swings. 2005a. Phylogeny and identification of *Enterococci* by atpA gene sequence analysis. *J. Clin. Microbiol.* **43:**2224–2230.

Naser, S. M., F. L. Thompson, B. Hoste, D. Gevers, P. Dawyndt, M. Vancanneyt, and J. Swings. 2005b. Application of multilocus sequence analysis (MLSA) for rapid identification of *Enterococcus* species based on rpoA and pheS genes. *Microbiology* **151:**2141–2150.

Pacini, F. 1854. Osservazione microscopiche e deduzioni patologiche sul cholera asiatico. *Gazette Medicale de Italiana Toscano Firenze* **6:**405–412.

Pedersen, K., L. Verdonck, B. Austin, D. A. Austin, A. R. Blanch, P. A. D. Grimont, J. Jofre, S. Koblavi, J. L. Larsen, T. Tiainen, Vigneulle M., and J. Swings. 1998. Taxonomic evidence that *Vibrio carchariae* Grimes *et al.* 1985 is a junior synonym of *Vibrio harveyi*(Johnson and Shunk 1936) Baumann *et al.* 1981. *Int. J. Syst. Bacteriol.* **48:**749–758.

Priest, F. G., M. Barker, L. W. Baillie, E. C. Holmes, and M. C. Maiden. 2004. Population structure and evolution of the *Bacillus cereus* group. *J. Bacteriol.* **186:**7959–7970.

Rademaker, J. L., B. Hoste, F. J. Louws, K. Kersters, J. Swings, L. Vauterin, P. Vauterin, and F. J. de Bruijn. 2000. Comparison of AFLP and rep-PCR genomic fingerprinting with DNA-DNA homology studies: *Xanthomonas* as a model system. *Int. J. Syst. Evol. Microbiol.* **50:**665–677.

Reichelt, J. L., P. Baumann, and L. Baumann. 1976. Study of genetic relationships among marine species of the genera *Beneckea* and *Photobacterium* by means of in vitro DNA/DNA hybridization. *Arch. Microbiol.* **110:**101–120.

Rossello-Mora, R., and R. Amann. 2001. The species concept for prokaryotes. *FEMS Microbiol. Rev.* **25:**39–67.

Sawabe, T., K. Hayashi, J. Moriwaki, F. L. Thompson, J. Swings, P. Potin, R. Christen, and Y. Ezura. 2004a. *Vibrio gallicus* sp. nov., isolated from the gut of the French abalone *Haliotis tuberculata. Int. J. Syst. Evol. Microbiol.* **54:**843–846.

Sawabe, T., K. Hayashi, J. Moriwaki, Y. Fukui, F. L. Thompson, J. Swings, and R. Christen. 2004b. *Vibrio neonatus* sp. nov. and *Vibrio ezurae* sp. nov. isolated from the gut of Japanese abalones. *Syst. Appl. Microbiol.* **27:**527–534.

Seo, H. J., S. S. Bae, S. H. Yang, J.-H. Lee, and S.-J. Kim. 2005a. *Photobacterium aplysiae* sp. nov., a lipolytic marine bacterium isolated from eggs of the sea hare *Aplysia kurodai. Int. J. Syst. Evol. Microbiol.* **55:**2293–2296.

Seo, H. J., S. S. Bae, J.-H. Lee, and S.-J. Kim. 2005b. *Photobacterium frigidiphilum* sp. nov., a psychrophilic, lipolytic bacterium isolated from deep-sea sediments of Edison Seamount. *Int. J. Syst. Evol. Microbiol.* **55:**1661–1666.

Skerman, V. B. D., McGowan V., and P. H. A. Sneath. 1980. Approved lists of bacterial names. *Int. J. Syst. Bacteriol.* **30:**225–420.

Smith, J. M., E. J. Feil, and N. H. Smith. 2000. Population structure and evolutionary dynamics of pathogenic bacteria. *BioEssays* **22:**1115–1122.

Sorokin, D. Y. 1992. *Catenococcus thiocyclus* gen. nov. sp. nov. a new facultatively anaerobic bacterium from a near-shore sulphidic hydrothermal area. *J. Gen. Microbiol.* **138:**2287–2292.

Stackebrandt, E., W. Frederiksen, G. M. Garrity, P. A. Grimont, P. Kampfer, M. C. Maiden, X. Nesme, R. Rossello-Mora, J. Swings, H. G. Truper, L. Vauterin, A. C. Ward, and W. B. Whitman. 2002. Report of the ad hoc committee for the re-evaluation of the species definition in bacteriology. *Int. J. Syst. Evol. Microbiol.* **52:**1043–1047.

Stackebrandt, E. 2003. The richness of prokaryotic diversity: there must be a species somewhere. *Food Technol. Biotechnol.* **41:**17–22.

Staley, J. T. 1997. Biodiversity: are microbial species threatened? *Curr. Opin. Biotechnol.* **8:**340–345.

Thompson, C. C., F. L. Thompson, K. Vandemeulebroecke, B. Hoste, P. Dawyndt, and J. Swings. 2004. Use of *recA* as an alternative phylogenetic marker in the family *Vibrionaceae. Int. J. Syst. Evol. Microbiol.* **54:**919–924.

Thompson, F. L., B. Hoste, K. Vendemeulebroecke, and J. Swings. 2001a. Genomic diversity amongst *Vibrio* isolates from different sources determined by fluorescent amplified fragment length polymorphism. *Syst. Appl. Microbiol.* **24:**520–538.

Thompson, F. L., B. Hoste, C. C. Thompson, Huys, G., and J. Swings. 2001b. The coral bleaching *Vibrio shiloi* Kushmaro et al. 2001 is a later synonym of *Vibrio mediterranei* Pujalte and Garay 1986. *Syst. Appl. Microbiol.* **24:**516–519.

Thompson, F. L., B. Hoste, K. Vandemeulebroecke, K. Engelbeen, R. Denys, and J. Swings. 2002a. *Vibrio trachuri* Iwamoto et al. 1996 is a junior synonym of *Vibrio harveyi* (Johnson and Shunk 1936) Baumann *et al.* 1981. *Int. J. Syst. Evol. Microbiol.* **52:**973–976.

Thompson, F., B. Hoste, C. C. Thompson, J. Goris, B. Gomez-Gil, L. Huys, P. De Vos, and J. Swings. 2002b. *Enterovibrio norvegicus* gen. nov., sp. nov., isolated from the gut of turbot (*Scophthalmus maximus*) larvae: a new member of the family *Vibrionaceae*. *Int. J. Syst. Evol. Microbiol.* **52:**2015–2022.

Thompson, F. L., T. Iida, and J. Swings. 2004. Biodiversity of vibrios. *Microbiol. Mol. Biol. Rev.* **38:**403–431.

Thompson, F. L., Y. Li, B. Gomez-Gil, C. C. Thompson, B. Hoste, K. Vandemeulebroecke, G. S. Rupp, A. Pereira, M. M. De Bem, P. Sorgeloos, and J. Swings. 2003a. *Vibrio neptunius* sp. nov., *V. brasiliensis* sp. nov. and *V. xuii* sp. nov., isolated from the marine aquaculture environment (bivalves, fish, rotifers and shrimps). *Int. J. Syst. Evol. Microbiol.* **53:**245–252.

Thompson, F. L., C. C. Thompson, Y. Li, B. Gomez-Gil, J. Vandenberghe, B. Hoste, and J. Swings. 2003b. Description of *Vibrio kanaloae* sp. nov., *Vibrio pomeroyi* sp. nov. and *Vibrio chagasii* sp. nov., from sea water and marine animals. *Int. J. Syst. Evol. Microbiol.* **53:**753–759.

Thompson, F. L., C. C. Thompson, and J. Swings. 2003c. *Vibrio tasmaniensis* sp. nov., isolated from Atlantic salmon (*Salmo salar* L.). *Syst. Appl. Microbiol.* **26:**65–69.

Thompson, F. L., C. C. Thompson, B. Hoste, K. Vandemeulebroecke, M. Gullian, and J. Swings. 2003d. Description of *Vibrio fortis* sp. nov., and *V. hepatarius* sp. nov., isolated from aquatic animals and the marine environment. *Int. J. Syst. Microbiol.* **53:**1495–1501.

Thompson, F. L., C. C. Thompson, S. Naser, B. Hoste, K. Vandemeulebroecke, C. Munn, D. Bourne, and J. Swings. 2005a. *Photobacterium rosenbergii* sp. nov. and *Enterovibrio coralii* sp. nov., vibrios associated with coral bleaching. *Int. J. Syst. Evol. Microbiol.* **55:**913–917.

Thompson, F. L., D. Gevers, C. C. Thompson, P. Dawyndt, S. Naser, B. Hoste, C. B. Munn, and J. Swings. 2005b. Phylogeny and molecular identification of vibrios on the basis of multilocus sequence analysis. *Appl. Environ. Microbiol.* **71:**5107–5115.

Thyssen, A., S. Van Eygen, L. Hauben, J. Goris, J. Swings, and F. Ollevier. 2000. Application of AFLP for taxonomic and epidemiological studies of *Photobacterium damselae* subsp. *piscicida*. *Int. J. Syst. Evol. Microbiol.* **50:**1013–1019.

Urwin, R., and M. C. J. Maiden. 2003. Multi-locus sequence typing: a tool for global epidemiology. *Trends Microbiol.* **11:**479–487.

Vandamme, P., B. Pot, M. Gillis, P. de Vos, K. Kersters, and J. Swings. 1996. Polyphasic taxonomy, a consensus approach to bacterial systematics. *Microbiol. Rev.* **60:**407–438.

Vandenberghe, J., L. Verdonck, R. Robles-Arozarena, G. Rivera, A. Bolland, M. Balladares, B. Gomez-Gil, J. Calderon, P. Sorgeloos, and J. Swings. 1999. Vibrios associated with *Litopenaeus vannamei* larvae, postlarvae, broodstock, and hatchery probionts. *Appl. Environ. Microbiol.* **65:**2592–2597.

Vandenberghe, J., F. L. Thompson, B. Gomez-Gil, and J. Swings. 2003. Phenotypic diversity amongst *Vibrio* isolates from marine aquaculture systems. *Aquaculture* **219:**9–20.

Vos, P., R. Hogers, M. Bleeker, M. Reijans, T. Vandelee, M. Hornes, A. Frijters, J. Pot, J. Peleman, M. Kuiper, and M. Zabeau. 1995. AFLP: a new technique for DNA-fingerprinting. *Nucleic Acids Res.* **23:**4407–4414.

Ward, D. M. 1998. A natural species concept for prokaryotes. *Curr. Opin. Microbiol.* **1:**271–277.

Woese, C. R., Olsen, G. J., Ibba, M., and D. Söll. 2000. Aminoacyl-tRNA synthetases, the genetic code, and the evolutionary process. *Microbiol. Mol. Biol. Rev.* **64:**202–236.

Woese, C. R. 2004. A new biology for a new century. *Microbiol. Mol. Biol. Rev.* **68:**173–186.

Xie, C. H., and A. Yokota. 2004. Transfer of *Hyphomicrobium indicum* to the genus *Photobacterium* as *Photobacterium indicum* comb. nov. *Int. J. Syst. Evol. Microbiol.* **54:**2113–2116.

Yoon J. H., J. K. Lee, Y. O. Kim, and T. K. Oh. 2005. *Photobacterium lipolyticum* sp. nov., a bacterium with lipolytic activity isolated from the Yellow Sea in Korea. *Int. J. Syst. Evol. Microbiol.* **55:**335–339.

Zeigler, D. R. 2003. Gene sequences useful for predicting relatedness of whole genomes in bacteria. *Int. J. Syst. Evol. Microbiol.* **53:**1893–1900.

The Biology of Vibrios
Edited by F. L. Thompson et al.
© 2006 ASM Press, Washington, D.C.

Chapter 4

Molecular Identification

MITSUAKI NISHIBUCHI

OVERVIEW

Biochemical characteristics are used to identify *Vibrio* species found in human infections. The vibrios are generally isolated in pure culture using direct plating of the clinical specimen onto a selective agar medium for *Vibrio*, e.g., thiosulfate citrate bile salt sucrose agar. However, the number of human pathogenic *Vibrio* species is limited, and only 12 species are of clinical significance, i.e., *V. cholerae*, *V. parahaemolyticus*, *V. mimicus*, *Grimontia* (*Vibrio*) *hollisae*, *V. fluvialis*, *V. furnissii*, *V. vulnificus*, *V. alginolyticus*, *Photobacterium* (*Vibrio*) *damselae*, *V. metschnikovii*, *V. cincinnatiensis*, and *V. harveyi*. The biochemical characteristics to differentiate them are summarized in Bergey's *Manual of Determinative Bacteriology*, 9th ed. (Holt et al., 1994). Commercial identification systems are available, and the biochemical tests can be used for most of these species; however, these systems may not always be accurate (O'Hara et al., 2003). Misidentification of strains of *Aeromonas*, which were isolated from clinical specimens, as *Vibrio* species using commercial systems has been reported (Abbott et al., 1998; Park et al., 2003). In addition, it is not easy to identify these species if the test strains come from environmental sources (e.g., seawater, sediments, and seafood). Samples are usually incubated in selective enrichment medium, e.g., alkaline peptone water, before plating onto a selective isolation medium. Nevertheless, various *Vibrio* species and related species may show similar biochemical characteristics. Holt et al. (1994) include 34 species in the genus *Vibrio*. A review published in 2004 listed 63 *Vibrio* species and 11 related species (F. L. Thompson et al., 2004). In short, it has become impossible to establish a comprehensive scheme to differentiate *Vibrio* species using only biochemical characteristics. Identification based on biochemical tests is not definitive, and the work is time-consuming and resource-intensive. For these reasons, workers have sought molecular genetic identification methods that are quicker and more definitive than biochemical tests. The genetic markers (e.g., a gene, a base[s] in a gene, a noncoding nucleotide sequence) are conserved only in a particular *Vibrio* species and can be selected as the target to identify the species. Genetic markers that are unique to a subpopulation of the species, e.g., the virulence-associated genes, can be used as targets to detect important subpopulations. The genetic targets that are potentially useful for this purpose are briefly reviewed here.

Simple and easy genetic techniques are useful for routine identification work. These techniques include gene probes and the PCR. The gene probe method detects target genes in the isolated strains. DNA fragments containing that gene or those derived from the internal portion of the gene or synthetic oligonucleotides complementary to the target gene sequences (referred to as DNA probes and oligonucleotide probes, respectively) are used as the probes in hybridization assays. Isotopically or nonisotopically labeled probes have been used successfully. Later, PCR methods were introduced that are more sensitive, more rapid, and simpler than the gene probe method. Therefore, the gene probe method has largely been replaced by PCR. The PCR allows for direct examination of the sample and does not require bacterial isolation. Quantitative PCR for enumeration of the target *Vibrio* species or subpopulation of the species has been developed. A nucleotide sequence amplification-based technique to quantitatively detect a particular mRNA is used that allows the detection of viable populations of a *Vibrio* species or to measure the expression level of the target gene. Medically important *Vibrio* species generally receive more attention than other *Vibrio* species: studies reporting the *Vibrio* species that are pathogenic to humans or aquatic animals are more frequently cited.

There are numerous reports on typing or differentiation of strains within the same species of *Vibrio*

Mitsuaki Nishibuchi • Center for Southeast Asian Studies, Kyoto University, Yoshida, Sakyo-ku, Kyoto 606-8501, Japan.

by various genetic fingerprinting methods. These include ribotyping, restriction fragment length polymorphism, amplified fragment length polymorphism, arbitrarily primed PCR (AP-PCR) or randomly amplified polymorphic DNA PCR, enterobacterial intergenic consensus sequence PCR, pulsed-field gel electrophoresis, and multilocus sequence typing. These techniques require more attention than the gene probe and the PCR techniques described above and are not suitable for routine examination. This aspect of molecular identification will not be discussed in this chapter.

GENE SEQUENCES POTENTIALLY USEFUL FOR TAXONOMY AND IDENTIFICATION

Genes Conserved in Bacteria and Other Organisms

Essential genes are conserved in bacteria and other organisms but the sequence may vary, and they are used for phylogenetic analysis. Particularly, the rRNA gene sequences and their intergenic regions are used frequently for this purpose. Several reports used the 16S rRNA gene sequence to study overall phylogenetic relationships of *Vibrionaceae* (Dorsch et al., 1992; Kita-Tsukamoto et al., 1993; Ruimy et al., 1994; Coelho et al., 1994) and fish-pathogenic *Vibrio* species (Wiik et al., 1995). Dorsch et al. (1992) proposed from the analysis of the 16S rRNA gene sequences of 10 *Vibrio* species that this gene could be used as the target for identification using oligonucleotide probes or PCR. However, considering the increasing number of described *Vibrio* species, the sequence variation in this gene is not large enough to be easily used as the target for differentiation of closely related *Vibrio* species or the subgroups within the same species. The 23S rRNA gene and the 16S to 23S rRNA intergenic spacer region may contain the regions where the sequences vary significantly and are more useful than the 16S gene. *V. vulnificus*-specific oligonucleotide probes that use the variable regions of the 23S rRNA have been reported (Aznar et al., 1994). Comparison of the nucleotide sequence of the 16S to 23S rRNA intergenic spacer region of *V. cholerae* indicates that there is a strain-to-strain variation (Lan and Reeves, 1998; Heidelberg et al., 2000). The sequence variation in the 16S to 23S rRNA intergenic spacer region was useful in differentiating strains belonging to the representative O serotypes of *V. cholerae*, and to distinguish *V. cholerae* from *V. mimicus* (Chun et al., 1999). Sparagano et al. (2002) reported that the 23S rRNA gene sequence is useful for phylogenetic analysis, and developed the probes and PCR methods for the identification of *Vibrio* species.

Various housekeeping genes have been evaluated. In particular, the *recA* gene is essential for genetic recombination. Stine et al. (2000) demonstrated divergence in the *recA* sequence among *V. cholerae* strains and related *Vibrio* species that suggest this gene may be useful for phylogenetic analysis of the genus *Vibrio*. C. C. Thompson et al. (2004) showed that the *recA* gene is more discriminatory than the 16S rRNA gene in the phylogenetic analysis of the family *Vibrionaceae*.

Kwok et al. (2002) show that a 600-bp partial sequence of the *hsp60* gene in 15 *Vibrio* species shared 71 to 82% sequence identity, whereas epidemiologically distinct strains within the same species had 96 to 100% sequence identity. This suggests that the *hsp60* gene sequence may be useful for phylogenetic analysis and identification of *Vibrio* species.

Genes coding the functions needed for DNA replication are also well conserved. In particular, the sequence variation of the *gyrB* gene coding for the B subunit of DNA gyrase is used for phylogenetic analysis of various organisms. *V. parahaemolyticus* and *V. alginolyticus* show very similar biochemical characteristics and sequence identity in their rRNA genes is >99%, whereas the *gyrB* sequence identity between these two species is only 86.8%. Venkateswaran et al. (1998) therefore claimed that the *gyrB* sequence is useful for genetic identification of vibrios. Moreover, comparison of the *gyrB* sequence was useful for phylogenetic analysis of *Vibrio splendidus* and related species (Le Roux et al., 2004a,b). It is noteworthy that the *gyrB* sequence served as the target to develop a PCR for the identification of *V. hollisae* (Vuddhakul et al., 2000b). The *gyrA* gene coding the A subunit of DNA gyrase and the *parC* gene coding for a subunit of the type II topoisomerase are also presumed to be conserved. Of relevance, the homology of the *gyrA* gene among bacterial species including *V. parahaemolyticus* was higher than that of the *parC* sequences (Okuda et al., 1999). Future nucleotide sequence analysis of these genes for species belonging to the *Vibrionaceae* will reveal whether these genes are useful for identification of the *Vibrionaceae* species.

The gene coding for the superoxide dismutase catalyzing the dismutation of the superoxide radical to hydrogen peroxide and oxygen is conserved in prokaryotes and eukaryotes. The *sodA* gene coding for manganese superoxide dismutase appears to be useful for identification of the *Vibrionaceae* species, and Shyu and Lin (1999) reported a *sodA*-targeted PCR for the identification of *V. parahaemolyticus*. Other workers have used the nucleotide sequence of the *sodA* gene from other *Vibrio* species (Kimoto et al., 2001).

The *groESL* gene coding for HSP and homologues of the *Escherichia coli rpoH* and *rpoS* genes coding, respectively, for the σ^{32} and σ^{S} subunits of RNA polymerase are associated with stress response; they were cloned from *V. cholerae,* and the nucleotide sequence was analyzed (Mizunoe et al., 1999; Sahu et al., 1997; Hiratsu et al., 1997). Other housekeeping genes were sequenced for phylogenetic use in strains belonging to *V. cholerae* and *V. parahaemolyticus* (Byun et al., 1999; Chowdhury et al., 2004b). Comparative analysis of these genes would reveal whether these genes could be used for identification of *Vibrionaceae* species.

The *lux* genes of various bacteria exhibiting bioluminescence appear to have come from a common ancestor. The nucleotide sequence of the *luxA* gene coding for luciferase was detected in a number of *V. cholerae* strains, including nonluminescent strains, whereas only about 10% of the *V. cholerae* strains exhibit a bioluminescence phenotype (Palmer and Colwell, 1991). The nucleotide sequence of the *luxA* gene was 99% identical among *V. cholerae* strains and 77% identical between *V. cholerae* and *V. harveyi.* This gene may be useful for the identification of luminescent species of the *Vibrionaceae.*

The *fur* gene coding for a regulator of an iron uptake system is detected in various bacterial species. Analysis of the *fur* sequences from *Vibrio* species so far characterized shows that this gene sequence may be useful for phylogenetic analysis of the genus *Vibrio* (Colquhoun and Sørum, 2002).

Genes Conserved Only or Predominantly in *Vibrionaceae*

The *toxR* gene was first discovered in *V. cholerae* as a positive transcriptional regulator for the *ctx* gene coding for the cholera toxin (Miller and Mekalanos, 1984). The *toxR* homologues coding for the regulators that stimulate the virulence gene expression in *V. parahaemolyticus* and *V. vulnificus* were also identified, and they were named the *toxR* gene (Lin et al., 1993; Lee et al., 2000). Subsequently, *V. hollisae,* *V. fischeri,* *V. (Listonella) anguillarum,* and a deepsea bacterium belonging to *Photobacterium* sp. were shown to carry the *toxR* sequences (Reich and Schoolnik, 1994; Vuddhakul et al., 2000b; Welch and Bartlett, 1998; Okuda et al., 2001). The nucleotide sequences homologous, at least in part, to the *toxR* gene were found in *V. fluvialis,* *V. alginolyticus,* *V. mimicus,* and two subspecies of *P. damselae* (Osorio and Klose, 2000). This suggests that the *toxR* gene is widely distributed in the family *Vibrionaceae,* including both pathogenic and nonpathogenic species. *toxR* is involved in the expression control of outer

membrane protein (OMP) in all *Vibrionaceae* species so far examined (Miller and Mekalanos, 1988; Crawford et al., 1998; Welch and Bartlett, 1998; Lee et al., 2000; Okuda et al., 2001). Therefore, the ancestral role of the *toxR* gene is the global control of genes involved in adaptation to environmental change (Okuda et al., 2001). The nucleotide sequence identity of the *toxR* gene sequences among *Vibrionaceae* species is relatively low. For example, *V. cholerae,* *V. parahaemolyticus,* and *V. vulnificus* are 52 to 63% identical (Miller and Mekalanos, 1984; Lin et al., 1993; Lee et al., 2000). The low interspecies homology values and universal distribution in the family *Vibrionaceae* make the *toxR* gene a good target for genetic identification of the *Vibrionaceae* species. Species-specific nucleotide bases in the *toxR* gene were used to develop PCR methods to identify *V. parahaemolyticus,* *V. hollisae,* *V. anguillarum,* and *V. cholerae* (Kim et al., 1999; Vuddhakul et al., 2000b; Okuda et al., 2001; Ghosh et al., 1997; Rivera et al., 2001). The nucleotide bases in the *toxR* gene are useful to distinguish newly emerging clones from other strains of *V. parahaemolyticus* (Matsumoto et al., 2000).

Vibrios are considered to play an important role in the turnover of chitin in the marine environment. The genes involved in chitin degradation may be conserved among a number of *Vibrionaceae* species. Two enzymes are usually required: a chitinase giving the disaccharide N, N'-diacethylchitobiose or $(GlcNAc)_2$ and a "chitobiase" giving GlcNAc from $(GlcNAc)_2$. The chitobiase genes of *V. vulnificus* and *V. harveyi* are evolutionarily closely related, but that of *V. parahaemolyticus* is not very similar (Somerville and Colwell, 1993; Wu and Laine, 1999). Future analysis of chitobiase genes from other vibrios may show the utility of this gene for identification.

THE GENE PROBE AND PCR TO IDENTIFY VIBRIOS

Some of the above-mentioned genes and other genes have been used as targets to identify and detect *Vibrionaceae* species using a DNA probe and PCR (Table 1). These are briefly explained for each species below.

V. cholerae

V. cholerae is widely distributed in estuarine and freshwater environments, and only a very small number of the *V. cholerae* strains in the natural environment have the ability to produce cholera toxin and cause the clinical symptoms of cholera (profuse, wa-

Table 1. Target genes for molecular identification of *Vibrionaceae*

Species	Target group	Target gene(s)	Detection method[a]	Reference(s)
P. damselae	Species	Capsular polysaccharide	PCR	Rajan et al., 2003
	Subsp. (*piscicida*)	16S rRNA	PCR	Kvitt et al., 2002; Osorio et al., 1999
	Subsp. (*damselae, piscicida*)	16S rRNA gene, *ureC* (urease)	Multiplex PCR	Osorio et al., 2000
P. leiognathi	Species	*luxA* (luciferase subunit)	DNA probe	Wimpee et al., 1991
P. phosphoreum	Species	*luxA* (luciferase subunit)	DNA probe	Wimpee et al., 1991
V. alginolyticus	Species	16S rRNA	PCR	Liu et al., 2004
V. anguillarum	Species	16S rRNA	Oligo probe	Rehnstam et al., 1989; Martinez-Picado et al., 1994
			PCR-DGG	Ji et al., 2004
		5S rRNA	Oligo probe	Ito et al., 1995
		vahl (hemolysin gene)	DNA probe	Hirono et al., 1996
			PCR	Hirono et al., 1996
		toxR (regulatory protein)	DNA probe	Okuda et al., 2001
			PCR	Okuda et al., 2001
			PCR	González et al., 2003
V. cholerae	Species	*rpoN* (alternative sigma factor)	Oligo probe	Jiang and Fu, 2001
		16S–23S rRNA intergenic space	PCR	Chun et al., 1999
		16S–23S rRNA intergenic space	PCR	Ghosh et al., 1997; Rivera et al., 2001
		toxR (regulatory protein)	PCR	Nandi et al., 2000
		ompW (outer membrane protein)	PCR	
		hylA (hemolysin)	Real-time PCR	Lyon, 2001
	Nonclassical strains	*rtxA* and *rtxC* (RTX toxin)	PCR	Chow et al., 2001
	Virulent strains	*ctx* (cholera toxin)	DNA probe	Kaper et al., 1989
			Oligo probe	Wright et al., 1992; Yoh et al., 1993
			PCR	Kobayashi et al., 1990; Shirai et al., 1991; Fields et al., 1992; Koch et al., 1993; Salles et al., 1993; Varela et al., 1993
		ompW, ctx	Nested PCR	Varela et al., 1994; Theron et al., 2000
		zot (zonula occludens toxin),	Multiplex PCR	Nandi et al., 2000
		ace (accessory cholera enterotoxin)	DNA probe	Kurazono et al., 1995; Novais et al., 1999; Rivera et al., 2001
		tcpA (pilin)	PCR	Keasler and Hall, 1993; Rivera et al., 2001
		ctx, tcp	Multiplex PCR	De et al., 2001
		O1 and O139 lipopolysaccharide	DNA probe	Waldor and Mekalanos, 1994; Yamasaki et al., 1999a,b
		ubeO and *rfb* (O1 and O139 lipopolysaccharide)	PCR	Lipp et al., 2003
		ctx, O1 and O139 lipopolysaccharide	Multiplex PCR	Hoshino et al., 1998
		ctxA, zot, ace, tcpA, ompU (outer membrane protein), *toxR*	Multiplex PCR	Singh et al., 2002
		hlyA for biotype differentiation	Oligo probe	Alim and Manning, 1990
		ctx, hlyA for biotype differentiation	Multiplex PCR	Shangkuan et al., 1995
		ctx, hlyA for biotype differentiation	Multiplex real-time PCR	Giglio et al., 2003
		pTLC plasmid	DNA probe	Rubin et al., 1998
		pTLC plasmid	PCR	Basu et al., 2000
		tdb (hemolysin)	DNA probe	Nishibuchi et al., 1990
			PCR	Tada et al., 1992b

Continued on following page

47

Table 1. *Continued*

Species	Target group	Target gene(s)	Detection method[a]	Reference(s)
		stn (heat-stable enterotoxin), *sto* (heat-stable enterotoxin)	PCR	Guglielmetti et al., 1994; Vicente et al., 1997; Rivera et al., 2001; Sarkar et al., 2002
V. costicola	Species	16S–23S rRNA intergenic space	PCR	Lee et al., 2002
V. diazotrophicus	Species	16S–23S rRNA intergenic space	PCR	Lee et al., 2002
V. fischeri	Species	*luxA* (luciferase subunit)	DNA probe	Wimpee et al., 1991
V. fluvialis	Species	16S–23S rRNA intergenic space	PCR	Lee et al., 2002
		16S rRNA	PCR-DGE	Ji et al., 2004
V. halioticoli	Species	*alyVG2* (alginate lyase gene)	In situ PCR	Sugimura et al., 2000
V. harveyi	Species	16S rRNA	PCR	Oakey et al., 2003
		toxR (regulatory protein)	PCR	Conejero and Hedreyda, 2003
		vhh (hemolysin)	PCR	Conejero and Hedreyda, 2004
		luxN (bioluminescence)	PCR	Hernandez and Olmos, 2004
		luxA (luciferase subunit)	DNA probe	Wimpee et al., 1991
V. hollisae	Species	*gyrB* (DNA gyrase)	PCR	Vuddhakul et al., 2000b
		toxR (regulatory protein)	PCR	Vuddhakul et al., 2000b
	Virulent strains	*tdh* (hemolysin)	DNA probe	Nishibuchi et al., 1985
			Oligo probe	Nishibuchi et al., 1986
			PCR	Nishibuchi et al., 1996
V. mimicus	Species	16S–23S rRNA intergenic space	PCR	Chun et al., 1999
		toxR (regulatory protein)	PCR	Bi et al., 2001
	Virulent strains	*vmh* (hemolysin)	PCR	Shinoda et al., 2004
		ctx (cholera toxin), *tcpP* (pilus production)	PCR	Shi et al., 1998; Bi et al., 2001
		ctxA, *zot* (zonula occludens toxin), *ace* (accessory cholera enterotoxin), *tcpA*, *ompU* (outer membrane protein), *toxR*	Multiplex PCR	Singh et al., 2002
		Heat-stable enterotoxin	Oligo probe	Ramamurthy et al., 1994
		Heat-stable enterotoxin	PCR	Vicente et al., 1997; Shinoda et al., 2004
		tdh (hemolysin)	DNA probe	Nishibuchi et al., 1990
		tdh	Oligo probe	Yamamoto et al., 1992
		tdh	PCR	Tada et al., 1992b; Uchimura et al., 1993
V. nigripulchritudo	Species	16S–23S rRNA intergenic space	PCR	Lee et al., 2002
V. ordalii	Species	5S rRNA	Oligo probe	Ito et al., 1995
V. parahaemolyticus	Species	*tlh* (hemolysin)	DNA probe	Taniguchi et al., 1986
		tlh	Oligo probe	McCarthy et al., 1999; Nordstrom and DePaola, 2003
		0.76-kb sequence in pR72H	DNA probe	Lee et al., 1995a
		0.76-kb sequence in pR72H	PCR	Lee et al., 1995b
		gyrB (DNA gyrase)	PCR	Venkateswaran et al., 1998
		toxR (regulatory protein)	PCR	Kim et al., 1999
		sodA (manganese superoxide dismutase)	PCR	Kim et al., 1999; Shyu and Lin, 1999

Organism	Specificity	Target	Method	Reference(s)
	Virulent strains	*tdh* (hemolysin)	DNA probe	Nishibuchi et al., 1985
			Real-time PCR	Blackstone et al., 2003
		trh (hemolysin)	DNA probe	Shirai et al., 1990; Kishishita et al., 1992
		tdh, *trh*	Oligo probe	Nishibuchi et al., 1986; Lee et al., 1992; Yamamoto et al., 1992; McCarthy et al., 2000
		tdh-specific mRNA, *trh*-specific mRNA	PCR	Tada et al., 1992a; Lee and Pan, 1993
			Real-time mRNA	Nakaguchi et al., 2004; Masuda et al., 2004
	Pandemic clone	*tlh*, *tdh*, *trh*	Multiplex PCR	Bej et al., 1999
		tlh, *tdh*, *trh*	Real-time PCR	Davis et al., 2004
		toxRS (regulatory proteins)	PCR	Matsumoto et al., 2000
		Filamentous phage sequence	PCR	Nasu et al., 2000; Iida et al., 2001; Laohaprertthisan et al., 2003
		Cloned sequence of unknown function	PCR	Okura et al., 2004
		toxRS, *tdh*	Multiplex PCR	Okura et al., 2003
V. penaeicida	Species	16S rRNA	PCR	Saulnier et al., 2000
V. proteolyticus	Species	16S rRNA	Oligo probe	Muniesa-Pere et al., 1996
V. salmonicida	Species	16S–23S rRNA intergenic space	PCR	Lee et al., 2002
V. splendidus	Species	16S–23S rRNA intergenic space	PCR	Lee et al., 2002
V. tubiashii	Species	16S–23S rRNA intergenic space	PCR	Lee et al., 2002
V. vulnificus	Species	16S–23S rRNA intergenic space	PCR	Lee et al., 2002
		16S rRNA	Oligo probe	Aznar et al., 1994; Cerda-Cuellar et al., 2000
		23S rRNA	Nested PCR	Arias et al., 1995
		vvh (hemolysin)	DNA probe	Wright et al., 1985; Morris et al., 1987
			Oligo probe	Wright et al., 1993; Banerjee et al., 2002
			PCR	Aono et al., 1997; Brauns et al., 1991; Coleman and Oliver, 1996; Hill et al., 1991; Lee et al., 1998; Coleman et al., 1996
	Clinically important strains	*vvh*-specific mRNA	Real-time PCR	Panicker et al., 2004b; Campbell and Wright, 2004
		toxR (regulatory protein)	Seminested RT-PCR	Fischer-Le Saux et al., 2002
		vvh, *viuB* (iron acquisition)	Real-time PCR	Takahashi et al., 2005
		Noncoding sequence	Multiplex real-time PCR	Panicker et al., 2004c
			PCR	Rosche et al., 2005

[a] Oligo probe, oligonucleotide probe; real-time mRNA, real-time mRNA monitoring; PCR-DGE, PCR followed by denaturing gel electrophoresis.

tery diarrhea and dehydration, leading to death). Gene probes and PCR methods to identify *V. cholerae* at the species level have been reported regardless of their ability to cause cholera symptoms. Jiang and Fu (2001) used an oligonucleotide probe for the 16S to 23S rRNA intergenic spacer region to identify colonies on a membrane in a hybridization assay to enumerate *V. cholerae* from coastal waters. PCR methods were used to target *V. cholerae*-specific bases in the 16S to 23S rRNA intergenic spacer region (Chun et al., 1999); the *toxR* gene (Ghosh et al., 1997; Rivera et al., 2001) or the *ompW* gene coding for an OMP was also used (Nandi et al., 2000). Identification of *V. cholerae* isolated from the environment by the outer membrane gene-targeted PCR was shown to be more accurate than biochemical test-based identification kits (Le Roux et al., 2004a,b). However, the *ompU* gene coding for another OMP, OmpU, a putative adherence factor and a bile resistance factor, was detected using a PCR method whereby only 77% of the *V. cholerae* strains isolated from the environment were found (Karunasagar et al., 2003). This outer membrane gene is not useful for identification at the species level. The *hylA* gene coding for a *V. cholerae* hemolysin is distributed in almost all strains of *V. cholerae* (Brown and Manning, 1985). Although this hemolysin does not appear to play a major role in pathogenesis, this gene was used as the target gene in developing a real-time PCR method for the rapid and quantitative detection of *V. cholerae* (Lyon, 2001). A multiplex PCR whereby the collagenase genes from three *Vibrio* species can be detected was used to identify *V. cholerae* (Table 2).

Medical microbiologists are interested in developing sensitive and definitive genetic methods to differentiate between virulent and avirulent strains of *V. cholerae*. The ability to produce cholera toxin is considered the major marker for virulent strains. Strains belonging to the O1 and O139 serotypes usually produce cholera toxin, and these serotypes are indirect markers for virulent strains in many clinical laboratories. To detect cholera toxin from *V. cholerae*, a latex agglutination-based immunology kit is commercially available. However, the genetic methods are more definitive because culture conditions affect toxin production. Therefore, the *ctx* gene coding for the cholera toxin and genetic markers associated with the *ctx*-bearing strains have been extensively studied. A DNA probe that specifically detects the *ctxA* gene coding for the A subunit of the cholera toxin was developed by Kaper et al. (1989). This probe was used in an epidemiological analysis of cholera cases in the United States, Hong Kong, and Australia (Kaper et al., 1982; Yam et al., 1989; Desmarchelier and Senn, 1989) and for the detection of environmental

and clinical strains of *V. cholerae* O1 in Japan (Minami et al., 1991). Examination of environmental strains of *V. cholerae* with this DNA probe found exceptional *V. cholerae* O1 strains lacking the *ctx* gene (Minami et al., 1991) and exceptional *V. cholerae* non-O1 strains carrying the *ctx* gene (Nair et al., 1988). The DNA probe labeled with an isotope was used as the *ctx*-specific DNA probe. A convenient nonisotopic hybridization assay using an alkaline phosphatase-labeled or -conjugated oligonucleotide probe was reported (Wright et al., 1992; Yoh et al., 1993). In the 1990s, PCRs to detect the *ctx* gene were reported by various groups (Kobayashi et al., 1990; Shirai et al., 1991; Fields et al., 1992; Koch et al., 1993; Salles et al., 1993; Varela et al., 1993). The *ctx*-specific PCR was useful for detecting *ctx*-bearing *V. cholerae* strains from fecal samples (Shirai et al., 1991; Ramamurthy et al., 1993; Miyagi et al., 1999) and from food samples (Koch et al., 1993; DePaola and Hwang, 1995; Karunasagar et al., 1995). Nested PCR to detect the *ctx* gene with high sensitivity was used to examine water samples and stool samples (Varela et al., 1994; Theron et al., 2000). Lipp et al. (2003) sequentially used PCR assays to detect the 16S to 23S intergenic spacer region, the *wbeO* and *rfb* genes (O1 and O139 antigens), and the *ctxA* gene from the DNA samples extracted from coastal waters and plankton to assess the presence of *ctx*-positive *V. cholerae* in these environmental samples. Nandi et al. (2000) reported a multiplex PCR to simultaneously detect the *ompW* and *ctx* genes to identify toxigenic strains of *V. cholerae* in a single step.

The *zot* and *ace* genes coding, respectively, for the zonula occludens toxin (Zot) and accessory cholera enterotoxin (Ace) and the *ctxAB* genes are located in the so-called "core region" (Baudry et al., 1992; Trucksis et al., 1993). The *zot* and *ace* genes are usually detected by specific DNA probes in *ctx*-bearing strains (Kurazono et al., 1995; Novais et al., 1999; Rivera et al., 2001). However, atypical non-O1/non-O139 strains carrying the *ctxA* gene, the *zot* gene, or both genes have been reported from the aquatic environment of California, where cholera epidemics are not known (Jiang et al., 2003).

The *tcpA* gene coding for the toxin-coregulated pilus is usually specific for *V. cholerae* O1 and O139 strains; the toxin-coregulated pilus is considered to play an important role in the intestinal colonization of these strains. The *tcpA* gene sequence varies to a degree between the classical and El Tor biotypes of O1 serotype (Iredell and Manning, 1994). The *tcpA* sequences of O139 strains are of the El Tor type (Rhine and Taylor, 1994). A multiplex PCR detecting the *ctx* and *tcp* genes and simultaneously differentiating the *tcpA* gene sequence of the classical and

Table 2. PCR methods for simultaneous detection of pathogenic species, including the genus *Vibrio*

PCR (reference)	Target	
	Species	Gene
Nested, uniplex (Wang et al., 1997)	*V. cholerae*	*ctx*
	V. parahaemolyticus	0.76-kb sequence in pR72H
	V. vulnificus	*vvh*
	10 other food-borne bacterial pathogens	
Multiplex (Brasher et al., 1998)	*V. cholerae*	*ctx* (cholera toxin)
	V. parahaemolyticus	*tlh* (hemolysin)
	V. vulnificus	*vvh* (hemolysin)
	Escherichia coli	*uidA* (β-glucuronidase)
	Salmonella enterica serovar Typhimurium	*invA* (invasion)
Multiplex (Kong et al., 2002)	*V. cholerae*	*epsM* (secretion protein)
	V. parahaemolyticus	16S–23S rRNA intergenic space
	7 other waterborne bacterial pathogens	
Multiplex (Lee et al., 2003)	*V. cholerae*	*ctx*
	V. parahaemolyticus	*tlh*
	V. vulnificus	*vvh*
	Total *Salmonella*	*hns* (histon-like protein)
	Pathogenic *Salmonella*	*spvB* (plasmid virulence)
Multiplex (Panicker et al., 2004a)	*Vibrio vulnificus*	*vvh*
		viuB (iron acquisition)
	V. cholerae	*ompU* (outer membrane protein)
		toxR (regulatory protein)
		tcpI (pilus production)
		hlyA (hemolysin)
	V. parahaemolyticus	*tlh*
		tdh (hemolysin)
		trh (hemolysin)
		Filamenous phage sequence
Multiplex (González et al., 2004)	*V. anguillarum*	*rpoN* (alternative sigma factor)
		fatA (iron transport)
	P. damselae subsp. *damselae*	*ureC* (urease)
		dly (phospholipase)
	V. parahaemolyticus	*gyrB* (DNA gyrase)
		toxR
	V. vulnificus	*vvh*
	Aeromonas salmonicida	*vap* (surface array protein)
		Plasmid sequence
Multiplex (Di Pinto et al., 2005)	*V. cholerae*	Collagenase
	V. parahaemolyticus	Collagenase
	V. alginolyticus	Collagenase
Universal (Hong et al., 2004)	*V. cholerae*	23S rRNA gene
	V. parahaemolyticus	23S rRNA gene
	12 other food-borne bacterial pathogens	
Multiplex reverse transcription (Morin et al., 2004)	*V. cholerae* O1	*rfbE* (lipopolysaccharide)
	E. coli O157:H7	*rfbE* (lipopolysaccharide)
		fliC (flagellin)
	Salmonella enterica serovar Typhi	*tyv* (lipopolysaccharide)
Duplex real-time (Fukushima et al., 2003)	*V. cholerae*	RTX toxin
		ctx
	V. parahaemolyticus	*tdh*
		trh
	V. vulnificus	*vvh*
	14 other food- and waterborne bacterial pathogens	

El Tor biotypes is available, but the PCR conditions require fine-tuning (De et al., 2001, 2004). The PCR methods demonstrate that the *tcpA* gene is well conserved in *V. cholerae* O1 strains (Keasler and Hall, 1993; Rivera et al., 2001), but atypical strains of *V. cholerae* were also found, e.g., *tcpA*-negative O1 strains isolated from clinical specimens (Vital Brazil et al., 2002); *ctx*-positive but *tcpA*-negative non-O1/non-O139 strains and *ctx*-negative but *tcpA*-positive strains of the non-O1/non-O139 serotypes were isolated from clinical and environmental sources (Ghosh et al., 1997; Novais et al., 1999). DNA sequences associated with the O1 and O139 antigen synthesis are useful as DNA probes to detect O1 and O139 strains (Waldor and Mekalanos, 1994; Nair et al., 1995; Yamasaki et al., 1999a,b). PCR was used to detect O1 serotype-specific genes in the clinically significant rough strains lacking the O1 antigen expression, and distribution of such strains in seafood was shown (De et al., 2004). A multiplex PCR examined the O1 and O139 antigens and the *ctx* gene in single PCR sets; it can substitute O1 and O139 serotyping as well as detection of cholera toxin production (Hoshino et al., 1998). A multiplex PCR one-step detection of six genes involved in virulence expression and regulation (*ctxA*, *zot*, *ace*, *tcpA*, *ompU*, and *toxR*) was also developed (Singh et al., 2002).

V. cholerae O1 strains are divided into classical and El Tor biotypes. The *hlyA* gene in O1 strains of the classical biotype has a deletion of 11-bp nucleotides. This causes an inability of the classical O1 strains to lyse sheep erythrocytes (Goldberg and Murphy, 1985; Alim et al., 1988; Alim and Manning, 1990) and is used as one of the criteria to differentiate between the two biotypes of *V. cholerae* O1. An oligonucleotide probe to specifically detect the 11-bp sequence in the *hlyA* gene of the El Tor strains could be used to differentiate the two biotypes (Alim and Manning, 1990). Rivera et al. (2001) reported a PCR to separately detect the *hly* genes of the different biotypes. Shangkuan et al. (1995) developed a multiplex PCR to differentiate the *hly* genes (three primers) and to detect the *ctx* gene (two primers) that enables differential detection of two biotypes of *V. cholerae* O1 at the same time. A real-time PCR technique was applied to this multiplex PCR system, but the results indicate that we have to be careful in designing the reaction conditions (Giglio et al., 2003).

The *ctx* core region is part of a temperate phage called CTXΦ (Waldor and Mekalanos, 1996); strains belonging to O1 and O139 serotypes appear to have acquired the CTXΦ. A plasmid named pTLC (toxin-linked cryptic plasmid) can exist as a covalently closed circular DNA and is tandem duplicated in the chromosome where a chromosomally integrated form

of pTLC is located adjacent to the CTX prophage. The pTLC may play a role in acquisition or replication of the CTXΦ phage (Rubin et al., 1998). The DNA probe or PCR targeting the pTLC plasmid sequence may be used as a marker for virulent strains, since the sequence was detected only in *ctx*-positive strains (Rubin et al., 1998). However, exceptional *ctx*-positive strains lacking the pTLC plasmid sequence have recently been reported (Chen et al., 2004). The RTX toxin (*repeats in toxin*: GD-rich repeats are conserved in this toxin family) gene cluster encoding cytotoxic activity to Hep-2 cells is genetically linked to the CTXΦ genome (Lin et al., 1999). However, PCR shows that the RTX toxin genes (*rtxA* and *rtxC*) are present in all strains of *V. cholerae* belonging to the O1, O139, and non-O1/O139 serotypes except in the O1 classical biotype (Chow et al., 2001).

Genetic examination of the *V. cholerae* strains shows rare strains carrying the toxin genes that are reported to be major virulence genes in other bacterial species. A study using a DNA probe to detect the *tdh* gene coding for the thermostable direct hemolysin (TDH) of *V. parahaemolyticus* showed that some strains of *V. cholerae* non-O1 isolated from patients with diarrhea carry the *tdh* gene (Baba et al., 1991). Some *V. cholerae* non-O1 strains isolated from patients with diarrhea produce a heat-stable enterotoxin (NAG-ST) that is related to the *E. coli* heat-stable enterotoxin. Studies with a DNA probe and an oligonucleotide probe specific to the NAG-ST gene showed some of the non-O1 strains from the environment, and some O1 strains isolated from both clinical and environmental sources carry the NAG-ST gene (Takeda et al., 1991; Pal et al., 1992; Mallard and Desmarchelier, 1995; Hoge et al., 1990). Subsequently, a PCR to detect the heat-stable enterotoxin genes of *V. cholerae* non-O1 and O1 (*stn* and *sto*, respectively) was reported (Guglielmetti et al., 1994; Vicente et al., 1997; Rivera et al., 2001; Sarkar et al., 2002). The *tdh* genes detected in various *Vibrio* species appear to be associated with transposon-like genetic units (Terai et al., 1991). The *stn* and *sto* genes are flanked by direct repeat sequences (Ogawa and Takeda, 1993), suggesting that *V. cholerae* probably acquired these genes from other bacteria during evolution.

V. parahaemolyticus

Biochemical identification of *V. parahaemolyticus* may be difficult when the strains are isolated from seafood or a marine environment. The samples may contain related *Vibrio* species or yet uncharacterized bacteria that may exhibit similar biochemical characteristics. Enumeration of the total number of *V. parahaemolyticus* cells in seafood is required to evaluate the safety of the seafood. The most probable number (MPN) determination method and identification method are usually combined to enumerate *V. parahaemolyticus*. Genetic identification methods that can replace biochemical tests and give a more definitive result can be very helpful in this regard. Taniguchi et al. (1986) cloned and characterized a gene coding for the thermolabile hemolysin (*tlh*). This gene was detected in all tested strains of *V. parahaemolyticus* isolated from clinical and environmental sources but was absent in other *Vibrio* species (Taniguchi et al., 1986). Nonisotopically labeled oligonucleotide probes to detect the *tlh* gene easily identify *V. parahaemolyticus* (McCarthy et al., 1999; Nordstrom and DePaola, 2003). To determine the MPN of *V. parahaemolyticus* in oysters, Gooch et al. (2001) used a hybridization method with these probes on a direct spread plate. Banerjee et al. (2002) also reported a nonisotopically labeled *tlh* gene probe to detect *V. parahaemolyticus* using a colony hybridization method. The *tlh* gene was included as a target gene to identify *V. parahaemolyticus* using a multiplex PCR and was selected as the target gene for quantitative detection of *V. parahaemolyticus* from environmental samples using a real-time PCR (Kaufman et al., 2004). Lee et al. (1995a) cloned a 0.76-kb nucleotide sequence that appears to be specific to *V. parahaemolyticus,* although its function is unknown. They named this clone pR72H and reported a hybridization method and a PCR assay targeting this sequence for specific identification of *V. parahaemolyticus* (Lee et al., 1995a,b). Identification of the strains isolated from seawater using this PCR was shown to be more specific than identification using biochemical tests (Robert-Pillot et al., 2002). This PCR procedure was used with a MPN approach to enumerate *V. parahaemolyticus* in mussels (Croci et al., 2002). Another line of thought for genetic identification of *V. parahaemolyticus* is the use of PCR to target *V. parahaemolyticus*-specific nucleotides within the genes conserved in *Vibrionaceae* or in higher taxons. These include the PCR using the *toxR* gene (Kim et al., 1999), the *gyrB* gene (Venkateswaran et al., 1998), and the *sodA* gene (Shyu and Lin, 1999). Stringent PCR conditions should be followed with these PCRs because these genes are also present in other *Vibrionaceae*. The multiplex PCR of Di-Pinto et al. (2005) detects the collagenase gene that may be used to identify *V. parahaemolyticus* (Table 2).

    V. parahaemolyticus strains are divided into virulent and avirulent strains based on the presence or absence of the *tdh* and/or the *trh* genes coding for the TDH and TDH-related hemolysin (TRH), respectively (Nishibuchi and Kaper, 1995). Strains possessing the *tdh* gene, the *trh* gene, or both genes are considered virulent strains; <2% of the environmental strains carry these virulence genes (Nishibuchi and Kaper, 1995). Therefore, methods to identify the virulent strains detecting these virulence genes have been reported. The hybridization method using the *tdh* gene probe was shown to be more definitive than the Kanagawa test detecting hemolytic activity of the TDH, because some of the *tdh*-positive strains show weak hemolytic activity (Nishibuchi et al., 1985). The amount of TRH produced from *trh*-bearing strains is much smaller than that of the TDH produced in Kanagawa test-positive strains (Shirai et al., 1990). The *trh* gene sequences in various strains vary but are close enough to detect either of two representative *trh* sequences, *trh1* and *trh2* (Kishishita et al., 1992). These two representative *trh* genes share ca. 80% sequence identity and are approximately 68% identical to the *tdh* gene. An enzyme-linked immunosorbent assay was reported to detect the *trh1* gene product (Shirai et al., 1990). It is not easy to detect the *trh2* gene product by hemolytic activity or by immunological methods (Kishishita et al., 1992). Urease-positive strains are rare, and TRH-producing strains are usually urease positive because the *ure* gene cluster is closely linked to the *trh* gene on the chromosome (Iida et al., 1997). Urease production therefore may be used as a marker for *trh*-bearing strains. Detection of the *trh* genes, however, is direct and more definitive than detecting the *trh* gene product or urease production. Hybridization assays using *trh1*- and *trh2*-specific DNA probes and *tdh*-specific probes show strong association of the *tdh* and/or *trh* genes with clinical strains (Shirai et al., 1990; Kishishita et al., 1992; Okuda et al., 1997b). Oligonucleotide probes to detect the *tdh* and *trh* genes were developed and applied to examine test strains or artificially contaminated oyster samples using the DNA colony hybridization and dot blot hybridization methods (Nishibuchi et al., 1986; Lee et al., 1992; McCarthy et al., 2000; Yamamoto et al., 1992). A PCR to detect the *tdh* and *trh* genes was developed by Tada et al. (1992b) and used for examination of strains isolated from clinical and environmental sources, artificially contaminated fecal specimens, and fish (Tada et al., 1992a; Okuda et al., 1997a,b; Matsumoto et al., 2000; Vuddhakul et al., 2000a; Islam et al., 2004; Hara-Kudo et al., 2003; Cabrera-Garcia et al., 2004). Lee and Pan (1993) also reported a *tdh*-targeted PCR and examined a contaminated fecal specimen. PCR protocols to detect the *tdh* and *trh* genes combined with a rapid automated fluorescent detection system were reported (Tada et al., 1992b). Blackstone et al.

(2003) developed a real-time PCR assay to detect the *tdh* gene and applied this assay to the examination of oyster samples. Automated assay systems to rapidly and quantitatively detect *tdh*- and *trh*-specific mRNA have been developed (Nakaguchi et al., 2004; Masuda et al., 2004). The assays are useful to detect and identify viable target strains actively producing TDH or TRH. A multiplex PCR to detect the *tlh*, *tdh*, and *trh* genes simultaneously was developed and used to examine shellfish samples and for an epidemiological study comparing clinical strains and oyster isolates (Bej et al., 1999; Kaufman et al., 2002). The *toxR*-targeted PCR to identify *V. parahaemolyticus* (Kim et al., 1999) and the above multiplex PCR were used to determine the MPN of total and virulent strains of *V. parahaemolyticus* in seawater and marine organic materials. The PCR-based MPN methods were shown to be more sensitive than the conventional culture-based MPN methods (Alam et al., 2002). Real-time PCR technology was recently applied to detect the *tlh*, *tdh*, and *trh* genes in food (Davis et al., 2004).

Methods to identify a genetically unique clone of virulent *V. parahaemolyticus* carrying the *tdh* gene have been reported. A new clone of *V. parahaemolyticus* causing a pandemic since 1996 was first suspected by the prevalence of identical serotypes (O3:K6) and by an AP-PCR technique (Okuda et al., 1997b; Matsumoto et al., 2000). Subsequently, a simple PCR to identify the clone was developed. This includes a PCR to detect the pandemic clone-specific nucleotides in the *toxRS* genes (Matsumoto et al., 2000), a pandemic clone-specific open reading frame in a lysogenic filamentous phage (Nasu et al., 2000; Iida et al., 2001; Laohaprertthisan et al., 2003), or a pandemic clone-specific nucleotide sequence originally detected by an AP-PCR method (Okura et al., 2004). A PCR to detect the possible pandemic O3:K6 strains was developed based on the enterobacterial intergenic consensus sequence-PCR, with which only a limited number of American strains were tested (Khan et al., 2002). Of these markers for the pandemic clone, the filamentous phage sequence appears to be the least specific (Okura et al., 2003; Chowdhury et al., 2004a). A multiplex PCR combining the detection of the pandemic clone-specific nucleotides in the *toxRS* genes and the *tdh* gene was reported (Okura et al., 2003).

V. vulnificus

V. vulnificus is classified into two subgroups termed biogroup 1 and biogroup 2. The two subgroups were initially differentiated using biochemical characteristics, especially indole production; sub-sequently, the difference in the O antigenicity was also found (Biosca et al., 1996). It is generally understood that biogroup 1 is a human pathogen whereas biogroup 2 was found to be an eel pathogen. Recently, an exceptional biogroup 2 strain was isolated from a human wound (Linkous and Oliver, 1999). Various 16S rRNA gene-targeted oligonucleotide probes to identify *V. vulnificus* have been reported (Aznar et al., 1994; Cerda-Cuellar et al., 2000). These include the biotype 1-specific probe and the probe specific for both biotypes 1 and 2 (Aznar et al., 1994). A sensitive nested PCR targeting the 23S rRNA gene of *V. vulnificus*, including biotypes 1 and 2, was developed to detect this bacterium in fish, sediments, and water (Arias et al., 1995). Randa et al. (2004) combined a nested PCR for the 23S rRNA gene, and the MPN estimation protocol was performed to enumerate *V. vulnificus* from water samples. A new subgroup named biogroup 3 was reported in an outbreak from a human infection (wound and bacteremia) (Bisharat et al., 1999). There are commercially available identification systems based on phenotypic characteristics that may misidentify biogroup 3 strains at the species level (Colodner et al., 2004). A recent study showed that the biogroup designations do not always correlate to the phylogenetic relationships shown by various genetic and enzymatic techniques (Gutacker et al., 2003). Therefore, subgrouping of *V. vulnificus* strains needs further evaluation.

In studies of the pathogenicity of *V. vulnificus*, a hemolysin termed cytotoxin-hemolysin received attention in the early stages of the investigation. Evidence now suggests that this hemolysin is not a virulence determinant; currently, the gene (*vvh*) coding for this hemolysin is studied as the marker to identify *V. vulnificus*. Hybridization studies using a 3.2-kb DNA fragment containing the *vvh* gene as the probe demonstrated specificity of the *vvh* gene; the sequence of the probe was present in *V. vulnificus* regardless of the source of isolation (clinical or environmental) and the biotype (biotypes 1 and 2) and in no other *Vibrio* species or other genera (Wright et al., 1985; Morris et al., 1987). A *vvh*-targeted alkaline phosphatase-labeled oligonucleotide probe developed by Wright et al. (1993) was shown to be more specific than a commercially available biochemical identification kit to identify *V. vulnificus* strains from the environment (Dalsgaard et al., 1996). The colony hybridization test using this oligonucleotide probe was found to enumerate *V. vulnificus* in various environmental samples, including water, oysters, plankton, and sediments (Wright et al., 1996; DePaola et al., 1997). Banerjee et al. (2002) report a nonisotopically labeled *vvh* gene probe for a colony hybridization assay. A PCR targeting the *vvh* gene was developed to examine isolated strains (Aono et al., 1997), the cells of the viable but nonculturable (VBNC) state (Brauns et al., 1991; Coleman and Oliver, 1996), environmental samples (Hill et al., 1991), and clinical specimens (Lee et al., 1998). Coleman et al. (1996) showed that the *vvh*-targeted PCR assay was specific to both biotype 1 and biotype 2. Real-time PCR assays targeting the *vvh* gene were found to quantitatively detect *V. vulnificus* in shellfish (Panicker et al., 2004b; Campbell and Wright, 2004). *vvh*-specific mRNA was detected using *V. vulnificus* VBNC with a seminested reverse transcription-PCR (Fischer-Le Saux et al., 2002). Quantitative detection of this species in the environment by various PCR methods, along with the culture-based identification methods, is useful for the study of *V. vulnificus* VBNC (Randa et al., 2004; Campbell and Wright, 2004).

Takahashi et al. (2005) looked at species-specific bases in the *toxR* gene to develop species-specific PCR primers, and the specificity was the same as that for the *vvh*-targeted PCR. They showed that the *toxR*-targeted PCR primers are useful in a real-time PCR system to enumerate *V. vulnificus* in seawater and seafood samples.

Some workers have attempted to find genetic markers useful for the identification of clinically important strains. The markers are primarily found for clinical strains and much less frequently for environmental strains. Nilsson et al. (2003) reported a statistically significant association between the nucleotide base change in the small subunit 16S rRNA gene and human clinical strains. Oligonucleotide probes or a PCR to detect this marker would be useful to detect clinically significant strains. Panicker et al. (2004c) showed that *viuB*, a gene involved in iron acquisition, was detectable by their PCR primers, established a multiplex PCR detecting the *vvh* and *viuB* genes, and used a real-time PCR system to detect these genes in shellfish. Rosche et al. (2005) found that the DNA sequence detected by a randomly amplified polymorphic DNA PCR assay was associated with the clinical strains and developed a PCR to detect this sequence.

Other Human-Pathogenic *Vibrio* Species

V. mimicus is isolated from patients with diarrhea and less frequently from extraintestinal infections and is widely distributed in the aquatic environment. This species is taxonomically closely related to *V. cholerae*; the two species are differentiated by the ability to ferment sucrose. Chun et al. (1999) proposed a PCR system to distinguish *V. cholerae* from *V. mimicus* based on a sequence variation in the 16S to 23S rRNA intergenic spacer region. Bi et al. (2001)

developed a *toxR*-targeted PCR to identify *V. mimicus*. Shinoda et al. (2004) report that the *vmh* gene coding for the *V. mimicus* hemolysin was detected in all strains of *V. mimicus* by using a PCR and suggest that PCR may be useful for identifying this species. Some strains of *V. mimicus* share important virulence genes with *V. cholerae* and less frequently with *V. parahaemolyticus*. The *ctx* and *tcpP* genes are detected in some strains of *V. mimicus* using PCR (Shi et al., 1998; Bi et al., 2001). An environmental strain of *V. mimicus* carries six virulence-associated genes (*ctxA*, *zot*, *ace*, *tcpA*, *ompU*, and *toxR*) of *V. cholerae* and was detected using a multiplex PCR (Singh et al., 2002). The NAG-ST gene was detected in some strains of *V. mimicus* using an oligonucleotide probe (Ramamurthy et al., 1994) and PCR (Vicente et al., 1997; Shinoda et al., 2004). The *tdh* gene of *V. parahaemolyticus* was detected in some strains using a DNA probe (Nishibuchi et al., 1985, 1990), an oligonucleotide probe method (Yamamoto et al., 1992), and a PCR method (Tada et al., 1992b; Uchimura et al., 1993).

V. hollisae is taxonomically distant from other pathogenic species in the genus *Vibrio*, and a new genus name, *Grimontia*, has been proposed (Thompson et al., 2003). *V. hollisae* has been isolated from patients with diarrhea and is rarely isolated from environmental samples. This species is biochemically inert; thus, it is difficult to develop a selective and differential isolation medium. This could be the reason for the rare isolation of *V. hollisae* from the environment. PCR assays targeting the *gyrB* and *toxR* genes were developed to identify *V. hollisae* to the species level (Vuddhakul et al., 2000b). The PCR could be useful not only to identify but also to screen for this species during isolation. A homologue of the *tdh* gene in *V. parahaemolyticus* was detected in all strains of *V. hollisae* by a DNA probe, suggesting that TDH may be an important virulence factor for *V. hollisae* (Nishibuchi et al., 1988, 1990, 1996; Yamasaki et al., 1991).

V. alginolyticus is listed as a human pathogen as well as a pathogen of aquatic organisms. *V. parahaemolyticus* and *Aeromonas* species are often misidentified as *V. alginolyticus* by phenotypic characteristics (Park et al., 1991; Robert-Pillot et al., 2002). Liu et al. (2004) used a PCR to target a specific sequence in the 16S rRNA gene to identify a strain isolated from diseased shrimp as *V. alginolyticus*. The collagenase gene-targeted multiplex PCR of Di Pinto et al. (2005) could be used to identify this species (Table 2).

Although genetic differentiation of *V. harveyi* from related species is not easy (Gauger and Gomez-Chiarri, 2002; Gomez-Gil et al., 2004), Oakey et al.

(2003) showed that a discriminative PCR targets the 16S rRNA gene for identification. The PCRs targeting the *toxR* gene, the *vhh* hemolysin gene, and the *luxN* (bioluminescence) gene may potentially be useful for this purpose (Conejero and Hedreyda, 2003, 2004; Hernandez and Olmos, 2004).

P. damselae consists of two subspecies, *P. damselae* subsp. *damselae* and *P. damselae* subsp. *piscicida*. *P. damselae* subsp. *damselae* was formerly *Vibrio damselae* and was considered a human pathogen. The nucleotide sequences of the rRNA genes and their intergenic sequences in the two subspecies are similar (Osorio et al., 2004, 2005). Rajan et al. (2003) developed a PCR targeting the capsular polysaccharide gene to identify *P. damselae* subsp. *piscicida*; however, both subspecies were detected using this PCR. A 16S rRNA gene-targeted PCR and nested PCR to identify *P. damselae* subsp. *piscicida* were proposed (Kvitt et al., 2002; Osorio et al., 1999). A multiplex PCR detecting the 16S rRNA sequence and the *ureC* gene could differentiate the two subspecies (Osorio et al., 2000).

Vibrio Species Pathogenic to Aquatic Animals

Some of the human-pathogenic *Vibrio* species described above are also known to be potentially pathogenic to aquatic animals. They are omitted in this section.

V. anguillarum identification is usually performed using biochemical tests and O serotyping. For a correct identification of this species, oligonucleotide probes for the 16S rRNA gene could be potentially useful (Rehnstam et al., 1989; Martinez-Picado et al., 1994). Ito et al. (1995) showed that oligonucleotide probes for the 5S rRNA gene could specifically differentiate and identify *V. anguillarum* and *V. ordalii*. Ji et al. (2004) reported amplification of the 16S rRNA sequence using PCR with universal primers followed by analysis of the amplicons by denaturing gel electrophoresis. They claim that six species of fish pathogens, including *V. anguillarum* and *V. fluvialis*, could be differentiated using this system. Detection of the hemolysin gene (*vah1*) using a DNA probe and PCR could be potentially useful in identifying *V. anguillarum* (Hirono et al., 1996). The DNA probe and PCR targeting the *toxR* gene (Okuda et al., 2001) and another PCR targeting the *rpoN* gene coding for an alternative sigma factor (González et al., 2003) could specifically identify *V. anguillarum*.

Lee et al. (2002) developed a PCR to target the 16S to 23S intergenic spacer region to identify eight *Vibrio* species that are potential pathogens of vertebrate and invertebrate aquatic animals: *Vibrio (Salinivibrio) costicola*, *V. diazotrophicus*, *V. fluvialis*, *V. ni-*

gripulchritudo, *V. proteolyticus*, *V. salmonicida*, *V. splendidus*, and *V. tubiashii*. Saulnier et al. (2000) reported a 16S rRNA gene-targeted PCR method that identified *V. penaeicida* as a pathogen of shrimp. An oligonucleotide probe for the 16S rRNA gene would be useful for identification of *V. proteolyticus* if it is used with a differential isolation medium (Muniesa-Pere et al., 1996).

Other *Vibrionaceae* Species

DNA probes to identify luminous members of the *Vibrionaceae* were reported. The DNA probes targeting the luciferase subunit genes (*luxA*) of *V. harveyi*, *V. fischeri*, *Photobacterium leiognathi*, and *Photobacterium phosphoreum* were shown to be specific at the species level when used under a high-stringency condition (Wimpee et al., 1991). Sugimura et al. (2000) developed an in situ PCR technique to detect the alginate lyase gene of *V. halioticoli*, a nonmotile member of the genus *Vibrio*.

Simultaneous Detection of Pathogenic Vibrios

V. cholerae, *V. parahaemolyticus*, and *V. vulnificus* are important food-borne pathogens. *V. cholerae* is an important waterborne pathogen. Workers have developed PCR-based systems to simultaneously detect these medically important vibrios and other food- and waterborne bacterial pathogens (Table 2). The most frequently used systems are based on the multiplex PCR formats. Alternative systems include individual PCRs using the same sample preparation and the same amplification conditions (Wang et al., 1997), a PCR using a universal primer (Hong et al., 2004), or a complicated real-time PCR system (Fukushima et al., 2003). The target genes employed were used for identification to the species level; however, the markers for virulent strains were included in some systems (Table 1). If the sensitivity and specificity of these systems suit the purpose of examination, the systems will be useful for examination of food and water samples.

CONCLUSIONS

Recently, many new species of the genus *Vibrio* have been proposed after due characterization of the strains isolated from aquatic organisms and natural environments. Although the differential characteristics include the 16S rRNA gene sequence, the 16S rRNA sequence may not always be useful for establishing molecular identification of a *Vibrionaceae* species at the species level due to its low discriminatory power. Alternative genes that are suitable for this purpose have been exploited and employed in identification systems for the *Vibrionaceae* species. Practical application of molecular identification techniques is limited mostly to medically important *Vibrio* species. This reflects the real need for rapid, easy, and reliable identification systems in clinical laboratories and for the fish farming industry. The target genes are not limited to those for species identification. The virulence genes have also been included as important targets to identify species in which differentiation between virulent and avirulent strains has been established. *V. cholerae* and *V. parahaemolyticus* are good examples of species for which the target genes have been extensively evaluated. Selection of suitable target genes and fine-tuning of their detection conditions are the basis for development of more sophisticated molecular identification systems, including multiplex PCR and real-time PCR.

REFERENCES

Abbott, S. L., L. S. Seli, M. Catino, Jr., M. A. Hartley, and J. M. Janda. 1998. Misidentification of unusual *Aeromonas* species as members of the genus *Vibrio*: a continuing problem. *J. Clin. Microbiol.* 36:1103–1104.

Alam, M. J., K. Tomochika, S. Miyoshi, and S. Shinoda. 2002. Environmental investigation of potentially pathogenic *Vibrio parahaemolyticus* in the Seto-Inland Sea, Japan. *FEMS Microbiol. Lett.* 208:83–87.

Alim, R. A., and P. A. Manning. 1990. Biotype-specific probe for *Vibrio cholerae* serogroup O1. *J. Clin. Microbiol.* 28:823–824.

Alim, R. A., U. H. Stroeher, and P. A. Manning. 1988. Extracellular proteins of *Vibrio cholerae*: nucleotide sequence of the structural gene (*hlyA*) for the haemolysin of the haemolytic El Tor strain O17 and characterization of the *hlyA* mutation in the non-haemolytic classical strain 569B. *Mol. Microbiol.* 2:481–488.

Aono, E., H. Sugita, J. Kawasaki, H. Sakakibara, T. Takahashi, K. Endo, and Y. Deguchi. 1997. Evaluation of the polymerase chain reaction method for identification of *Vibrio vulnificus* isolated from marine environments. *J. Food Prot.* 60:81–83.

Arias, C. R., E. Garay, and R. Aznar. 1995. Nested PCR method for rapid and sensitive detection of *Vibrio vulnificus* in fish, sediments, and water. *Appl. Environ. Microbiol.* 61:3476–3478.

Aznar, R., W. Ludwig, R. I. Amann, and K. H. Schleifer. 1994. Sequence determination of rRNA genes of pathogenic *Vibrio* species and whole-cell identification of *Vibrio vulnificus* with rRNA-targeted oligonucleotide probes. *Int. J. Syst. Bacteriol.* 44:330–337.

Baba, K., H. Shirai, A. Terai, K. Kumagai, Y. Takeda, and M. Nishibuchi. 1991. Similarity of the *tdh* gene-bearing plasmids of *Vibrio cholerae* non-01 and *Vibrio parahaemolyticus*. *Microb. Pathog.* 10:61–70.

Banerjee, S. K., S. Pandian, E. C. Todd, and J. M. Farber. 2002. A rapid and improved method for the detection of *Vibrio parahaemolyticus* and *Vibrio vulnificus* strains grown on hydrophobic grid membrane filters. *J. Food Prot.* 65:1049–1053.

Basu, A., A. K. Mukhopadhyay, P. Garg, S. Chakraborty, T. Ramamurthy, S. Yamasaki, Y. Takeda, and G. B. Nair. 2000. Diversity in the arrangement of the CTX prophages in classical strains of *Vibrio cholerae* O1. *FEMS. Microbiol. Lett.* 182:35–40.

Baudry, B., A. Fasano, J. Ketley, and J. B. Kaper. 1992. Cloning of a gene (*zot*) encoding a new toxin produced by *Vibrio cholerae*. *Infect. Immun.* 60:428–434.

Bej, A. K., D. P. Patterson, C. W. Brasher, M. C. Vickery, D. D. Jones, and C. A. Kaysner. 1999. Detection of total and hemolysin-producing *Vibrio parahaemolyticus* in shellfish using multiplex PCR amplification of *tlh*, *tdh*, and *trh*. *J. Microbiol. Methods* 36:215–225.

Bi, K., S. I. Miyoshi, K. I. Tomochika, and S. Shinoda. 2001. Detection of virulence associated genes in clinical strains of *Vibrio mimicus*. *Microbiol. Immunol.* 45:613–616.

Biosca, E. G., J. D. Oliver, and C. Amaro. 1996. Phenotypic characterization of *Vibrio vulnificus* biotype 2, a lipopolysaccharide-based homogeneous O serogroup within *Vibrio vulnificus*. *Appl. Environ. Microbiol.* 62:918–927.

Bisharat, N., V. Agmon, R. Finkelstein, R. Raz, G. Ben-Dror, L. Lerner, S. Soboh, R. Colodner, D. N. Cameron, D. L. Wykstra, D. L. Swerdlow, and J. J. Farmer III. 1999. Clinical, epidemiological, and microbiological features of *Vibrio vulnificus* biogroup 3 causing outbreaks of wound infection and bacteraemia in Israel. *Lancet* 354:1421–1424.

Blackstone, G. M., J. L. Nordstrom, M. C. Vickery, M. D. Bowen, R. F. Meyer, and A. DePaola. 2003. Detection of pathogenic *Vibrio parahaemolyticus* in oyster enrichments by real time PCR. *J. Microbiol. Methods* 53:149–155.

Brasher, C. W., A. DePaola, D. D. Jones, and A. K. Bej. 1998. Detection of microbial pathogens in shellfish with multiplex PCR. *Curr. Microbiol.* 37:101–117.

Brauns, L. A., M. C. Hudson, and J. D. Oliver. 1991. Use of the polymerase chain reaction in detection of culturable and nonculturable *Vibrio vulnificus* cells. *Appl. Environ. Microbiol.* 57:2651–2655.

Brown, M. H., and P. A. Manning. 1985. Haemolysin genes of *Vibrio cholerae*: presence of homologous DNA in non-haemolytic O1 and haemolytic non-O1 strains. *FEMS Microbiol. Lett.* 30:197–201.

Byun, R., B. R. Elbourne, R. Lan, and P. R. Reeves. 1999. Evolutionary relationships of pathogenic clones of *Vibrio cholerae* by sequence analysis of four housekeeping genes. *Infect. Immun.* 7:1116–11124.

Cabrera-García, M. E., C. Vázquez-Salinas, and E. I. Quiñones-Ramírez. 2004. Serologic and molecular characterization of *Vibrio parahaemolyticus* strains isolated from seawater and fish products of the Gulf of Mexico. *Appl. Environ. Microbiol.* 70:6401–6406.

Campbell, M. S., and A. C. Wright. 2004. Real-time PCR analysis of *Vibrio vulnificus* from oysters. *Appl. Environ. Microbiol.* 69:7137–7144.

Cerda-Cuellar, M., J. Jofre, and A. R. Blanch. 2000. A selective medium and a specific probe for detection of *Vibrio vulnificus*. *Appl. Environ. Microbiol.* 66:855–859.

Chen, C.-H., T. Shimada, N. Elhadi, S. Radu, and M. Nishibuchi. 2004. Phenotypic and genotypic characteristics of *ctx*$^+$ strains of *Vibrio cholerae* isolated from seafood in Malaysia and the significance of the strains in epidemiology. *Appl. Environ. Microbiol.* 70:1964–1972.

Chow, K. H., T. K. Ng, K. Y. Yuen, and W. C. Yam. 2001. Detection of RTX toxin gene in *Vibrio cholerae* by PCR. *J. Clin. Microbiol.* 39:2594–2597.

Chowdhury, A., M. Ishibashi, V. D. Thiem, D. T. Tuyet, T. V. Tung, B. T. Chien, L. von Seidlein, D. G. Canh, J. Clemens, D. D. Trach, and M. Nishibuchi. 2004a. Emergence and serovar transition of *Vibrio parahaemolyticus* pandemic strains isolated during a diarrhea outbreak in Vietnam between 1997 and 1999. *Microbiol. Immunol.* 48:319–327.

Chowdhury, N. R., O. C. Stine, J. G. Morris, and G. B. Nair. 2004b. Assessment of evolution of pandemic *Vibrio parahaemo-*

lyticus by multilocus sequence typing. *J. Clin. Microbiol.* **42:** 1280–1282.

Chun J., A. Huq, and R. R. Colwell. 1999. Analysis of 16S-23S rRNA intergenic spacer regions of *Vibrio cholerae* and *Vibrio mimicus*. *Appl. Environ. Microbiol.* **65:**2202–2208.

Coelho, A., H. Momen, A. C. Vicente, and C. A. Salles. 1994. An analysis of the V1 and V2 regions of *Vibrio cholerae* and *Vibrio mimicus* 16S rRNA. *Res. Microbiol.* **145:**151–156.

Coleman, S. S., D. M. Melanson, E. G. Biosca, and J. D. Oliver. 1996. Detection of *Vibrio vulnificus* biotypes 1 and 2 in eels and oysters by PCR amplification. *Appl. Environ. Microbiol.* **62:**1378–1382.

Coleman, S. S., and J. D. Oliver. 1996. Optimization of conditions for the polymerase chain reaction amplification of DNA from culturable and nonculturable cells of *Vibrio vulnificus*. *FEMS Microbiol. Ecol.* **19:**127–132.

Colodner, R., R. Raz, I. Meir, T. Lazarovich, L. Lerner, J. Kopelowitz, Y. Keness, W. Sakran, S. Ken-Dror, and N. Bisharat. 2004. Identification of the emerging pathogen *Vibrio vulnificus* biotype 3 by commercially available phenotypic methods. *J. Clin. Microbiol.* **42:**4137–4140.

Colquhoun, D. J., and H. Sørum. 2002. Cloning, characterisation and phylogenetic analysis of the *fur* gene in *Vibrio salmonicida* and *Vibrio logei*. *Gene* **296:**213–220.

Conejero, M. J., and C. T. Hedreyda. 2003. Isolation of partial *toxR* gene of *Vibrio harveyi* and design of *toxR*-targeted PCR primers for species detection. *J. Appl. Microbiol.* **95:**602–611.

Conejero, M. J., and C. T. Hedreyda. 2004. PCR detection of hemolysin (*vhh*) gene in *Vibrio harveyi*. *J. Gen. Appl. Microbiol.* **50:**137–142.

Crawford, J. A., J. B. Kaper, and V. J. DiRita. 1998. Analysis of ToxR-dependent transcription activation of *ompU*, the gene encoding a major envelope protein in *Vibrio cholerae*. *Mol. Microbiol.* **29:**235–246.

Croci, L., E. Suffredini, L. Cozzi, and L. Toti. 2002. Effects of depuration of molluscs experimentally contaminated with *Escherichia coli*, *Vibrio cholerae* O1, and *Vibrio parahaemolyticus*. *J. Appl. Microbiol.* **92:**460–465.

Dalsgaard, A., I. Dalsgaard, L. Hoi, and J. L. Larsen. 1996. Comparison of a commercial biochemical kit and an oligonucleotide probe for identification of environmental isolates of *Vibrio vulnificus*. *Lett. Appl. Microbiol.* **22:**184–188.

Davis, C. R., L. C. Heller, K. K. Peak, D. L. Wingfield, C. L. Goldstein-Hart, D. W. Bodager, A. C. Cannons, P. T. Amuso, and J. Cattanii. 2004. Real-time PCR detection of the thermostable direct hemolysin and thermolabile hemolysin genes in a *Vibrio parahaemolyticus* cultured from mussels and mussel homogenate associated with a foodborne outbreak. *J. Food Prot.* **67:**1005–1008.

De, K., T. Ramamurthy, S. M. Faruque, S. Yamasaki, Y. Takeda, G. B. Nair, and R. K. Nandy. 2004. Molecular characterisation of rough strains of *Vibrio cholerae* isolated from diarrhoeal cases in India and their comparison to smooth strains. *FEMS Microbiol. Lett.* **232:**23–30.

De, K., T. Ramamurthy, A. C. Ghose, M. S. Islam, Y. Takeda, G. B. Nair, and R. K. Nandy. 2001. Modification of the multiplex PCR for unambiguous differentiation of the El Tor and classical biotypes of *Vibrio cholerae* O1. *Indian J. Med. Res.* **114:**77–82.

DePaola, A., and G. C. Hwang. 1995. Effect of dilution, incubation time, and temperature of enrichment on cultural and PCR detection of *Vibrio cholerae* obtained from the oyster *Crassostrea virginica*. *Mol. Cell. Probes* **9:**75–81.

DePaola, A., M. L. Motes, D. W. Cook, J. Veazey, W. E. Garthright, and R. Blodgett. 1997. Evaluation of an alkaline phosphatase-labeled DNA probe for enumeration of *Vibrio vulnificus* in Gulf Coast oysters. *J. Microbiol. Methods* **29:**115–120.

Desmarchelier, P. M., and C. R. Senn. 1989. A molecular epidemiological study of *Vibrio cholerae* in Australia. *Med. J. Aust.* **150:** 631–634.

Di Pinto, A., G. Ciccarese, G. Tantillo, D. Catalano, and V. T. Forte. 2005. A collagenase-targeted multiplex PCR assay for identification of *Vibrio alginolyticus*, *Vibrio cholerae*, and *Vibrio parahaemolyticus*. *J. Food Prot.* **68:**150–153.

Dorsch, M., D. Lane, and E. Stackebrandt. 1992. Towards a phylogeny of the genus *Vibrio* based on 16S rRNA sequences. *Int. J. Syst. Bacteriol.* **42:**58–63.

Fields, P. I., T. Popovic, K. Wachsmuth, and O. Olsik. 1992. Use of the polymerase chain reaction for detection of toxigenic *Vibrio cholerae* O1 strains from the Latin American cholera epidemic. *J. Clin. Microbiol.* **30:**2118–2121.

Fischer-Le Saux, M., D. Hervio-Heath, S. Loaec, R. R. Colwell, and M. Pommepuy. 2002. Detection of cytotoxin-hemolysin mRNA in nonculturable populations of environmental and clinical *Vibrio vulnificus* strains in artificial seawater. *Appl. Environ. Microbiol.* **68:**5641–5646.

Fukushima, H., Y. Tsunomori, and R. Seki. 2003. Duplex realtime SYBR green PCR assays for detection of 17 species of foodor waterborne pathogens in stools. *J. Clin. Microbiol.* **41:**5134–5146.

Gauger, E. J., and M. Gomez-Chiarri. 2002. 16S ribosomal DNA sequencing confirms the synonymy of *Vibrio harveyi* and *V. carchariae*. *Dis. Aquat. Organ.* **52:**39–46.

Ghosh, C., R. K. Nandy, S. K. Dasgupta, G. B. Nair, R. H. Hall, and A. C. Ghose. 1997. A search for cholera toxin (CT), toxin coregulated pilus (TCP), the regulatory element ToxR and other virulence factors in non-O1/non-O139 *Vibrio cholerae*. *Microb. Pathog.* **22:**199–208.

Giglio, S., P. T. Monis, and C. P. Saint. 2003. Demonstration of preferential binding of SYBR Green I to specific DNA fragments in real-time multiplex PCR. *Nucleic Acids Res.* **31:**e136.

Goldberg, S. L., and J. R. Murphy. 1985. Cloning and characterization of the hemolysin determinants from *Vibrio cholerae* RV79 (Hly⁺), RV79(Hly⁻), and 569B. *J. Bacteriol.* **162:**35–41.

Gomez-Gil, B., S. Soto-Rodriguez, A. Garcia-Gasca, A. Roque, R. Vazquez-Juarez, F. L. Thompson, and J. Swings. 2004. Molecular identification of *Vibrio harveyi*-related isolates associated with diseased aquatic organisms. *Microbiology* **150:**1769–1777.

González, S. F., M. J. Krug, M. E. Nielsen, Y. Santos, and D. R. Call. 2004. Simultaneous detection of marine fish pathogens by using multiplex PCR and a DNA microarray. *J. Clin. Microbiol.* **42:**1414–1419.

González, S. F., C. R. Osorio, and Y. Santos. 2003. Development of a PCR-based method for the detection of *Listonella anguillarum* in fish tissues and blood samples. *Dis. Aquat. Organ.* **55:**109–115.

Gooch, J. A., A. DePaola, C. A. Kaysner, and D. L. Marshall. 2001. Evaluation of two direct plating methods using nonradioactive probes for enumeration of *Vibrio parahaemolyticus* in oysters. *Appl. Environ. Microbiol.* **67:**721–724.

Guglielmetti, P., L. Bravo, A. Zanchi, R. Monte, G. Lombardi, and G. M. Rossolini. 1994. Detection of the *Vibrio cholerae* heatstable enterotoxin gene by polymerase chain reaction. *Mol. Cell. Probes* **8:**39–44.

Gutacker, M., N. Conza, C. Benagli, A. Pedroli, M. V. Bernasconi, L. Permin, R. Aznar, and J. C. Piffaretti. 2003. Population genetics of *Vibrio vulnificus*: identification of two divisions and a distinct eel-pathogenic clone. *Appl. Environ. Microbiol.* **69:** 3203–3212.

Hara-Kudo, Y., K. Sugiyama, M. Nishibuchi, A. Chowdhury, J. Yatsuyanagi, Y. Ohtomo, A. Saito, H. Nagano, T. Nishina, H. Nakagawa, H. Konuma, M. Miyahara, and S. Kumagai. 2003. Prevalence of pandemic thermostable direct hemolysin-producing *Vibrio parahaemolyticus* O3:K6 in seafood and the coastal environment in Japan. *Appl. Environ. Microbiol.* **69:** 3883–3891.

Heidelberg, J. F., J. A. Eisen, W. C. Nelson, R. A. Clayton, M. L. Gwinn, R. J. Dodson, D. H. Haft, E. K. Hickey, J. D. Peterson, L. Umayam, S. R. Gill, K. E. Nelson, T. D. Read, H. Tettelin, D. Richardson, M. D. Eermolaeva, J. Vamathevan, S. Bass, H. Quin, I. Dragoi, P. Sellers, L. McDonald, T. Utterback, R. D. Fleishmann, W. C. Nierman, O. White, S. L. Salzberg, H. O. Smith, R. R. Colwell, J. J. Mekalanos, J. C. Venter, and C. M. Fraser. 2000. DNA sequence of both chromosomes of the cholera pathogen *Vibrio cholerae*. *Nature* **406:**477–484.

Hernandez, G., and J. Olmos. 2004. Molecular identification of pathogenic and nonpathogenic strains of *Vibrio harveyi* using PCR and RAPD. *Appl. Microbiol. Biotechnol.* **63:**722–727.

Hill, W. E., S. P. Keasler, M. W. Trucksess, P. Feng, C. A. Kaysner, and K. A. Lampel. 1991. Polymerase chain reaction identification of *Vibrio vulnificus* in artificially contaminated oysters. *Appl. Environ. Microbiol.* **57:**707–711.

Hiratsu, K., K. Yamamoto, and K. Makino. 1997. Molecular cloning and functional analysis of an *rpoS* homologue gene from *Vibrio cholerae* N86. *Genes Genet. Syst.* **72:**115–118.

Hirono, I., T. Masuda, and T. Aoki. 1996. Cloning and detection of the hemolysin gene of *Vibrio anguillarum*. *Microb. Pathog.* **21:**173–182.

Hoge, C. W., O. Sethabutr, L. Bodhidatta, P. Echeverria, D. C. Robertson, and J. G. Morris, Jr. 1990. Use of a synthetic oligonucleotide probe to detect strains of non-serovar O1 *Vibrio cholerae* carrying the gene for heat-stable enterotoxin (NAG-ST). *J. Clin. Microbiol.* **28:**1473–1476.

Holt, J. G., N. R. Krieg, P. H. A. Sneath, J. T. Staley, and S. T. Williams (ed.). 1994. *Bergey's Manual of Determinative Bacteriology*, 9th ed. Lippincott Williams & Wilkins, Baltimore, Md.

Hong, B. X., L. F. Jiang, Y. S. Hu, D. Y. Fang, and H. Y. Guo. 2004. Application of oligonucleotide array technology for the rapid detection of pathogenic bacteria of foodborne infections. *J. Microbiol. Methods* **58:**403–411.

Hoshino, K., S. Yamasaki, A. K. Mukhopadhyay, S. Chakraborty, A. Basu, S. K. Bhattacharya, G. B. Nair, T. Shimada, and Y. Takeda. 1998. Development and evaluation of a multiplex PCR assay for rapid detection of toxigenic *Vibrio cholerae* O1 and O139. *FEMS Immunol. Med. Microbiol.* **20:** 201–207.

Iida, T., A. Hattori, K. Tagomori, H. Nasu, R. Naim, and T. Honda. 2001. Filamentous phage associated with recent pandemic strains of *Vibrio parahaemolyticus*. *Emerg. Infect. Dis.* **7:**477–478.

Iida, T., O. Suthienkul, K. S. Park, G. Q. Tang, R. K. Yamamoto, M. Ishibashi, K. Yamamoto, and T. Honda. 1997. Evidence for genetic linkage between the *ure* and *trh* genes in *Vibrio parahaemolyticus*. *J. Med. Microbiol.* **46:**639–645.

Iredell, J. R., and P. A. Manning. 1994. Biotype-specific *tcpA* genes in *Vibrio cholerae*. *FEMS Microbiol. Lett.* **121:**47–54.

Islam, M. S., R. Tasmin, S. I. Khan, H. B. Bakht, Z. H. Mahmood, M. Z. Rahman, N. A. Bhuiyan, M. Nishibuchi, G. B. Nair, R. B. Sack, A. Huq, R. R. Colwell, and D. A. Sack. 2004. Pandemic strains of O3:K6 *Vibrio parahaemolyticus* in the aquatic environment of Bangladesh. *Can. J. Microbiol.* **50:**827–834.

Ito, H., H. Ito, I. Uchida, T. Sekizaki, and N. Terakado. 1995. A specific oligonucleotide probe based on 5S rRNA sequences for identification of *Vibrio anguillarum* and *Vibrio ordalii*. *Vet. Microbiol.* **43:**167–171.

Ji, N., B. Peng, G. Wang, S. Wang, and X. Peng. 2004. Universal primer PCR with DGGE for rapid detection of bacterial pathogens. *J. Microbiol. Methods* **57:**409–413.

Jiang, S., W. Chu, and W. Fu. 2003. Prevalence of cholera toxin genes (*ctxA* and *zot*) among non-O1/O139 *Vibrio cholerae*

strains from Newport Bay, California. *Appl. Environ. Microbiol.* **69:**7541–7544.

Jiang, S. C., and W. Fu. 2001. Seasonal abundance and distribution of *Vibrio cholerae* in coastal waters quantified by a 16S-23S intergenic spacer probe. *Microb. Ecol.* **42:**540–548.

Kaper, J. B., H. B. Bradford, N. C. Roberts, and S. Falkow. 1982. Molecular epidemiology of *Vibrio cholerae* in the U. S. Gulf Coast. *J. Clin. Microbiol.* **16:**129–134.

Kaper, J. B., J. G. Morris, Jr., and M. Nishibuchi. 1989. DNA probes for pathogenic *Vibrio* species, p. 65–77 *In* F. C. Tenover (ed.), *DNA Probes for Infectious Diseases*. CRC Press, Boca Raton, Fla.

Karunasagar, I., I. Rivera, B. Joseph, B. Kennedy, V. R. Shetty, A. Huq, I. Karunasagar, and R. R. Colwell. 2003. *ompU* genes in non-toxigenic *Vibrio cholerae* associated with aquaculture. *J. Appl. Microbiol.* **95:**338–343.

Karunasagar, I., G. Sugumar, and I. Karunasagar. 1995. Rapid detection of *Vibrio cholerae* contamination of seafood by polymerase chain reaction. *Mol. Mar. Biol. Biotechnol.* **4:**365–368.

Kaufman, G. E., G. M. Blackstone, M. C. Vickery, A. K. Bej, J. Bowers, M. D. Bowen, R. F. Meyer, and A. DePaola. 2004. Real-time PCR quantification of *Vibrio parahaemolyticus* in oysters using an alternative matrix. *J. Food Prot.* **67:**2424–2429.

Kaufman, G. E., M. L. Myers, C. L. Pass, A. K. Bej, and C. A. Kaysner. 2002. Molecular analysis of *Vibrio parahaemolyticus* isolated from human patients and shellfish during US Pacific north-west outbreaks. *Lett. Appl. Microbiol.* **34:**155–161.

Keasler, S. P., and R. H. Hall. 1993. Detection and biotyping *Vibrio cholerae* O1 with multiplex polymerase chain reaction. *Lancet* **341:**1661.

Khan, A. A., S. McCarthy, R. F. Wang, and C. E. Cerniglia. 2002. Characterization of United States outbreak isolates of *Vibrio parahaemolyticus* using enterobacterial repetitive intergenic consensus (ERIC) PCR and development of a rapid PCR method for detection of O3:K6 isolates. *FEMS Microbiol. Lett.* **206:** 209–214.

Kim, Y. B., J. Okuda, C. Matsumoto, N. Takahashi, S. Hashimoto, and M. Nishibuchi. 1999. Identification of *Vibrio parahaemolyticus* at the species level by PCR targeted to the *toxR* gene. *J. Clin. Microbiol.* **37:**1173–1177.

Kimoto, R., T. Funahashi, N. Yamamoto, S. Miyoshi, S. Narimatsu, and S. Yamamoto. 2001. Identification and characterization of the *sodA* genes encoding manganese superoxide dismutases in *Vibrio parahaemolyticus*, *Vibrio mimicus*, and *Vibrio vulnificus*. *Microbiol. Immunol.* **45:**135–142.

Kishishita, M., N. Matsuoka, K. Kumagai, S. Yamasaki, Y. Takeda, and M. Nishibuchi. 1992. Sequence variation in the thermostable direct hemolysin-related hemolysin (*trh*) gene of *Vibrio parahaemolyticus*. *Appl. Environ. Microbiol.* **58:**2449–2457.

Kita-Tsukamoto, K., H. Oyaizu, K. Nanba, and U. Simidu. 1993. Phylogenetic relationships of marine bacteria, mainly members of the family *Vibrionaceae*, determined on the basis of 16S rRNA sequences. *Int. J. Syst. Bacteriol.* **43:**8–19.

Kobayashi, K., K. Seto, S. Akasaka, and M. Makino. 1990. Detection of toxigenic *Vibrio cholerae* O1 using polymerase chain reaction for amplifying the cholera enterotoxin gene. *J. Jpn. Assoc. Infect. Dis.* **64:**1323–1329.

Koch, W. H., W. L. Payne, B. A. Wentz, and T. A. Cebula. 1993. Rapid polymerase chain reaction method for detection of *Vibrio cholerae* in foods. *Appl. Environ. Microbiol.* **59:**556–560.

Kong, R. Y., S. K. Lee, T. W. Law, S. H. Law, and R. S. Wu. 2002. Rapid detection of six types of bacterial pathogens in marine waters by multiplex PCR. *Water Res.* **36:**2802–2812.

Kurazono, H., A. Pal, P. K. Bag, G. B. Nair, T. Karasawa, T. Mihara, and Y. Takeda. 1995. Distribution of genes encoding cholera toxin, zonula occludens toxin, accessory cholera toxin, and El Tor hemolysin in *Vibrio cholerae* of diverse origins. *Microb. Pathog.* **18:**231–235.

Kvitt, H., M. Ucko, A. Colorni, C. Batargias, A. Zlotkin, and W. Knibb. 2002. *Photobacterium damselae* ssp. *piscicida*: detection by direct amplification of 16S rRNA gene sequences and genotypic variation as determined by amplified fragment length polymorphism (AFLP). *Dis. Aquat. Organ.* **48:**187–195.

Kwok, A. Y., J. T. Wilson, M. Coulthart, L. K. Ng, L. Mutharia, and A. W. Chow. 2002. Phylogenetic study and identification of human pathogenic *Vibrio* species based on partial *hsp60* gene sequences. *Can. J. Microbiol.* **48:**903–910.

Lan, R., and P. R. Reeves. 1998. Recombination between rRNA operons created most of the ribotype variation observed in the seventh pandemic clone of *Vibrio cholerae*. *Microbiology* **144:**1213–1221.

Laohaprertthisan, V., A. Chowdhury, U. Kongmuang, S. Kalnauwakul, M. Ishibashi, C. Matsumoto, and M. Nishibuchi. 2003. Prevalence and serodiversity of the pandemic clone among the clinical strains of *Vibrio parahaemolyticus* isolated in southern Thailand. *Epidemiol. Infect.* **130:**1–12.

Lee, C., and S. F. Pan. 1993. Rapid and specific detection of the thermostable direct haemolysin gene in *Vibrio parahaemolyticus* by the polymerase chain reaction. *J. Gen. Microbiol.* **139:**3225–3231.

Lee, C. Y., C. H. Chen, and Y. W. Chou. 1995a. Characterization of a cloned pR72H probe for *Vibrio parahaemolyticus* detection and development of a nonisotopic colony hybridization assay. *Microbiol. Immunol.* **39:**177–183.

Lee, C. Y., S. F. Pan, and C. H. Chen. 1995b. Sequence of a cloned pR72H fragment and its use for detection of *Vibrio parahaemolyticus* in shellfish with the PCR. *Appl. Environ. Microbiol.* **61:**1311–1317.

Lee, C. Y., L. H. Chen, M. L. Liu, and Y. C. Su. 1992. Use of an oligonucleotide probe to detect *Vibrio parahaemolyticus* in artificially contaminated oysters. *Appl. Environ. Microbiol.* **58:**3419–3422.

Lee, C. Y., G. Panicker, and A. K. Bej. 2003. Detection of pathogenic bacteria in shellfish using multiplex PCR followed by CovaLink NH microwell plate sandwich hybridization. *J. Microbiol. Methods* **53:**199–209.

Lee, S. E., S. H. Shin, S. Y. Kim, Y. R. Kim, D. H. Shin, S. S. Chung, Z. H. Lee, J. Y. Lee, K. C. Leong, S. H. Choi, and J. H. Rhee. 2000. *Vibrio vulnificus* has the transmembrane transcription activator *toxRS* stimulating the expression of the hemolysin gene *vvhA*. *J. Bacteriol.* **182:**3405–3415.

Lee, S. E., S. Y. Kim, S. J. Kim, H. S. Kim, J. H. Shin, S. H. Choi, S. S. Chung, and J. H. Rhee. 1998. Direct identification of *Vibrio vulnificus* in clinical specimens by nested PCR. *J. Clin. Microbiol.* **36:**2887–2892.

Lee, S. K., H. Z. Wang, S. H. Law, R. S. Wu, and R. Y. Kong. 2002. Analysis of the 16S-23S rDNA intergenic spacers (IGSs) of marine vibrios for species-specific signature DNA sequences. *Mar. Pollut. Bull.* **44:**412–420.

Le Roux, F., M. Gay, C. Lambert, J. L. Nicolas, M. Gouy, and F. Berthe. 2004a. Phylogenetic study and identification of *Vibrio splendidus*-related strains based on *gyrB* gene sequences. *Dis. Aquat. Organ.* **58:**143–150.

Le Roux, W. J., D. Masoabi, C. M. de Wet, and S. N. Venter. 2004b. Evaluation of a rapid polymerase chain reaction based identification technique for *Vibrio cholerae* isolates. *Water Sci. Technol.* **50:**229–232.

Lin, W., K. J. Fullner, R. Clayton, J. A. Sexton, M. B. Rogers, K. E. Calia, S. B. Calderwood, C. Fraser, and J. J. Mekalanos. 1999. Identification of a *Vibrio cholerae* RTX toxin gene cluster that is tightly linked to the cholera toxin prophage. *Proc. Natl. Acad. Sci. USA* **96:**1071–1076.

Lin, Z., K. Kumagai, K. Baba, J. J. Mekalanos, and M. Nishibuchi. 1993. *Vibrio parahaemolyticus* has a homolog of the *Vibrio cholerae toxRS* operon that mediates environmentally induced regulation of the thermostable direct hemolysin gene. *J. Bacteriol.* **175:**3844–3855.

Linkous, D. A., and J. D. Oliver. 1999. Pathogenesis of *Vibrio vulnificus*. *FEMS Microbiol. Lett.* **174:**207–214.

Lipp, E. K., I. N. Rivera, A. I. Gil, E. M. Espeland, N. Choopun, V. R. Louis, E. Russek-Cohen, A. Huq, and R. R. Colwell. 2003. Direct detection of *Vibrio cholerae* and *ctxA* in Peruvian coastal water and plankton by PCR. *Appl. Environ. Microbiol.* **69:**3676–3680.

Liu, C. H., W. Cheng, J. P. Hsu, and J. C. Chen. 2004. *Vibrio alginolyticus* infection in the white shrimp *Litopenaeus vannamei* confirmed by polymerase chain reaction and 16S rDNA sequencing. *Dis. Aquat. Organ.* **61:**169–174.

Lyon, W. J. 2001. TaqMan PCR for detection of *Vibrio cholerae* O1, O139, non-O1, and non-O139 in pure cultures, raw oysters, and synthetic seawater. *Appl. Environ. Microbiol.* **67:**4685–4693.

Mallard, K. E., and P. M. Desmarchelier. 1995. Detection of heat-stable enterotoxin genes among Australian *Vibrio cholerae* O1 strains. *FEMS Microbiol. Lett.* **127:**111–115.

Martinez-Picado, J., A. R. Blanch, and J. Jofre. 1994. Rapid detection and identification of *Vibrio anguillarum* by using a specific oligonucleotide probe complementary to 16S rRNA. *Appl. Environ. Microbiol.* **60:**732–737.

Masuda, N., K. Yasukawa, Y. Isawa, R. Horie, J. Saito, T. Ishiguro, Y. Nakaguchi, M. Nishibuchi, and T. Hayashi. 2004. Rapid detection of *tdh* and *trh* mRNAs of *Vibrio parahaemolyticus* by the transcription-reverse transcription concerted (TRC) method. *J. Biosci. Bioeng.* **98:**236–243.

Matsumoto, C., J. Okuda, M. Ishibashi, M. Iwanaga, P. Garg, T. Rammamurthy, H.-C. Wong, A. DePaola, Y. B. Kim, M. J. Albert, and M. Nishibuchi. 2000. Pandemic spread of an O3:K6 clone of *Vibrio parahaemolyticus* and emergence of related strains evidenced by arbitrarily primed PCR and *toxRS* sequence analyses. *J. Clin. Microbiol.* **38:**578–585.

McCarthy, S. A., A. DePaola, D. W. Cook, C. A. Kaysner, and W. E. Hill. 1999. Evaluation of alkaline phosphatase- and digoxigenin-labelled probes for detection of the thermolabile hemolysin (*tlh*) gene of *Vibrio parahaemolyticus*. *Lett. Appl. Microbiol.* **28:**66–70.

McCarthy, S. A., A. DePaola, C. A. Kaysner, W. E. Hill, and D. W. Cook. 2000. Evaluation of nonisotopic DNA hybridization methods for detection of the *tdh* gene of *Vibrio parahaemolyticus*. *J. Food. Prot.* **63:**1660–1664.

Miller, V. L., and J. J. Mekalanos. 1984. Synthesis of cholera toxin is positively regulated at the transcriptional level by *toxR*. *Proc. Natl. Acad. Sci. USA* **81:**3471–3475.

Miller, V. L., and J. J. Mekalanos. 1988. A novel suicide vector and its use in construction of insertion mutations: osmoregulation of outer membrane proteins and virulence determinants in *Vibrio cholerae* requires *toxR*. *J. Bacteriol.* **170:**2575–2583.

Minami, A., S. Hashimoto, H. Abe, M. Arita, T. Taniguchi, T. Honda, T. Miwatani, and M. Nishibuchi. 1991. Cholera enterotoxin production in *Vibrio cholerae* O1 strains isolated from the environment and from humans in Japan. *Appl. Environ. Microbiol.* **57:**2152–2157.

Miyagi, K., K. Sano, C. Morita, S. Imura, S. Morimatsu, T. Goto, Y. Nakano, Y. Omura, Y. Matsumoto, K. Maeda, S. Hashimoto, and T. Honda. 1999. An improved method for detecting faecal *Vibrio cholerae* by PCR of the toxin A gene. *J. Med. Microbiol.* **48:**883–889.

Mizunoe, Y., S. N. Wai, K. Umene, T. Kokubo, S. Kawabata, and S. Yoshida. 1999. Cloning, sequencing, and functional expression in *Escherichia coli* of chaperonin (*groESL*) genes from *Vibrio cholerae*. *Microbiol. Immunol.* **43:**513–520.

Morin, N. J., Z. Gong, and X. F. Li. 2004. Reverse transcription-multiplex PCR assay for simultaneous detection of *Escherichia coli* O157:H7, *Vibrio cholerae* O1, and *Salmonella* Typhi. *Clin. Chem.* **50:**2037–2044.

Morris, J. G., Jr., A. C. Wright, D. M. Roberts, P. K. Wood, L. M. Simpson, and J. D. Oliver. 1987. Identification of environmental *Vibrio vulnificus* isolates with a DNA probe for the cytotoxin-hemolysin gene. *Appl. Environ. Microbiol.* **53:**193–195.

Muniesa-Pere, M., J. Jofre, and A. R. Blanch. 1996. Identification of *Vibrio proteolyticus* with a differential medium and a specific probe. *Appl. Environ. Microbiol.* **62:**2673–2675.

Nair, G. B., P. K. Bag, T. Shimada, T. Ramamurthy, T. Takeda, S. Yamamoto, H. Kurazono, and Y. Takeda. 1995. Evaluation of DNA probes for specific detection of *Vibrio cholerae* O139 Bengal. *J. Clin. Microbiol.* **33:**2186–2187.

Nair, G. B., Y. Oku, Y. Takeda, A. Ghosh, R. K. Ghosh, S. Chattopadhyyay, S. C. Pal, J. B. Kaper, and T. Takeda. 1988. Toxin profiles of *Vibrio cholerae* non-O1 from environmental sources in Calcutta, India. *Appl. Environ. Microbiol.* **54:**3180–3182.

Nakaguchi, Y., T. Ishizuka, S. Ohnaka, T. Hayashi, K. Yasukawa, T. Ishiguro, and M. Nishibuchi. 2004. Rapid and specific detection of *tdh*, *trh*1, and *trh*2 mRNA of *Vibrio parahaemolyticus* using transcription-reverse transcription concerted (TRC) reaction with an automated system. *J. Clin. Microbiol.* **42:**4284–4292.

Nandi, B., R. K. Nandy, S. Mukhopadhyay, G. B. Nair, T. Shimada, and A. C. Ghose. 2000. Rapid method for species-specific identification of *Vibrio cholerae* using primers targeted to the gene of outer membrane protein OmpW. *J. Clin. Microbiol.* **38:**4145–4151.

Nasu, H., T. Iida, T. Sugahara, Y. Yamaguchi, K.-S. Park, K. Yokoyama, K. Makino, H. Shinagawa, and T. Honda. 2000. A filamentous phage associated with recent pandemic *Vibrio parahaemolyticus* O3:K6 strains. *J. Clin. Microbiol.* **8:**2156–2161.

Nilsson, W. B., R. N. Paranjpye, A. DePaola, and M. S. Strom. 2003. Sequence polymorphism of the 16S rRNA gene of *Vibrio vulnificus* is a possible indicator of strain virulence. *J. Clin. Microbiol.* **41:**442–446.

Nishibuchi, M., S. Doke, S. Toizumi, T. Umeda, M. Yoh, and T. Miwatani. 1988. Isolation from a coastal fish of *Vibrio hollisae* capable of producing a hemolysin similar to the thermostable direct hemolysin of *Vibrio parahaemolyticus*. *Appl. Environ. Microbiol.* **54:**2144–2146.

Nishibuchi, M., W. E. Hill, G. Zon, W. L. Payne, and J. B. Kaper. 1986. Synthetic oligodeoxyribonucleotide probes to detect Kanagawa phenomenon-positive *Vibrio parahaemolyticus*. *J. Clin. Microbiol.* **23:**1091–1095.

Nishibuchi, M., M. Ishibashi, Y. Takeda, and J. B. Kaper. 1985. Detection of the thermostable direct hemolysin gene and related DNA sequence in *Vibrio parahaemolyticus* and other *Vibrio* species by the DNA colony hybridization test. *Infect. Immun.* **49:**481–486.

Nishibuchi, M., J. M. Janda, and T. Ezaki. 1996. The thermostable direct hemolysin gene (*tdh*) of *Vibrio hollisae* is dissimilar in prevalence to and phylogenetically distant from the *tdh* genes of other vibrios: implications in the horizontal transfer of the *tdh* gene. *Microbiol. Immunol.* **40:**59–65.

Nishibuchi, M., and J. B. Kaper. 1995. Thermostable direct hemolysin gene of *Vibrio parahaemolyticus*: a virulence gene acquired by a marine bacterium. *Infect. Immun.* **63:**2093–2099.

Nishibuchi, M., V. Khaeomanee-iam, T. Honda, J. B. Kaper, and T. Miwatani. 1990. Comparative analysis of the hemolysin genes of *Vibrio cholerae* non-O1, *V. mimicus*, and *V. hollisae* that are similar to the *tdh* gene of *V. parahaemolyticus*. *FEMS Microbiol. Lett.* **55:**251–256.

Nordstrom, J. L., and A. DePaola. 2003. Improved recovery of pathogenic *Vibrio parahaemolyticus* from oysters using colony hybridization following enrichment. *J. Microbiol. Methods* **52:**273–277.

Novais, R. C., A. Coelho, C. A. Salles, and A. C. P. Vincente. 1999. Toxin co-regulated pilus cluster in non-O1, non-toxigenic *Vibrio cholerae*: evidence of a third allele of pilin gene. *FEMS Microbiol. Lett.* **171:**49–55.

Oakey, H. J., N. Levy, D. G. Bourne, B. Cullen, and A. Thomas. 2003. The use of PCR to aid in the rapid identification of *Vibrio harveyi* isolates. *J. Appl. Microbiol.* **95:**1293–1303.

Ogawa, A., and T. Takeda. 1993. The gene encoding the heat-stable enterotoxin of *Vibrio cholerae* is flanked by 123-base pair direct repeats. *Microbiol. Immunol.* **37:**607–616.

O'Hara, C. M., E. G. Sowers, C. A. Bopp, S. B. Duda, and N. A. Strockbine. 2003. Accuracy of six commercially available systems for identification of members of the family *Vibrionaceae*. *J. Clin. Microbiol.* **41:**5654–5659.

Okuda, J., E. Hayakawa, M. Nishibuchi, and T. Nishino. 1999. Sequence analysis of the *gyrA* and *parC* homologues of a wild-type strain of *Vibrio parahaemolyticus* and its fluoroquinolone-resistant mutants. *Antimicrob. Agents Chemother.* **43:**1156–1162.

Okuda, J., M. Ishibashi, S. Abbott, J. M. Janda, and M. Nishibuchi. 1997a. Analysis of the thermostable direct hemolysin (*tdh*) gene and the *tdh*-related hemolysin (*trh*) genes in urease-positive strains of *Vibrio parahaemolyticus* isolated on the west coast of United States. *J. Clin. Microbiol.* **35:**1965–1971.

Okuda, J., M. Ishibashi, E. Hayakawa, T. Nishino, Y. Takeda, A. Mukhopadhyay, S. Garg, S. K. Bhattacharya, G. B. Nair, and M. Nishibuchi. 1997b. Emergence of a unique O3:K6 clone of *Vibrio parahaemolyticus* in Calcutta, India, and isolation of strains from the same clonal group from Southeast Asian travelers arriving in Japan. *J. Clin. Microbiol.* **35:**3150–3155.

Okuda, J., T. Nakai, P. S. Chang, T. Oh, T. Nishino, T. Koitabashi, and M. Nishibuchi. 2001. The *toxR* gene of *Vibrio* (*Listonella*) *anguillarum* controls expression of the major outer membrane proteins but not virulence in a natural host model. *Infect. Immun.* **69:**6091–6101.

Okura, M., R. Osawa, A. Iguchi, E. Arakawa, J. Terajima, and H. Watanabe. 2003. Genotypic analyses of *Vibrio parahaemolyticus* and development of a pandemic group-specific multiplex PCR assay. *J. Clin. Microbiol.* **41:**4676–4682.

Okura, M., R. Osawa, A. Iguchi, M. Takagi, E. Arakawa, J. Terajima, and H. Watanabe. 2004. PCR-based identification of pandemic group *Vibrio parahaemolyticus* with a novel group-specific primer pair. *Microbiol. Immunol.* **48:**787–790.

Osorio, C. R., M. D. Collins, J. L. Romalde, and A. E. Toranzo. 2004. Characterization of the 23S and 5S rRNA genes and 23S-5S intergenic spacer region (ITS-2) of *Photobacterium damselae*. *Dis. Aquat. Organ.* **61:**33–39.

Osorio, C. R., M. D. Collins, J. L. Romalde, and A. E. Toranzo. 2005. Variation in 16S-23S rRNA intergenic spacer regions in *Photobacterium damselae*: a mosaic-like structure. *Appl. Environ. Microbiol.* **71:**636–645.

Osorio, C. R., M. D. Collins, A. E. Toranzo, J. L. Barja, and J. L. Romalde. 1999. 16S rRNA gene sequence analysis of *Photobacterium damselae* and nested PCR method for rapid detection of the causative agent of fish pasteurellosis. *Appl. Environ. Microbiol.* **65:**2942–2946.

Osorio, C. R., and K. E. Klose. 2000. A region of the transmembrane regulatory protein ToxR that tethers the transcriptional activation domain to the cytoplasmic membrane displays wide divergence among *Vibrio* species. *J. Bacteriol.* **182:**526–528.

Osorio, C. R., A. E. Toranzo, J. L. Romalde, and J. L. Barja. 2000. Multiplex PCR assay for *ureC* and 16S rRNA genes clearly discriminates between both subspecies of *Photobacterium damselae*. *Dis. Aquat. Organ.* **40:**177–183.

Pal, A., T. Ramamurthy, R. K. Bhadra, T. Takeda, T. Shimada, Y. Takeda, G. B. Nair, S. C. Pal, and S. Chakrabarti. 1992. Reassessment of the prevalence of heat-stable enterotoxin (NAG-ST) among environmental *Vibrio cholerae* non-O1 strains isolated from Calcutta, India, by using a NAG-ST DNA probe. *Appl. Environ. Microbiol.* **58:**2485–2489.

Palmer, L. M., and R. R. Colwell. 1991. Detection of luciferase gene sequence in nonluminescent *Vibrio cholerae* by colony hybridization and polymerase chain reaction. *Appl. Environ. Microbiol.* **57:**1286–1293.

Panicker, G., D. R. Call, M. J. Krug, and A. K. Bej. 2004a. Detection of pathogenic *Vibrio* spp. in shellfish by using multiplex PCR and DNA microarrays. *Appl. Environ. Microbiol.* **70:**7436–7444.

Panicker, G., M. L. Myers, and A. K. Bej. 2004b. Rapid detection of *Vibrio vulnificus* in shellfish and Gulf of Mexico water by real-time PCR. *Appl. Environ. Microbiol.* **70:**498–507.

Panicker, G., M. C. Vickery, and A. K. Bej. 2004c. Multiplex PCR detection of clinical and environmental strains of *Vibrio vulnificus* in shellfish. *Can. J. Microbiol.* **50:**911–922.

Park, S. D., H. S. Shon, and N. J. Joh. 1991. *Vibrio vulnificus* septicemia in Korea: clinical and epidemiologic findings in 70 patients. *J. Am. Acad. Dermatol.* **24:**397–403.

Park, T. S., S. H. Oh, E. Y. Lee, T. K. Lee, K. H. Park, M. J. Figueras, and C. L. Chang. 2003. Misidentification of *Aeromonas veronii* biovar sobria as *Vibrio alginolyticus* by the Vitek system. *Lett. Appl. Microbiol.* **37:**349–353.

Rajan, P. R., J. H. Lin, M. S. Ho, and H. L. Yang. 2003. Simple and rapid detection of *Photobacterium damselae* ssp. *piscicida* by a PCR technique and plating method. *J. Appl. Microbiol.* **95:**1375–1380.

Ramamurthy, T., M. J. Albert, A. Huq, R. R. Colwell, Y. Takeda, T. Takeda, T. Shimada, B. K. Mandal, and G. B. Nair. 1994. *Vibrio mimicus* with multiple toxin types isolated from human and environmental sources. *J. Med. Microbiol.* **40:**194–196.

Ramamurthy, T., A. Pal, P. K. Bag, S. K. Bhattacharya, G. B. Nair, H. Kurozano, S. Yamasaki, H. Shirai, T. Takeda, and Y. Uesaka. 1993. Detection of cholera toxin gene in stool specimens by polymerase chain reaction: comparison with bead enzyme-linked immunosorbent assay and culture method for laboratory diagnosis of cholera. *J. Clin. Microbiol.* **31:**3068–3070.

Randa, M. A., M. F. Polz, and E. Lim. 2004. Effects of temperature and salinity on *Vibrio vulnificus* population dynamics as assessed by quantitative PCR. *Appl. Environ. Microbiol.* **70:**5469–5476.

Rehnstam, A. S., A. Norqvist, H. Wolf-Watz, and A. Hagström. 1989. Identification of *Vibrio anguillarum* in fish by using partial 16S rRNA sequences and a specific 16S rRNA oligonucleotide probe. *Appl. Environ. Microbiol.* **55:**1907–1910.

Reich, K. A., and G. K. Schoolnik. 1994. The light organ symbiont *Vibrio fischeri* possesses a homolog of the *Vibrio cholerae* transmembrane transcriptional activator ToxR. *J. Bacteriol.* **176:**3085–3088.

Rhine, J. A., and R. K. Taylor. 1994. TcpA pilin sequences and colonization requirements for O1 and O139 *Vibrio cholerae*. *Mol. Microbiol.* **13:**1013–1020.

Rivera, T. G., J. Chun, A. Huq, R. B. Sack, and R. R. Colwell. 2001. Genotypes associated with virulence in environmental isolates of *Vibrio cholerae*. *Appl. Environ. Microbiol.* **67:**2421–2429.

Robert-Pillot, A., A. Guenole, and J. M. Fournier. 2002. Usefulness of R72H PCR assay for differentiation between *Vibrio parahaemolyticus* and *Vibrio alginolyticus* species: validation by DNA-DNA hybridization. *FEMS Microbiol. Lett.* **215:**1–6.

Rosche, T. M., Y. Yano, and J. D. Oliver. 2005. A rapid and simple PCR analysis indicates there are two subgroups of *Vibrio vulnificus* which correlate with clinical or environmental isolation. *Microbiol. Immunol.***49:**381–389.

Rubin, E. J., W. Lin, J. J. Mekalanos, and M. K. Waldor. 1998. Replication and integration of a *Vibrio cholerae* cryptic plasmid linked to the CTX prophage. *Mol. Microbiol.* **28:**1247–1254.

Ruimy, R., V. Breittmayer, P. Elbaze, B. Lafay, O. Boussemart, M. Gauthier, and R. Christine. 1994. Phylogenetic analysis and assessment of the genera *Vibrio*, *Photobacterium*, *Aeromonas*, and *Plesiomonas* deduced from small-subunit rRNA sequences. *Int. J. Syst. Bacteriol.* **44:**416–426.

Sahu, G. K., R. Chowdhury, and J. Das. 1997. The *rpoH* gene encoding σ³² homolog of *Vibrio cholerae*. *Gene* **189:**203–207.

Salles, C. A., H. Momen, A. C. P. Vicente, and A. Coelho. 1993. *Vibrio cholerae* in South America: polymerase chain reaction and zymovar analysis. *Trans. R. Soc. Trop. Med. Hyg.* **87:**272.

Sarkar, B., T. Bhattacharya, T. Ramamurthy, T. Shimada, Y. Takeda, and G. B. Nair. 2002. Preferential association of the heat-stable enterotoxin gene (*stn*) with environmental strains of *Vibrio cholerae* belonging to the O14 serogroup. *Epidemiol. Infect.* **129:**245–251.

Saulnier, D., J. C. Avarre, G. Le Moullac, D. Ansquer, P. Levy, and V. Vonau. 2000. Rapid and sensitive PCR detection of *Vibrio penaeicida*, the putative etiological agent of syndrome 93 in New Caledonia. *Dis. Aquat. Organ.* **40:**109–115.

Shangkuan, Y. H., Y. S. Show, and T. M. Wang. 1995. Multiplex polymerase chain reaction to detect toxigenic *Vibrio cholerae* and to biotype *Vibrio cholerae* O1. *J. Appl. Bacteriol.* **79:**264–273.

Shi, L., S. Miyoshi, M. Hiura, K. Tomochika, T. Shimada, and S. Shinoda. 1998. Detection of genes encoding cholera toxin (CT), zonula occludens toxin (ZOT), accessory cholera enterotoxin (ACE) and heat-stable enterotoxin (ST) in *Vibrio mimicus* clinical strains. *Microbiol. Immunol.* **42:**823–828.

Shinoda, S., T. Nakagawa, L. Shi, K. Bi, Y. Kanoh, K. Tomochika, S. Miyoshi, and T. Shimada. 2004. Distribution of virulence-associated genes in *Vibrio mimicus* isolates from clinical and environmental origins. *Microbiol. Immunol.* **48:**547–551.

Shirai, H., H. Ito, T. Hirayama, Y. Nakamoto, N. Nakabayashi, K. Kumagai, Y. Takeda, and M. Nishibuchi. 1990. Molecular epidemiologic evidence for association of thermostable direct hemolysin (TDH) and TDH-related hemolysin of *Vibrio parahaemolyticus* with gastroenteritis. *Infect. Immun.* **58:**3568–3573.

Shirai, H., M. Nishibuchi, T. Ramamurthy, S. K. Bhattacharya, S. C. Pal, and Y. Takeda. 1991. Polymerase chain reaction for detection of the cholera enterotoxin operon of *Vibrio cholerae*. *J. Clin. Microbiol.* **29:**2517–2521.

Shyu, Y. C., and F. P. Lin. 1999. Cloning and characterization of manganese superoxide dismutase gene from *Vibrio parahaemolyticus* and application to preliminary identification of *Vibrio* strains. *IUBMB Life* **48:**345–352.

Singh, D. V., S. R. Isac, and R. R. Colwell. 2002. Development of a hexaplex PCR assay for rapid detection of virulence and regulatory genes in *Vibrio cholerae* and *Vibrio mimicus*. *J. Clin. Microbiol.* **40:**4321–4324.

Somerville, C. C., and R. R. Colwell. 1993. Sequence analysis of the β-N-acetylhexosaminidase gene of *Vibrio vulnificus*: evidence for a common evolutionary origin of hexosaminidases. *Proc. Natl. Acad. Sci, USA* **90:**6751–6755.

Sparagano, O. A., P. A. Robertson, I. Purdom, J. McInnes, Y. Li, D. H. Yu, Z. J. Du, H. S. Xu, and B. Austin. 2002. PCR and molecular detection for differentiating *Vibrio* species. *Ann. N. Y. Acad. Sci.* **969:**60–65.

Stine, O. C., S. Sozhamannan, Q. Gou, S. Zheng, J. G. Morris, Jr., and J. A. Johnson. 2000. Phylogeny of *Vibrio cholerae* based on *recA* sequence. *Infect. Immun.* **68:**7180–7185.

Sugimura, I. I., T. Sawabe, and Y. Ezura. 2000. In situ polymerase chain reaction visualization of *Vibrio halioticoli* using alginate lyase gene *alyVG2*. *Mar. Biotechnol.* (New York) **2**:74–79.

Tada, J., T. Ohashi, N. Nishimura, H. Ozaki, S. Fukushima, J. Takano, M. Nishibuchi, and Y. Takeda. 1992a. Non-isotopic microtitre plate-based assay for detecting products of polymerase chain reaction amplification: application to detection of the *tdh* gene of *Vibrio parahaemolyticus*. *Mol. Cell. Probes* **6**: 489–494.

Tada, J., T. Ohashi, N. Nishimura, Y. Shirasaki, H. Ozaki, S. Fukushima, J. Takano, M. Nishibuchi, and Y. Takeda. 1992b. Detection of thermostable direct hemolysin gene (*tdh*) and the thermostable direct hemolysin-related hemolysin gene (*trh*) of *Vibrio parahaemolyticus* by polymerase chain reaction. *Mol. Cell. Probes* **6**:477–487.

Takahashi, H., Y. Hara-Kudo, J. Miyasaka, S. Kumagai, and H. Konuma. 2005. Development of a quantitative real-time polymerase chain reaction targeted to the *toxR* for detection of *Vibrio vulnificus*. *J. Microbiol. Methods* **61**:77–85.

Takeda, T., Y. Peina, A. Ogawa, S. Dohi, H. Abe, G. B. Nair, and S. C. Pal. 1991. Detection of heat-stable enterotoxin in a cholera toxin gene-positive strain of *Vibrio cholerae* O1. *FEMS Microbiol. Lett.* **64**:23–27.

Taniguchi, H., H. Hirano, S. Kubomura, K. Higashi, and Y. Mizuguchi. 1986. Comparison of the nucleotide sequences of the genes for the thermostable direct hemolysin from *Vibrio parahaemolyticus*. *Microb. Pathog.* **1**:425–432.

Terai, A., K. Baba, H. Shirai, O. Yoshida, Y. Takeda, and M. Nishibuchi. 1991. Evidence for insertion-sequence-mediated spread of the thermostable direct hemolysin gene among *Vibrio* species. *J. Bacteriol.* **173**:5036–5046.

Theron, J., J. Cilliers, M. Du Preez, V. S. Brozel, and S. N. Venter. 2000. Detection of toxigenic *Vibrio cholerae* from environmental water samples by an enrichment broth cultivation-pit-stop semi-nested PCR procedure. *J. Appl. Microbiol.* **89**: 539–546.

Thompson, C. C., F. L. Thompson, K. Vandemeulebroecke, B. Hoste, P. Dawyndt, and J. Swings. 2004. Use of *recA* as an alternative phylogenetic marker in the family *Vibrionaceae*. *Int. J. Syst. Evol. Microbiol.* **54**:919–924.

Thompson, F. L., B. Hoste, K. Vandemeulebroecke, and J. Swings. 2003. Reclassification of *Vibrio hollisae* as *Grimontia hollisae* gen. nov., comb. nov. *Int. J. Syst. Evol. Microbiol.* **53**:1615–1617.

Thompson, F. L., T. Iida, and J. Swings. 2004. Biodiversity of vibrios. *Microbiol. Mol. Biol. Rev.* **68**:403–431.

Trucksis, M., J. E. Galen, J. Michalski, A. Fasano, and J. B. Kaper. 1993. Accessory cholera enterotoxin (Ace), the third toxin of a *Vibrio cholerae* virulence cassette. *Proc. Natl. Acad. Sci. USA* **90**:5267–5271.

Uchimura, M., K. Koiwai, Y. Tsuruoka, and H. Tanaka. 1993. High prevalence of thermostable direct hemolysin (TDH)-like toxin in *Vibrio mimicus* strains isolated from diarrhoeal patients. *Epidemiol. Infect.* **111**:49–53.

Varela, P., G. D. Pollevick, M. Rivas, I. Chinen, N. Binsztein, A. C. Frasch, and R. A. Ugalde. 1994. Direct detection of *Vibrio cholerae* in stool samples. *J. Clin. Microbiol.* **32**:1246–1248.

Varela, P., M. Rivas, N. Binsztein, M. L. Cremona, P. Herrmann, O. Burrone, R. A. Ugalde, and A. C. Frasch. 1993. Identification of toxigenic *Vibrio cholerae* from the Argentine outbreak by PCR for ctx A1 and ctx A2-B. *FEBS Lett.* **315**: 74–76.

Venkateswaran, K., N. Dohmoto, and S. Harayama. 1998. Cloning and nucleotide sequence of the *gyrB* gene of *Vibrio parahaemolyticus* and its application in detection of this pathogen in shrimp. *Appl. Environ. Microbiol.* **64**:681–687.

Vicente, A. C. P., A. Coelho, and C. A. Salles. 1997. Detection of *Vibrio cholerae* and *V. mimicus* heat-stable toxin gene sequence by PCR. *J. Med. Microbiol.* **46**:1–5.

Vital Brazil, J. M., R. M. Alves, I. N. Rivera, D. P. Rodrigues, D. K. Karaolis, and L. C. Campos. 2002. Prevalence of virulence-associated genes in clinical and environmental *Vibrio cholerae* strains isolated in Brazil between 1991 and 1999. *FEMS Microbiol. Lett.* **215**:15–21.

Vuddhakul, V., A. Chowdhury, V. Laohaprertthisan, P. Pungrasamee, N. Patararungrong, P. Thianmontri, M. Ishibashi, C. Matsumoto, and M. Nishibuchi. 2000a. Isolation of *Vibrio parahaemolyticus* strains belonging to a pandemic O3:K6 clone from environmental and clinical sources in Thailand. *Appl. Environ. Microbiol.* **66**:2685–2689.

Vuddhakul, V., T. Nakai, C. Matsumoto, T. Oh, T. Nishino, C.-H. Chen, M. Nishibuchi, and J. Okuda. 2000b. Analysis of *gyrB* and *toxR* gene sequences of *Vibrio hollisae* and development of *gyrB*- and *toxR*-targeted PCR methods for isolation of *Vibrio hollisae* from the environment and its identification. *Appl. Environ. Microbiol.* **66**:3506–3514.

Waldor, M. K., and J. J. Mekalanos. 1994. *Vibrio cholerae* O139 specific gene sequences. *Lancet* **343**:1366.

Waldor, M. K., and J. J. Mekalanos. 1996. Lysogenic conversion by a filamentous phage encoding cholera toxin. *Science* **272**: 1910–1914.

Wang, R. F., W. W. Cao, and C. E. Cerniglia. 1997. A universal protocol for PCR detection of 13 species of foodborne pathogens in foods. *J. Appl. Microbiol.* **83**:727–736.

Welch, T. J., and D. H. Bartlett. 1998. Identification of a regulatory protein required for pressure-responsive gene expression in the deep-sea bacterium *Photobacterium* species strain SS9. *Mol. Microbiol.* **27**:977–985.

Wiik, R., E. Stackebrandt, O. Valle, F. L. Daae, O. M. Rodseth, and K. Andersen. 1995. Classification of fish-pathogenic vibrios based on comparative 16S rRNA analysis. *Int. J. Syst. Bacteriol.* **45**:421–428.

Wimpee, C. F., T. L. Nadeau, and K. H. Nealson. 1991. Development of species-specific hybridization probes for marine luminous bacteria by using in vitro DNA amplification. *Appl. Environ. Microbiol.* **57**:1319–1324.

Wright A., J. G. Morris, Jr., D. R. Maneval, Jr., K. Richardson, and J. B. Kaper. 1985. Cloning of the cytotoxin-hemolysin gene of *Vibrio vulnificus*. *Infect. Immun.* **50**:922–924.

Wright, A. C., Y. Guo, J. A. Johnson, J. P. Nataro, and J. G. Morris, Jr. 1992. Development and testing of a nonradioactive DNA oligonucleotide probe that is specific for *Vibrio cholerae* cholera toxin. *J. Clin. Microbiol.* **30**:2302–2306.

Wright, A. C., R. T. Hill, J. A. Johnson, M.-C. Roghman, R. R. Colwell, and J. G. Morris, Jr. 1996. Distribution of *Vibrio vulnificus* in the Chesapeake Bay. *Appl. Environ. Microbiol.* **62**:717–724.

Wright, A. C., G. A. Miceli, W. L. Landry, J. B. Christy, W. D. Watkins, and J. G. Morris, Jr. 1993. Rapid identification of *Vibrio vulnificus* on nonselective media with an alkaline phosphatase-labeled oligonucleotide probe. *Appl. Environ. Microbiol.* **50**:541–546.

Wu, M. H., and R. A. Laine. 1999. Sequence of the *V. parahaemolyticus* gene for cytoplasmic N, N′-diacetylchitobiase and homology with related enzymes. *J. Biochem.* (Tokyo) **125**:1086–1093.

Yam, W. C., M. L. Lung, K. Y. Ng, and M. H. Ng. 1989. Molecular epidemiology of *Vibrio cholerae* in Hong Kong. *J. Clin. Microbiol.* **27**:1900–1902.

Yamamoto, K., T. Honda, T. Miwatani, S. Tamatsukuri, and S. Shibata. 1992. Enzyme-labeled oligonucleotide probes for detection of the genes for thermostable direct hemolysin (TDH) and TDH-related hemolysin (TRH) of *Vibrio parahaemolyticus*. *Can. J. Microbiol.* **38**:410–416.

Yamasaki, S., S. Garg, G. B. Nair, and Y. Takeda. 1999a. Distrib-ution of *Vibrio cholerae* O1 antigen biosynthesis genes among O139 and other non-O1 serogroups of *Vibrio cholerae*. *FEMS Microbiol. Lett.* **179**:115–121.

Yamasaki, S., T. Shimizu, K. Hoshino, S. T. Ho, T. Shimada, G. B. Nair, and Y. Takeda. 1999b. The genes responsible for O-anti-gen synthesis of *Vibrio cholerae* O139 are closely related to those of *Vibrio cholerae* O22. *Gene* **237**:321–332

Yamasaki, S., H. Shirai, Y. Takeda, and M. Nishibuchi. 1991. Analysis of the gene of *Vibrio hollisae* encoding the hemolysin similar to the thermostable direct hemolysin of *Vibrio para-haemolyticus. FEMS Microbiol. Lett.* **80**:259–264.

Yoh, M., K. Miyagi, Y. Matsumoto, K. Hayashi, Y. Takarada, K. Yamamoto, and T. Honda. 1993. Development of an en-zyme-labeled oligonucleotide probe for the cholera toxin gene. *J. Clin. Microbiol.* **31**:1312–1314.

IV. GENOME EVOLUTION

The Biology of Vibrios
Edited by F. L. Thompson et al.
© 2006 ASM Press, Washington, D.C.

Chapter 5

Comparative Genomics: Genome Configuration and the Driving Forces in the Evolution of Vibrios

Tetsuya Iida and Ken Kurokawa

There has been important recent progress in the study of genus *Vibrio* bacteria (Thompson et al., 2004), with the discovery of the two-chromosome configuration of their genomes and the completion of whole-genome sequencing of several species. *Vibrio cholerae* was the first for which the whole genome sequence was reported (Heidelberg et al., 2000). The genome sequences of *Vibrio parahaemolyticus* (Makino et al., 2003) and two strains of *Vibrio vulnificus*, strains YJ016 (Chen et al., 2003) and CMCP6 (unpublished data; GenBank accession nos. AE016795 and AE016796), followed this breakthrough. Genome sequencing projects for a strain of *Vibrio fischeri* (http://ergointegratedgenomics.com/Genomes/VFI/), *Vibrio lentus* (F. Le Roux and D. Mazel, personal communication), and *Photobacterium profundum* (unpublished data; GenBank accession nos. CR354531 and CR354532) are now under way. In this chapter, we outline recent progress in the study of the *Vibrio* genomes, including information from the genome sequences covered to date.

TWO-CHROMOSOME CONFIGURATION

Discovery of the Two-Chromosome Configuration

A Japanese group and an American group independently reported the possession of two circular chromosomes for vibrios. Yamaichi and his colleagues constructed a physical map of the genomic DNA for *V. parahaemolyticus* strain AQ4673 (Yamaichi et al., 1999). Unexpectedly, the results showed two circular replicons of 3.2 and 1.9 Mb. Pulsed-field gel electrophoresis (PFGE) of undigested genomic DNA revealed two bands of corresponding sizes. Analysis both by NotI digestion and by Southern

blotting of the two isolated bands confirmed the existence of two replicons. The presence of housekeeping genes on both the replicons indicated that they are chromosomes rather than megaplasmids.

Two similar bands were also detected after PFGE of undigested genomic DNA of *V. parahaemolyticus* strains other than AQ4673, and of strains belonging to other *Vibrio* species, such as *V. vulnificus*, *V. fluvialis*, and various serovars and biovars of *V. cholerae*. From these data, Yamaichi et al. (1999) suggested that a two-replicon structure is common among *Vibrio* species. Almost at the same time, Trucksis et al. (1998) documented the physical map of *V. cholerae* 395, a classical Ogawa strain, and demonstrated that the genome is composed of two unique circular chromosomes (2.4 and 1.6 Mb).

Subsequent Reports

The possession of two distinct circular chromosomes for vibrios was confirmed by genome sequencing. Thus far, the genomes of three human-pathogenic *Vibrio* species—*V. cholerae* (Heidelberg et al., 2000), *V. parahaemolyticus* (Makino et al., 2003), and *V. vulnificus* (Chen et al., 2003; unpublished data, GenBank accession nos. AE016795 and AE016796)—have been completely sequenced. In all the cases, sequencing revealed the presence of two circular chromosomes in each strain: one large and one small.

Recently, Okada et al. (2005) reported extensive investigations of the genomic structure of vibrios, including pathogenic and nonpathogenic species, and of closely related species. They examined the prevalence of the two-chromosome configuration in 34 species of vibrios and closely related genera, such as *Photobacterium* and *Salinivibrio*. PFGE of undigested genomic

T. Iida • Department of Bacterial Infections, Research Institute for Microbial Diseases, Osaka University, 3-1 Yamadaoka, Suita, Osaka 565-0871, Japan. **Ken Kurokawa** • Laboratory of Comparative Genomics, Graduate School of Information Science, Nara Institute of Science and Technology, 8916-5 Takayamacho, Ikoma 630-0192, Japan.

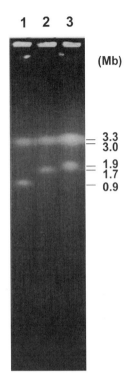

Figure 1. PFGE of undigested genomic DNA of vibrios. 1, *V. metschnikovii*; 2, *V. proteolyticus*; 3, *V. parahaemolyticus*. Two distinct bands corresponding to each chromosome are apparent.

DNA (Fig. 1) suggested that all the *Vibrio* strains have two chromosomes. Moreover, their results suggested that this feature is not limited to the genus *Vibrio*, but is common among the family *Vibrionaceae*. In contrast, strains of the genera *Aeromonas* and *Plesiomonas*, which were included in the *Vibrionaceae* until recently but have now been reclassified into the families *Aeromonadaceae* and *Enterobacteriaceae*, respectively (Ruimy et al., 1994), possess only one chromosome. Thus, possession of two chromosomes, a common feature among the *Vibrionaceae*, may not extend outside this family (Okada et al., 2005). Retrospectively, it is surprising that classical taxonomy has precisely predicted a grouping of bacteria that have a unique genome structure.

Reports of the possession of multiple chromosomes (or mega-sized replicons) by bacteria are not limited to the family *Vibrionaceae* (Kolsto, 1999; Okada et al., 2005). Eubacteria with multiple chromosomes are found in the alpha-proteobacteria (genera *Agrobacterium, Sinorhizobium, Rhizobium, Rhodobacter,* and *Brucella*) and beta-proteobacteria (genera *Burkholderia, Ralstonia,* and others) (Suwanto and Kaplan, 1989; Sobral et al., 1991; Jumas-Bilak et al., 1998a; Kolsto, 1999). However, some features of the chromosomes of these bacteria differ from those of the vibrios (Okada et al., 2005).

Okada et al. (2005) clearly demonstrated the features of the vibrio genomes. First, all the vibrios examined possess two chromosomes: no vibrios with only one chromosome were found. Second, the size of the large chromosome is relatively constant among the vibrios (see below). Finally, no vibrios possess large replicons (>0.3 Mb) apart from their two chromosomes. Thus, the genome structure of vibrios appears to be more constant and more stable than those of other bacteria with multiple chromosomes. This fact, along with the relatively large size of the smaller chromosomes (20 to 40% of the genome [Okada et al., 2005]) and the presence of essential genes on both the chromosomes (Heidelberg et al., 2000; Makino et al., 2003), suggests that the small chromosome is indispensable for these bacteria.

Why Do Vibrios Have Two Chromosomes?

Why do vibrios have two chromosomes? Why has the smaller not been integrated into the larger? Yamaichi et al. (1999) suggested that the split of the genome into two replicons would be advantageous for DNA replication. This hypothesis would partly account for the fast growth rate of some species, such as *V. parahaemolyticus*, which has a doubling time of only 8 to 9 min (Joseph et al., 1982). On the other hand, Heidelberg et al. (2000) suggested that the small chromosome might play an important specialized function that applies evolutionary selective pressure against integration. They hypothesized that, under certain conditions, there may be differences in copy numbers of the larger and smaller chromosomes, potentially increasing the effective level of gene expression on the more numerous chromosome, to the organism's advantage. The extreme in chromosome copy number asymmetry would be the loss of one chromosome, and nonreplicating, single-chromosome cells ("drone" cells), within a nutrient-stressed population of normal two-chromosome cells, might contribute to the survival of the community by continuing the secretion of enzymes that break down molecules in the surrounding microenvironment. Heidelberg et al. (2000) proposed that the resulting nutrients might be used by normal members of a community at risk, thereby promoting survival of the species as a whole. Although there is no direct evidence for such a change in stoichiometry of the two chromosomes, splitting the genome into two replicons might be advantageous for those vibrios that experience varied environments in a natural setting (Colwell, 1996; Schoolnik and Yildiz, 2000).

The distribution of genes of known function between the large and small chromosomes of vibrios provides tantalizing clues about how the two-chromosome configuration of the *Vibrionaceae* might confer an evolutionary advantage. The large chromosome contains most of the genes that are required for growth (Heidelberg et al., 2000; Makino et al., 2003), while the small chromosome contains more genes for bacterial adaptation to environmental change (Makino et al., 2003). This implies that the two chromosomes may play different roles. The global expression pattern of the genes on the two chromosomes of *V. cholerae* has been analyzed using a whole-genome microarray (Merrell et al., 2002; Xu et al., 2003; Bina et al., 2003) and other methods (Hang et al., 2003). In the rabbit ileal loop model, Xu et al. (2003) compared the global transcriptional pattern of in vivo-grown cells with those grown to mid-exponential phase in rich medium under aerobic conditions. Under both conditions, the genes showing the highest levels of expression resided primarily on the large chromosome. However, a shift occurred in vivo that resulted in many more small-chromosome genes being expressed during growth in the intestine. This again implies that the two chromosomes may play different roles.

Origin of the Small Chromosome

How did the two-chromosome configuration of vibrios develop? One possible explanation is that the small chromosome may have arisen by excision from a single, large ancestral genome (Waldor and Raychaudhuri, 2000). Such an example has been reported for species other than those in the family *Vibrionaceae* (Jumas-Bilak et al., 1998b). In contrast with the above explanation, Heidelberg et al. (2000) hypothesized that the small chromosome of *V. cholerae* may be a megaplasmid acquired by an ancestral vibrio. Several lines of evidence support this hypothesis. Phylogenetic analysis of the *parA* gene homologues located near the putative origin of replication of each chromosome showed that the large chromosome *parA* tends to group with other chromosomal *parA*s, whereas the *parA* from the small chromosome tends to group with plasmid, phage, and megaplasmid *parA*s. In general, genes on the small chromosome, with an apparently identical functioning copy on the large chromosome, appeared less similar to orthologues present in other gamma-proteobacteria species. In addition, the large chromosome contained all the ribosomal RNA operons and at least one copy of all the transfer RNAs. Also, an integron, an element often found on plasmids (Mazel et al., 1998), is located on the small chromosome in the genome of *V. cholerae* (Heidelberg et al., 2000). On a side note, in *V. parahaemolyticus* and *V. vulnificus*, the integron (a superintegron) is present on the large chromosome (Tagomori et al., 2002; Makino et al., 2003; Chen

et al., 2003). This difference in the location of the superintegron in species belonging to the same genus indicates that the chromosomal location of the superintegron cannot be cited as evidence to support the megaplasmid hypothesis. Finally, the bias in the housekeeping gene content is more easily explained if the small chromosome was originally a megaplasmid.

Recently, Egan and Waldor (2003) identified the origins of replication (*ori*) of the two *V. cholerae* chromosomes. Their work revealed that the replication origin of the large chromosome ($oriCI_{VC}$) largely conforms to the basic features of *Escherichia coli* *oriC*, which has been thought to define the chromosomal origins of replication in gamma-proteobacteria. Although the replication origin of the small chromosome ($oriCII_{VC}$) shares certain features with *E. coli oriC*, several unusual features for a bacterial chromosome are also present. Thus, $oriCII_{VC}$-based replication displays several features that characterize certain plasmid replicons. The authors suggested that these similarities might support the proposal (Heidelberg et al., 2000) that the small chromosome was originally acquired as a plasmid. In any case, because the two *Vibrio* chromosomes appear to have coexisted throughout *Vibrionaceae* speciation (Okada et al., 2005; Egan and Waldor, 2003), the generation of the two-chromosome configuration must have occurred before the diversification of this family.

Replication and Segregation of the Two Chromosomes

It is interesting to consider whether the mechanisms of replication for the two chromosomes of vibrios are common. Whereas prokaryotic and eukaryotic chromosomes replicate once per cell cycle, plasmids replicate autonomously and independently of the host chromosome. Egan and Waldor (2003) analyzed the origins of replication of each chromosome in *V. cholerae* and revealed that the origin of replication of the large chromosome, $oriCI_{VC}$, has significant sequence homology and functional similarity to the origin of replication of the *E. coli* chromosome (*oriC*). In contrast, $oriCII_{VC}$ is unusual, containing a combination of features characteristic of both chromosomal and plasmid replicons. The two chromosomes have both shared and distinct mechanisms to control replication initiation (Egan and Waldor, 2003). Evidence now indicates that the two chromosomes of *V. cholerae* initiate replication synchronously, once per cell cycle (Egan et al., 2004). Thus, whatever the origin of the small chromosome of vibrios, stable maintenance of genomes with multiple chromosomes might have required the evolution of shared mechanisms to control replication (Egan et al., 2004).

Another interesting issue concerns the mechanism that enables the bacteria to properly segregate and distribute the two chromosomes to daughter cells during replication and cell division. Using static and time-lapse fluorescence microscopy, Fogel and Waldor (2005) visualized the localization and segregation of the origins of replication of the *V. cholerae* chromosomes. At all stages of the cell cycle, the two origins localized to distinct subcellular locations. In newborn cells, *oriCI*$_{VC}$ was located near the cell pole, and *oriCII*$_{VC}$ was at the cell center. Segregation of *oriCI*$_{VC}$ occurred asymmetrically from a polar position, with one duplicated origin traversing the length of the cell toward the opposite pole and the other remaining relatively fixed. In contrast, *oriCII*$_{VC}$ segregated later in the cell cycle than *oriCI*$_{VC}$, and the two duplicated *oriCII*$_{VC}$ regions became repositioned to the new cell centers. The differences in localization and timing of segregation of *oriCI*$_{VC}$ and *oriCII*$_{VC}$ suggest that distinct mechanisms govern the localization and segregation of the two *V. cholerae* chromosomes. However, the molecular mechanisms that account for the different patterns of localization of the two chromosomes of vibrios are yet to be elucidated.

GENOME COMPARISON

Comparison of Genome Structure Among Species

Genome sequencing of three species of vibrios enabled us to precisely compare the genome structures of the strains (Color Plate 2). A comparison of the genomes of *V. cholerae*, *V. parahaemolyticus*, and *V. vulnificus* revealed that, although the large chromosome does not differ greatly in size among the genomes (3.0 to 3.4 Mb), the small chromosome is much larger in *V. parahaemolyticus* and *V. vulnificus* (1.9 Mb) than in *V. cholerae* (1.1 Mb) (Heidelberg et al., 2000; Makino et al., 2003; Chen et al., 2003; unpublished data, GenBank accession nos. AE016795 and AE016796). The smaller chromosomes seem to have higher proportions of genes unique to each *Vibrio* species (Makino et al., 2003; Chen et al., 2003). These size differences and the discrepancy in the number of unique genes suggest that the smaller chromosomes of vibrios are more diverse in structure and gene content than the larger.

Examination of more species from the *Vibrionaceae* also demonstrated that, whereas the sizes of the large chromosome do not differ greatly, those of the smaller chromosome are variable (Okada et al., 2005). The sizes of the large chromosomes of various vibrios, except for a few strains, clustered in a relatively narrow range of 3.0 to 3.3 Mb, whereas the sizes of the small chromosomes varied considerably, from 0.8 to 2.4 Mb (Fig. 2) (Okada et al., 2005). Thus, the smaller chromosomes of vibrios seem to be structurally more flexible than the larger ones. A comparison of the relative position of the conserved genes in *V. cholerae* and *V. parahaemolyticus* demonstrated that extensive genome rearrangements have occurred within and between the larger and smaller chromosomes during evolution (Tagomori et al., 2002; Makino et al., 2003). Of the 2,293 conserved genes on the larger chromosome of *V. cholerae*, 2,076 (90.5%) were also found on the larger chromosome in *V. parahaemolyticus*, and 539 (85.0%) of 634 conserved genes found on the smaller chromosome of *V. cholerae* were also found on the *V. parahaemolyticus* smaller chromosome (Makino et al., 2003). These results suggest that, despite extensive genome re-

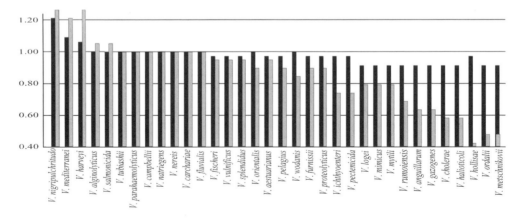

Figure 2. Comparison of the size of the large and small chromosomes. The vertical axis is the relative size of each chromosome from various vibrios; the chromosomes of *V. parahaemolyticus* were used as a standard. The size differences of the chromosomes of each vibrio are presented by dividing the size of the large or small chromosome of each strain by the size of the corresponding chromosome of *V. parahaemolyticus*. Black bars, large chromosomes; gray bars, small chromosomes.

arrangement, the location of most of the conserved genes on either the larger or smaller chromosome is conserved between the two species of *Vibrio*. This was also the case when *V. vulnificus* was included in the comparison (Chen et al., 2003). This implies that interchromosomal rearrangements were less frequent than intrachromosomal rearrangements in the evolution of vibrios.

Regarding the genome-sequenced vibrios, most of the essential genes required for growth and viability occur on the larger chromosomes, although some are definitely present only on the smaller ones (Heidelberg et al., 2000; Makino et al., 2003). The smaller chromosomes seem to contain more genes related to transcriptional regulation and the transport of various substrates than do the larger ones. Genes classified in these categories have a role in bacterial adaptation to environmental changes. The features of the large and small chromosomes of these vibrios are likely to be similar in other representatives of the genus. We speculate that the ancestral vibrio diversified into various species with most essential genes maintaining on the large chromosome. This view is supported by the above-mentioned low frequency of interchromosomal rearrangements (Makino et al., 2003). It also explains the greater structural flexibility of the smaller chromosomes.

Another characteristic feature in the genome rearrangement of vibrios is the apparent symmetric translocation of orthologous genes around the replication origin and terminus during evolution (Color Plate 2) (Makino et al., 2003). In other words, most of the intrachromosomal rearrangements, especially in the larger chromosome, may have occurred without changing their relative distance from the origin of replication. Such symmetrical rearrangements have been observed in other bacterial chromosomes (Liu and Sanderson, 1996; Stibitz and Yang, 1997; Eisen et al., 2000). What could cause such rearrangements that pivot around the origin and terminus of the genome to occur more frequently than others? One possibility is that many rearrangements occur, but there is selection against those that change the distance of a gene from the origin or terminus. There are two possible explanations. The gene dosage hypothesis postulates that genes at different distances from the origin of replication have different gene dosages and hence different gene expression, so that during evolution genes become adapted to their specific location. By contrast, the genomic balance hypothesis states that the origin and terminus of replication must be separated by 180°. These hypotheses, however, need further investigation (Liu and Sanderson, 1996). Alternatively, the symmetrical rearrangement events could be linked to the bidirectional chromosome replication (Eisen et al., 2000). Whatever the mechanisms, such symmetrical rearrangements seem to be a common feature of bacterial evolution (Eisen et al., 2000).

Comparisons within Species

One useful way to analyze the diversity of genomes of strains within a certain species is to compare the whole gene contents of the strains. Comparative genomic hybridization using microarrays is a feasible approach for this and has been applied to *V. cholerae* (Dziejman et al., 2002). The results revealed a surprisingly high degree of conservation among the strains tested. The work also identified genes unique to all pandemic strains, as well as genes specific for the seventh pandemic El Tor and related O139 serogroup strains (Dziejman et al., 2002). Similar analysis is now under way with *V. parahaemolyticus* strains (T. Iida, unpublished data). This approach is surely useful to the analysis of diversity within a certain species and could be applicable for the comparison of the gene content among phylogenetically closely related species.

DRIVING FORCES IN EVOLUTION

Genomic study of the vibrios, especially comparative genome analysis between *V. cholerae*, *V. parahaemolyticus*, and *V. vulnificus*, has demonstrated evidence of a variety of genetic events that occurred during the evolution of the organisms (Makino et al., 2003; Chen et al., 2003). Those include mutations, chromosomal rearrangements, loss of genes by decay or deletion, and gene acquisitions through duplication or lateral transfer. All of these events could have been driving forces in the evolution and diversification of bacteria (Hacker and Carniel, 2001; Hacker et al., 2003; Ochman et al., 2000; Ochman and Moran, 2001). Among them, lateral gene transfer could have been an efficient mechanism to introduce new phenotypes into bacterial genomes. When compared with the genome of, for example, pathogenic *E. coli*, which shows numerous traces of lateral gene transfer (Ohnishi et al., 2001), vibrios seem to have fewer mobile genetic elements—transposons and phages and DNA regions with a G+C content that differs from the whole-genome average (which are indicative of recent lateral gene transfer) (Heidelberg et al., 2000; Makino et al., 2003). Nevertheless, it has been demonstrated by recent studies, including genome sequencing, that lateral gene transfer has contributed to several important characteristics of vibrios, such as pathogenicity.

Here, we briefly introduce some recent topics on horizontal gene transfer in vibrios, mainly in relation to the acquisition of the genes for pathogenicity. More general and detailed explanations on lateral gene transfer, bacteriophages, and integrons will appear in the following chapters.

Bacteriophages

The genes for the cholera toxin, the most important virulence factor of *V. cholerae*, have long been believed to be encoded in the chromosome of the bacterium. Waldor and Mekalanos (1996) reported that the genes are actually encoded in the genome of a newly identified bacteriophage, CTXϕ, and that the phage genome is integrated into the bacterial chromosome as a prophage (Waldor and Mekalanos, 1996). Unlike other toxin-converting phages, most of which are double-stranded DNA phages, CTXϕ is a filamentous phage with a single-stranded DNA genome, and it shares some features with the well-known filamentous phage, M13. Although the M13 phage does not integrate into the host chromosome, CTXϕ is able to do so. CTXϕ infects host cholera vibrios through pili, using the toxin-coregulated pili as its receptor (Waldor and Mekalanos, 1996). Phylogenetically related filamentous phages have been identified for *V. cholerae* and other *Vibrio* species (Kar et al., 1996; Honma et al., 1997; Ikema and Honma, 1998; Chang et al., 1998; Nasu et al., 2000; Campos et al., 2003). The genomes of this group of filamentous phages are composed of both conserved and distinctive regions. The distinctive regions are considered foreign DNA sequences, often containing genes unique to each phage. Examples of this are the cholera toxin genes of CTXϕ. Thus, the phages can play a role in genetic transmission between bacterial strains. Recently, it has been revealed that this group of phages targets *dif*-like sites as their integration site (Huber and Waldor, 2002; Iida et al., 2002; Campos et al., 2003). The *dif* site is located in the replication terminus region of bacterial chromosomes, where it serves to resolve dimeric chromosomes formed during replication.

Superintegrons

One of the characteristics of vibrio genomes is the presence of superintegrons. Classically described integrons are natural cloning and expression systems that incorporate open reading frames and convert them into functional genes. Integrons have been widely identified as the constituents of transferable elements responsible for the evolution of multidrug resistance: multiresistance integrons (Rowe-Magnus

et al., 2002a). The integron platform codes for an integrase (*intI*) that mediates recombination between a proximal primary recombination site (*attI*) and a target recombination sequence, called an *attC* site (59 bp). The *attC* site is usually found associated with a single open reading frame in a circularized structure, termed a gene cassette. Insertion of the gene cassette at the *attI* site drives expression of the encoded proteins. Superintegrons constitute a large ancestral integron, which was first discovered in the *V. cholerae* genome (Clark et al., 1997; Mazel et al., 1998). This superintegron spans 126 kb and harbors 179 cassettes of mainly unassigned function (Heidelberg et al., 2000). To date, superintegrons have been identified in a number of *Vibrio* species (Rowe-Magnus et al., 2001, 2002c). A comparison of the superintegrons of *V. cholerae* and *V. parahaemolyticus* revealed that the contents of the gene cassettes differed substantially between the two (Makino et al., 2003). This suggests that the genes making up the superintegron are highly diverse between species. Considering that *Vibrionaceae* is one of the dominant families of marine microorganisms, clearly the superintegrons in various vibrios as a whole constitute a vast gene pool in natural environments. Recently, it was reported that chromosomal superintegrons of vibrios might be a genetic source for the evolution of resistance to clinically relevant antibiotics through integron-mediated recombination events (Rowe-Magnus et al., 2002b).

Pathogenicity Islands

Pathogenicity islands (PAIs) were first described in the genomes of pathogenic *E. coli* (Blum et al., 1994). Subsequently, they were found in other pathogens, where they form specific entities associated with bacterial pathogenicity (Groisman and Ochman, 1996; Hacker and Carniel, 2001). PAIs are DNA regions in bacterial genomes, generally somewhere between 10 and 200 kb in length, which have features characteristic of transferred elements (different G+C content and codon usage compared with the rest of the genome, and the presence of mobility genes such as insertion elements and parts of phages). PAIs often accommodate large clusters of genes contributing a particular virulence phenotype and are distributed exclusively in pathogenic strains, but not nonpathogenic strains, in a set of related organisms. It is now accepted that the generation of PAIs often starts with the integration of plasmids, phages, or conjugative transposons into specific target genes, preferentially on the chromosomes (Kaper and Hacker, 1999). These targets are often tRNA genes. After integration into the bacterial genome, the integrated elements experience multiple genetic events, such as mutations, deletions, and insertions of genes under specific selective pressures, before resulting in the formation of PAIs (Kaper and Hacker, 1999).

In *V. cholerae*, a cluster of genes for the biogenesis of toxin-coregulated pili, which play an important role in human intestinal colonization by the pathogen, are encoded in a PAI designated VPI (Karaolis et al., 1998). The presence of PAIs has also been suggested for the genome of *V. parahaemolyticus* strains pathogenic to humans (Suthienkul et al., 1995; Iida et al., 1997, 1998; Park et al., 2000). Sequencing the genome of *V. parahaemolyticus* RIMD 2210633 demonstrated that, on the smaller chromosome of the strain, there was a PAI of about 80 kb (Makino et al., 2003). This PAI encoded the genes for hemolysins, toxins, enzymes, possible membrane proteins, and the type III secretion system (TTSS) (Hueck, 1998). There may well be involvement of these newly identified genes in the pathogenicity of *V. parahaemolyticus* (Park et al., 2004a,b; T. Iida et al., unpublished data).

Sequencing of entire bacterial genomes revealed that PAI-like DNA regions are much more widespread than previously thought and represent a paradigm of more general entities that are present in the genomes of many bacterial species: genomic islands (Hacker and Carniel, 2001). Functions encoded in genomic islands are not restricted to pathogenicity but can include many other aspects, such as antibiotic resistance, symbiosis, metabolism, degradation, and secretion, that increase bacterial fitness in certain environments. The evolutionary advantage of genomic islands over mutations and other smaller insertions is that a large number of genes (e.g., operons and gene clusters encoding related functions) may be transferred and incorporated en bloc into the recipient genome. This transfer may lead to dramatic changes in the behavior of the organism, resulting ultimately in "evolution in quantum leaps" (Groisman and Ochman, 1996). Obviously, there must have been a great contribution of genomic islands to bacterial evolution.

CONCLUDING REMARKS

To date, full genome sequencing has been completed for at least three species in the genus *Vibrio*. The sequencing and following analysis revealed a number of common and diverse genome features among the species. One of the most notable common aspects of vibrio genomes is the two-chromosome configuration. Vibrios include more than 60 species, with various features and lifestyles. Nevertheless, they have conserved this unique genomic structure during their evolution. It would be interesting to know why vibrios have diversified while maintaining the two-chromosome configuration. This might be related to the lifestyle unique to vibrios. There have been a number of reports on bacteria other than vibrios with multiple chromosomes (or mega-sized replicons), and comparison of the lifestyles of those bacteria with that of vibrios could help elucidate this.

Genome sequencing and comparative analysis also give novel insights into the features unique to each organism. An example is the discovery of functional TTSS genes in the genome of *V. parahaemolyticus* (Makino et al., 2003; Park et al., 2004b), which are not found in the genomes of *V. cholerae* and *V. vulnificus*. The TTSS is a bacterial protein secretion apparatus by which bacteria intimately interact with eukaryotic cells (Hueck, 1998). The possession of such secretion systems by *V. parahaemolyticus* implies that, in the life cycle of this organism in natural environments, there should be a stage or phase in which this organism intimately interacts with certain eukaryotic organisms. As with *V. parahaemolyticus*, further genome sequencing and comparative analysis of more vibrios should give us exciting new knowledge about vibrios exhibiting a variety of lifestyles.

Acknowledgment. We thank Kazuhisa Okada for assistance in preparation of the figures in this chapter.

REFERENCES

Bina, J., J. Zhu, M. Dziejman, S. Faruque, S. Calderwood, and J. Mekalanos. 2003. ToxR regulon of *Vibrio cholerae* and its expression in vibrios shed by cholera patients. *Proc. Natl. Acad. Sci. USA* **100**:2801–2806.

Blum, G., M. Ott, A. Lischewski, A. Ritter, H. Imrich, H. Tschape, and J. Hacker. 1994. Excision of large DNA regions termed pathogenicity island from tRNA-specific loci in the chromosome of an *Escherichia coli* wild-type pathogen. *Infect. Immun.* **62**:606–614.

Campos, J., E. Martinez, E. Suzarte, B. L. Rodriguez, K. Marrero, Y. Silva, T. Ledon, R. del Sol, and R. Fando. 2003. VGJø, a novel filamentous phage of *Vibrio cholerae*, integrates into the same chromosomal site as CTXφ. *J. Bacteriol.* **185**:5685–5696.

Chang, B., H. Taniguchi, H. Miyamoto, and S. Yoshida. 1998. Filamentous bacteriophages of *Vibrio parahaemolyticus* as a possible clue to genetic transmission. *J. Bacteriol.* **180**:5094–5101.

Chen, C. Y., K. M. Wu, Y. C. Chang, C. H. Chang, H. C. Tsai, T. L. Liao, Y. M. Liu, H. J. Chen, A. B. Shen, J. C. Li, T. L. Su, C. P. Shao, C. T. Lee, L. I. Hor, and S. F. Tsai. 2003. Comparative genome analysis of *Vibrio vulnificus*, a marine pathogen. *Genome Res.* **13**:2577–2587.

Clark, C. A., L. Purins, P. Kaewrakon, and P. A. Manning. 1997. VCR repetitive sequence elements in the *Vibrio cholerae* chromosome constitute a mega-integron. *Mol. Microbiol.* **26**:1137–1138.

Colwell, R. R. 1996. Global climate and infectious disease: the cholera paradigm. *Science* **274**:2025–2031.

Dziejman, M., E. Balon, D. Boyd, C. M. Fraser, J. F. Heidelberg, and J. J. Mekalanos. 2002. Comparative genomic analysis of

Vibrio cholerae: genes that correlate with cholera endemic and pandemic disease. *Proc. Natl. Acad. Sci. USA* **99**:1556–1561.

Egan, E. S., and M. K. Waldor. 2003. Distinct replication requirements for the two *Vibrio* chromosomes. *Cell* **114**:521–530.

Egan, E. S., A. Løbner-Olesen, and M. K. Waldor. 2004. Synchronous replication initiation of the two *Vibrio cholerae* chromosomes. *Curr. Biol.* **14**:R501–R502.

Eisen, J. A., J. F. Heidelberg, O. White, and S. L. Salzberg. 2000. Evidence for symmetric chromosomal inversions around the replication origin in bacteria. *Genome Biol.* **1**:RESEARCH0011.1–0011.9.

Fogel, M. A., and M. K. Waldor. 2005. Distinct segregation dynamics of the two *Vibrio cholerae* chromosomes. *Mol. Microbiol.* **55**:125–136.

Groisman, E. A., and H. Ochman. 1996. Pathogenicity islands: bacterial evolution in quantum leaps. *Cell* **87**:791–794.

Hacker, J., and E. Carniel. 2001. Ecological fitness, genomic islands and bacterial pathogenicity. *EMBO Rep.* **2**:376–381.

Hacker, J., U. Hentschel, and U. Dobrindt. 2003. Prokaryotic chromosomes and disease. *Science* **301**:790–793.

Hang, L., M. John, M. Asaduzzaman, E. A. Bridges, C. Vanderspurt, T. J. Kirn, R. K. Taylor, J. D. Hillman, A. Progulske-Fox, M. Handfield, E. T. Ryan, and S. B. Calderwood. 2003. Use of *in vivo*-induced antigen technology (IVIAT) to identify genes uniquely expressed during human infection with *Vibrio cholerae*. *Proc. Natl. Acad. Sci. USA* **100**:8508–8513.

Heidelberg, J. F., J. A. Eisen, W. C. Nelson, R. A. Clayton, M. L. Gwinn, R. J. Dodson, D. H. Haft, E. K. Hickey, J. D. Peterson, L. Umayam, S. R. Gill, K. E. Nelson, T. D. Read, H. Tettelin, D. Richardson, M. D. Ermolaeva, J. Vamathevan, S. Bass, H. Qin, I. Dragoi, P. Sellers, L. McDonald, T. Utterback, R. D. Fleishmann, W. C. Nierman, O. White, S. L. Salzberg, H. O. Smith, R. R. Colwell, J. J. Mekalanos, J. C. Venter, and C. M. Fraser. 2000. DNA sequence of both chromosomes of the cholera pathogen *Vibrio cholerae*. *Nature* **406**:477–483.

Honma, Y., M. Ikema, C. Toma, M. Ehara, and M. Iwanaga. 1997. Molecular analysis of a filamentous phage (fs1) of *Vibrio cholerae* O139. *Biochim. Biophys. Acta* **1362**:109–115.

Huber, K. E., and M. K. Waldor. 2002. Filamentous phage integration requires the host recombinases XerC and XerD. *Nature* **417**:656–659.

Hueck, C. J. 1998. Type III protein secretion systems in bacterial pathogens of animals and plants. *Microbiol. Mol. Biol. Rev.* **62**:379–433.

Iida, T., K. Makino, H. Nasu, K. Yokoyama, K. Tagomori, A. Hattori, H. Okuno, H. Shinagawa, and T. Honda. 2002. Filamentous bacteriophages of vibrios are integrated into the dif-like site of the host chromosome. *J. Bacteriol.* **184**:4933–4935.

Iida, T., K.-S. Park, O. Suthienkul, J. Kozawa, Y. Yamaichi, K. Yamamoto, and T. Honda. 1998. Close proximity of the *tdh*, *trh* and *ure* genes on the chromosome of *Vibrio parahaemolyticus*. *Microbiology* **144**:2517–2523.

Iida, T., O. Suthienkul, K.-S. Park, G.-Q. Tang, R. K. Yamamoto, M. Ishibashi, K. Yamamoto, and T. Honda. 1997. Evidence for genetic linkage between the *ure* and *trh* genes in *Vibrio parahaemolyticus*. *J. Med. Microbiol.* **46**:639–645.

Ikema, M., and Y. Honma. 1998. A novel filamentous phage, fs2, of *Vibrio cholerae* O139. *Microbiology* **144**:1901–1906.

Joseph, S. W., R. R. Colwell, and J. B. Kaper. 1982. *Vibrio parahaemolyticus* and related halophilic vibrios. *Crit. Rev. Microbiol.* **10**:77–124.

Jumas-Bilak. E., S. Michaux-Charachon, G. Bourg, M. Ramuz, and A. Allardet-Servent. 1998a. Unconventional genomic organization in the alpha subgroup of the Proteobacteria. *J. Bacteriol.* **180**:2749–2755.

Jumas-Bilak, E., S. Michaux-Charachon, G. Bourg, D. O'Callaghan, and M. Ramuz. 1998b. Differences in chromosome number and genome rearrangements in the genus *Brucella*. *Mol. Microbiol.* **27**:99–106.

Kaper, J. B., and J. Hacker. 1999. *Pathogenicity Islands and Other Mobile Virulence Elements*. ASM Press, Washington, D.C.

Kar, S., R. K. Ghosh, A. N. Ghosh, and A. Ghosh. 1996. Integration of the DNA of a novel filamentous bacteriophage VSK from *Vibrio cholerae* O139 into the host chromosomal DNA. *FEMS Microbiol. Lett.* **145**:17–22.

Karaolis, D. K., J. A. Johnson, C. C. Bailey, E. C. Boedeker, J. B. Kaper, and P. R. Reeves. 1998. A *Vibrio cholerae* pathogenicity island associated with epidemic and pandemic strains. *Proc. Natl. Acad. Sci. USA* **95**:3134–3139.

Kolsto, A. B. 1999. Time for a fresh look at the bacterial chromosome. *Trends Microbiol.* **7**:223–226.

Liu, S.-L., and K. E. Sanderson. 1996. Highly plastic chromosomal organization in *Salmonella typhi*. *Proc. Natl. Acad. Sci. USA* **93**:10303–10308.

Makino, K., K. Oshima, K. Kurokawa, K. Yokoyama, T. Uda, K. Tagomori, Y. Iijima, M. Najima, M. Nakano, A. Yamashita, Y. Kubota, S. Kimura, T. Yasunaga, T. Honda, H. Shinagawa, M. Hattori, and T. Iida. 2003. Genome sequence of *Vibrio parahaemolyticus*: a pathogenic mechanism distinct from that of *V. cholerae*. *Lancet* **361**:743–749.

Mazel, D., B. Dychinco, V. A. Webb, and J. Davies. 1998. A distinctive class of integron in the *Vibrio cholerae* genome. *Science* **280**:605–608.

Merrell, D. S., S. M. Butler, F. Qadri, N. A. Dolganov, A. Alam, M. B. Cohen, S. B. Calderwood, G. K. Schoolnik, and A. Camilli. 2002. Host-induced epidemic spread of the cholera bacterium. *Nature* **417**:642–645.

Nasu, H., T. Iida, T. Sugahara, Y. Yamaichi, K.-S. Park, K. Yokoyama, K. Makino, H. Shinagawa, and T. Honda. 2000. A filamentous phage associated with recent pandemic *Vibrio parahaemolyticus* O3:K6 strains. *J. Clin. Microbiol.* **38**:2156–2161.

Ochman, H., J. G. Lawrence, and E. A. Groisman. 2000. Lateral gene transfer and the nature of bacterial innovation. *Nature* **405**:299–304.

Ochman, H., and N. A. Moran. 2001. Genes lost and genes found: evolution of bacterial pathogenesis and symbiosis. *Science* **292**:1096–1098.

Ohnishi, M., K. Kurokawa, and T. Hayashi. 2001. Diversification of *Escherichia coli* genomes: are bacteriophages the major contributors? *Trends Microbiol.* **9**:481–485.

Okada, K., T. Iida, K. Kita-Tsukamoto, and T. Honda. 2005. Vibrios commonly possess two chromosomes. *J. Bacteriol.* **187**:752–757.

Park, K.-S., T. Iida, Y. Yamaichi, T. Oyagi, K. Yamamoto, and T. Honda. 2000. Genetic characterization of DNA region containing the *trh* and *ure* genes of *Vibrio parahaemolyticus*. *Infect. Immun.* **68**:5742–5748.

Park, K.-S., T. Ono, M. Rokuda, M.-H. Jang, T. Iida, and T. Honda. 2004a. Cytotoxicity and enterotoxicity of the thermostable direct hemolysin-deletion mutants of *Vibrio parahaemolyticus*. *Microbiol. Immunol.* **48**:313–318.

Park, K.-S., T. Ono, M. Rokuda, M.-H. Jang, K. Okada, T. Iida, and T. Honda. 2004b. Functional characterization of two type III secretion systems of *Vibrio parahaemolyticus*. *Infect. Immun.* **48**:313–318.

Rowe-Magnus, D. A., A.-M. Guetout, P. Ploncard, B. Dychinco, J. Davies, and D. Mazel. 2001. The evolutionary history of chromosomal super-integrons provides an ancestry for multiresistant integrons. *Proc. Natl. Acad. Sci. USA* **98**:652–657.

Rowe-Magnus, D. A., J. Davies, and D. Mazel. 2002a. Impact of integrons and transposons on the evolution of resistance and virulence. *Curr. Top. Microbiol. Immunol.* **264:**167–188.

Rowe-Magnus, D. A., A.-M. Guetout, and D. Mazel. 2002b. Bacterial resistance evolution by recruitment of super-integron gene cassettes. *Mol. Microbiol.* **43:**1657–1669.

Rowe-Magnus, D. A., A.-M. Guetout, L. Biskri, P. Bouige, and D. Mazel. 2002c. Comparative analysis of superintegrons: engineering extensive genetic diversity in the *Vibrionaceae. Genome Res.* **13:**428–442.

Ruimy, R., V. Breittmayer, P. Elbaze, B. Lafay, O. Boussemart, M. Gauthier, and R. Christen. 1994. Phylogenetic analysis and assessment of the genera *Vibrio, Photobacterium, Aeromonas,* and *Plesiomonas* deduced from small-subunit rRNA sequences. *Int. J. Syst. Bacteriol.* **44:**416–426.

Schoolnik, G. K., and F. H. Yildiz. 2000. The complete genome sequence of *Vibrio cholerae*: a tale of two chromosomes and of two lifestyles. *Genome Biol.* **1:**1016.1–1016.3.

Sobral, B. W., R. J. Honeycutt, A. G. Atherly, and M. McClelland. 1991. Electrophoretic separation of the three *Rhizobium meliloti* replicons. *J. Bacteriol.* **173:**5173–5180.

Stibitz, S., and M.-S. Yang. 1997. Genomic fluidity of *Bordetella pertussis* assessed by a new method for chromosomal mapping. *J. Bacteriol.* **179:**5820–5826.

Suthienkul, O., M. Ishibashi, T. Iida, N. Nettip, S. Supavej, B. Eampokalap, M. Makino, and T. Honda. 1995. Urease production correlates with possession of the *trh* gene in *Vibrio para-haemolyticus* strains isolated in Thailand. *J. Infect. Dis.* **172:**1405–1408.

Suwanto, A., and S. Kaplan. 1989. Physical and genetic mapping of the *Rhodobacter sphaeroides* 2.4.1 genome: presence of two unique circular chromosomes. *J. Bacteriol.* **171:**5850–5859.

Tagomori, K., T. Iida, and T. Honda. 2002. Comparison of genome structures of vibrios, bacteria possessing two chromosomes. *J. Bacteriol.* **184:**4351–4358.

Thompson, F. L., T. Iida, and J. Swings. 2004. Biodiversity of vibrios. *Microbiol. Mol. Biol. Rev.* **68:**403–431.

Trucksis, M., J. Michalski, Y. K. Deng, and J. B. Kaper. 1998. The *Vibrio cholerae* genome contains two unique circular chromosomes. *Proc. Natl. Acad. Sci. USA* **95:**14464–14469.

Waldor, M. K., and J. J. Mekalanos. 1996. Lysogenic conversion by a filamentous phage encoding cholera toxin. *Science* **272:**1910–1914.

Waldor, M. K., and D. Raychaudhuri. 2000. Treasure trove for cholera research. *Nature* **406:**469–470.

Xu, Q., M. Dziejman, and J. J. Mekalanos. 2003. Determination of the transcriptome of *Vibrio cholerae* during intraintestinal growth and midexponential phase *in vitro. Proc. Natl. Acad. Sci. USA* **100:**1286–1291.

Yamaichi, Y., T. Iida, K.-S. Park, K. Yamamoto, and T. Honda. 1999. Physical and genetic map of the genome of *Vibrio para-haemolyticus*: presence of two chromosomes in *Vibrio* species. *Mol. Microbiol.* **31:**1513–1521.

The Biology of Vibrios
Edited by F. L. Thompson et al.
© 2006 ASM Press, Washington, D.C.

Chapter 6

Gene Duplicates in Vibrio Genomes

DIRK GEVERS AND YVES VAN DE PEER

INTRODUCTION

Completely sequenced genomes provide us with an opportunity to study the evolution of the genome at a comprehensive level. During the last decade, the increasing number of sequenced genomes has made it clear that prokaryotic genomes should not be regarded as static structures but rather as relatively variable and flexible structures. Evolutionary processes, including gene duplication, horizontal gene transfer (HGT), gene loss, and chromosomal rearrangements, act to optimize the genome organization, allowing the best adaptive response of the cell within its natural environment (Gogarten et al., 2002; Snel et al., 2002; Kunin and Ouzounis, 2003; Coenye et al., 2005). By comparing related genomes and inferring ancestral ones, it becomes possible to study the processes of genome evolution that shape bacterial genomes and to reconstruct ancient evolutionary relationships.

Gene duplication is considered a mechanistic antecedent of gene innovation, and consequently of genetic novelty, that has facilitated adaptation to changing environments and exploitation of new niches (Hooper and Berg, 2003). The role of gene duplications in evolution has been studied extensively (Tekaia and Dujon, 1999; Jordan et al., 2001; Snel et al., 2002; Kunin and Ouzounis, 2003; Gevers et al., 2004). Examples of evolutionary response have been observed in bacteria exposed to different selection pressures, such as starvation conditions and thermal stress (Yamanaka et al., 1998; Riehle et al., 2001; Caporale, 2003). When the selective pressure is removed, the duplicates can be rapidly lost, thereby forming a reversible adaptive mutation that alters gene dosage without really altering genetic information. In addition to this short-term evolutionary advantage of paralogs, gene duplication and consequent functional divergence are considered an important evolutionary step toward diversity in the functional

repertoire of an organism, presumably enabling the organism to adapt to varying environmental conditions and broadening the phenotypes (Jordan et al., 2001; Hooper and Berg, 2003). Prokaryotic genomes have a considerable fraction of genes that are homologous to other genes within the same genome (Tekaia and Dujon, 1999; Coissac et al., 1997). But there are basically two ways by which such intragenome homologs can arise: (i) through duplication (both gene copies are then called paralogs) and (ii) by acquiring similar genes from outside sources via HGT (both gene copies are then called xenologs). The term "synologs" was proposed as an agnostic name for homologs within a genome arising from either process (Lerat et al., 2005). The work described in this chapter was not based on a comparison between gene phylogenies and organism phylogeny, but rather on a taxonomical distribution of homologous genes per gene family, and therefore is not intended to make a distinction between xenologs and paralogs.

Gene families are defined as those groups of genes which share a common evolutionary history. In practice, protein family construction is based on two steps. Within a given set of protein sequences, one has to (i) detect homologous genes based on a significant similarity measure and (ii) group homologous genes together by sequence clustering. Similarity measures are obtained by pairwise amino acid sequence comparison, e.g., using BLAST. Sequence homology can be reliably inferred from statistically significant similarity. To select "suitable" homologous genes, one can apply absolute criteria such as an E-value or a similarity score cutoff. In many cases, however, sequences have diverged to the extent that their common origin is untraceable by a direct sequence comparison. It becomes very difficult to correctly detect homology for pairs with a pairwise sequence identity between 20 and 30%, the so-called twilight zone (Rost, 1999). Rost (1999) proposed an empirical for-

Dirk Gevers • Laboratory of Microbiology, Ghent University, K.L. Ledeganckstraat 35, B-9000 Ghent, Belgium. **Yves Van de Peer** • BioInformatics and Evolutionary Genomics, Ghent University/VIB Technologiepark 927, B-9052 Ghent, Belgium.

mula (HSSP curve) based on the sequence similarities between proteins with known structures to distinguish between true and false positives for low levels of similarity (Rost, 1999). A second difficulty occurs when high identity value is obtained when a short protein shares one or more domains with a longer protein, the "multi-domain problem." Consequently, relying on significant similarities as such will result in families of possible nonhomologous proteins sharing the same domain. If possible, genes should be classified in one family only if they have highly similar domain architecture. This means that a high level of sequence identity does not confer homology unless alignable length versus total protein length is taken into account. To cope with this, additional parameters to distinguish global homology from local homology, such as coverage of the alignable region on both potentially homologous genes, can be used (Li et al., 2001). For the second step, grouping homologous genes in families, the single linkage algorithm is used, i.e., if protein A hits protein B and B hits C, then proteins A, B, and C are put in the same cluster, regardless of whether protein A hits protein C or not (transitive homology). In this case, stringent pairwise detection of homologous genes is required to construct reliable gene families, which is essential to avoid noise in the detection of gene duplications.

Here, we discuss the collection of gene duplicates (the paranome) in relation to the whole proteome, the functional composition, and organization for all currently available *Vibrionaceae* genomes (as of March 2005), i.e., *Vibrio cholerae*, *Vibrio vulnificus*, *Vibrio parahaemolyticus*, and *Photobacterium profundum*.

GENOME EXPANSION BY DUPLICATION

Whereas in the pregenomic era, it was suggested that bacterial genomes may have evolved from small to large genomes by several genome duplications (Kunisawa, 1995), the analysis of the complete genomes of *Haemophilus influenzae* and *Mycoplasma genital-*

ium did not support this idea (Kolsto, 1997). A recent analysis of the prevalence and genomic organization of paralogs in bacterial genomes showed that most of the duplicated genes in bacteria seem to have been created by small gene duplication events (Gevers et al., 2004). Evidence for large-scale gene duplication events such as those observed in eukaryotic genomes (Van de Peer, 2004) has not been detected in any of the bacterial genomes investigated so far. Nevertheless, gene duplicates constitute a significant fraction of the bacterial genome coding capacity (see Table 1 for data on vibrio genomes). Previous work on a broader range of taxa had determined that the fraction of paralogs is strongly correlated with genome size in a linear regression ($R^2 = 0.94$) and that the fraction in gamma-proteobacteria ranges from 6.9% (*Buchnera aphidicola*) to 43.7% (*Escherichia coli* O157:H7 strain EDL933) (Coissac et al., 1997; Tekaia and Dujon, 1999; Jordan et al., 2001; Gevers et al., 2004; Pushker et al., 2004). A closer look at the *Vibrio* genomes shows that their data fit that trend (Table 1). The average family size (expansion level) increases with genome size and thus confirms that gene duplicates have a critical role in the expansion of genome size. However, *V. parahaemolyticus* seems to be an exception, as the expansion level is relatively low, compensated by a relatively large fraction of orphans (genes without any homolog) and a higher number of families. *V. vulnificus* CMCP6 shows quite an opposite picture, as the fraction of orphans is relatively low and the fraction of duplicated genes is higher. Thus, we should conclude that genome expansion is partly due to the expansion of existing gene families but is also due to an increase in the number of families and the number of orphans, and their ratio is organism specific. The vibrio genomes are organized into two replicons, a main chromosome and an auxiliary chromosome, the latter characterized by less gene synteny than the former (Gomez-Gil et al., 2005). We quantified the fractions of duplicates on the different replichores and found that the distribution of the duplicates is proportional to the distri-

Table 1. Properties of the vibrio genomes

Species	Genes	Families[a]	Exp. level[b]	Orphans[c] (%)	Singletons[d] (%)	Duplicates (%)
V. cholerae	3,828	355	3.05	592 (15.46)	2,157 (56.35)	1,079 (28.19)
V. vulnificus CMCP6	4,537	451	3.17	360 (7.03)	2,751 (60.63)	1,426 (31.43)
V. parahaemolyticus	4,832	484	3.04	825 (17.07)	2,538 (52.52)	1,469 (30.40)
V. vulnificus YJ016	5,028	469	3.16	761 (15.14)	2,786 (55.41)	1,481 (29.46)
P. profundum	5,474	554	3.63	1,151 (21.03)	2,316 (42.31)	2,007 (36.66)

[a] Group of homologs within a genome (intragenomic).
[b] Expansion level, or the average number of duplicates per family.
[c] Unique genes, i.e, no homologs in the *Proteobacteria* genomes.
[d] Genes with homologs among *Proteobacteria*, but in this genome, present in only one copy.

bution of the genes on the replichores. No indications for a preferential expansion of the auxiliary replichores, as shown in alpha-proteobacteria (Boussau et al., 2004), was found.

EVOLUTIONARY SCENARIOS FOR GENOME EVOLUTION

To quantify the flux of genes along the lineage of *Vibrionaceae* and to determine the evolutionary history of the genomes, ancestral proteomes were inferred and the numbers of gene losses, duplications, and genesis events along each branch of the tree were estimated (Fig. 1). Parsimonious scenarios for individual gene families and ancestral character states on a given species tree were constructed as reported before (Snel et al., 2002; Mirkin et al., 2003; Boussau et al., 2004). Gene families were constructed from the combined proteome of 74 available *Proteobacteria* genomes, and the species tree was constructed from 16S rRNA gene sequence. We consider three elementary evolutionary events: (i) gene loss, (ii) gene genesis, i.e., introduction of a new gene, and (iii) expansion of existing gene or family by duplication or horizontal acquisition. An evolutionary scenario is any combination of elementary events that leads to the observed phyletic pattern of the gene family, given

the topology of the species tree. The scenario that includes the minimum number of events is the (most) parsimonious scenario, i.e., the scenario that is most consistent with the topology of the species tree. The parsimonious scenario is not necessarily unique, as there may be multiple scenarios with the same minimal number of events. Despite several efforts, and because of inconsistent results, the relative contributions of each genetic force remain controversial (Kunin and Ouzounis, 2003; Daubin et al., 2003; Hong et al., 2004; Lerat et al., 2005). The essence of the problem is that any phylogenetic evidence of HGT can also be explained via a combination of gene duplication and lineage-specific gene loss events. Thus, this analysis should be considered a crude attempt at constructing evolutionary scenarios using comparative genomic data and is only one plausible picture of genome evolution.

The analysis of gene content alterations at the branches of the tree revealed three major trends that are observed irrespective of different penalties for acquisition, deletions, and gene genesis used (Fig. 1; Table 2). First, the ancestral *Vibrionaceae* proteome was estimated to contain 3,894 proteins and 303 families. Second, there has been a large increase of genes in the evolution of the *Photobacterium* lineage. This may be linked to the ability to respond rapidly to favorable changes in growth conditions (Vezzi et al.,

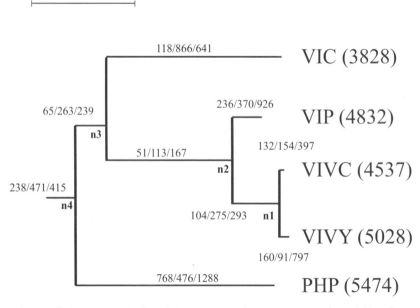

Figure 1. Parsimonious evolutionary scenario for vibrio proteomes. A reconstruction of acquisition, loss, and gene genesis events was based on the species tree (constructed from 16S rRNA gene sequence) and on *Proteobacteria* protein families as described in the introduction. Numbers along the branches refer to the number of acquisitions, losses, and genesis, respectively. Each node has a number that refers to data in Table 2. The number between brackets is the total number of genes for that particular genome. Inference of ancestral gene contents was made by parsimony analysis in PAUP with penalties for acquisition, deletion, and gene genesis set to 1, 1, and 5, respectively.

Table 2. Properties of the vibrio ancestral nodes[a]

Node	Genes	Families	Exp. level	Singletons (%)	Duplicates (%)
Node 1	4,162	399	3.17	2,896 (69.58)	1,266 (30.42)
Node 2	4,040	369	3.14	2,882 (71.34)	1,158 (28.66)
Node 3	3,935	344	3.18	2,839 (72.15)	1,096 (27.85)
Node 4	3,894	303	3.28	2,900 (74.47)	994 (25.53)

[a]See Table 1.

2005). The expansion involved both acquisition and genesis, with the latter being roughly twice as frequent. The role of acquisition is also reflected in an increased expansion level from 3.28 at the ancestral node to 3.63 for *P. profundum*, whereas the role of genesis is reflected in a strong increase in the number of new families (*n* = +251). Third, whereas *V. parahaemolyticus* and *V. vulnificus* have increased their gene content with acquisitions and/or genesis (genesis being roughly three to five times more frequent), *V. cholerae* is the only branch that has a net gene content reduction. This size difference among *Vibrio* genomes has arisen not only through a reduced frequency of acquisition (expansion level decreased from 3.18 to 3.05) and genesis (only 11 new families compared to ancestral *Vibrio* node) in *V. cholerae*, but mainly through a more frequent gene decay or deletion in *V. cholerae*. This is in agreement with the earlier finding that *V. cholerae* is more dependent on a human host than are other vibrios (Gomez-Gil et al.,

2005) and confirms a general trend in bacterial pathogens that have become dependent on the functions of their hosts (Klasson and Andersson, 2004).

SEQUENCE DIVERGENCE WITHIN FAMILIES

Because gene duplication might precede a functional diversification between duplicates, it would be of interest to look at the divergence level among duplicates. For each of the studied genomes, the average sequence identity for all the members of a multigene family was calculated and plotted versus the family size. In general, we can conclude that large families are more divergent and thus likely to be older (Fig. 2). Smaller families cover a range from those with very similar members to those with more divergent ones. The latter probably represent either old families in which further expansion does not yield a selective advantage, or families composed of paralogs and/or xenologs, resulting in a lower similarity due to a dissimilar evolutionary history. A large number of exceptions are found in *P. profundum* SS9, which has larger gene families with unusually high average sequence similarity. Most of them were families of transposases or phage-related families, known to be found at a higher frequency in this genome (Vezzi et al., 2005). In other genomes, a few exceptions of large families containing hypothetical proteins with a high average sequence similarity were found, most of which were identified as recent lineage-specific expansions.

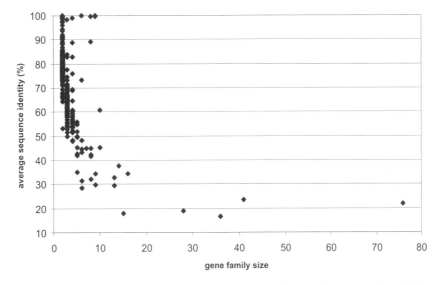

Figure 2. Sequence divergence within families versus family size. For each family of the *V. vulnificus* CMCP6 proteome, the average amino acid sequence identity of pairwise comparisons among the members of the family is determined and plotted against the family size.

STRAIN-SPECIFIC EXPANSIONS

Comparative genome analyses were applied to the *Vibrionaceae* genomes for delineation and examination of strain- and/or lineage-specific expansions. Strain-specific expansions are groups of paralogous genes generated subsequent to the divergence of the prokaryotic strain analyzed. In our analysis, a strain-specific expansion (SSE) refers to a gene family that is expanded in one particular strain only, resulting in a phylogenetic profile with at least two paralogous genes in one particular strain and a maximum of one gene per strain for all other *Proteobacteria*. By definition, such SSEs resemble relatively recently duplicated genes. In the *Vibrio* species, SSEs account for 2.5% of the predicted proteome, whereas in *P. profundum* it is as high as 6.9%. The average expansion level (i.e., the number of genes per family) of an SSE in *V. cholerae* or both *V. vulnificus* genomes is 2.7; in *V. parahaemolyticus* it is 2.1, and in *P. profundum* it is 3.4. This means that the majority of the detected SSE families consist of only two to four genes, with only very few exceptions. The general trends for the SSEs are similar with respect to the expansion trends that were reported for the complete paranome, i.e., an increased expansion in *P. profundum* and a relatively smaller expansion in *V. parahaemolyticus*. The SSEs include genes with higher pairwise amino acid sequence identity, on average >82%, thereby confirming that these are recent duplications. For comparison, the average identity level between orthologs (homologs in different genomes) typically lies at ~30% (Grishin et al., 2000). Table 3 gives for each of the included genomes a number of examples of SSEs. The majority (>77%) of the SSEs consist of hypothetical proteins or proteins with unknown functions, of which most have no homologous genes outside the *Vibrionaceae* (not shown in Table 3). Other SSEs have a diversity of functions.

THE FUNCTIONAL LANDSCAPE OF THE PARANOME

Previous studies have shown a preferential enrichment of certain functional classes in prokaryotic genomes (Pushker et al., 2004; Gevers et al., 2004). To perform a similar analysis for the *Vibrionaceae*, gene function descriptions and functional class designations were retrieved for the *V. cholerae* proteome using the functional annotation provided for the so-called COGs (clusters of orthologous groups) at NCBI (http://www.ncbi.nlm.nih.gov/COG) (Tatusov et al., 2003). Approximately 74% of the *V. cholerae* proteome is COG annotated. The four unclassified proteomes were checked in BLAST against all the proteins in the COG data set. An in-house implementation of the COGNITOR software (Tatusov et al., 2000) was used to automatically assign the genes in our data set to existing COGs and functional categories. Using this approach, approximately 64% of the unclassified proteomes could be assigned COGs. An overview of the functional landscape of the paranome for the *Vibrionaceae* is presented in Color Plate 3, which allows us to determine whether duplicate retention is biased toward specific functional classes for each of the bacterial strains. It appears that the preferentially retained duplicated genes mainly belong to the functional classes that are associated with amino acid metabolism (class E) and transcription (class K). Such genes are directly or indirectly (via regulation) involved in adapting to a constantly changing environment, showing the importance of gene duplicates for biological evolution. For class E, the preferentially retained expansion can partly (~30%) be explained by the amplification of transport proteins, which are known to be the largest families that contain the greatest numbers of paralogs in a single organism (Saier and Paulsen, 1999). Other known large families include the transcriptional regulatory proteins and response regulators, which are members of class K (Babu and Teichmann, 2003). In addition, more organism-specific cases of duplicate retention can be deduced from our analysis. For example, in the paranome of *P. profundum*, three functional classes with an excess of retained gene duplicates are prominent, namely, carbohydrate transport and metabolism (class G, Color Plate 3B), energy production and conversion (class C, Color Plate 3A), and replication, recombination, and repair (class L, Color Plate 3A). Complex carbohydrates are an important carbon source in oceanic abyssal environments, the environment from which *P. profundum* was isolated, and the expansion of class G might be an indication of adaptation to the available carbon sources (Vezzi et al., 2005). The expansion of class L, on the other hand, is mainly (78%) explained by the presence of transposon- and/or phage-related genes in multiple copies. In the *V. cholerae* genome there is a preferential enrichment of inorganic ion transport and metabolism proteins (class P, Color Plate 3B), with 77% of these proteins being transport proteins with peptides, metals, and phosphate/sulfate as the main substrate categories. As *V. cholerae* has a diverse natural habitat, it is no surprise that this organism maintains a large repertoire of transport proteins with broad substrate specificity (Heidelberg et al., 2000). A last notable expansion is class X in *V. vulnificus* CMCP6 (Color Plate 3B), containing several expanded families that are unique to this organism and described as hypothetical proteins.

Table 3. Examples of strain-specific expansions

Organism	Gene family	Profile[a]	Seq ID[b]	Func.[c]	Function description
V. cholerae	24013	-2---	69.32	X	Hemolysin-related protein
	10057	-(14)1--	77.56	X	Hypothetical protein
	524	-2111	77.86	R	Zn-dependent hydrolase
	22606	-2---	94.88	R	Predicted Fe-S protein
	9011	-2-11	100	KR	Acetyltransferase
V. vulnificus CMCP6	14375	11121	64.49	M	Outer membrane biogenesis
	2648	11121	66.24	E	NAD-specific glutamate dehydrogenase
	2378	11121	66.67	KL	DNA or RNA helicase of superfamily II
	17654	11121	66.67	M	Outer membrane lipoprotein-sorting protein
	5084	-1121	66.67	X	Outer membrane protein
	22445	--121	80.2	X	Permease of the major facilitator superfamily
	24008	---2-	97.66	X	Cation transport ATPase
	15651	---2-	99.93	V	Penicillin-binding protein
	2633	---9-	100	X	Cell wall-associated hydrolase
V. vulnificus YJ016	22811	---12	85.13	X	Flop pilus assembly protein
	11283	1-112	89.31	X	Ribose 5-phosphate isomerase
	6737	---12	91.51	X	Drug/metabolite transporter
	9615	--1-4	99.21	X	PAS factor
	9355	---12	99.51	X	NTP pyrophosphohydrolase
	13887	----2	100	KT	SOS-response transcriptional repressors
V. parahaemolyticus	57	11311	65.8	V	Na$^+$-driven multidrug efflux pump
	1317	11211	75.57	I	4-Diphosphocytidyl-2-methyl-D-erythritol synthase
	7176	11211	75.59	GEPR	Permease of the major facilitator superfamily
	3865	1-211	76.04	M	Outer membrane phospholipase A
	7649	11211	77.72	X	Putative virulence-associated protein VacB/Rnase R
	27988	1-2--	78.12	X	Putative membrane lipoprotein
	13293	1-211	78.29	X	Putative outer membrane lipoprotein
	12191	--2--	80.8	K	Transcriptional regulator
	8727	1-2-1	81.13	T	Signal transduction histidine kinase
	5406	11211	84.97	P	Carbonic anhydrase
	4653	11211	87.24	ER	Hydrolase of the PHP family
	1503	11211	90.71	I	Acetyl-coA carboxylase beta subunit
P. profundum	15056	2----	74.95	M	Glycosyltransferases
	676	3-111	75.11	E	Dipeptide/tripeptide permease
	6855	21111	77.15	G	Galactose mutarotase and related enzymes
	9860	21111	77.71	V	Na$^+$-driven multidrug efflux pump
	21605	2-111	78.22	K	Transcriptional regulator
	1224	2-1-1	78.48	C	Anaerobic dehydrogenase
	13739	21111	79.38	C	Nitrate reductase cytochrome c-type subunit
	20904	21111	81.71	I	Lysophospholipase
	5564	21111	85.41	C	Malate synthase
	2574	(67)----	87.11	X	Hypothetical protein
	6417	21111	92.27	M	Predicted UDP-glucose 6-dehydrogenase
	9520	3----	93.84	V	Restriction endonuclease

[a] The number of genes found in each genome, in this order: P. profundum/V. cholerae/V.parahaemolyticus/V. vulnificus CMCP6/V. vulnificus YJ016. A hyphen (-) is shown when no member of this family was found in the proteome.
[b] Average amino acid sequence identity of pairwise comparisons among the members of the family.
[c] Functional class according to the annotation for COGs (see Color Plate 3).

As discussed above, the overall distribution of gene duplicates did not reveal a preferential expansion on one of the two replicons (data not shown). Looking at the functional landscape of the paranome does reveal an unequal distribution. All *Vibrionaceae* gene duplicates of class E are mainly located on chromosome 1, whereas those of class K are spread over both replicons. The gene duplicates of classes G and L in *P. profundum* are mainly located on chromosome 2, and the gene du-

plicates of class P in *V. cholerae* are on chromosome 1. It is known that both replicons are essential for growth and viability, because genes involved in essential metabolic pathways are spread over both replicons, but at the same time it is suggested that chromosome 2 takes the main role of adaptation to environmental change (Hayashi et al., 2001). The results described here, however, suggest that both replicons play a role in bacterial adaptation by gene family expansion.

ORGANIZATION OF THE PARANOME WITHIN THE GENOME

Gene duplications can occur on a gene-by-gene basis, resulting in tandem duplicates, or they can result from the duplication of larger regions. Rearrangements after duplication and acquisition of homologs via HGT result in duplicates' being dispersed over the genome. Using a tool for detection of conserved gene content and order, i.e., colinearity (such as the ADHoRe software [Vandepoele et al., 2002]), within 106 bacterial genomes revealed that overall 15% of the paranome consists of tandem duplicates, and 9.5% is located in duplicated segments (containing multiple genes). The latter suggests that these genes have been duplicated en bloc. The majority of these segmental duplications are three to four genes long.

A similar analysis was performed for the *Vibrionaceae* genomes (Table 4). Tandem duplicates, here defined as two homologous genes with a maximum of two intervening genes, account for a fraction ranging between 6.2% of the paranome in *V. cholerae* and 13.8% in *V. vulnificus* CMCP6. The fraction of duplicated genes in segments ranges between 3.3% for *P. profundum* and 6.9% in *V. vulnificus* CMCP6. The latter genome contains a segment of five to six genes that is present in nine copies and is unique among the *Proteobacteria* (Color Plate 4). This observation reflects a remarkable difference between the two *V. vulnificus* strains. All genes in this segment are annotated as hypothetical proteins, except for one that is labeled with cell wall-associated protein, and thus no explanation of the significance of this strain-specific duplication has been formulated so far. This observation corresponds to the fact that class X (Color Plate 3B) contains relatively more gene duplicates. The majority of the other gene duplicates located in segments are functionally biased toward proteins associated with amino acid transport and metabolism (class E) (Color Plate 3B). This is in contrast to the duplicates of class K that are preferentially not duplicated en bloc or, alternatively, are rearranged and dispersed after duplication.

Table 4. Organization of the *Vibrionaceae* paranome

Species	Tandem dupl.[a]	Block dupl.[a]	Segments	Avg. segment size[b]
V. cholerae	6.2	3.4	9	4.3
V. vulnificus CMCP6	13.8	6.9	19	5.5
V. vulnificus YJ016	9.3	4.7	14	5.6
V. parahaemolyticus	7.6	3.9	13	4.8
P. profundum	10.6	3.3	18	3.9

[a] Expressed as percentage of the paranome.
[b] Expressed as number of genes.

The average size of a segment among the *Vibrionaceae* is 4.84 genes, resembling the typical bacterial operon size, thereby indicating a putative mechanism for operon duplication and evolution. Operons are the principal form of gene organization and regulation in prokaryotes and often encode functionally linked proteins. Previously, comparative genomics of prokaryotic genomes has shown that only a few operons are conserved across large evolutionary distances (Mushegian and Koonin, 1996). So far, it is not clear whether the segmental duplications are under strong selective pressure that preserves gene coexpression and coregulation in duplicate or, alternatively, whether they represent recently duplicated gene strings that have been spared from rearrangements and disruption. Studies on genome rearrangements have shown that relative gene position is not essential for the gene to function (Wolf et al., 2001). Furthermore, not only the number of amino acid substitutions but also the degree of genome rearrangements constantly increases as a function of the time of divergence (Suyama and Bork, 2001). Analysis of divergence in gene content and organization among duplicated segments often shows evolution by gene loss, tandem duplications, gene fusion/fission, inversion, and gene acquisition.

Placing the obtained intragenomic results in a phylogenetic context, i.e., an intergenome comparison, provides an extra evolutionary dimension of the conservation of the segmental duplication event over larger evolutionary distances (Color Plate 4). Two clusters of homologous segments, called multiplicons, are present in all *Vibrionaceae* analyzed and are even widely spread among other *Proteobacteria*. These are two operons of four and five genes, respectively, both encoding a peptide ABC transport system, with different operons encoding slightly different substrate specificity. On the other hand, there are also multiplicons with homologous segments in only one particular *Vibrionaceae* genome and some distantly related *Proteobacteria*. For example, homologous operons of four genes encoding a c4-dicarboxylate transport system are present in *P. profundum* SS9, *Salmonella enterica* serovar Typhimurium LT2, and *Mannheimia succiniciproducens* (strain MBEL55E) each time in three copies. Homologous proteins are present in other *Vibrionaceae* but did not retain the operon organization found in *P. profundum*.

REFERENCES

Babu, M. M., and S. A. Teichmann. 2003. Evolution of transcription factors and the gene regulatory network in *Escherichia coli*. *Nucleic Acids Res.* **31**:1234–1244.

Boussau, B., E. O. Karlberg, A. C. Frank, B. A. Legault, and S. G. Andersson. 2004. Computational inference of scenarios for alpha-proteobacterial genome evolution. *Proc. Natl. Acad. Sci. USA* **101**:9722–9727.

Caporale, L. H. 2003. Natural selection and the emergence of a mutation phenotype: an update of the evolutionary synthesis considering mechanisms that affect genome variation. *Annu. Rev. Microbiol.* **57**:467–485.

Coenye, T., D. Gevers, Y. Van de Peer, P. Vandamme, and J. Swings. 2005. Towards a prokaryotic genomic taxonomy. *FEMS Microbiol. Rev.* **29**:147–167.

Coissac, E., E. Maillier. and P. Netter. 1997. A comparative study of duplications in bacteria and eukaryotes: the importance of telomeres. *Mol. Biol. Evol.* **14**:1062–1074.

Daubin, V., N. A. Moran, and H. Ochman. 2003. Phylogenetics and the cohesion of bacterial genomes. *Science* **301**:829–832.

Gevers, D., K. Vandepoele, C. Simillon, and Y. Van de Peer. 2004. Gene duplication and biased functional retention of paralogs in bacterial genomes. *Trends Microbiol.* **12**:148–154.

Gogarten, J. P., W. F. Doolittle, and J. G. Lawrence. 2002. Prokaryotic evolution in light of gene transfer. *Mol. Biol. Evol.* **19**:2226–2238.

Gomez-Gil, B., F. L. Thompson, T. Sawabe, A. Vicente, A. Coelho, D. Gevers, and J. Swings. 2005. The genus *Vibrio. In* M. Dworkin et al., (ed.), *The Prokaryotes: an Evolving Electronic Resource for the Microbiological Community*, 3rd ed. Springer-Verlag, New York, N.Y.

Grishin, N. V., Y. I. Wolf, and E. V. Koonin. 2000. From complete genomes to measures of substitution rate variability within and between proteins. *Genome Res.* **10**:991–1000.

Hayashi, T., K. Makino, M. Ohnishi, K. Kurokawa, K. Ishii, K. Yokoyama, C. G. Han, E. Ohtsubo, K. Nakayama, T. Murata, M. Tanaka, T. Tobe, T. Iida, H. Takami, T. Honda, C. Sasakawa, N. Ogasawara, T. Yasunaga, S. Kuhara, T. Shiba, M. Hattori, and H. Shinagawa. 2001. Complete genome sequence of enterohemorrhagic *Escherichia coli* O157:H7 and genomic comparison with a laboratory strain K-12. *DNA Res.* **8**:11–22.

Heidelberg, J. F., J. A. Eisen, W. C. Nelson, R. A. Clayton, M. L. Gwinn, R. J. Dodson, D. H. Haft, E. K. Hickey, J. D. Peterson, L. Umayam, S. R. Gill, K. E. Nelson, T. D. Read, H. Tettelin, D. Richardson, M. D. Ermolaeva, J. Vamathevan, S. Bass, H. Qin, I. Dragoi, P. Sellers, L. McDonald, T. Utterback, R. D. Fleishmann, W. C. Nierman, O. White, S. L. Salzberg, H. O. Smith, R. R. Colwell, J. J. Mekalanos, J. C. Venter, and C. M. Fraser. 2000. DNA sequence of both chromosomes of the cholera pathogen *Vibrio cholerae. Nature* **406**:477–483.

Hong, S. H., T. Y. Kim, and S. Y. Lee. 2004. Phylogenetic analysis based on genome-scale metabolic pathway reaction content. *Appl. Microbiol. Biotechnol.* **65**:203–210.

Hooper, S. D., and O. G. Berg. 2003. On the nature of gene innovation: duplication patterns in microbial genomes. *Mol. Biol. Evol.* **20**:945–954.

Jordan, I. K., K. S. Makarova, J. L. Spouge, Y. I. Wolf, and E. V. Koonin. 2001. Lineage-specific gene expansions in bacterial and archaeal genomes. *Genome Res.* **11**:555–565.

Klasson, L., and S. G. Andersson. 2004. Evolution of minimal-gene-sets in host-dependent bacteria. *Trends Microbiol.* **12**:37–43.

Kolsto, A. B. 1997. Dynamic bacterial genome organization. *Mol. Microbiol.* **24**:241–248.

Kunin, V., and C. A. Ouzounis. 2003. The balance of driving forces during genome evolution in prokaryotes. *Genome Res.* **13**:1589–1594.

Kunisawa, T. 1995. Identification and chromosomal distribution of DNA sequence segments conserved since divergence of *Escherichia coli* and *Bacillus subtilis. J. Mol. Evol.* **40**:585–593.

Lerat, E., V. Daubin, H. Ochman, and N. A. Moran. 2005. Evolutionary origins of genomic repertoires in bacteria. *PLoS Biol.* **3**:e130. (First published 5 April 2005; 10.1371/journal.pbio.0030130.)

Li, W. H., Z. Gu, H. Wang, and A. Nekrutenko. 2001. Evolutionary analyses of the human genome. *Nature* **409**:847–849.

Mirkin, B. G., T. I. Fenner, M. Y. Galperin, and E. V. Koonin. 2003. Algorithms for computing parsimonious evolutionary scenarios for genome evolution, the last universal common ancestor and dominance of horizontal gene transfer in the evolution of prokaryotes. *BMC Evol. Biol.* **3**:2. [Online.] http://www.biomedcentral.com/1471-2148/3/2.

Mushegian, A. R., and E. V. Koonin. 1996. Gene order is not conserved in bacterial evolution. *Trends Genet.* **12**: 289–290.

Pushker, R., A. Mira, and F. Rodriguez-Valera. 2004. Comparative genomics of gene-family size in closely related bacteria. *Genome Biol.* **5**:R27.

Riehle, M. M., A. F. Bennett, and A. D. Long. 2001. Genetic architecture of thermal adaptation in *Escherichia coli. Proc. Natl. Acad. Sci. USA* **98**:525–530.

Rost, B. 1999. Twilight zone of protein sequence alignments. *Protein Eng.* **12**:85–94.

Saier, M. H., and I. T. Paulsen. 1999. Paralogous genes encoding transport proteins in microbial genomes. *Res. Microbiol.* **150**:689–699.

Snel, B., P. Bork, and M. A. Huynen. 2002. Genomes in flux: the evolution of archaeal and proteobacterial gene content. *Genome Res.* **12**:17–25.

Suyama, M., and P. Bork. 2001. Evolution of prokaryotic gene order: genome rearrangements in closely related species. *Trends Genet.* **17**:10–13.

Tatusov, R. L., N. D. Fedorova, J. D. Jackson, A. R. Jacobs, B. Kiryutin, E. V. Koonin, D. M. Krylov, R. Mazumder, S. L. Mekhedov, A. N. Nikolskaya, B. S. Rao, S. Smirnov, A. V. Sverdlov, S. Vasudevan, Y. I. Wolf, J. J. Yin, and D. A. Natale. 2003. BMC database: an updated version includes eukaryotes. *BMC Bioinformatics* **4**:41. [Online.] http://www.biomedcentral.com/1471-2105/4/41.

Tatusov, R. L., M. Y. Galperin, D. A. Natale, and E. V. Koonin. 2000. The Cog database: a tool for genome-scale analysis of protein functions and evolution. *Nucleic Acids Res.* **28**:33–36.

Tekaia, F., and B. Dujon. 1999. Pervasiveness of gene conservation and persistence of duplicates in cellular genomes. *J. Mol. Evol.* **49**:591–600.

Van de Peer, Y. 2004. Computational approaches to unveiling ancient genome duplications. *Nat. Rev. Genet.* **5**:752–763.

Vandepoele, K., Y. Saeys, C. Simillion, J. Raes, and Y. Van de Peer. 2002. The automatic detection of homologous regions (ADHoRe) and its application to microcolinearity between arabidopsis and rice. *Genome Res.* **12**:1792–1801.

Vezzi, A., S. Campanaro, M. D'Angelo, F. Simonato, N. Vitulo, et al. 2005. Life at depth: *Photobacterium profundum* genome sequence and expression analysis. *Science* **307**:1459–1461.

Wolf, Y. I., I. B. Rogozin, A. S. Kondrashov, and E. V. Koonin. 2001. Genome alignment, evolution of prokaryotic genome organization, and prediction of gene function using genomic context. *Genome Res.* **11**:356–372.

Yamanaka, K., L. Fang, and M. Inouye. 1998. The Cspa family in *Escherichia coli*: multiple gene duplication for stress adaptation. *Mol. Microbiol.* **27**:247–255.

The Biology of Vibrios
Edited by F. L. Thompson et al.
© 2006 ASM Press, Washington, D.C.

Chapter 7

The Roles of Lateral Gene Transfer and Vertical Descent in Vibrio Evolution

YAN BOUCHER AND HATCH W. STOKES

INTRODUCTION TO LATERAL GENE TRANSFER

A microbial group is defined as much by the genes shared by its representatives as by those they exchange among one another. In vibrios, the latter process might be particularly important, as these bacteria display many specialized mechanisms that facilitate the transfer of genetic material. In this chapter, we will look at the detection of lateral gene transfer (LGT), from the level of individual genes up to whole-genome and multiple locus analyses. The genetic elements most frequently involved in these gene transfers and the genomic hot spots for such events will also be described.

It is very important to start by defining LGT accurately, since the term usually encompasses a range of different phenomena, all having in common that some DNA fragment from an exogenous source is assimilated in the genome of an organism. It is termed lateral (or horizontal) gene transfer, as opposed to the vertical transmission of genetic material from a parent cell. Laterally acquired DNA can be integrated in a genome through homologous (legitimate) or illegitimate recombination. Homologous recombination includes all processes requiring a segment of at least 20 identical base pairs between the recipient DNA and the recombining fragment (such as RecA-mediated homologous recombination) (Smith, 1988). The frequency of homologous recombination falls off rapidly as sequence identity decreases between the two DNA molecules to be recombined (Majewski et al., 2000). Nevertheless, homologous recombination can occur between genes of rather distantly related genomes (even different "species"), especially if they are slowly evolving genes (e.g., RNA polymerase or ribosomal RNA genes). Genes brought in on stably maintained plasmids by conjugation or fragments integrated by illegitimate recombination (e.g., topoisomerase errors or nicking at a plasmid's origin of replication) are not required to have any similarity to the host's DNA (Michel, 1999). Other processes only require very short stretches of similarity, which are not necessarily part of the acquired gene itself (e.g., inverted repeats from transposable elements, *attI* sites of integrons, integration sites of temperate bacteriophages, or spontaneous rearrangements in chromosomes occurring in short homologous sequences of 3 to 20 bp long).

MODES OF DNA TRANSFER

Lateral DNA transfer is known to occur through three main modes. Conjugation is the transfer of DNA involving cell-to-cell contact (which can be achieved through a protein complex such as a pilus or by direct cell fusion). Transformation is the uptake of free DNA by a host (either by direct penetration through a porous cell membrane or using uptake machinery). Transduction is phage- or vesicle-mediated transfer of DNA between two cells (no cell-to-cell contact is required). Transformation is unlikely to be of importance in *Vibrio*, as there is no representative of this group known to be naturally competent. Conversely, there are many conjugative plasmids and phages associated with various *Vibrio* species, making it likely that conjugation and transduction are major modes of LGT in this genus.

Detection Methods

Many methods are used to detect LGT events. Each targets specific types of events (occurring across large or short phylogenetic distances, recent or an-

Yan Boucher and Hatch W. Stokes • Department of Chemistry and Biomolecular Sciences, Macquarie University, Sydney, NSW 2109, Australia.

cient, involving a particular process or genetic structure). Here is a short description of the most common methods, along with some of their main characteristics. For more substantial descriptions see Ragan (2001) and Koonin et al. (2001).

Unusual ranking of similarity among homologs

Similarity between a query gene and sequences in public databases is often determined using a BLAST-based tool (Altschul et al., 1997). When searching public databases with BLAST, if a protein or DNA sequence query "hits" a homolog from a distant taxon with a higher score than it hits homologs from close relatives, there is a possibility that this gene was acquired by LGT. This is a very quick method, but there are several potential problems with it. It is very difficult to determine a BLAST score threshold to retrieve only orthologs of the query (and even harder to find a threshold that will work for multiple gene families). Also, large-scale tests indicate that as many as 40% of BLASTp best matches may target sequences other than the phylogenetically nearest neighbor (Koski and Golding, 2001). This method is, therefore, not very sensitive and is only effective to detect transfers across large phylogenetic distances (orders or domains). It has been applied most recently by Koonin et al. (2001) on completely sequenced prokaryotic genomes, which included *Vibrio cholerae*.

Atypical composition

Anomalous nucleotide composition (codon usage bias, G+C content, trinucleotide frequencies) is widely used to detect LGT events but is only applicable to recent transfers (Lawrence and Ochman, 1997). Indeed, genes acquired from other organisms with a different genomic nucleotide composition will be gradually modified to match the nucleotide composition of their new host (over a time scale of a few million to a few hundred million years). Genes can also have compositional biases for reasons other than an exogenous origin (e.g., expression levels and strand biases). Furthermore, genes from compositionally similar donors will certainly be missed. Detection of LGT by atypical composition is applied to completely sequenced genomes (which include two *Vibrio vulnificus* and *V. cholerae*) in the horizontal gene transfer database (Garcia-Vallve et al., 2003).

Presence of elements associated with particular LGT processes

The identification of site-specific phage insertion sequences, transposon insertion elements, gene cas-

sette *attC* sites, or plasmid remnants flanking a gene is very direct evidence for an exogenous origin. These elements, however, are quite frequently highly mutated and can be difficult to identify. Completely sequenced genomes have been screened for prophage presence using automated scripts (Canchaya et al., 2004) and an integrated approach (Canchaya et al., 2003), which identified the well-known CTX-psi prophage of *V. cholerae* O1 El Tor and a prophage-like element in *V. vulnificus* CMCP6. A similar approach has also been used to locate genomic islands integrated in tRNA or tmRNA genes, allowing the detection of several islands in *Vibrio parahaemolyticus* RIMD2210633 and *V. vulnificus* CMCP6 (Mantri and Williams, 2004). Extrachromosomal elements, such as plasmids, can also be included in this category and are easily detected by simple plasmid isolation and profiling techniques. Plasmid profiling is a commonly used technique to type disease-causing aquatic vibrios (Le Chevalier et al., 2003).

Difference in gene content among the genomes of close relatives

When two closely related prokaryotes differ greatly in their gene content, it is likely that many of the genes found in only one of these neighbors were acquired through LGT. One of the most striking examples is *Escherichia coli* O157:H7, which contains 1,387 genes absent from *E. coli* K-12, which in turn contains 528 genes absent from O157:H7 (Perna et al., 2001). Basic gene content comparison between prokaryotic genomes can be performed using a reciprocal BLASTp best-match approach (Charlebois and Doolittle, 2004).

Incongruence among phylogenetic trees

LGT can result in the anomalous placement of a particular taxon in reference to an "organismal" tree. For example, a typically bacterial gene grouping strongly among archaeal homologs, well away from its bacterial counterparts, is likely to have been acquired by its bacterial host through LGT from an archaeon. Phylogenetic evidence can be very convincing if it comes from a statistically well-supported tree of a single orthologous family. However, methodological artifacts and paralogy cannot always be ruled out and can make the interpretation of a tree fairly difficult (Smith et al., 1992). Approaches have been developed to obtain the "phylome" (the assembled phylogenetic trees of all genes in a given genome) of prokaryotic genomes (Sicheritz-Ponten and Andersson, 2001; Frickey and Lupas, 2004), but they have yet to be applied to vibrios. Several anecdotal cases of LGT detected using a phylogenetic approach are presented here.

Hybridization techniques

An organism for which the complete genome sequence is available can be compared with other strains or species by microarray hybridization (Kim et al., 2002). The genes of the organism that had its genome sequenced are amplified by PCR and spotted as an array. This microarray is then hybridized with genomic DNA from the strains being compared to the arrayed genome. It is also possible to focus on extracting the DNA regions that differ between two organisms by suppressive subtractive hybridization (Nesbø et al., 2002). This technique involves the hybridization of DNA from two genomes for the removal of common sequences and yields strain-specific clones that can then be sequenced. The microarray hybridization approach has been used to detect *V. cholerae* O1 El Tor genes missing from other strains of *V. cholerae* (Dziejman et al., 2002).

Multilocus sequence analysis

When one or more loci are sequenced from a number of closely related strains (usually of the same species), homologous recombination events between these organisms can be detected by a number of methods. If a single orthologous family is being studied (one locus), recombination can be suggested by the presence of a network in a split decomposition analysis (Bryant and Moulton, 2004) or a nonrandom clustering of polymorphic nucleotide sites (Stephens, 1985). Direct evidence for recombination can be obtained when multiple loci are compared, which is often termed multilocus sequence analysis or multilocus

sequence typing (Feil et al., 2001). A lack of congruence among the phylogenetic trees of the different genes analyzed (i.e., congruence is no better than what is observed between individual gene trees and trees of random topology) indicates frequent recombination at one or more loci (Feil et al., 2001). With a large enough data set, the ratio of nucleotide substitutions caused by recombination versus point mutation can also be calculated. A substantial multilocus sequence analysis has been done with strains of *V. cholerae* O139 (Garg et al., 2003), and smaller-scale analyses have been performed with other *V. cholerae* (Farfan et al., 2002) and *V. parahaemolyticus* (Chowdhury et al., 2004).

DETECTION OF LGT EVENTS IN VIBRIOS

Anecdotal Evidence for LGT at a Single Locus

LGT of specific vibrio genes has been observed over a wide range of phylogenetic distances. For example, phylogenetic analysis of the HMG-coenzyme A reductase gene (*mvaA*) suggests the acquisition of this mevalonate biosynthesis/degradation gene by an ancestral *Vibrio* from an archaeal donor (interdomain transfer) (Table 1). Homologous recombination events have been shown to occur at the intraspecies (*recA*, *dnaE*), interspecies (*asd*), and intergenera (*gmd*) levels through phylogenetic analyses and nonrandom clustering of polymorphic nucleotide sites (Table 1). These genes code for proteins performing a variety of different metabolic and housekeeping functions. Such individual gene studies, however, cannot estimate the importance of LGT for the evolution of vibrios as a whole. This requires studies on a larger scale, such as whole-genome analysis or multilocus sequence analysis of a large strain collection.

Table 1. Representative cases of *Vibrio* genes that have been laterally transferred

Gene(s)	Function	Donor/recipient	Type of LGT[a]	Detection method[b]	Reference
mvaA	Lipid metabolism	Archaea/ancestral *Vibrio*	GA	IP	Boucher and Doolittle, 2000
tdh	Hemolysin	?/*V. parahaemolyticus*	GA	IP	Nishibuchi and Kaper, 1995
recA	Homologous recombination	Environmental *V. cholerae*/6th pandemic *V. cholerae*	HR	NP	Byun et al., 1999
dnaE	DNA polymerase	Environmental *V. cholerae*/U.S. Gulf Coast *V. cholerae*	HR	NP	Byun et al., 1999
asd	Amino acid biosynthesis	*V. mimicus*/several non-O1 *V. cholerae*	HR	NP	Karaolis et al., 1995
wbf/*otn*	Polysaccharide synthesis	*V. cholerae* O22/*V. cholerae* O139	OR	HB	Dumontier and Berche, 1998
gmd	Amino acid biosynthesis	*E. coli*/*V. cholerae* O139	HR	SM	Garg et al., 2003

[a] GA, gene acquisition; HR, homologous recombination; OR, orthologous replacement.
[b] IP, incongruence among phylogenetic trees; SM, overall sequence similarity; NP, nonrandom clustering of polymorphic nucleotide sites; HB, hybridization.

Detection of Laterally Acquired Genes in Completely Sequenced Genomes

Unusual ranking of similarity among homologs is often used to estimate the proportion of a genome that could have been acquired by LGT. Such similarity-based methods are limited to detecting events that occurred over large phylogenetic distances (see "Detection Methods"). Using the BLASTp program to compare *V. cholerae* open reading frames (ORFs) with the nonredundant NCBI protein sequence database, Koonin et al. (2001) estimated that 0.7% of N16961 genes had been acquired by interdomain lateral transfer events and 5.6% had been acquired by interphylum events. Each of these candidate LGT events was confirmed by a phylogenetic analysis to avoid false positives.

The horizontal gene transfer database uses a different method to detect lateral (horizontal) transfer events in microbial genomes (Garcia-Vallve et al., 2003). Their method assumes codon usage and G+C content are distinct and global features of each prokaryotic genome. Under this assumption, laterally acquired genes would have an atypical nucleotide composition diverging from the genomic average (see "Detection Methods"). Using this method, both *V. vulnificus* genomes and *V. cholerae* N16961 are estimated to have acquired 11 to 12% of their genes by LGT (Table 2). Only *V. cholerae* shows a strong bias for laterally acquired genes to be present on the smaller of its two chromosomes. This bias is most likely due to the presence of a *V. cholerae* 126-kb integron gene cassette array on its small chromosome, as opposed to this highly modular element being found on the large chromosome for both *V. vulnificus* strains.

The two methods described above for estimating foreign gene content in a genome give different estimates because they detect different laterally acquired gene subsets. Unusual ranking of similarity among homologs is biased toward transfers over large phylogenetic distances, and atypical composition is biased toward recent transfers. These two gene subsets do not overlap substantially (Ragan, 2001), which suggests a total estimate of laterally acquired genes rating above either individual estimate but below their combined total (~15 to 20%).

Variation of Gene Content Among *Vibrio* Species

Microarray analysis is an efficient technique to compare a completed microbial genome to closely related strains. Such a microarray was constructed for *V. cholerae* O1 El Tor N16961 (seventh pandemic) genome. This array, representing 93% of the N16961 genome, was hybridized with genomic DNA from a variety of pandemic and nonpandemic *V. cholerae* strains to compare their gene content (Dziejman et al., 2002). These nine strains, differing in biotype, serogroup, and year and site of isolation, lacked about 1% of N16961 genes. The regions where N16961 genes were missing in other strains included two new *Vibrio* pathogenicity islands (VSP-I and VSP-II), the repeat toxin gene cluster, the region encoding enzymes for capsule and O-antigen biosynthesis, the chromosomal integron, and the CTX-psi phage region. This suggests that these regions are among the fastest evolving in *V. cholerae*, a claim that is also supported by studies focusing on the evolution of these individual loci (Faruque and Mekalanos, 2003). Besides these regions, the microarray analysis suggested little variation in genome content among *V. cholerae* strains. This type of analysis, however, cannot determine if the strains compared with N16961 had acquired genes not present in this reference genome. Gene acquisition, which can occur by gene duplication or LGT, certainly played an important role in the evolution of vibrios. Comparison of the four completely sequenced *Vibrio* genomes reveals important differences in gene content. The two *V. vulnificus* strains for which complete genome data are available (YJ016 and CMCP6) reveal extensive differences when compared by a reciprocal BLASTp best-match approach (see "Detection Methods"). They share only 3,882 ORFs when each contains 4,959 and 4,537 identifiable ORFs, respectively. This represents a difference of 15 to 20% of gene content between two strains of the same species. Gene content difference is obviously even higher between *Vibrio* species, reaching 30 to 45% between *V. cholerae* and *V. parahaemolyticus*.

Detection of Homologous Recombination Through Multilocus Sequence Analysis

Homologous recombination is often the process allowing for the integration of foreign DNA in a host genome. At the very least, it is an important source

Table 2. Laterally acquired genes detected in *Vibrio* genomes through atypical nucleotide composition analysis (Garcia-Vallve et al., 2003)

Organism	Chromosome I[a]	Chromosome II	Total
V. cholerae O1 El Tor N16961	234/2,742 (8.5%)	204/1,093 (18.7%)	438/3,835 (11.4%)
V. vulnificus YJ016	375/3,262 (11.5%)	207/1,697 (12.2%)	582/4,959 (11.7%)
V. vulnificus CMCP6	320/2,972 (10.8%)	199/1,565 (12.7%)	519/4,537 (11.4%)

[a] No. of laterally acquired ORFs/total no. of ORFs (percentage of laterally acquired ORFs).

of nucleotide substitutions for bacterial genomes. Indeed, multilocus sequence analysis has shown that recombination can produce as much as or more genetic variation than point mutations in some *Neisseria* and *Streptococcus* species (Feil et al., 2001). Although vibrios have not been investigated in as much depth, some studies have suggested that homologous recombination could also play an important role in the evolution of this group of bacteria.

Relatively few multilocus sequence analyses have been carried out on vibrios, and most have specifically targeted *V. cholerae*. The most extensive study was done with 96 strains of *V. cholerae* O139, each of which was sequenced at seven housekeeping loci, one locus that carries the gene for cholera toxin, and another locus close to the insertion sequence within the O139 *wbf* region (Garg et al., 2003). Putative cases of LGT were found for all nine loci, with a minimal estimate of 26 recombination events among the 96 strains. These LGT events, although less frequent than the occurrence of point mutations, have introduced three times more nucleotide changes in their host genomes (120 versus 38 nucleotide substitutions). In most cases, the source of the acquired DNA was within the vibrios, but in one case, the *gmd* allele from an O139 isolate was more similar to an *E. coli* allele than to its vibrio homologs (Table 1).

Another study has looked at four housekeeping loci from 33 *V. cholerae* isolates (Byun et al., 1999). These isolates were relatively diverse, representing geographically distant clinical and environmental sources. Recombination events between environmental and clinical isolates were detected for three of the four loci. This illustrates that recombination occurs between distantly related environmental and clinical *V. cholerae* strains as well as within groups of pathogenic isolates (e.g., O139).

Besides *V. cholerae*, substantial multilocus sequence analysis has only been performed for *V. parahaemolyticus* (Chowdhury et al., 2004), in which no recombination was detected. However, the *V. parahaemolyticus* O3:K6 isolates studied were much less diverse than a similar clinical sample of *V. cholerae* O139, 94% of 54 strains being identical at the four loci sequenced. Studies of more divergent *V. parahaemolyticus* strains would be required to determine the importance of recombination in this species. The analyses performed on *V. cholerae* do, however, suggest that, as with *Streptococcus* and *Neisseria*, recombination can be responsible for a large part of the genetic variation in vibrios and that any gene can be altered through this process. As the level of clonality can vary between species (Feil et al., 2001), multilocus sequence analysis of *Vibrio* species other than *V. cholerae* is required to determine if recombina-

tion is as frequent in other representatives of the genus.

GENETIC ELEMENTS OF VIBRIOS INVOLVED IN LGT

Genomic Islands

Genomic islands are large DNA regions that have been acquired by LGT and inserted in the host chromosome. Their exact characteristics can vary from one island to another, but usual features include insertion near a tRNA gene, G+C content and codon bias that differs from the rest of the genome, the presence of insertion or prophage-related elements, flanking by direct repeats, and the presence of a functional integrase gene. Most genomic islands that have been identified so far in vibrios contain clusters of genes related to virulence and are often called pathogenicity islands (Dobrindt et al., 2004). Among these is the first *Vibrio* genomic island to be discovered, which was termed vibrio pathogenicity island (VPI), as it contained a gene cluster encoding the cholera toxin phage receptor and an essential colonization factor as well as a regulator of virulence genes (Karaolis et al., 1998). This 40-kb *V. cholerae* O1 El Tor N16961 island was flanked by putative 20-bp *att*-like attachment sequences and contained integrase and transposase genes. The presence of this island in clinical non-O1/non-O139 strains in addition to epidemic strains also suggests that it was transferred within *V. cholerae* (Karaolis et al., 1998). Since then, more genomic islands have been identified in *V. cholerae* strains. Microarray analysis comparing *V. cholerae* O1 El Tor N16961 to other *V. cholerae* strains identified two clusters of genes (16 kb and 7.5 to 13.5 kb) that were present only in seventh pandemic strains (Dziejman et al., 2002). The G+C content of these seventh pandemic islands (VSP-I and VSP-II) was lower than the genome average, suggesting lateral transfer. Another genomic island with G+C content differing from the host genome average was identified in numerous toxigenic *V. cholerae* O1 strains and was absent from all nontoxigenic strains examined (Jermyn and Boyd, 2002). This 57.3-kb island was named VPI-2, as it contained genes encoding for neuraminidase (*nanH*) and amino sugar metabolism.

Genomic islands have also been identified in other species of *Vibrio* besides *V. cholerae*. The complete genome sequence of *V. parahaemolyticus* RIMD 2210633 revealed an 80-kb region that had a much lower G+C content than the genome average (Makino et al., 2003). This island contains a thermostable direct hemolysin (*tdh*) gene (which is only associated with pathogenic strains), genes coding for the type III secretion system, homologs to the *E. coli* cytotoxic necrotizing factor, and a large gene also present in the *V. cholerae* VPI. A search directed at genomic islands inserted in tRNA and tmRNA genes identified four more islands of smaller size in RIMD2210633 (24, 16.5, 6, and 6 kb) (Mantri and Williams, 2004). This search also located two islands in *V. vulnificus* CMCP6 (42 and 27 kb).

Genomic islands can represent a significant proportion of the host genome (a total of at least 125 kb, or 3%, in *V. cholerae* O1 El Tor N16961) that has been acquired by LGT. This is likely to be an underestimate, however, since the detection of these islands mostly relies on their G+C content diverging from the genome average; therefore, only transfer events of recent evolutionary origin are likely to be detected. Older genomic islands, which have undergone a process of amelioration to match the host genome G+C content (Lawrence and Ochman, 1997), would be much harder to identify with certainty. Also, most efforts have been directed to locate genomic regions related to pathogenicity and fortuitously lead to the discovery of genomic islands. As we now know that these islands can carry functions that are not related to virulence determinants (Dobrindt et al., 2004), many other laterally acquired clusters could yet be found in vibrios.

Although the capsular and O-antigen biosynthesis genomic region of vibrios does not have some of the most common characteristics of genomic islands, such as a diverging G+C content, it is a large DNA region which has most certainly been transferred laterally among vibrios. The *V. cholerae* O139 Bengal strain, responsible for recent epidemics in the Indian subcontinent, was most likely created by the replacement of the 22-kb capsule and O-antigen biosynthesis region of *V. cholerae* O1 El Tor epidemic strain by a 40-kb O139-specific DNA fragment (Dumontier and Berche, 1998). This region is linked to the insertion sequence IS*1538*, which encodes a transposase and could have been involved in the acquisition of these new polysaccharide biosynthesis genes by O139 Bengal (Dumontier et al., 1998). The capsule and O-antigen biosynthesis region of the *V. cholerae* O37 Sudan strain is also linked to an insertion element, IS*1004*. This strain is closely related to *V. cholerae* O1 classical strains, and it is likely that lateral transfer of this region occurred between the two groups (although it is hard to determine the direction of this transfer) (Bik et al., 1996). Molecular typing of various strains harboring the O1 antigen revealed that these strains did not form a monophyletic group, suggesting that the genes synthesizing this antigen have been laterally transferred among *V. cholerae* (Karaolis et al., 1995). This region in the *V. cholerae* chromosome therefore seems to be a hot spot for lateral transfer.

Plasmids and Integrative/Conjugative Elements

Plasmids are so frequent and diverse in vibrios that they have been used in the differentiation of strains within a species, a technique known as plasmid profiling. *Vibrio* plasmids are found in a wide range of sizes (0.8 to 290 kb), and up to seven can be present in a single strain (Hoi et al., 1998). The types and number of plasmids can vary widely, even within a single serogroup of a specific species (Table 3). Some of these plasmids are conjugative (Chen et al., 2003) and others can be mobilized when a conjugative genetic element is present (Hochhut et al., 2000). In addition to their variable distribution, vibrio plasmids can show considerable microheterogeneity. pJM1, for example, is a well-studied 65-kb plasmid frequently found in *V. anguillarum*. This plasmid contains insertion elements and transposon-like structures, which could account for the variable restriction endonuclease patterns observed for different isolates as well as modification of the expression levels of some siderophore biosynthesis genes (Di Lorenzo et al., 2003).

Table 3. Plasmid profile studies in *Vibrio* species

Species	No. of strains (% with plasmids)	No. of plasmid types (sizes)	No. of profiles	Reference
V. salmonicida	32 (97%)	4 (2.6–24 MDa)	4	Wiik et al., 1989
V. tapetis	9 (100%)	4 (60–100 kb)	3	Le Chevalier et al., 2003
V. anguillarum serogroup O2	129 (46%)	17 (3.8–200 kb)	18	Tiainen et al., 1995
V. vulnificus	85 (98%)	10 (3.9–135 kb)	11	Hoi et al., 1998
Vibrio spp.	30 (80%)	15 (0.8–84 kb)	11	Molina-Aja et al., 2002
V. vulnificus	36 (46%)	10 (1.4–9.7 MDa)	8	Radu et al., 1998
V. anguillarum	291 (89%)	21 (2.6–260 kb)	22	Pedersen et al., 1999
P. damselae	31 (81%)	18 (3.0–190 kb)	14	Pedersen et al., 1997

A conjugative element that can integrate in its host chromosome but also has a circular nonreplicative extrachromosomal form was recently discovered. This element, termed SXT constin, is found in virtually all clinical isolates of *V. cholerae* from the Indian subcontinent and carries resistance determinants to multiple antibiotics (Beaber et al., 2002). Its large size (100 kb) and capacity to integrate in the chromosome imply that it can bring substantial novelty to its host genome. This genetic element can also mediate the transfer of chromosomal DNA in a fashion similar to that of the *E. coli* F plasmid (Hochhut et al., 2000), making possible the lateral transfer of large genomic DNA regions.

Given the widespread presence of many large plasmids and an integrative/conjugative element like SXT, a substantial amount of genetic material has the potential to be laterally transferred by conjugation in vibrios. The presence of highly selectable traits on some of these elements, such as iron transport or antibiotic resistance, further facilitates their spread and/or maintenance.

Vibriophages and Prophages

Phages are very abundant in marine ecosystems: seawater samples containing 10^4 to 10^7 phage particles/ml are routinely isolated from across the globe (Wommack and Colwell, 2000). Direct counts have allowed an estimate of about 3 to 10 virus-like particles for every cell in the marine environment (Bergh et al., 1989). Temperate phages are also present in great numbers inside bacterial cells, as lysogeny (phage-induced lysis of the host after exposure to an inducing agent such as UV light or mitomycin C) is very frequent in marine bacteria. A study of 116 bacterial isolates from oligotrophic waters revealed that 40% were lysogens (Jiang and Paul, 1998b). A more specific survey of *V. cholerae* O1 strains by electron microscopy revealed that 60% of isolates were lysogenic (i.e., contained at least one prophage) (Shimodori et al., 1984). Phages specific to vibrios (vibriophages) can be readily isolated from marine environments or animals (Baross et al., 1978; DePaola et al., 1997). A study by Moebus and Nattkemper (1983) found that 280 of 366 phage-sensitive bacteria isolated from the Atlantic belonged to the genus *Vibrio*. Even phages that infect a specific species, such as *V. parahaemolyticus*, could be isolated from 37% of 375 enrichments from marine samples (Baross et al., 1978).

Transduction could therefore be a major vector for LGT in vibrios. It can be of a generalized nature, wherein random host DNA is mistakenly packaged into the capsid during the production of phage particles, or specialized, wherein a temperate bacteriophage inserted in the host chromosome packages genes located close to its insertion site. The phage genes themselves, once inserted in the host chromosome, can give the cell a selective advantage (e.g., cholera toxin of CTX-psi).

Frequency of transduction in natural marine communities has been estimated (Jiang and Paul, 1998a) using *Pseudomonas* bacteriophage–host systems isolated from coastal waters, values ranging from 1.6×10^{-8} to 3.7×10^{-8} transductants per PFU. Even with low transduction frequencies predicted using this system, which are likely underestimates, a simple mathematical model predicts a significant impact of transduction on LGT (up to 100 transductants per day in each liter of water in the Tampa Bay estuary) (Jiang and Paul, 1998a). Other phage–host systems could generate different predictions; for example, higher transduction frequencies were detected in a vibriophage transduction system in a laboratory setting (Ichige et al., 1989).

Direct evidence for LGT through transduction is mostly limited to temperate phages. These phages can insert in their host's chromosome and can be readily detected in complete genome sequences using scripts looking for characteristics typical of these insertion sites (Canchaya et al., 2004). The best-studied example is the CTX-psi phage of *V. cholerae*. This phage, which bears the genes coding for the cholera toxin, is widely distributed among *V. cholerae*. A study screened 300 strains of epidemic and nonepidemic *V. cholerae* and found that 12 strains of varying serogroup and origin were positive for CTX-psi (Li et al., 2003). Other vibrio phages, such as the *V. parahaemolyticus* f237 phage, are also known to integrate in their host's chromosome (Table 4). The latter also has a widespread distribution: 53 of 96 clinical *V. parahaemolyticus* strains representing various serogroups tested positive for this phage (Iida et al., 2001). These types of studies, along with evidence that a large proportion of environmental vibrios are lysogenic, suggest that specialized transduction is responsible for a significant amount of gene transfer in vibrio populations. The predictions on the impact of transduction in marine populations of bacteria also add to the potential significance of LGT in natural *Vibrio* populations, although studies more specific to this group of bacteria are required to make more accurate estimates.

Integrons and Gene Cassette Arrays

The integron/gene cassette system of vibrios is noteworthy here, given the substantial contribution it likely makes to LGT. Since integrons are specialized systems for the acquisition, rearrangement, and expression of genes (packaged as gene cassettes),

Table 4. Completely sequenced vibriophage genomes

Phage	Phage family	Host range	Genome size (kb)	Found in host chromosome?	Reference
f237	*Inoviridae*	Several *V. parahaemolyticus* serogroups	8.8	Yes	Nasu et al., 2000
K139	*Myoviridae*	Several *V. cholerae* serogroups	33	Yes	Kapfhammer et al., 2002
VP16T/VP16C	*Siphoviridae*	*V. parahaemolyticus* strain 16	50	Uncertain	Seguritan et al., 2003
CTX	*Inoviridae*	*V. cholerae* and *V. mimicus* strains	7.0	Yes	Waldor and Mekalanos, 1996
KVP40	*Myoviridae*	Eight *Vibrio* species	245	No	Miller et al., 2003
Vf12/Vf33	*Inoviridae*	*V. parahaemolyticus* and *P. damselae*	8.0	Yes	Chang et al., 1998
VHML	*Myoviridae*	Mostly *V. harveyi*, some other *Vibrio* spp.	43	Likely	Oakey et al., 2002
VpV262	*Podoviridae*	*V. parahaemolyticus* strain 94Z944	46	No	Hardies et al., 2003

they are a hot spot for LGT in their host genome (Holmes et al., 2003). What makes integrons particularly relevant in vibrios is their potential ubiquity and the large size of their associated gene cassette arrays. Indeed, integrons appear to be very common in vibrios, having been found in at least a dozen species of the genus by either Southern analysis, PCR screening, or direct sequencing (Clark et al., 2000; Rowe-Magnus et al., 2003). The cassette arrays associated with integrons found on vibrio chromosomes range from 50 kb (*V. parahaemolyticus* RIMD2210633) to 150 kb (*V. vulnificus* CMCP6), numbering ~70 to 220 gene cassettes and representing 1 to 3% of their genomes (Chen et al., 2003; Makino et al., 2003; Rowe-Magnus et al., 2003). In addition, other integrons with smaller cassette arrays (5 to 10 gene cassettes), usually containing antibiotic resistance genes, are frequently found on vibrio plasmids or other genetic elements (such as the SXT constin) (Dalsgaard et al., 2000; Hochhut et al., 2001).

Comparison of the gene content of the integron gene cassettes arrays of two *V. vulnificus* strains, CMCP6 and YJ016, revealed that only 29 of 129 unique ORFs were shared between the arrays of the two strains (Chen et al., 2003). This represents a difference of 75% in gene content, much greater than the 15 to 20% difference observed in their total genomic gene contents. Even fewer ORFs are conserved when the cassette arrays of two different vibrio species are compared (6 to 12 ORFs when comparing species that had their genomes completely sequenced). This variability supports the idea that integron gene cassette arrays evolve much faster than the rest of the genome by being involved in frequent transfer of gene cassettes. The arrays themselves are possibly mobilizable, as they are often associated or flanked by multiple transposase genes. The *V. cholerae* O1 El Tor N16961 cassette array, being located on the small chromosome (as opposed to the larger one for *V. vulnificus* and *V. parahaemolyticus*), has likely been transferred between chromosomes through transposon activity.

PERSPECTIVES ON THE IMPORTANCE OF LATERAL AND VERTICAL INHERITANCE IN THE EVOLUTION OF VIBRIOS

Most efforts in studying the evolution of vibrios have been devoted to pathogenic species. From these studies, many novel mobile genetic elements have been discovered (new phage types, SXT constins, large integron/gene cassette array systems, genomic islands, large plasmids). Several of these elements have been subsequently found in nonpathogenic environmental vibrios, marking them as general tools for vibrio evolution.

On a qualitative level, there is no doubt about the importance of transduction and conjugation for lateral gene transfer in vibrios. Most studies on vibriophages have been done in marine environments, in which these phages are present in great numbers and can easily be isolated. The high proportion of lysogenic marine bacteria in general and vibrio cells in particular, as well as the frequent identification of prophage DNA in vibrio chromosomes, confirms the propensity of transduction in this genus. Plasmid profiling studies of marine vibrios pathogenic to aquatic animals have shown how widespread large plasmids are in these microbes. The simultaneous presence of many plasmids in one cell also seems very frequent, enhancing chances of transfer of mobilizable plasmids along with their conjugative counterparts. An integrative/conjugative element such as the SXT constin found in *V. cholerae*, which can mediate the transfer chromosomal DNA as well as plasmids, further enhances the potential of conjugation as a mode of LGT. Another genetic element that allows the integration of novel DNA in vibrio genomes is the integron/gene cassette system. Integrons associated with large gene cassette arrays seem to be a ubiquitous feature in this group of microbes and represent a nonnegligible proportion of most vibrio genomes (1 to 3%).

Much progress has recently been made in trying to quantify laterally acquired DNA in the genomes of vibrios, although many more taxa need to be examined for accurate representation. If we take the genome of *V. cholerae* O1 El Tor N16961 as an example, 7% of it is composed of recently acquired or obviously mobile DNA (281 kb from a total genome size of 4,034 kb: 126-kb integron/gene cassette array system, 7-kb CTX-psi prophage, 22-kb capsular and O-antigen biosynthesis region, 40-kb VPI-1, 57-kb VPI-2, 16-kb VSP-1, and 13-kb VSP-2). N16961 does not seem to be an exceptional vibrio genome in terms of LGT, but it has definitely been scrutinized in more detail than others for genetic material of foreign origin. The 48-kb plasmid and 138-kb integron of *V. vulnificus* YJ016 alone represent 3.5% of its genome. The 100-kb SXT constin, 40-kb polysaccharide biosynthesis region, and 120-kb integron of *V. cholerae* 139 Bengal represent 6.5% of its genome. These figures suggest that ~5 to 10% of most vibrio genomes are fast-evolving regions involved in LGT. This value would be a subset of the 15 to 20% estimate obtained by BLAST-based and atypical nucleotide composition methods, as these regions are very likely to be detected by at least one of these methods.

On a smaller evolutionary scale, intraspecific and interspecific homologous recombination takes place between vibrios. The number of nucleotide substitutions caused by recombination versus point mutation has only been calculated for *V. cholerae* O139, in which it was 3:1. It does not mean that this value can be translated to all other vibrio species, but it does suggest a very important, possibly even predominant, role for recombination in the microevolution of vibrio genomes.

To get a better idea of the role of LGT in the evolution of vibrio genomes, more extensive studies are required. More vibrio genome sequences and microarray comparisons would help define the variability in gene content between and within vibrio species and therefore determine which genes are part of a stable core and which genes are transient. Phylomes (phylogenetic trees of all genes found in a genome) of vibrio genomes would allow an estimate of their content in laterally acquired genes that could be compared to values obtained through other types of whole-genome analyses. Multilocus sequence analysis of many more strains and species of vibrios using as many loci as possible would allow for more accurate estimates of recombination rates in this taxonomical group. Taken together, these different techniques should allow for a global picture of *Vibrio* evolution, from the microevolution of conserved housekeeping genes by recombination to mobile genes traveling rapidly between species.

REFERENCES

Altschul, S. F., T. L. Madden, A. A. Schaffer, J. Zhang, Z. Zhang, W. Miller, and D. J. Lipman. 1997. Gapped BLAST and PSI-BLAST: a new generation of protein database search programs. *Nucleic Acids Res.* **25**:3389–3402.

Baross, J. A., J. Liston, and R. Y. Morita. 1978. Incidence of *Vibrio parahaemolyticus* bacteriophages and other Vibrio bacteriophages in marine samples. *Appl. Environ. Microbiol.* **36**:492–499.

Beaber, J. W., B. Hochhut, and M. K. Waldor. 2002. Genomic and functional analyses of SXT, an integrating antibiotic resistance gene transfer element derived from *Vibrio cholerae*. *J. Bacteriol.* **184**:4259–4269.

Bergh, Ø., K. Y. Borsheim, G. Bratbak, and M. Heldal. 1989. High abundance of viruses found in aquatic environments. *Nature* **340**:467–468.

Bik, E. M., R. D. Gouw, and F. R. Mooi. 1996. DNA fingerprinting of *Vibrio cholerae* strains with a novel insertion sequence element: a tool to identify epidemic strains. *J. Clin. Microbiol.* **34**:1453–1461.

Boucher, Y., and W. F. Doolittle. 2000. The role of lateral gene transfer in the evolution of isoprenoid biosynthesis pathways. *Mol. Microbiol.* **37**:703–716.

Bryant, D., and V. Moulton. 2004. Neighbor-net: an agglomerative method for the construction of phylogenetic networks. *Mol. Biol. Evol.* **21**:255–265.

Byun, R., L. D. Elbourne, R. Lan, and P. R. Reeves. 1999. Evolutionary relationships of pathogenic clones of *Vibrio cholerae* by sequence analysis of four housekeeping genes. *Infect. Immun.* **67**:1116–1124.

Canchaya, C., G. Fournous, and H. Brussow. 2004. The impact of prophages on bacterial chromosomes. *Mol. Microbiol.* **53**:9–18.

Canchaya, C., C. Proux, G. Fournous, A. Bruttin, and H. Brussow. 2003. Prophage genomics. *Microbiol. Mol. Biol. Rev.* **67**:238–276.

Chang, B., H. Taniguchi, H. Miyamoto, and S. Yoshida. 1998. Filamentous bacteriophages of *Vibrio parahaemolyticus* as a possible clue to genetic transmission. *J. Bacteriol.* **180**:5094–5101.

Charlebois, R. L., and W. F. Doolittle. 2004. Computing prokaryotic gene ubiquity: rescuing the core from extinction. *Genome Res.* **14**:2469–2477.

Chen, C. Y., K. M. Wu, Y. C. Chang, C. H. Chang, H. C. Tsai, T. L. Liao, Y. M. Liu, H. J. Chen, A. B. Shen, J. C. Li, T. L. Su, C. P. Shao, C. T. Lee, L. I. Hor, and S. F. Tsai. 2003. Comparative genome analysis of *Vibrio vulnificus*, a marine pathogen. *Genome Res.* **13**:2577–2587.

Chowdhury, N. R., O. C. Stine, J. G. Morris, and G. B. Nair. 2004. Assessment of evolution of pandemic *Vibrio parahaemolyticus* by multilocus sequence typing. *J. Clin. Microbiol.* **42**:1280–1282.

Clark, C. A., L. Purins, P. Kaewrakon, T. Focareta, and P. A. Manning. 2000. The *Vibrio cholerae* O1 chromosomal integron. *Microbiology* **146**:2605–2612.

Dalsgaard, A., A. Forslund, O. Serichantalergs, and D. Sandvang. 2000. Distribution and content of class 1 integrons in different *Vibrio cholerae* O-serotype strains isolated in Thailand. *Antimicrob. Agents Chemother.* **44**:1315–1321.

DePaola, A., S. McLeroy, and G. McManus. 1997. Distribution of *Vibrio vulnificus* phage in oyster tissues and other estuarine habitats. *Appl. Environ. Microbiol.* **63**:2464–2467.

Di Lorenzo, M., M. Stork, M. E. Tolmasky, L. A. Actis, D. Farrell, T. J. Welch, L. M. Crosa, A. M. Wertheimer, Q. Chen, P. Salinas, L. Waldbeser, and J. H. Crosa. 2003. Complete sequence of virulence plasmid pJM1 from the marine fish pathogen *Vibrio anguillarum* strain 775. *J. Bacteriol.* **185:**5822-5830.

Dobrindt, U., B. Hocchut, U. Hentschel, and J. Hacker. 2004. Genomic islands in pathogenic and environmental microorganisms. *Nature Rev.* **2:**414–424.

Dumontier, S., and P. Berche. 1998. *Vibrio cholerae* O22 might be a putative source of exogenous DNA resulting in the emergence of the new strain of *Vibrio cholerae* O139. *FEMS Microbiol. Lett.* **164:**91–98.

Dumontier, S., P. Trieu-Cuot, and P. Berche. 1998. Structural and functional characterization of IS1358 from *Vibrio cholerae*. *J. Bacteriol.* **180:**6101–6106.

Dziejman, M., E. Balon, D. Boyd, C. M. Fraser, J. F. Heidelberg, and J. J. Mekalanos. 2002. Comparative genomic analysis of *Vibrio cholerae*: genes that correlate with cholera endemic and pandemic disease. *Proc. Natl. Acad. Sci. USA* **99:**1556–1561.

Farfan, M., D. Minana-Galbis, M. C. Fuste, and J. G. Loren. 2002. Allelic diversity and population structure in *Vibrio cholerae* O139 Bengal based on nucleotide sequence analysis. *J. Bacteriol.* **184:**1304–1313.

Faruque, S. M., and J. J. Mekalanos. 2003. Pathogenicity islands and phages in *Vibrio cholerae* evolution. *Trends Microbiol.* **11:**505–510.

Feil, E. J., E. C. Holmes, D. E. Bessen, M. S. Chan, N. P. Day, M. C. Enright, R. Goldstein, D. W. Hood, A. Kalia, C. E. Moore, J. Zhou, and B. G. Spratt. 2001. Recombination within natural populations of pathogenic bacteria: short-term empirical estimates and long-term phylogenetic consequences. *Proc. Natl. Acad. Sci. USA* **98:**182–187.

Frickey, T., and A. N. Lupas. 2004. PhyloGenie: automated phylome generation and analysis. *Nucleic Acids Res.* **32:**5231–5238.

Garcia-Vallve, S., E. Guzman, M. A. Montero, and A. Romeu. 2003. HGT-DB: a database of putative horizontally transferred genes in prokaryotic complete genomes. *Nucleic Acids Res.* **31:**187–189.

Garg, P. G., A. Aydanian, D. Smith, J. G. Morris, G. B. Nair, and O. C. Stine. 2003. Molecular epidemiology of O139 *Vibrio cholerae*: mutation, lateral gene transfer, and founder flush. *Emerg. Infect. Dis.* **9:**810–814.

Hardies, S. C., A. M. Comeau, P. Serwer, and C. A. Suttle. 2003. The complete sequence of marine bacteriophage VpV262 infecting *Vibrio parahaemolyticus* indicates that an ancestral component of a T7 viral supergroup is widespread in the marine environment. *Virology* **310:**359–371.

Hochhut, B., Y. Lotfi, D. Mazel, S. M. Faruque, R. Woodgate, and M. K. Waldor. 2001. Molecular analysis of antibiotic resistance gene clusters in *Vibrio cholerae* O139 and O1 SXT constins. *Antimicrob. Agents Chemother.* **45:**2991–3000.

Hochhut, B., J. Marrero, and M. K. Waldor. 2000. Mobilization of plasmids and chromosomal DNA mediated by the SXT element, a constin found in *Vibrio cholerae* O139. *J. Bacteriol.* **182:**2043–2047.

Hoi, L., I. Dalsgaard, A. DePaola, R. J. Siebeling, and A. Dalsgaard. 1998. Heterogeneity among isolates of *Vibrio vulnificus* recovered from eels (*Anguilla anguilla*) in Denmark. *Appl. Environ. Microbiol.* **64:**4676–4682.

Holmes, A. J., M. R. Gillings, B. S. Nield, B. C. Mabbutt, K. M. Nevalainen, and H. W. Stokes. 2003. The gene cassette metagenome is a basic resource for bacterial genome evolution. *Environ. Microbiol.* **5:**383–394.

Ichige, A., S. Matsutani, K. Oishi, and S. Mizushima. 1989. Establishment of gene transfer systems for and construction of the genetic map of a marine *Vibrio* strain. *J. Bacteriol.* **171:**1825–1834.

Iida, T., A. Hattori, K. Tagomori, H. Nasu, R. Naim, and T. Honda. 2001. Filamentous phage associated with recent pandemic strains of *Vibrio parahaemolyticus*. *Emerg. Infect. Dis.* **7:**477–478.

Jermyn, W. S., and E. F. Boyd. 2002. Characterization of a novel *Vibrio* pathogenicity island (VPI-2) encoding neuraminidase (nanH) among toxigenic *Vibrio cholerae* isolates. *Microbiology* **148:**3681–3693.

Jiang, S. C., and J. H. Paul. 1998a. Gene transfer by transduction in the marine environment. *Appl. Environ. Microbiol.* **64:**2780–2787.

Jiang, S. C., and J. H. Paul. 1998b. Significance of lysogeny in the marine environment: studies with isolates and a model of lysogenic phage production. *Microb. Ecol.* **35:**235–243.

Kapfhammer, D., J. Blass, S. Evers, and J. Reidl. 2002. *Vibrio cholerae* phage K139: complete genome sequence and comparative genomics of related phages. *J. Bacteriol.* **184:**6592–6601.

Karaolis, D. K., J. A. Johnson, C. C. Bailey, E. C. Boedeker, J. B. Kaper, and P. R. Reeves. 1998. A *Vibrio cholerae* pathogenicity island associated with epidemic and pandemic strains. *Proc. Natl. Acad. Sci. USA* **95:**3134–3139.

Karaolis, D. K., R. Lan, and P. R. Reeves. 1995. The sixth and seventh cholera pandemics are due to independent clones separately derived from environmental, nontoxigenic, non-O1 *Vibrio cholerae*. *J. Bacteriol.* **177:**3191–3198.

Kim, C. C., E. A. Joyce, K. Chan, and S. Falkow. 2002. Improved analytical methods for microarray-based genome-composition analysis. *Genome Biol.* **3:**1–17.

Koonin, E. V., K. S. Makarova, and L. Aravind. 2001. Horizontal gene transfer in prokaryotes: quantification and classification. *Annu. Rev. Microbiol.* **55:**709–742.

Koski, L. B., and G. B. Golding. 2001. The closest BLAST hit is often not the nearest neighbor. *J. Mol. Evol.* **52:**540–542.

Lawrence, J. G., and H. Ochman. 1997. Amelioration of bacterial genomes: rates of change and exchange. *J. Mol. Evol.* **44:**383–397.

Le Chevalier, P., C. Le Boulay, and C. Paillard. 2003. Characterization by restriction fragment length polymorphism and plasmid profiling of *Vibrio tapetis* strains. *J. Basic Microbiol.* **43:**414–422.

Li, M., M. Kotetishvili, Y. Chen, and S. Sozhamannan. 2003. Comparative genomic analyses of the vibrio pathogenicity island and cholera toxin prophage regions in nonepidemic serogroup strains of *Vibrio cholerae*. *Appl. Environ. Microbiol.* **69:**1728–1738.

Majewski, J., P. Zawadzki, P. Pickerill, F. M. Cohan, and C. G. Dowson. 2000. Barriers to genetic exchange between bacterial species: *Streptococcus pneumoniae* transformation. *J. Bacteriol.* **182:**1016–1023.

Makino, K., K. Oshima, K. Kurokawa, K. Yokoyama, T. Uda, K. Tagomori, Y. Iijima, M. Najima, M. Nakano, A. Yamashita, Y. Kubota, S. Kimura, T. Yasunaga, T. Honda, H. Shinagawa, M. Hattori, and T. Iida. 2003. Genome sequence of *Vibrio parahaemolyticus*: a pathogenic mechanism distinct from that of *V. cholerae*. *Lancet* **361:**743–749.

Mantri, Y., and K. P. Williams. 2004. Islander: a database of integrative islands in prokaryotic genomes, the associated integrases and their DNA site specificities. *Nucleic Acids Res.* **32:**55–58.

Michel, B. 1999. Illegitimate recombination in bacteria, p. 129–150. *In* R. Charlesbois (ed.), *Organization of the Prokaryotic Genome*. ASM Press, Washington, D.C.

Miller, E. S., J. F. Heidelberg, J. A. Eisen, W. C. Nelson, A. S. Durkin, A. Ciecko, T. V. Feldblyum, O. White, I. T. Paulsen, W. C. Nierman, J. Lee, B. Szczypinski, and C. M. Fraser. 2003. Complete genome sequence of the broad-host-range vibriophage KVP40: comparative genomics of a T4-related bacteriophage. *J. Bacteriol.* **185:**5220–5233.

Moebus, K., and H. Nattkemper. 1983. Taxonomic investigation of bacteriophage sensitive bacteria isolated from marine waters. *Helgoland Mar. Res.* **36**:357–373.

Molina-Aja, A., A. Garcia-Gasca, A. Abreu-Grobois, C. Bolan-Mejia, A. Roque, and B. Gomez-Gil. 2002. Plasmid profiling and antibiotic resistance of *Vibrio* strains isolated from cultured penaeid shrimp. *FEMS Microbiol. Lett.* **213**:7–12.

Nasu, H., T. Iida, T. Sugahara, Y. Yamaichi, K. S. Park, K. Yokoyama, K. Makino, H. Shinagawa, and T. Honda. 2000. A filamentous phage associated with recent pandemic *Vibrio parahaemolyticus* O3:K6 strains. *J. Clin. Microbiol.* **38**:2156–2161.

Nesbø, C. L., K. E. Nelson, and W. F. Doolittle. 2002. Suppressive subtractive hybridization detects extensive genomic diversity in *Thermotoga maritima*. *J. Bacteriol.* **184**:4475–4488.

Nishibuchi, M., and J. B. Kaper. 1995. Thermostable direct hemolysin gene of *Vibrio parahaemolyticus*: a virulence gene acquired by a marine bacterium. *Infect. Immun.* **63**:2093–2099.

Oakey, H. J., B. R. Cullen, and L. Owens. 2002. The complete nucleotide sequence of the *Vibrio harveyi* bacteriophage VHML. *J. Appl. Microbiol.* **93**:1089–1098.

Pedersen, K., I. Dalsgaard, and J. L. Larsen. 1997. *Vibrio damsela* associated with diseased fish in Denmark. *Appl. Environ. Microbiol.* **63**: 3711–3715.

Pedersen, K., I. Kuhn, J. Seppanen, A. Hellstrom, T. Tiainen, E. Rimaila-Parnanen, and J. L. Larsen. 1999. Clonality of *Vibrio anguillarum* strains isolated from fish from the Scandinavian countries, Sweden, Finland and Denmark. *J. Appl. Microbiol.* **86**:337–347.

Perna, N. T., G. Plunkett III, V. Burland, B. Mau, J. D. Glasner, D. J. Rose, G. F. Mayhew, P. S. Evans, J. Gregor, H. A. Kirkpatrick, G. Posfai, J. Hackett, S. Klink, A. Boutin, Y. Shao, L. Miller, E. J. Grotbeck, N. W. Davis, A. Lim, E. T. Dimalanta, K. D. Potamousis, J. Apodaca, T. S. Anantharaman, J. Lin, G. Yen, D. C. Schwartz, R. A. Welch, and F. R. Blattner. 2001. Genome sequence of enterohaemorrhagic *Escherichia coli* O157:H7. *Nature* **409**:529–533.

Radu, S., N. Elhadi, Z. Hassan, G. Rusul, S. Lihan, N. Fifadara, Yuherman, and E. Purwati. 1998. Characterization of *Vibrio vulnificus* isolated from cockles (*Anadara granosa*): antimicrobial resistance, plasmid profiles and random amplification of polymorphic DNA analysis. *FEMS Microbiol. Lett.* **165**: 139–143.

Ragan, M. A. 2001. On surrogate methods for detecting lateral gene transfer. *FEMS Microbiol. Lett.* **201**:187–191.

Rowe-Magnus, D. A., A. M. Guerout, L. Biskri, P. Bouige, and D. Mazel. 2003. Comparative analysis of superintegrons: engineering extensive genetic diversity in the Vibrionaceae. *Genome Res.* **13**:428–442.

Seguritan, V., I. W. Feng, F. Rohwer, M. Swift, and A. M. Segall. 2003. Genome sequences of two closely related *Vibrio parahaemolyticus* phages, VP16T and VP16C. *J. Bacteriol.* **185**: 6434–6447.

Shimodori, A., K. Takeya, and A. Takade. 1984. Lysogenicity and prophage type of the strains of *V. cholerae* O1 isolated mainly from the natural environment. *Am. J. Epidemiol.* **120**:759–768.

Sicheritz-Ponten, T., and S. G. Andersson. 2001. A phylogenomic approach to microbial evolution. *Nucleic Acids Res.* **29**:545–552.

Smith, G. R. 1988. Homologous recombination in procaryotes. *Microbiol. Rev.* **52**:1–28.

Smith, M. W., D. F. Feng, and R. F. Doolittle. 1992. Evolution by acquisition: the case for horizontal gene transfers. *Trends Biochem. Sci.* **17**:489–493.

Stephens, J. C. 1985. Statistical methods of DNA sequence analysis: detection of intragenic recombination or gene conversion. *Mol. Biol. Evol.* **2**:539–556.

Tiainen, T., K. Pedersen, and J. L. Larsen. 1995. Ribotyping and plasmid profiling of Vibrio anguillarum serovar O2 and *Vibrio ordalii*. *J. Appl. Bacteriol.* **79**:384–392.

Waldor, M. K., and J. J. Mekalanos. 1996. Lysogenic conversion by a filamentous phage encoding cholera toxin. *Science* **272**: 1910–1914.

Wiik, R., K. Andersen, F. L. Daae, and K. A. Hoff. 1989. Virulence studies based on plasmid profiles of the fish pathogen *Vibrio salmonicida*. *Appl. Environ. Microbiol.* **55**:819–825.

Wommack, K. E., and R. R. Colwell. 2000. Virioplankton: viruses in aquatic ecosystems. *Microbiol. Mol. Biol. Rev.* **64**:69–114.

The Biology of Vibrios
Edited by F. L. Thompson et al.
© 2006 ASM Press, Washington, D.C.

Chapter 8

The Adaptive Genetic Arsenal of Pathogenic *Vibrio* Species: the Role of Integrons

Dean A. Rowe-Magnus, Mohammed Zouine, and Didier Mazel

Microbiology entered the genomic age with considerable success. As of today, >290 bacterial genomes have been completely sequenced, and their analysis has indisputably demonstrated the roles of the different types of mobile DNA elements in the overall plasticity of bacterial genomes. The finding that many accessory functions, such as detoxification, pathogenicity, and virulence, are embedded in mobile elements solidifies their importance as an unlimited resource of adaptive potential for bacteria. For example, phages and phage-like elements play a leading role in the spread of pathogenicity genes, and many pathogenicity islands (Hacker et al., 1992; Faruque and Mekalanos, 2003) correspond to such elements. Our review is focused on integrons of both the mobile and superintegron types. We discuss their specific roles in the adaptive capacity of the *Vibrionaceae*, with emphasis on pathogenic species and antibiotic resistance.

THE PATHOGENIC VIBRIOS

The *Vibrionaceae* is a large family of marine gammaproteobacteria, the majority of which are nonpathogenic members of the genus *Vibrio*. However, the genus includes several pathogenic *Vibrio* species, a small number of which cause human diseases (Janda et al., 1988). Infections caused by these organisms are usually associated with ingestion of raw shellfish or exposure of wounds to seawater. The clinical presentation and severity of infections are wide-ranging. The most common presentation is self-limiting gastroenteritis, but soft tissue infections and septicemia do occur, and their morbidity and mortality are high, particularly in patients with liver disease. The pathology of cholera, caused by *Vibrio cholerae* O1 and O139 strains, *Vibrio parahaemolyticus* food

poisoning, and *Vibrio vulnificus* septicemia has garnered the most attention, and these species are well-documented human pathogens. The diarrheal symptoms caused by cholera and *V. parahaemolyticus* food poisoning include profuse purging of watery stools, vomiting, and dehydration. The diseases are usually self-limiting, and the mainstay of therapy for patients is oral or intravenous rehydration salt solution. *V. vulnificus* infections are categorized as skin and soft tissue infections and primary septicemia. Mortality can reach 75% in septic patients, with most deaths occurring within 48 h of hospital admission. *Vibrio hollisae*, *Vibrio fluvialis*, and *Vibrio fetus* have also been associated with human disease (Ullmann, 1969; Abbott and Janda, 1994; Ahmed et al., 2004). Antibiotics such as tetracycline, doxycycline, norfloxacin, ciprofloxacin, and furazolidone may be used as an adjunct to rehydration therapy and are critical in the treatment of septicemic patients (Lima, 2001; Bhattacharya, 2003; Chiang and Chuang, 2003). Resistance to many of these drugs has emerged in these pathogens and is a matter of major concern, particularly in the case of *V. vulnificus* and *V. cholerae*. Reports delineating the origins of antibiotic resistance genes in gram-negative clinical and environmental isolates abound in the literature. These resistance genes may be chromosomally encoded (for example, multidrug efflux pumps), carried on conjugative plasmids, or residing within integrons.

THE PATHS OF RESISTANCE

Chromosomal Multidrug Efflux Pumps

Even though many reports have demonstrated that the presence of antibiotic resistance genes in plas-

Dean A. Rowe-Magnus • Department of Microbiology, Sunnybrook & Women's College Health Sciences Centre, Toronto, Ontario, M4N 3N5, Canada. **Mohammed Zouine and Didier Mazel** • Unité postulante "Plasticité du Génome Bactérien" CNRS URA 2171, Dept. Structure et Dynamique des Génomes, Institut Pasteur, 75724 Paris, France.

mids or integrons in *V. cholerae* was the cause of resistance to antimicrobial agents, the mechanism of resistance in other cases was unknown. Studies by Colmer et al. (1998) and Huda et al. (2003) to elucidate the multidrug resistance systems in non-O1 *V. cholerae* strains led to the characterization of the multidrug efflux pump, VceAB, which belongs to the major facilitator superfamily, and VcaM, an ABC multidrug efflux pump. More recently, two Na^+-driven multidrug efflux pumps, VcmA and VcrM, from non-O1 *V. cholerae* strains (Huda et al., 2001, 2003) and the Na^+-driven multidrug efflux pump, NorM, from *V. parahaemolyticus* (Morita et al., 1998) have also been identified. Norfloxacin, ciprofloxacin, kanamycin, streptomycin, tetracycline, and fluoroquinolones are substrates of VcaM. Because these antibiotics are commonly used for the treatment of vibrio infections, the overexpression of these pumps would be a serious threat to patient treatment protocols.

Integrative and Conjugative Elements (ICEs)

ICEs are a diverse class of mobile elements found in both gram-positive and gram-negative bacteria (Beaber et al., 2002a; Burrus et al., 2002a). Like prophages, ICEs integrate into the chromosome of their hosts. Recombination between two specific sequences, *attL* and *attR*, that flank the element leads to its excision from the chromosome in the form of a covalently closed circular molecule that is usually nonreplicative. Like conjugative plasmids, ICEs encode conjugation systems that can transfer the excised DNA to a new host, where it can integrate into the chromosome by site-specific recombination between the *attP* site generated by excision and a target sequence, *attB* in the host chromosome. Different ICEs integrate into a variety of sites and encode diverse recombination, conjugation, and regulation systems. They also carry genes encoding a variety of functions, including catabolic pathways, antibiotic resistances, nitrogen fixation, and phage resistance mechanisms (Rice, 1998; Sullivan and Ronson, 1998; Burrus et al., 2002b; Whittle et al., 2002; van der Meer and Sentchilo, 2003).

The SXT element is a 99.5-kb ICE that was originally identified in a *V. cholerae* serogroup O139 clinical isolate (Waldor et al., 1996; Hochhut and Waldor, 1999; Beaber et al., 2002b). It encodes resistance to sulfamethoxazole, trimethoprim, and streptomycin. SXTMO10-related elements have been identified in *V. cholerae* O1 and O139 clinical isolates from Asia (Amita et al., 2003) and Africa (Dalsgaard et al., 2001). These SXT-related elements may contain different antibiotic resistance genes or may even lack antibiotic resistance genes.

R391 is an ICE originally isolated from *Providencia rettgeri* and also identified in *Vibrio* species. It mediates resistance to kanamycin and mercury (Murphy and Pembroke, 1995; Boltner et al., 2002). R391 and several related IncJ elements integrate into *prfC*, the same chromosomal locus used by SXT (Hochhut et al., 2001a). Comparison of the nucleotide sequences of these two ICEs showed that they consist of a conserved set of genes that mediate regulation, excision/integration, and conjugative transfer of the respective ICEs (Beaber et al., 2002a). Insertions into this shared common backbone confer element-specific properties, such as resistance to particular antibiotics. Cells can harbor both SXT and R391 elements (Hochhut et al., 2001a).

The Rise of Antibiotic Resistance: from Conjugative Plasmids to Transposons to Integrons

The emergence and spread of antibiotic resistance among human pathogens is certainly the most striking evolution that has arisen in bacteria within the past 6 decades. Indeed, resistance has been encountered as an impediment to antibiotic therapy for as long as antibiotics have been used. Single resistance phenotypes were not entirely unforeseen, as early laboratory studies demonstrated that, for example, penicillin-resistant point mutants were readily selected. In contrast, multidrug resistance was never anticipated, since the coappearance of multiple mutations conferring such phenotypes was considered to be beyond the evolutionary potential of a given bacterial population. However, only 6 years after the introduction and massive production of streptomycin, tetracycline, and chloramphenicol in Japan, *Shigella dysenteriae*, which was simultaneously resistant to these three antibiotics and sulfonamide, was isolated (Mitsuhashi et al., 1961). It became clear that the emergence of multiply resistant strains could not be attributed to mutation alone. This was actually the time when transposons, and with them integrons, appeared in the human gram-negative pathogen and commensal bacterial population as factors in resistance gene dissemination through hitchhiking on conjugative plasmids and interspecies transfer. Nevertheless, integrons were only formally identified as agents of resistance gene recruitment in the late 1980s through the observation that loci in transposons and resistance plasmids expressing different antibiotic resistance spectra shared the same genetic backbone and differed only in the resistance genes they harbored (Stokes and Hall, 1989). It is clear, however, that they were definitely part of the first multidrug resistance outbreaks in the 1950s, as attested by the involvement of Tn*21*, an integron-containing transposon, in the resistance phenotype propagated by plasmid NR1 (R100) in the very first events in Japan (Leon and Roy, 2003). In light of the time scale for the development of multiresistant *Shigella* strains described above, it was hardly disputable that bacteria were already prepared to face such a challenge and had already evolved the appropriate genetic tools. Integrons likely correspond to one of the most refined tools selected by bacteria, as suggested by the data collected during the last 15 years. These data have shown how powerful the integron recombination machinery is, though its functional platform is fairly simple.

INTEGRONS

Toward a Minimal Definition

Integrons are natural genetic engineering platforms that incorporate open reading frames (ORFs) and convert them to functional genes by ensuring their correct expression. All integrons characterized so far are composed of three key elements necessary for the procurement of exogenous genes: (i) a gene coding for an integrase of the tyrosine recombinase family (*intI*), (ii) a primary recombination site (*attI*), and (iii) a strong, outward-oriented promoter (Pc). Integron integrases can recombine, in a *recA*-independent manner, discrete units of circularized DNA known as gene cassettes downstream of the resident Pc promoter at the proximal *attI* site, permitting ex-

pression of their encoded proteins (Fig. 1). All integron-inserted cassettes identified share specific structural characteristics. The integrated cassettes generally include a single gene and an imperfect inverted repeat located at the 3' end of the gene called an *attC* site or "59-base element." The *attC* sites are a diverse family of sequences that function as recognition sites for the site-specific integrase. The *attC* sites vary from 57 to 141 bp in length, and their nucleotide sequence similarities are primarily restricted to their boundaries; the boundaries correspond to the inverse core site or R″ sequence (RYYYAAC, where R is a purine and Y is a pyrimidine) and the core site or R′ sequence (G/TTRRRY, where the shill indicates the recombination point; see Fig. 1).

In its first description, the definition included the fact that the integron platform was itself a mobile DNA element (Stokes and Hall, 1989). This last point has been the source of misunderstanding, as it suggested that the integron integrase was also able to move the whole platform. This assumption was due to the fact that the first characterized integrons were carried by transposons, but transposition does not depend on the activity of the integron integrase. The integron integrase only mobilizes the gene cassettes within integrons. With the discovery of other types of integrons, either carried by different transposons or as sedentary components of the genomes of various species, the definition of an integron has evolved toward the minimal set presented above. The neces-

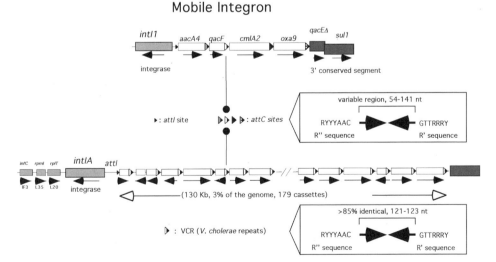

Mobile Integron

Chromosomal Superintegron (*V. cholerae*)

Figure 1. Structural comparison of a "classical" mobile integron and the *V. cholerae* N16961 superintegron. (Top) Schematic representation of In40; the various resistance genes are associated with different *attC* sites (see text). Antibiotic resistance cassettes confer resistance to the following compounds: *aacA4*, aminoglycosides; *qac*, quaternary ammonium compounds; *cmlA2*, chloramphenicol; *oxa9*, beta-lactams. The *sul* gene, which provides resistance to sulfonamides, is not a gene cassette. (Bottom) The ORFs are separated by highly homologous sequences, the VCRs. See text for details.

sity and the legitimacy of dividing the integrons into two subsets, the mobile integrons (MI), primarily involved in the spread of resistance genes, and the superintegrons (SI), are still disputed and will be discussed in the section about SIs.

Mobile Integrons

At present, there are five different classes of MI involved in the dissemination of antibiotic resistance genes. The classes have been defined on the basis of the divergence of the integrase genes. All five are physically linked to mobile DNA structures, either through their association with insertion sequences, transposons, and/or conjugative plasmids, all of which can serve as vehicles for the intra- and interspecies transmission of genetic material. Among these MIs are the three "historical" classes involved in the phenotype of multiple antibiotic resistance (Hall, 1997). Class 1 integrons are found associated with functional transposons such as Tn*21* (Liebert et al., 1999). Class 2 integrons are exclusively found inside Tn7 derivatives (Sundstrom et al., 1991; Radstrom et al., 1994), whereas class 3 integrons have been found on mainly uncharacterized plasmids (Arakawa et al., 1995; Correia et al., 2003; Shibata et al., 2003). The two other classes of MI are known for their involvement in the development of trimethoprim resistance in *Vibrio* species; one is a component of a subset of SXT elements found in *V. cholerae* (Hochhut et al., 2001b), and the other has been identified by Henning Sørum and colleagues (Norwegian School of Veterinary Science, Oslo) in a compound transposon carried on a plasmid in *V. salmonicida* (unpublished data; GenBank no. AJ277063). Each class of integron is considered to be able to share and acquire the same gene cassettes. Class 1 integrons are found extensively in clinical isolates, and most of the known resistance gene cassettes have been characterized in class 1 integrons. As of today, even considering only those cassettes that differ in nucleotide sequence by more than 5%, over 80 different cassettes from class 1 integrons have been described that confer resistance to all β-lactams, all aminoglycosides, chloramphenicol, trimethoprim, streptothricin, rifampin, erythromycin, and antiseptics of the quaternary ammonium compound family (reviewed by Rowe-Magnus and Mazel, 2002; Fluit and Schmitz, 2004). The second most prevalent class of integrons carrying resistance genes is class 2, but only five different resistance cassettes have been found in these structures to date. This is likely due to the observation that the integrase of class 2 integrons has been inactivated by a stop codon in position 179 in the gene, yielding a truncated, nonfunctional protein (Hansson et al., 2002). Several cassettes have

been identified in different classes of integrons, and the class 1 integrase, IntI1, has been demonstrated to recombine several structurally diverse *att*C sites (Collis et al., 2001). The IntI integrases belong to the catalytic family of the tyrosine (Y) recombinases that are involved in the movement of numerous bacteriophages through site-specific recombination (such as the λ phage integrase, λInt) or in fundamental cellular processes such as chromosome dimer resolution in cell division (XerC/D). Despite these relationships, the integron integrases have a unique characteristic among the Y recombinases—they are able to recombine sequences that are only poorly related.

The proficiency of the partnership of integrons and mobile DNA elements is confirmed by the marked differences in codon usage among cassettes within the same MI, indicating that the genes are of diverse origins. With this system, bacteria are capable of stockpiling exogenous resistance genes to establish an appreciable antimicrobial armamentarium; MIs harboring up to eight different resistance cassettes have been characterized (Naas et al., 2001). MIs carrying a plethora of resistance gene cassettes have been identified in diverse gram-negative bacteria. With regard to the *Vibrionaceae*, MIs have been found in *V. cholerae* O1, O139, non-O1/non-O139, and *V. fluvialis* strains (Falbo et al., 1999; Dalsgaard et al., 2000, 2001; Thungapathra et al., 2002; Ahmed et al., 2004; Ehara et al., 2004; Iwanaga et al., 2004). Furthermore, integrons are no longer restricted to the gram-negative bacteria. Thus, class 1 integrons have also been found in the gram-positive bacterial genera *Mycobacterium* (Martin et al., 1990), *Corynebacterium* (Nesvera et al., 1998; Tauch et al., 2002), *Aerococcus*, *Brevibacterium*, and *Staphylococcus* (Nandi et al., 2004).

Another Type of Integron: the Chromosomal Superintegrons

In the late 1990s, studies examining the relationship between MI gene cassette arrays and a cluster of repeated sequences identified in the *V. cholerae* genome (called VCRs for *V. cholerae* repeats) (Barker et al., 1994) led to the discovery of a distinct type of integron in chromosome 2 (Mazel et al., 1998; Heidelberg et al., 2000). This integron, which possesses a specific integrase, VchIntIA, that is related to the integrases of MI, has two structural characteristics that distinguish it from known MIs: the large number of cassettes that are gathered and the high homology observed between the *att*C sites of these cassettes (the VCRs in the case of *V. cholerae*; see Fig. 1). Finally, the functional platform (IntIA + *att*I site) of this structure did not appear to be mobile; it was located on the chromosome and was not associated with mobile DNA elements. These key features define SIs.

The division of integrons into MI and SI has been disputed by Hall and Stokes (2004), who argue that (i) MI could be located on the chromosome, (ii) chromosomal integrons such as the one found in *Shewanella oneidensis* only carried a few cassettes, and (iii) gene cassettes conferring antibiotic resistance could be found inside SI. The authors recommend using the single term integron to describe all types of integron structures, supporting this suggestion with the fact that the different integrons use the same recombination processes and machinery. However, this point is not steadfast, particularly in the light of a recent study of the recombination reaction catalyzed by Vch IntIA, the integrase associated with the *V. cholerae* SI (Biskri et al., 2005). We do agree with several of the points made by Hall and Stokes, and we now propose to use the terms MI to describe integrons that are carried on mobile elements such as Tn or plasmids, and SI to describe integrons that have the key features listed above, as it is not yet confirmed that MI and SI truly function in the same way.

The superintegron is now known to be an integral component of many gammaproteobacterial genomes, and such SI structures have been identified among the *Vibrionaceae*, their close relatives *Xanthomonas*, and a branch of *Pseudomonas* (Table 1). SIs share the same general characteristics; i.e., they are large and carry >20 cassettes, and they exhibit extensive homology between their endogenous cassette *attC* sites. SIs clearly predate the antibiotic era, as they are present in isolates from the 19th century, as demonstrated through the study of a *Vibrio metschnikovii* isolate from 1888 (Mazel et al., 1998). The SIs carried in the genomes of the four *Vibrio* species that have been sequenced all show a large number of cassettes, ranging from 72 in *V. parahaemolyticus* (Makino et al., 2003) to more than 200 in *V. vulnificus* (Chen et al., 2003). In the case of *Vibrio fischeri* strain ES114, a preliminary analysis we made of the recently published complete genome sequence (Ruby et al., 2005) showed that the SI cassette array was split into two segments, one associated with the Vfi*intIA* gene on chromosome 2, and a second, smaller array on chromosome 1 (unpublished data). A preliminary analysis of the *Photobacterium profundum* SS9 genome sequence (accession numbers NC_006370 and NC_006371) (Vezzi et al., 2005) also shows a splitting of the array into four different segments along with a few isolated cassettes on both chromosomes (unpublished data). Incidentally, *P. profundum*'s *intIA* gene has been inactivated by a deletion covering approximately two-thirds of the C-terminal part of the gene (unpublished data). Such a

splitting also likely occurred for the cassette array of the *Pseudomonas alcaligenes* SI (Vaisvila et al., 2001). The SI of the *V. cholerae* strain El Tor N16961 gathers at least 216 mostly unidentified ORFs in an array of 179 cassettes that starts from the Vch*intIA* gene and occupies about 3% of the genome (Rowe-Magnus et al., 2003). Structural analysis of this SI cassette array showed that the *attC* sites carried by 149 of the 179 cassettes differed by less than 10% over their entire length of 122 to 124 nucleotides (Rowe-Magnus et al., 2003). Similar analyses performed on subsets of the SI cassette arrays from other species (i.e., *V. fischeri*, *V. metschnikovii*, *P. alcaligenes*, *Pseudomonas stutzeri*, and *Xanthomonas campestris*) have shown conservation of this characteristic (Vaisvila et al., 2001; Rowe-Magnus et al., 2001, 2003; Holmes et al., 2003b). The high level of identity shared by the majority of the *attC* sites in each SI, and in a species-specific manner, suggests that gene cassettes are assembled in the species harboring the SI through the physical association of an *attC* site with an incoming DNA fragment. The mechanics of this process are still unknown.

Several integrons have been characterized in the genomes of different bacterial species that do not share all the characteristics of the typical SI described above. Such is the case for the SI found in the genome of two *Shewanella* species, which only gather a handful of cassettes with structurally heterogeneous *attC* sites, a situation reminiscent of the resistance cassette arrays. The integron found in the genome of *Nitrosomonas europaea* is devoid of any cassette. However, the integrases and cognate *attI* sites from both species have been shown to be functional (Drouin et al., 2002; Leon and Roy, 2003). Furthermore, integron integrase-like genes have also been identified in the genomes of other proteobacteria from the δ and β subgroups, and in bacterial species belonging to other taxa such as the marine planctomycete *Rhodopirellula baltica* SH 1 (also known as *Pirellula* sp. strain 1), but they have not been further characterized (Table 1). A structure having all the typical SI features has also been found in the genome of the spirochete *Treponema denticola* ATCC 35405 (Coleman et al., 2004). This integron displays a yet unseen variation in its organization, as its *intIA* gene is inversely oriented compared to the canonical structure. Indeed, its *attI* site is located adjacent to the 3' end of the *intIA* gene, followed by an array of 45 cassettes carrying highly related *attC* sites (Coleman et al., 2004).

The ubiquitous nature of integrons in environmental bacteria has been confirmed by the detection of a number of integron integrase genes and cassettes from markedly different environmental DNA samples

Table 1. Bacterial species harboring chromosomal integrons and superintegrons

Radiation	Species	Characteristic	IntIA accession no.
Gammaproteobacteria	*Vibrionaceae* and close relatives		
	Vibrio cholerae	Etiological agent of cholera in humans	AAC38424
	V. mimicus	Certain serogroups are enterotoxic human pathogens	AAD55407
	V. metschnikovii	Certain serogroups are enterotoxic human pathogens	AAK02074
	V. parahaemolyticus	Certain strains can cause seafood-borne gastroenteritis in humans	AAK02076
	V. splendidus	Strains are mutualistic, opportunistic, and pathogenic for marine animals	
	V. harveyi	Pathogen of black tiger prawns	
	V. natriegens		AAO38263
	Grimontia hollisae	Certain strains can cause gastroenteritis in humans	
	V. salmonicida	A marine fish pathogen	
	V. fischeri	Nonpathogenic, luminescent bacterium	AAK02079
	V. vulnificus	Human and animal pathogen causing septicemia	Strain CIP 75.4, AAN33109; strain CMCP6, AAO10775; strain YJ016, BAC94705
	V. anguillarum	A marine fish pathogen	AAM95157
	Listonella pelagia	A marine bacterium that produces tetrodotoxin	AAK02082
	Alteromonas macleodii	Marine bacterium	
	Photobacterium phosphoreum SS9	Luminescent bacterium	CR378678 (translation)
	Moritella marina	Psychrophilic marine bacterium	
	Shewanella	A diverse genus of bacteria well known for their ability to utilize dissimilar compounds as electron acceptors; they figure prominently in fish spoilage and hull oxidation in the marine industry	
	S. oneidenis		AAN55084
	S. putrefaciens		AAK01408
	Xanthomonads	A gram-negative phytopathogenic species responsible for disease in virtually all major taxa of plant life	
	Xanthamonas campestris pv. *campestris*		AAK07444
	X. campestris pv. *badrii*		AAK07443
	Xanthamonas sp. 102397		AAK07447
	Xanthamonas sp. 102336		
	Xanthamonas sp. 102338		
	Xanthamonas sp. 105155		
	X. oryzae		
	Pseudomonads	Include opportunistic pathogens	
	P. pseudoalcaligenes		
	P. alcaligenes		AAK73287
	P. mendocina		
	P. stutzeri		Strain BAM, AAN16071, strain Q, AAN16061
	Pseudomonas sp. NEB 376		
	Various species		
	Microbulbifer degradans	Marine bacterium capable of degrading insoluble complex polysaccharides	ZP_00318025
	Dechloromonas aromatica	Capable of anaerobic benzene oxidation	ZP_00149963
Betaproteobacteria	*Nitrosomonas europaea*	Soil bacterium	CAD84361, CAD86100
	Thiobacillus denitrificans	Capable of autotrophic denitrification	ZP_00335637
	Azoarcus sp. EbN1	Anaerobic aromatic-degrading denitrifying bacterium	YP_157034

Continued on following page

Table 1. *Continued*

Radiation	Species	Characteristics	IntIA accession no.
	Methylobacillus flagellatus KT	Obligate methylotroph	ZP_00172930
	Dechloromonas aromatica RCB	Capable of anaerobic benzene oxidation	ZP_00149963
	Rubrivivax gelatinosus PM1	Facultative phototrophic nonsulfur bacterium	ZP_00244730
Deltaproteobacteria	*Geobacter metallireducens* GS-15	Fe(III)-reducing bacterium	ZP_00299196
	Geobacter sulfurreducens	Fe(III)-reducing bacterium	NP-953513
Planctomycetes	*Rhodopirellula baltica* SH1	Marine strain	NP_865348
Spirochaetales	*Treponema denticola*	A cause of periodontal disease	NP_972448

that ranged in origin from national park soil to heavily metal-contaminated soil. Using PCR primers directed against conserved regions of the integron integrase genes and *attC* sites, the groups of Stokes and Schmidt were able to amplify 19 new integron integrases and hundreds of cassettes (Nield et al., 2001; Stokes et al., 2001; Holmes et al., 2003a; Nemergut et al., 2004). Unfortunately, in most cases the protocol used does not permit determination of the source of these integrons, be it the endogenous SI of a soil bacterium or an integron located on a mobile structure. However, these findings support the hypothesis developed by the discovery of SIs: that integrons are widespread among bacterial populations, either as components of mobile DNA elements or the chromosome and that they are not confined to pathogenic or multidrug-resistant bacteria.

The Phylogeny of the Integron Integrases Indicates That Integrons Are Ancient Evolutionary Apparatuses

Comparative analysis of all integron integrases, be they from characterized integrons or identified from genomic analyses, shows that they clearly group together and form a specific clade within the Y-recombinase family (Rowe-Magnus et al., 2001; Rowe-Magnus and Mazel, 2001; Nemergut et al., 2004) (Fig. 2). Furthermore, it has been observed that all integron integrases contain a specific stretch of 16 amino acids (Messier and Roy, 2001; Nield et al., 2001) located between conserved patches II and III of the tyrosine recombinase family (Nunes-Duby et al., 1998). The integron platform is undoubtedly ancient, as attested to by the species-specific clustering of the respective SI integrase genes in a pattern that adheres, in several cases, to the line of descent among the bacterial species in which they are found (such as in the *Vibrio* radiation). Thus, the establishment of SIs likely predates speciation within the respective genera, indicating that integrons are ancient structures that have been steering the evolution of

bacterial genomes for hundreds of millions of years. Even if SI platforms are not carried by an identifiable mobile DNA element, it remains possible that horizontal transfer of either a part or all of an SI occurred during such a long period of evolution. This could be the origin of the discrepancies observed between the SI integrase and 16S rRNA gene trees for *V. fischeri*. This is also certainly the case for many of the *intI*-like genes identified in the different genome sequences of δ and β proteobacteria; further analysis may reveal that these genes are part of mobile elements. For example, the *intI* gene of *Azoarcus* strain EbN1 is located inside a phage-like structure that has its own integrase (accession no. NC_006513) (Rabus et al., 2005).

It is interesting to note that the phylogenetic congruency in the tree topologies for the IntI and the 16S rRNA sequences also recovers the respective niches (Fig. 2). Indeed, integron integrases from marine bacteria all group together, while those from freshwater or soil bacteria, be they beta-, delta-, or gammaproteobacteria, are more closely related to one another. This probably reflects the extent of DNA exchange among species sharing similar environments.

In the case of the *Vibrionaceae* radiation, it is of note that SIs have been found in all genera examined (i.e., *Vibrio*, *Listonella*, and *Photobacterium*). Interestingly, a recent study demonstrated that the division of the genome into two circular chromosomes (Trucksis et al., 1998; Yamaichi et al., 1999; Tagomori et al., 2002) was not limited to *Vibrio* species but was a common characteristic of all the *Vibrionaceae*, including *Listonella*, *Photobacterium*, and *Salinivibrio* species (Okada et al., 2005). It is not known yet if the *Salinivibrio* species also carry an SI in their genome, but if this is the case, it would be tempting to speculate that acquisition of the SI and the evolution of a two-chromosome genome are somehow linked.

Cassette Functions

Apart from the antibiotic resistance cassettes found in MIs, a precise inventory of the functions encoded by integron gene cassettes remains to be estab-

Figure 2. Phylogenetic relationship of the integron *intI* genes among the proteobacteria. The dendrogram is based on known *intI* gene sequences. The tree was rooted using XerC and XerD from *E. coli* (Eco) and *Thiobacillus denitrificans* ATCC 25259 (Thd). The integrases from the five classes of MI are boxed. Abbreviations for the organism in which the integron integrases are found are as follows: *Azoarcus* sp. EbN1 (Azo), *Dechloromonas aromatica* (Dar), *E. coli* (Eco), *Geobacter metallireducens* (Gme), *Listonella pelagia* (Lpe), *Vibrio anguillarum* (Lan), *Methylobacillus flagellatus* (Mef), *Microbulbifer degradans* (Mid), *Nitrosomonas europaea* (Neu), *P. alcaligenes* (Pal), *P. stutzeri* BAM (PstBAM), *P. stutzeri* Q (PstQ), *Rubrivivax gelatinosus* (Rug), *S. oneidensis* (Son), *Shewanella putrefaciens* (Spu), *T. denitrificans* (Thd), *Treponema denticola* (Tde), *V. cholerae* (Vch), *V. metschnikovii* (Vme), *V. mimicus* (Vmi), *V. parahaemolyticus* (Vpa), *Vibrio splendidus* (Vsp), *V. fischeri* (Vfi), *Xanthomonas campestris* (Xca), *X. oryzae* (Xor), and *Xanthomonas* species (Xsp). The sources of *IntI6*, *IntI7*, and *IntI8* are unknown. Branch lengths were drawn proportional to the amount of evolution based on genetic distances. Accession numbers (when available) can be found in Table 1.

lished, especially for the hundreds of SI gene cassettes. The majority of the SI cassettes examined so far appear to be unique to the host species, and a majority of their encoded genes have no counterparts in the database, or the sole homologues are ORFs of unassigned function. Nevertheless, the reservoir of adaptive functions residing within SIs includes cassettes

with significant homology to known antibiotic resistance genes. Although a known antibiotic resistance gene cassette identified in a clinical isolate has not yet been identified within a SI, several potential progenitor cassettes with significant homology to aminoglycoside, phosphinotricin, fosfomycin, and streptothricin resistance genes are present (Rowe-Magnus et al.,

1999). Recently, two such genes from environmental cassettes were expressed as recombinant proteins and assayed for the appropriate enzyme activity (Nield et al., 2004). One of these, which showed similarity to an aminoglycoside phosphotransferase, displayed an ATPase activity that was consistent with the presence of characteristic Mg^{2+}-binding residues. However, this activity was not enhanced by the aminoglycosides hygromycin B or kanamycin, and its substrate is still unknown. The product of the second gene tested showed sequence motifs of the RNA methyltransferase superfamily, and the recombinant version showed methyltransferase activity with RNA. Furthermore, three genes showing activity against clinically relevant antibiotics have been characterized in the SI of several *V. cholerae* isolates (see below).

Preliminary studies indicated that SI cassettes also encode adaptive functions that extend beyond antibiotic resistance. In *V. cholerae*, two pathogenicity genes, the heat-stable toxin gene (*sto*) and the mannose-fucose-resistant hemagglutinin gene (*mrhA*), as well as a lipoprotein gene, have been found to be cassette encoded (Ogawa and Takeda, 1993; Abbott and Janda, 1994; Barker and Manning, 1997). In *V. vulnificus*, a gene identified through a transposon mutagenesis screen as essential for the expression of its major virulence determinant, the capsular polysaccharide, has also been found to be cassette encoded (Smith and Siebeling, 2003). We have determined the metabolic function of a sulfate-binding protein in a *V. cholerae* SI cassette, a psychrophilic lipase in a *Moritella marina* SI cassette, and a restriction enzyme (XbaI) and its cognate methylase in a *Xanthomonas campestris* pathovar *badrii* SI cassette (Rowe-Magnus et al., 2001). Genes with homology to DNA methylases, immunity proteins, restriction endonucleases, dNTP triphosphohydrolases, lipases, and 8-oxoguanine triphosphatases (MutT), among others, have been found.

The determination of the metabolic activities of several SI cassettes whose activities are not related to antibiotic resistance or virulence confirms that integrons operate as a general gene capture system in bacterial adaptation. If each bacterial species harboring an SI has its own cassette pool, the resource in terms of gene cassette availability will be immense. It is noteworthy that the proportion of proteins predicted to contain transmembrane segments, which is suggestive of a location at the cell periphery and interfacing with the environment, is higher in the cassette-encoded proteins than in the proteins encoded by the rest of the genome (Vaisvila et al., 2001; Rowe-Magnus et al., 2003). The crystal structure of one of these gene cassette products, Bal32a, was solved to 1.8Å resolution and was found to be a new member of the

adaptable $\alpha + \beta$ barrel family of transport proteins and enzymes (Robinson et al., 2005). These studies demonstrate that the analysis of integrons can lead to the discovery of novel proteins, and the functions of the encoded genes have fantastic potential from both genetic and biotechnological standpoints.

Interspecies Cassette Content Variations

A comparison of the gene cassette contents between different *Vibrio* species indicated that the large majority of the cassettes were unique to their host species (Rowe-Magnus et al., 2003). For example, comparison of the *V. cholerae* N16961 and *V. parahaemolyticus* RIMD2210633 SI contents revealed that only three cassettes were highly related between them (Makino et al., 2003), whereas the *V. cholerae* and *V. vulnificus* YJ016 SI had no cassettes in common (Chen et al., 2003). Also, we made a comparison of the *V. fischeri* and *P. profundum* cassette contents and found that they do not share any cassettes (unpublished data). These observations clearly indicate that the process of cassette genesis is constant and efficient in these species. Furthermore, even if many of the cassette-encoded genes have no counterparts in the database, those that have homologues of bacterial origin do not show any bias for a specific group, which suggests that both gram-positive and gram-negative bacterial species are sources for the recruitment of gene cassettes (Rowe-Magnus et al., 2003).

Intraspecies Cassette Content Variations

The activity of integron cassettes offers a fast track to bacterial innovation. The size of SIs and the ancient and dynamic nature of the system are a reminder that the cassettes that currently occupy the SI represent only a fraction of those that may have participated in the evolution of the host, because the cassettes will presumably be subject to episodic selection. The >165 different O serotypes of *V. cholerae* are represented by species of ecological, geographical, and temporal diversity. Thus, comparison of SI organization from recent and earlier isolates as well as between recent isolates from different geographical locations and ecological niches may yield valuable information. Clark and colleagues (2000) examined the global SI organization of 65 different *V. cholerae* O serotypes by PCR and Southern hybridization. Extensive restriction polymorphism was observed even among closely related isolates, suggesting an appreciable plasticity for these structures and their microevolution through integrase-mediated gene acquisition and gene loss as well as via cassette rearrangement events. A recent microarray-based comparative genomic analysis of different *V. cholerae* isolates that included non-O1/non-O139 strains showed that, whereas the O1 and O139 isolates shared >95% of the integron cassettes found in the SI of N16961, this dropped to only 50 to 75% when the non-O1/non-O139 isolates were compared with N16961 (Dziejman et al., 2005). The SI organization of two *V. cholerae* strains suggests that cassettes can be mobilized in clusters; the cassettes in positions 1 to 4 of the SI of strain 569B are found in the same order in the SI of strain N16961, but they occupy positions 79 to 82 (a displacement of more than 40 kb). This has also been seen for other clusters of cassettes (A. M. Guerout and D. Mazel, unpublished results). However, it is not known if these differences represent a true group mobilization event or simply temporal differences in cassette acquisition. In addition, there is evidence that not all of the SI cassette *attC* sites are equally functional; some are known to contain mutations or deletions within the core site that could render them nonfunctional (D. S. Rowe-Magnus, A. M. Guerout, and D. Mazel, unpublished results). Therefore, their movement would have to be coordinated with those of other cassettes. Collis and Hall (1992) demonstrated that integron gene cassettes are excised as covalently closed circles and observed differences in the resulting recombination products. Some cassettes could be mobilized as individual units, whereas others were only excised in tandem with another cassette. Whether such cassette hitchhiking is by design to ensure simultaneous transmission of genes or is just a matter of happenstance is not known.

The cassette content of SI appears to vary extensively among isolates from the same species for other vibrios as well. Comparison of the IntIAs from both the *V. fischeri* type strain CIP 103206 and the sequenced ES114 strain showed that they share 98% amino acid identity. However, a comparison of the *V. fischeri* ES114 SI cassette contents with the 19 unique cassettes identified through the partial characterization of the SI of the CIP 103206 type strain showed that these two strains only share 2 of the 19 cassettes. This suggests that the cassette content of the SI between these *V. fischeri* strains is extremely variable (Rowe-Magnus et al., 2003).

The SIs found in the two different isolates of *V. vulnificus*, which have been sequenced, show a similar number of cassettes: 188 in the SI of strain YJ016 and 211 in the SI of strain CMCP6 (Chen et al., 2003). Interestingly, while the two IntIAs only differ by a single amino acid substitution, Chen and collaborators noticed that the two SIs only shared about 20% of their cassettes, with only six of them being identical. Furthermore, our analysis (Color Plate 5) revealed that the cassettes nearest the inte-

grase genes, particularly those in CMPC6 (with a single exception), were not found in YJ016. This arrangement suggested that these cassettes might correspond to more recent cassette acquisition or genesis events, and as such are unique to the different isolates.

Are *Vibrio* SIs at the Origin of MI and Their Resistance Cassettes?

MIs and SIs share an identical structural organization, and the antiquity of SIs suggests that they are the likely ancestors of MI. These relationships led to the proposal that MI evolved from SI through the entrapment of *intI* genes and their cognate *attI* sites by mobile structures, such as easily and randomly assembled compound transposons. Furthermore, we noticed that 12 different resistance cassettes carried an *attC* site almost identical to those specifically found in, and characteristic of, *Xanthomonas* and *Vibrio* SIs (Rowe-Magnus et al., 2001). Once they were mobile, one can imagine that the subsequent harvesting of cassettes from various SI sources then led to the establishment of contemporary MIs, as reflected by the great diversity of the *attC* sites associated with the gene cassettes of MIs.

The hypothetical evolution of MI from the SI has received support from two recent observations. The first came from the discovery made by Sørum of a plasmid associated with trimethoprim resistance in *V. salmonicida*. Sørum and collaborators found that trimethoprim resistance was due to a *dfrA1* cassette carried in a novel integron class, according to its *intI* gene (unpublished; GenBank no. AJ277063). The *dfrA1* cassette, which is commonly found in both class 1 and class 2 MIs, was in the second position of eight cassettes. This new integron has two features that corroborate our model. First, its integrase gene clearly originated from a *Vibrio* species or closely related genus, as attested to by its 74% similarity to the *V. cholerae* and *V. mimicus* IntIA proteins. Second, the *attC* sites of three of the seven other cassettes are structurally homogeneous, a characteristic until now only found in the SI cassette arrays. As this integron is surrounded by insertion sequences, it is very tempting to imagine that it corresponds to an intermediate in the evolution of an MI of the type found in clinical isolates. The creation of an MI that imparts a multiresistant phenotype could be achieved by the harvesting of other resistance cassettes using the mobility of the assembled compound transposon and multiple transfers.

The second piece of supporting evidence came from our demonstration that an MI could directly recruit the cassettes of an SI and confer a resistance phenotype. The recruitment of SI gene cassettes was shown to be random (Rowe-Magnus et al., 2002). By then applying selective pressure for the development of antibiotic resistance, a novel chloramphenicol acetyltransferase gene cassette, *catB9*, was discovered in the *V. cholerae* SI (Rowe-Magnus et al., 2002). Along with previous results demonstrating that SI cassettes are substrates for the integrase of class 1 integrons when present on a high-copy-number plasmid (Mazel et al., 1998), these results clearly demonstrated that environmental conditions, such as antibiotic selection, dictate which of the randomly recruited cassettes are retained within the MI of clinical isolates.

Two other resistance cassettes, coding for the novel carbenicillinases CARB-7 and CARB-9, have been identified in the SI of two non-O1, non-O139 *V. cholerae* environmental isolates (Melano et al., 2002; Petroni et al., 2004). These cassettes, like the *catB9* cassette, are associated with *attC* sites that are canonical VCRs. Interestingly, they confer different levels of β-lactam resistance to their host, with the MIC of CARB-9 being one-fourth to one-eighth that of CARB-7, depending on the β-lactam used (Petroni et al., 2004). These carbenicillinases belong to a subgroup of the class A β-lactamase family, the RSG carbenicillinases (Lim et al., 2001), which are all cassette encoded and associated with either a VCR-like *attC* site 123 bp in length or a 103-bp-long *attC* site. With the exception of CARB-7 and -9, these RSG carbenicillinase cassettes have been identified in class 1 integrons from a variety of bacterial hosts (*Pseudomonas aeruginosa, Salmonella enterica* serovar Typhimurium, *Proteus mirabilis,* and *V. cholerae*). Noticeably, phylogenetic analysis showed that the most ancestral cassettes were those associated with a VCR-like *attC* site (Petroni et al., 2004). It is thus very tempting to speculate that the ancestral RSG carbenicillinase originated in *V. cholerae*. This would corroborate our hypothesis that the cassettes found in contemporary MIs were recruited from various SI sources.

It is very likely that the assembly of complex MIs (i.e., MIs carrying more than two resistance cassettes) occurred through recombination between different MIs, rather than through successive direct recruitment events from the SI cassette arrays of different environmental species. Niches exist which could favor the exchange of cassettes between MIs on a large scale. The recent characterization of a remarkable collection of cassettes, integrons, and plasmids circulating in a single wastewater treatment plant (Tennstedt et al., 2003), as well as the recent and unpredicted demonstration that aerobic gram-positive *Corynebacterium* spp. can be a reservoir of class 1 MI in poultry litter (Nandi et al., 2004), clearly indicates that environments other than the obvious clinical settings exist in which cassette exchange between MIs may be fairly common.

EVIDENCE FOR A NOVEL RECOMBINATION PATHWAY IN INTEGRONS

As discussed previously, the integron platform consists of an integrase gene and an associated primary recombination target called the *attI* site. The IntI integrases belong to the λ integrase family of tyrosine (Y) recombinases (for a review, see Azaro and Landy, 2002). The IntI integrases mediate recombination involving two types of sites: their specific *attI* site and the cassette-associated *attC* site. Contrary to the integrases of the same family that have been studied (λ, Cre, FLP, XerC/XerD, FimB/FimE), those from integrons are able to recombine distantly related DNA sequences. This last point is reflected in the large structural disparity that is observed between the *attC* sites from different cassettes.

The recombination activity of the integrases from the class 1, 2, and 3 MIs, and of several SIs, has been studied to different extents. The paradigm for integron recombination is undoubtedly the class 1 integrase, IntI1. Indeed, since its initial characterization in 1990 (Martinez and de la Cruz, 1990), IntI1-catalyzed recombination has been extensively studied. The ability of IntI1 to recombine distantly related *attC* sites has been demonstrated in different studies, and five types of recombination reactions have been established. Three types correspond to the different recombinations possible between the normal sites (*attI* × *attC*, *attC* × *attC*, and *attI* × *attI*) and the other two correspond to inadvertent recombination between either an *attI* or an *attC* site and what has been called a secondary site (Francia et al., 1993, 1997; Recchia et al., 1994; Recchia and Hall, 1995; Francia and Garcia Lobo, 1996; Hansson et al., 1997). The recombination events involving two *attI* sites are significantly less efficient than the reactions involving two *attC* sites, and the *attI* × *attC* reaction is preferred over all others (Collis et al., 2001). Although the integron integrases are able to recombine structurally different *attC* sites, they are specific for recombination with their own *attI* site. However, IntI1 and IntINeu have been shown to be able to delete cassettes in first position of heterologous *attI* sites, albeit at low frequencies (Hansson et al., 2002; Leon and Roy, 2003).

The cassette deletion and integration activities of IntISon, from *S. oneidensis*, and IntINeu, from *Nitrosomonas europaea*, which are associated with a small number of cassettes or no cassettes at all, have been demonstrated but not quantified (Drouin et al., 2002; Leon and Roy, 2003). In addition, the integrative recombination (*attI* × *attC* sites) catalyzed by the *P. stutzeri* SI integrase (IntIPstQ) has been demonstrated (Holmes et al., 2003b), while the deletion activity through $attC_{aadB}$ × VCR recombination was previously established for the *V. cholerae* SI integrase, VchIntIA (formerly IntI4) (Rowe-Magnus et al., 2001). With the exception of VchIntIA, all of these integrases were able to catalyze cointegrate formation through recombination of their cognate *attI* sites and at least one resistance cassette *attC* site.

As mentioned above, within each SI the cassette *attC* sites are extremely homogenous and species-specific, whereas the *attC* sites of MI cassettes are highly variable in length and sequence. A hypothesis to explain these structural differences is that recombination activities of the MI and SI integrases differ. Indeed, one could imagine that the narrow range of structural heterogeneity shown by the SI cassette *attC* sites directly reflects the recognition spectrum of the SI integrase. This question has been addressed through a comparative study of the recombination activities of IntI1 and VchIntIA, which are the integrases of the class 1 MI and the *V. cholerae* SI, respectively (Biskri et al., 2005). This study showed that, although the structural range of the *attC* sites recombined by VchIntIA was narrower than that for IntI1, recombination was not limited to *attC* sites that were related to the VCR. Other structural families of *attC* sites, such as the short *attC* site of the *aadA7* cassette, were also substrates for recombination by VchIntIA. This suggested that the nearly exclusive presence of cassettes carrying a VCR in the *V. cholerae* SI was not due to the restricted spectrum of IntIA site recognition (Biskri et al., 2005). Another puzzling observation was that, in *Escherichia coli*, an *attI1* × VCR recombination catalyzed by IntI1 was far more efficient than an *attI*Vch × VCR recombination catalyzed by VchIntIA. Cointegrate formation, using the same substrates and both integrases, was also tested in *V. cholerae*. In this organism, the *attI*Vch × VCR recombination by VchIntIA increased 2,000-fold compared to what was measured in *E. coli*, while an *attI1* × VCR recombination by IntI1 was identical to that measured in *E. coli*. These results were unexpected, as the VCR and the *attI*Vch recombination sites are the natural substrates of VchIntIA. Taken together, these results indicate that the substrate recognition and recombination reactions catalyzed by VchIntIA might differ from the class 1 MI paradigm.

Until now, all integron integrases tested were able to catalyze recombination between various *attC* sites and their own *attI* site in *E. coli* (Martinez and de la Cruz, 1990; Hall et al., 1999; Collis et al., 2002; Drouin et al., 2002; Hansson et al., 2002; Holmes et al., 2003b; Leon and Roy, 2003); however, in most cases the efficiency of the recombination was not determined. It was assumed that recombination by these IntI integrases did not require any accessory protein

for site-specific recombination. Furthermore, no known recognition sites for the DNA-binding protein required in other tyrosine recombinase recombination systems, such as IHF (integration host factor) FIS (factor for inversion stimulation) (Landy, 1989; Azam and Ishihama, 1999), have been identified adjacent to any IntI recombination site. However, since an *in vitro* recombination assay has yet to be successfully developed, this presumption is still questionable. Therefore, the simplest explanation for the observed low level of integration in a reconstituted system in *E. coli* is that VchIntIA requires, at least for the integration process, an accessory protein which is either absent or too divergent in *E. coli* to sustain the VchIntIA-mediated integration process. In the case of IntI1, if such an accessory protein is required for IntI1 integration at its *attI1* site, the evolutionary constraints exerted on MI systems, which are carried on mobile elements and then selected to be operational in multiple hosts, may explain why it is able to recombine at the same rates in *E. coli* and *V. cholerae*.

It is also possible that a lower intracellular concentration of VchIntIA in *E. coli*, due to either a lower synthesis or a lower stability of the protein, would not be restrictive for intramolecular deletion reactions but would become limiting for intermolecular recombination reactions. However, VchIntIA-catalyzed deletion of a cassette in the first position ($attI$Vch-$lacI^q$-VCR), which is also an intramolecular deletion, was not detected, suggesting that the $attI$Vch recombination complex is, for reasons unknown, not functional in *E. coli* (Biskri et al., 2005).

Another interesting discovery is linked to the nature of the recombination substrates. A unique trait of the integron recombination system resides in the structure of the recombination sites. Typical Y-recombinase core recombination sites consist of a pair of highly conserved 9- to 13-bp inverted binding sites separated by a 6- to 8-bp central region. The *attI* sites differ from this canonical organization, as one of the putative binding sites within the core site is always extremely degenerate, and the central region differs greatly between the different *attI* sites. However, in vitro experiments have shown binding of IntI to four regions of the double-stranded *attI* site, two of which correspond to the core site. The other two form direct repeats located 5′ to the core site (see Fig. 1) (Collis et al., 1998; Gravel et al., 1998). The structure of the *attC* site is more complex. It consists of two potential core sites, R″-L″ and L′-R′ (called 1L-2L and 2R-1R, respectively, by Stokes et al. [1997]) (Fig. 1), separated by a central region. Only L′-R′ is recombinogenic, and if the central region can be highly variable in sequence and size, all structurally different *attC* sites can potentially form cruciform structures

(Hall et al., 1991; Stokes et al., 1997; Rowe-Magnus et al., 2003) and are efficiently recombined by IntI1. In 1999, Francia and collaborators (1999) demonstrated that purified IntI1, the integrase of class 1 integrons, bound specifically to the bottom strand (bs) of single-stranded $attC_{aadA1}$ DNA but not to a double-stranded (ds) $attC_{aadA1}$ site. Sundstrom and colleagues (Johansson et al., 2004) confirmed this seminal observation and identified several key elements in the $attC_{aadA1}$ sequence that act as recognition determinants for IntI1 binding and play important roles in the potential secondary structure of the $attC$ site.

In most circularized cassettes, self-pairing on the same single strand (ss) can be structurally extended up to the R' and R" sequences, which usually show a stretch of 9 to 11 consecutive complementary nucleotides (Hansson et al., 1997; Rowe-Magnus et al., 2003). Such a self-paired stem could be seen as an almost canonical core site consisting of the L"-L' duplex, an unpaired central region followed by an R"-R' duplex. Recognition and recombination by the IntI integrase of such a structure with a canonical ds-*attI* site would lead to a Holliday junction intermediate that may be resolved by a replication step. This hypothetical model has received recent support from the in vivo demonstration that only one $attC$ site strand was recombined by IntI1. Indeed, using an in vivo recombination assay based on two different conjugation systems as a mechanism to independently deliver either the top strand (ts) or bs of the different integron recombination sites in ss form into a recipient strain, Bouvier et al. found that recombination rates of the different substrates varied over a wide range (Bouvier et al., 2005). They determined that the rate of recombination following transfer of the $attC$-bs with a ds-*attI* site carried on a plasmid in the recipient was 1,000-fold higher than the rate after conjugation of the $attC$-ts. Furthermore, the recombination rate measured after delivery of the $attC$-bs was found to be identical to that obtained in a classical assay using ds-*attC* and ds-*attI* sites carried on plasmids and comaintained in bacterial cells expressing IntI1. Conversely, only the $attC$ sites appeared to recombine under ss form, since recombination following conjugative transfer of either the ts or bs strands of the *attI* site was found to occur at a rate 1,000-fold lower than when measured in a classical assay using ds-*attC* and ds-*attI* sites carried on plasmids and comaintained in bacterial cells expressing IntI1. Altogether, these results strongly support a recombination model for the insertion of integron cassettes at the ds-*attI* site that involves only the $attC$-bs strand folded in a stem-and-loop structure and a resolution of the generated Holliday junction intermediate through replication.

The selective advantages that have led to the development of these ss recombination sites and processes are still elusive. However, this might be linked to the phenomenon of gene dissemination by horizontal transfer, which in many cases goes through an ss stage, as demonstrated for conjugation and for natural transformation in bacteria.

CONCLUSIONS

Recruitment of exogenous genes is the most rapid adaptation against antimicrobial compounds, and the integron functional platform provides a gene capture system perfectly suited to face the challenges of multiple antibiotic treatment regimens. With the discovery of SI and of the thousands of cassettes entrapped in the integrons of environmental species, especially in the *Vibrionaceae*, the role of integrons in evolution clearly extends beyond the phenomenon of antibiotic resistance. We now have a better view of the immense resource of this system. The integron gene cassettes for which an activity has been experimentally demonstrated, be they from SI arrays or from soil DNA, encode proteins related to simple enzymatic functions; their recruitment is seen as providing the bacterial host with an adaptive advantage. Both experimental and phylogenetic data suggest that SIs are the source of the MI and resistance gene cassettes observed within clinical isolates. Nevertheless, several questions of importance about this system are still without satisfactory answers. In particular, there is a need to elucidate the specifics of the recombination process and of the process of cassette genesis, as well as to study the dynamics of cassette exchange in complex bacterial populations.

Acknowledgments. Work in the Mazel laboratory was supported by the Institut Pasteur, the CNRS, and the Programme de Recherche en Microbiologie from the MENESR and the IFREMER. Work in the Rowe-Magnus laboratory was supported by the Canadian Institutes of Health Research (CIHR) and the Sunnybrook Trust.

REFERENCES

Abbott S. L., and J. M. Janda. 1994. Severe gastroenteritis associated with *Vibrio hollisae* infection: report of two cases and review. *Clin. Infect. Dis.* **18:**310–312.

Ahmed, A. M., T. Nakagawa, E. Arakawa, T. Ramamurthy, S. Shinoda, and T. Shimamoto. 2004. New aminoglycoside acetyltransferase gene, aac(3)-Id, in a class 1 integron from a multiresistant strain of *Vibrio fluvialis* isolated from an infant aged 6 months. *J. Antimicrob. Chemother.* **53:**947–951.

Amita, S. R. Chowdhury, M. Thungapathra, T. Ramamurthy, G. B. Nair, and A. Ghosh. 2003. Class I integrons and SXT elements in El Tor strains isolated before and after 1992 *Vibrio cholerae* O139 outbreak, Calcutta, India. *Emerg. Infect. Dis.* **9:**500–502.

Arakawa, Y., M. Murakami, K. Suzuki, H. Ito, R. Wacharotayankun, S. Ohsuka, N. Kato, and M. Ohta. 1995. A novel integron-like element carrying the metallo-beta-lactamase gene *blaIMP*. *Antimicrob. Agents Chemother.* **39:**1612–1615.

Azam, T. A., and A. Ishihama. 1999. Twelve species of the nucleoid-associated protein from *Escherichia coli*. Sequence recognition specificity and DNA binding affinity. *J. Biol. Chem.* **274:** 33105–33113.

Azaro, M. A., and A. Landy. 2002. λ integrase and the λ Int family, p. 118–148. *In* N. L. Craig, R. Craigie, M. Gellert, and A. M. Lambowitz (ed.), *Mobile DNA II.* ASM Press, Washington, D.C.

Barker, A., C. A. Clark, and P. A. Manning. 1994. Identification of VCR, a repeated sequence associated with a locus encoding a hemagglutinin in *Vibrio cholerae* O1. *J. Bacteriol.* **176:**5450–5458.

Barker, A., and P. A. Manning. 1997. VlpA of *Vibrio cholerae* O1: the first bacterial member of the alpha 2-microglobulin lipocalin superfamily. *Microbiology* **143:**1805–1813.

Beaber, J. W., V. Burrus, B. Hochhut, and M. K. Waldor. 2002a. Comparison of SXT and R391, two conjugative integrating elements: definition of a genetic backbone for the mobilization of resistance determinants. *Cell. Mol. Life Sci.* **59:**2065–2070.

Beaber, J. W., B. Hochhut, and M. K. Waldor. 2002b. Genomic and functional analyses of SXT, an integrating antibiotic resistance gene transfer element derived from *Vibrio cholerae*. *J. Bacteriol.* **184:**4259–4269.

Bhattacharya, S. K. 2003. An evaluation of current cholera treatment. *Expert Opin. Pharmacother.* **4:**141–146.

Biskri, L., M. Bouvier, A. M. Guerout, S. Boisnard, and D. Mazel. 2005. Comparative study of class 1 integron and *Vibrio cholerae* superintegron integrase activities. *J. Bacteriol.* **187:**1740–1750.

Boltner, D., C. MacMahon, J. T. Pembroke, P. Strike, and A. M. Osborn. 2002. R391: a conjugative integrating mosaic comprised of phage, plasmid, and transposon elements. *J. Bacteriol.* **184:**5158–5169.

Bouvier, M., G. Demarre, and D. Mazel. 2005. Integron cassette insertion: a recombination process involving a folded single strand substrate. *EMBO J.* **24:**4356–4367.

Burrus, V., G. Pavlovic, B. Decaris, and G. Guedon. 2002a. Conjugative transposons: the tip of the iceberg. *Mol. Microbiol.* **46:** 601–610.

Burrus, V., G. Pavlovic, B. Decaris, and G. Guedon. 2002. The ICESt1 element of *Streptococcus thermophilus* belongs to a large family of integrative and conjugative elements that exchange modules and change their specificity of integration. *Plasmid* **48:**77–97.

Chen, C. Y., K. M. Wu, Y. C. Chang, C. H. Chang, H. C. Tsai, T. L. Liao, Y. M. Liu, H. J. Chen, A. B. Shen, J. C. Li, T. L. Su, C. P. Shao, C. T. Lee, L. I. Hor, and S. F. Tsai. 2003. Comparative genome analysis of *Vibrio vulnificus*, a marine pathogen. *Genome Res.* **13:**2577–2587.

Chiang, S. R., and Y. C. Chuang. 2003. *Vibrio vulnificus* infection: clinical manifestations, pathogenesis, and antimicrobial therapy. *J. Microbiol. Immun. Infect.* **36:**81–88.

Clark, C. A., L. Purins, P. Kaewrakon, T. Focareta, and P. A. Manning. 2000. The *Vibrio cholerae* O1 chromosomal integron. *Microbiology* **146:**2605–2612.

Coleman, N., S. Tetu, N. Wilson, and A. Holmes. 2004. An unusual integron in *Treponema denticola*. *Microbiology* **150:**3524–3526.

Collis, C. M., and R. M. Hall. 1992. Gene cassettes from the insert region of integrons are excised as covalently closed circles. *Mol. Microbiol.* **16:**2875–2885.

Collis, C. M., M. J. Kim, S. R. Partridge, H. W. Stokes, and R. M. Hall. 2002. Characterization of the class 3 integron and the site-specific recombination system it determines. *J. Bacteriol.* **184:** 3017–3026.

Collis, C. M., M. J. Kim, H. W. Stokes, and R. M. Hall. 1998. Binding of the purified integron DNA integrase IntI1 to integron-

and cassette-associated recombination sites. *Mol. Microbiol.* **29:** 477–490.

Collis, C. M., G. D. Recchia, M. J. Kim, H. W. Stokes, and R. M. Hall. 2001. Efficiency of recombination reactions catalyzed by class 1 integron integrase IntI1. *J. Bacteriol.* **183:** 2535–2542.

Colmer, J. A., J. A. Fralick, and A. N. Hamood. 1998. Isolation and characterization of a putative multidrug resistance pump from *Vibrio cholerae*. *Mol. Microbiol.* **27:**63–72.

Correia, M., F. Boavida, F. Grosso, M. J. Salgado, L. M. Lito, J. M. Cristino, S. Mendo, and A. Duarte. 2003. Molecular characterization of a new class 3 integron in *Klebsiella pneumoniae*. *Antimicrob. Agents Chemother.* **47:**2838–2843.

Dalsgaard, A., A. Forslund, A. Petersen, D. J. Brown, F. Dias, S. Monteiro, K. Molbak, P. Aaby, A. Rodrigues, and A. Sandstrom. 2000. Class 1 integron-borne, multiple-antibiotic resistance encoded by a 150-kilobase conjugative plasmid in epidemic *Vibrio cholerae* O1 strains isolated in guinea-bissau. *J. Clin. Microbiol.* **38:**3774–3779.

Dalsgaard, A., A. Forslund, D. Sandvang, L. Arntzen, and K. Keddy. 2001. *Vibrio cholerae* O1 outbreak isolates in Mozambique and South Africa in 1998 are multiple-drug resistant, contain the SXT element and the aadA2 gene located on class 1 integrons. *J. Antimicrob. Chemother.* **48:**827–838.

Drouin, F., J. Melancon, and P. H. Roy. 2002. The IntI-like tyrosine recombinase of *Shewanella oneidensis* is active as an integron integrase. *J. Bacteriol.* **184:**1811–1815.

Dziejman, M., D. Serruto, V. C. Tam, D. Sturtevant, P. Diraphat, S. M. Faruque, M. H. Rahman, J. F. Heidelberg, J. Decker, L. Li, K. T. Montgomery, G. Grills, R. Kucherlapati, and J. J. Mekalanos. 2005. Genomic characterization of non-O1, non-O139 *Vibrio cholerae* reveals genes for a type III secretion system. *Proc. Natl. Acad. Sci. USA* **102:**3465–3470.

Ehara, M., B. M. Nguyen, D. T. Nguyen, C. Toma, N. Higa, and M. Iwanaga. 2004. Drug susceptibility and its genetic basis in epidemic *Vibrio cholerae* O1 in Vietnam. *Epidemiol. Infect.* **132:** 595–600.

Falbo, V., A. Carattoli, F. Tosini, C. Pezzella, A. M. Dionisi, and I. Luzzi. 1999. Antibiotic resistance conferred by a conjugative plasmid and a class I integron in *Vibrio cholerae* O1 El Tor strains isolated in Albania and Italy. *Antimicrob. Agents Chemother.* **43:** 693–696.

Faruque, S. M., and J. J. Mekalanos. 2003. Pathogenicity islands and phages in *Vibrio cholerae* evolution. *Trends Microbiol.* **11:** 505–510.

Fluit, A. C., and F. J. Schmitz. 2004. Resistance integrons and super-integrons. *Clin. Microbiol. Infect.* **10:**272–288.

Francia, M. V., P. Avila, F. de la Cruz, and J. M. Garcia Lobo. 1997. A hot spot in plasmid F for site-specific recombination mediated by Tn21 integron integrase. *J. Bacteriol.* **179:**4419–4425.

Francia, M. V., F. de la Cruz, and J. M. Garcia Lobo. 1993. Secondary-sites for integration mediated by the Tn21 integrase. *Mol. Microbiol.* **10:**823–828.

Francia, M. V., and J. M. Garcia Lobo. 1996. Gene integration in the *Escherichia coli* chromosome mediated by Tn21 integrase (Int21). *J. Bacteriol.* **178:**894–898.

Francia, M. V., J. C. Zabala, F. de la Cruz, and J. M. Garcia-Lobo. 1999. The IntI1 integron integrase preferentially binds single-stranded DNA of the *attC* site. *J. Bacteriol.* **181:**6844–6849.

Gravel, A., B. Fournier, and P. H. Roy. 1998. DNA complexes obtained with the integron integrase IntI1 at the attI1 site. *Nucleic Acids Res.* **26:**4347–4355.

Hacker, J., M. Ott, G. Blum, R. Marre, J. Heesemann, H. Tschape, and W. Goebel. 1992. Genetics of *Escherichia coli* uropathogenicity: analysis of the O6:K15:H31 isolate 536. *Zentbl. Bakteriol.* **276:**165–175.

Hall, R. M. 1997. Mobile gene cassettes and integrons: moving antibiotic resistance genes in gram-negative bacteria. *Ciba Found. Symp.* **207**:192–202.

Hall, R. M., D. E. Brookes, and H. W. Stokes. 1991. Site-specific insertion of genes into integrons: role of the 59-base element and determination of the recombination cross-over point. *Mol. Microbiol.* **5**:1941–1959.

Hall, R. M., C. M. Collis, M. J. Kim, S. R. Partridge, G. D. Recchia, and H. W. Stokes. 1999. Mobile gene cassettes and integrons in evolution. *Ann. N. Y. Acad. Sci.* **870**:68–80.

Hall, R. M., and H. W. Stokes. 2004. Integrons or super integrons? *Microbiology* **150**:3–4.

Hansson, K., O. Skold, and L. Sundstrom. 1997. Non-palindromic attI sites of integrons are capable of site-specific recombination with one another and with secondary targets. *Mol. Microbiol.* **26**:441–453.

Hansson, K., L. Sundstrom, A. Pelletier, and P. H. Roy. 2002. IntI2 integron integrase in Tn7. *J. Bacteriol.* **184**:1712–1721.

Heidelberg, J. F., J. A. Eisen, W. C. Nelson, R. A. Clayton, M. L. Gwinn, R. J. Dodson, D. H. Haft, E. K. Hickey, J. D. Peterson, L. Umayam, S. R. Gill, K. E. Nelson, T. D. Read, H. Tettelin, D. Richardson, M. D. Ermolaeva, J. Vamathevan, S. Bass, H. Qin, I. Dragoi, P. Sellers, L. McDonald, T. Utterback, R. D. Fleishmann, W. C. Nierman, O. White, S. L. Salzberg, H. O. Smith, R. R. Colwell, J. J. Mekalanos, J. C. Venter, and C. M. Fraser. 2000. DNA sequence of both chromosomes of the cholera pathogen *Vibrio cholerae*. *Nature* **406**:477–483.

Hochhut, B., J. W. Beaber, R. Woodgate, and M. K. Waldor. 2001a. Formation of chromosomal tandem arrays of the SXT element and R391, two conjugative chromosomally integrating elements that share an attachment site. *J. Bacteriol.* **183**:1124–1132.

Hochhut, B., Y. Lotfi, D. Mazel, S. M. Faruque, R. Woodgate, and M. K. Waldor. 2001b. Molecular analysis of antibiotic resistance gene clusters in *Vibrio cholerae* O139 and O1 SXT constins. *Antimicrob. Agents Chemother.* **45**:2991–3000.

Hochhut, B., and M. K. Waldor. 1999. Site-specific integration of the conjugal *Vibrio cholerae* SXT element into prfC. *Mol. Microbiol.* **32**:99–110.

Holmes, A. J., M. R. Gillings, B. S. Nield, B. C. Mabbutt, K. M. Nevalainen, and H. W. Stokes. 2003a. The gene cassette metagenome is a basic resource for bacterial genome evolution. *Environ. Microbiol.* **5**:383–394.

Holmes, A. J., M. P. Holley, A. Mahon, B. Nield, M. Gillings, and H. W. Stokes. 2003b. Recombination activity of a distinctive integron-gene cassette system associated with *Pseudomonas stutzeri* populations in soil. *J. Bacteriol.* **185**:918–928.

Huda, M. N., J. Chen, Y. Morita, T. Kuroda, T. Mizushima, and T. Tsuchiya. 2003. Gene cloning and characterization of VcrM, a Na+-coupled multidrug efflux pump, from *Vibrio cholerae* non-O1. *Microbiol. Immunol.* **47**:419–427.

Huda, M. N., Y. Morita, T. Kuroda, T. Mizushima, and T. Tsuchiya. 2001. Na+-driven multidrug efflux pump VcmA from *Vibrio cholerae* non-O1, a non-halophilic bacterium. *FEMS Microbiol. Lett.* **203**:235–239.

Iwanaga, M., C. Toma, T. Miyazato, S. Insisiengmay, N. Nakasone, and M. Ehara. 2004. Antibiotic resistance conferred by a class I integron and SXT constin in *Vibrio cholerae* O1 strains isolated in Laos. *Antimicrob. Agents Chemother.* **48**:2364–2369.

Janda, J. M., C. Powers, R. G. Bryant, and S. L. Abbott. 1988. Current perspectives on the epidemiology and pathogenesis of clinically significant *Vibrio* spp. *Clin. Microbiol. Rev.* **1**:245–267.

Johansson, C., M. Kamali-Moghaddam, and L. Sundstrom. 2004. Integron integrase binds to bulged hairpin DNA. *Nucleic Acids Res.* **32**:4033–4043.

Landy, A. 1989. Dynamic, structural, and regulatory aspects of lambda site-specific recombination. *Annu. Rev. Biochem.* **58**:913–949.

Leon, G., and P. H. Roy. 2003. Excision and integration of cassettes by an integron integrase of *Nitrosomonas europaea*. *J. Bacteriol.* **185**:2036–2041.

Liebert, C. A., R. M. Hall, and A. O. Summers. 1999. Transposon Tn21, flagship of the floating genome. *Microbiol. Mol. Biol. Rev.* **63**:507–522.

Lim, D., F. Sanschagrin, L. Passmore, L. De Castro, R. C. Levesque, and N. C. Strynadka. 2001. Insights into the molecular basis for the carbenicillinase activity of PSE-4 beta-lactamase from crystallographic and kinetic studies. *Biochemistry* **40**:395–402.

Lima, A. A. 2001. Tropical diarrhoea: new developments in traveller's diarrhoea. *Curr. Opin. Infect. Dis.* **14**:547–552.

Makino, K., K. Oshima, K. Kurokawa, K. Yokoyama, T. Uda, K. Tagomori, Y. Iijima, M. Najima, M. Nakano, A. Yamashita, Y. Kubota, S. Kimura, T. Yasunaga, T. Honda, H. Shinagawa, M. Hattori, and T. Iida. 2003. Genome sequence of *Vibrio parahaemolyticus*: a pathogenic mechanism distinct from that of *V. cholerae*. *Lancet* **361**:743–749.

Martin, C., J. Timm, J. Rauzier, R. Gomez-Lus, J. Davies, and B. Gicquel. 1990. Transposition of an antibiotic resistance element in mycobacteria. *Nature* **345**:739–743.

Martinez, E., and F. de la Cruz. 1990. Genetic elements involved in Tn21 site-specific integration, a novel mechanism for the dissemination of antibiotic resistance genes. *EMBO J.* **9**:1275–1281.

Mazel, D., B. Dychinco, V. A. Webb, and J. Davies. 1998. A distinctive class of integron in the *Vibrio cholerae* genome. *Science* **280**:605–608.

Melano, R., A. Petroni, A. Garutti, H. A. Saka, L. Mange, F. Pasteran, M. Rapoport, A. Rossi, and M. Galas. 2002. New carbenicillin-hydrolyzing beta-lactamase (CARB-7) from *Vibrio cholerae* non-O1, non-O139 strains encoded by the VCR region of the *V. cholerae* genome. *Antimicrob. Agents Chemother.* **46**:2162–2168.

Messier, N., and P. H. Roy. 2001. Integron integrases possess a unique additional domain necessary for activity. *J. Bacteriol.* **183**:6699–6706.

Mitsuhashi, S., K. Harada, H. Hashimoto, and R. Egawa. 1961. On the drug-resistance of enteric bacteria. *Jpn. J. Exp. Med.* **31**:47–52.

Morita, Y., K. Kodama, S. Shiota, T. Mine, A. Kataoka, T. Mizushima, and T. Tsuchiya. 1998. NorM, a putative multidrug efflux protein, of *Vibrio parahaemolyticus* and its homolog in *Escherichia coli*. *Antimicrob. Agents Chemother.* **42**:1778–1782.

Murphy, D. B., and J. T. Pembroke. 1995. Transfer of the IncJ plasmid R391 to recombination deficient *Escherichia coli* K12: evidence that R391 behaves as a conjugal transposon. *FEMS Microbiol. Lett.* **134**:153–158.

Naas, T., Y. Mikami, T. Imai, L. Poirel, and P. Nordmann. 2001. Characterization of In53, a class 1 plasmid- and composite transposon-located integron of *Escherichia coli* which carries an unusual array of gene cassettes. *J. Bacteriol.* **183**:235–249.

Nandi, S., J. J. Maurer, C. Hofacre, and A. O. Summers. 2004. Gram-positive bacteria are a major reservoir of class 1 antibiotic resistance integrons in poultry litter. *Proc. Natl. Acad. Sci. USA* **101**:7118–7122.

Nemergut, D. R., A. P. Martin, and S. K. Schmidt. 2004. Integron diversity in heavy-metal-contaminated mine tailings and inferences about integron evolution. *Appl. Environ. Microbiol.* **70**:1160–1168.

Nesvera, J., J. Hochmannova, and M. Patek. 1998. An integron of class 1 is present on the plasmid pCG4 from gram-positive bacterium *Corynebacterium glutamicum*. *FEMS Microbiol. Lett.* **169**:391–395.

Nield, B. S., A. J. Holmes, M. R. Gillings, G. D. Recchia, B. C. Mabbutt, K. M. Nevalainen, and H. W. Stokes. 2001. Recovery of new integron classes from environmental DNA. *FEMS Microbiol. Lett.* **195**:59–65.

Nield, B. S., R. D. Willows, A. E. Torda, M. R. Gillings, A. J. Holmes, K. M. Nevalainen, H. W. Stokes, and B. C. Mabbutt. 2004. New enzymes from environmental cassette arrays: functional attributes of a phosphotransferase and an RNA-methyltransferase. *Protein Sci.* **13**:1651–1659.

Nunes-Duby, S. E., H. J. Kwon, R. S. Tirumalai, T. Ellenberger, and A. Landy. 1998. Similarities and differences among 105 members of the Int family of site-specific recombinases. *Nucleic Acids Res.* **26**:391–406.

Ogawa, A., and T. Takeda. 1993. The gene encoding the heat-stable enterotoxin of *Vibrio cholerae* is flanked by 123-base pair direct repeats. *Microbiol. Immunol.* **37**:607–616.

Okada, K., T. Iida, K. Kita-Tsukamoto, and T. Honda. 2005. Vibrios commonly possess two chromosomes. *J. Bacteriol.* **187**:752–757.

Petroni, A., R. G. Melano, H. A. Saka, A. Garutti, L. Mange, F. Pasteran, M. Rapoport, M. Miranda, D. Faccone, A. Rossi, P. S. Hoffman, and M. F. Galas. 2004. CARB-9, a carbenicillinase encoded in the VCR region of *Vibrio cholerae* non-O1, non-O139 belongs to a family of cassette-encoded beta-lactamases. *Antimicrob. Agents Chemother.* **48**:4042–4046.

Rabus, R., M. Kube, J. Heider, A. Beck, K. Heitmann, F. Widdel, and R. Reinhardt. 2005. The genome sequence of an anaerobic aromatic-degrading denitrifying bacterium, strain EbN1. *Arch. Microbiol.* **183**:27–36.

Radstrom, P., O. Skold, G. Swedberg, J. Flensburg, P. H. Roy, and L. Sundstrom. 1994. Transposon Tn5090 of plasmid R751, which carries an integron, is related to Tn7, Mu, and the retroelements. *J. Bacteriol.* **176**:3257–3268.

Recchia, G. D., and R. M. Hall. 1995. Plasmid evolution by acquisition of mobile gene cassettes: plasmid pIE723 contains the aadB gene cassette precisely inserted at a secondary site in the incQ plasmid RSF1010. *Mol. Microbiol.* **15**:179–187.

Recchia, G. D., H. W. Stokes, and R. M. Hall. 1994. Characterisation of specific and secondary recombination sites recognised by the integron DNA integrase. *Nucleic Acids Res.* **22**:2071–2078.

Rice, L. B. 1998. Tn916 family conjugative transposons and dissemination of antimicrobial resistance determinants. *Antimicrob. Agents Chemother.* **42**:1871–1877.

Robinson, A., P. S. Wu, S. J. Harrop, P. M. Schaeffer, Z. Dosztanyi, M. R. Gillings, A. J. Holmes, K. M. Helena Nevalainen, H. W. Stokes, G. Otting, N. E. Dixon, P. M. Curmi, and B. C. Mabbutt. 2005. Integron-associated mobile gene cassettes code for folded proteins: the structure of Bal32a, a new member of the adaptable alpha+beta barrel family. *J. Mol. Biol.* **346**:1229–1241.

Rowe-Magnus, D. A., A.-M. Guerout, and D. Mazel. 1999. Superintegrons. *Res. Microbiol.* **150**:641–651.

Rowe-Magnus, D. A., A.-M. Guerout, P. Ploncard, B. Dychinco, J. Davies, and D. Mazel. 2001. The evolutionary history of chromosomal super-integrons provides an ancestry for multi-resistant integrons. *Proc. Natl. Acad. Sci. USA* **98**:652–657.

Rowe-Magnus, D. A., A. M. Guerout, L. Biskri, P. Bouige, and D. Mazel. 2003. Comparative analysis of superintegrons: engineering extensive genetic diversity in the Vibrionaceae. *Genome Res.* **13**:428–442.

Rowe-Magnus, D. A., A. M. Guerout, and D. Mazel. 2002. Bacterial resistance evolution by recruitment of super-integron gene cassettes. *Mol. Microbiol.* **43**:1657–1669.

Rowe-Magnus, D. A., and D. Mazel. 2001. Integrons: natural tools for bacterial genome evolution. *Curr. Opin. Microbiol.* **4**:565–569.

Rowe-Magnus, D. A., and D. Mazel. 2002. The role of integrons in antibiotic resistance gene capture. *Int. J. Med. Microbiol.* **292**:115–125.

Ruby, E. G., M. Urbanowski, J. Campbell, A. Dunn, M. Faini, R. Gunsalus, P. Lostroh, C. Lupp, J. McCann, D. Millikan, A. Schaefer, E. Stabb, A. Stevens, K. Visick, C. Whistler, and E. P. Greenberg. 2005. Complete genome sequence of *Vibrio fischeri*: a symbiotic bacterium with pathogenic congeners. *Proc. Natl. Acad. Sci. USA* **102**:3004–3009.

Shibata, N., Y. Doi, K. Yamane, T. Yagi, H. Kurokawa, K. Shibayama, H. Kato, K. Kai, and Y. Arakawa. 2003. PCR typing of genetic determinants for metallo-beta-lactamases and integrases carried by gram-negative bacteria isolated in Japan, with focus on the class 3 integron. *J. Clin. Microbiol.* **41**:5407–5413.

Smith, A. B., and R. J. Siebeling. 2003. Identification of genetic loci required for capsular expression in *Vibrio vulnificus*. *Infect. Immun.* **71**:1091–1097.

Stokes, H. W., and R. M. Hall. 1989. A novel family of potentially mobile DNA elements encoding site-specific gene-integration functions: integrons. *Mol. Microbiol.* **3**:1669–1683.

Stokes, H. W., A. J. Holmes, B. S. Nield, M. P. Holley, K. M. Nevalainen, B. C. Mabbutt, and M. R. Gillings. 2001. Gene cassette PCR: sequence-independent recovery of entire genes from environmental DNA. *Appl. Environ. Microbiol.* **67**:5240–5246.

Stokes, H. W., D. B. O'Gorman, G. D. Recchia, M. Parsekhian, and R. M. Hall. 1997. Structure and function of 59-base element recombination sites associated with mobile gene cassettes. *Mol. Microbiol.* **26**:731–745.

Sullivan, J. T., and C. W. Ronson. 1998. Evolution of rhizobia by acquisition of a 500-kb symbiosis island that integrates into a phe-tRNA gene. *Proc. Natl. Acad. Sci. USA* **95**:5145–5149.

Sundstrom, L., P. H. Roy, and O. Skold. 1991. Site-specific insertion of three structural gene cassettes in transposon Tn7. *J. Bacteriol.* **173**:3025–3028.

Tagomori, K., T. Iida, and T. Honda. 2002. Comparison of genome structures of vibrios, bacteria possessing two chromosomes. *J. Bacteriol.* **184**:4351–4358.

Tauch, A., S. Gotker, A. Puhler, J. Kalinowski, and G. Thierbach. 2002. The 27.8-kb R-plasmid pTET3 from *Corynebacterium glutamicum* encodes the aminoglycoside adenyltransferase gene cassette aadA9 and the regulated tetracycline efflux system Tet 33 flanked by active copies of the widespread insertion sequence IS6100. *Plasmid* **48**:117–129.

Tennstedt, T., R. Szczepanowski, S. Braun, A. Puhler, and A. Schluter. 2003. Occurrence of integron-associated resistance gene cassettes located on antibiotic resistance plasmids isolated from a wastewater treatment plant. *FEMS Microbiol. Ecol.* **45**:239–252.

Thungapathra, M., Amita, K. K. Sinha, S. R. Chaudhuri, P. Garg, T. Ramamurthy, G. B. Nair, and A. Ghosh. 2002. Occurrence of antibiotic resistance gene cassettes aac(6′)-Ib, dfrA5, dfrA12, and ereA2 in class I integrons in non-O1, non-O139 *Vibrio cholerae* strains in India. *Antimicrob. Agents Chemother.* **46**:2948–2955.

Trucksis, M., J. Michalski, Y. K. Deng, and J. B. Kaper. 1998. The *Vibrio cholerae* genome contains two unique circular chromosomes. *Proc. Natl. Acad. Sci. USA* **95**:14464–14469.

Ullmann, U. 1969. [Vibrio fetus as cause of disease in man]. [In German.] *Dtsch. Med. Wochenschr.* **94**:2399–2402.

Vaisvila, R., R. D. Morgan, J. Posfai, and E. A. Raleigh. 2001. Discovery and distribution of super-integrons among pseudomonads. *Mol. Microbiol.* **42**:587–601.

van der Meer, J. R., and V. Sentchilo. 2003. Genomic islands and the evolution of catabolic pathways in bacteria. *Curr. Opin. Biotechnol.* **14**:248–254.

Vezzi, A., S. Campanaro, M. D'Angelo, F. Simonato, N. Vitulo, F. M. Lauro, A. Cestaro, G. Malacrida, B. Simionati, N. Cannata, C. Romualdi, D. H. Bartlett, and G. Valle. 2005. Life at depth: *Photobacterium profundum* genome sequence and expression analysis. *Science* 307:1459–1461.

Waldor, M. K., H. Tschape, and J. J. Mekalanos. 1996. A new type of conjugative transposon encodes resistance to sulfamethoxazole, trimethoprim, and streptomycin in *Vibrio cholerae* O139. *J. Bacteriol.* 178:4157–4165.

Whittle, G., N. B. Shoemaker, and A. A. Salyers. 2002. The role of *Bacteroides* conjugative transposons in the dissemination of antibiotic resistance genes. *Cell Mol. Life Sci.* 59:2044–2054.

Yamaichi, Y., T. Iida, K. S. Park, K. Yamamoto, and T. Honda. 1999. Physical and genetic map of the genome of *Vibrio parahaemolyticus*: presence of two chromosomes in *Vibrio* species. *Mol. Microbiol.* 31:1513–1521.

V. PHYSIOLOGY

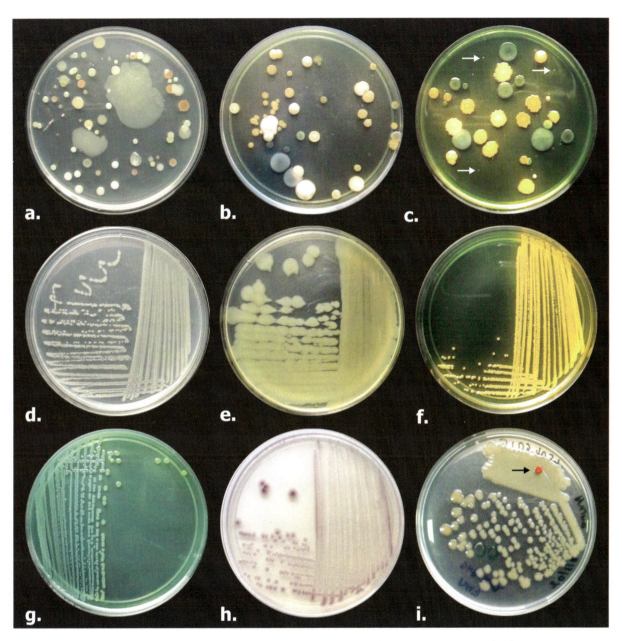

Color Plate 1 (Chapter 2). Colonial morphology of vibrios in different media. (a) Estuarine water plated on marine agar. (b) Estuarine water plated on TCBS agar after 24 h of incubation. (c) Vibrios on TCBS agar; note non-vibrios forming very small colonies (arrows). (d) *V. alginolyticus* on marine agar. (e) *V. alginolyticus* on TSA agar; note spreading on this medium. (f) *V. alginolyticus* on TCBS agar, a sucrose-positive colony. (g) *V. parahaemolyticus* on TCBS agar, a sucrose-negative colony. (h) *V. parahaemolyticus* on CHROMagar chromogenic agar; note the violet color of the colonies. (i) Cryopreservation glass bead (arrow) streaked onto TSA.

Large chromosome

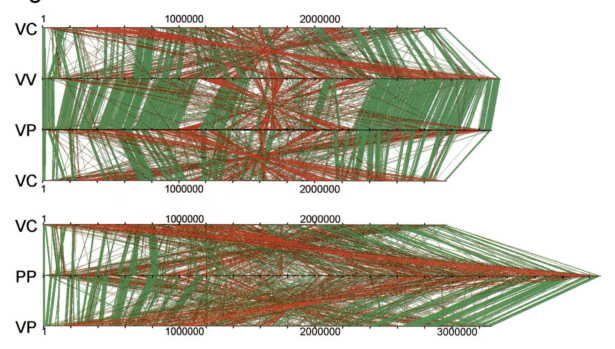

Small chromosome

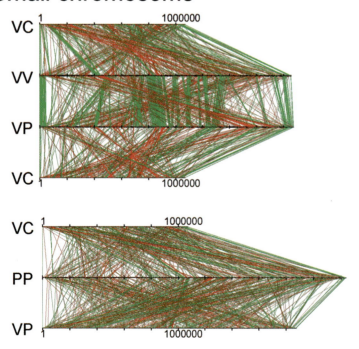

Color Plate 2 (Chapter 5). Comparison of relative positions of conserved genes among *Vibrio* and *Photobacterium* species. Gene pairs were generated by BLASTp analysis of predicted genes from each genome. Lines connect conserved genes among the organisms. The chromosomes are depicted with the origins of replication at the left end. VC, *V. cholerae* (Heidelberg et al., 2000); VV, *V. vulnificus* (Chen et al., 2003); VP, *V. parahaemolyticus* (Makino et al., 2003); PP, *P. profundum* (GenBank accession nos. CR354531 and CR354532).

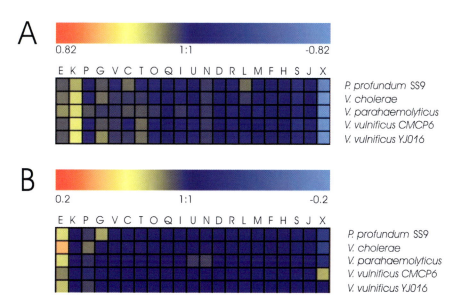

Color Plate 3 (Chapter 6). The functional landscape of the paranome. Preferentially retained duplicated genes according to their functional classes are visualized for all *Vibrionaceae* genomes. The functional distribution of the paranome in terms of percentages is subtracted from the functional distribution of the single-copy genes in terms of percentages, revealing the functional classes that have relatively more (yellow to red) or less (gradient of blue) paralogs compared to single-copy genes. The functional classes have been ordered according to relative average contribution. The paranome was divided into the fraction that is not in duplicated segments (A) and the fraction that is in duplicated segments (B). The different functional classes are abbreviated as follows: [C] energy production and conversion, [D] cell cycle control, cell division, chromosome partitioning, [E] amino acid transport and metabolism, [F] nucleotide transport and metabolism, [G] carbohydrate transport and metabolism, [H] coenzyme transport and metabolism, [I] lipid transport and metabolism, [J] translation, ribosomal structure, and biogenesis, [K] transcription, [L] replication, recombination, and repair, [M] cell wall/membrane/envelope biogenesis, [N] cell motility, [O] posttranslational modification, protein turnover, chaperones, [P] inorganic ion transport and metabolism, [Q] secondary metabolite biosynthesis, transport, and catabolism, [R] general function prediction only, [S] function unknown, [U] intracellular trafficking, secretion, and vesicular transport, [V] defense mechanisms, [X] none of the above. Note that all genes or open reading frames annotated as being transposon or phage related have been removed from this analysis.

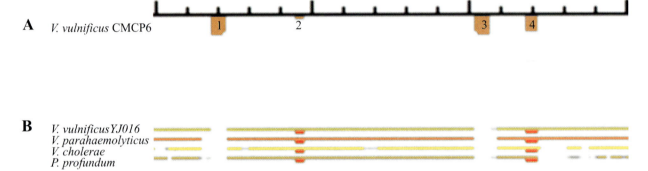

Color Plate 4 (Chapter 6). Visualization of segmental duplication and conservation among *Vibrionaceae*. (Top) Segmental duplication in a genomic fragment of the *V. vulnificus* CMCP6 chromosome I (gene 900 to 1050). Each block is a segmental duplication where the width represents the number of genes in the segment, and the height is the number of homologous copies found in the genome. (Bottom) Colinearity between the *V. vulnificus* CMCP6 genomic fragment and other *Vibrionaceae*. Conservation of gene content and order is represented by the colored lines. Conservation of content without order is represented by gray lines. When a colinear region is also duplicated in segments, a red line is drawn. The scale in the ruler is 10 genes between two lines. Blocks labeled with 1 and 3 are homologous segments unique in *V. vulnificus* CMCP6, whereas blocks 2 and 4 are conserved in all *Vibrionaceae*.

V. vulnificus YJ016

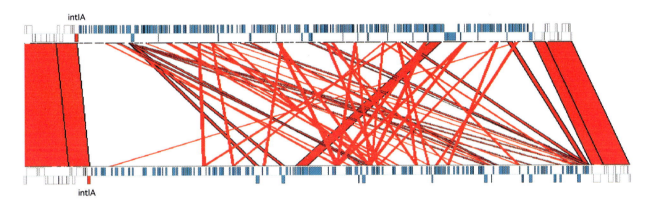

intIA

intIA

V. vulnificus CMCP6

Color Plate 5 (Chapter 8). Comparison of relative positions of conserved ORFs between *V. vulnificus* YJ016 and CMCP6 superintegron regions. ORF pairs were generated by BLASTn analysis of predicted genes from each genome. Pairs with a BLASTp score ≥600 are shown. Red lines connect both conserved genes between the two organisms. Cassette-encoded ORFs are symbolized by blue boxes, the *intIA* gene is shown as a red box, and white boxes correspond to ORFs located outside the SI; the relative orientation of the ORFs is also shown.

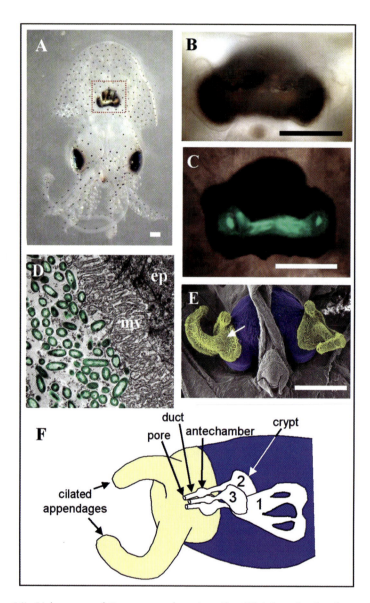

Color Plate 6 (Chapter 14). Light organs of *Euprymna scolopes* juveniles. (A) A juvenile placed ventral side up and lit from above. Red box indicates location of light organ. Gold color within light organ is reflective tissue. (B) Backlit light organ. Translucent symbiotic tissue sits above the darker ink sac. (C) Similar to panel B, except epifluorescence microscopy reveals the location of green fluorescent protein-labeled *V. fischeri* (green) within the light organ. (D) Transmission electron microscopy of bacterial symbionts (green) colonizing microvillous surface (mv) of a crypt epithelial cell (ep). (E) Scanning electron microscopy of a light organ colored to show ciliated fields on the light organ surface (yellow) and other light organ tissues (blue). The white arrow indicates the approximate location of three 10- to 15-μm-diameter pores. (F) Cartoon depiction of a light organ, with yellow and blue coloration as in panel E. White regions indicate integral internal symbiotic tissues, described in the text, with numbers labeling crypts 1, 2, and 3, respectively. Solid bars in panels A, B, C, and E are ~250 μm.

Color Plate 7 (Chapter 16). A colony of *O. patagonica* showing bleached and healthy tissues.

Color Plate 8 (Chapter 16). Photograph of the fireworm *H. carunculata* feeding on a coral colony.

Color Plate 9 (Chapter 16). Photographs of a bacterium-bleached *P. damicornis* coral (a) and a healthy one (b).

Color Plate 10 (Chapter 17). The two historical entrance routes of cholera to South America (heavy arrows). Major dissemination route of cholera in Brazil in the 1991 pandemic (light arrows). Foci of distinct nonepidemic *V. cholerae* lineages in the South American pandemic (1991): purple, Tucumán strain; orange, Amazonia strain; and blue, sucrose late-fermenting strain.

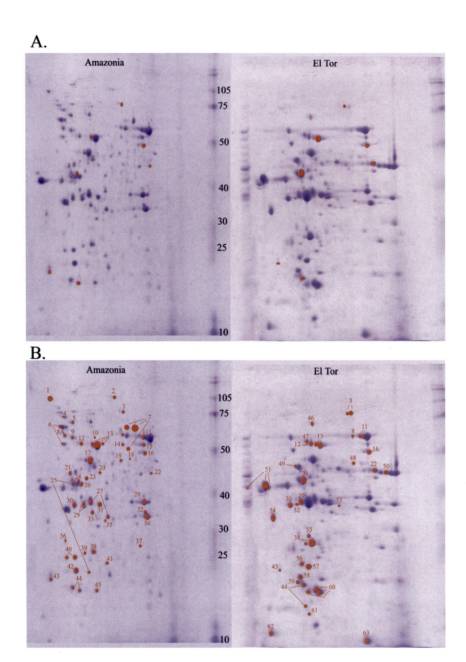

Color Plate 11 (Chapter 17). Comparison of the proteins present in supernatant-enriched fractions of the Amazonia strain 3509 and the El Tor strain N16961, in the pH range of 4 to 7 (left to right), and M_r from 10 to 110 kDa. The 2D gels were stained with Coomassie blue. (A) 2D gels for each strain. The seven pairs of orange spots in this panel correspond to proteins with the same identification and position in both gels. They were used as reference spots for the alignment of the rest of the gels. (B) The same gels with superimposed orange spots for the identified proteins. The numbers correspond to those in Table 1. Spots were numbered from left to right and from top to bottom. Proteins with the same identification received the same number in both gels.

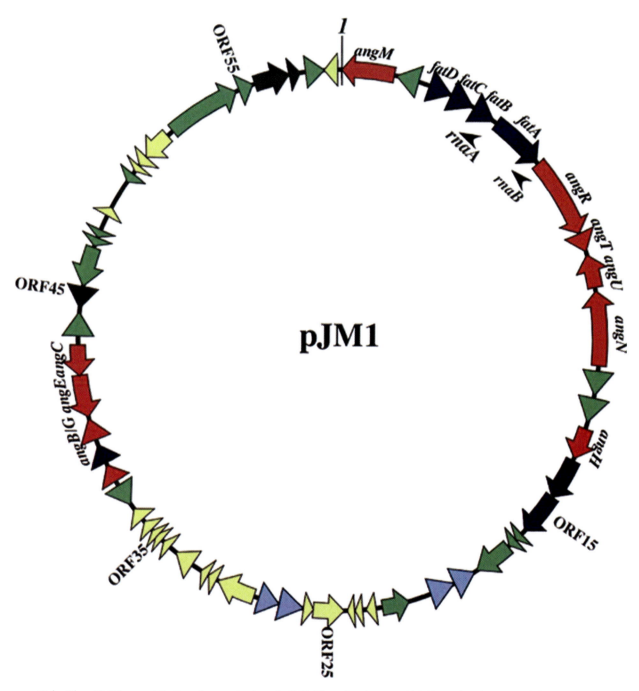

Color Plate 12 (Chapter 18). Genetic organization of pJM1. The colored arrows identify predicted ORFs and their direction of transcription. Red, ORFs related to siderophore biosynthesis; blue, ORFs related to siderophore transport; green, ORFs related to insertion elements and composite transposon; cyan, ORFs related to replication and partitioning; yellow, conserved hypothetical ORFs and ORFs with no known functions; black, ORFs with functions that do not fall in any of the above categories. The black arrowheads represent genes encoding antisense RNAs: *rnaA*, antisense RNAα, and *rnaB*, antisense RNAβ.

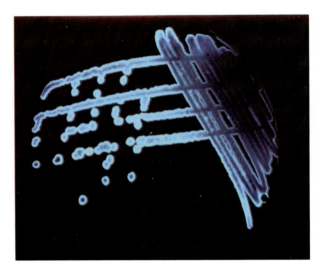

Color Plate 13 (Chapter 19). Bioluminescent *V. harveyi* ISO7 isolated from seawater by Phillip Burza. Photograph by Dr. Jane Oakey, James Cook University.

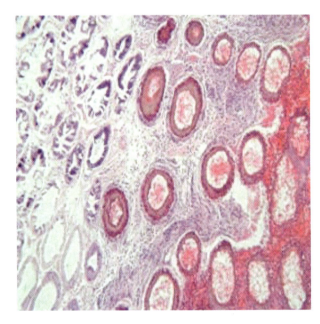

Color Plate 14 (Chapter 19). Septic hepatopancreatic tubules from broodstock of *P. monodon* naturally infected with *V. harveyi* from northern Australia. Photograph by Leigh Owens.

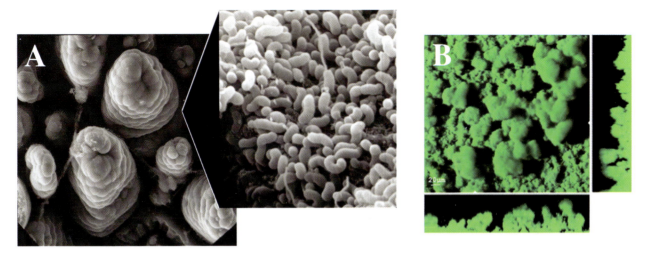

Color Plate 15 (Chapter 23). (A) Micrograph of *V. cholerae* colonizing the suckling mouse intestine. First image shows villous architecture within the small intestine; inset is a magnification of crypts within the first image detailing *V. cholerae* cells in intimate contact with epithelial cells. Image courtesy of M. Neal Guentzel. (B) Biofilm structures of green fluorescent protein-expressing *V. cholerae* as imaged by confocal scanning laser microscopy. Large box is x-y scanned image; sidebars show x-z reconstruction. Image courtesy of F. Yildiz. Reprinted from Yildiz et al. (2004) with permission of Blackwell Publishing.

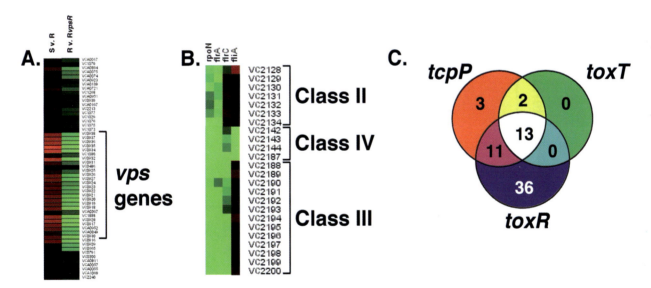

Color Plate 16 (Chapter 23). Microarray analysis of whole-genome transcription profiling of *V. cholerae*. In cluster analysis, green corresponds to downregulated expression, and red corresponds to upregulated expression. (A) Cluster analysis of smooth-rugose and rugose-rugose *vpsR* strains, detailing changes in *vps* transcription. Image courtesy of F. Yildiz. Reprinted from Yildiz et al. (2001) with permission of Blackwell Publishing. (B) Cluster analysis of *rpoN*, *flrA*, *flrC*, and *fliA* strains, demonstrating class II, class III, and class IV patterns of flagellar transcription. Image courtesy of K. Klose and F. Yildiz. (C) Venn diagram representing *toxR*-, *tcpP*-, and *toxT*-dependent genes, as determined by microarray analysis. Image courtesy of J. Bina and J. Mekalanos. Reprinted from Bina et al. (2003) with permission of *Proceedings of the National Academy of Sciences USA*.

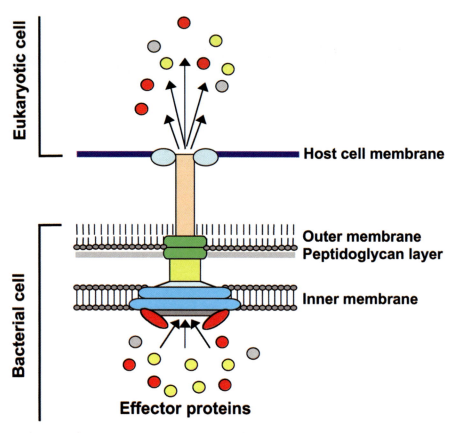

Color Plate 17 (Chapter 24). Type III secretion system (TTSS). The type III secretion system is a protein secretion apparatus of gram-negative bacteria. Through this machinery, bacteria can inject their own proteins (effectors) into the cytosol of target eukaryotic cells (Hueck, 1998). Bacterial pathogens that cause disease by intimate interactions with eukaryotic cells, such as *Salmonella*, *Shigella*, *Yersinia*, and plant pathogens, have this system.

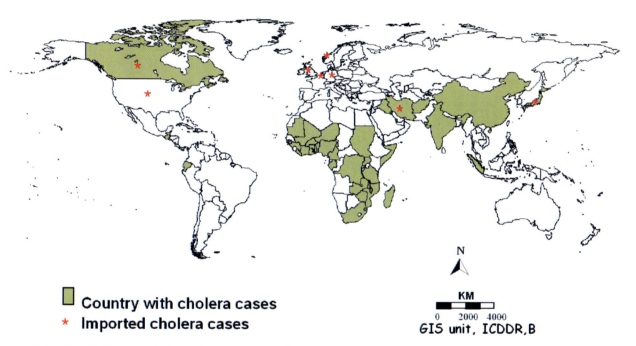

□ **Country with cholera cases**
* **Imported cholera cases**

KM

0   2000   4000

GIS unit, ICDDR,B

Color Plate 18 (Chapter 27). Countries/areas reporting cholera cases in 2003. This map was adapted from the map published in the *Weekly Epidemiological Record*, 30 July 2004: World Health Organization (2004). Cholera, 2003. *Wkly. Epidemiol. Rec.* **79**:281–288.

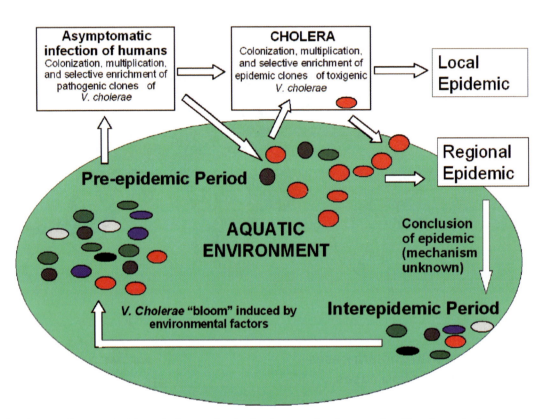

Color Plate 19 (Chapter 27). Schematic diagram showing involvement of environmental and host factors in the occurrence of seasonal epidemics in an area of endemic cholera. Following blooms of diverse vibrios due to environmental factors, enrichment of pathogenic *V. cholerae* strains occurs in humans with asymptomatic infection prior to a seasonal epidemic. Red circles represent *V. cholerae* strains with epidemic potential, whereas other circles represent diverse environmental *V. cholerae* populations.

The Biology of Vibrios
Edited by F. L. Thompson et al.
© 2006 ASM Press, Washington, D.C.

Chapter 9

Motility and Chemotaxis

LINDA L. MCCARTER

INTRODUCTION: SWIMMING AND SWARMING

Flagella, working as helical propellers, provide bacteria with a highly effective means of locomotion. *Vibrio* species are particularly fast swimmers. For example, the average swimming speed of *V. alginolyticus* is 116 μm/s, and cells have been clocked as high as 147 μm/s (the average cell body size is ~2 μm in length [Magariyama et al., 1995]). Amazingly fast motors spin the flagella (~5 μm in length) at rates as high as 100,000 rpm (Magariyama et al., 1994). The numbers and arrangement of the propeller vary, but the mode of insertion is of two major types, i.e., polar or peritrichous (Baumann and Baumann, 1981). The flagellar patterns of some members of the *Vibrionaceae* are listed in Table 1.

Environmental conditions influence the type of flagellation. When grown planktonically, all *Vibrio* species display polar flagella. Some species (e.g., *V. cholerae*) possess a single polar flagellum (monotrichous flagellation), whereas others can have multiple flagella inserted at one pole (lophotrichous) (e.g., *V. fischeri*). These flagella are usually sheathed by what appears to be an extension of the cell outer membrane (Follett and Gordon, 1963). When grown on solid medium, some members of the *Vibrionaceae* alter their flagellation, and these organisms produce both polar and peritrichous (usually called lateral) flagella (e.g., *V. alginolyticus* and *V. parahaemolyticus*). The lateral flagella are highly abundant (>100/cell), unsheathed, and used for movement over surfaces and through viscous environments. Figure 1 contrasts the flagellation patterns of liquid- and plate-grown *V. parahaemolyticus*. Environmental conditions have also been shown to affect the number of polar flagella. Specifically, the magnesium ion concentration influences the number of polar flagella of *V. fischeri* and has a significant effect on swimming motility (O'Shea et al., 2005).

To possess a functioning flagellar motility system, three things are requisite: a propeller, a motor, and a system of navigation. In addition, there is a dedicated pathway for assembly of the complex flagellar organelle. As a consequence, the number of genes necessary to constitute a flagellar system is quite large, often >50. Like all flagellar systems that have been studied (reviewed by Macnab, 1996; Aldridge and Hughes, 2002), expression of these many flagellar genes in *Vibrio* spp. is carefully regulated and tied to morphogenesis of the organelle. For the *Vibrio* species that have polar and lateral systems, there are no shared structural parts. The genes encoding each system are distinct, as is their regulation. In addition, the energy source powering flagellar rotation differs for the polar and lateral motors. In this chapter, the swimming and swarming systems will be considered separately. There is a central processing system that is shared by both flagellar systems, i.e., chemotaxis, and it allows the bacteria to detect signals in their environment and respond by modifying their movement.

SWIMMING: THE POLAR FLAGELLAR SYSTEM

Polar Flagellar Genes

The polar flagellar genes are listed in Table 2. The products of many of these flagellar genes can be described much like those of an engine (Berg, 2003). The stator contains MotA and MotB, which participate in torque generation. There is a rotor, containing FliF, FliG, and the C-ring components FliM and FliN. The drive shaft, or flagellar rod, contains FlgB, FlgC, FlgF, and FlgG. FlgH and FlgI form bushings, called the L and P rings, that are found in the outer membrane and peptidoglycan layer, respectively. The bushings surround the drive shaft and keep it in place. There is a flexible universal joint called the hook (FlgE) that joins the drive shaft to the propeller via

Linda L. McCarter • Microbiology Department, The University of Iowa, Iowa City, IA 52242.

Table 1. Flagellation patterns of Vibrio species[a]

Organism	Sheathed polar flagella[b]	Unsheathed peritrichous flagella[c]
V. alginolyticus	Monotrichous	+
V. anguillarum	Monotrichous	−
V. cholerae	Monotrichous	−
V. fischeri	Lophotrichous (2–8)	−
V. harveyi	Monotrichous	+
V. parahaemolyticus	Monotrichous	+
V. vulnificus	Monotrichous	−

[a] Determined by electron microscopy studies of Allen and Baumann (1971) and Baumann and Baumann (1981).
[b] Determined when grown in liquid medium. Other Vibrio species reported to have multiple, polarly inserted flagella include V. logei, V. leiognathi, V. angustum, and V. phosphoreum.
[c] Determined when grown on solid medium. A plus sign indicates that the majority of strains examined (>80%) possessed lateral flagella; a minus sign indicates that none possessed lateral flagella. Other Vibrio species reported to have lateral flagella include V. proteolyticus, V. furnissii, V. fluvialis, V. mediterranei, and V. splendidus (Allen and Baumann, 1971; Baumann and Baumann, 1981; Shinoda et al., 1992).

two adaptor proteins (or hook-associated proteins, HAP1 [FlgK] and HAP3 [FlgL]). The propeller, or flagellar filament, is polymerized from multiple flagellin subunit monomers called flagellins. At the tip of the filament is a cap, called FliD or HAP2.

The flagellum is assembled sequentially, initiating with the membrane-embedded basal body (rotor, drive shaft, and bushings), followed by addition of the hook and finally the filament. A dedicated type III secretion apparatus, containing FlhA, FlhB, FliO, FliP, FliQ, FliR, FliH, FliI, and FliJ, directs morphogenesis of the organelle (reviewed by Macnab, 2003, 2004). The above-mentioned genes are common to all flagellar systems that have been studied (reviewed by Berg, 2003). In addition, there are genes that seem peculiar to some but not all flagellar systems. One pair of genes includes flhF and flhG. FlhF, first discovered in Bacillus subtilis (Carpenter et al., 1992), shows homology to FtsY, which is a GTP-binding protein involved in the signal recognition particle targeting pathway. Immediately downstream of flhF is flhG, which encodes a protein that shows homology to MinD. MinD is a membrane ATPase involved in septum site determination (Levin et al., 1998; Marston et al., 1998). FlhF and FlhG seem to work as checkpointing regulators. Disruption of flhF in Pseudomonas putida leads to random arrangement of flagellar insertion (Pandza et al., 2000). Disruption of flhG in Pseudomonas aeruginosa leads to multiple polar flagella (Dasgupta et al., 2000); moreover, in this organism FlhG has been demonstrated to bind to the master flagellar regulator (equivalent to FlrA in Vibrio) and act as an antiactivator (Dasgupta and Ram-

phal, 2001). In V. cholerae, disruption of flhF influences expression of class 3 and class 4 flagellar genes (encoding basal body, hook, filament, and motor parts) whereas disruption of FlhG increases expression of all flagellar genes including the master regulator FlrA (Correa et al., 2005).

The Propeller

Flagellar filaments, which act as propellers, consist of self-assembling protein subunits (flagellin) arranged in a helix and forming a hollow tube (reviewed by Namba and Vonderviszt, 1997). Subunits move down the hollow core and are polymerized at the tip of the flagellum. The Vibrio polar organelles are complex flagella, heteropolymers of multiple flagellin subunits. Five genes encode the subunits in V. anguillarum and V. cholerae; these are found in two loci: flaAC and flaFBA (McGee et al., 1996; Klose and Mekalanos, 1998a). Six flagellin genes are found in V. fischeri: five are linked (flaABCDE), and one is alone (flaF) (Millikan and Ruby, 2003, 2004). The six flagellin genes of V. parahaemolyticus are organized as flaCDE and flaFBA (McCarter, 1995; Kim and McCarter, 2000). The flagellin genes in the genome sequence of V. vulnificus are annotated similar to V. parahaemolyticus. Analysis of gene expression and protein composition of purified flagella from wild type and strains with mutations in flagellin genes suggests that all (or most) of the flagellins are incorporated into the organelle (McCarter, 1995; McGee et al., 1996; Klose and Mekalanos, 1998a; Kim and McCarter, 2000; Millikan and Ruby, 2004). Because the flagellins are immunogenically cross-reactive, nothing is known with respect to their spatial arrangement in the flagellum. In V. fischeri, a mutation in flaA considerably reduces motility (and numbers of polar flagella), whereas a mutation in flaC causes no swimming defect (Millikan and Ruby, 2004). In V. parahaemolyticus, all of the flagellins are dispensable: mutants with single flagellin gene defects show no defect in swimming (McCarter, 1995; Kim and McCarter, 2000). However, for V. anguillarum and V. cholerae, one of the flagellin genes (flaA) is critical, and loss of function severely impairs swimming motility (Milton et al., 1996; Klose and Mekalanos, 1998a). In V. cholerae, expression of flaA is uniquely dependent on σ^{54}, whereas the other flagellin genes require the flagellar-specific sigma factor σ^{28} (Klose and Mekalanos, 1998b). Transcription of V. fischeri flaA also requires σ^{54}, and potential σ^{28}-dependent promoters are found in the upstream sequences for flaB, flaC, flaD, and flaF (Millikan and Ruby, 2004). The critical flagellin FlaA is most equivalent with respect to gene location

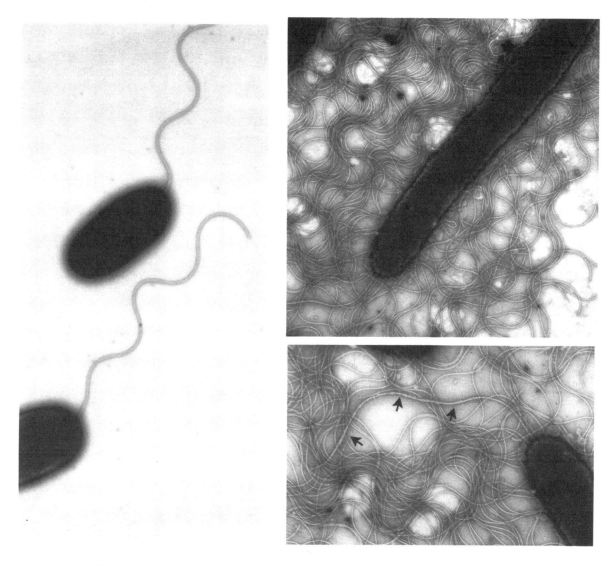

Figure 1. Flagellation patterns of liquid- and plate-grown *V. parahaemolyticus*. The *V. parahaemolyticus* swimmer cells (left panel) possess single, sheathed polar flagella. This electron micrograph is of planktonically grown cells, stained with uranyl acetate. Cell proportions are ~1 × 2 μm. The swarmer cells (right panels) are elongated and display numerous lateral flagella in addition to the single, sheathed polar flagellum. *V. parahaemolyticus* swarmer cells were harvested from a plate and stained with phosphotungstic acid. The arrows point to the partially dissolved sheath of polar flagellum. The diameter of the sheathed polar flagellum is ~30 nm, and the diameter of the unsheathed lateral flagellum is ~15 nm.

as well as predicted protein sequence to *V. parahaemolyticus* FlaC. Although not an essential (or dominant) flagellin gene in *V. parahaemolyticus*, *flaC*$_{vp}$ is regulated differently from the other flagellin genes, which are σ^{28}-dependent (Kim and McCarter, 2000).

It is not clear why these organisms possess such large numbers of flagellins. The similarity of the gene products and the dispensability of most of the genes suggest that there may be no special structural requirements. However, filament structure/function could be more complex and adapted to specific circumstances than our simple laboratory motility tests can reveal. For instance, the *flaA*$_{vf}$ defect has pro-

found consequences for *V. fischeri* host colonization (Millikan and Ruby, 2004); *V. anguillarum* strains with *flaD*$_{va}$ or *flaE*$_{va}$ mutations show significantly decreased virulence although a swimming speed defect was undetected (McGee et al., 1996); and the *flaC*$_{vp}$ gene is expressed under unique environmental conditions (Kim and McCarter, 2000). Immune system evasion might also account for some of the diversity. Bacteria are known to modulate the antigenicity of their flagellar filaments by expressing different flagellin genes or by recombination and rearrangement of flagellin genes (reviewed by Wilson and Beveridge, 1993). The multiplicity of flagellin genes suggests a

Table 2. Polar flagellar and linked chemotaxis genes

Gene name[a]	Alternative gene name	VP locus tag	VC locus tag	VF locus tag	Predicted function
Region 1					
flgN		VP0770	VC2205	VF1882	Chaperone
flgM		VP0771	VC2204	VF1881	Anti-σ^{28} factor
flgA		VP0772	VC2203	VF1880	P-ring chaperone
cheV	cheV-3	VP0773	VC2202	VF1879	CheW/CheY hybrid
cheR	cheR-2	VP0774	VC2201	VF1878	Chemotaxis methyl transferase
flgB		VP0775	VC2200	VF1877	Proximal rod
flgC		VP0776	VC2199	VF1876	Proximal rod
flgD		VP0777	VC2198	VF1875	Hook cap
flgE		VP0778	VC2197	VF1874	Hook
flgF		VP0780	VC2196	VF1873	Proximal rod
flgG		VP0781	VC2195	VF1872	Distal rod
flgH		VP0782	VC2194	VF1871	L ring
flgI		VP0783	VC2193	VF1870	P ring
flgJ		VP0784	VC2192	VF1869	Peptidoglycan hydrolase; rod cap
flgK		VP0785	VC2191	VF1868	HAP1 (hook-filament junction)
flgL		VP0786	VC2190	VF1867	HAP3 (hook-filament junction)
flaX[b]		VP0788	VC2189	VF1866	Flagellin
flaX		VP0790	VC2187	VF1865	Flagellin
flaX		VP0791		VF1864	Flagellin
Region 2				VF1863	Flagellin
flaX[c]		VP2261	VC2144	VF1862	Flagellin
flaX		VP2259	VC2143	VF2079	Flagellin (VF2079 is found in a different locus)
flaX		VP2258	VC2142		Flagellin
flaG		VP2257	VC2141	VF1861	Unknown; axial protein?
fliD	flaH	VP2256	VC2140	VF1860	HAP2; filament cap
flaI		VP2255	VC2139	VF1859	Unknown
fliS	flaJ	VP2254	VC2138	VF1858	Flagellin chaperone
flrA	flaK	VP2253	VC2137	VF1856	σ^{54}-dependent regulator
flrB	flaL	VP2252	VC2136	VF1855	Two-component sensor kinase
flrC	flaM	VP2251	VC2135	VF1854	Two-component response regulator
fliE		VP2250	VC2134	VF1853	Hook-basal body component; MS ring-rod junction
fliF		VP2249	VC2133	VF1852	MS ring
fliG		VP2248	VC2132	VF1851	Rotor/switch component
fliH		VP2247	VC2131	VF1850	Fla export; negative regulator of FliI
fliI		VP2246	VC2130	VF1849	Fla export; ATPase
fliJ		VP2245	VC2129	VF1848	Fla export
fliK		VP2244	VC2128	VF1847	Hook-length control
fliL		VP2243	VC2127	VF1846	Unknown
fliM		VP2242	VC2126	VF1845	C ring/switch component
fliN		VP2241	VC2125	VF1844	C ring/switch component
fliO		VP2240	VC2124	VF1843	Fla export
fliP		VP2239	VC2123	VF1842	Fla export
fliQ		VP2238	VC2122	VF1841	Fla export
fliR		VP2237	VC2121	VF1840	Fla export
flhB		VP2236	VC2120	VF1849	Fla export
flhA		VP2235	VC2069	VF1837	Fla export
flhF		VP2234	VC2068	VF1836	Flagellar regulator; GTP-binding protein; homologous to FtsY
flhG		VP2233	VC2067	VF1835	Conserved domain COG0455 (ATPase involved in cell division and chromosome binding); flagellar synthesis regulator
fliA		VP2232	VC2066	VF1834	RNA polymerase σ^{28} factor
cheY	cheY-3	VP2231	VC2065	VF1833	Response regulator; causes change in direction of flagellar rotation
cheZ		VP2230	VC2064	VF1832	CheY phosphatase
cheA	cheA-2	VP2229	VC2063	VF1831	Histidine autokinase
cheB	cheB-2	VP2228	VC2062	VF1830	Chemotaxis methyl esterase

Continued on following page

Table 2. *Continued*

Gene name[a]	Alternative gene name	VP locus tag	VC locus tag	VF locus tag	Predicted function
ORF 1		VP2227	VC2061	VF1829	Unknown; Soj-like and other chromosome partitioning ATPase proteins
ORF 2		VP2226	VC2060	VF1828	Unknown (370 aa)
cheW	*cheW-1*	VP2225	VC2059	VF1827 VF1826	Chemotaxis coupling protein
ORF 3		VP2224	VC2058	VF1825	Unknown (163 aa)
Motor genes					
motA	*pomA*	VP0689	VC0892	VF0714 VFA0186	Na$^+$ motor (torque generation)
motB	*pomB*	VP0690	VC0893	VF0715 VFA0187	Na$^+$ motor (torque generation)
motX		VP2811	VC2601	VF2317	Na$^+$ motor component
motY		VP2111	VC1008	VF0926	Na$^+$ motor component

[a] All polar flagellar genes are found on chromosome 1. The physical organization of the genes is highly conserved between *Vibrio* species with the exception that the *V. cholerae* and *V. parahaemolyticus* flagellar genes are found in two regions, whereas the *V. fischeri* flagellar genes are found contiguously.

[b] There is a variation in the number, arrangement, and gene designation of the flagellin genes at this position. *V. parahaemolyticus* and *V. vulnificus* genes are designated *flaC*, *flaD*, and *flaE*; *V. anguillarum* and *V. cholerae* genes are designated *flaA* and *flaC*; and *V. fischeri* genes are designated *flaA*, *flaB*, *flaC*, *flaD*, and *flaE*.

[c] There is variation in the number, arrangement, and gene designation of the flagellin genes at this position. *V. parahaemolyticus* and *V. vulnificus* genes are designated *flaF*, *flaB*, and *flaA*; *V. anguillarum* and *V. cholerae* genes are designated *flaE*, *flaD*, and *flaB*. The other *V. fischeri* flagellin, designated *flaF* (VF2079), is not linked with the other flagellar genes.

significant reservoir for antigenic or phase variation. Electron microscopy suggests that the sheath may be fragile (Allen and Baumann, 1971; Sjoblad et al., 1983), and so the sheath may not afford adequate protection against the immune response of a host.

The Polar Flagellar Sheath

The polar flagellum is sheathed by an apparent extension of the cell membrane (Follett and Gordon, 1963; Allen and Baumann, 1971). Very little is known about the composition, formation, or function of flagellar sheaths (reviewed by Sjoblad et al., 1983). Evidence suggests that the sheath contains both lipopolysaccharide and proteins, and it may exist as a stable membrane domain distinct from the outer membrane (Hranitzky et al., 1980; Fuerst and Perry, 1988). How the sheath is formed remains essentially uninvestigated. It has been postulated that the sheath forms concomitantly with the elongation of the flagellar filament (Fuerst, 1980). However, it is provocative to note that "tubules" or "cell wall extension" structures that appear to be empty sheaths lacking filament have been observed, which suggests a possibility of uncoupling of flagellar core and sheath assembly (Allen and Baumann, 1971). The mechanism of how a sheathed flagellum rotates has also not been elucidated. Potentially, the flagellar filament could rotate within the sheath, or the two could rotate as a unit (Fuerst, 1980). One of three major sheath proteins of *V. alginolyticus* has been characterized. Ge-

netic and biochemical evidence suggests that it is a lipoprotein (Furuno et al., 2000). Mutants of *V. anguillarum* lacking a major flagellar sheath antigen, which is probably polysaccharide, are avirulent (Norqvist and Wolf-Watz, 1993). Such data suggest that the sheath may be quite important for interaction with the environment.

Polar Propulsion: the Sodium-Driven Flagellar Motor

The flagellar filament is turned by a reversible rotary motor that is embedded in the membrane (reviewed by Berg, 2003; Blair, 2003; Kojima and Blair, 2004). Energy to power flagellar rotation is derived from the transmembrane electrochemical potential of specific ions. Rotation is tightly coupled to the flow of ions through the motor. The coupling ions can be protons or sodium ions (Imae et al., 1986; Macnab, 1986). The polar flagellum of *Vibrio* is sodium-driven. Sodium channel-blocking drugs, particularly amiloride and phenamil, specifically inhibit sodium-driven motility (Sugiyama et al., 1988). Phenamil prevents swimming motility in *V. alginolyticus*, *V. anguillarum*, *V. cholerae*, and *V. parahaemolyticus* (Kawagishi et al., 1996; Muramoto et al., 1996; Asai et al., 1997; Kojima et al., 1999c; Okabe et al., 2001; Larsen et al., 2004).

Four genes (located on chromosome 1) have been described that are required for sodium-dependent flagellar rotation: *motA* (a.k.a. *pomA*), *motB* (a.k.a. *pomB*), *motX*, and *motY* (McCarter, 1994a,b; Okunishi et al., 1996; Asai et al., 1997; Gosink and

Hase, 2000; Boles and McCarter, 2000; Okabe et al., 2001). Loss of function of any one of the four motor genes completely abolishes motility but does not prevent flagellar assembly. Interestingly, genome sequencing of *V. fischeri* has revealed an extra set of *motA* and *motB* genes on chromosome 2 (Ruby et al., 2005). Much intensive work has been done on the sodium-driven motor of *V. alginolyticus*, and a detailed molecular understanding of sodium-type motor function is being developed (reviewed by Yorimitsu and Homma, 2001).

The MotA-MotB complex: the stator

With respect to membrane topology and function, the *Vibrio* sodium-type proteins resemble MotA and MotB of the proton-type motor (Asai et al., 1997). MotA contains four transmembrane domains, whereas MotB contains one transmembrane domain and a C-terminal peptidoglycan-interacting domain. The C-terminal domain of MotB is essential for motor function (Yakushi et al., 2005). Much data suggest that these proteins form the Na^+-conducting channel. Initial evidence implicating *Vibrio* MotA and MotB in Na^+ translocation was provided by the isolation of mutants that could swim in the presence of the sodium-channel inhibitor phenamil (Jaques et al., 1999; Kojima et al., 1999a). Some of the mutations conferring phenamil resistance also altered the ion specificity of the motor, resulting, for example, in increased or decreased swimming rates in the presence of lithium ions (Jaques et al., 1999). Reconstitution experiments in liposomes provided strong proof that purified MotA-MotB complexes catalyzed Na^+ flux (Sato and Homma, 2000a). To further probe the stoichiometry of the torque generator, a chimeric construct was designed to produce a fused dimer of MotA (Sato and Homma, 2000b). Such a dimer was found to support motility; furthermore, mutational inactivation of either half of the fusion dimer caused complete loss of swimming motility and loss of ability to interact with MotB. This evidence suggests that dual MotA components are essential for motor function. Furthermore, purification studies indicated that MotA and MotB physically interact, can be purified as a complex of two MotB and four MotA subunits, and perhaps may form larger assemblies (Sato and Homma, 2000a; Yakushi et al., 2004a; Yorimitsu et al., 1999, 2004). Intermolecular cross-linking studies using cysteine substitutions have demonstrated that adjacent MotA molecules are physically close within the motor (Yorimitsu et al., 2000) and that MotA and MotB molecules are also proximal to each other (Yakushi et al., 2004b; Yorimitsu et al., 2004).

Torque generation and ion specificity

Current models for torque generation (H^+ or Na^+ driven) suggest that conformational changes in the stator (MotA-MotB) are induced upon ion influx, and these changes are transmitted from a cytoplasmic region of MotA to FliG of the rotor, via electrostatic interactions. Random and site-directed mutagenesis has been used to probe motor structure and function (Kojima et al., 1999b, 2000; Asai et al., 2000b; Yorimitsu et al., 2002). These studies suggest that there may be some differences in the rotor/stator interface of the H^+ and the Na^+ motor. For example, a single electrostatic interaction between a positively and negatively charged amino acid can be critical for torque generation in *E. coli* (Zhou et al., 1998) whereas multiple (and different) charge interactions may be at work in *V. alginolyticus* (Yorimitsu et al., 2001, 2003; Fukuoka et al., 2004). Perhaps more interactions are required for the sodium-type motor, which spins much faster than the proton-type motor.

Nevertheless, experiments using chimeras of mixed sodium and proton motor parts provide strong evidence that the general mechanism of torque generation is the same in the two kinds of motors. *Escherichia coli* motor proteins can weakly substitute to power motility using the proton motive force in a *V. cholerae* strain eviscerated for the native sodium motor genes (Gosink and Hase, 2000). Sodium-dependent motility could be restored to a MotA-deficient *V. alginolyticus* strain on provision of the proton-type MotA from *Rhodobacter sphaeroides* (Asai et al., 1999). Although MotB$_{rs}$ was not functional with *V. alginolyticus* motor parts, MotB chimeras containing portions of the N-terminal transmembrane domain of *R. sphaeroides* and the C-terminal, linker-peptidoglycan binding domains of *V. alginolyticus* were able to reconstitute Na^+-driven motility (Asai et al., 2000a). Thus, chimeras have also provided information on coupling ion specificity. Most informative was the successful construction of a *Vibrio/E. coli* hybrid MotB that functions as a sodium motor solely with MotA$_{va}$ without a requirement for MotX or MotY (Asai et al., 2003).

Unique motor components

The MotX and MotY proteins are also required components of the *Vibrio* polar motor, and yet their specific roles are not known. The genes encoding these proteins are not linked. Loss of function of either *motX* or *motY* produces a paralyzed mutant completely defective for swimming but competent for flagellar assembly (McCarter, 1994a, b; Okunishi et al., 1996; Gosink and Hase, 2000; Okabe et al., 2001). The C terminus of MotY contains a peptidoglycan-interaction domain that shows striking homology to a number of open reading frames known to interact with peptidoglycan, e.g., OmpA and peptidoglycan-associated lipoproteins (McCarter, 1994b). The simplest hypothesis for the role of MotY is that the polar flagellar motor possesses two elements for anchoring the force generator. The role of MotX is even less clear, although it is known that MotX and MotY interact (McCarter, 1994a). MotX and MotY have been shown to colocalize to the outer membrane in *V. alginolyticus*, which is a unique location for motor components (Okabe et al., 2002). Originally, *motX* and *motY* were believed to be components specific to the sodium-type motor: however, essential roles for additional MotY-like proteins have now been identified for proton-type motors (specifically, the lateral flagellar motor of *V. parahaemolyticus* and one of the flagellar motors of *P. aeruginosa*) (Stewart and McCarter, 2003; Doyle et al., 2004). Furthermore, MotB chimeras functioning solely with MotA without MotX and MotY as sodium motors directly show that there is no absolute requirement for MotX and MotY contributions to sodium-driven motility (Asai et al., 2003). So, although essential in *Vibrio* for motor function, the role of MotX and MotY remains mysterious.

Regulation of Polar Flagellar Gene Expression

As can be seen in Table 2, a considerable number of genes are required to encode a flagellar motility system; therefore, maintenance of flagellation is a sizable investment with respect to cellular economy. As a result, flagellar systems are highly regulated. In systems in which it has been studied, the general scheme of gene control represents a hierarchical cascade of regulation that couples sequential expression of specific classes of genes to assembly of the organelle. Genes in each temporal class must be functional for expression of the subsequent class to occur. This also seems to be the case in the *Vibrionaceae*; the potential hierarchy is outlined in Fig. 2. Two specialized sigma factors are required for flagellar gene expression: RpoN, or σ^{54}, and FliA, or σ^{28} (Kawagishi et al., 1997; O'Toole et al., 1997; Klose and Mekalanos, 1998b; Kim and McCarter, 2000; Stewart and McCarter, 2003; Wolfe et al., 2004). The hierarchy seems generally consistent in all *Vibrio* spp. that have been studied; it has been delineated by using mutant, sequence, and transcriptional fusion analyses in combination with primer extension mapping of promoter sequences (Kim and McCarter, 2000; Prouty et al., 2001; Millikan and Ruby, 2003, 2004).

Class 1: a σ^{54}-dependent master flagellar regulatory protein

FlrA (a.k.a. FlaK) appears to be the master transcriptional regulator of polar genes. The central portion of FlrA contains a σ^{54}-interaction domain. FlrA and σ^{54} are required for expression of class 2 flagellar genes. Loss of function of *flrA* renders *V. fischeri* completely nonmotile (Millikan and Ruby, 2003), causes a profound (but not absolute) motility defect in *V. cholerae* (Klose and Mekalanos, 1998b), and has little effect on swimming motility in *V. parahaemolyticus* (Stewart and McCarter, 1996). The difference in the *V. parahaemolyticus* phenotype can be accounted for by the lateral flagellar regulator LafK, which compensates for loss of FlaK (FlrA) (Kim and McCarter, 2004). There is some evidence in *V. cholerae* and *V. fischeri* that FlrA may also regulate nonflagellar genes (Klose and Mekalanos, 1998b; Millikan and Ruby, 2003).

Class 2

Genes in class 2 encode the type 3 flagellar assembly apparatus, regulators FlhF and FlhG, chemotaxis proteins, σ^{28} and its anti-sigma factor FlgM, a sensor kinase FlrB, and an additional σ^{54}-dependent regulator FlrC (Prouty et al., 2001). FlrC is a two-component response regulator that can be phosphorylated by FlrB (Correa et al., 2000). FlhF and FlhG each possess nucleotide binding domains and, although they are not transcription factors, participate in regulating flagellar placement and number.

Class 3

FlrC is absolutely required for swimming motility and directs σ^{54}-dependent expression of the class 3 genes encoding the hook and basal body components, the critical flagellin (FlaA), filament adaptor proteins (or HAPs), and one motor protein (Prouty et al., 2001). FlrC may regulate other, nonflagellar genes, as *flrC V. cholerae* mutants display a more profound virulence defect than mutants with flagellar structural defects (Correa et al., 2000).

Class 4

The final proteins required to complete flagellar biogenesis are produced only after the hook-basal body structure is completed. This timing of gene expression is coupled to morphogenesis of the organelle via a mechanism conserved in flagellar systems (reviewed by Aldridge and Hughes, 2002). The transcription of class 4 genes is directed by the special-

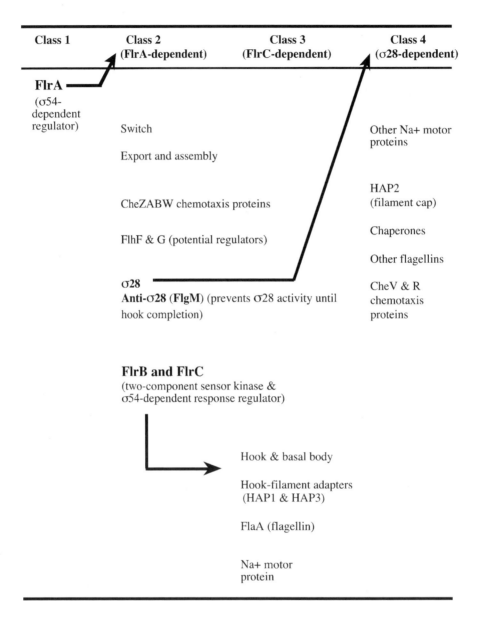

Figure 2. Regulatory hierarchy and morphogenetic pathway for polar flagellar biogenesis. Flagellar assembly is an ordered process, initiating with integral membrane proteins forming a scaffold for the export and assembly of the remainder of the organelle and culminating with polymerization of the filament (reviewed by Macnab, 2004). Gene expression is strictly regulated and tied into morphogenesis of the organelle (reviewed by Aldridge and Hughes, 2002). Early *Vibrio* polar gene expression requires σ^{54} and late gene expression σ^{28}. The scheme of control seems conserved for the *Vibrio* sp. in which it has been studied, and this figure, which depicts a general pathway for gene expression and organelle biogenesis, represents a summary of what is currently known (Kim and McCarter, 2000; Prouty et al., 2001; Millikan and Ruby, 2003, 2004). The master flagellar regulator FlrA (a.k.a. FlaK) directs transcription of class 2 genes, which encode the export and assembly apparatus and other regulators that act sequentially. FlhF and FlhG regulate flagellar number and placement by an unknown mechanism, but one that also influences transcription (Correa et al., 2005). Expression of class 3 genes is controlled by the activity of a two-component regulator, FlrC (a.k.a. FlaM) and results in the assembly of the hook and basal body. Completion of the hook allows export of the anti-sigma factor FlgM, which frees the σ^{28} to direct transcription of class 4 genes (Correa et al., 2004).

ized σ^{28} factor, and its activity is kept in check by the anti-sigma factor FlgM. On completion of the hook-basal body, FlgM is secreted through the flagellar export apparatus and σ^{28} is thereby freed to transcribe late gene expression (Correa et al., 2004). Thus, although the flagellar sheath can act as a barrier to trap flagellin subunits (McCarter, 1995, 2001; Nishioka et al., 1998), FlgM (and perhaps other proteins) can be effectively secreted through the sheathed flagellum (Correa et al., 2004).

SWARMING: THE LATERAL FLAGELLAR SYSTEM

In liquid environments, the swimming speed of the marine *Vibrios* is approximately 100 μm/s. However, as viscosity increases, the polar flagellum is not an effective propulsive organelle: increasing viscosity slows swimming (Atsumi et al., 1996; Kawagishi et al., 1996). In compensation, the lateral flagella system is induced by growth on surfaces (reviewed by McCarter, 2004). These peritrichous flagella are quite functional in viscous environments and enable the bacterium to move over and colonize surfaces (Shinoda and Okamoto, 1977; McCarter, 1999).

Lateral Flagellar Genes

To date, the genome of only a single organism with a lateral system has been sequenced. The genetics of the system has been dissected by mutational and transcriptional analyses (McCarter and Wright, 1993; Stewart and McCarter, 2003; Jaques, 2004). Like the polar system, the lateral genes are named with the common designations used in other flagellar systems when identifiable counterparts exist; the subscripts P and L are used to distinguish polar and lateral genes, respectively. Lateral genes were originally assigned names that were permutations of the standard flagellar nomenclature (e.g., *lfg* for *flg*); these names are still used in other bacteria (e.g., *Aeromonas* species and *E. coli* [Kirov, 2003; Ren et al., 2005]). Whereas polar flagellar genes are found exclusively on the large chromosome, all of the lateral flagellar genes are on the small chromosome (which is ~1.9 Mb). There is no general bias in the G+C content, which is ~46% for lateral as well as polar genes (compared to 45% for the entire genome). There is a full set of lateral flagellar genes, distinct from the polar set, encoding structural, export, and assembly as well as regulatory components (e.g., the lateral system has its own flagellar-specific σ^{28} as well as a distinct type 3 flagellar secretion system). The genes and their organization are presented in Table 3.

Some notable differences can be found between the complement of *V. parahaemolyticus* lateral flagellar genes and its polar genes or the genes of *E. coli*. The lateral, but not the polar, system lacks an *fliO* homolog. FliO is a membrane component of the flagellar export apparatus, but its precise role in export is not defined (Macnab, 2004). Although there are six polar flagellin genes (*flaA, B, C, D, E,* and *F*), only a single gene (*lafA*) encodes subunits for the lateral filament (McCarter and Wright, 1993). The lateral flagellar system has no *flhF* and *flhG* homologs. However, it does have some novel genes: one *is flgO*. It is essential for motility (Jaques, 2004), and its predicted product is a lytic transglycosylase. The *V. parahaemolyticus flgJ*$_L$ gene is unusual. *E. coli* FlgJ functions as a peptidoglycan hydrolase and a rod capping protein (Macnab, 2004). Lateral FlgJ$_{vp}$ is truncated, lacking the muramidase domain. Two genes in the lateral system of *V. parahaemolyticus* may encode the dual functions of FlgJ$_{EC}$. There are also a few open reading frames, found within lateral flagellar gene clusters, of unknown function.

Lateral Propulsion: the Proton-Driven Flagellar Motor

Although *V. alginolyticus* and *V. parahaemolyticus* utilize sodium-driven motors for swimming in dilute liquid environments, the energy force driving rotation of the lateral flagella is derived from the proton motive force (Atsumi et al., 1992; Kawagishi et al., 1995). Three genes are required for the lateral, proton-type motor function: *motA, motB,* and *motY* (Stewart and McCarter, 2003); whereas four genes are required for the polar, sodium-type motor function: *motA, motB, motX,* and *motY* (and only two genes for the *E. coli* proton-driven motor). Each of these motor genes is specific for its dedicated motility system; there is no functional overlap in polar and lateral motor parts.

Regulation of Lateral Flagellar Gene Expression

The hierarchy of lateral flagellar gene expression and organelle biogenesis is depicted in Fig. 3. In some respects, lateral regulation is more similar to the polar system than it is to the regulatory hierarchy of peritrichous, proton-driven flagella employed for swimming and swarming by *E. coli*, *Salmonella enterica* serovar Typhimurium, or *Proteus mirabilis*. Specifically, early gene transcription requires σ^{54} and a σ^{54}-dependent transcription factor, LafK, which is similar to the polar regulator FlaK (Stewart and McCarter, 2003). Late lateral gene expression requires a lateral flagellar-specific σ^{28} (McCarter and Wright, 1993).

Table 3. *V. parahaemolyticus* lateral flagellar genes[a]

Gene name	Alternative gene name	VP locus tag	Predicted function
Region 1			
flgO	*lfgO*	VPA0260	Murein transglycosylase
flgN	*lfgN*	VPA0261	Chaperone for FlgK and FlgL
flgM	*lfgM*	VPA0262	Anti-σ^{28} factor
flgA	*lfgA*	VPA0263	P-ring chaperone
flgB	*lfgB*	VPA0264	Proximal rod
flgC	*lfgC*	VPA0265	Proximal rod
flgD	*lfgD*	VPA0266	Hook cap
flgE	*lfgE, lafX*	VPA0267	Hook
flgF	*lfgF*	VPA0268	Proximal rod
flgG	*lfgG*	VPA0269	Distal rod
flgH	*lfgH*	VPA0270	L ring
flgI	*lfgI*	VPA0271	P ring
flgJ	*lfgJ*	VPA0272	Peptidoglycan hydrolase; rod cap
flgK	*lfgK*	VPA0273	HAP1; hook-filament junction
flgL	*lfgL*	VPA0274	HAP3; hook-filament junction
flgU	*lfgU*	VPA0275	Unknown but required; axial component?
Region 2			
fliJ	*lfiJ*	VPA1532	Laf export
fliI	*lfiI*	VPA1533	Laf export; ATPase
fliH	*lfiH*	VPA1534	Laf export; negative regulator of FliI
fliG	*lfiG*	VPA1535	Rotor/switch component
fliF	*lfiF*	VPA1536	MS ring
fliE	*lfiE*	VPA1537	Hook-basal body component; MS ring-rod junction
lafK		VPA1538	σ^{54}-dependent regulator
motY	*lafY*	VPA1539	H^{+} motor component
fliM	*lfiM*	VPA1540	C ring/switch component
fliN	*lfiN*	VPA1541	C ring/switch component
fliP	*lfiP*	VPA1542	Laf export
fliQ	*lfiQ*	VPA1543	Laf export
fliR	*lfiR*	VPA1544	Laf export
flhB	*lfhB*	VPA1545	Laf export
flhA	*lfhA*	VPA1546	Laf export
ORF		VPA1547	Unknown
lafA		VPA1548	Flagellin
ORF		VPA1549	Unknown
fliD	*lafB*	VPA1550	HAP2; hook-associated protein 2, filament cap
fliS	*lafC*	VPA1551	Chaperone?
fliT	*lafD*	VPA1552	Chaperone?
fliK	*lafK*	VPA1553	Hook-length control
fliL	*lafL*	VPA1554	Unknown
fliA	*lafS*	VPA1555	RNA polymerase σ^{28} factor
motA	*lafT*	VPA1556	H^{+} motor component
motB	*lafU*	VPA1557	H^{+} motor component

[a] All lateral flagellar genes are found on chromosome 2 of *V. parahaemolyticus*.

In fact, there may be some cross-regulation of the two flagellar systems operating through LafK (Kim and McCarter, 2004).

FUNCTIONAL DIFFERENCES BETWEEN POLAR AND LATERAL FLAGELLA

Why are there two flagellar systems? The two flagellar systems operate to propel the bacteria under different circumstances. The polar flagellum, which is shared by all *Vibrio* spp., is produced continuously and is used for swimming. The lateral flagella, which are found in only some species, are used for swarming. The polar filament, composed of multiple flagellin subunits, is sheathed by membrane and can rotate very fast, using energy derived from the sodium membrane potential. However, this motor seems to be optimized to rotate exceedingly fast in low viscosity, and it performs quite poorly in environments of increased viscosity (Magariyama et al., 1995; Atsumi et al., 1996). The unsheathed lateral filaments, composed of a single flagellin, can also rotate, albeit not as quickly as the polar filaments, using energy derived from the proton motive force. Lateral flagella are synthesized only under conditions that do not favor polar flagellar rotation, i.e., growth on surfaces or viscous environments (Belas et al., 1986; McCarter and Silverman, 1990). Lateral flagella perform well under such conditions to allow highly effective movement over and colonization of surfaces (Belas and Colwell, 1982; Atsumi et al., 1996). Thus, a polar flagellum seems to provide maximal swimming proficiency, whereas numerous lateral flagella support the most vigorous swarming.

CHEMOTAXIS

Being able to move is not enough—navigation is also important. Chemotaxis integrates environmental signaling to modulate behavior by biasing movement toward more favorable conditions or away from unfavorable environments. One consequence is that not just motility, but also chemotaxis is important for survival and colonization. For example, motility and chemotaxis have been shown to play a role in the virulence of *V. anguillarum* (Graf et al., 1994; O'Toole et al., 1996; Larsen and Boesen, 2001). In *V. cholerae*, a number of studies show differential expression of various motility and chemotaxis genes under in vivo and in vitro conditions (Das et al., 2000; Lee et al., 2001; Banerjee et al., 2002; Merrell et al., 2002; Hang et al., 2003; Xu et al., 2003). For a more thorough review focusing specifically on *V. cholerae* chemotaxis, the reader is referred to Boin et al. (2004).

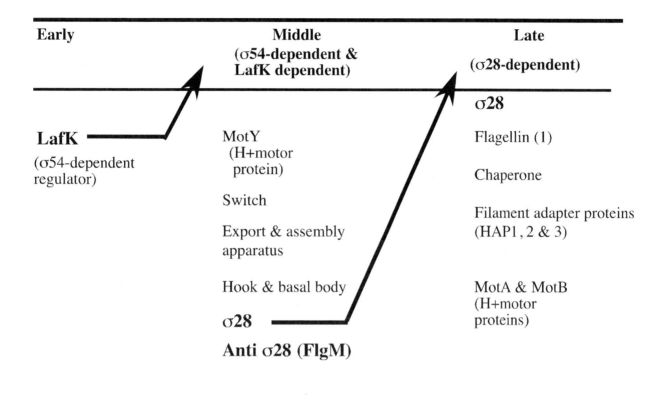

Figure 3. Regulatory hierarchy and morphogenetic pathway for lateral flagellar biogenesis. Although it is a distinct flagellar system, with its own complete set of flagellar genes, the lateral flagellar hierarchy of *V. parahaemolyticus* resembles the polar hierarchy in that a σ^{54}-dependent transcriptional regulator directs expression of intermediate genes, and σ^{28} controls late flagellar gene expression (Stewart and McCarter, 2003). The master lateral regulator LafK has also been shown to affect polar gene transcription (Kim and McCarter, 2004).

Behavior

Bacterial chemotaxis has been most extensively studied in organisms with a few (~6 to 10) peritrichously arranged flagella, e.g., *E. coli*, *Salmonella enterica* serovar Typhimurium, and *B. subtilis* (reviewed by Blair, 1995; Szurmant and Ordal, 2004; Wadhams and Armitage, 2004). Seen by light microscopy, the cells move in alternating periods of smooth, forward trajectories of swimming (called running) and chaotic movement (called tumbling). Forward translation occurs when the flagellar motor rotates in a direction (e.g., counterclockwise) that results in propagation of the semirigid helical wave of the flagellum, which acts like a propeller and exerts a pushing motion on the cell. Wave propagation proceeds from the cell proximal to the distal end of the flagellum, and the peritrichous flagella coalesce to rotate synchronously and form a propulsive bundle (reviewed by Macnab,

2003). When the filaments are rotated in the opposite direction (clockwise), structural changes are induced within the filament, the bundle dissociates, and the cell tumbles, which results in random reorientation. In the absence of chemotaxis, bacteria such as *E. coli* move randomly in the alternating pattern of smooth swimming and tumbling. In the presence of chemotactic stimuli, the time spent in one mode is biased; e.g., in the presence of a gradient of attractant, the probability of tumbling is decreased, thereby prolonging the smooth period of swimming. Thus, chemotaxis controls the frequency of switching of the direction of flagellar rotation and hence the frequency of tumbling.

In the light microscope, *Vibrio* cells do not run and tumble. Rather, the bacteria swim and reverse. Movement is in smooth, slightly curved lines, punctuated by periods of directional change. Sometimes there is a reversal of direction so that the cells back up, while at other times, a small circular path of swimming or an abrupt change of direction can be observed. Thus, the mechanism of taxis is not the same for polarly flagellated bacteria: reversal of the direction of flagellar rotation produces forward and backward swimming cells with the flagellum pushing or pulling the cell body (Magariyama et al., 2001). Chemotaxis mutants are biased toward swimming in one direction (forward or backward) (Homma et al., 1996). Although back-and-forth reversals of direction are not precise and may provide random directional alteration (similar to a tumble), additional mechanics may contribute to polar chemotaxis. The backward swimming speed is ~1.5 times faster than the forward swimming speed; such changes in swimming speed may provide a fundamental contribution to *Vibrio* taxis (Magariyama et al., 2001). Since differences in swimming speed have significant consequences with respect to attachment to glass (Kogure et al., 1998), this forward and backward speed may be of particular relevance within a host organism. Thus, it seems quite interesting that *che* genes are differentially regulated within hosts (Das et al., 2000; Lee et al., 2001; Banerjee et al., 2002; Merrell et al., 2002; Hang et al., 2003; Xu et al., 2003) and, moreover, that *che* mutants with different rotational biases show profoundly different colonization phenotypes in the host (Larsen and Boesen, 2001; Butler and Camilli, 2004). Another highly interesting observation is that the backward swimming trajectory, which is straight and similar to the forward trajectory in dilute liquid environments, becomes curved as the bacterium approaches a surface (Kudo et al., 2005; Magariyama et al., 2005). Such an observation suggests the intriguing potential for an alternative, or modified, chemotaxis strategy when the bacterium is near a surface.

Chemoattractants and Repellents

Numerous *Vibrio* species have been reported to undergo chemotaxis to a variety of amino acids and carbohydrates (Freter and O'Brien, 1981; Sar et al., 1990). Serine is a particularly good chemoattractant, whereas aspartate chemotaxis has not been observed. Many pathogenic *Vibrio* species display movement toward mucus. For example, *V. anguillarum* and *V. alginolyticus* undergo chemotaxis to mucus collected from fish skin and intestines (Bordas et al., 1998); *V. shiloi*, a coral pathogen, migrates toward coral mucus (Banin et al., 2001), and *V. cholerae* moves into intestinal mucus (Freter et al., 1981). There can be broad specificity in the attraction to mucus; e.g., the human intestinal pathogen *V. cholerae* will migrate toward porcine and fish mucus, and *V. anguil-

larum will move toward human and porcine mucus (O'Toole et al., 1999). However, species specificity to components of mucus has also been observed. *V. fischeri* displays unique chemotaxis to *N*-acetylneuraminic acid, a component of the mucus derived from the Hawaiian squid light organ (DeLoney-Marino et al., 2003). Chemotaxis to soluble chitin oligosaccharides has been particularly well characterized in *V. furnissii* (Bassler et al., 1991; Yu et al., 1993). Algal products also serve as *Vibrio* chemoattractants (Sjoblad and Mitchell, 1979). One most interesting and unusual set of chemoattractants has been reported for *V. fischeri*: nucleosides (DeLoney-Marino et al., 2003). One repellent that has been studied is phenol, which elicits a rapidly reversing phenotype (back-and-forth movement) (Homma et al., 1996).

Chemoreceptors

It seems likely that the repertoire of chemoattractants for the *Vibrionaceae* is very large. The genomes of the sequenced *Vibrio* species reveal a plethora of potential chemoreceptors, and they are found on both chromosomes. *V. cholerae*, *V. vulnificus*, and *V. fischeri* each contain ~40 or more potential methyl-accepting chemotaxis protein (MCP) genes, and *V. parahaemolyticus* contains ~30. Although these genes are annotated as potential MCPs, it is not clear whether they all participate in coordinating movement. This large number of MCPs nevertheless suggests a great capacity for sensing and responding to signals in the environment. Elucidating the specific signals (and the signal transduction pathways) for each MCP promises to be a complex task. Some MCP localization has been performed, using antibodies directed to an *E. coli* chemoreceptor. In swimmer cells, the MCPs localize to both poles; in the swarmer cells, MCPs are found at the poles and at intervals along the cell body (Gestwicki et al., 2000).

Central Cytoplasmic Chemotaxis Genes

A general scheme for chemotaxis (reviewed by Szurmant and Ordal, 2004; Wadhams and Armitage, 2004) has been found for most flagellated bacteria. It involves signal reception by the MCP and a conformational change in the receptor modulating a phosphorelay cascade that works through the coupling protein CheW, the kinase CheA, and CheY. Phosphorylated CheY (CheY~P) is the output regulator that interacts with the flagellar motor and causes clockwise rotation. Another chemotaxis protein, the phosphatase CheZ, destroys CheY~P. Adaptation, or receptor sensitivity, is determined by the methylating and demethylating enzymes CheR and

CheB, respectively. CheB activity is controlled by phosphate transfer from CheA~P. Some organisms, such as *B. subtilis* (Szurmant and Ordal, 2004), have additional Che components: CheD and CheV. CheD acts to modify (deamidate) residues in the MCP, resulting in greater receptor sensitivity. CheV is a CheW/CheY hybrid.

Multiple copies of the central chemotaxis genes can be found in *Vibrio* genomes (Table 4). For example, *V. cholerae* has three clusters of *che*-like genes, as well as some orphan *che* paralogs scattered about the chromosome (Boin et al., 2004). However, evidence is emerging that not all of the gene sets participate in a chemosensory signal transduction cascade controlling flagellar rotation and cell motility (or at least they do not operate to control motility in laboratory conditions). The three *cheA* genes of *V. cholerae* were deleted, and only one (*cheA2*) affected swimming motility (Gosink et al., 2002); similarly, of the five potential CheY proteins in *V. cholerae*, only one interacts with the flagellar motor (Hyakutake et al., 2005). Therefore, in *V. cholerae* only the genes in chemotaxis cluster 2 have been functionally implicated in chemotaxis. Significantly, this is the cluster that is found in *V. parahaemolyticus and V. fischeri*. Although these latter two organisms have only one complete set of central chemotaxis genes, they do possess multiple copies of the MCP-coupling proteins CheW and CheV. So, just as there is a great reservoir of sensory transducers, some *Vibrio* species have multiple potential chemotaxis-like signal transduction cascades (genome sequences show three in *V. cholerae* and two in *V. vulnificus*); however, present evidence only implicates one gene set as controlling motility.

One Chemotaxis System Coordinates Swimming and Swarming

The navigation system is one point of overlap in the *Vibrio* species with dual flagellar systems. By using mutant strains that were proficient in only one of the two motility systems (i.e., Polar$^+$ Lateral$^-$ and Polar$^-$ Lateral$^+$), effects of chemotaxis defects could be assessed separately for swimming and swarming (Sar et al., 1990). Insertions in the *cheYZABW* or *cheVR* locus of *V. parahaemolyticus* affected both polar and lateral motility. Although the central chemotaxis genes are shared, some differences between the polar and lateral chemotactic responses have been observed, specifically with respect to adaptation to the repellent phenol (Homma et al., 1996). This might be a consequence of having one versus hundreds of flagellar motors, or there may be specific chemotaxis components operating in one flagellar system and not the other. Although it is clear that chemotaxis coordinates lateral flagella and swarming motility, it is somewhat provocative to imagine how the chemotaxis systems effectively exerts control on so many flagellar motors (see Fig. 1).

NONFLAGELLAR MOTILITY

Although this chapter has focused on flagellar-mediated motility, the capacity for other forms of *Vibrio* motility should not be discounted. One novel form of surface motility that has been discovered, but is not understood, is a flagellum-independent spreading motility observed in *V. cholerae* (Brown and Hase,

Table 4. Multiple copies of potential chemotaxis genes are encoded in *Vibrio* genomes

Gene	Predicted function	No. of potential genes[a]			
		V. cholerae El Tor N16961	*V. parahaemolyticus* RIMD	*V. vulnificus*[b]	*V. fischeri* ES114
cheA	Sensor kinase	3(1)[c]	1	2	1
cheB	Methyl esterase	2	1	2	1
cheD	Deaminase	1		1	
cheR	Methyl transferase	3	1	3	1
cheV	CheW/CheY hybrid	4	4	4	3
cheW	Chemotaxis adapter protein	4	1	4	2
cheY	Response regulator tumble effector	5(1)[d]	2(1)[e]	2	1
cheZ	CheY phosphatase	1	1	1	1
MCP	Methyl-accepting chemotaxis (and aerotaxis) receptors	~45	~29	~45	~39

[a] Annotated as chemotaxis or putative chemotaxis genes at http://www.tigr.org and http://www.ergo-light.com.
[b] Numbers of genes are the same for *V. vulnificus* strains CMCP6 and YJ016.
[c] One *cheA* gene has been shown to be essential for chemotaxis (Gosnik et al., 2002).
[d] One *cheY* gene has been shown to be essential for chemotaxis (Hyakutake et al., 2005).
[e] One *cheY* gene has been shown to be essential for chemotaxis (unpublished observation).

2001). Also, genome sequence suggests that there may be an undiscovered capacity for twitching motility, as type IV pili genes can be found in *Vibrio* genomes.

SURFACE SENSING: POLAR FLAGELLAR PERFORMANCE AFFECTS GENE EXPRESSION

Evidence suggests that, in addition to its role as a propulsive organelle, the polar flagellum can act as a sensor informing the cell of contact with surfaces or viscous environments. As a result, polar flagellar performance influences gene expression. In the case of *V. parahaemolyticus*, contact with surfaces induces swarming, whereas in the case of *V. cholerae*, the status of the polar flagellum seems to influence the production of extracellular polysaccharide. Thus, under difficult conditions for swimming (i.e., on a surface), *Vibrio* spp. respond by inducing alternative strategies for survival and colonization.

Lateral Flagella

In *V. parahaemolyticus*, mutations that have been isolated that affect polar flagellar performance cause expression of lateral flagellar genes in liquid. These include mutations that affect assembly and rotation (McCarter et al., 1988; Boles and McCarter, 2000; McCarter, 2001). Mutations in the central chemotaxis genes (Che⁻) do not perturb lateral flagellar gene expression (unpublished observations). Not only is there genetic linkage between the polar and the lateral flagellar systems, but there is also a functional linkage. Physical conditions that impede polar flagellar performance, such as high viscosity or antibody agglutination, induce lateral flagellar gene expression (reviewed by McCarter and Silverman, 1990). Most important, polar rotation can be directly manipulated by using the sodium channel-blocking drug phenamil. Increasing concentrations of phenamil slow polar flagellar rotation and concomitantly induce lateral flagellar gene expression (Kawagishi et al., 1996). Thus, the flagellum seems to act as a mechanosensor: interference with rotation signals swarmer cell gene expression and lateral flagella production (McCarter et al., 1988; McCarter and Silverman, 1990).

Extracellular Polysaccharide

Flagellar mutants of *V. cholerae* O139 have rugose colony morphologies due to overproduction of extracellular polysaccharide. These include mutants with defects in the critical flagellin gene *flaA* and mutants with defects in the flagellar regulators *flrA* and *flrC*; however, motor mutations do not have such an effect on colony morphology (Watnick et al., 2001). Thus, it seems that rather than flagellar performance per se, *V. cholerae* may be sensing the presence or absence of a flagellum (Watnick et al., 2001). Alternatively, it has been proposed that the flagellar motor itself directly participates in a signal transduction cascade influencing expression of *Vibrio* polysaccharide (*vps*) genes (Lauriano et al., 2004). The result is that polar flagellar function influences expression of cell surface polysaccharide, which has important consequences for biofilm formation and host colonization (Watnick et al., 2001; Lauriano et al., 2004).

REFERENCES

Aldridge, P., and K. T. Hughes. 2002. Regulation of flagellar assembly. *Curr. Opin. Microbiol.* **5:**160–165.

Allen, R. D., and P. Baumann. 1971. Structure and arrangement of flagella in species of the genus *Beneckea* and *Photobacterium fischeri*. *J. Bacteriol.* **107:**295–302.

Asai, Y., I. Kawagishi, R. E. Sockett, and M. Homma. 2000a. Coupling ion specificity of chimeras between H(+)- and Na(+)-driven motor proteins, MotB and PomB, in *Vibrio* polar flagella. *EMBO J.* **19:**3639–3648.

Asai, Y., T. Shoji, I. Kawagishi, and M. Homma. 2000b. Cysteine-scanning mutagenesis of the periplasmic loop regions of PomA, a putative channel component of the sodium-driven flagellar motor in *Vibrio alginolyticus*. *J. Bacteriol.* **182:**1001–1007.

Asai, Y., I. Kawagishi, R. E. Sockett, and M. Homma. 1999. Hybrid motor with H(+)- and Na(+)-driven components can rotate *Vibrio* polar flagella by using sodium ions. *J. Bacteriol.* **181:**6332–6338.

Asai, Y., S. Kojima, H. Kato, N. Nishioka, I. Kawagishi, and M. Homma. 1997. Putative channel components for the fast-rotating sodium-driven flagellar motor of a marine bacterium. *J. Bacteriol.* **179:**5104–5110.

Asai, Y., T. Yakushi, I. Kawagishi, and M. Homma. 2003. Ion-coupling determinants of Na+-driven and H+-driven flagellar motors. *J. Mol. Biol.* **327:**453–463.

Atsumi, T., Y. Maekawa, T. Yamada, I. Kawagishi, Y. Imae, and M. Homma. 1996. Effect of viscosity on swimming by the lateral and polar flagella of *Vibrio alginolyticus*. *J. Bacteriol.* **178:**5024–5026.

Atsumi, T., L. McCarter, and Y. Imae. 1992. Polar and lateral flagellar motors of marine *Vibrio* are driven by different ion-motive forces. *Nature* **355:**182–184.

Banerjee, R., S. Das, K. Mukhopadhyay, S. Nag, A. Chakrabortty, and K. Chaudhuri. 2002. Involvement of in vivo induced *cheY-4* gene of *Vibrio cholerae* in motility, early adherence to intestinal epithelial cells and regulation of virulence factors. *FEBS Lett.* **532:**221–226.

Banin, E., T. Israely, M. Fine, Y. Loya, and E. Rosenberg. 2001. Role of endosymbiotic zooxanthellae and coral mucus in the adhesion of the coral-bleaching pathogen *Vibrio shiloi* to its host. *FEMS Microbiol. Lett.* **199:**33–37.

Bassler, B. L., P. J. Gibbons, C. Yu, and S. Roseman. 1991. Chitin utilization by marine bacteria. Chemotaxis to chitin oligosaccharides by *Vibrio furnissii*. *J. Biol. Chem.* **266:**24268–24275.

Baumann, P., and L. Baumann. 1981. The marine Gram-negative eubacteria: genera *Photobacterium*, *Beneckea*, *Alteromonas*, and *Alcaligenes*, p. 1302–1331. *In* M. Starr, H. Stolp, H. Truper, A.

Balows, and H. Schlegel (ed.), *The Prokaryotes*. Springer-Verlag, New York, N.Y.

Belas, M. R., and R. R. Colwell. 1982. Adsorption kinetics of laterally and polarly flagellated *Vibrio*. *J. Bacteriol.* **151:**1568–1580.

Belas, R., M. Simon, and M. Silverman. 1986. Regulation of lateral flagella gene transcription in *Vibrio parahaemolyticus*. *J. Bacteriol.* **167:**210–218.

Berg, H. C. 2003. The rotary motor of bacterial flagella. *Annu. Rev. Biochem.* **72:**19–54.

Blair, D. F. 2003. Flagellar movement driven by proton translocation. *FEBS Lett.* **545:**86–95.

Blair, D. F. 1995. How bacteria sense and swim. *Annu. Rev. Microbiol.* **49:**489–522.

Boin, M. A., M. J. Austin, and C. C. Hase. 2004. Chemotaxis in *Vibrio cholerae*. *FEMS Microbiol. Lett.* **239:**1–8.

Boles, B. R., and L. L. McCarter. 2000. Insertional inactivation of genes encoding components of the sodium-type flagellar motor and switch of *Vibrio parahaemolyticus*. *J. Bacteriol.* **182:**1035–1045.

Bordas, M. A., M. C. Balebona, J. M. Rodriguez-Maroto, J. J. Borrego, and M. A. Morinigo. 1998. Chemotaxis of pathogenic *Vibrio* strains towards mucus surfaces of gilt-head sea bream (*Sparus aurata L.*). *Appl. Environ. Microbiol.* **64:**1573–1575.

Brown, I. I., and C. C. Hase. 2001. Flagellum-independent surface migration of *Vibrio cholerae* and *Escherichia coli*. *J. Bacteriol.* **183:**3784–3790.

Butler, S. M., and A. Camilli. 2004. Both chemotaxis and net motility greatly influence the infectivity of *Vibrio cholerae*. *Proc. Natl. Acad. Sci. USA* **101:**5018–5023.

Carpenter, P. B., D. W. Hanlon, and G. W. Ordal. 1992. *flhF*, a *Bacillus subtilis* flagellar gene that encodes a putative GTP-binding protein. *Mol. Microbiol.* **6:**2705–2713.

Correa, N. E., J. R. Barker, and K. E. Klose. 2004. The *Vibrio cholerae* FlgM homologue is an anti-sigma28 factor that is secreted through the sheathed polar flagellum. *J. Bacteriol.* **186:**4613–4619.

Correa, N. E., C. M. Lauriano, R. McGee, and K. E. Klose. 2000. Phosphorylation of the flagellar regulatory protein FlrC is necessary for *Vibrio cholerae* motility and enhanced colonization. *Mol. Microbiol.* **35:**743–755.

Correa, N. E., F. Peng, and K. E. Klose. 2005. Roles of the regulatory proteins FlhF and FlhG in the *Vibrio cholerae* flagellar transcription hierarchy. *J. Bacteriol.* **187:**6324–6332.

Das, S., A. Chakrabortty, R. Banerjee, S. Roychoudhury, and K. Chaudhuri. 2000. Comparison of global transcription responses allows identification of *Vibrio cholerae* genes differentially expressed following infection. *FEMS Microbiol. Lett.* **190:**87–91.

Dasgupta, N., S. K. Arora, and R. Ramphal. 2000. fleN, a gene that regulates flagellar number in *Pseudomonas aeruginosa*. *J. Bacteriol.* **182:**357–364.

Dasgupta, N., and R. Ramphal. 2001. Interaction of the antiactivator FleN with the transcriptional activator FleQ regulates flagellar number in *Pseudomonas aeruginosa*. *J. Bacteriol.* **183:**6636–6644.

DeLoney-Marino, C. R., A. J. Wolfe, and K. L. Visick. 2003. Chemoattraction of *Vibrio fischeri* to serine, nucleosides, and N-acetylneuraminic acid, a component of squid light-organ mucus. *Appl. Environ. Microbiol.* **69:**7527–7530.

Doyle, T. B., A. C. Hawkins, and L. L. McCarter. 2004. The complex flagellar torque generator of *Pseudomonas aeruginosa*. *J. Bacteriol.* **186:**6341–6350.

Follett, E. A., and J. Gordon. 1963. An electron microscope study of *Vibrio* flagella. *J. Gen. Microbiol.* **32:**235–239.

Freter, R., B. Allweiss, P. C. O'Brien, S. A. Halstead, and M. S. Macsai. 1981. Role of chemotaxis in the association of motile bacteria with intestinal mucosa: in vitro studies. *Infect. Immun.* **34:**241–249.

Freter, R., and P. C. O'Brien. 1981. Role of chemotaxis in the association of motile bacteria with intestinal mucosa: chemotactic responses of *Vibrio cholerae* and description of motile nonchemotactic mutants. *Infect. Immun.* **34:**215–221.

Fuerst, J. A. 1980. Bacterial sheathed flagella and the rotary motor model for the mechanism of bacterial motility. *J. Theor. Biol.* **4:**761–774.

Fuerst, J. A., and J. W. Perry. 1988. Demonstration of lipopolysaccharide on sheathed flagella of *Vibrio cholerae* O:1 by protein A-gold immunoelectron microscopy. *J. Bacteriol.* **170:**1488–1494.

Fukuoka, H., T. Yakushi, and M. Homma. 2004. Concerted effects of amino acid substitutions in conserved charged residues and other residues in the cytoplasmic domain of PomA, a stator component of Na+-driven flagella. *J. Bacteriol.* **186:**6749–6758.

Furuno, M., K. Sato, I. Kawagishi, and M. Homma. 2000. Characterization of a flagellar sheath component, PF60, and its structural gene in marine Vibrio. *J. Biochem.* (Tokyo) **127:**29–36.

Gestwicki, J. E., A. C. Lamanna, R. M. Harshey, L. L. McCarter, L. L. Kiessling, and J. Adler. 2000. Evolutionary conservation of methyl-accepting chemotaxis protein location in Bacteria and Archaea. *J. Bacteriol.* **182:**6499–6502.

Gosink, K. K., and C. C. Hase. 2000. Requirements for conversion of the Na(+)-driven flagellar motor of *Vibrio cholerae* to the H(+)-driven motor of *Escherichia coli*. *J. Bacteriol.* **182:**4234–4240.

Gosink, K. K., R. Kobayashi, I. Kawagishi, and C. C. Hase. 2002. Analyses of the roles of the three cheA homologs in chemotaxis of *Vibrio cholerae*. *J. Bacteriol.* **184:**1767–1771.

Graf, J., P. V. Dunlap, and E. G. Ruby. 1994. Effect of transposon-induced motility mutations on colonization of the host light organ by *Vibrio fischeri*. *J. Bacteriol.* **176:**6986–6991.

Hang, L., M. John, M. Asaduzzaman, E. A. Bridges, C. Vanderspurt, T. J. Kirn, R. K. Taylor, J. D. Hillman, A. Progulske-Fox, M. Handfield, E. T. Ryan, and S. B. Calderwood. 2003. Use of in vivo-induced antigen technology (IVIAT) to identify genes uniquely expressed during human infection with *Vibrio cholerae*. *Proc. Natl. Acad. Sci. USA* **100:**8508–8513.

Homma, M., H. Oota, S. Kojima, I. Kawagishi, and Y. Imae. 1996. Chemotactic responses to an attractant and a repellent by the polar and lateral flagellar systems of *Vibrio alginolyticus*. *Microbiology* **142:**2777–2783.

Hranitzky, K. W., A. Mulholland, A. D. Larson, E. R. Eubanks, and L. T. Hart. 1980. Characterization of a flagellar sheath protein of *Vibrio cholerae*. *Infect. Immun.* **27:**597–603.

Hyakutake, A., M. Homma, M. J. Austin, M. A. Boin, C. C. Hase, and I. Kawagishi. 2005. Only one of the five CheY homologs of *Vibro cholerae* is involved in chemotaxis. *J. Bacteriol.* **187:**8403–8410.

Imae, Y., H. Matsukura, and S. Kobayashi. 1986. Sodium-driven flagellar motors motors of alkalophilic *Bacillus*. *Methods Enzymol.* **125:**582–592.

Jaques, S. 2004. *Regulation of Swarming in* Vibrio parahaemolyticus. University of Iowa Press, Iowa City, Iowa.

Jaques, S., Y. K. Kim, and L. L. McCarter. 1999. Mutations conferring resistance to phenamil and amiloride, inhibitors of sodium-driven motility of *Vibrio parahaemolyticus*. *Proc. Natl. Acad. Sci. USA* **96:**5740–5745.

Kawagishi, I., M. Imagawa, Y. Imae, L. McCarter, and M. Homma. 1996. The sodium-driven polar flagellar motor of marine *Vibrio* as the mechanosensor that regulates lateral flagellar expression. *Mol. Microbiol.* **20:**693–699.

Kawagishi, I., Y. Maekawa, T. Atsumi, M. Homma, and Y. Imae. 1995. Isolation of the polar and lateral flagellum-defective mutants in *Vibrio alginolyticus* and identification of their flagellar driving energy sources. *J. Bacteriol.* **177:**5158–5160.

Kawagishi, I., M. Nakada, N. Nishioka, and M. Homma. 1997. Cloning of a *Vibrio alginolyticus rpoN* gene that is required for polar flagellar formation. *J. Bacteriol.* **179:**6851–6854.

Kim, Y. K., and L. L. McCarter. 2000. Analysis of the polar flagellar gene system of *Vibrio parahaemolyticus. J. Bacteriol.* **182:** 3693–3704.

Kim, Y. K., and L. L. McCarter. 2004. Cross-regulation in *Vibrio parahaemolyticus*: compensatory activation of polar flagellar genes by the lateral flagellar regulator LafK. *J. Bacteriol.* **186:**4014–4018.

Kirov, S. M. 2003. Bacteria that express lateral flagella enable dissection of the multifunctional roles of flagella in pathogenesis. *FEMS Microbiol. Lett.* **224:**151–159.

Klose, K. E., and J. J. Mekalanos. 1998a. Differential regulation of multiple flagellins in *Vibrio cholerae. J. Bacteriol.* **180:**303–316.

Klose, K. E., and J. J. Mekalanos. 1998b. Distinct roles of an alternative sigma factor during both free-swimming and colonizing phases of the *Vibrio cholerae* pathogenic cycle. *Mol. Microbiol.* **28:**501–520.

Kogure, K., E. Ikemoto, and H. Morisaki. 1998. Attachment of *Vibrio alginolyticus* to glass surfaces is dependent on swimming speed. *J. Bacteriol.* **180:**932–937.

Kojima, S., Y. Asai, T. Atsumi, I. Kawagishi, and M. Homma. 1999a. Na+-driven flagellar motor resistant to phenamil, an amiloride analog, caused by mutations in putative channel components. *J. Mol. Biol.* **285:**1537–1547.

Kojima, S., M. Kuroda, I. Kawagishi, and M. Homma. 1999b. Random mutagenesis of the *pomA* gene encoding a putative channel component of the Na(+)-driven polar flagellar motor of *Vibrio alginolyticus. Microbiology* **145:**1759–1767.

Kojima, S., K. Yamamoto, I. Kawagishi, and M. Homma. 1999c. The polar flagellar motor of *Vibrio cholerae* is driven by an Na+ motive force. *J. Bacteriol.* **181:**1927–1930.

Kojima, S., and D. F. Blair. 2004. The bacterial flagellar motor: structure and function of a complex molecular machine. *Int. Rev. Cytol.* **233:**93–134.

Kojima, S., T. Shoji, Y. Asai, I. Kawagishi, and M. Homma. 2000. A slow-motility phenotype caused by substitutions at residue Asp31 in the PomA channel component of a sodium-driven flagellar motor. *J. Bacteriol.* **182:**3314–3318.

Kudo, S., N. Imai, M. Nishitoba, S. Sugiyama, and Y. Magariyama. 2005. Asymmetric swimming pattern of *Vibrio alginolyticus* cells with single polar flagella. *FEMS Microbiol. Lett.* **242:**221–225.

Larsen, M. H., N. Blackburn, J. L. Larsen, and J. E. Olsen. 2004. Influences of temperature, salinity and starvation on the motility and chemotactic response of *Vibrio anguillarum. Microbiology* **150:**1283–1290.

Larsen, M. H., and H. T. Boesen. 2001. Role of flagellum and chemotactic motility of *Vibrio anguillarum* for phagocytosis by and intracellular survival in fish macrophages. *FEMS Microbiol. Lett.* **203:**149–152.

Lauriano, C. M., C. Ghosh, N. E. Correa, and K. E. Klose. 2004. The sodium-driven flagellar motor controls exopolysaccharide expression in *Vibrio cholerae. J. Bacteriol.* **186:**4864–4874.

Lee, S. H., S. M. Butler, and A. Camilli. 2001. Selection for in vivo regulators of bacterial virulence. *Proc. Natl. Acad. Sci. USA* **98:** 6889–6894.

Levin, P. A., J. J. Shim, and A. D. Grossman. 1998. Effect of *minCD* on FtsZ ring position and polar septation in *Bacillus subtilis. J. Bacteriol.* **180:**6048–6051.

Macnab, R. M. 1996. Flagella and motility, p. 123–146. *In* F. C. Neidhardt, R. Curtiss III, J. L. Ingraham, E. C. C. Lin, K. B. Low, B. Magasanik, W. S. Reznikoff, M. Riley, M. Schaechter, and H. E. Umbarger (ed.), Escherichia coli *and* Salmonella *Cellular and Molecular Biology*, 2nd ed. ASM Press, Washington, D.C.

Macnab, R. M. 2003. How bacteria assemble flagella. *Annu. Rev. Microbiol.* **57:**77–100.

Macnab, R. M. 1986. Proton-driven bacterial flagellar motor. *Methods Enzymol.* **125:**563–579.

Macnab, R. M. 2004. Type III flagellar protein export and flagellar assembly. *Biochim. Biophys. Acta* **1694:**207–217.

Magariyama, Y., M. Ichiba, K. Nakata, K. Baba, T. Ohtani, S. Kudo, and T. Goto. 2005. Difference in bacterial motion between forward and backward swimming caused by the wall effect. *Biophys. J.* **88:**3648–3658.

Magariyama, Y., S. Masuda, Y. Takano, T. Ohtani, and S. Kudo. 2001. Difference between forward and backward swimming speeds of the single polar-flagellated bacterium, *Vibrio alginolyticus. FEMS Microbiol. Lett.* **205:**343–347.

Magariyama, Y., S. Sugiyama, K. Muramoto, I. Kawagishi, Y. Imae, and S. Kudo. 1995. Simultaneous measurement of bacterial flagellar rotation rate and swimming speed. *Biophys. J.* **69:**2154–2162.

Magariyama, Y., S. Sugiyama, K. Muramoto, Y. Maekawa, I. Kawagishi, Y. Imae, and S. Kudo. 1994. Very fast flagellar rotation. *Nature* **371:**752.

Marston, A. L., H. B. Thomaides, D. H. Edwards, M. E. Sharpe, and J. Errington. 1998. Polar localization of the MinD protein of *Bacillus subtilis* and its role in selection of the mid-cell division site. *Genes Dev.* **12:**3419–3430.

McCarter, L. L. 1994a. MotX, the channel component of the sodium-type flagellar motor. *J. Bacteriol.* **176:**5988–5998.

McCarter, L. L. 1994b. MotY, a component of the sodium-type flagellar motor. *J. Bacteriol.* **176:**4219–4225.

McCarter, L. L. 1995. Genetic and molecular characterization of the polar flagellum of *Vibrio parahaemolyticus. J. Bacteriol.* **177:**1595–1609.

McCarter, L. 1999. The multiple identities of *Vibrio parahaemolyticus. J. Mol. Microbiol. Biotechnol.* **1:**51–57.

McCarter, L. L. 2001. Polar flagellar motility of the *Vibrionaceae. Microbiol. Mol. Biol. Rev.* **65:**445–462.

McCarter, L. L. 2004. Dual flagellar systems enable motility under different circumstances. *J. Mol. Microbiol. Biotechnol.* **7:**18–29.

McCarter, L., M. Hilmen, and M. Silverman. 1988. Flagellar dynamometer controls swarmer cell differentiation of *V. parahaemolyticus. Cell* **54:**345–351.

McCarter, L., and M. Silverman. 1990. Surface-induced swarmer cell differentiation of *Vibrio parahaemolyticus. Mol. Microbiol.* **4:**1057–1062.

McCarter, L. L., and M. E. Wright. 1993. Identification of genes encoding components of the swarmer cell flagellar motor and propeller and a sigma factor controlling differentiation of *Vibrio parahaemolyticus. J. Bacteriol.* **175:**3361–3371.

McGee, K., P. Horstedt, and D. L. Milton. 1996. Identification and characterization of additional flagellin genes from *Vibrio anguillarum. J. Bacteriol.* **178:**5188–5198.

Merrell, D. S., S. M. Butler, F. Qadri, N. A. Dolganov, A. Alam, M. B. Cohen, S. B. Calderwood, G. K. Schoolnik, and A. Camilli. 2002. Host-induced epidemic spread of the cholera bacterium. *Nature* **417:**642–645.

Millikan, D. S., and E. G. Ruby. 2003. FlrA, a sigma54-dependent transcriptional activator in *Vibrio fischeri*, is required for motility and symbiotic light-organ colonization. *J. Bacteriol.* **185:**3547–3557.

Millikan, D. S., and E. G. Ruby. 2004. *Vibrio fischeri* flagellin A is essential for normal motility and for symbiotic competence during initial squid light organ colonization. *J. Bacteriol.* **186:**4315–4325.

Milton, D., R. O'Toole, P. Horstedt, and H. Wolf-Watz. 1996. Flagellin A is essential for virulence of *Vibrio anguillarum. J. Bacteriol.* **178:**1310–1319.

Muramoto, K., Y. Magariyama, M. Homma, I. Kawagishi, S. Sugiyama, Y. Imae, and S. Kudo. 1996. Rotational fluctuation of the sodium-driven flagellar motor of *Vibrio alginolyticus* induced by binding of inhibitors. *J. Mol. Biol.* **259**:687–695.

Namba, K., and F. Vonderviszt. 1997. Molecular architecture of bacterial flagellum. *Q. Rev. Biophys.* **30**:1–65.

Nishioka, N., M. Furuno, I. Kawagishi, and M. Homma. 1998. Flagellin-containing membrane vesicles excreted from *Vibrio alginolyticus* mutants lacking a polar-flagellar filament. *J. Biochem.* (Tokyo) **123**:1169–1173.

Norqvist, A., and H. Wolf-Watz. 1993. Characterization of a novel chromosomal virulence locus involved in expression of a major surface flagellar sheath antigen of the fish pathogen *Vibrio anguillarum*. *Infect. Immun.* **61**:2434–2444.

Okabe, M., T. Yakushi, Y. Asai, and M. Homma. 2001. Cloning and characterization of *motX*, a *Vibrio alginolyticus* sodium-driven flagellar motor gene. *J. Biochem.* (Tokyo) **130**:879–884.

Okabe, M., T. Yakushi, M. Kojima, and M. Homma. 2002. MotX and MotY, specific components of the sodium-driven flagellar motor, colocalize to the outer membrane in *Vibrio alginolyticus*. *Mol. Microbiol.* **46**:125–134.

Okunishi, I., I. Kawagishi, and M. Homma. 1996. Cloning and characterization of *motY*, a gene coding for a component of the sodium-driven flagellar motor in *Vibrio alginolyticus*. *J. Bacteriol.* **178**:2409–2415.

O'Shea, T. M., C. R. DeLoney-Marino, S. Shibata, S. Aizawa, A. J. Wolfe, and K. L. Visick. 2005. Magnesium promotes flagellation of *Vibrio fischeri*. *J. Bacteriol.* **187**:2058–2065.

O'Toole, R., S. Lundberg, S. A. Fredriksson, A. Jansson, B. Nilsson, and H. Wolf-Watz. 1999. The chemotactic response of *Vibrio anguillarum* to fish intestinal mucus is mediated by a combination of multiple mucus components. *J. Bacteriol.* **181**:4308–4317.

O'Toole, R., D. L. Milton, P. Horstedt, and H. Wolf-Watz. 1997. RpoN of the fish pathogen *Vibrio (Listonella) anguillarum* is essential for flagellum production and virulence by the waterborne but not intraperitoneal route of inoculation. *Microbiology* **43**:3849–3859.

O'Toole, R., D. L. Milton, and H. Wolf-Watz. 1996. Chemotactic motility is required for invasion of the host by the fish pathogen *Vibrio anguillarum*. *Mol. Microbiol.* **19**:625–637.

Pandza, S., M. Baetens, C. H. Park, T. Au, M. Keyhan, and A. Matin. 2000. The G-protein FlhF as a role in polar flagellar placement and general stress response induction in *Pseudomonas putida*. *Mol. Microbiol.* **36**:414–423.

Prouty, M., N. Correa, and K. Klose. 2001. The novel σ^{54}- and σ^{28}- dependent flagellar gene transcription hierarchy of *Vibrio cholerae*. *Mol. Microbiol.* **39**:1595–1609.

Ren, C. P., S. A. Beatson, J. Parkhill, and M. J. Pallen. 2005. The Flag-2 locus, an ancestral gene cluster, is potentially associated with a novel flagellar system from *Escherichia coli*. *J. Bacteriol.* **187**:1430–1440.

Ruby, E. G., M. Urbanowski, J. Campbell, A. Dunn, M. Faini, R. Gunsalus, P. Lostroh, C. Lupp, J. McCann, D. Millikan, A. Schaefer, E. Stabb, A. Stevens, K. Visick, C. Whistler, and E. P. Greenberg. 2005. Complete genome sequence of *Vibrio fischeri*: a symbiotic bacterium with pathogenic congeners. *Proc. Natl. Acad. Sci. USA.* **102**:3004–3009.

Sar, N., L. McCarter, M. Simon, and M. Silverman. 1990. Chemotactic control of the two flagellar systems of *Vibrio parahaemolyticus*. *J. Bacteriol.* **172**:334–341.

Sato, K., and M. Homma. 2000a. Functional reconstitution of the Na(+)-driven polar flagellar motor component of *Vibrio alginolyticus*. *J. Biol. Chem.* **275**:5718–5722.

Sato, K., and M. Homma. 2000b. Multimeric structure of PomA, a component of the Na+-driven polar flagellar motor of *Vibrio alginolyticus*. *J. Biol. Chem.* **275**:20223–20228.

Shinoda, S., and K. Okamoto. 1977. Formation and function of *Vibrio parahaemolyticus* lateral flagella. *J. Bacteriol.* **129**:1266–1271.

Shinoda, S., I. Yakiyama, S. Yasui, Y. M. Kim, B. Ono, and S. Nakagami. 1992. Lateral flagella of vibrios: serological classification and genetical similarity. *Microbiol. Immunol.* **36**:303–309.

Sjoblad, R. D., C. W. Emala, and R. N. Doetsch. 1983. Bacterial sheaths: structures in search of function. *Cell Motility* **3**:93–103.

Sjoblad, R. D., and R. Mitchell. 1979. Chemotactic responses of *Vibrio alginolyticus* to algal extracellular products. *Can. J. Microbiol.* **25**:964–967.

Stewart, B. J., and L. L. McCarter. 2003. Lateral flagellar gene system of *Vibrio parahaemolyticus*. *J. Bacteriol.* **185**:4508–4518.

Stewart, B. J., and L. L. McCarter. 1996. *Vibrio parahaemolyticus* FlaJ, a homologue of FliS, is required for production of a flagellin. *Mol. Microbiol.* **20**:137–149.

Sugiyama, S. J., E. J. Cragoe, and Y. Imae. 1988. Amiloride, a specific inhibitor for the Na+-driven flagellar motors of alkalophilic *Bacillus*. *J. Biol. Chem.* **263**:8215–8219.

Szurmant, H., and G. W. Ordal. 2004. Diversity in chemotaxis mechanisms among the bacteria and archaea. *Microbiol. Mol. Biol. Rev.* **68**:301–319.

Wadhams, G. H., and J. P. Armitage. 2004. Making sense of it all: bacterial chemotaxis. *Nat. Rev. Mol. Cell Biol.* **5**:1024–1037.

Watnick, P. I., C. M. Lauriano, K. E. Klose, L. Croal, and R. Kolter. 2001. The absence of a flagellum leads to altered colony morphology, biofilm development and virulence in *Vibrio cholerae* O139. *Mol. Microbiol.* **39**:223–235.

Wilson, D. R., and T. J. Beveridge. 1993. Bacterial flagellar filaments and their component flagellins. *Can. J. Microbiol.* **39**:451–472.

Wolfe, A. J., D. S. Millikan, J. M. Campbell, and K. L. Visick. 2004. *Vibrio fischeri* sigma54 controls motility, biofilm formation, luminescence, and colonization. *Appl. Environ. Microbiol.* **70**:2520–2524.

Xu, Q., M. Dziejman, and J. J. Mekalanos. 2003. Determination of the transcriptome of *Vibrio cholerae* during intraintestinal growth and midexponential phase *in vitro*. *Proc. Natl. Acad. Sci. USA* **100**:1286–1291.

Yakushi, T., N. Hattori, and M. Homma. 2005. Deletion analysis of the carboxyl-terminal region of the PomB component of the *Vibrio alginolyticus* polar flagellar motor. *J. Bacteriol.* **187**:778–784.

Yakushi, T., M. Kojima, and M. Homma. 2004a. Isolation of *Vibrio alginolyticus* sodium-driven flagellar motor complex composed of PomA and PomB solubilized by sucrose monocaprate. *Microbiology* **150**:911–920.

Yakushi, T., S. Maki, and M. Homma. 2004b. Interaction of PomB with the third transmembrane segment of PomA in the Na+-driven polar flagellum of *Vibrio alginolyticus*. *J. Bacteriol.* **186**:5281–5291.

Yorimitsu, T., Y. Asai, K. Sato, and M. Homma. 2000. Intermolecular cross-linking between the periplasmic Loop3-4 regions of PomA, a component of the Na+-driven flagellar motor of *Vibrio alginolyticus*. *J. Biol. Chem.* **275**:31387–31391.

Yorimitsu, T., and M. Homma. 2001. Na(+)-driven flagellar motor of *Vibrio*. *Biochim. Biophys. Acta* **1505**:82–93.

Yorimitsu, T., M. Kojima, T. Yakushi, and M. Homma. 2004. Multimeric structure of the PomA/PomB channel complex in the Na+-driven flagellar motor of *Vibrio alginolyticus*. *J. Biochem.* (Tokyo) **135**:43–51.

Yorimitsu, T., A. Mimaki, T. Yakushi, and M. Homma. 2003. The conserved charged residues of the C-terminal region of FliG, a rotor component of the Na+-driven flagellar motor. *J. Mol. Biol.* **334**:567–583.

Yorimitsu, T., K. Sato, Y. Asai, I. Kawagishi, and M. Homma. 1999. Functional interaction between PomA and PomB, the Na(+)-driven flagellar motor components of *Vibrio alginolyticus. J. Bacteriol.* **181:**5103–5106.

Yorimitsu, T., Y. Sowa, A. Ishijima, T. Yakushi, and M. Homma. 2002. The systematic substitutions around the conserved charged residues of the cytoplasmic loop of Na+-driven flagellar motor component PomA. *J. Mol. Biol.* **320:**403–413.

Yu, C., B. L. Bassler, and S. Roseman. 1993. Chemotaxis of the marine bacterium *Vibrio furnissii* to sugars. A potential mechanism for initiating the chitin catabolic cascade. *J. Biol. Chem.* **268:**9405–9409.

Zhou, J., S. A. Lloyd, and D. F. Blair. 1998. Electrostatic interactions between rotor and stator in the bacterial flagellar motor. *Proc. Natl. Acad. Sci. USA* **95:**6436–6441.

The Biology of Vibrios
Edited by F. L. Thompson et al.
© 2006 ASM Press, Washington, D.C.

Chapter 10

Adaptive Responses of Vibrios

DIANE MCDOUGALD AND STAFFAN KJELLEBERG

INTRODUCTION

Bacterial survival depends on efficient adaptation of cells to changes in environmental conditions. Members of the genus *Vibrio* are able to successfully proliferate in areas of high substrate availability and cell density (e.g., biofilms), as well as to persist as free-living pelagic cells. Vibrios have been found associated with higher organisms and with inanimate surfaces and can constitute a significant fraction of the gastrointestinal tract of fish and invertebrates. Thus, life cycles of individual cells may be complex and contain intervals of nongrowth with intermittent periods of unbalanced growth with long generation times. The ability to survive under these changing conditions indicates that vibrios have a diversity of adaptive responses. In this chapter we will discuss several aspects of adaptation that have been suggested to play a role in the survival of vibrios, in particular, starvation adaptation, the viable but nonculturable (VBNC) response, and biofilm formation. In addition, quorum sensing, which has been shown to control many phenotypes associated with survival under different conditions, will be discussed.

STARVATION ADAPTATION IN VIBRIOS

Two main nonbiological factors limiting bacterial growth in the marine environment are water temperature and the availability of nutrients (Moriarty and Bell, 1993), with carbon concentrations in aquatic environments varying from 80 μM in surface waters to 40 μM in deep waters (Williams, 2000). A significant portion of this carbon may be in particulate form and thus is largely unavailable for immediate utilization by microorganisms. Therefore, in nature, brief periods of rapid growth interspersed with long periods of nongrowth are common for a range of heterotrophic bacteria. Many *Vibrio* spp. have very fast doubling times during feast conditions, which allows them to quickly take advantage of nutrient availability and outcompete other species (Giovannoni and Rappé, 2000). In one study, North Sea bacterioplankton was subjected to enrichment conditions often used to isolate pelagic bacteria, and the population of *Vibrio* spp. was enriched more dramatically than any other group: it went from nearly undetectable numbers at the start of the treatment to 65% of the total bacterial population after 40 h of incubation (Eilers et al., 2000). Thus, *Vibrio* spp. maintain the potential to react quickly to changes in growth conditions even during extended periods of nongrowth. In this section we will explore some of the mechanisms and responses of *Vibrio* spp. to adaptation to nongrowth.

Ultramicrobacteria

Marine waters contain large numbers of ultramicrocells, a subpopulation of which have been demonstrated to be starved cells. In an early study, it was observed that several species isolated from filtered seawater were no longer filterable after laboratory cultivation, prompting speculation that the small forms could be "reproductive elements" or could possibly reflect a change in the size of the parent cell (Anderson and Heffernan, 1965). The formation of "round bodies" was later described for aged cultures of *Vibrio marinus* (Felter et al., 1969).

These observations prompted a series of studies on responses of pure cultures to starvation conditions. Upon starvation of Ant-300, a marine vibrio, there was an initial increase in cell number (Novitsky and Morita, 1978), and after 3 weeks 50% of the population was able to pass through a 0.4-μm-pore-size filter (Novitsky and Morita, 1976). These cells changed shape from rods to cocci while maintaining

Diane McDougald and Staffan Kjelleberg • School of Biotechnology and Biomolecular Sciences, Centre for Marine Biofouling and Bio-Innovation, University of New South Wales, Sydney 2052, Australia.

normal cellular structures and were demonstrated to remain viable for at least 2.5 years (Amy et al., 1983) under starvation conditions. When these cells were inoculated into nutrient-rich medium, they increased in size and resumed a rod shape. Likewise, water samples from a Gulf Coast estuary were passed through a 0.2-μm filter and incubated with dilute nutrient broth for 21 days, after which normally sized *Vibrio*, *Pseudomonas*, *Aeromonas*, and *Alcaligenes* spp. were isolated (MacDonell and Hood, 1982). Thus, the development of ultramicrocells appears to be a general response of vibrios to starvation conditions. The decrease in cell size provides the cell with an increased surface-to-volume ratio and thus a more efficient substrate-scavenging capacity. It has also been shown that small cells are better able to escape predation than larger cells (Gonzalez et al., 1992; Matz et al., 2002a). Consequently, the formation of small cells by starving bacteria may be a multifactorial response enhancing nutrient uptake and escape from predation.

Starvation-Induced Differentiation

It is now appreciated that nonsporulating gram-negative bacteria are nondifferentiating but exhibit an elaborate and highly developed starvation adaptation program that involves alterations in gene expression as well as physiological changes (Hengge-Aronis, 1993, 1996, 2002; Kjelleberg et al., 1993; Kolter et al., 1993; Matin et al., 1989; Siegele and Kolter, 1992). Starvation differentiation has been well studied in *Vibrio* spp. and is typically characterized by such changes as reductions in cell volume, DNA and ribosome content, and the rate of protein synthesis (Ostling et al., 1993). The best-studied system for starvation adaptation is that in *Vibrio angustum* S14 (Fig. 1), where three distinct stages of the starvation adaptation program have been identified.

The first stage of the starvation adaptation response has been defined as the stringent response phase and is governed by the temporary accumulation of guanosine 3'-diphosphate 5'-diphosphate (ppGpp) (Cashel et al., 1996; Flärdh et al., 1994). This phase is exemplified by the shutdown of macromolecular synthesis with an increase in the rate of protein degradation allowing for extensive reorganization of cellular components. Studies on *Vibrio* sp. Ant-300 (Novitsky and Morita, 1977) and *Vibrio cholerae* (Hood et al., 1986) provided information of a similar kind.

It has been shown that degradation of rRNA (~80% of the total cellular RNA) begins immediately when *Escherichia coli* cells are placed in a carbon-limited medium (Davis et al., 1986), leading to the proposal that the major cause of death during prolonged starvation is excessive loss of ribosomes. However, in *V. angustum*, it was found that ribosomes exist in large excess over the requirement for synthesis during carbon starvation (Flärdh et al., 1992). Using two-dimensional polyacrylamide gel electrophoresis (2D-PAGE), it was also shown that phase 1 of starvation includes differential gene expression. Proteins that are produced during the exponential phase of growth are quickly degraded, and new starvation-specific proteins are synthesized in the early stages of starvation; these newly synthesized proteins are important for subsequent resistance and survival.

In the second stage of the starvation adaptation program (0.5 to 6 h), there is a decrease in ppGpp with a concomitant increase in macromolecular synthesis. Reorganizations may include shifts in the fatty acid composition of the membranes (Malmcrona-Friberg et al., 1986; Oliver and Stringer, 1984), degradation of reserve material (Holmquist and Kjelleberg, 1993; Malmcrona-Friberg et al., 1986), and onset of development of resistance against a variety of stress conditions (Nyström et al., 1992), as shown for *V. angustum* S14, *Vibrio* sp. Ant-300 (Novitsky and Morita, 1978; Preyer and Oliver, 1993), *Vibrio* sp. DW1 (Jouper-Jaan et al., 1992), *Vibrio vulnificus* (Oliver et al., 1991; Paludan-Müller et al., 1996; Weichart and Kjelleberg, 1996), *Vibrio anguillarum* (Nelson et al., 1997), and *Vibrio parahaemolyticus* (Koga and Takumi, 1995; Wong and Wang, 2004).

In the third phase of starvation adaptation, there is a gradual decline in macromolecular synthesis and metabolic activities, such as endogenous respiration. These changes prepare the cell for successful survival under continued stress conditions and allow for outgrowth should conditions become favorable for regrowth. It has been demonstrated that in *V. angustum* S14, de novo RNA synthesis is not required for the initial phases of outgrowth, indicating that rRNA and ribosomes needed for growth are already present in the starving cell (Albertson et al., 1990a; Flärdh et al., 1992). An increase in the mean mRNA half-life brought about by global mRNA stabilization and the presence of long-lived starvation-specific messages allow for continued protein synthesis in long-term-starved cells (Albertson et al., 1990b; Takayama and Kjelleberg, 2000).

2D-PAGE demonstrated that up to 18 specific proteins are synthesized immediately (within 3 min) after glucose addition to *V. angustum* S14 cells that had been starved for 48 h (Marouga and Kjelleberg, 1996). These immediate upshift proteins (Iups) are expressed at 10-fold-higher levels during upshift when compared to exponentially growing cells. Twelve of the 18 Iups are synthesized in the presence of the transcriptional inhibitor rifampin, indicating

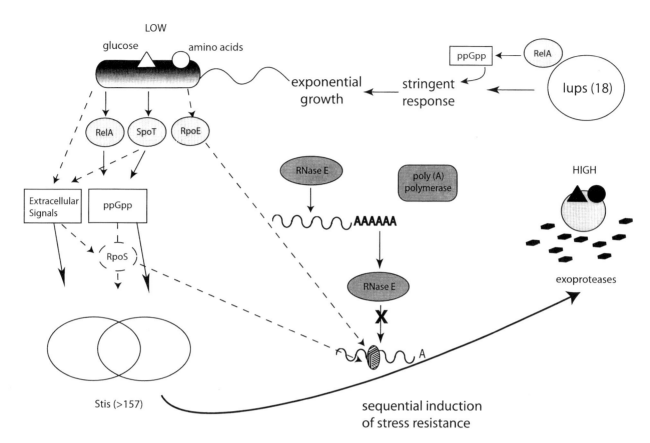

Figure 1. Model for the entry into and exit from carbon starvation in *Vibrio* S14 (shaded rod and coccoid forms). Ovals represent regulators (i.e., RelA, SpoT, RpoE, RpoS). Rectangles denote signaling molecules (i.e., ppGpp, quorum-sensing autoinducer molecules). Dotted lines refer to postulated regulators and pathways. Large shaded ovals depict Sti and Iup protein responders. Overlap between Sti ovals indicates protein responders, which are induced by both SpoT and RelA. One hundred fifty-seven Stis and 18 Iups have been mapped by 2D-PAGE, and their respective rates of synthesis and modes of regulation have been characterized. mRNA (~~~) modification and stability during starvation are proposed to be modulated by RNase E (large oval), mRNA-binding protein(s) (small hatched oval), and polyadenylation (polyA tail). Triangles and circles on the bacterial cells represent low- and high-affinity uptake for glucose and amino acids, respectively.

that the transcripts exist during starvation and that translation is induced upon nutrient upshift. Thus, when starved cells are presented with a suitable energy source, the machinery needed to utilize the substrate is immediately available.

Expression of Proteins in Response to Starvation

Most starvation-induced (Sti) proteins are synthesized early during starvation and seem to be the most important for survival, as the addition of inhibitors of protein synthesis for a brief time at the onset of starvation significantly affects long-term starvation survival (Nyström et al., 1992). Many of these proteins play specific roles in protection against external stresses, e.g., heat, osmotic stress, and oxidation (Dukan and Nyström, 1999; Matin, 1991). This results in starved cells' being resistant to a variety of stresses and is referred to as starvation-induced cross

protection (Jenkins et al., 1988, 1990). This phenomenon is controlled by several regulators, including the sigma factor RpoS, in many species (Lange and Hengge-Aronis, 1991).

By employing high-resolution 2D-PAGE, it was observed that of the 760 analyzed proteins, the relative rate of protein synthesis was increased for 157 proteins and decreased for 144 proteins after 1 h of carbon starvation in *V. angustum* S14 (Ostling et al., 1996, 1997). The identification of temporal classes of proteins induced during starvation indicates that the response to starvation is an ordered progression of a programmed sequence of events that allows for survival during stress (Nyström et al., 1988, 1990).

The possibility exists that protein turnover during starvation targets specific polypeptides rather than being random. Nyström (2004) proposed that the oxidation of specific proteins may make them more susceptible to degradation. In starving cells, specific proteins related to peptide chain elongation, protein folding, central carbon metabolism, and stress resistance become oxidized (Dukan and Nystrom, 1998, 1999). This increase in oxidation of proteins is associated with the generation of aberrant proteins produced owing to misincorporation of amino acids (Ballesteros et al., 2001). The reduction in translation accuracy is caused by ribosomes that are starved for charged tRNAs and results in the selective degradation of a subset of proteins in the starved cell (Nyström, 2004). Whether or not the sensitivity of some proteins to oxidation has a molecular basis, or whether it is simply related to cellular localization of proteins or some other unidentified factor is not currently understood.

Effect of Limiting Nutrients on Starvation Response

Many of the early studies on starvation survival employed total nutrient and energy limitation, while subsequent investigations explored the response of cells to starvation for individual nutrients. Carbon-starved cells of *V. angustum* S14 were shown to survive significantly better than nitrogen- or phosphorus-starved cells (Nyström et al., 1992). Cells starved simultaneously for all three nutrients had viability similar to that of the carbon-starved cultures. Phosphorus-starved cells of *V. angustum* S14 showed an initial increase in optical density that was probably due to the accumulation of poly(3-hydroxybutyrate) (Mårdén et al., 1985; Nyström et al., 1992). These cells exhibited only transient development of stress resistance and failed to survive long-term phosphorus starvation. The response of *V. angustum* S14 to nitrogen starvation was similar to its response to phosphorus limitation. Thus, the response to carbon starvation, which results in both nutrient and energy limitation, appears to provoke a response very different from that of phosphorus or nitrogen.

Nitrogen starvation does not induce a strict stationary phase such as the cessation of growth seen during carbon starvation. This is mainly due to the continued growth resulting from the use of intracellular reserves of nitrogenous polymers (Mason and Egli, 1993). Similar utilization of internal reserves of phosphorus will delay the onset of phosphorus limitation. Starvation for phosphorus results in the *spoT*-dependent accumulation of ppGpp (Spira, 1995). Low phosphorus levels are detected by PhoR, which then activates PhoB, the primary phosphate regulator, which in turn regulates the expression of about 30 genes (Makino, 1989). Polyphosphates, linear polymers of inorganic phosphate, have been found in all organisms examined. Inorganic polyphosphate has been

shown to be essential for adaptation to stress and survival in stationary phase (Rao and Kornberg, 1996). It was shown in *E. coli* that there was massive accumulation of polyphosphate when cells were starved for amino acids and phosphate; when this accumulation was prevented by a mutation in the polyphosphate synthase gene, cells failed to survive in stationary phase. These results led to the speculation that the induction of ppGpp may be an important factor in the development of cross protection, as carbon and phosphorus limitation results in its accumulation (Rao and Kornberg, 1996). Interestingly, in *V. vulnificus*, carbon and phosphorus starvation allows for maintenance of culturability at low temperatures (Paludan-Müller et al., 1996).

Role of Oxidative Stress in Starvation Survival in *V. angustum*

More recent studies with *V. angustum* S14 have led to a model for the starvation survival response (Fig. 1) in which the main regulators of the stringent control, RelA and SpoT, as well as the signal molecule ppGpp, regulate the expression of starvation-specific protein responders (Ostling et al., 1996, 1997). While RelA is dispensable for noncompetitive survival (in laboratory cultures) during carbon starvation, SpoT is an essential regulator of the carbon starvation response and the successful formation of competent ultramicrocells (Ostling et al., 1996). The inability of carbon-starved *spoT* mutants to maintain culturability appears to be directly related to difficulties in the development of starvation-induced protection against oxidative stress (McDougald et al., 2002). Survival of the *spoT* mutant is fully maintained during carbon starvation in the dark, but not when exposed to cool fluorescent light (Gong et al., 2002). By using various UV cutoff filters, it was determined that light-dependent lethality is mainly due to low-intensity UVA radiation. The proposal that the lethal effect of UVA radiation was due to reactive oxygen species (ROS) caused by photo-oxidation was supported by the finding that the *spoT* mutant displayed increased sensitivity to oxidative stress during starvation when compared to the wild type. In addition, the *spoT* mutant strain shows decreased catalase activity and increased oxidation levels of lipids and proteins compared with the parent strain. Furthermore, the addition of exogenous catalase to the carbon-starved culture reverses most of the loss of culturability of the *spoT* mutant provoked by light exposure. It was also determined that RpoE plays a similar role in stationary-phase survival of *V. angustum*, as an *rpoE* mutant was not able to survive when exposed to near-UV irradiation during carbon starvation. The potential role for oxidative stress in the VBNC response is discussed below.

A role for extracellular signaling molecules in the regulation of starvation and stress adaptation has also been demonstrated. In *V. angustum* S14, addition of signal molecule-containing supernatants induces stationary phase prematurely in growing cells and up-regulates a subset of the carbon starvation regulon (Srinivasan et al., 1998). Furthermore, quorum sensing has been demonstrated to regulate starvation survival and oxidative stress resistance in *V. vulnificus* (McDougald et al., 2001, 2002).

From these studies, it is clear that bacteria have evolved complex mechanisms to deal with conditions that are routinely encountered in the natural environment. Such adaptive responses are characterized by changes in gene expression, physiology, and morphology. Studies on starvation and stress responses of heterotrophic bacteria have elucidated key regulatory pathways controlling expression or repression of genes that ultimately regulate adaptation to unfavorable growth conditions.

THE VIABLE BUT NONCULTURABLE RESPONSE (VBNC)

When *Vibrio* spp. are exposed to certain long-term stresses, for example, incubation of *V. vulnificus* in the estuarine environment during winter (Oliver et al., 1995), the culturable cell count decreases while the total cell count remains constant. One hypothesis for this discrepancy is that cells may lose culturability while maintaining viability and are thus in a VBNC state (Barer et al., 1993; Colwell et al., 1985; Oliver, 1993a). The existence of this VBNC state has been debated for over 20 years, as others argue that nonculturable cells are by definition dead cells (Bogosian et al., 1996, 1998; Kell et al., 1998). Indeed, little evidence has been reported for the genetic regulation of a programmed response resulting in VBNC cells, and thus the significance of the VBNC state is still a subject of debate. However, recent studies have suggested that ROS play a central role in the VBNC state (Cuny et al., 2005; Desnues et al., 2003; Mizunoe et al., 1999, 2000). The potential for nonculturable cells to resuscitate, irrespective of whether the VBNC state is truly a protective strategy or is simply a consequence of stress leading to death, does have implications for the survival of vibrios in the natural environment. The latter statement may also be applied to environmental and medical consequences as well, since most monitoring is based on plate counts. In this section we present some of the current theories on factors driving the generation and resuscitation of nonculturable cells.

The VBNC Hypothesis

The first reports on VBNC cells came from studies by Colwell and colleagues (Colwell et al., 1985; Xu et al., 1982), which demonstrated that cells that had become nonplateable on medium normally used for culturing still possessed indications of metabolic activity. A VBNC cell was later defined as "a cell which can be demonstrated to be metabolically active, while being incapable of undergoing the sustained cellular division required for growth in or on a medium normally supporting growth of that cell" (Oliver, 1993a). In addition to most (if not all) *Vibrio* spp., many genera of bacteria have been suggested to form VBNC cells. Several environmental stresses have been shown to induce this state, including starvation, salinity, visible light, and temperature. However, while there have been many reports of bacteria exhibiting a loss of culturability under stress conditions, the lack of reproducible resuscitation or rescue of such cells has only fueled the debate over whether VBNC cells are truly alive or are simply dying. Certainly, resuscitation must occur for the VBNC response to be a survival strategy.

Indirect Assays of Viability

Several assays have been used to assess metabolic activity of VBNC cells, including direct viable count (Kogure et al., 1979), microautoradiography (Rahman et al., 1994), and reduction of tetrazolium salts (Rodriguez et al., 1992) (reviewed by McDougald et al., 1998). More recently, randomly amplified polymorphic DNA (Warner and Oliver, 1998) and PCR have been used successfully in some cases for the detection of VBNC cells (Campbell and Wright, 2003; Gil et al., 2004; Hasan et al., 1994; Lee and Ruby, 1995), whereas others have shown that increased amounts of DNA of VBNC cells are required for amplification to occur (Brauns et al., 1991; Coleman and Oliver, 1996). This may be due to condensation of DNA upon entry into the VBNC state by DNA-binding proteins or to thickening of the cell wall, which may affect the efficiency of DNA extraction, making PCR less efficient in VBNC cells when compared to culturable cells. A major disadvantage of these DNA-based assays is the potential to detect DNA from dead cells (Josephson et al., 1993). A more recent report evaluated current viability markers for detection of heat- and UV-killed *E. coli* cells (Villarino et al., 2000); it was concluded that immediately after UV irradiation, all killed cells were detected by fluorescent stains (4′,6′-diamidino-2-phenylindole, fluorescence in situ hybridization, Live/Dead BacLight, or 5-cyano-2,3-ditolyl tetrazolium chloride). For the UV-killed cells, residual activities were detected for 48 h at 20°C, whereas heat-killed cells were detected for several hours. Of all the fluorescent markers used, only a combination of direct viable count and fluorescence in situ hybridization was negative for both UV- and heat-killed cells.

The detection of mRNA has been suggested to be a good viability marker owing to its short half-life. Results have shown that RpoS and 16SrRNA mRNA can be detected in VBNC cultures of *V. parahaemolyticus* using reverse transcription-PCR (Coutard et al., 2005) and that cytolysin mRNA was maintained in VBNC *V. vulnificus* cells for up to 4.5 months (Fischer-Le Saux et al., 2002). The authors reported a loss of the RT-PCR product after boiling to kill VBNC cells, indicating that the product was indeed from viable cells. We have shown that *V. vulnificus* cells maintain intact pools of RNA and DNA for a period of time after becoming nonculturable but that prolonged incubation results in a gradual loss of RNA followed by DNA (Weichart et al., 1997), further indicating that mRNA would be a useful target for detection.

Resuscitation of VBNC Cells

Many workers have proposed that the VBNC response is analogous to a stress response and is therefore a survival strategy (Colwell et al., 1985; McDougald et al., 1998, 1999; Oliver, 1993a,b; Rice et al., 2000), while others argue that VBNC cells are moribund and would eventually succumb to the stress only to be returned to the pool of carbon and nitrogen whence they came (Bogosian et al., 1996, 1998; Kell et al., 1998). A universally agreed upon benchmark for a true VBNC response is the demonstration of resuscitation. If VBNC cells can be demonstrated to resuscitate, it would suggest that those nonculturable cells were still viable and hence would be evidence that the VBNC state truly represents an adaptive response. Even death of the majority of cells with a minority being able to take advantage of the pool of carbon for subsequent resuscitation would still be a benefit to the population.

There have been numerous reports of resuscitation induced by mechanisms such as temperature upshift (Nilsson et al., 1991; Oliver and Bockian, 1995; Oliver et al., 1995; Whitesides and Oliver, 1997) and heat shock (Ravel et al., 1995; Wai et al., 1996). Reports of in vivo resuscitation in animal models (Colwell et al., 1985; Oliver and Bockian, 1995), tissue

culture (Oliver, 1993a), or humans (Colwell et al., 1996) suggest that VBNC cells may maintain pathogenicity. In most cases, resuscitation is only possible for those cells that have been VBNC for short periods of time; cells that have been VBNC for longer times appear to have lost infectivity. Thus, there is likely to be a small window of time during which VBNC cells may be capable of resuscitation. Clearly, the resuscitation of nonculturable cells in vivo has major implications for human health and for the regulation of bacterial numbers in water and foods (e.g. oysters).

The methods used to determine resuscitation have left some room for the possibility that there was regrowth of a few undetected culturable cells remaining in the VBNC samples (Bogosian et al., 1996, 1998; Weichart et al., 1992). For example, Bogosian et al. (1996) developed a mixed culture method that showed that reactivation of several strains was due to the regrowth of culturable cells and was not due to the resuscitation of VBNC cells. Such reports highlight the need to separate possible regrowth from bona fide resuscitation. This is an especially important issue so that the conditions and factors required for resuscitation can be investigated and identified; however, there are no experimental systems that seem to be accepted as giving unambiguous results concerning resuscitation of VBNC cells. This highlights a need for accepted methodologies and theoretical frameworks that can be used to better study this phenomenon. Obviously, the identification of genes or regulatory pathways involved in VBNC formation would provide a much-needed understanding of the biology of the VBNC phenomenon.

One particularly convincing model of a VBNC phenotype is that of the coral-bleaching pathogen *Vibrio shiloi* (Rosenberg and Ben-Haim, 2002; Sussman et al., 2003). This organism adheres to host coral only at elevated seawater temperatures and penetrates the epidermal layer of the coral. The bacterial cells can be seen to multiply inside the host cells but are not recovered by plating (Israely et al., 2001). In addition to dividing, these cells are able to produce toxins that cause bleaching of the coral while not being culturable. If the water temperature falls below 20°C, the bacterial cells rapidly lyse, indicating that *V. shiloi* does not use the coral as a winter reservoir. In fact, the bacterium can be found in high numbers in the marine fireworm *Hermodice carunculata* (Sussman et al., 2003), but again, >99.9% of the cells are VBNC. Whether these cells are dependent on host factors for resuscitation or growth remains to be determined. These studies are especially intriguing, as resuscitation is not necessary for this organism to be able to invade, divide, and produce toxins.

The Oxidative Injury Hypothesis

One current hypothesis that has lent an experimental framework to the VBNC phenomenon is the generation of VBNC cells in relation to oxidative injury in growth-arrested cells. Stasis caused by a variety of conditions induces the expression of regulators involved in the prevention and repair of oxidative damage to cellular components. The growth arrest observed in VBNC cells may be phenotypically similar to that of stasis-induced growth arrest and may require similar repair processes, although different inducers may be involved in its onset. ROS can damage DNA and lipid membranes and proteins and have been implicated as a cause of cellular aging due to nongrowth conditions (Dukan and Nystrom, 1998, 1999; Nyström, 1998, 1999). Whereas many of the recent reports on bacterial stasis and oxidative damage use *E. coli* as a model organism, the same effects are likely to play a role in damage to VBNC cells of *Vibrio* spp.

The addition of increasing concentrations of hydrogen peroxide—one of the photoproducts formed in natural aquatic systems—to *E. coli* cultures gave rise to the formation of VBNC cells (Arana et al., 1992). Conversely, in illuminated microcosms, the addition of compounds that eliminate hydrogen peroxide (i.e., catalase, sodium pyruvate, and thioglycolate) had a protective effect, as the colony counts on minimal medium and on recuperation medium were significantly higher than those detected in the absence of these compounds. Mutants of *V. vulnificus* (Kong et al., 2004) or *E. coli* (Dukan and Nyström, 1999) that lack the oxidative stress regulator OxyR lose plateability on nutrient plates but can be recovered by plating anaerobically or with high levels of catalase added to plates, indicating the involvement of oxidative deterioration in loss of culturability. Likewise, VBNC cells of *V. parahaemolyticus* were resuscitated on media containing catalase or sodium pyruvate (Mizunoe et al., 2000). These data indicate that oxidative damage is at least one inducer of a nonculturable phenotype and that VBNC cells may be oxidatively damaged cells. As most of these experiments involve direct plating of VBNC cells rather than incubation of broth cultures for extended periods, the pitfalls of other resuscitation studies, in which regrowth of a few cells may occur, are avoided, highlighting the possibility that under the correct physiological conditions, resuscitation of previously nonculturable cells is possible. One could infer from these results that the damage experienced from the oxidative burst that occurs in resuscitating cells upon the resumption of metabolism prevents the recovery of a subpopulation of the nonculturable cells in the population. Another hypothesis for the cause of increased protein oxidation is that there is an increased concentration of substrates available for oxidative attack owing to aberrant and misfolded proteins (Dukan et al., 2000), which result from mistranslation due to reductions in transcriptional or translational fidelity in growth-arrested cells (Ballesteros et al., 2001).

Population Differentiation

Density-dependent cell sorting was utilized to separate a stationary-phase culture of *V. parahaemolyticus* into fractions that differed in culturability (ranging from 11.5 to 96.2%) and resistance to low-temperature stress (Nishino et al., 2003). Using an ultracentrifugation separation technique, a VBNC population was isolated from a 48-h *E. coli* culture (Desnues et al., 2003). The VBNC population exhibited a decrease in superoxide dismutase activity, resulting in an increase in oxidative damage and induction of stress regulons, such as those regulated by RpoS and RpoE. Taken together, these results indicate that the loss of culturability is due to ROS and resultant oxidative stress in nongrowing cells.

In a subsequent study, a culture was separated after 10 h into populations that differed in the rate of VBNC formation while having virtually the same induction ratio of proteins involved in defense against ROS and similar oxidative damage (based on carbonyl content) (Cuny et al., 2005). These experiments indicate that there are physiological changes that occur in these populations that are not directly related to defenses against ROS. Thus, there is a subpopulation that is more susceptible to becoming VBNC after stress, and it is possible to separate this subpopulation before loss of culturability occurs. It is suggested that the presence of pre-VBNC cells may be a population strategy for long-term survival in stationary phase. The pre-VBNC cells have a higher death rate and could thus provide nutrients for the remaining culturable cells. This population could be analogous to the population of cells that undergo cell death in biofilms, thereby providing nutrients for reactivation of dispersal cells (Webb et al., 2003).

Conclusions

Until a genetically programmed pathway is shown to control VBNC formation, the concept of VBNC as a survival strategy remains difficult. In addition, the lack of inarguable cases of true resuscitation continues to fuel debate about a VBNC state. Discussions on semantics have further hampered agreement on nonculturable cells and their impact in the environment. It remains to be definitively established whether VBNC cells of pathogenic bacteria pose any threat to human health, but the possibility remains that cells may become undetectable while still capable of causing disease, even if for only short periods of time. The fact that *Vibrio* spp. become largely undetectable during the winter months but reemerge when the water temperature rises indicates that there is at least a small fraction of cells that survive. However, a recent report suggests that this inability to detect *V. vulnificus* during winter stems from the disappearance of the cells from the water column, as they were unable to detect cells by PCR even when using techniques adapted for VBNC cells (Randa et al., 2004). Perhaps the most convincing case for VBNC cells' being alive is the fact that *V. shiloi* can infect, invade, and divide and produce toxins while not being culturable.

Despite some of the uncertainties in this intriguing field, it seems certain that any culture entering nongrowth or starvation-induced stasis is composed of at least three populations of cells: cells retaining culturability and stress resistance, cells that are metabolically active but not readily culturable, and cells showing no detectable biological activity. While the role of these respective populations in the environment remains to be demonstrated, researchers are beginning to unravel the causes of induction of nonculturability, one of which is probably related to protein oxidation due to increased mistranslation, and damage caused by ROS during stasis and during regrowth. Despite the caveats presented above and the difficulty in generating unambiguous examples of resuscitation, it is clear that a VBNC phenotype is manifested by a range of species in different habitats. The potential of VBNC cells to cause disease in different hosts makes this area of research highly relevant.

THE ROLE OF QUORUM SENSING IN VIBRIOS

Bacteria communicate through the secretion and uptake of small diffusible autoinducer (AI) molecules that induce coordination of phenotypic expression and provide a selective advantage in the natural environment. Autoinducer circuits, in which the AI accumulates during growth until a threshold concentration is attained, at which point the system is switched on, are considered to be cell density-sensing systems. However, it should be stated that, theoretically, the density of signaling molecules could also be increased by limiting the space around the cells or by altering diffusion of the AI, rather than by increasing the cell number, as has been recently discussed by Redfield (2002). Since the AI accumulates extracellularly and increases in concentration as a function of population density, a coordinated response will occur only when a critical cell density is reached (i.e., when

enough AI accumulates). Thus, it has been suggested that the regulation of virulence determinants by AI systems in pathogenic bacteria functions to evade host defenses (Bassler, 1999). The premature expression of bacterial toxins might be a poor strategy, as it could alert the host and elicit defensive responses. Quorum sensing (QS) could function to delay virulence factor production until cell numbers are high enough to result in a productive infection. In support of this hypothesis, transgenic tobacco plants that express the native acylated homoserine lactone (AHL) synthase from *Erwinia carotovora* were more resistant to infection by wild-type *E. carotovora* than the wild-type plant (Mae et al., 2001). Similarly, exogenous addition of AHL also reduced disease of the wild-type plant to 10% even though the AHL system acts to induce virulence factor expression. This indicates that early expression of virulence factors at low cell density may indeed disadvantage the bacterial community and its ability to successfully colonize a host (Mae et al., 2001).

In vibrios, a picture is emerging whereby QS systems act as coordinate regulators of a wide range of phenotypes at different cell densities; i.e., some are active at low cell density and others are active at high cell density. These studies highlight the fact that, at least in *Vibrio* spp., there is a complex program of induction/repression of genes that is cell density-dependent rather than high cell density-dependent. Recent evidence also suggests that QS systems may play a more central role in the physiology of bacteria by regulating starvation adaptation pathways and entry into stationary phase. In *V. vulnificus*, the QS system regulates starvation adaptation (McDougald et al., 2001) as well as resistance to oxidative stress (McDougald et al., 2002). Furthermore, extracellular signals in *V. angustum* serve to affect starvation adaptation by increasing the stability of mRNAs, preparing the cells for a rapid response to nutrient addition (Takayama and Kjelleberg, 2000).

Quorum Sensing in *V. fischeri*

The study of intercellular signaling in marine *Vibrio* spp. focused for many years on the symbiotic marine bacterium *Vibrio fischeri*, which colonizes the light organ of the Hawaiian bobtail squid, *Euprymna scolopes*. Early experiments showed that bioluminescence was regulated in a cell density-dependent manner by an AHL (Eberhard et al., 1981), synthesized by the *luxI* gene product (AI synthase), and detected by the *luxR*-encoded receptor/transcriptional activator. When the concentration of AHL (*N*-3-oxo-hexanoyl homoserine lactone) in the cells reaches a certain level, such as in the light organ, a signal transduction cascade is initiated that leads to the production of luciferase (Engebrecht et al., 1983) owing to the association of the AHL-LuxR complex with the *lux* operon promoter and transcription of the *lux* operon (Engebrecht and Silverman, 1984). The LuxR complex binds to a region in the *lux* operator called the *lux* box (Stevens et al., 1994; Stevens and Greenberg, 1997) to stimulate expression. The transcriptional regulator LitR, which further enhances the LuxR/I system (Fidopiastis et al., 2002), induces transcription of *luxR*.

luxI is transcribed at a low basal rate, allowing for a constant, low concentration of AI production, which at low cell density diffuses away from the cells. At high cell density, the AHL-LuxR complex stimulates transcription of *luxI*, which leads to even more AHL production, thus forming a positive feedback loop termed autoinduction (Nealson, 1977). This autoinduction circuit allows for a rapid increase in signal production and therefore leads to a rapid induction of phenotypic expression by members of the population.

V. fischeri produces a second AHL (*N*-octanoylhomoserine lactone) encoded by *ainS* (Gilson et al., 1995; Kuo et al., 1994) that has no similarities to the LuxI family of AI synthases but does show some similarity to the LuxM protein of *Vibrio harveyi* (see below). Transcription of *ainS* is considerably decreased in the absence of the signal produced by AinS, suggesting an autoregulatory mechanism as with *luxI* (Lupp and Ruby, 2004). This autoinduction loop is either directly or indirectly dependent on LitR. The AinS signal interacts with LuxR as well as with a second regulator/sensor, AinR (Gilson et al., 1995). In addition to the LuxR/I and AinR/S systems, a two-component phosphorelay cascade affects gene expression via LuxU through a response regulator, the LuxO protein (Miyamoto et al., 2000). At low cell density (in the absence of signals), the response regulators autophosphorylate and transfer phosphate to the LuxU protein, which in turn phosphorylates LuxO (Freeman and Bassler, 1999a,b; Freeman et al., 2000). Phosphorylated LuxO is active and interacts with RpoN to activate transcription of a repressor. Thus, at low cell density, luminescence is repressed by LuxO but becomes partially induced at moderate cell densities (10^8 to 10^9 cells/ml) owing to inactivation of LuxO by the AinR/S system, resulting in derepression of the *lux* operon (Miyamoto et al., 2000). Inactivation of LuxO also results in increased transcription of *litR*, the gene encoding the positive regulator of *luxR* transcription (Fidopiastis et al., 2002). Furthermore, at these cell densities, low levels of induction of the *lux* operon occur owing to the binding of AinS to LuxR (Lupp et al., 2003). The system becomes fully induced at high cell densities (10^{10}, as would occur in the squid light organ) by the functioning of both the Ain and Lux systems. AinS is important not only for luminescence regulation, especially at cell densities preceding induction of the LuxR/I system, but also for successful colonization of the squid host and other putative functions, one of which has been determined to be motility (Lupp and Ruby, 2005). *V. fischeri* QS systems play sequential roles in the symbiotic colonization of the squid, as AinS controls early *lux*-independent colonization at lower cell densities, while LuxI controls late *lux*-dependent colonization at higher cell densities.

The presence of the AI-2 QS system, first described in *V. harveyi* (Schauder et al., 2001; Surette et al., 1999), has been identified in *V. fischeri* (Lupp and Ruby, 2004). This signaling system exists in both gram-negative and gram-positive bacteria, suggesting that AI-2-mediated signal systems arose before the AHL system in gram-negative bacteria or the peptide signaling system in gram-positive bacteria. In the AI-2 system, *luxS* encodes for an enzyme that synthesizes a signal named autoinducer 2 (AI-2), a furanosyl borate diester (Chen et al., 2002). In *V. fischeri*, the LuxS signal regulates luminescence through inactivation of LuxO (Lupp and Ruby, 2004). Both the AinS and LuxS signals affect gene expression through the LuxO cascade, but while *ainS* transcription is regulated by a positive feedback loop (by LitR), *luxS* transcription is constant. A mutation in *luxS* does not affect colonization of the squid light organ, but a *luxS ainS* double mutant shows decreased colonization when compared to the *ainS* mutant, suggesting that the two systems regulate gene expression synergistically (Lupp et al., 2003). Thus, it appears that the quantitative contribution of LuxS is small when compared to the AinS contribution.

In addition to autoregulation of the QS genes, other regulatory systems are known to influence QS regulation. For example, conditions known to increase *lux* operon expression include reduced levels of nutrient, iron and oxygen, and the presence of toxins or DNA-damaging agents (Nealson and Hastings, 1979). Cyclic AMP and cyclic AMP receptor protein are required for transcription of *luxR* in *V. fischeri* (Dunlap and Greenberg, 1988). Control of QS regulation by cyclic AMP might serve to channel information about the nutritional status of the cell into the QS mechanism and delay stimulation of the response when nutritional levels and growth rates are high. In addition, the AI-2 QS system has been suggested to be a metabolic signal rather than a QS signal per se (DeLisa et al., 2001; Winzer et al., 2002). The occurrence of multiple signaling systems may function to integrate such information as nutritional status and cell density of microbial populations and may be a general regulatory mechanism utilized by *Vibrio* spp.

Quorum Sensing in *V. harveyi*

The free-living marine bacterium *V. harveyi* possesses three autoinducer-response systems that function to control bioluminescence (Bassler et al., 1994), type III secretion (Henke and Bassler, 2004a), metalloprotease production (Mok et al., 2003), siderophore production, and colony morphology (Lilley and Bassler, 2000). Signaling system 1 is composed of a homologue of *ainS*, an autoinducer synthase (*luxM*) that produces HAI-1 (N-3-hydroxybutanoyl-homoserine lactone) (Cao and Meighen, 1989), and sensor 1 (*luxN*) (Bassler et al., 1993) (see Table 1 for homologues). The response regulator/sensor LuxN, which responds to and binds the AHL, is homologous to AinR. Signaling system 2 is composed of the autoinducer synthase 2, LuxS (which produces AI-2), and sensor 2 (LuxPQ), another two-component regulatory pair (Bassler et al., 1994). Information from sensor 1 and sensor 2 is relayed to LuxO via the phosphorelay protein LuxU (Freeman and Bassler, 1999a). LuxU is responsible for integrating signaling events from LuxN and LuxQ to the common response regulator. At low cell density (in the absence of signals), LuxN and LuxQ autophosphorylate and transfer phosphate to the shared LuxU protein, which in turn phosphorylates LuxO (Freeman and Bassler, 1999a,b; Freeman et al., 2000). Phosphorylated LuxO is active and interacts with RpoN to activate transcription of genes encoding small regulatory RNAs (sRNAs) that, together with the RNA chaperone Hfq, destabilize the mRNA encoding for LuxR (Lenz et al., 2004; Lilley and Bassler, 2000). LuxR is required for expression of the target genes and, thus, no light is produced (Showalter et al., 1990). The concentration of autoinducers accumulates as the cell density increases and binding of the AIs to LuxN and LuxP causes LuxN and LuxQ to switch from kinases to phosphatases. Unphosphorylated LuxO is inactive and the sRNAs are not expressed, leading to translation of

luxR mRNA (Lenz et al., 2004). The target genes are transcriptionally activated by LuxR, resulting in the production of light.

LuxR functions to induce bioluminescence and metalloprotease production as well as to repress other genes. In both *V. harveyi* and *V. parahaemolyticus*, the type III secretion system is repressed by QS; this effect is dependent on LuxR, although whether LuxR directly represses gene expression or exerts its effect indirectly is not known (Henke and Bassler, 2004a). Furthermore, siderophore production and rugose colony morphology are repressed by QS (Lilley and Bassler, 2000).

The third QS system in *V. harveyi* is homologous to signaling system 1 in *V. cholerae* (Miller et al., 2002) (see below). This system is composed of the autoinducer synthase CqsA, which produces the as yet unidentified CAI-1 signal, and CqsS, which is the cognate sensor (Henke and Bassler, 2004b). All three autoinducers (HAI-1, AI-2, and CqsA) act synergistically to induce bioluminescence in *V. harveyi*; HAI-1 has a stronger influence than AI-2, which has a stronger influence than CAI-1. All three systems act through the LuxU and LuxO phosphorelay system. The CqsS sensor switches from a kinase to a phosphatase at very low cell densities, much earlier than the LuxN and LuxQ sensors. Under natural conditions, the input from each system could therefore vary depending on the environmental conditions. For example, the contribution of system 3 to luminescence is greater than that of system 2 when cells are grown on agar, but the reverse is true for broth cultures (Henke and Bassler, 2004b).

Quorum Sensing in *V. cholerae*

V. cholerae possesses three QS systems (Fig. 2) that also operate through the LuxO protein (Miller et al., 2002) and HapR, a homologue of the *V. harveyi* LuxR transcriptional regulator (Jobling and

Table 1. Quorum-sensing homologues in *Vibrio* spp.; system components are grouped together

V. fischeri	*V. harveyi*	*V. cholerae*	*V. anguillarum*	*V. parahaemolyticus*	*V. vulnificus*
LuxI			VanI		
LuxR			VanR		
AinS	LuxM		VanM	LuxM	
AinR	LuxN		VanN	LuxN	
	CqsA	CqsA		CqsA	
	CqsS	CqsS		CqsS	
LuxS	LuxS	LuxS	VanS	LuxS	LuxS
LuxPQ	LuxPQ	LuxPQ	VanPQ	LuxPQ	LuxPQ
LuxU	LuxU	LuxU	VanU	LuxU	LuxU
LuxO	LuxO	LuxO	VanO	LuxO	LuxO
LitR	LuxR	HapR	VanT	OpaR	SmcR

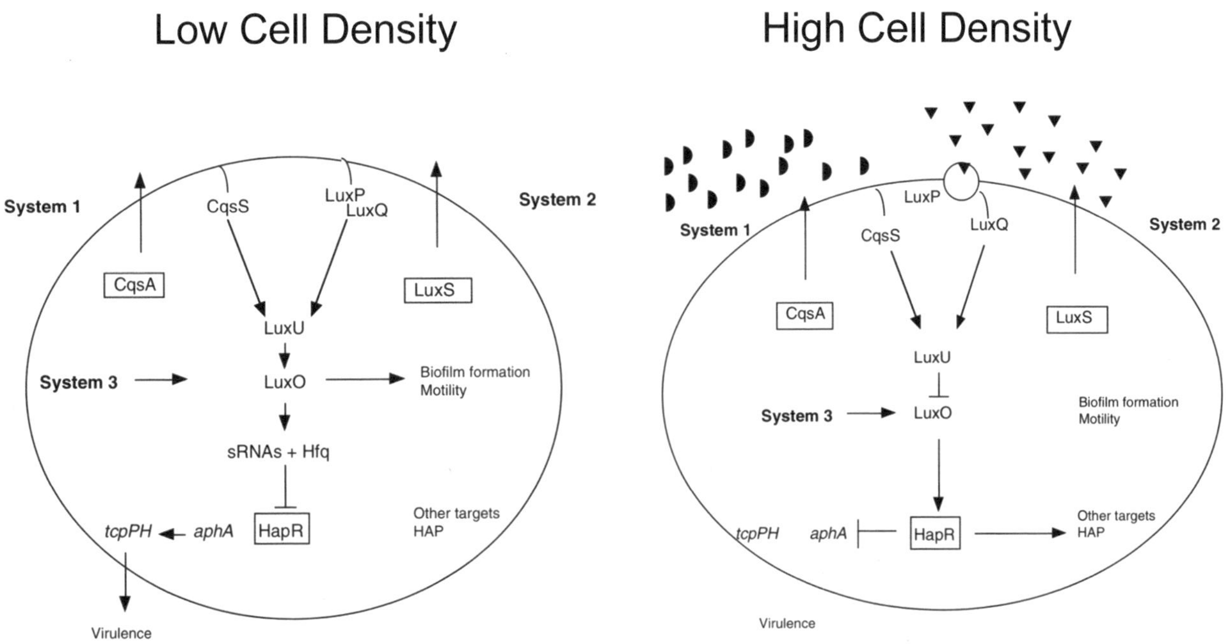

Figure 2. *V. cholerae* QS systems. Three systems regulate biofilm formation, virulence, and HA/protease production. Details of the signal transduction pathways are in the text.

Holmes, 1997). Signaling system 1 is homologous to the CqsA/S system described above but was first identified in *V. cholerae* as the cholera quorum-sensing system (Miller et al., 2002). System 2 is composed of the LuxS-dependent autoinducer AI-2 and its cognate sensor LuxPQ (Schauder et al., 2001). The *V. cholerae* and *V. harveyi* QS systems operate similarly, except that the response to CqsA is greater than the response to AI-2 in *V. cholerae*, which is in contrast to the system in *V. harveyi*. Furthermore, the switch of CqsS from kinase to phosphatase mode occurs at a much higher cell density in *V. cholerae* than it does in *V. harveyi*. This more than likely reflects the fine-tuning of signals for optimal survival in different environments.

A putative system 3 has been proposed to occur in *V. cholerae*, as inactivation of both signaling systems 1 (Cqs) and 2 (AI-2) does not abolish density-dependent gene expression. Signaling system 3 channels its sensory information to the LuxO, but data indicate that LuxU is not involved in this signal relay. Thus, while systems 1 and 2 feed signal information to LuxO via LuxU, system 3 appears to feed information directly to LuxO. Furthermore, the system 3 signal is proposed to be intracellular rather than extracellular, as are the CAI-1 and AI-2 signals.

In *V. cholerae*, virulence factor expression is QS-regulated. A critical step in the *V. cholerae* ToxR virulence cascade is activation of *tcpPH* expression by AphA and AphB (Kovacikova and Skorupski, 1999). HapR directly represses *aphA* expression, thereby in-

hibiting virulence gene expression (Kovacikova and Skorupski, 2002). At low cell density, the sensors act as kinases and transfer phosphate via LuxU to LuxO; active LuxO together with σ^{54} induces transcription of sRNAs, which, together with Hfq, destabilizes *hapR* mRNA (Lenz et al., 2004). This relieves HapR repression of AphA and AphB and allows for ToxR-regulated virulence genes (Kovacikova and Skorupski, 2002). At high cell density, the sensors act as phosphatases, resulting in dephosphorylation and inactivation of LuxO, allowing HapR to activate transcription of the HA/P metalloprotease (Jobling and Holmes, 1997) or to repress expression of the AphA, thereby preventing ToxR-regulated virulence gene expression. Thus, HapR acts at an early stage of growth to repress virulence factor production (Kovacikova and Skorupski, 2002; Miller et al., 2002; Zhu et al., 2002), but it remains unclear how HapR regulates virulence gene expression in vivo.

In addition to regulating virulence gene expression, HapR functions to regulate rugose colony morphology by repression of the *vps* operons that are responsible for exopolysaccharide (EPS) production and thereby also represses biofilm formation (see below) (Hammer and Bassler, 2003; Zhu and Mekalanos, 2003). In *V. cholerae*, QS-controlled virulence gene expression and colonization occur at low cell density, in the absence of autoinducers, and these phenotypes are repressed at high cell density when the signal molecules have accumulated. This implies that it is beneficial to be biofilm/adhesion competent and to express virulence under low-cell-density conditions early in infections. After the bacterial population has grown to high cell density, the induction of the QS system results in repression of virulence factor production and induces the protease, which acts as a detachase, resulting in the return of *V. cholerae* to the environment.

Quorum Sensing in Other *Vibrio* Species

The salmonid fish pathogen *V. anguillarum* contains multiple QS systems homologous to the *V. fischeri luxI/R* system (*vanI/R*) (Milton et al., 1997) and to the *V. harveyi luxM/N* system (*vanM/N*) (Milton et al., 2001). In *V. anguillarum*, a homologue of *V. harveyi* LuxR, VanT, positively regulates extracellular protease activity, pigment production, and biofilm formation (Croxatto et al., 2002). Unlike other vibrios, the VanT in *V. anguillarum* is produced at low cell density, and the amount does not differ during growth (Croxatto et al., 2004). VanO does repress expression of *vanT*, as in other Vibrios, but less significantly. The level of *vanT* expression is also limited by autorepression. VanU plays a role in the activation of *vanT* expression, a response that is unique to this species. Furthermore, VanT interacts with the *vanOU* promoter, further inducing expression during late growth.

V. parahaemolyticus exists planktonically or attached to particulate matter in the marine or estuarine environment and is a major cause of gastroenteritis. This organism has been shown to possess two of the three QS systems found in *V. harveyi*, the LuxM/N system as well as the LuxS/PQ system, in addition to the LuxU/LuxO phosphorelay proteins (Henke and Bassler, 2004a). Furthermore, it has been demonstrated that at high cell density (in the presence of autoinducers) QS represses the type III section system similarly to the situation in *V. harveyi* (Henke and Bassler, 2004a). In addition, the LuxR homologue OpaR positively regulates colony opacity (McCarter, 1998).

The human pathogen *V. vulnificus* has been shown to possess the AI-2 QS system (McDougald et al., 2000; Shao and Hor, 2001), which was demonstrated to regulate a number of virulence factors as well as core response phenotypes of starvation adaptation and stress resistance. Specifically, the LuxR homologue SmcR positively regulates metalloprotease expression and negatively regulates hemolysin expression, but an SmcR mutant strain showed no difference in virulence in a mouse model of infection (Shao and Hor, 2001). In contrast, LuxS was shown to regulate cytotoxicity of the organism to HeLa cells and virulence to mice (Kim et al., 2003). We demon-

strated that, in contrast to metalloprotease production, pili production, motility, and biofilm formation were all repressed by QS in *V. vulnificus* (McDougald et al., 2001). In addition to regulating virulence-related phenotypes, the QS system in this bacterium regulates starvation adaptation as well as resistance to stresses such as oxidative stress (McDougald et al., 2001, 2003). These results expand the role of QS systems to include global regulation of nongrowth physiology.

Although a regulatory circuit involving LuxO appears to be conserved in *Vibrio* species, the relative importance of each of the signals may vary widely among species. CAI-1 may also be restricted to use within certain members of the genus *Vibrio*. Database analysis indicates that, while *V. cholerae* and *V. parahaemolyticus* possess *cqsA*, *V. vulnificus* and *V. fischeri* do not (Henke and Bassler, 2004b). *V. anguillarum*, *V. alginolyticus*, and *V. furnissii* supernatants possess CAI-1 activity, indicating the presence of *cqsA*. The variety of QS systems found in *Vibrio* spp. and their complex multichannel networks may have evolved to regulate the switch from various environments (for example, from a nonpathogenic to a pathogenic lifestyle) that is critical for survival of these organisms in different niches. The presence of multiple systems may protect the system from noise due to similar AIs or to signal degradation by other organisms. In addition, the individual systems may respond to different environmental cues and could thus allow for the integration of multiple cues in the regulation of signaling phenotypes. It has also been proposed that these different QS systems may function to allow populations of *Vibrio* spp. to respond to signals from either conspecific or nonconspecific bacteria (Bassler et al., 1997). The use of QS systems to control behaviors such as biofilm formation, symbiosis, and virulence factor expression indicates that QS regulates phenotypes that modulate the association of bacteria with higher organisms or surfaces. Thus, the ability of bacteria to distinguish self from nonself could be a fundamental property of QS systems. Clearly, this makes QS systems an attractive target for inhibition of virulence or biofilm formation (Rice et al., 2005). Indeed, the use of QS antagonists for these purposes is a fast-growing and innovative field.

BIOFILM FORMATION AS A SURVIVAL STRATEGY

It is now assumed that a majority of bacterial cells in nature exist in biofilms, which are assemblages of bacteria on a surface encased in an extracellular matrix, rather than as free-swimming entities (Costerton et al., 1978). In fact, it has been proposed that prokaryotic cellular communities, including the assembly of the first bacterial and archael cells, develop preferentially on surfaces and that the planktonic cell phenotype has evolved as a mechanism of dispersal (Stoodley et al., 2002). Biofilm formation is recognized as a developmental process that requires a series of discrete and well-regulated steps. The bacteria within a biofilm have been shown to have increased resistance, compared to their planktonic counterparts, to a variety of stresses, including UV light, acid conditions, dehydration, oxidative environments, and antimicrobial agents. Biofilm communities also offer increased metabolic efficiency for the population (Jefferson, 2004). It is therefore of significant interest to understand the factors and regulators used by bacteria in the formation of biofilms in order to elucidate mechanisms of adaptation and to develop strategies for the control of biofilms in situations where they represent a problem (e.g., medically, industrially, or environmentally). As could be expected, the regulation of biofilm formation is complex, involving a range of physical and biological factors, the influences of which vary between species. This section reviews biofilm formation by *Vibrio* spp., with a strong emphasis on *V. cholerae*, for which the most data are available.

Several studies have suggested that biofilm-mediated attachment to abiotic and biotic surfaces may be important for the survival of *Vibrio* spp., especially *V. cholerae*, in the environment (Dumontet et al., 1996; Hood and Winter, 1997; Kumazawa et al., 1991; Montanari et al., 1999; Watnick and Kolter, 1999; Watnick et al., 2001). For example, *Vibrio* species have been shown to attach to plants, filamentous green algae, copepods, crustaceans, and insects (Dumontet et al., 1996; Hood and Winter, 1997; Kumazawa et al., 1991; Montanari et al., 1999). The eel pathogen *V. vulnificus* serovar E has been reported to form biofilms on the epidermal cells of eels (Marco-Noales et al., 2001). Biofilm formation on biotic surfaces has implications for the outbreak of disease. It has been proposed that outbreaks of vibriosis on eel farms were due to stress conditions, such as overcrowding, allowing the cells within the biofilm to cause opportunistic infections in the eels. Similarly, biofilm formation on biotic surfaces has implications for human disease. Because zooplankton and phytoplankton blooms precede cholera outbreaks, the association of *V. cholerae* with these surfaces has been implicated as playing a role in the epidemiology of cholera (Colwell, 1996). The phytoplankton provide food for zooplankton, resulting in blooms of zooplankton, including copepods on which *V. cholerae* has been shown to

form biofilms (Colwell, 2002). Significantly, this relationship can be used to predict when and where cholera outbreaks will occur, using remote sensing to determine when a phytoplankton bloom is occurring (Gil et al., 2004). It remains to be shown that biofilms predominate in many disease settings, but based on biofilm formation on the epithelia of eels and colonization of the light organ of squid by *V. fischeri*, it is highly likely that biofilm formation is a significant contributor to a variety of pathological conditions, especially in chronic infections such as cholera.

Protective Effects of Biofilms

The processes of survival and persistence are especially relevant to bacteria that appear sporadically and that have significant impact on humans. The environmental reservoir for *V. cholerae* is unknown but is presumed to be mediated by attachment and biofilm formation on zooplankton. The removal of zooplankton from contaminated water by filtration is 48% effective in reducing infection by *V. cholerae* (Colwell et al., 2003). To colonize the human host, *V. cholerae* must survive passage through the stomach, leading to the hypothesis that biofilm cells of *V. cholerae* may be more resistant to acid stress than nonbiofilm cells (Zhu and Mekalanos, 2003). It was demonstrated that planktonic cells were readily killed after exposure to pH 4.5 for 30 min, while biofilm-associated cells were 1,000-fold more resistant.

Protozoan grazing has been identified as one of the key environmental pressures faced by bacteria; hence, the survival and persistence of bacteria depend on their ability to adapt to this pressure. Protozoans are ubiquitous in aquatic and soil environments and have been shown to have a major impact on bacterial community composition (Jurgens et al., 1999). In response to protozoan grazing, bacterial communities can develop inedible phenotypes, referred to as grazing-resistant bacteria, leading to pronounced changes in the structural and taxonomic composition of the community. A frequent protective mechanism is the development of grazing-resistant microcolonies, or flocs (Hahn et al., 2000; Matz et al., 2002b). For example, the presence of the flagellate *Rhynchomonas nasuta* results in the formation of grazing-resistant microcolonies in *Pseudomonas aeruginosa* biofilms (Matz et al., 2004). In addition to microcolony formation, the production of antiprotozoal factors was identified as a second protective mechanism.

We have recently demonstrated that planktonic *V. cholerae* cells are rapidly eliminated by grazing, while biofilms of both rugose and smooth variants were unaffected (Matz et al., 2005). Grazing on planktonic *V. cholerae* cells resulted in significantly enhanced biofilm formation by both variants, with the rugose variant exhibiting the greatest increase in biofilm biomass (Fig. 3). When biofilms were pregrown and then exposed to grazers, there was a decrease in flagellate cell numbers, indicating that the biofilm but not the planktonic phase had an antiprotozoal effect. In contrast, a *hapR* mutant biofilm was rapidly eliminated, indicating that the antiprotozoal activity is QS regulated.

It is likely that protozoan grazing has been a selective force in the evolution of pathogens. One hypothesis for the evolution of pathogenicity is that virulence factors first arose as mechanisms of protection in the environment from pressures such as grazing by predatory protozoans. Thus, protozoans have played an important role in the evolution of human pathogens. Understanding the factors involved in resistance to grazing, and for biofilm formation more generally, will be important for determining how and where *V. cholerae* strains persist in the environment and how they may lead to novel targets for the effective control of these important pathogens.

Role of Temperature and Media in Biofilm Formation

While many *Vibrio* species have been shown to form biofilms, the specifics of biofilm formation differ, a fact that may relate to differences in environmental niches occupied by these species. Therefore, various studies have compared biofilm formation at different temperatures, to reflect either environmental or infection conditions, and under different salinity conditions for organisms that exist in estuarine or marine habitats in attempts to understand if the organisms prefer attached or planktonic growth in the different environments. Attachment of *V. parahaemolyticus* was greater when cells grown in defined minimal salts media were compared to attachment of cells grown in Luria-Bertani (Wong et al., 2002); addition of glucose and other sugars inhibited attachment. Using strains of *V. vulnificus*, we have demonstrated similar results (D. McDougald, in preparation). The observation of reduced biofilm formation when grown in complex medium agrees with results from other organisms and therefore may reflect a general trend in biofilm formation. In addition, we ob-

V. cholerae Strains

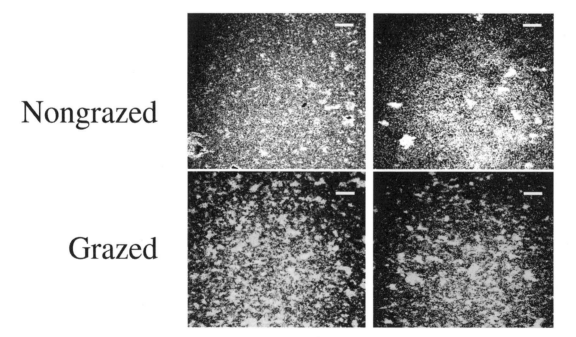

Figure 3. Confocal laser scanning microscope images of biofilms of *V. cholerae* strains under nongrazed (top of each panel) and grazed (bottom of each panel) treatments. Biofilms were stained with Live/Dead viability probe, and images were taken using a confocal laser scanning microscope. Magnification, ×200. Scale bar, 50 μm.

served that there was no significant difference in attachment ability of strains incubated in Luria-Bertani containing either 1% or 3.5% NaCl or when incubated at 20 or 37°C. In contrast, a study by Joseph and Wright (2004) showed that more biofilm was formed when *V. vulnificus* was incubated at 30 or 37°C when compared to 25°C, and when incubated in 1% NaCl compared to 2% NaCl. These studies used similar conditions, with the exception of gentle agitation of microtiter plates in our study, in contrast to static incubation. It is probable that the differences seen are due to strain differences and not agitation. Strain differences may account for much of the disparity of results, a concept that has been demonstrated through comparison of El Tor and O139 strains of *V. cholerae* (see below); hence, it remains unclear whether temperature and salinity play dominant roles in the formation of biofilms by *Vibrio* spp.

Role of Exopolysaccharide (EPS) in Biofilm Formation

The production of capsule and vibrio EPS has been well studied for *Vibrio* spp. in terms of its relevance to infection and the resistance to environmental stresses (Wai et al., 1996; Yildiz and Schoolnik, 1999). Such studies have been used to interpret the significance of strains that have encapsulated (opaque) and unencapsulated (translucent) or smooth and rugose (EPS producer) forms. These cells have the ability to switch between opaque and translucent, or smooth and rugose. It has been shown previously that opaque or rugose cells are more resistant and more infectious. Recently, studies have begun to focus on the role of the capsule in biofilm formation. Joseph and Wright (2004) demonstrated that unencapsulated (translucent) strains of *V. vulnificus* biotype 1 attached better and formed thicker biofilms than the encapsulated (opaque) strains in a microtiter plate assay. *V. parahaemolyticus* opaque and translucent strains both form biofilms, but the opaque biofilm is loose structurally and lacks discrete microcolonies as well as the three-dimensional structures produced by the translucent strain (Enos-Berlage et al., 2005). Unlike the El Tor biotype, the O139 biotype of *V. cholerae* possesses a capsule (Comstock et al., 1995). Spontaneous unencapsulated mutants of *V. cholerae* O139 also exhibit increased surface association (Kierek and Watnick, 2003b). These studies led to the conclusion that the formation of the translucent morphotype may be important for the formation of biofilms in the environment. Thus, the ability to switch from encapsulated (opaque) to unencapsulated (translucent) morphotypes may be due to the role of the different morphotypes in different environmental

niches. For example, it has been suggested that the translucent strains of *V. vulnificus* may be better adapted for attachment to surfaces such as algae, while opaque strains may be selected for in the oyster, as they are able to avoid phagocytic oyster hemocytes (Joseph and Wright, 2004).

In addition to the genes necessary for the production of capsule, *V. cholerae* O139 possesses genes responsible for the production of EPS that are also involved in biofilm formation. This biotype thus forms *vps*-independent and *vps*-dependent biofilms that differ in architecture and are activated by different environmental factors (Kierek and Watnick, 2003a). The activators (monosaccharides) of the *vps*-dependent biofilm are present in fresh water, while the activator (calcium) of *vps*-independent biofilm is present in seawater. It was shown that the O139 O antigen as well as the capsule mediate *vps*-independent biofilm development, where calcium interacts directly with the O-antigen polysaccharide to maintain biofilm structure (Kierek and Watnick, 2003b). In addition to O139 strains, other serogroups of *V. cholerae*, as well as *V. alginolyticus*, *V. fluvialis*, and *V. parahaemolyticus*, also exhibit dissolution of biofilms in the absence of calcium. The authors suggest that dissolution of calcium-dependent biofilms may have implications for the epidemiology of cholera. In this scenario, mixing of seawater with fresh water, which occurs during monsoon seasons, would cause *V. cholerae* biofilms to disperse, thereby increasing planktonic bacterial counts and hence human exposure. It will be particularly exciting to establish which conditions and environments are relevant to these distinct pathways for biofilm formation.

In contrast, EPS production appears to be essential for biofilm formation by *V. cholerae* O1 El Tor, and a *vps*-independent pathway has not been identified (Wai et al., 1998; Yildiz and Schoolnik, 1999). The production of EPS regulates the switch from the smooth to the rugose phase variant (Yildiz and Schoolnik, 1999). This phase transition increases the resistance of the organism to osmotic, acid, and oxidative stress and enhances its ability to form a biofilm (Wai et al., 1998; Watnick et al., 2001; Yildiz and Schoolnik, 1999). Thus, EPS production in El Tor strains results in biofilm phenotypes that are similar to those described above for other *Vibrio* spp. Recent results indicate that rugose *V. cholerae* biofilms have a high percentage of smooth morphotypes in the biofilm effluent (D. McDougald, unpublished), indicating that the smooth form may be the dispersal phenotype. It could account for the predominance of the smooth strains in previous studies, which looked at planktonic cells. In this model, the opaque or rugose cells would represent the environmentally resistant forms, capable of causing infection due to the presence of the capsule which helps to evade phagocytosis, and the smooth form might be a dispersal stage to aid in the colonization of new sites, e.g., when nutrients become rich and cell numbers increase rapidly or when the salinity of the water decreases during monsoons.

Quorum-Sensing Regulation of Biofilm Formation

It has been demonstrated that QS regulates biofilm formation in a number of *Vibrio* species (Croxatto et al., 2002; Enos-Berlage et al., 2005; Hammer and Bassler, 2003; McDougald et al., 2001; Zhu and Mekalanos, 2003). However, most of the work on biofilms and QS has focused on *V. cholerae*, and this is reflected in the information presented here. The QS systems found in *V. cholerae* are widely found in other vibrios, and, hence, it is anticipated that the QS system will behave similarly in species other than *V. cholerae*. In contrast to other bacterial pathogens, such as *P. aeruginosa*, that induce virulence and biofilm production at high cell density in the presence of AIs, *V. cholerae* represses these behaviors at high cell density. A *V. cholerae* mutant that constitutively expresses HapR is deficient in biofilm formation on abiotic surfaces, while a HapR-deficient mutant is increased in its ability to form a biofilm (Zhu et al., 2002) and shows decreased detachment from biofilms when compared with the wild type (Zhu and Mekalanos, 2003). The increased biofilm formation by a HapR-deficient strain is most likely related to increased production of extracellular matrix material, as HapR-deficient biofilm cells were shown to overexpress *vps* genes, which, as described above, play a significant role in biofilm formation (Watnick and Kolter, 1999; Watnick et al., 2001; Yildiz and Schoolnik, 1999; Yildiz et al., 2004). Interestingly, long-term incubation of biofilm-deficient *luxO* mutant strains, which constitutively express *hapR*, has reduced *vps* expression, accumulated mutations in *hapR* that led to overproduction of EPS, and, therefore, enhanced biofilm formation (Hammer and Bassler, 2003). Furthermore, deletions of genes within the *vps* operon completely abolished biofilm formation in both wild type and *hapR* mutants, confirming the roles of *vps* genes in biofilm formation in the *hapR* mutant strain (Zhu and Mekalanos, 2003). (It should be pointed out that the experiments described above utilized O1 El Tor strains of *V. cholerae*, which do not seem to form *vps*-independent biofilms, in contrast to the O139 strains). *vps* expression was undetectable in wild-type planktonic cells but readily detectable in wild-type biofilms.

QS is also involved in the control of attachment to biotic surfaces. While the HapR mutant was able to colonize infant mice at wild-type levels, a LuxO mutant was defective in colonization, probably owing to a reduction in pili production (Zhu et al., 2002). These studies indicated that HapR is involved in detachment, probably owing to the activation of the HA/P protease, and that a HapR-deficient strain may remain attached to the epithelium longer than wild-type strains. However, QS may not be involved in long-term colonization, as it was found that the planktonic wild-type, *hapR*, and *vps* mutant bacteria could all be recovered 11 days postinfection in similar numbers (Zhu and Mekalanos, 2003). These results, especially those obtained with the *vps* mutants and the *hapR* mutants, should be compared with those on attachment and biofilm formation on biotic surfaces to determine if the pathways and genes involved in both are similar or involve separate gene sets. Moreover, studies of the role of QS in biofilm formation in other *Vibrio* spp. would be useful to establish if *V. cholerae* serves as an appropriate model for the genus as a whole or whether there are important differences that may reflect various life styles and environments of the various members.

Role of Non-Quorum-Sensing Regulators and Surface Structures in Biofilm Formation

HapR has also been shown to repress the expression of *vpsR*, encoding for a response regulator, which has homology with σ^{54}-dependent activators (Yildiz et al., 2004). VpsR has been identified as a positive regulator of *vps* gene transcription (Yildiz and Schoolnik, 1999; Yildiz et al., 2001). Disruption of *vpsR* in a rugose strain results in smooth colony morphology, prevents *vps* gene expression, and abolishes three-dimensional biofilm formation. In addition, VieA, a two-component response regulator (Tischler and Camilli, 2004), represses VpsR transcription. A second, positive regulator of *vps* genes (VpsT) has also been identified and determined to be necessary for mature biofilm formation (Casper-Lindley and Yildiz, 2004); it was shown that both VpsR and VpsT are necessary for full induction of *vps* genes and that both of these regulators positively influence their own expression. It has been suggested that these two response regulators respond to different environmental stimuli to separately control biofilm formation.

A CytR homologue has been identified as a repressor of *vps* gene transcription and biofilm development (Haugo and Watnick, 2002). In *E. coli*, CytR represses nucleoside uptake and catabolism when nucleosides are scarce. It was shown that in *V. cholerae*, for both planktonic and biofilm-associated cells, CytR regulates EPS synthesis at the level of *vps* gene transcription. The presence of a HapR binding motif at the *cytR* promoter indicates that HapR may repress *vps* expression by increasing expression of CytR, as well as by direct repression of *vps* (Yildiz et al., 2004). These data indicate that elevated intracellular cytidine levels may lead to an increase in EPS production and thus an increase in biofilm formation.

In a screen for biofilm-deficient transposon mutants of *V. cholerae* O1 El Tor, it was determined that flagellar motility and the synthesis of EPS were essential for the development of a biofilm (Watnick and Kolter, 1999; Watnick et al., 2001). Similarly, *V. parahaemolyticus* flagellar mutants were arrested at the monolayer stage (Enos-Berlage et al., 2005). In contrast, the loss of flagella does not greatly affect attachment and subsequent biofilm formation in *V. cholerae* O139 (Watnick et al., 2001), whereas the sodium-driven motor is necessary for *vps* expression as well as biofilm formation, as nonmotile flagellated strains have reduced EPS (Lauriano et al., 2004). It would appear that the sodium-driven motor is a mechanosensor that is used for regulation of *V. cholerae* biofilm development, similar to *V. parahaemolyticus* and *V. alginolyticus*, where it regulates the swimming-to-swarming transition by responding to changes in torque on the flagellum (Belas et al., 1986; McCarter et al., 1988).

Thus, at least three distinct pathways regulate EPS production in *V. cholerae*: QS repression by HapR, flagellum-regulated repression of EPS, and increased EPS expression as seen in the rugose morphotypes. The presence of multiple signaling pathways for EPS regulation and biofilm formation may indicate that different pathways will be operational in different environments or that selection of different strains occurs under certain conditions (Heithoff and Mahan, 2004). Furthermore, the observation that O1 and O139 have distinct pathways for biofilm formation may ultimately provide insights into differences between their habitats/lifestyles that account for their reservoirs between infections.

In addition to the above regulators, in a screen of a *V. parahaemolyticus* transposon library, H-NS mutants were found to be the most severely impaired for biofilm formation and were also defective for initial attachment to surfaces (Enos-Berlage et al., 2005). This is likely due to the fact that these mutants fail to make polar flagella as well as the capsule, a theme reflected in the regulation of biofilm formation for *V. cholerae*. Mutants in capsule production in both opaque and translucent strains were able to attach but were not able to form microcolonies. The mannose-sensitive hemagglutinin (MSHA) pilus aided attachment to certain surfaces but was not required for three-dimensional biofilm maturation.

V. cholerae O1 El Tor uses the MSHA pilus for attachment to abiotic surfaces as well as for attachment to cellulose (biotic surfaces such as algae); however, attachment to chitin was shown to occur independent of the MSHA pilus (Watnick et al., 1999). Instead, the toxin-coregulated pilus (TCP), which is required for intestinal colonization and for acquisition of cholera toxin, mediates bacterial interactions necessary for biofilm formation on chitinous surfaces (Reguera and Kolter, 2005). Furthermore, a 53-kDa protein has been implicated in the promotion of attachment to chitin, copepods, and intestinal epithelial cells in *V. cholerae* (Zampini et al., 2005). It is likely that *Vibrio* spp. have evolved similar/overlapping mechanisms regulating attachment to chitin and other surfaces in seawater that also allow for host cell colonization. Thus, evolution of phenotypic traits enhancing survival and persistence of these bacteria in the environment may have surreptitiously increased their ability to invade host organisms.

CONCLUSIONS

These are exciting times in microbial ecology as we improve our understanding of the life cycles of microorganisms in the environment. Rapid progress is being made in the identification of factors affecting persistence of bacteria in nature. Vibrios are found in a broad range of environments and are able to persist partly because of their ability to survive cycles of feast and famine. The starvation adaptation pathway protects cells against a number of stresses and prepares them for subsequent outgrowth when nutrients become plentiful. Members of this genus have been shown to lose culturability, while retaining measures of viability, a state termed "viable but nonculturable." The resuscitation of cells from this state may account for the reemergence of bacteria after periods of apparent absence and has significant implications for human health. Another adaptive response is the ability to form biofilms, which protects the cells from a variety of stresses, such as protozoan grazing. All of these behaviors need to be regulated for them to be effective strategies for survival. One mechanism that many species in this genus utilize is quorum sensing, which has been shown to regulate virulence, biofilm formation, luminescence, and host colonization. Thus, these organisms can serve as models for a range of processes involved in environmental adaptations and survival.

Acknowledgments. The authors thank Scott Rice for comments and critical reading of the manuscript. We also thank Pui Yi Young for confocal images. This work was supported by funding from the Centre for Marine Biofouling and Bio-Innovation and National Health and Medical Research Council project grant number 222847.

REFERENCES

Albertson, N. H., T. Nyström, and S. Kjelleberg. 1990a. Macromolecular synthesis during recovery of the marine *Vibrio* sp. S14 from starvation. *J. Gen. Microbiol.* **136**:2201–2207.

Albertson, N. H., T. Nyström, and S. Kjelleberg. 1990b. Functional mRNA half-lives in the marine *Vibrio* sp. S14 during starvation and recovery. *J. Gen. Microbiol.* **136**:2195–2199.

Amy, P. S., C. Pauling, and R. Y. Morita. 1983. Starvation-survival processes of a marine Vibrio. *Appl. Environ. Microbiol.* **45**:1041–1048.

Anderson, J. I. W., and W. P. Heffernan. 1965. Isolation and characterization of filterable marine bacteria. *J. Bacteriol.* **90**:1713–1718.

Arana, I., A. Muela, J. Iriberri, L. Egea, and I. Baracina. 1992. Role of hydrogen peroxide in loss of culturability mediated by visible light in *Escherichia coli* in a freshwater system. *Appl. Environ. Microbiol.* **58**:3903–3907.

Ballesteros, M., A. Fredriksson, J. Henriksson, and T. Nystrom. 2001. Bacterial senescence: protein oxidation in non-proliferating cells is dictated by the accuracy of the ribosomes. *EMBO J.* **20**:5280–5289.

Barer, M. R., L. T. Gribbon, C. R. Harwood, and C. E. Nwoguh. 1993. The viable but non-culturable hypothesis and medical microbiology. *Rev. Med. Microbiol.* **4**:183–191.

Bassler, B. L., M. Wright, R. E. Showalter, and M. R. Silverman. 1993. Intercellular signalling in *Vibrio harveyi*: sequence and function of genes regulating expression of luminescence. *Mol. Microbiol.* **9**:773–786.

Bassler, B. L., M. Wright, and M. R. Silverman. 1994. Multiple signalling systems controlling expression of luminescence in *Vibrio harveyi*: sequence and function of genes encoding a second sensory pathway. *Mol. Microbiol.* **13**:273–286.

Bassler, B. L., E. P. Greenberg, and A. M. Stevens. 1997. Cross-species induction of luminescence in the quorum-sensing bacterium *Vibrio harveyi*. *J. Bacteriol.* **179**:4043–4045.

Bassler, B. L. 1999. How bacteria talk to each other: regulation of gene expression by quorum sensing. *Curr. Opin. Microbiol.* **2**:582–587.

Belas, R., M. Simon, and M. Silverman. 1986. Regulation of lateral flagella gene transcription in *Vibrio parahaemolyticus*. *J. Bacteriol.* **167**:210–218.

Bogosian, G., L. E. Sammons, P. J. L. Morris, J. P. O'Neil, M. A. Heitkamp, and D. B. Weber. 1996. Death of the *Escherichia coli* K-12 strain W3110 in soil and water. *Appl. Environ. Microbiol.* **62**:4114–4120.

Bogosian, G., P. J. L. Morris, and J. P. O'Neil. 1998. A mixed culture recovery method indicates that enteric bacteria do not enter the viable but nonculturable state. *Appl. Environ. Microbiol.* **64**:1736–1742.

Brauns, L. A., M. C. Hudson, and J. D. Oliver. 1991. Use of the polymerase chain reaction in detection of culturable and nonculturable *Vibrio vulnificus* cells. *Appl. Environ. Microbiol.* **57**:2651–2655.

Campbell, M. S., and A. C. Wright. 2003. Real-time PCR analysis of *Vibrio vulnificus* from oysters. *Appl. Environ. Microbiol.* **69**:7137–7144.

Cao, J.-G., and E. A. Meighen. 1989. Purification and structural identification of an autoinducer for the luminescence system of *Vibrio harveyi*. *J. Biol. Chem.* **264**:21670–21676.

Cashel, M., D. R. Gentry, V. J. Hernandez, and D. Vinella. 1996. The stringent response, p. 1458–1496. *In* F. C. Neidhardt, R. Curtiss III, J. L. Ingraham, E. C. C. Lin, K. B. Low, B. Magasanik, W. S. Reznikoff, M. Riley, M. Schaechter, and H. E. Umbarger (ed.), Escherichia coli *and* Salmonella typhimurium: Cellular and Molecular Biology, 2nd ed. ASM Press, Washington, D.C.

Casper-Lindley, C., and F. H. Yildiz. 2004. VpsT is a transcriptional regulator required for expression of *vps* biosynthesis genes and the development of rugose colonial morphology in *Vibrio cholerae* O1 El Tor. *J. Bacteriol.* **186**:1574–1578.

Chen, X., S. Schauder, N. Potier, A. Van Dorsselaer, I. Pelczer, B. L. Bassler, and F. M. Hughson. 2002. Structural identification of a bacterial quorum-sensing signal containing boron. *Nature* **415**:545–549.

Coleman, S. S., and J. D. Oliver. 1996. Optimization of conditions for the polymerase chain reaction amplification of DNA from culturable and nonculturable cells of *Vibrio vulnificus*. *FEMS Microbiol. Ecol.* **19**:127–132.

Colwell, R. R. 1996. Global climate and infectious disease: the cholera paradigm. *Science* **274**:2025–2031.

Colwell, R. R. 2002. A voyage of discovery: cholera, climate and complexity. *Environ. Microbiol.* **4**:67–69.

Colwell, R. R., P. R. Brayton, D. J. Grimes, D. B. Roszak, S. A. Huq, and L. M. Palmer. 1985. Viable but non-culturable *Vibrio cholerae* and related pathogens in the environment: implications for release of genetically engineered microorganisms. *Bio/Technology* **3**:817–820.

Colwell, R. R., P. Brayton, D. Herrington, B. Tall, A. Huq, and M. M. Levine. 1996. Viable but non-culturable *Vibrio cholerae* O1 revert to a culturable state in the human intestine. *World J. Microbiol. Biotechnol.* **12**:28–31.

Colwell, R. R., A. Huq, M. S. Islam, et al. 2003. Reduction of cholera in Bangladeshi villages by simple filtration. *Proc. Natl. Acad. Sci. USA* **100**:1051–1055.

Comstock, L. E., D. Maneval, Jr., P. Panigrahi, A. Joseph, M. M. Levine, J. B. Kaper, J. G. Morris, Jr., and J. A. Johnson. 1995. The capsule and O antigen in *Vibrio cholerae* O139 Bengal are associated with a genetic region not present in *Vibrio cholerae* O1. *Infect. Immun.* **63**:317–323.

Costerton, J. W., G. G. Geesey, and K. J. Cheng. 1978. How bacteria stick. *Sci. Amer.* **238**:86–95.

Coutard, F., M. Pommepuy, S. Loaec, and D. Hervio-Heath. 2005. mRNA detection by reverse transcription-PCR for monitoring viability and potential virulence in a pathogenic strain of *Vibrio parahaemolyticus* in viable but nonculturable state. *J. Appl. Microbiol.* **98**:951–961.

Croxatto, A., V. J. Chalker, J. Lauritz, J. Jass, A. Hardman, P. Williams, M. Camara, and D. L. Milton. 2002. VanT, a homologue of *Vibrio harveyi* LuxR, regulates serine, metalloprotease, pigment, and biofilm production in *Vibrio anguillarum*. *J. Bacteriol.* **184**:1617–1629.

Croxatto, A., J. Pride, A. Hardman, P. Williams, M. Camara, and D. L. Milton. 2004. A distinctive dual-channel quorum-sensing system operates in *Vibrio anguillarum*. *Mol. Microbiol.* **52**:1677–1689.

Cuny, C., L. Dukan, L. Fraysse, M. Ballesteros, and S. Dukan. 2005. Investigation of the first events leading to loss of culturability during *Escherichia coli* starvation: future nonculturable bacteria form a subpopulation. *J. Bacteriol.* **187**:2244–2248.

Davis, B. D., S. M. Luger, and P. C. Tai. 1986. Role of ribosome degradation in the death of starved *Escherichia coli* cells. *J. Bacteriol.* **166**:439–445.

DeLisa, M. P., J. J. Valdes, and W. E. Bentley. 2001. Quorum signaling via Al-2 communicates the "metabolic burden" associated with heterologous protein production in *Escherichia coli*. *Biotechnol. Bioeng.* **75**:439–450.

Desnues, B., G. Gregori, S. Dukan, H. Aguilaniu, and T. Nyström. 2003. Differential oxidative damage and expression of stress regulons in culturable and nonculturable cells of *Escherichia coli*. *EMBO Rep.* **4**:400–404.

Dukan, S., and T. Nystrom. 1998. Bacterial senescence: stasis results in increased and differential oxidation of cytoplasmic proteins leading to developmental induction of the heat shock regulon. *Genes Dev.* **12:**3431–3441.

Dukan, S., and T. Nystrom. 1999. Oxidative stress defense and deterioration of growth-arrested *Escherichia coli* cells. *J. Biol. Chem.* **274:**26027–26032.

Dukan, S., A. Farewell, M. Ballesteros, F. Taddei, M. Radman, and T. Nystrom. 2000. Protein oxidation in response to increased transcriptional or translational errors. *Proc. Natl. Acad. Sci. USA* **97:**5746–5749.

Dumontet, S., K. Krovacek, S. B. Baloda, R. Grottoli, V. Pasquale, and S. Vanucci. 1996. Ecological relationship between *Aeromonas* and *Vibrio* spp. and planktonic copepods in the coastal marine environment in Southern Italy. *Comp. Immunol. Microbiol. Infect. Dis.* **19:**245–254.

Dunlap, P. V., and E. P. Greenberg. 1988. Control of *Vibrio fischeri lux* gene transcription by a cyclic AMP receptor protein-LuxR protein regulatory circuit. *J. Bacteriol.* **170:**4040–4046.

Eberhard, A., A. L. Burlingame, C. Eberhard, G. L. Kenyon, K. H. Nealson, and N. J. Oppenheimer. 1981. Structural identification of autoinducer of *Photobacterium fischeri* luciferase. *Biochemistry* **28:**2444–2449.

Eilers, H., J. Pernthaler, and R. Amann. 2000. Succession of pelagic marine bacteria during enrichment: a close look at cultivation-induced shifts. *Appl. Environ. Microbiol.* **66:**4634–4640.

Engebrecht, J., K. Nealson, and M. Silverman. 1983. Bacterial bioluminescence: isolation and genetic analysis of functions from *Vibrio fischeri. Cell* **32:**773–781.

Engebrecht, J., and M. Silverman. 1984. Identification of genes and gene products necessary for bacterial bioluminescence. *Proc. Natl. Acad. Sci. USA* **81:**4154–4158.

Enos-Berlage, J. L., Z. T. Guvener, C. E. Keenan, and L. L. McCarter. 2005. Genetic determinants of biofilm development of opaque and translucent *Vibrio parahaemolyticus. Mol. Microbiol.* **55:**1160–1182.

Felter, R. A., R. R. Colwell, and G. B. Chapman. 1969. Morphology and round body formation in *Vibrio marinus. J. Bacteriol.* **99:**326–335.

Fidopiastis, P. M., C. M. Miyamoto, M. G. Jobling, E. A. Meighen, and E. G. Ruby. 2002. LitR, a new transcriptional activator in *Vibrio fischeri*, regulates luminescence and symbiotic light organ colonization. *Mol. Microbiol.* **45:**131–143.

Fischer-Le Saux, M., D. Hervio-Heath, S. Loaec, R. R. Colwell, and M. Pommepuy. 2002. Detection of cytotoxin-hemolysin mRNA in nonculturable populations of environmental and clinical *Vibrio vulnificus* strains in artificial seawater. *Appl. Environ. Microbiol.* **68:**5641–5646.

Flärdh, K., P. S. Cohen, and S. Kjelleberg. 1992. Ribosomes exist in large excess over the apparent demand for protein synthesis during carbon starvation in marine *Vibrio* sp. strain CCUG 15956. *J. Bacteriol.* **174:**6780–6788.

Flärdh, K., T. Axberg, N. Albertson, and S. Kjelleberg. 1994. Stringent control during carbon starvation of marine *Vibrio* sp. strain S14: molecular cloning, nucleotide sequence, and deletion of the *relA* gene. *J. Bacteriol.* **176:**5949–5957.

Freeman, J. A., and B. L. Bassler. 1999a. Sequence and function of LuxU: a two-component phosphorelay protein that regulates quorum sensing in *Vibrio harveyi. J. Bacteriol.* **181:**899–906.

Freeman, J. A., and B. L. Bassler. 1999b. A genetic analysis of the function of LuxO, a two-component response regulator involved in quorum sensing in *Vibrio harveyi. Mol. Microbiol.* **31:**665–677.

Freeman, J. A., B. N. Lilley, and B. L. Bassler. 2000. A genetic analysis of the functions of LuxN: a two-component hybrid sensor kinase that regulates quorum sensing in *Vibrio harveyi. Mol. Microbiol.* **35:**139–149.

Gil, A. I., V. R. Louis, I. N. G. Rivera, et al. 2004. Occurrence and distribution of *Vibrio cholerae* in the coastal environment of Peru. *Environ. Microbiol.* **6:**699–706.

Gilson, L., A. Kuo, and P. V. Dunlap. 1995. AinS and a new family of autoinducer synthesis proteins. *J. Bacteriol.* **177:**6946–6951.

Giovannoni, S., and M. Rappé. 2000. Evolution, diversity, and molecular ecology of marine prokaryotes, p. 47–84. *In* D. Kirchman (ed.), *Microbial Ecology of the Oceans.* Wiley-Liss, Inc., New York, N.Y.

Gong, L., K. Takayama, and S. Kjelleberg. 2002. Role of spoT-dependent ppGpp accumulation in the survival of light-exposed starved bacteria. *Microbiology* **148:**559–570.

Gonzalez, J. M., J. Iriberri, L. Egea, and I. Barcina. 1992. Characterization of culturability, protistan grazing, and death of enteric bacteria in aquatic ecosystems. *Appl. Environ. Microbiol.* **58:**998–1004.

Hahn, M. W., E. R. B. Moore, and M. G. Hofle. 2000. Role of microcolony formation in the protistan grazing defense of the aquatic bacterium *Pseudomonas* sp. MWH1. *Microb. Ecol.* **39:**175–185.

Hammer, B. K., and B. L. Bassler. 2003. Quorum sensing controls biofilm formation in *Vibrio cholerae. Mol. Microbiol.* **50:**101–104.

Hasan, J. A. K., M. A. R. Chowdhury, M. Shahabuddin, A. Huq, L. Loomis, and R. R. Colwell. 1994. Cholera toxin gene polymerase chain reaction for detection of non-culturable *Vibrio cholerae* O1. *World J. Microbiol. Biotechnol.* **10:**568–571.

Haugo, A. J., and P. I. Watnick. 2002. *Vibrio cholerae* CytR is a repressor of biofilm development. *Mol. Microbiol.* **45:**471–483.

Heithoff, D. M., and M. J. Mahan. 2004. *Vibrio cholerae* biofilms: stuck between a rock and a hard place. *J. Bacteriol.* **186:**4835–4837.

Hengge-Aronis, R. 1993. Survival of hunger and stress: the role of *rpoS* in early stationary phase gene regulation in *E. coli. Cell* **72:**165–168.

Hengge-Aronis, R. 1996. Regulation of gene expression during entry into stationary phase, p. 1497–1512. *In* F. C. Neidhardt, R. Curtiss III, J. L. Ingraham, E. C. C. Lin, K. B. Low, B. Magasanik, W. S. Reznikoff, M. Riley, M. Schaechter, and H. E. Umbarger (ed.), Escherichia coli *and* Salmonella typhimurium: *Cellular and Molecular Biology*, 2nd ed. ASM Press, Washington, D.C.

Hengge-Aronis, R. 2002. Signal transduction and regulatory mechanisms involved in control of the σ(S) (RpoS) subunit of RNA polymerase. *Microbiol. Mol. Biol. Rev.* **66:**373–395.

Henke, J. M., and B. L. Bassler. 2004a. Quorum sensing regulates type III secretion in *Vibrio harveyi* and *Vibrio parahaemolyticus. J. Bacteriol.* **186:**3794–3805.

Henke, J. M., and B. L. Bassler. 2004b. Three parallel quorum-sensing systems regulate gene expression in *Vibrio harveyi. J. Bacteriol.* **186:**6902–6914.

Holmquist, L., and S. Kjelleberg. 1993. Changes in viability, respiratory activity and morphology of the marine *Vibrio* sp. strain S14 during starvation of individual nutrients and subsequent recovery. *FEMS Microbiol. Ecol.* **12:**215–224.

Hood, M. A., J. B. Guckert, D. C. White, and F. Deck. 1986. Effect of nutrient deprivation on lipid, carbohydrate, DNA, RNA and protein levels in *Vibrio cholerae. Appl. Environ. Microbiol.* **52:**788–793.

Hood, M. A., and P. A. Winter. 1997. Attachment of *Vibrio cholerae* under various environmental conditions and to selected substrates. *FEMS Microbiol. Ecol.* **22:**215–223.

Israely, T., E. Banin, and E. Rosenberg. 2001. Growth, differentiation and death of *Vibrio shiloi* in coral tissue as a function of seawater temperature. *Aquat. Microb. Ecol.* **24:**1–8.

Jefferson, K. K. 2004. What drives bacteria to produce a biofilm? *FEMS Microbiol. Lett.* **236**:163–173.

Jenkins, D. E., J. E. Schultz, and A. Matin. 1988. Starvation-induced cross protection against heat or H2O2 challenge in *Escherichia coli. J. Bacteriol.* **170**:3910–3914.

Jenkins, D. E., S. A. Chaisson, and A. Matin. 1990. Starvation-induced cross protection against osmotic challenge in *Escherichia coli. J. Bacteriol.* **172**:2779–2781.

Jobling, M. G., and R. K. Holmes. 1997. Characterization of *hapR*, a positive regulator of the *Vibrio cholerae* HA/protease gene *hap*, and its identification as a functional homologue of the *Vibrio harveyi luxR* gene. *Mol. Microbiol.* **26**:1023–1034.

Joseph, L. A., and A. C. Wright. 2004. Expression of *Vibrio vulnificus* capsular polysaccharide inhibits biofilm formation. *J. Bacteriol.* **186**:889–893.

Josephson, K. L., C. P. Gerba, and I. L. Pepper. 1993. Polymerase chain reaction detection of nonviable bacterial pathogens. *Appl. Environ. Microbiol.* **59**:3513–3515.

Jouper-Jaan, Å., A. E. Goodman, and S. Kjelleberg. 1992. Bacteria starved for prolonged periods develop increased protection against lethal temperatures. *FEMS Microbiol. Ecol.* **101**:229–236.

Jurgens, K., J. Pernthaler, S. Schalla, and R. Amann. 1999. Morphological and compositional changes in a planktonic bacterial community in response to enhanced protozoan grazing. *Appl. Environ. Microbiol.* **65**:1241–1250.

Kell, D. B., A. S. Kaprelyants, D. H. Weichart, C. R. Harwood, and M. R. Bare. 1998. Viability and activity in readily culturable bacteria: a review and discussion of the practical issues. *Antonie Leeuwenhoek* **73**:169–187.

Kierek, K., and P. I. Watnick. 2003a. Environmental determinants of *Vibrio cholerae* biofilm development. *Appl. Environ. Microbiol.* **69**:5079–5088.

Kierek, K., and P. I. Watnick. 2003b. The *Vibrio cholerae* O139 O-antigen polysaccharide is essential for Ca2+-dependent biofilm development in sea water. *Proc. Natl. Acad. Sci. USA* **100**:14357–14362.

Kim, S. Y., S. E. Lee, Y. R. Kim, C. M. Kim, P. Y. Ryu, H. E. Choy, S. S. Chung, and J. H. Rhee. 2003. Regulation of *Vibrio vulnificus* virulence by the LuxS quorum-sensing system. *Mol. Microbiol.* **48**:1647–1664.

Kjelleberg, S., N. Albertson, K. Flärdh, L. Holmquist, Å. Jouper-Jaan, R. Marouga, J. Östling, B. Svenblad, and D. Weichart. 1993. How do non-differentiating bacteria adapt to starvation? *Antonie Leeuwenhoek* **63**:333–341.

Koga, T., and K. Takumi. 1995. Nutrient starvation induces cross protection against heat, osmotic or H$_2$O$_2$ challenge in *Vibrio parahaemolyticus. Microbiol. Immunol.* **39**:213–215.

Kogure, K., U. Simidu, and N. Taga. 1979. A tentative direct microscopic method for counting living marine bacteria. *Can. J. Microbiol.* **25**:415–420.

Kolter, R., D. A. Siegele, and A. Tormo. 1993. The stationary phase of the bacterial life cycle. *Annu. Rev. Microbiol.* **47**:855–874.

Kong, I.-S., T. C. Bates, A. Hulsmann, H. Hassan, B. E. Smith, and J. D. Oliver. 2004. Role of catalase and *oxyR* in the viable but nonculturable state of *Vibrio vulnificus. FEMS Microbiol. Ecol.* **50**:133–142.

Kovacikova, G., and K. Skorupski. 1999. A *Vibrio cholerae* LysR homolog, AphB, cooperates with AphA at the *tcpPH* promoter to activate expression of the ToxR virulence cascade. *J. Bacteriol.* **181**:4250–4256.

Kovacikova, G., and K. Skorupski. 2002. Regulation of virulence gene expression in *Vibrio cholerae* by quorum sensing: HapR functions at the *aphA* promoter. *Mol. Microbiol.* **46**:1135–1147.

Kumazawa, N. H., N. Fukuma, and Y. Komoda. 1991. Attachment of *Vibrio parahaemolyticus* strains to estuarine algae. *J. Vet. Med. Sci.* **53**:201–205.

Kuo, A., N. V. Blough, and P. V. Dunlap. 1994. Multiple N-acyl-L-homoserine lactone autoinducers of luminescence in the marine symbiotic bacterium *Vibrio fischeri. J. Bacteriol.* **176**:7558–7565.

Lange, R., and R. Hengge-Aronis. 1991. Identification of a central regulator of stationary-phase gene expression in *Escherichia coli. Mol. Microbiol.* **5**:49–59.

Lauriano, C. M., C. Ghosh, N. E. Correa, and K. E. Klose. 2004. The sodium-driven flagellar motor controls exopolysaccharide expression in *Vibrio cholerae. J. Bacteriol.* **186**:4864–4874.

Lee, K.-H., and E. G. Ruby. 1995. Symbiotic role of the viable but nonculturable state of *Vibrio fischeri* in Hawaiian coastal seawater. *Appl. Environ. Microbiol.* **61**:278–283.

Lenz, D. H., K. C. Mok, B. N. Lilley, R. V. Kulkarni, N. S. Wingreen, and B. L. Bassler. 2004. The small RNA chaperone Hfq and multiple small RNAs control quorum sensing in *Vibrio harveyi* and *Vibrio cholerae. Cell* **118**:69–82.

Lilley, B. N., and B. L. Bassler. 2000. Regulation of quorum sensing in *Vibrio harveyi* by LuxO and sigma 54. *Mol. Microbiol.* **36**:940–954.

Lupp, C., M. Urbanowski, E. P. Greenberg, and E. G. Ruby. 2003. The *Vibrio fischeri* quorum-sensing systems *ain* and *lux* sequentially induce luminescence gene expression and are important for persistence in the squid host. *Mol. Microbiol.* **50**:319–331.

Lupp, C., and E. G. Ruby. 2004. *Vibrio fischeri* LuxS and AinS: comparative study of two signal synthases. *J. Bacteriol.* **186**:3873–3881.

Lupp, C., and E. G. Ruby. 2005. *Vibrio fischeri* uses two quorum-sensing systems for the regulation of early and late colonization factors. *J. Bacteriol.* **187**:3620–3629.

MacDonell, M. T., and M. A. Hood. 1982. Isolation and characterization of ultramicrobacteria from a gulf coast estuary. *Appl. Environ. Microbiol.* **43**:566–571.

Mae, A., M. Montesano, V. Koiv, and E. T. Palva. 2001. Transgenic plants producing the bacterial pheromone N-acyl-homoserine lactone exhibit enhanced resistance to the bacterial phytopathogen *Erwinia carotovora. Mol. Plant-Microbe Interact.* **14**:1035–1042.

Makino, K., H. Shinagawa, M. Amemura, T. Kawamoto, M. Yamada, and A. Nakata. 1989. Signal transduction in the phosphate regulon of Escherichia coli involves phosphotransfer between PhoR and PhoB proteins. *J. Mol. Biol.* **210**:551–559.

Malmcrona-Friberg, K., A. Tunlid, P. Mårdén, S. Kjelleberg, and G. Odham. 1986. Chemical changes in cell envelope and poly-β-hydroxybutyrate during short term starvation of a marine bacterial isolate. *Arch. Microbiol.* **144**:340–345.

Marco-Noales, E., M. Milan, B. Fouz, E. Sanjuan, and C. Amaro. 2001. Transmission to eels, portals of entry, and putative reservoirs of *Vibrio vulnificus* serovar E (biotype 2). *Appl. Environ. Microbiol.* **67**:4717–4725.

Mårdén, P., A. Tunlid, K. Malmcrona-Friberg, G. Odham, and S. Kjelleberg. 1985. Physiological and morphological changes during short term starvation of marine bacterial isolates. *Arch. Microbiol.* **142**:326–332.

Marouga, R., and S. Kjelleberg. 1996. Synthesis of immediate upshift (Iup) proteins during recovery of marine *Vibrio* sp. strain S14 subjected to long-term carbon starvation. *J. Bacteriol.* **178**:817–822.

Mason, C. A., and T. Egli. 1993. Dynamics of microbial growth in the decelerating and stationary phase of batch culture, p. 81–102. *In* S. Kjelleberg (ed.), *Starvation in Bacteria.* Plenum Press, New York, N.Y.

Matin, A., E. A. Auger, P. H. Blum, and J. E. Schultz. 1989. Genetic basis of starvation survival in nondifferentiating bacteria. *Annu. Rev. Microbiol.* 43:293–316.

Matin, A. 1991. The molecular basis of carbon-starvation-induced general resistance in *Escherichia coli. Mol. Microbiol.* 5:3–10.

Matz, C., J. Boenigk, H. Arndt, and K. Jurgens. 2002a. Role of bacterial phenotypic traits in selective feeding of the heterotrophic nanoflagellate *Spumella* sp. *Aquat. Microb. Ecol.* 27:137–148.

Matz, C., P. Deines, and K. Jurgens. 2002b. Phenotypic variation in *Pseudomonas* sp. CM10 determines microcolony formation and survival under protozoan grazing. *FEMS Microbiol. Ecol.* 39:57–65.

Matz, C., T. Bergfeld, S. A. Rice, and S. Kjelleberg. 2004. Microcolonies, quorum sensing and cytotoxicity determine the survival of *Pseudomonas aeruginosa* biofilms exposed to protozoan grazing. *Environ. Microbiol.* 6:218–226.

Matz, C., D. McDougald, A. M. Moreno, P. Y. Yung, F. H. Yildiz, and S. Kjelleberg. 2005. Biofilm formation and phenotypic variation enhance predation-driven persistence of *Vibrio cholerae. Proc. Natl. Acad. Sci. USA* 102:16819–16824.

McCarter, L. L., M. Hilmen, and M. Silverman. 1988. Flagellar dynamometer controls swarmer cell differentiation of *Vibrio parahaemolyticus. Cell* 54:345–351.

McCarter, L. L. 1998. OpaR, a homolog of *Vibrio harveyi* LuxR, controls opacity of *Vibrio parahaemolyticus. J. Bacteriol.* 180:3166–3173.

McDougald, D., S. A. Rice, D. Weichart, and S. Kjelleberg. 1998. Nonculturability: adaptation or debilitation? *FEMS Microbiol. Ecol.* 25:1–9.

McDougald, D., S. A. Rice, and S. Kjelleberg. 1999. New perspectives on the viable but nonculturable response. *Biologia* 54:617–623.

McDougald, D., S. A. Rice, and S. Kjelleberg. 2001. SmcR-dependent regulation of adaptive responses in *Vibrio vulnificus. J. Bacteriol.* 183:758–762.

McDougald, D., L. Gong, S. Srinivasan, E. Hild, L. Thompson, K. Takayama, S. A. Rice, and S. Kjelleberg. 2002. Defences against oxidative stress during starvation in bacteria. *Antonie Leeuwenhoek* 81:3–13.

McDougald, D., S. Srinivasan, S. A. Rice, and S. Kjelleberg. 2003. Signal-mediated cross-talk regulates stress adaptation in Vibrio species. *Microbiology* 149:1923–1933.

Miller, M. B., K. Skorupski, D. H. Lenz, R. K. Taylor, and B. L. Bassler. 2002. Parallel quorum sensing systems converge to regulate virulence in *Vibrio cholerae. Cell* 110:303–314.

Milton, D. L., A. Hardman, M. Camara, S. R. Chhabra, B. W. Bycroft, G. S. Stewart, and P. Williams. 1997. Quorum sensing in *Vibrio anguillarum*: characterization of the *vanI/vanR* locus and identification of the autoinducer N-(3-oxodecanoyl)-L-homoserine lactone. *J. Bacteriol.* 179:3004–3012.

Milton, D. L., V. J. Chalker, D. Kirke, A. Hardman, M. Camara, and P. Williams. 2001. The LuxM homologue VanM from *Vibrio anguillarum* directs the synthesis of N-(3-hydroxyhexanoyl)homoserine lactone and N-hexanoylhomoserine lactone. *J. Bacteriol.* 183:3537–3547.

Miyamoto, C. M., Y. H. Lin, and E. A. Meighen. 2000. Control of bioluminescence in *Vibrio fischeri* by the LuxO signal response regulator. *Mol. Microbiol.* 36:594–607.

Mizunoe, Y., S. N. Wai, A. Takade, and S.-I. Yoshida. 1999. Restoration of culturability of starvation-stressed and low-temperature-stressed *Escherichia coli* O157 cells by using H_2O_2-degrading compounds. *Arch. Microbiol.* 172:63–67.

Mizunoe, Y., S. N. Wai, T. Ishikawa, A. Takade, and S.-I. Yoshida. 2000. Resuscitation of viable but nonculturable cells of *Vibrio parahaemolyticus* induced at low temperature under starvation. *FEMS Microbiol. Lett.* 186:115–120.

Mok, K. C., N. S. Wingreen, and B. L. Bassler. 2003. Vibrio harveyi quorum sensing: a coincidence detector for two autoinducers controls gene expression. *EMBO J.* 22:870–881.

Montanari, M. P., C. Pruzzo, L. Pane, and R. R. Colwell. 1999. Vibrios associated with plankton in a coastal zone of the Adriatic Sea (Italy). *FEMS Microbiol. Ecol.* 29:241–247.

Moriarty, D. J. W., and R. T. Bell. 1993. Bacterial growth and starvation in aquatic environments, p. 25–53. *In* S. Kjelleberg (ed.), *Starvation in Bacteria.* Plenum Press, New York, N.Y.

Nealson, K. H. 1977. Autoinduction of bacterial luciferase: occurrence, mechanism and significance. *Arch. Microbiol.* 112:73–79.

Nealson, K. H., and J. W. Hastings. 1979. Bacterial bioluminescence: its control and ecological significance. *Microbiol. Rev.* 43:496–518.

Nelson, D. R., Y. Sadlowski, M. Eguchi, and S. Kjelleberg. 1997. The starvation-stress response of *Vibrio (Listonella) anguillarum. Microbiology* 143:2305–2312.

Nilsson, L., J. D. Oliver, and S. Kjelleberg. 1991. Resuscitation of *Vibrio vulnificus* from the viable by nonculturable state. *J. Bacteriol.* 173:5054–5059.

Nishino, T., B. B. Nayak, and K. Kogure. 2003. Density-dependent sorting of physiologically different cells of *Vibrio parahaemolyticus. Appl. Environ. Microbiol.* 69:3569–3572.

Novitsky, J. A., and R. Y. Morita. 1976. Morphological characterization of small cells resulting from nutrient starvation of a psychrophilic marine *Vibrio. Appl. Environ. Microbiol.* 32:617–622.

Novitsky, J. A., and R. Y. Morita. 1977. Survival of a psychrophilic marine vibrio under long-term nutrient starvation. *Appl. Environ. Microbiol.* 33:635–641.

Novitsky, J. A., and R. Y. Morita. 1978. Possible strategy for the survival of marine bacteria under starvation conditions. *Mar. Biol.* 48:289–295.

Nyström, T., N. Albertson, and S. Kjelleberg. 1988. Synthesis of membrane and periplasmic proteins during starvation of a marine *Vibrio* sp. *J. Gen. Microbiol.* 134:1645–1651.

Nyström, T., K. Flärdh, and S. Kjelleberg. 1990. Responses to multiple-nutrient starvation in marine *Vibrio* sp. strain CCUG 15956. *J. Bacteriol.* 172:7085–7097.

Nyström, T., R. M. Olsson, and S. Kjelleberg. 1992. Survival, stress resistance, and alteration in protein expression in the marine *Vibrio* sp. strain S14 during starvation for different individual nutrients. *Appl. Environ. Microbiol.* 58:55–65.

Nyström, T. 1998. To be or not to be: the ultimate decision of the growth-arrested bacterial cell. *FEMS Microbiol. Rev.* 21:283–290.

Nyström, T. 1999. Starvation, cessation of growth and bacterial aging. *Curr. Opin. Microbiol.* 2:214–219.

Nyström, T. 2004. Stationary-phase physiology. *Annu. Rev. Microbiol.* 58:161–181.

Oliver, J. D., and W. F. Stringer. 1984. Lipid composition of a psychrophilic marine *Vibrio* sp. during starvation-induced morphogenesis. *Appl. Environ. Microbiol.* 47:461–466.

Oliver, J. D., L. Nilsson, and S. Kjelleberg. 1991. The formation of nonculturable *Vibrio vulnificus* cells and its relationship to the starvation state. *Appl. Environ. Microbiol.* 57:2640–2644.

Oliver, J. D. 1993a. Formation of viable but nonculturable cells, p. 239–272. *In* S. Kjelleberg (ed.), *Starvation in Bacteria.* Plenum Press, New York, N.Y.

Oliver, J. D. 1993b. Nonculturability and resuscitation of *Vibrio vulnificus*, p. 187–191. *In* R Guerrero and C Pedrós-Alió (ed.), *Trends in Microbial Ecology.* Spanish Society for Microbiology, Madrid, Spain.

Oliver, J. D., and R. Bockian. 1995. In vivo resuscitation, and virulence towards mice, of viable but nonculturable cells of *Vibrio vulnificus. Appl. Environ. Microbiol.* 61:2620–2623.

Oliver, J. D., F. Hite, D. McDougald, N. L. Andon, and L. M. Simpson. 1995. Entry into, and resuscitation from, the viable but nonculturable state by *Vibrio vulnificus* in an estuarine environment. *Appl. Environ. Microbiol.* **61:**2624–2630.

Ostling, J., L. Holmquist, K. Flärdh, B. Svenblad, Å. Jouper-Jaan, and S. Kjelleberg. 1993. Starvation and recovery of *Vibrio*, p. 103–127. *In* S. Kjelleberg (ed.), *Starvation in Bacteria.* Plenum Press, New York, N.Y.

Ostling, J., L. Holmquist, and S. Kjelleberg. 1996. Global analysis of the carbon starvation response of a marine *Vibrio* species with disruptions in genes homologous to *relA* and *SpoT. J. Bacteriol.* **178:**4901–4908.

Ostling, J., D. McDougald, R. Marouga, and S. Kjelleberg. 1997. Global analysis of physiological responses in marine bacteria. *Electrophoresis* **18:**1441–1450.

Paludan-Müller, C., D. Weichart, D. McDougald, and S. Kjelleberg. 1996. Analysis of starvation conditions that allow for prolonged culturability of *Vibrio vulnificus* at low temperature. *Microbiology* **142:**1675–1684.

Preyer, J. M., and J. D. Oliver. 1993. Starvation-induced thermal tolerance as a survival mechanism in a psychrophilic marine bacterium. *Appl. Environ. Microbiol.* **59:**2653–2656.

Rahman, I., M. Shahamat, P. A. Kirchman, E. Russek-Cohen, and R. R. Colwell. 1994. Methionine uptake and cytopathogenicity of viable but nonculturable *Shigella dysenteriae* type 1. *Appl. Environ. Microbiol.* **60:**3573–3578.

Randa, M. A., M. F. Polz, and E. Lim. 2004. Effects of temperature and salinity on *Vibrio vulnificus* population dynamics as assessed by quantitative PCR. *Appl. Environ. Microbiol.* **70:**5469–5476.

Rao, N. N., and A. Kornberg. 1996. Inorganic phosphate supports resistance and survival of stationary-phase *Escherichia coli. J. Bacteriol.* **178:**1394–1400.

Ravel, J., I. T. Knight, C. E. Monahan, R. T. Hill, and R. R. Colwell. 1995. Temperature induced recovery of *Vibrio cholerae* from the viable but nonculturable state: growth or resuscitation? *Microbiology* **141:**377–383.

Redfield, R. J. 2002. Is quorum sensing a side effect of diffusion sensing? *Trends Microbiol.* **10:**365–370.

Reguera, G., and R. Kolter. 2005. Virulence and the environment: a novel role for *Vibrio cholerae* toxin-coregulated pili in biofilm formation on chitin. *J. Bacteriol.* **187:**3551–3555.

Rice, S. A., D. McDougald, and S. Kjelleberg. 2000. *Vibrio vulnificus:* a physiological and genetic approach to the viable but nonculturable response. *J. Infect. Chemother.* **6:**115–120.

Rice, S. A., D. McDougald, N. Kumar, and S. Kjelleberg. 2005. The use of quorum sensing blockers as therapeutic agents for the control of biofilm associated infections. *Curr. Opin. Invest. Drugs* **6:** 178–184.

Rodriguez, G. G., D. Phipps, K. Ishiguro, and H. F. Ridgeway. 1992. Use of a fluorescent redox probe for direct visualization of actively respiring bacteria. *Appl. Environ. Microbiol.* **58:**1801–1808.

Rosenberg, E., and Y. Ben-Haim. 2002. Microbial diseases of corals and global warming. *Environ. Microbiol.* **4:**318–326.

Schauder, S., K. Shokat, M. G. Surette, and B. L. Bassler. 2001. The LuxS family of bacterial autoinducers: biosynthesis of a novel quorum-sensing signal molecule. *Mol. Microbiol.* **41:**463–476.

Shao, C.-P., and L.-I. Hor. 2001. Regulation of metalloprotease gene expression in *Vibrio vulnificus* by a *Vibrio harveyi* LuxR homologue. *J. Bacteriol.* **183:**1369–1375.

Showalter, R. E., M. O. Martin, and M. R. Silverman. 1990. Cloning and nucleotide sequence of *luxR*, a regulatory gene controlling bioluminescence in *Vibrio harveyi. J. Bacteriol.* **172:** 2946–2954.

Siegele, D. A., and R. Kolter. 1992. Life after log. *J. Bacteriol.* **174:** 345–348.

Spira, B., N. Silberstein, and E. Yagil. 1995. Guanosine 3′,5′-bispyrophosphate (ppGpp) synthesis in cells of *Escherichia coli* starved for Pi. *J. Bacteriol.* **177:**4053–4058.

Srinivasan, S., J. Ostling, T. Charlton, R. deNys, K. Takayama, and S. Kjelleberg. 1998. Extracellular signal molecule(s) involved in the carbon starvation response of marine *Vibrio* sp. strain S14. *J. Bacteriol.* **180:**201–209.

Stevens, A. M., K. M. Dolan, and E. P. Greenberg. 1994. Synergistic binding of the *Vibrio fischeri* LuxR transcriptional activator domain and RNA polymerase to the *lux* promoter region. *Proc. Natl. Acad. Sci. USA* **91:**12619–12623.

Stevens, A. M., and E. P. Greenberg. 1997. Quorum sensing in *Vibrio fischeri:* essential elements for activation of the luminescence genes. *J. Bacteriol.* **179:**557–562.

Stoodley, P., K. Sauer, D. G. Davies, and J. W. Costerton. 2002. Biofilms as complex differentiated communities. *Annu. Rev. Microbiol.* **56:**187–209.

Surette, M. G., M. B. Miller, and B. L. Bassler. 1999. Quorum sensing in *Escherichia coli, Salmonella typhimurium,* and *Vibrio harveyi:* a new family of genes responsible for autoinducer production. *Proc. Natl. Acad. Sci. USA* **96:**1639–1644.

Sussman, M., Y. Loya, M. Fine, and E. Rosenberg. 2003. The marine fireworm *Hermodice carunculata* is a winter reservoir and spring-summer vector for the coral-bleaching pathogen *Vibrio shiloi. Environ. Microbiol.* **5:**250–255.

Takayama, K., and S. Kjelleberg. 2000. The role of RNA stability during bacterial stress responses and starvation. *Environ. Microbiol.* **2:**355–365.

Tischler, A. D., and A. Camilli. 2004. Cyclic diguanylate (c-di-GMP) regulates *Vibrio cholerae* biofilm formation. *Mol. Microbiol.* **53:** 857–869.

Villarino, A., O. M. Bouvet, B. Regnault, S. Martin-Delautre, and P. A. D. Grimont. 2000. Exploring the frontier between life and death in *Escherichia coli:* evaluation of different viability markers in live and heat- or UV-killed cells. *Res. Microbiol.* **151:**755–768.

Wai, S. N., T. Moriya, K. Kondo, H. Misumi, and K. Amako. 1996. Resuscitation of *Vibrio cholerae* O1 strain TSI-4 from a viable but nonculturable state by heat shock. *FEMS Microbiol. Lett.* **136:**187–191.

Wai, S. N., Y. Mizunoe, A. Takade, S.-L. Kawabata, and S.-I. Yoshida. 1998. *Vibrio cholerae* O1 strain TSI-4 produces the exopolysaccharide materials that determine colony morphology, stress resistance, and biofilm formation. *Appl. Environ. Microbiol.* **64:**3648–3655.

Warner, J. M., and J. D. Oliver. 1998. Randomly amplified polymorphic DNA analysis of starved and viable nonculturable *Vibrio vulnificus* cells. *Appl. Environ. Microbiol.* **64:**3025–3028.

Watnick, P., K. J. Fullner, and R. Kolter. 1999. A role for the mannose-sensitive hemagglutinin in biofilm formation by *Vibrio cholerae* El Tor. *J. Bacteriol.* **181:**3606–3609.

Watnick, P. I., and R. Kolter. 1999. Steps in the development of a *Vibrio cholerae* El Tor biofilm. *Mol. Microbiol.* **34:**586–595.

Watnick, P. I., C. M. Lauriano, K. E. Klose, L. Croal, and R. Kolter. 2001. The absence of a flagellum leads to altered colony morphology, biofilm development and virulence in *Vibrio cholerae* O139. *Mol. Microbiol.* **39:**223–235.

Webb, J. S., L. S. Thompson, S. James, T. Charlton, T. Tolker-Nielsen, B. Koch, M. Givskov, and S. Kjelleberg. 2003. Cell death in *Pseudomonas aeruginosa* biofilm development. *J. Bacteriol.* **185:**4585–4592.

Weichart, D., J. D. Oliver, and S. Kjelleberg. 1992. Low temperature induced non-culturability and killing of *Vibrio vulnificus. FEMS Microbiol. Lett.* **100:**205–210.

Weichart, D., and S. Kjelleberg. 1996. Stress resistance and recovery potential of culturable and viable but nonculturable cells of *Vibrio vulnificus. Microbiology* **142:**845–853.

Weichart, D., D. McDougald, D. Jacobs, and S. Kjelleberg. 1997. *In situ* analysis of nucleic acids in cold-induced nonculturable *Vibrio vulnificus. Appl. Environ. Microbiol.* **63:**2754–2758.

Whitesides, M. D., and J. D. Oliver. 1997. Resuscitation of *Vibrio vulnificus* from the viable but nonculturable state. *Appl. Environ. Microbiol.* **63:**1002–1115.

Williams, P. 2000. Heterotrophic bacteria and the dynamics of dissolved organic material, p. 153–200. *In* D. L. Kirchman (ed.), *Microbial Ecology of the Oceans.* Wiley-Liss, Inc, New York, N.Y.

Winzer, K., K. R. Hardie, N. Burgess, et al. 2002. LuxS: its role in central metabolism and the *in vitro* synthesis of 4-hydroxy-5-methyl-3(2H)-furanone. *Microbiology* **148:**909–922.

Wong, H.-C., Y.-C. Chung, and J.-A. Yu. 2002. Attachment and inactivation of *Vibrio parahaemolyticus* on stainless steel and glass surface. *Food Microbiol.* **19:**341–350.

Wong, H. C., and P. Wang. 2004. Induction of viable but nonculturable state in *Vibrio parahaemolyticus* and its susceptibility to environmental stresses. *J. Appl. Microbiol.* **96:**359–366.

Xu, H.-S., N. Roberts, F. L. Singleton, R. W. Attwell, D. J. Grimes, and R. R. Colwell. 1982. Survival and viability of nonculturable *Escherichia coli* and *Vibrio cholerae* in the estuarine and marine environment. *Microb. Ecol.* **8:**313–323.

Yildiz, F. H., and G. K. Schoolnik. 1999. *Vibrio cholerae* O1 El Tor: identification of a gene cluster required for the rugose colony type, exopolysaccharide production, chlorine resistance, and biofilm formation. *Proc. Natl. Acad. Sci. USA* **96:**4028–4033.

Yildiz, F. H., N. A. Dolganov, and G. K. Schoolnik. 2001. VpsR, a member of the response regulators of the two-component regulatory systems, is required for expression of *vps* biosynthesis genes and EPS^{ETr}-associated phenotypes in *Vibrio cholerae* O1 El Tor. *J. Bacteriol.* **183:**1716–1726.

Yildiz, F. H., X. S. Liu, A. Heydorn, and G. K. Schoolnik. 2004. Molecular analysis of rugosity in a *Vibrio cholerae* O1 El Tor phase variant. *Mol. Microbiol.* **53:**497–515.

Zampini, M., C. Pruzzo, V. P. Bondre, R. Tarsi, M. Cosmo, A. Bacciaglia, A. Chhabra, R. Srivastava, and B. S. Srivastava. 2005. *Vibrio cholerae* persistence in aquatic environments and colonization of intestinal cells: involvement of a common adhesion mechanism. *FEMS Microbiol. Lett.* **244:**267.

Zhu, J., M. B. Miller, R. E. Vance, M. Dziejman, B. L. Bassler, and J. J. Mekalanos. 2002. Quorum-sensing regulators control virulence gene expression in *Vibrio cholerae. Proc. Natl. Acad. Sci. USA* **99:**3129–3134.

Zhu, J., and J. J. Mekalanos. 2003. Quorum sensing-dependent biofilms enhance colonization in *Vibrio cholerae. Dev. Cell* **5:**647–656.

The Biology of Vibrios
Edited by F. L. Thompson et al.
© 2006 ASM Press, Washington, D.C.

Chapter 11

Extremophilic *Vibrionaceae*

DOUGLAS H. BARTLETT

INTRODUCTION

In this chapter, particular attention will be paid to some of the more "eccentric" relatives within the family *Vibrionaceae*. This is meant to indicate those members which flourish under conditions in which most microbial growth is inhibited. In a sense, all species can be considered "eccentric" or "extremophilic," as each species is genetically programmed for a specific subset of nutritional and physiochemical niches. However, in keeping with generally accepted notions of what is and is not extreme, physiological specializations will not be covered, but rather attention will be given to adaptation to physiochemical limits. Many members of the *Vibrionaceae* are noteworthy for their ability to tolerate conditions perceived in this way as extreme. There exist species which grow under conditions of low and high osmolarity, low temperature, and/or elevated hydrostatic pressure. As described below, they range from antarctic and arctic polar environments to abyssal and hadal habitats, and from saturated brines to hydrothermal vent regions. Considering that most of the habitable portions of Earth are salty, deep, and permanently below 5°C (at least outside of the deep-subsurface biosphere [Fredrickson and Onstott, 1996]), many of the environments in question are not at all unusual when viewed as a fraction of the total volume of the biosphere.

Currently, eight genera are included in the *Vibrionaceae*. According to *Bergey's Manual of Systematic Bacteriology* (http://dx.doi.org/10.1007/bergeysoutline 200210), these genera are *Allomonas*, *Catenococcus*, *Enhydrobacter*, *Grimontia*, *Listonella*, *Photobacterium*, *Salinivibrio*, and *Vibrio*. It is worth noting that Farmer has provided arguments against the genus designations *Allomonas* and *Listonella* and the placement of *Enhydrobacter* in this family (Farmer, 1999). Focus here is given to *Photobacterium*, *Salinivibrio*, and *Vibrio*.

CATEGORIES OF EXTREMOPHILIC *VIBRIONACEAE*

Psychrophiles/Psychrotolerants

Psychrophiles have been defined as being able to grow at or below 0°C, having an optimum growth temperature of <15°C, and possessing a maximum temperature of <20°C (Morita, 1975; Russell and Hamamoto, 1998; Feller and Gerday, 2003). Psychrotolerants (previously called psychrotrophs) are capable of growth at or close to 0°C but have optimum and maximum temperatures of >20°C (Morita, 1975; Russell and Hamamoto, 1998). Care must be taken when characterizing the thermal growth ranges for many bacteria, including members of the *Vibrionaceae*, as substrate utilization can be strongly influenced by temperature (Rueger, 1988).

Vibrionaceae, which are capable of growth at low temperature (≤5°C), are indicated in Table 1. These include *Photobacterium angustum*, *Photobacterium iliopiscarium*, *Photobacterium phosphoreum*, *Photobacterium profundum*, *Vibrio kanaloae*, *Vibrio logei*, *Vibrio pomeroyi*, *Vibrio splendidus*, *Vibrio tapetis*, and *Vibrio wodanis*. It should be noted that a large proportion of *Vibrionaceae* isolates do not match with their related type strains in terms of lower temperature growth limits. This has been documented for strains isolated from a variety of environments (Grimes et al., 1993; Martinkearley and Gow, 1994). Another example of this phenomenon is that different subtypes of *P. phosphoreum* (comprising both nonluminous and luminous strains) often exhibit different growth abilities at 0 to 15°C (Dalgaard et al., 1997b). The bases of such strain-to-strain differences are unknown and deserve further study. One dramatic example of the influence of culture temperature on a *Vibrionaceae* phenotype was the isolation of *Vibrio fischeri* strains that exhibited striking yellow

Douglas H. Bartlett • Marine Biology Research Division (0202), Scripps Institution of Oceanography, University of California, San Diego, La Jolla, CA 92093-0202.

Table 1. Psychrotolerant or psychrophylic members of the *Vibrionaceae*

Species	Environmental characteristics	Reference(s)
P. angustum	Seawater	Baumann et al., 1971; Radjasa et al., 2001; Reichelt et al., 1976; Urakawa et al., 1999c
P iliopiscarium	Cold seawater fish intestines	Onarheim et al., 1994; Urakawa et al., 1999b
P. phosphoreum	Light organs, surface and digestive tracts of many fish families, fish spoilage at low temperature	Budsberg et al., 2003; Farmer III and Hickman-Brenner, 1992; Ruby and Morin, 1978; Ruby et al., 1980
P. profundum[a]	Deep-sea amphipods, marine sediments, seawater	Bowman et al., 2003; Campanaro et al., in preparation; De Long et al., 1997; GenBank accession number AY849798; Nogi et al., 1998a; Radjasa et al., 2001
V. kanaloae	Diseased oyster in France	Thompson et al., 2003
V. logei	Light organs of sepiolid squids, cold seawater environments, including the Arctic, exoskeleton lesions of tanner crabs, fish intestinal contents, scallops, and marine sediments	Baumann, 1981; Farmer III and Hickman-Brenner, 1992; Fidopiastis et al., 1998; Urakawa et al., 1999c
V. ordalii	Fish pathogen, also isolated from giant scallops	Martinkearley and Gow, 1994; Martikearley et al., 1994; Schiewie et al., 1981
V. pomeroyi	Bivalve larvae in southern Brazil	Thompson et al., 2003
V. salmonicida[a]	Cold water vibriosis in salmonids	Colquhoun et al., 2002; Urakawa et al., 1999c
V. splendidus	Coastal waters, deep-sea waters, fish, oysters, and shrimp	Batacidos et al., 1990; Gatesoupe et al., 1999; Jensen et al., 2003; Lacoste et al., 2001; Radjasa et al., 2001; Urukawa et al., 1999a,c
V. tapetis	Brown ring disease in clams, also isolated from halibut	Borrego et al., 1996; Pinhassi et al., 1999; Reid et al., 2003
V. wodanis	"Winter ulcer" disease of salmonids	Lunder et al., 2000

[a]Many but not all members of these species are psychrophilic.

bioluminescence (Ruby and Nealson, 1977). This only occurred at low temperature (18°C). At higher temperatures, the strains exhibited more typical blue-green luminescence due to the dissociation of a yellow fluorescent protein from the luciferase (M. Haygood, personal communication).

Five examples of low-temperature-adapted *Vibrionaceae* are listed below.

1. *P. phosphoreum* (psychrotolerant). It has long been known that one source for the bioluminescent bacterium *P. phosphoreum* is the light organs of bathyal fish from the family *Macrouridae*, whose ambient environment is 2 to 10°C (Ruby and Morin, 1978). *P. phosphoreum* is also noted for the fact that its optimal luminescence is not affected by deep-sea temperatures. Surprisingly, *P. phosphoreum* has been isolated from migrating salmon in Alaska >1,000 km from seawater (Budsberg et al., 2003). A fairly specific enrichment for *P. phosphoreum*, as described by Baumann and Baumann (1977), consists of incubating fish, squid, or octopus half submerged in seawater at 10 to 15°C. Any luminous spots appearing after overnight incubation generally belong to *P. phosphoreum*. This organism is an important source of spoilage of modified atmosphere-packed fish products stored at low temperatures, where it respires trimethylamine oxide and produces trimethylamine (Dalgaard et al., 1997a). Efforts are under way to in-

hibit the growth of this *Photobacterium* sp. in refrigerated fish (Dalgaard et al., 1998). *P. phosphoreum* can also convert histidine to histamine, resulting in histamine fish poisoning in people, particularly in fish samples stored below 15°C (Kanki et al., 2004).

2. *P. profundum* (psychrophile/psychrotolerant). *P. profundum* has been isolated from a variety of mostly low-temperature, deep-sea Pacific Ocean samples. These include a 5.1-km deep-sea Ryuku Trench sediment (Nogi et al., 1998a), an Ocean Drilling Program Leg 201 sediment obtained from the slope of the Peru Trench at a water depth of 5.1 km (GenBank accession number AY849798), a shallow-water sediment obtained in San Diego Bay (S. Campanaro et al., unpublished data), a 2.5-km deep-sea Sulu Sea amphipod (DeLong et al., 1997), and deep-sea seawater samples obtained in the northwestern Pacific Ocean (Radjasa et al., 2001) and from a continental shelf area located off eastern Antarctica (Bowman et al., 2003).

Depending on the strain, *P. profundum* can be psychrotolerant or psychrophilic. All *P. profundum* strains grow well near 0°C and the one isolate from San Diego Bay grows at the highest known temperature for the species, 25°C. Some, but not all, strains are moderately piezophilic, with some strains growing at pressures as high as 90 MPa.

3. *V. logei* (psychrotolerant). *V. logei* is a bioluminescent vibrio related to *Vibrio salmonicida* and

V. wodanis, and strains closely related to *V. logei* share with their relatives the ability to associate with salmonids reared at low temperatures (Benediktsdottir et al., 1998, 2000). *V. logei* also resides in the light organs of several species of Mediterranean sepiolid squids; at low temperatures it is the dominant symbiont, whereas at higher temperatures (26°C) it is outcompeted by *V. fischeri* (Fidopiastis et al., 1998; Nishiguchi, 2000). Curiously, *V. logei* can also be found in tropical waters (Esiobu and Yamazaki, 2003).

4. *V. salmonicida* (psychrophile). The disease coldwater vibriosis is caused by the psychrophile *V. salmonicida*, whose optimal growth temperature is in the range of ~10°C in liquid media (Colquhoun et al., 2002; Holm et al., 1985). The upper temperature range for coldwater vibriosis corresponds to this temperature optimum (Colquhoun et al., 2002). Direct counts of *V. salmonicida* in Norwegian fish farm waters are generally highest in the winter months (Enger et al., 1991).

5. *V. splendidus* (psychrotolerant). *V. splendidus* is noteworthy among the listed psychrotolerant bacteria because it is so abundant in various environments (Urakawa et al., 1999a,c). Using restriction fragment length polymorphism analysis of psychrophilic and psychrotolerant *Vibrio* and *Photobacterium* species, Urakawa observed that *V. splendidus* was dominant in the Otsuchi Bay in the winter (Urakawa et al., 1999c). It has also been observed that *V. splendidus* is abundant in shallow and deep-sea waters from the northwestern Pacific Ocean (Radjasa et al., 2001). High numbers of *V. splendidus* in coastal waters have also previously been reported (Ortigosa et al., 1994; Pinhassi et al., 1999).

A relative of *V. splendidus*, *Vibrio orientalis*, along with all the *Photobacterium* species, has the ability to accumulate poly(3-hydroxybutyrate) (PHB). *P. profundum* accumulates PHB and its oligomers at low temperature and at high pressure (Martin et al., 2002). The ability to utilize PHB has been correlated with resistance to a variety of stresses (Kadouri et al., 2003), but its relationship to cold (or pressure) adaptation is unknown.

Piezophiles

Microbes that grow best at pressures in excess of atmospheric pressure are termed piezophiles (previously known as barophiles) (Yayanos, 1995). While a number of members of the family have been isolated from deep-sea high-pressure environments, only one species, *P. profundum*, has been obtained which displays piezophilic growth. In general, it has often been observed that *Photobacterium* species are abundant in cold deep-sea environments, whereas certain *Vibrio* species are more abundant in cold surface waters (e.g., see Urakawa et al., 1999a).

Although no piezophilic or thermophilic vent isolates have yet been reported, the *Vibrionaceae* have also been isolated from deep-sea as well as shallow-water hydrothermal vent environments. *Vibrio diabolicus* was isolated from the deep-sea hydrothermal vent polychaete *Alvinella pompejana* (Raguénès et al., 1997). This microbe is capable of growth at temperatures as high as 45°C and produces a polysaccharide of potential biotechnological value (Colliec-Jouault et al., 2004). *Catenococcus thiocyclus* is a thiosulfate oxidizer from a shallow marine hydrothermal vent which grows as a chain-forming coccus (Sorokin, 1992).

It seems likely that many piezophilic *Vibrionaceae* exist, as they have been isolated from a variety of deep-sea environments. *P. phosphoreum*, while not including any piezophilic strains, has been isolated in the Puerto Rico Trench at 50 cells/ml at 1,000 m depth (Ruby et al., 1980). An omega-3 polyunsaturated fatty acid (docosahexaenoic acid)-producing *Vibrio* sp. was obtained from a deep-sea fish (Yano et al., 1994). A strain of *V. alginolyticus* was isolated which caused epidermal lesions and mortality in deep-sea Bahamian echinoids (Bauer and Young, 2000). Also, culture-independent analyses of deep-sea sediment microbial diversity have frequently indicated the presence of *Vibrionaceae* (Li et al., 1999a,b).

Halophiles/Haloversatiles

A halophile is an organism that requires NaCl for life; extreme halophiles grow at salt concentrations of up to 5.2 M, and haloversatile organisms grow over wide ranges of salt concentrations (Bartlett and Roberts, 2000). *Salinivibrio costicola* is a particularly fascinating haloversatilic microbe because of the tremendous range of water activities it will tolerate (Ventosa et al., 1998), ranging from close to fresh water to close to saturated NaCl. It is also one of the few *Vibrionaceae* species that can be isolated outside the marine environment. *S. costicola* subsp. *vallismortis* has been isolated from a hypersaline pond in Death Valley, Calif., and can grow at temperatures ranging from 5 to 50°C (Huang et al., 2000).

ACCLIMATIZATION AND ADAPTATION TO LOW TEMPERATURE

One of the challenges in describing low-temperature adaptation in the *Vibrionaceae* is that many of the model systems that have been used for these studies have undergone taxonomic reclassification. *Vibrio*

ANT-300, *Vibrio* sp. strain 5710, and *Vibrio marinus* are all now considered to be members of the genus *Moritella* (Dalluge et al., 1997; Saito and Nakayama, 2004; Urakawa et al., 1999c), and *Vibrio* ABE-1 has been reassigned to *Colwellia maris* (Yumoto et al., 1998). Some of the results from studies of these microbes will be described despite their "defection" from the family *Vibrionaceae*.

The Cold Shock Response

The Arrhenius equation has been used to define the effect of temperature on the growth properties of many microorganisms. Beyond the boundaries where a straight-line relationship exists between the logarithm of the growth rate and the reciprocal of the absolute temperature, deleterious changes occur in the topology and flexibility of membranes, proteins, and nucleic acids (Ingraham and Marr, 1996). When *Escherichia coli* is exposed to a substantial temperature decrease, growth stops for an acclimatization period of several hours followed by resumption of growth at a rate slower than predicted from purely physical-chemical considerations.

During the acclimatization period, the translation rate of ~30 proteins, termed cold shock (CSP) or cold-induced proteins, increases (e.g., see Gualerzi et al., 2003), and several dozen more genes have recently been reported to be low temperature inducible by transcriptome profiling (Phadtare and Inouye, 2004; Polissi et al., 2003). It has been suggested that immediately after low-temperature stress, there is direct inhibition of most translation to prevent protein miscoding (Ermolenko and Makhatadze, 2002). Indeed, antibiotics that decrease translational fidelity induce CSPs (VanBogelen and Neidhardt, 1990). Thus, part of the cold shock response might be produced to block translation while the remaining components help adapt the cells to lower temperatures.

Many of the CSPs possess nucleic acid-binding properties that are presumed to help the cell deal with unfavorable RNA and DNA secondary structures formed at lower temperatures. These include several of the CspA family of proteins (see below), RNA helicases Dead (CsdA) and RbfA, translation initiation factors IF1-3, transcription elongation factor NusA, RNase, PNPase, and the DNA-binding proteins DNA replication protein DnaA, DNA recombination protein RecA, the α subunit of DNA gyrase, and the nucleoid-associated proteins H-NS and the β subunit of protein HU.

Factors associated with protein stability that are cold induced include the RpoE sigma factor and its anti-sigma factor RseA, OtsA/B involved in trehalose synthesis, trigger factor, and the Hsp70-type chaperone and cochaperone Hsc66 and HscB. Among the CSPs, certain subsets of the CspA family along with PNPase, trigger factor, OtsA/B, Dead, RbfA, and H-NS have all been demonstrated by gene knockout studies to play a role in cell growth or survival at low temperature (Dersch et al., 1994; Jones et al., 1996; Kandror and Goldberg, 1997; Xia et al., 2001, 2003; Kandror et al., 2002).

While the bulk of these studies have been performed using *E. coli*, genome sequence analyses (F. M. Lauro and D. H. Bartlett, unpublished results) and some physiological studies indicate extensive overlap in the cold shock program of *E. coli* with those present in the *Vibrionaceae*. Cold shock responses have been characterized for *Vibrio cholerae*, *Vibrio vulnificus*, and *Vibrio parahaemolyticus* (Bryan et al., 1999; Datta and Bhadra, 2003; Huels et al., 2003; Lin et al., 2004; McGovern and Oliver, 1995). *V. parahaemolyticus* and *V. vulnificus* cold shock enhances survival at lower temperatures in a protein synthesis-dependent manner (Bryan et al., 1999; Lin et al., 2004). As with *E. coli*, *V. cholerae* has a cold-inducible H-NS-related gene (Tendeng et al., 2000). A cold shock response has also been documented for the psychrophilic *Vibrio* sp. strain ANT-300 (now reclassified as a *Moritella* species) following a shift down in temperature from 13°C to 0°C. In this case it was suggested that many of the newly made proteins are for adaptation of the protein synthesis machinery (Araki, 1991a,b).

Nucleic Acid Function at Low Temperature

Among the CSP nucleic acid-binding proteins, the CspA family proteins have received the most attention. *E. coli* contains nine genes encoding homologous proteins of the CspA family. These gene products overlap in function, and deletion of at least four of these genes (*cspA, B, G,* and *E*) is required to produce a cold-sensitive strain (Xia et al., 2001). The CspA protein is the major CSP. Its levels transiently reach 100 μM after a temperature downshift (Goldstein et al., 1990). CspA family proteins bind single-stranded nucleic acids and reduce secondary structure (reviewed by Ermolenko and Makhatadze, 2002; Phadtare et al., 2004). By acting as nucleic acid chaperones, they facilitate transcription and translation at low temperature. Structurally, they are small proteins, rich in β-sheet structure, which contain two well-conserved RNA-binding motifs. Aromatic residues presumably intercalate between nucleic acid bases and contribute to "melting" (Phadtare et al., 2004). Genes encoding proteins that contain CspA-like cold shock domains have been identified in other organisms, including psychrophilic *Archaea* (Saunders et al., 2003).

The regulation of the cold shock response occurs at multiple levels. Transcription can be modulated. For example, the small RNA molecule DsrA accumulates at low temperature and increases RpoS sigma factor translation, which in turn is employed for the activation of transcription of the genes for trehalose biosynthesis (Lease and Belfort, 2000; Wassarman, 2002). CspA family proteins also regulate gene expression. Deletion of the *cspA, B, G,* and *E* genes results in a loss of many cold-inducible genes (Phadtare and Inouye, 2004). CspA protein induction itself has been extensively studied and found to result primarily from posttranscriptional and translational regulation (reviewed by Ermolenko and Makhatadze, 2002; Gualerzi et al., 2003). *cspA* mRNA is greatly stabilized during the initial phase of cold shock, until the CSPs PNPase and CsdA accumulate and promote *cspA* mRNA degradation. There is also preferential translation of many cold shock gene mRNAs. The initial 5-codon sequence of *cspA* enhances its translation initiation, and there is evidence that the accumulation of initiation factor 3 at low temperature also promotes preferential *csp* translation. The *cspA* gene's untranslated region and translation-enhancing-element sequence have been used commercially to prepare vectors for massive overproduction of heterologous proteins of interest at low temperature in *E. coli* (Qing et al., 2004). Curiously, at least some vibrios must possess a different mode of CspA family gene regulation. The cold-inducible *cspV* gene of *V. cholerae* lacks a long 5′ untranslated region (Datta and Bhadra, 2003).

Cold stress results in the inactivation of a large fraction of the ribosomes. Two cold-induced ribosome-associated RNA helicases help to restore ribosome function. CsdA facilitates translation initiation/elongation, ribosome assembly/maturation, and activity of the cold shock degradosome responsible for RNA turnover (Jones et al., 1996; Prud'homme-Genereux et al., 2004). Also, RbfA is needed for 16S rRNA processing (Bylund et al., 1998; Luttinger et al., 1996; Xia et al., 2003). RNA modification is also important for tRNA function at low temperature. Psychrophilic bacteria have been reported to contain 40 to 70% more dihydrouridine in their tRNA molecules than *E. coli* tRNAs (Dalluge et al., 1997). The species examined included *Vibrio* 29-6, *Moritella* ANT-300, and *Moritella* 5710. Posttranscriptional tRNA modification has also been inferred to be the case for optimizing tRNA flexibility in two cold-adapted *Archaea* (Saunders et al., 2003).

DNA structure is also sensitive to low temperature. The nucleoid-binding protein H-NS (Dersch et al., 1994) and the ratio of the HU β and α subunit (Giangrossi et al., 2002) have been implicated in low-temperature adaptation. DNA gyrase also appears to perform a critical role in temperature adaptation. Mutations in the genes for DNA gyrase can have a marked effect on the temperature range for bacterial growth (Droffner and Yamamoto, 1991).

Membrane Alteration at Low Temperature

The fatty acid component of membrane lipids generally changes to an increased extent of unsaturation with a decrease in growth temperature. In the case of *E. coli*, the composition of glycerophospholipids shifts toward greater unsaturation at lower temperatures owing to increased β-ketoacyl-ACP synthase II (FabF) activity (Garwin et al., 1980), and the palmitoleoyl (16:1) fatty acid content of the lipid A component of lipopolysaccharide likewise increases owing to palmitoleoyltransferase mRNA accumulation (Vorachek-Warren et al., 2002). The gram-positive organism *Bacillus subtilis* adopts a somewhat different low temperature adaptation strategy and utilizes a desaturase enzyme to introduce double bonds into preexisting fatty acid lipid components at colder temperatures. The abundance of this enzyme is regulated by a signal transduction pathway responding to changes in the physical state of the membrane (Aguilar et al., 2001).

The value of incorporating membrane-fluidizing unsaturated fatty acids into the membrane is most often considered in the context of homeoviscous or homeophasic adaptation (McElhaney, 1982; Hochachka and Somero, 2002). However, other theories have been proposed for the functional significance of membrane fatty acid modulation with temperature, including the maintenance of ion permeability for bioenergetic purposes (van de Vossenberg et al., 1995) and the adjustment of membrane curvature elastic stress (Attard et al., 2000). Only a few studies of membrane fatty acid compositional changes in the *Vibrionaceae* have been performed. It has been observed that the psychrophilic *Vibrio* sp. strain 5710 (now reclassified as a *Moritella* species) increases the amount of its omega-3 polyunsaturated fatty acid docosahexaenoic acid (22:6) and decreases the amount of the saturated fatty acid palmitic acid (16:0) following a switch from 10°C to 0°C (Hamamoto et al., 1994). Psychrophilic vibrios and related bacteria often synthesize *trans*-unsaturated fatty acids (Henderson et al., 1993). These behave more like saturated fatty acids than *cis*-unsaturated fatty acids in terms of their physical effects on the membrane, and are often present as a counterbalance to polyunsaturated fatty acids.

Another component of the membrane that is likely to undergo modification of function in psychrophilic members of the *Vibrionaceae* is the protein export apparatus. *E. coli* protein export is inherently cold sensitive (Pogliano and Beckwith, 1993).

Psychrophile Enzymes

Excellent reviews on psychrophilic enzymes can be found in the articles by Glansdorff and Xu (2002) and Feller and Gerday (2003). As a general but far from absolute rule, enzymes from psychrophiles are more thermolabile (and also cold labile!) than their mesophile counterparts. Adaptation to low temperature is accompanied by a decrease in the amount of enthalpy-driven interactions required during activity, resulting in greater flexibility for the parts of the enzyme involved and increased conformational entropy for the enzyme in the E-S ground state. In some cases, the adaptation to low-temperature catalysis is accompanied by extreme overall flexibility. For example, amylase from the antarctic bacterium *Alteromonas haloplanktis* appears to undergo cold denaturation at $-12°C$ (Feller et al., 1999). These results are all the more intriguing because $-12°C$ has been suggested to be close to the lower temperature limit for life: cells have been observed to retain liquid water in a supercooled state down to about this temperature (Russell, 1990). Thus, more than the physical state of water might determine the lower temperature limit of microorganisms. Psychrophile enzymes generally exhibit more solvent interactions and are less compact than their counterparts from mesophiles. This is accomplished by reducing the number of proline and arginine residues, which restrict flexibility and can form multiple salt bridges and hydrogen bonds, respectively. In addition, fewer weak interactions exist, and the interior core is less hydrophobic. Laboratory evolution experiments demonstrate that it is possible to enhance enzyme activity at low temperature without compromising stability, although random mutations that improve activity and stability are rare (see, for example, Wintrode and Arnold, 2000). Table 2 indicates enzymes from psychrotolerant and psychrophilic current and former (reclassified) *Vibrio* species that have been examined.

The isocitrate dehydrogenase (ICDH) isozymes produced by strain ABE-1 (now classified as *C. maris*) are particularly fascinating examples of psychrophile enzyme adaptation to low temperature. ICDH is a key enzyme of the tricarboxylic acid cycle. Most microorganisms contain a single ICDH gene, whereas ABE-1 contains two ICDH genes: *icdhI*, which encodes an enzyme with thermostability properties similar to those of mesophile enzymes, and *icdII*, which encodes a thermolabile enzyme (Ochiai et al., 1979; Ishii et al., 1987). It has previously been suggested that microbes may expand their temperature range for growth by producing isozymes for key sets of enzymes (Wiegel, 1990). Kinetic studies of the purified enzymes indicate that the catalytic efficiency of the

icdII product is much higher than that of *icdI* (Ochiai et al., 1979). The transcription of *icdII* is also induced at low temperature. Expression of *icdII* in *E. coli* stimulates growth at low temperature (Sahara et al., 1999), suggesting that at least under some conditions ICDH activity can limit bacterial growth at low temperature.

Protein Folding at Low Temperature

While mostly appreciated for their role in protein folding and stabilization at high temperature, protein refolding chaperonins also govern the growth and survival of bacteria at low temperature. Among the three major chaperonins in *E. coli*, if the genes for the DnaK and trigger factor proteins are deleted, the resulting cells are still capable of growth at low temperatures (Genevaux et al., 2004). However, mutants impaired in the gene for the ribosome-associated protein trigger factor do exhibit more rapid death at 4°C (Kandror and Goldberg, 1997). Mutants in the synthesis of the chemical chaperonin trehalose also exhibit reduced viability at 4°C (Kandror et al., 2002). However, the GroEL chaperonin seems to be critical for low-temperature growth. Expression in *E. coli* of the *groEL*, *groES* homologs from the psychrophilic antarctic bacterium *Oleispira antarctica* decreased its theoretical lower-temperature growth limit from 7.5°C to $-13.7°C$ (Ferrer et al., 2003). In vitro analyses of wild type of mutant *O. antarctica* GroEL indicated that its dissociation from a tetradecameric doublering structure to a single heptamer ring was critical to optimal refolding activity at low temperature (Ferrer et al., 2004).

ACCLIMATIZATION AND ADAPTATION TO HIGH PRESSURE

Diversity of Piezophiles

Thermophilic and hyperthermophilic piezophiles have been isolated from both the *Euryarchaeota* and *Crenarchaeota* kingdoms of the *Archaea* obtained from hydrothermal vent regions (Bernhardt et al., 1988; Marteinsson et al., 1997, 1999a; Miller et al., 1988) and from members of the gammaproteobacteria of the domain *Bacteria* in cold, deep-sea environments (Bale et al., 1997; DeLong et al., 1997; Deming et al., 1988; Kato et al., 1995, 1998; Liesack et al., 1991; Nakayama et al., 1994; Nogi et al., 1998a,b, 2002, 2004; Saito and Nakayama, 2004; Xu et al., 2003a,b; Yano et al., 1994, 1997; Yayanos et al., 1979, 1981; Yayanos, 1986, 1995). The latter include species of *Colwellia, Moritella, Photobacterium, Psy-*

Table 2. Examples of enzymes from cold-adapted *Vibrio* species that have been studied

Enzyme	Bacterial species	Reference(s)
Alkaline phosphatase	*Vibrio* sp.	Asgeirsson and Andresson, 2001; Hauksson et al., 2000
Aspartate carbamoyl transferase	*Vibrio* sp. strain 2693	Xu et al., 1998
Chitinase	*Vibrio* sp. strain Fi:7	Bendt et al., 2001
Isocitrate dehydrogenase	*Vibrio* sp. strain ABE-1 (now classified as *C. maris*)	Ochiai et al., 1979; Yoneta et al., 2004
Isocitrate lyase	*Vibrio* sp. strain ABE-1 (now classified as *C. maris*)	Watanabe et al., 2001, 2003
Malate dehydrogenase	*Vibrio* sp. strain 5710 (now classified as *Moritella* sp. strain 5710)	Saito and Nakayama, 2004
Malate synthase	*Vibrio* sp. strain ABE-1 (now classified as *C. marinus*)	Watanabe et al., 2001
Subtilisin-like serine protease	*Vibrio* sp. strain PA44	Kristjansson et al., 1999; Arnorsdottir et al., 2002
Isopropylmalate dehydrogenase	*Vibrio* sp. strain 15	Wallon et al., 1997; Svingor et al., 2001
Triosphosphate isomerase	*Vibrio* sp. strain ANT-300 (now classified as *Moritella* sp. strain ANT-300)	Adler and Knowles, 1995

chromonas, and *Shewanella*. The only exceptions to this clustering at this time are the thermophilic bacterium *Marinitoga piezophila*, obtained from a hydrothermal vent (Alain et al., 2002), and *Desulfovibrio profundus*, obtained from a deep subsurface environment (Bale et al., 1997). About 75% of the isolates have been obtained from various deep Pacific Ocean locations. Seawater, shallow sediment, fish, amphipods, and holothurians have all been used as source material for piezophiles, which have been collected using free vehicles, manned submersibles, and remote-operated vehicles with or without decompression during sample recovery. There are no obligately piezophilic *Archaea* as is the case for several of the bacterial isolates. Culture-independent studies indicate that a subgroup within the *Crenarchaeota* is widespread in deep ocean environments (DeLong et al., 1994; Fuhrman et al., 1992; Karner et al., 2001) as well as *Thermoplasma*-related marine group II within the *Euryarchaeota* (Lopez-Garcia et al., 2001a,b), and small members of the *Eukarya* related to the alveolates (Lopez-Garcia et al., 2001c).

The one piezophilic member of the *Vibrionaceae* is *P. profundum*. As mentioned above, strains of *P. profundum* have been isolated from amphipods, sediments, and water samples from depths as great as 5.1 km (DeLong, 1986; Nogi et al., 1998a; Bowman et al., 2003; Radjasa et al., 2001).

Pressure Effects on Mesophilic Organisms

Pressures of the magnitude present in the marine environment inhibit a variety of cell processes in mesophilic bacteria such as *E. coli*. These include motility (10 MPa), the transport of various nutrients (26 MPa), cell division and FtsZ ring formation (20

to 50 MPa), growth (50 MPa), DNA replication (50 MPa), translation (60 MPa), transcription (77 MPa), and viability (100 MPa) (Gross et al., 1993; Gross and Jaenicke, 1994; Ishii et al., 2004; Landau, 1967; Meganathan and Marquis, 1973; Pagan and Mackey, 2000; Yayanos and Pollard, 1969; ZoBell and Cobet, 1950, 1962, 1963; ZoBell, 1970). *E. coli* responds to moderate pressure stress by inducing a large number of both heat shock and cold shock proteins (Gross et al., 1994; Welch et al., 1993). Transcriptomic analysis of *E. coli* gene expression as a function of pressure has recently been reported (Ishii et al., 2005). Little induction of heat shock and cold shock gene expression is observed, suggesting that the increase in the abundance of the products of these genes seen via proteomic approaches is either transient or largely due to posttranscriptional regulation.

E. coli mutants with defects in the nucleoid-associated protein H-NS display enhanced pressure sensitivity (Ishii et al., 2005), as do mutants with defects in *cydD*, which is required for cytochrome *bd* complex assembly (Kato et al., 1996).

Membrane Alteration at High Pressure

As with decreased temperature, increased pressure causes fatty acid chains to pack together more tightly and decrease gel state (L_β) to liquid crystalline state (L_α) transitions of membranes (Bartlett, 1999, 2002; Bartlett and Bidle, 1999; Suutari and Laakso, 1994). For many phospholipids and natural membranes, the transition slope in a temperature, pressure-phase diagram for the L_β-L_α main transition is approximately 20°C/100 MPa at pressures below 100 MPa. Based on this relationship, the combined effect of pressure and temperature on the phase state of a membrane from a deep-sea bacterium existing at 100 MPa and 2°C is equivalent to an identical membrane at atmospheric pressure and −18°C (possibly lower than the temperature limit for life). Deep-sea organisms cope with the combined effects of low temperature and high pressure on their membranes by modifying their *anteiso* to *iso* branching, *cis-trans* isomerization, acyl chain length, and extent and nature of unsaturation (Bartlett, 2002; Kamimura et al., 1993; Russell and Hamamoto, 1998).

Deep-sea bacterial unsaturated fatty acids can be either monounsaturated or polyunsaturated. It is generally rare for bacteria to produce polyunsaturated fatty acids, and there is considerable biotechnological interest in those microbes, including the psychrophiles and piezophiles, which produce omega-3 polyunsaturated fatty acids. This is because of the biomedical and biotechnological value of these fatty acids in human health and development and their use as essential dietary supplements in aquaculture and poultry farming (Simopoulos, 1991; Valentine and Valentine, 2004). It has been suggested that omega-3 polyunsaturated fatty acids function in the facilitation of proton-driven respiration and block or disrupt islands of gel-phase lipids (Valentine and Valentine, 2004).

Curiously, in *P. profundum* SS9, polyunsaturated fatty acids expression is dispensable, whereas the 16:1 and 18:1 monounsaturated fatty acids appear to be critical to growth at high pressure, at least under fermentative growth conditions (Allen et al., 1999; Allen and Bartlett, 2000). Mutants impaired in β-ketoacyl-acyl carrier protein synthase II (*fabF*) produce little 18:1 and are no longer capable of piezophilic growth (Allen and Bartlett, 2000). Mutants impaired in β-ketoacyl-acyl carrier protein synthase I (*fabB*) have an even more dramatic phenotype. They produce little 16:1 or 18:1 and grow well at 15°C and atmospheric pressure but are incapable of growth at 9°C and 28 MPa pressure (Fig. 1).

Piezophile Enzymes

Studies of enzymes which function at high pressure and low temperature have mostly been performed with deep-sea animals (Somero, 1990, 1992). A malate dehydrogenase enzyme has been kinetically characterized from the obligate piezophile *Moritella* sp. strain 2D2 (Saito and Nakayama, 2004). In comparison with an enzyme obtained from a closely related psychrophilic *Moritella* strain, it was found to display increased thermostability and, surprisingly, K_m and k_{cat} values for NADH which were stimulated by a pressure of 62 MPa by two- to threefold. Curiously, the malate dehydrogenase from the piezophile *P. profundum* strain SS9 shares with the 2D2 malate

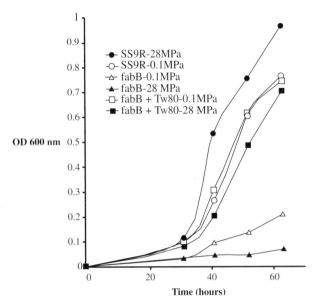

Figure 1. Growth characteristics of the *P. profundum* strain SS9 *fabB* (β-ketoacyl-acyl carrier protein synthase I) mutant at low pressure (0.1 MPa, 9°C) and high pressure (28 MPa, 9°C) compared with the parental SS9R strain. Tween 80 (18:1) was supplemented at a final concentration of 0.025% (vol/vol) (E. E. Allen and D. H. Bartlett, unpublished results).

dehydrogenase a basic amino substitution in the same protein domain involved in subunit interaction (Welch and Bartlett, 1997).

Enhanced quaternary stability could be a hallmark feature of enzymes from piezophiles. Single-stranded DNA-binding protein obtained from more piezophilic *Shewanella* species exhibits increased resistance to pressure-induced subunit dissociation (Chilukuri and Bartlett, 1997). And RNA polymerase obtained from the piezophile *Shewanella violacea* DSS12 is more pressure resistant than the *E. coli* enzyme as long as its sigma 70 subunit is associated with the core enzyme (Kawano et al., 2004).

These few promising examples clearly indicate that there is a need for more examination of the enzymatic and structural adaptations of proteins from piezophiles.

P. profundum Strain SS9 Genetics/Genomics

P. profundum strain SS9 has been a useful model system for studies of adaptation to high pressure because it is amenable to genetic manipulation (Allen and Bartlett, 2000, 2001; Allen et al., 1999; Bartlett and Chi, 1994; Bidle and Bartlett, 1999; Chi and Bartlett, 1993, 1995; Welch and Bartlett, 1996, 1998). Genetic studies have identified the membrane-localized transcription factor ToxR, ubiquitous among the

Vibrionaceae, as a pressure signal transducer (Welch and Bartlett, 1998). The basis of SS9 ToxR pressure sensing, in contrast to *V. cholerae* ToxR, which responds to changes in osmolarity, pH, and the extracellular levels of certain amino acids (Gardel and Mekalanos, 1994), is unknown but could localize to the periplasmic domains where the greatest protein sequence divergence exists. A few SS9 genes have been identified whose products are particularly important for high-pressure growth. These include the aforementioned *fab* genes, whose disruption alters monounsaturated fatty acid biosynthesis, as well as genes facilitating membrane protein assembly and cell division (Allen and Bartlett, 2000; Bidle and Bartlett, 1999; Chi and Bartlett, 1995).

A major breakthrough in studies of the molecular basis of high-pressure adaptation has been the recent report of the complete SS9 genome sequence (Vezzi et al., 2005). It is 25 to 50% larger than that of many other members of the *Vibrionaceae* and contains a large number of gene duplications. These include the largest number of ribosomal operons identified thus far in a bacterial genome (15), two complete operons for F_1F_0ATP synthase, two complete flagellar gene systems, and three complete sets of *cbb3* cytochrome oxidase genes. All of these duplications are noteworthy in the context of pressure adaptation, as ribosome function, proton translocation, motility, and electron transport are all documented to be pressure sensitive in mesophilic bacteria (Gross et al., 1993; Marquis and Bender, 1987; Matsumura and Marquis, 1977; Meganathan and Marquis, 1973; Schwarz and Landau, 1972; Tamegai et al., 1998). Genome-enabled transcriptome studies have also been performed and have revealed hundreds of pressure-regulated genes (Vezzi et al., 2005). At low pressure, the protein-folding genes *htpG*, *dnaK*, *dnaJ*, and *groEL* are upregulated, suggesting that in contrast to mesophiles, the proteins of piezophiles such as SS9 require increased chaperone function to fold correctly at atmospheric pressure. Consistent with these results, the piezophile *Thermococcus barophilus* induces a heat shock-like protein following decompression (Marteinsson et al., 1999b).

OSMOTIC ADAPTATION

The internal osmolyte levels of microbial cells are in balance with those of their environment. In environments of low osmolarity (high water activity), the challenge is to preserve internal pH, ionic strength, metabolite concentration, and the levels of specific ions, whereas at high osmolarity the need is to supplement the cellular milieu with solutes least in-

hibitory to cellular processes (Bartlett and Roberts, 2000). All members of the *Vibrionaceae* are halophiles, but most are only moderately halophilic. When the NaCl concentration exceeds 200 mM, *V. cholerae* experiences a growth delay during which time it accumulates small, uncharged compatible solutes such as ectoine (by synthesis) or glycine betaine (by transport) (Pflughoeft et al., 2003). De novo ectoine biosynthesis may be unique to the genus *Vibrio* among the *Vibrionaceae* (Schmitz and Galinski, 1996). Glutamate is also of general importance. *V. cholerae* preferentially transports glutamate at high osmolarity (Adams et al., 1987; Munro and Gauthier, 1994). Also, under conditions of external proline, osmotic adaptation in *V. vulnificus* follows the osmotic induction of proline transport and its conversion to glutamate (Lee et al., 2003). Trehalose is also an important compatible solute, and among the *Vibrionaceae*, *V. fischeri* is known to be capable of its synthesis (Schmitz and Galinski, 1996). At low osmolarity, *Vibrio* species may increase putrescine content (Kamekura et al., 1986; Yamamoto et al., 1986, 1989), perhaps as a replacement for the decrease in potassium as a counter ion to the phosphate backbone of nucleic acids (Csonka and Epstein, 1996). Membrane lipid and fatty acid composition are also subject to modification in response to osmotic changes (Fukunaga et al., 1995). Finally, it has been postulated that polyunsaturated fatty acids are present in many psychrotolerant and psychrophilic marine bacteria, in part to help cope with a halophilic lifestyle. The hyperfluid membrane resulting from polyunsaturated fatty acid incorporation could enhance the removal of toxic sodium ions from cells (Valentine and Valentine, 2004).

Compatible solutes might be required not just for adaptation to increased osmotic pressure but also for adaptation to increased hydrostatic pressure. Deep-sea fish, skates, and crustaceans increase their levels of the organic osmolyte trimethylamine-*N*-oxide with depth, and trimethylamine-*N*-oxide has been shown to decrease pressure effects on enzyme substrate binding (Gillett et al., 1997; Kelly and Yancey, 1999). *P. profundum* strain SS9 accumulates increased amounts of β-hydroxybutyrate and its oligomers at both high pressure and high osmolarity (Martin et al., 2002); these compounds could perform a similar role to that played by trimethylamine-*N*-oxide in deep-sea animals. Also, SS9 has been found to upregulate at high pressure the expression of genes for the reductive deamination of glycine, sarcosine, and betaine via the Stickland reaction (Vezzi et al., 2005). It has been suggested that the function of this process in SS9 may be to operate in reverse for osmolyte production (Galperin, 2005).

CONNECTION BETWEEN EXTREME ENVIRONMENTS AND THE VIABLE BUT NONCULTURABLE STATE

In response to starvation conditions at low temperature (i.e., 5°C) many bacteria enter into a physiological state in which they cannot be propagated on culture media but remain metabolically active. Such cells have been referred to as having entered the viable but nonculturable (VBNC) state. Cold stress is often a critical aspect of this phenomenon, although shifting cells gradually to lower temperatures and allowing the induction of cold shock proteins may prevent the formation of this physiological state (Bryan et al., 1999). It has been suggested that the induction of the VBNC state is a potential cold adaptation (Chattopadhyay, 2000). Whereas the VBNC state has mostly been documented for some prominent human pathogenic members of the *Vibrionaceae*—*V. cholerae*, *V. vulnificus*, and *V. parahaemolyticus* (Chaiyanan et al., 2001; Kong et al., 2004; Wong et al., 2004)—even the psychrophiles *V. salmonicida* and *Moritella* ANT-300 enter into VBNC states, indicating that cold stress is not a prerequisite for the induction of this condition (Enger, 1990; Preyer and Oliver, 1993). Also, since starvation has been demonstrated to increase resistance to supraoptimal temperature and pressure in *Moritella* ANT-300 (Novitsky and Morita, 1978; Preyer and Oliver, 1993), starvation adaptation could influence the survivability of microbes transiently transported under oligotrophic conditions to certain extreme environments.

Acknowledgment. I gratefully acknowledge the support of the National Science Foundation (MCB 02-37059).

REFERENCES

Adams, R., B. J., M. Kogut, and N. J. Russell. 1987. The role of osmotic effects in haloadaptation of *Vibrio costicola*. *J. Gen. Microbiol.* **133**:1861–1870.

Adler, E., and J. A. Knowles. 1995. Thermolabile triosephosphate isomerase from the psychrophile *Vibrio* sp. strain ANT-300. *Arch. Biochem. Biophys.* **321**:137–139.

Aguilar, P. S., A. M. Hernandez-Arriaga, L. E. Cybulski, A. C. Erazo, and D. de Mendoza. 2001. Molecular basis of thermosensing: a two-component signal transduction thermometer in *Bacillus subtilis*. *EMBO J.* **20**:1681–1691.

Alain, K., V. G. Marteinsson, M. L. Miroshnichenko, E. A. Bonch-Osmolovskaya, D. Prieur, and J.-L. Birrien. 2002. *Marinitoga piezophila* sp. nov., a rod-shaped, thermo-piezophilic bacterium isolated under high hydrostatic pressure from a deep-sea hydrothermal vent. *Int. J. Syst. Evol. Microbiol.* **52**:1331–1339.

Allen, E. E., and D. H. Bartlett. 2000. FabF is required for piezoregulation of cis-vaccenic acid levels and piezophilic growth of the deep-sea bacterium *Photobacterium profundum* strain SS9. *J. Bacteriol.* **182**:1264–1271.

Allen, E. E., and D. H. Bartlett. 2001. Structure and regulation of the omega-3 polyunsaturated fatty acid synthase genes from the deep-sea bacterium *Photobacterium profundum* strain SS9. *Microbiology* **148**:1903–1913.

Allen, E. E., D. Facciotti, and D. H. Bartlett. 1999. Monounsaturated but not polyunsaturated fatty acids are required for growth at high pressure and low temperature in the deep-sea bacterium *Photobacterium profundum* strain SS9. *Appl. Environ. Microbiol.* **65**:1710–1720.

Araki, T. 1991a. Changes in rates of synthesis of individual proteins in a psychrophilic bacterium after a shift in temperature. *Can. J. Microbiol.* **37**:840–847.

Araki, T. 1991b. The effect of temperature shifts on protein synthesis by the psychrophilic bacterium *Vibrio* sp. strain ANT-300. *J. Gen. Microbiol.* **137**:817–826.

Arnorsdottir, J., R. B. Smaradottir, O. T. Magnusson, S. H. Thorbjarnardottir, G. Eggertsson, and M. M. Kristjansson. 2002. Characterization of a cloned subtilisin-like serine proteinase from a psychrotrophic *Vibrio* species. *Eur. J. Biochem.* **269**:5536–5546.

Asgeirsson, B., and O. S. Andresson. 2001. Primary structure of cold-adapted alkaline phosphatase from a *Vibrio* sp. as deduced from the nucleotide gene sequence. *Biochim. Biophys. Acta* **1549**:99–111.

Attard, G. S., R. H. Templer, W. S. Smith, A. N. Hunt, and S. Jackowski. 2000. Modulation of CTP:phosphocholine cytidylyltransferase by membrane curvature elastic stress. *Proc. Natl. Acad. Sci. USA* **97**:9032–9036.

Bale, S. J., K. Goodman, P. A. Rochelle, J. R. Marchesi, J. C. Fry, A. J. Weightman, and R. J. Parkes. 1997. *Desulfovibrio profundus* sp nov, a novel barophilic sulfate-reducing bacterium from deep sediment layers in the Japan Sea. *Int. J. Syst. Bacteriol.* **47**:515–521.

Bartlett, D. H. 1999. Microbial adaptations to the psychrosphere/piezosphere. *J. Mol. Microbiol. Biotechnol.* **1**:93–100.

Bartlett, D. H. 2002. Pressure effects on in vivo microbial processes. *Biochim. Biophys. Acta* **1595**:367–381.

Bartlett, D. H., and K. A. Bidle. 1999. Membrane-based adaptations of deep-sea piezophiles, p. 501–512. *In* J. Seckbach (ed.), *Enigmatic Microorganisms and Life in Extreme Environments.* Kluwer Publishing Co., Dordrecht, The Netherlands.

Bartlett, D. H., and E. Chi. 1994. Genetic characterization of *ompH* mutants in the deep-sea bacterium *Photobacterium* species strain SS9. *Arch. Microbiol.* **162**:323–328.

Bartlett, D. H., and M. F. Roberts. 2000. Osmotic stress, p. 502–516. *In* J. Lederberg (ed.), *Encyclopedia of Microbiology.* Academic Press, Inc., San Diego, Calif.

Batacidos, M. C. L., C. R. LaVilla-Pitogo, E. R. Cruz-Lacierda, L. D. De La Pena, and N. A. Sunaz. 1990. Studies on the chemical control of luminous bacteria *Vibrio harveyi* and *Vibrio splendidus* isolated from diseased *Penaeus mondon* larvae and rearing water. *Dis. Aquat. Org.* **9**:133–140.

Bauer, J. C., and C. M. Young. 2000. Epidermal lesions and mortality caused by vibriosis in deep-sea Bahamian echinoids: a laboratory study. *Dis. Aquat. Org.* **39**:193–199.

Baumann, P., L. Baumann, and M. Mandel. 1971. Taxonomy of marine bacteria: the genus *Benekea*. *J. Bacteriol.* **107**:268–294.

Baumann, P., and L. Baumann. 1977. Biology of the marine enterobacteria: genera *Benekea* and *Photobacterium*. *Annu. Rev. Microbiol.* **31**:39–61.

Baumann, P., and L. Baumann. 1981. The marine gram-negative eubacteria: genera *Photobacterium*, *Beneckea*, *Alteromonas*, *Pseudomonas*, and *Alcaligenes*, p. 1302–1331. *In* H. S. M. P. Starr, H. G. Trüper, A. Balows, and H. G. Schlegel (ed.), *The Prokaryotes*. Springer-Verlag, Berlin, Germany.

Bendt, A., H. Huller, U. Kammel, E. Helmke, and T. Schweder. 2001. Cloning, expression, and characterization of a chitinase gene from the antarctic psychrotolerant bacterium *Vibrio* sp. strain Fi:7. *Extremophiles* **5**:119–126.

Benediktsdottir, E., S. Helgason, and H. Sigurjonsdottir. 1998. *Vibrio* spp. isolated from salmonids with shallow skin lesions and reared at low temperature *J. Fish Dis.* **21**:19–28.

Benediktsdottir, E., L. Verdonck, C. Sproer, S. Helgason, and J. Swings. 2000. Characterization of *Vibrio viscosus* and *Vibrio wodanis* isolated at different geographical locations: a proposal for reclassification of *Vibrio viscosus* as *Moritella viscosa* comb. nov. *Int. J. Syst. Evol. Microbiol.* **50**:479–488.

Bernhardt, G., R. Jaenicke, H.-D. Ludemann, H. Koning, and K. O. Stetter. 1988. High pressure enhances the growth rate of the thermophilic archaebacterium *Methanococcus thermolithotrophicus* without extending its temperature range. *Appl. Environ. Microbiol.* **54**:1258–1261.

Bidle, K. A., and D. H. Bartlett. 1999. RecD function is required for high pressure growth in a deep-sea bacterium. *J. Bacteriol.* **181**:2330–2337.

Borrego, J. J., D. Castro, A. Luque, C. Paillard, P. Maes, M. T. Garcia, and A. Ventosa. 1996. *Vibrio tapetis* sp. nov., the causative agent of the brown ring disease affecting cultured clams. *Int. J. Syst. Bacteriol.* **46**:480–484.

Bowman, J. P., S. A. McCammon, J. A. Gibson, L. Robertson, and P. D. Nichols. 2003. Prokaryotic metabolic activity and community structure in antarctic continental shelf sediments. *Appl. Environ. Microbiol.* **69**:2448–2462.

Bryan, P. J., R. J. Steffan, A. DePaola, J. W. Foster, and A. K. Bej. 1999. Adaptive response to cold temperatures in *Vibrio vulnificus. Curr. Microbiol.* **38**:168–175.

Budsberg, K. J., C. F. Wimpee, and J. F. Braddock. 2003. Isolation and identification of *Photobacterium phosphoreum* from an unexpected niche: migrating salmon. *Appl. Environ. Microbiol.* **69**:6938–6942.

Bylund, G. O., L. C. Wipemo, L. A. C. Lundberg, and P. M. Wikstrom. 1998. RimM and RbfA are essential for efficient processing of 16S rRNA in *Escherichia coli. J. Bacteriol.* **180**:73–82.

Chaiyanan, S., S. Chaiyanan, A. Huq, T. Maugel, and R. R. Colwell. 2001. Viability of the nonculturable *Vibrio cholerae* O1 and O139. *Syst. Appl. Microbiol.* **24**:331–341.

Chattopadhyay, M. K. 2000. Cold-adaptation of antarctic microorganisms—possible involvement of viable but nonculturable state. *Polar Biol.* **23**:223–224.

Chi, E., and D. H. Bartlett. 1993. Use of a reporter gene to follow high pressure signal transduction in the deep-sea bacterium *Photobacterium* SS9. *J. Bacteriol.* **175**:7533–7540.

Chi, E., and D. H. Bartlett. 1995. An *rpoE*-like locus controls outer membrane protein synthesis and growth at cold temperatures and high pressures in the deep-sea bacterium *Photobacterium* SS9. *Mol. Microbiol.* **17**:713–726.

Chilukuri, L. N., and D. H. Bartlett. 1997. Isolation and characterization of the gene encoding single-stranded-DNA-binding protein (SSB) from four marine *Shewanella* strains that differ in their temperature and pressure optima for growth. *Microbiology* **143**:1163–1174.

Colliec-Jouault, S., P. Zanchetta, D. Helley, J. Ratiskol, C. Sinquin, A. M. Fischer, and J. Guezennec. 2004. [Microbial polysaccharides of marine origin and their potential in human therapeutics] [In French.] *Pathol. Biol.* (Paris) **52**:127–130.

Colquhoun, D. J., K. Alvheim, K. Dommarsnes, C. Syvertsen, and H. Sorum. 2002. Relevance of incubation temperature for *Vibrio salmonicida* vaccine production. *J. Appl. Microbiol.* **92**:1087–1096.

Csonka, L. N., and W. Epstein. 1996. Osmoregulation, p. 1210–1223. *In* F. C. Neidhardt, R. Curtiss III, J. L. Ingraham, E. C. C. Lin, K. B. Low, B. Magasanik, W. S. Reznikoff, M. Riley, M. Schaechter, and H. E. Umbarger (ed.), Escherichia coli *and* Salmonella typhimurium: *Cellular and Molecular Biology*, 2nd ed. ASM Press, Washington, D.C.

Dalgaard, P., L. Garcia Munoz, and O. Mejlholm. 1998. Specific inhibition of *photobacterium phosphoreum* extends the shelf life of modified-atmosphere-packed cod fillets. *J. Food Prot.* **61**:1191–1194.

Dalgaard, P., G. P. Manfio, and M. Goodfellow. 1997a. Classification of photobacteria associated with spoilage of fish products by numerical taxonomy and pyrolysis mass spectrometry. *Zentbl. Bakteriol.* **285**:157–168.

Dalgaard, P., O. Mejlholm, T. J. Christiansen, and H. H. Huss. 1997b. Importance of *Photobacterium phosphoreum* in relation to spoilage of modified atmosphere-packed fish products. *Lett. Appl. Microbiol.* **24**:373–378.

Dalluge, J. J., T. Hamamoto, K. Horikoshi, R. Y. Morita, K. O. Stetter, and J. A. Mccloskey. 1997. Posttranscriptional modification of tRNA in psychrophilic bacteria. *J. Bacteriol.* **179**:1918–1923.

Datta, P. P., and R. K. Bhadra. 2003. Cold shock response and major cold shock proteins of *Vibrio cholerae. Appl. Environ. Microbiol.* **69**:6361–6369.

DeLong, E. F. 1986. *Adaptations of Deep-Sea Bacteria to the Abyssal Environment*. Ph.D. thesis. University of California, San Diego.

DeLong, E. F., D. G. Franks, and A. A. Yayanos. 1997. Evolutionary relationships of cultivated psychrophilic and barophilic deep-sea bacteria. *Appl. Environ. Microbiol.* **63**:2105–2108.

DeLong, E. F., K. Y. Wu, B. B. Prezelin, and R. V. M. Jovine. 1994. High abundance of archaea in antarctic marine picoplankton. *Nature* **371**:695–696.

Deming, J. W., L. K. Somers, W. L. Straube, D. G. Swartz, and M. T. Macdonell. 1988. Isolation of an obligately barophilic bacterium and description of a new genus, *Colwellia* gen. nov. *Syst. Appl. Microbiol.* **10**:152–160.

Dersch, P., S. Kneip, and E. Bremer. 1994. The nucleoid-associated DNA binding protein H-NS is required for the efficient adaptation of *Escherichia coli* K-12 to a cold environment. *Mol. Gen. Genet.* **245**:255–259.

Droffner, M. L., and N. Yamamoto. 1991. Prolonged environmental stress via a two step process selects mutants of *Escherichia, Salmonella* and *Pseudomonas* that grow at 54 degrees C. *Arch. Microbiol.* **156**:307–311.

Enger, Ø. 1990. Starvation survival of the fish pathogenic bacteria *Vibrio anguillarum* and *Vibrio salmonicida* in marine environments. *FEMS Microbiol. Ecol.* **74**:215–220.

Enger, Ø., B. Husevag, and J. Goksoyr. 1991. Seasonal variation in presence of *Vibrio salmonicida* and total bacterial counts in Norwegian fish-farm water. *Can. J. Microbiol.* **37**:618–623.

Ermolenko, D. N., and G. I. Makhatadze. 2002. Bacterial cold-shock proteins. *Cell. Mol. Life Sci.* **59**:1902–1913.

Esiobu, N., and K. Yamazaki. 2003. Analysis of bacteria associated with the gut of healthy wild penaeid shrimps: a step towards effective probiotics in aquaculture. *J. Aquacult. Tropics* **18**:275–286.

Farmer, J. J., III. 1999. The family Vibrionaceae. *In The Prokaryotes: an Evolving Electronic Resource for the Microbiological Community*. Springer-Verlag, New York, N.Y. http://link.springer-ny.com/link/service/books/10125/.

Farmer, J. J., III, and F. W. Hickman-Brenner. 1992. The genera *Vibrio* and *Photobacterium*, p. 2952–3011. *In* A. Balow, H. G. Truper, W. Harder, and K.-H. Schleifer (ed.), *The Prokaryotes: a Handbook on the Biology of Bacteria: Ecophysiology, Isolation, Identification, Applications*. Springer-Verlag, New York, N.Y.

Feller, G., D. d'Amico, and C. Gerday. 1999. Thermodynamic stability of a cold-active alpha-amylase from the antarctic bacterium *Alteromonas haloplanctis. Biochemistry* **38**:4613–4619.

Feller, G., and C. Gerday. 2003. Psychrophilic enzymes: hot topics in cold adaptation. *Nat. Rev. Microbiol.* **1**:200–208.

Ferrer, M., T. N. Chernikova, M. M. Yakimov, P. Golyshin, and K. N. Timmis. 2003. Chaperonins govern growth of *Escherichia coli* at low temperatures. *Nat. Biotechnol.* 21:1266–1267.

Ferrer, M., H. Lunsdorf, T. N. Chernikova, M. Yakimov, K. N. Timmis, and P. N. Golyshin. 2004. Functional consequences of single:double ring transitions in chaperonins: life in the cold. *Mol. Microbiol.* 53:167–182.

Fidopiastis, P. M., S. von Boletzky, and E. G. Ruby. 1998. A new niche for *Vibrio logei*, the predominant light organ symbiont of squids in the genus *Sepiola*. *J. Bacteriol.* 180:59–64.

Fredrickson, J. K., and T. C. Onstott. 1996. Microbes deep inside the earth. *Sci. Am.* 275:68–73.

Fuhrman, J. A., K. McCallum, and A. A. Davis. 1992. Novel major archaebacterial group from marine plankton. *Nature* 356:148–149.

Fukunaga, N., M. Wada, M. Honjo, Y. Setaishi, N. Hayashinaka, Y. Takada, and J. Nishikawa. 1995. Effects of temperature and salt on lipid and fatty acid compositions of a bacterium isolated from the bottom layer of Lake Vanda, Antarctica. *J. Gen. Appl. Microbiol.* 41:191–205.

Galperin, M. Y. 2005. The vibrio that sheds light. *Environ. Microbiol.* 7:757–760.

Gardel, C. L., and J. J. Mekalanos. 1994. Regulation of cholera toxin by temperature, pH, and osmolarity. *Methods Enzymol.* 235:517–527.

Garwin, J. L., A. L. Klages, and J. E. Cronan. 1980. Structural, enzymatic, and genetic studies of β-ketoacyl-acyl carrier protein synthases I and II of *Escherichia coli*. *J. Biol. Chem.* 255:11949–11956.

Gatesoupe, F. J., C. Lambert, and J. L. Nicolas. 1999. Pathogenicity of *Vibrio splendidus* strains associated with turbot larvae, *Scophthalmus maximus*. *J. Appl. Microbiol.* 87:757–763.

Genevaux, P., F. Keppel, F. Schwager, P. S. Langendijk-Genevaux, F. U. Hartl, and C. Georgopoulos. 2004. *In vivo* analysis of the overlapping functions of DnaK and trigger factor. *EMBO Rep.* 5:195–200.

Giangrossi, M., A. M. Giuliodori, C. O. Gualerzi, and C. L. Pon. 2002. Selective expression of the beta-subunit of nucleoid-associated protein HU during cold shock in *Escherichia coli*. *Mol. Microbiol.* 44:205–216.

Gillett, M. B., J. R. Suko, F. O. Santoso, and P. H. Yancey. 1997. Elevated levels of trimethylamine oxide in muscles of deep-sea gadiform teleosts: a high-pressure adaptation? *J. Exp. Zool.* 279:386–391.

Glansdorff, N., and Y. Xu. 2002. Microbial life at low temperatures: mechanisms of adaptation and extreme biotopes. Implications for exobiology and the origin of life. *Recent Res. Dev. Microbiol.* 6:1–21.

Goldstein, J., N. S. Pollitt, and M. Inouye. 1990. Major cold shock protein of *Escherichia coli*. *Proc. Natl. Acad. Sci. USA* 87:283–287.

Grimes, D. J., D. Jacobs, D. G. Swartz, P. R. Brayton, and R. R. Colwell. 1993. Numerical taxonomy of gram-negative, oxidase-positive rods from carcharhinid sharks. *Int. J. Syst. Bacteriol.* 43:88–98.

Gross, M., I. J. Kosmowsky, R. Lorenz, H. P. Molitoris, and R. Jaenicke. 1994. Response of bacteria and fungi to high-pressure stress as investigated by two-dimensional polyacrylamide gel electrophoresis. *Electrophoresis* 15:1559–1565.

Gross, M., and R. Jaenicke. 1994. Proteins under pressure—the influence of high hydrostatic pressure on structure, function and assembly of protein complexes. *Eur. J. Biochem.* 221:617–630.

Gross, M., K. Lehle, R. Jaenicke, and K. H. Nierhaus. 1993. Pressure-induced dissociation of ribosomes and elongation cycle intermediates. Stabilizing conditions and identification of the most sensitive functional state. *Eur. J. Biochem.* 218:463–468.

Gualerzi, C. O., A. M. Giuliodori, and C. L. Pon. 2003. Transcriptional and post-transcriptional control of cold-shock genes. *J. Mol. Biol.* 331:527–539.

Hamamoto, T., N. Takata, T. Kudo, and K. Horikoshi. 1994. Effect of temperature and growth phase on fatty acid composition of the psychrophilic *Vibrio* sp. strain 5710. *FEMS Microbiol. Lett.* 119:77–82.

Hauksson, J. B., O. S. Andresson, and B. Asgeirsson. 2000. Heat-labile bacterial alkaline phosphatase from a marine *Vibrio* sp. *Enzyme Microb. Technol.* 27:66–73.

Henderson, R. J., R.-M. Millar, J. R. Sargent, and J.-P. Jostensen. 1993. Trans-monoenoic and polyunsaturated fatty acids in phospholipids of a *Vibrio* species of bacterium in relation to growth conditions. *Lipids* 28:389–396.

Hochachka, P. W., and G. N. Somero. 2002. *Biochemical Adaptation. Mechanism and Process in Physiological Evolution*. Oxford University Press, Oxford, England.

Holm, K. O., E. Strom, K. Stensvag, J. Raa, and T. Jorgensen. 1985. Characteristics of a *Vibrio* sp. associated with the "Hitra disease" of Atlantic salmon in Norwegian fish farms. *Fish Pathol.* 20:125–130.

Huang, C. Y., J. L. Garcia, B. K. Patel, J. L. Cayol, L. Baresi, and R. A. Mah. 2000. *Salinivibrio costicola* subsp. *vallismortis* subsp. nov., a halotolerant facultative anaerobe from Death Valley, and emended description of *Salinivibrio costicola*. *Int. J. Syst. Evol. Microbiol.* 50:615–622.

Huels, K. L., Y. J. Brady, M. A. Delaney, and J. A. Bader. 2003. Evidence of a cold shock response in *Vibrio vulnificus*, a human pathogen transmitted via raw eastern oysters, *Crassostrea virginica*, from the Gulf of Mexico. *J. Shellfish Res.* 22:336.

Ingraham, J. L., and A. G. Marr. 1996. Effect of temperature, pressure, pH, and osmotic stress on growth, p. 1570–1578. *In* F. C. Neidhardt, R. Curtiss III, J. L. Ingraham, E. C. C. Lin, K. B. Low, B. Magasanik, W. S. Reznikoff, M. Riley, M. Schaechter, and H. E. Umbarger (ed.), Escherichia coli *and* Salmonella typhimurium: *Cellular and Molecular Biology*, 2nd ed. ASM Press, Washington, D.C.

Ishii, A., S. Imagawa, N. Fukunaga, S. Sasaki, O. Minowa, Y. Mizuno, and H. Shiokawa. 1987. Isozymes of isocitrate dehydrogenase from an obligately psychrophilic bacterium, *Vibrio* sp. strain ABE-1: purification and modulation of activities by growth conditions. *J. Biochem.* 102:1489–1498.

Ishii, A., T. Oshima, T. Sato, K. Nakasone, H. Mori, and C. Kato. 2005. Analysis of hydrostatic pressure effects on transcription in *Escherichia coli* by DNA microarray procedure. *Extremophiles* 9:65–73.

Ishii, A., T. Sato, M. Wachi, K. Nagai, and C. Kato. 2004. Effects of high hydrostatic pressure on bacterial cytoskeleton FtsZ polymers *in vivo* and *in vitro*. *Microbiology* 150:1965–1972.

Jensen, S., O. B. Samuelsen, K. Andersen, L. Torkildsen, C. Lambert, G. Choquet, C. Paillard, and O. Bergh. 2003. Characterization of strains of *Vibrio splendidus* and *V. tapetis* isolated from corkwing wrasse *Symphodus melops* suffering vibriosis. *Dis. Aquat. Org.* 53:25–31.

Jones, P. G., M. Mitta, Y. Kim, W. Jiang, and M. Inouye. 1996. Cold shock induces a major ribosomal-associated protein that unwinds double-stranded RNA in *Escherichia coli*. *Proc. Natl. Acad. Sci. USA* 93:76–80.

Kadouri, D., E. Jurkevitch, and Y. Okon. 2003. Poly beta-hydroxybutyrate depolymerase (PhaZ) in *Azospirillum brasilense* and characterization of a *phaZ* mutant. *Arch. Microbiol.* 180:309–318.

Kamekura, M., S. Bardocz, P. Anderson, R. Wallace, and D. J. Kushner. 1986. Polyamines in moderately and extremely halophilic bacteria. *Biochim. Biophys. Acta* 880:204–208.

Kamimura, K., H. Fuse, O. Takimura, and Y. Yamaoka. 1993. Effects of growth pressure and temperature on fatty acid composition of a barotolerant deep-sea bacterium. *Appl. Environ. Microbiol.* **59:**924–926.

Kandror, O., A. DeLeon, and A. L. Goldberg. 2002. Trehalose synthesis is induced upon exposure of *Escherichia coli* to cold and is essential for viability at low temperatures. *Proc. Natl. Acad. Sci. USA* **99:**9727–9732.

Kandror, O., and A. L. Goldberg. 1997. Trigger factor is induced upon cold shock and enhances viability of *Escherichia coli* at low temperatures. *Proc. Natl. Acad. Sci. USA* **94:**4978–4981.

Kanki, M., T. Yoda, M. Ishibashi, and T. Tsukamoto. 2004. *Photobacterium phosphoreum* caused a histamine fish poisoning incident. *Int. J. Food Microbiol.* **92:**79–87.

Karner, M. B., E. F. DeLong, and D. M. Karl. 2001. Archaeal dominance in the mesopelagic zone of the Pacific Ocean. *Nature* **409:**507–510.

Kato, C., L. Li, Y. Nogi, Y. Nakamura, J. Tamaoka, and K. Horikoshi. 1998. Extremely barophilic bacteria isolated from the Mariana Trench, Challenger Deep, at a depth of 11,000 meters. *Appl. Environ. Microbiol.* **64:**1510–1513.

Kato, C., T. Sato, and K. Horikoshi. 1995. Isolation and properties of barophilic and barotolerant bacteria from deep-sea mud samples. *Biodivers. Conserv.* **4:**1–9.

Kato, C., H. Tamegai, A. Ikegami, R. Usami, and K. Horikoshi. 1996. Open reading frame 3 of the barotolerant bacterium strain DSS12 is complementary with *cydD* in *Escherichia coli*: *cydD* functions are required for cell stability at high pressure. *J. Biochem.* (Tokyo) **120:**301–305.

Kawano, H., K. Nakasone, M. Matsumoto, Y. Yoshida, R. Usami, C. Kato, and F. Abe. 2004. Differential pressure resistance in the activity of RNA polymerase isolated from *Shewanella violacea* and *Escherichia coli*. *Extremophiles* **8:**367–375.

Kelly, R. H., and P. H. Yancey. 1999. High contents of trimethylamine oxide correlating with depth in deep-sea teleost fishes, skates, and decapod crustaceans. *Biol. Bull.* **196:**18–25.

Kong, I.-S., T. C. Bates, A. Hulsmann, H. Hassan, B. E. Smith, and J. D. Oliver. 2004. Role of catalase and *oxyR* in the viable but nonculturable state of *Vibrio vulnificus*. *FEMS Microbiol. Ecol.* **50:**133–142.

Kristjansson, M. M., O. T. Magnusson, H. M. Gudmundsson, G. A. Alfredsson, and H. Matsuzawa. 1999. Properties of a subtilisin-like proteinase from a psychrotrophic *Vibrio* species—comparison with proteinase K and aqualysin I. *Biochemistry* **260:**752–760.

Lacoste, A., F. Jalabert, S. Malham, A. Cueff, F. Gelebart, C. Cordevant, M. Lange, and S. A. Poulet. 2001. A *Vibrio splendidus* strain is associated with summer mortality of juvenile oysters *Crassostrea gigas* in the Bay of Morlaix (North Brittany, France). *Dis. Aquat. Org.* **46:**139–145.

Landau, J. V. 1967. Induction, transcription, and translation in *Escherichia coli*: a hydrostatic pressure study. *Biochim. Biophys. Acta* **149:**506–512.

Lease, R. A., and M. Belfort. 2000. Riboregulation by DsrA RNA: trans-actions for global economy. *Mol. Microbiol.* **38:**667–672.

Lee, J. H., N. Y. Park, M. H. Lee, and S. H. Choi. 2003. Characterization of the *Vibrio vulnificus putAP* operon, encoding proline dehydrogenase and proline permease, and its differential expression in response to osmotic stress. *J. Bacteriol.* **185:**3842–3852.

Li, L., J. Guezennec, P. Nichols, P. Henry, M. Yanagibayashi, and C. Kato. 1999a. Microbial diversity in Nankai Trough sediments at a depth of 3,843 m. *J. Oceanography* **55:**635–642.

Li, L. N., C. Kato, and K. Horikoshi. 1999b. Bacterial diversity in deep-sea sediments from different depths. *Biodivers. Conserv.* **8:**659–677.

Liesack, W., H. Weyland, and E. Stackebrandt. 1991. Potential risks of gene amplification by PCR as determined by 16S rDNA

analysis of a mixed-culture of strict barophilic bacteria. *Microb. Ecol.* **21:**191–198.

Lin, C., R.-C. Yu, and C.-C. Chou. 2004. Susceptibility of *Vibrio parahaemolyticus* to various environmental stresses after cold shock treatment. *Int. J. Food Microbiol.* **92:**207–215.

Lopez-Garcia, P., A. Lopez-Lopez, D. Moreira, and F. Rodriguez-Valera. 2001a. Diversity of free-living prokaryotes from a deep-sea site at the Antarctic Polar Front. *FEMS Microbiol. Ecol.* **36:**193–202.

Lopez-Garcia, P., D. Moreira, A. Lopez-Lopez, and F. Rodríguez-Valera. 2001b. A novel haloarchaeal-related lineage is widely distributed in deep oceanic regions. *Environ. Microbiol.* **3:**72–78.

Lopez-Garcia, P., F. Rodriguez-Valera, C. Pedros-Alio, and D. Moreira. 2001c. Unexpected diversity of small eukaryotes in deep-sea antarctic plankton. *Nature* **409:**603–607.

Lunder, T., H. Sorum, G. Holstad, A. G. Steigerwalt, P. Mowinckel, and D. J. Brenner, 2000. Phenotypic and genotypic characterization of *Vibrio viscosus* sp. nov. and *Vibrio wodanis* sp. nov. isolated from Atlantic salmon (*Salmo salar*) with "winter ulcer." *Int. J. Syst. Bacteriol.* **50:**427–450.

Luttinger, A., J. Hahn, and D. Dubnau. 1996. Polynucleotide phosphorylase in necessary for competence development in *Bacillus subtilis*. *Mol. Microbiol.* **19:**343–356.

Makemson, J. C. 1973. Control of in vivo luminescence in psychrophilic marine photobacterium. *Arch. Microbiol.* **93:**347–358.

Marquis, R. E., and G. R. Bender. 1987. *Barophysiology of Prokaryotes and Proton-Translocating ATPases.* Academic Press, Inc., London, England.

Marteinsson, V. T., J. L. Birrien, A. L. Reysenbach, M. Vernet, D. Marie, A. Gambacorta, P. Messner, U. Sleytr, and D. Prieur. 1999a. *Thermococcus barophilus* sp. nov., a new barophilic and hyperthermophilic archaeon isolated under high hydrostatic pressure from a deep-sea hydrothermal vent. *Int. J. Syst. Bacteriol.* **49:**351–359.

Marteinsson, V. T., A.-L. Reysenbach, J.-L. Birrien, and D. Prieur. 1999b. A stress protein is induced in the deep-sea barophilic hyperthermophile *Thermococcus barophilus* when grown under atmospheric pressure. *Extremophiles* **3:**277–282.

Marteinsson, V. T., P. Moulin, J.-L. Birrien, A. Gambacorta, M. Vernet, and D. Prieur. 1997. Physiological responses to stress conditions and barophilic behavior of the hyperthermophilic vent archaeon *Pyrococcus abyssi*. *Appl. Environ. Microbiol.* **63:**1230–1236.

Martin, D. D., D. H. Bartlett, and M. F. Roberts. 2002. Solute accumulation in the deep-sea bacterium *Photobacterium profundum*. *Extremophiles* **6:**507–514.

Martinkearley, J., and J. A. Gow. 1994. Numerical taxonomy of *Vibrionaceae* from Newfoundland coastal waters. *Can. J. Microbiol.* **40:**355–361.

Martinkearley, J., J. A. Gow, M. Peloquin, and C. W. Greer. 1994. Numerical analysis and the application of random amplified polymorphic DNA polymerase chain reaction to the differentiation of *Vibrio* strains from a seasonally cold ocean. *Can. J. Microbiol.* **40:**446–455.

Matsumura, P., and R. E. Marquis. 1977. Energetics of streptococcal growth inhibition by hydrostatic pressure. *Appl. Environ. Microbiol.* **33:**885–892.

McElhaney, R. N. 1982. Effects of membrane lipids on transport and enzymic activities, p. 317–380. *In* S. Razin and S. Rottem (ed.), *Current Topics in Membranes and Transport.* Academic Press, Inc., New York, N.Y.

McGovern, V. P., and J. D. Oliver. 1995. Induction of cold-responsive proteins in *Vibrio vulnificus*. *J. Bacteriol.* **177:**4131–4133.

Meganathan, R., and R. E. Marquis. 1973. Loss of bacterial motility under pressure. *Nature* **246:**526–527.

Miller, J. F., N. N. Shah, C. M. Nelson, J. M. Lulow, and D. S. Clark. 1988. Pressure and temperature effects on growth and methane production of the extreme thermophile *Methanococcus jannaschii*. *Appl. Environ. Microbiol.* **54**:3039–3042.

Morita, R. Y. 1975. Psychrophilic bacteria. *Bacteriol. Rev.* **39**:144–167.

Munro, P. M., and M. J. Gauthier. 1994. Uptake of glutamate by *Vibrio cholerae* in media of low and high osmolarity, and in seawater. *Lett. Appl. Microbiol.* **18**:197–199.

Nakayama, A., Y. Yano, and K. Yoshida. 1994. New method for isolating barophiles from intestinal contents of deep-sea fishes retrieved from the abyssal zone. *Appl. Environ. Microbiol.* **60**:4210–4212.

Nishiguchi, M. K. 2000. Temperature affects species distribution in symbiotic populations of *Vibrio* spp. *Appl. Environ. Microbiol.* **66**:3550–3555.

Nogi, Y., S. Hosoya, C. Kato, and K. Horikoshi. 2004. *Colwellia piezophila* sp. nov., a novel piezophilic species from deep-sea sediments of the Japan Trench. *Int. J. Syst. Evol. Microbiol.* **54**:1627–1631.

Nogi, Y., C. Kato, and K. Horikoshi. 2002. *Psychromonas kaikoae* sp. nov., a novel piezophilic bacterium from the deepest cold-seep sediments in the Japan Trench. *Int. J. Syst. Evol. Microbiol.* **52**:1527–1532.

Nogi, Y., N. Masui, and C. Kato. 1998a. *Photobacterium profundum* sp. nov., a new moderately barophilic bacterial species isolated from a deep-sea sediment. *Extremophiles* **2**:1–7.

Nogi, Y., N. Masui, and C. Kato. 1998b. Taxonomic studies of deep-sea barophilic *Shewanella* species, and *Shewanella violacea* sp. nov., a new moderately barophilic bacterial species. *Arch. Microbiol.* **170**:331–338.

Novitsky, J. A., and R. Y. Morita. 1978. Starvation-induced barotolerance as a survival mechanism of a psychrophilic marine *Vibrio* in the waters of the Atlantic convergence. *Mar. Biol.* **49**:7–10.

Ochiai, T., N. Fukunaga, and S. Sasaki. 1979. Purification and some properties of two NADP-specific isocitrate dehydrogenases from an obligately psychrophilic marine bacterium, *Vibrio* sp. strain ABE-1. *J. Biochem.* **86**:377–384.

Onarheim, A. M., R. Wiik, J. Burghardt, and E. Stackebrandt. 1994. Characterization and identification of two *Vibrio* species indigenous to the intestine of fish in cold sea water—description of *Vibrio iliopiscarius* sp nov. *Syst. Appl. Microbiol.* **17**:370–379.

Ortigosa, M., E. Garay, and M.-J. Pujalte. 1994. Numerical taxonomy of *Vibrionaceae* isolated from oysters and seawater along an annual cycle. *Syst. Appl. Microbiol.* **17**:216–225.

Pagan, R., and B. Mackey. 2000. Relationship between membrane damage and cell death in pressure-treated *Escherichia coli* cells: differences between exponential- and stationary-phase cells and variation among strains. *Appl. Environ. Microbiol.* **66**:2829–2834.

Pflughoeft, K. J., K. Kierek, and P. I. Watnick. 2003. Role of ectoine in *Vibrio cholerae* osmoadaptation. *Appl. Environ. Microbiol.* **69**:5919–5927.

Phadtare, S., and M. Inouye. 2004. Genome-wide transcriptional analysis of the cold shock response in wild-type and cold-sensitive, quadruple-*csp*-deletion strains of *Escherichia coli*. *J. Bacteriol.* **186**:7007–7014.

Phadtare, S., M. Inouye, and K. Severinov. 2004. The mechanism of nucleic acid melting by a CspA family protein. *J. Mol. Biol.* **337**:147–155.

Pinhassi, J., F. Azam, J. Hemphälä, R. Long, J. Martinez, U. Zweifel, and Å. Hagström. 1999. Coupling between bacterioplankton species composition, population dynamics, and organic matter degradation. *Aquat. Microb. Ecol.* **17**:13–26.

Pogliano, K. J., and J. Beckwith. 1993. The Cs sec mutants of *Escherichia coli* reflect the cold sensitivity of protein export itself. *Genetics* **133**:763–773.

Polissi, A., W. De Laurentis, S. Zangrossi, F. Briani, V. Longhi, G. Pesole, and G. Deho. 2003. Changes in *Escherichia coli* transcriptome during acclimatization at low temperature. *Res. Microbiology* **154**:573–580.

Preyer, J. M., and J. D. Oliver. 1993. Starvation-induced thermal tolerance as a survival mechanism in a psychrophilic marine bacterium. *Appl. Environ. Microbiol.* **59**:2653–2656.

Prud'homme-Genereux, A., R. K. Beran, I. Lost, C. S. Ramey, G. A. Mackie, and R. W. Simons. 2004. Physical and functional interactions among RNase E, polynucleotide phosphorylase and the cold-shock protein, CsdA: evidence for a "cold shock degradosome." *Mol. Microbiol.* **54**:1409–1421.

Qing, G., L.-C. Ma, A. Khorchid, G. V. T. Swapna, T. K. Mal, M. M. Takayama, B. Xia, S. Phadtare, H. Ke, T. Acton, G. T. Montelione, M. Ikura, and M. Inouye. 2004. Cold-shock induced high-yield protein production in *Escherichia coli*. *Nat. Biotechnol.* **22**:877–882.

Radjasa, O. K., H. Urakawa, K. Kita-Tsukamoto, and K. Ohwada. 2001. Characterization of psychrotrophic bacteria in the surface and deep-sea waters from the northwestern Pacific Ocean based on 16S ribosomal DNA analysis. *Mar. Biotechnol.* **3**:454–462.

Raguénès, G., R. Christen, J. Guezennec, P. Pignet, and G. Barbier. 1997. *Vibrio diabolicus* sp. nov., a new polysaccharide-secreting organism isolated from a deep-sea hydrothermal vent polychaete annelid, *Alvinella pompejana*. *Int. J. Syst. Bacteriol.* **47**:989–995.

Reichelt, J. L., P. Baumann, and L. Baumann. 1976. Study of genetic relationships among marine species of the genera *Beneckea* and *Photobacterium* by means of in vitro DNA/DNA hybridization. *Arch. Microbiol.* **110**:101–120.

Reid, H. I., H. L. Duncan, L. A. Laidler, D. Hunter, and T. H. Birkbeck. 2003. Isolation of *Vibrio tapetis* from cultivated Atlantic halibut (*Hippoglossus hippoglossus* L.). *Aquaculture* **221**:65–74.

Ruby, E. G., E. P. Greenberg, and J. W. Hastings. 1980. Planktonic marine luminous bacteria: species distribution in the water column. *Appl. Environ. Microbiol.* **39**:302–306.

Ruby, E. G., and J. G. Morin. 1978. Specificity of symbiosis between deep-sea fishes and psychrotrophic luminous bacteria. *Deep-Sea Res.* **25**:161–167.

Ruby, E. G., and K. H. Nealson. 1977. A luminous bacterium that emits yellow light. *Science* **196**:432–434.

Rueger, H.-J. 1988. Substrate-dependent cold adaptations in some deep-sea sediment bacteria. *Syst. Appl. Microbiol.* **11**:90–93.

Russell, N. J. 1990. Cold adaptation of microorganisms. *Phil. Trans. R. Soc. London B* **326**:595–611.

Russell, N. J., and T. Hamamoto. 1998. Psychrophiles, p. 25–45. *In* K. Horikoshi and W. D. Grant (ed.), *Extremophiles: Microbial Life in Extreme Environments*. Wiley-Liss, Inc., New York, N.Y.

Sahara, T., M. Suzuki, J.-I. Tsuruha, Y. Takada, and N. Fukunaga. 1999. cis-Acting elements responsible for low-temperature-inducible expression of the gene coding for the thermolabile isocitrate dehydrogenase isozyme of a psychrophilic bacterium, *Vibrio* sp. strain ABE-1. *J. Bacteriol.* **181**:2602–2611.

Saito, R., and A. Nakayama. 2004. Differences in malate dehydrogenases from obligately piezophilic deep-sea bacterium *Moritella* sp. strain 2d2 and the psychrophilic bacterium *Moritella* sp. strain 5710. *FEMS Microbiol. Lett.* **233**:165–172.

Saunders, N. F. W., T. Thomas, P. M. G. Curmi, J. S. Mattick, E. Kuczek, R. Slade, J. Davis, P. D. Franzmann, D. Boone, K. Rusterholtz, R. Feldman, C. Gates, S. Bench, K. R. Sowers, K. Kadner, A. Aerts, P. Dehal, C. Detter, T. Glavina, S. Lucas, P. M. Richardson, F. Larimer, L. Hauser, M. Land, and R. Cavicchioli. 2003. Mechanisms of thermal adaptation revealed from the genomes of the antarctic Archaea *Methanogenium frigidum* and *Methanococcoides burtonii*. *Genome Res.* **13**:1580–1588.

Schiewe, M. H., T. J. Trust, and J. H. Crosa. 1981. *Vibrio ordalii* sp. nov.: a causative agent of vibriosis in fish. *Curr. Microbiol.* **6**:343–348.

Schmitz, R. P. H., and E. A. Galinski. 1996. Compatible solutes in luminescent bacteria of the genera *Vibrio, Photobacterium* and *Photorhabdus (Xenorhabdus)*: occurrence of ectoine, betaine and glutamate. *FEMS Microbiol. Lett.* **142**:195–201.

Schwarz, J. R., and J. V. Landau. 1972. Inhibition of cell-free protein synthesis by hydrostatic pressure. *J. Bacteriol.* **112**:1222–1227.

Simopoulos, A. P. 1991. Omega-3 fatty acids in health and disease and in growth and development. *Am. J. Clin. Nutr.* **54**:438–463.

Somero, G. N. 1990. Life at low volume change: hydrostatic pressure as a selective factor in the aquatic environment. *Am. Zool.* **30**:123–135.

Somero, G. N. 1992. Biochemical ecology of deep-sea animals. *Experientia* **48**:537–543.

Sorokin, D. Y. 1992. *Catenococcus thiocyclus* gen. nov.—a new facultatively anaerobic bacterium from a near-shore sulphidic hydrothermal area. *J. Gen. Microbiol.* **138**:2287–2292.

Suutari, M., and S. Laakso. 1994. Microbial fatty acids and thermal adaptation. *Crit. Rev. Microbiol.* **20**:285–328.

Svingor, A., J. Kardos, I. Hajdu, A. Nemeth, and P. Zavodszky. 2001. A better enzyme to cope with cold—comparative flexibility studies on psychrotrophic, mesophilic, and thermophilic IP-MDHS. *J. Biol. Chem.* **276**:28121–28125.

Tamegai, H., C. Kato, and K. Horikoshi. 1998. Pressure-regulated respiratory system in barotolerant bacterium *Shewanella* sp. strain DSS12. *J. Biochem. Mol. Biol. Biophys.* **1**:213–220.

Tendeng, C., C. Badaut, E. Krin, P. Gounon, S. Ngo, A. Danchin, S. Rimsky, and P. Bertin. 2000. Isolation and characterization of *vicH*, encoding a new pleiotropic regulator in *Vibrio cholerae. J. Bacteriol.* **182**:2026–2032.

Thompson, F. L., C. C. Thompson, Y. Li, B. Gomez-Gil, J. Vandenberghe, B. Hoste, and J. J. Swings. 2003. *Vibrio kanaloae* sp. nov., *Vibrio pomeroyi* sp. nov. and *Vibrio chagasii* sp. nov., from sea water and marine animals. *Int. J. Syst. Bacteriol.* **53**:753–759.

Urakawa, H., K. Kita-Tsukamoto and K. Ohwada. 1999a. 16S rDNA restriction fragment length polymorphism analysis of psychrotrophic vibrios from Japanese coastal water. *Can. J. Microbiol.* **45**:1001–1007.

Urakawa, H., K. Kita-Tsukamoto, and K. Ohwada. 1999b. Reassessment of the taxonomic position of *Vibrio iliopiscarius* (Onarheim et al. 1994) and proposal for *Photobacterium iliopiscarium* comb. nov. *Int. J. Syst. Bacteriol.* **49**:257–260.

Urakawa, H., K. Kita-Tsukamoto, and K. Ohwada. 1999c. Restriction fragment length polymorphism analysis of psychrophilic and psychrotrophic *Vibrio* and *Photobacterium* from the northwestern Pacific Ocean and Otsuchi Bay, Japan. *Can. J. Microbiol.* **45**:67–76.

Valentine, R. C., and D. L. Valentine. 2004. Omega-3 fatty acids in cellular membranes: a unified concept. *Prog. Lipid Res.* **43**:383–402.

VanBogelen, R. A., and F. C. Neidhardt. 1990. Ribosomes as sensors of heat and cold shock in *Escherichia coli. Proc. Natl. Acad. Sci. USA* **87**:5589–5593.

van de Vossenberg, J. L., T. Ubbink-Kok, M. G. Elferink, A. J. Driessen, and W. N. Konings. 1995. Ion permeability of the cytoplasmic membrane limits the maximum growth temperature of bacteria and archaea. *Mol. Microbiol.* **18**:925–932.

Ventosa, A., J. J. Nieto, and A. Oren. 1998. Biology of moderately halophilic aerobic bacteria. *Microbiol. Mol. Biol. Rev.* **62**:504–544.

Vezzi, A., S. Campanaro, M. D'Angelo, F. Simonato, N. Vitulo, F. M. Lauro, A. Cestaro, G. Malacrida, B. Simionati, N. Can-

nata, C. Romualdi, D. H. Bartlett, and G. Valle. 2005. Life at depth: *Photobacterium profundum* genome sequence and expression analysis. *Science* **307**:1459–1461.

Vorachek-Warren, M. K., S. M. Carty, S. Lin, R. J. Cotter, and C. R. H. Raetz. 2002. An *Escherichia coli* mutant lacking the cold shock-induced palmitoleoyltransferase of lipid A biosynthesis. *J. Biol. Chem.* **277**:14186–14193.

Wallon, G., S. T. Lovett, C. Magyar, A. Svingor, A. Szilagyi, P. Zavodszky, D. Ringe, and G. A. Petsko. 1997. Sequence and homology model of 3-isopropylmalate dehydrogenase from the psychrotrophic bacterium *Vibrio* sp I5 suggest reasons for thermal instability. *Protein Eng.* **10**:665–672.

Wassarman, K. M. 2002. Small RNAs in bacteria: diverse regulators of gene expression in response to environmental changes. *Cell* **109**:141–144.

Watanabe, S., Y. Takada, and N. Fukunaga. 2001. Purification and characterization of a cold-adapted isocitrate lyase and a malate synthase from *Colwellia maris*, a psychrophilic bacterium. *Biosci. Biotechnol. Biochem.* **65**:1095–1103.

Watanabe, S., N. Yamaoka, Y. Takada, and N. Fukunaga. 2003. The cold-inducible *icl* gene encoding thermolabile isocitrate lyase of a psychrophilic bacterium, *Colwellia maris. Microbiology* **148**:2579–2589.

Welch, T. J., and D. H. Bartlett. 1996. Isolation and characterization of the structural gene for OmpL, a pressure-regulated porin-like protein from the deep-sea bacterium *Photobacterium* species strain SS9. *J. Bacteriol.* **178**:5027–5031.

Welch, T. J., and D. H. Bartlett. 1997. Cloning, sequencing and overexpression of the gene encoding malate dehydrogenase from the deep-sea bacterium *Photobacterium* species strain SS9. *Biochim. Biophys. Acta* **1350**:41–46.

Welch, T. J., and D. H. Bartlett. 1998. Identification of a regulatory protein required for pressure-responsive gene expression in the deep-sea bacterium *Photobacterium* species strain SS9. *Mol. Microbiol.* **27**:977–985.

Welch, T. J., A. Farewell, F. C. Neidhardt, and D. H. Bartlett. 1993. Stress response in *Escherichia coli* induced by elevated hydrostatic pressure. *J. Bacteriol.* **175**:7170–7177.

Wiegel, J. 1990. Temperature spans for growth: hypothesis and discussion. *FEMS Microbiol. Rev.* **75**:155–170.

Wintrode, P. L., and F. H. Arnold. 2000. Temperature adaptation of enzymes: lessons from laboratory evolution. *Adv. Protein Chem.* **55**:161–225.

Wong, H.-C., C.-T. Shen, C.-N. Chang, Y.-S. Lee, and J. D. Oliver. 2004. Biochemical and virulence characterization of viable but nonculturable cells of *Vibrio parahaemolyticus. J. Food Prot.* **67**:2430–2435.

Xia, B., H. Ke, and M. Inouye. 2001. Acquirement of cold sensitivity by quadruple deletion of the *cspA* family and its suppression by PNPase S1 domain in *Escherichia coli. Mol. Microbiol.* **40**:179–188.

Xia, B., H. Ke, U. Shinde, and M. Inouye. 2003. The role of RbfA in 16 S rRNA processing and cell growth at low temperature in *Escherichia coli. J. Mol. Biol.* **332**:575–584.

Xu, Y., Y. Nogi, C. Kato, Z. Liang, H.-J. Rueger, D. De Kegel, and N. Glansdorff. 2003a. *Moritella profunda* sp. nov. and *Moritella abyssi* sp. nov., two psychropiezophilic organisms isolated from deep Atlantic sediments. *Int. J. Syst. Evol. Microbiol.* **53**:533–538.

Xu, Y., Y. Nogi, C. Kato, Z. Liang, H.-J. Rueger, D. De Kegel, and N. Glansdorff. 2003b. *Psychromonas profunda* sp. nov., a psychropiezophilic bacterium from deep Atlantic sediments. *Int. J. Syst. Evol. Microbiol.* **53**:527–532.

Xu, Y., Y. F. Zhang, Z. Y. Liang, M. Vandecasteele, C. Legrain, and N. Glansdorff. 1998. Aspartate carbamoyltransferase from a psychrophilic deep-sea bacterium, *Vibrio* strain 2693—properties of the enzyme, genetic organization and synthesis in *Escherichia coli. Microbiology* **144**:1435–1441.

Yamamoto, S., K. Yamasaki, K. Takashina, T. Katsu, and S. Shinoda. 1989. Characterization of putrescine production in nongrowing *Vibrio-parahaemolyticus* cells in response to external osmolality. *Microbiol. Immunol.* **33**:11–22.

Yamamoto, S., M. Yoshida, H. Nakao, M. Koyama, Y. Hashimoto, and S. Shinoda. 1986. Variations in cellular polyamine compositions and contents of *Vibrio* species during growth in media with various sodium chloride concentrations. *Chem. Pharm. Bull.* **34**:3038–3042.

Yano, Y., A. Nakayama, H. Saito, and K. Ishihara. 1994. Production of docosahexaenoic acid by marine bacteria isolated from deep sea fish. *Lipids* **29**:527–528.

Yano, Y., A. Nakayama, and K. Yoshida. 1997. Distribution of polyunsaturated fatty acids in bacteria present in intestines of deep-sea fish and shallow-sea poikilothermic animals. *Appl. Environ. Microbiol.* **63**:2572–2577.

Yayanos, A. A. 1986. Evolutional and ecological implications of the properties of deep-sea barophilic bacteria. *Proc. Natl. Acad. Sci. USA* **83**:9542–9546.

Yayanos, A. A. 1995. Microbiology to 10,500 meters in the deep sea. *Annu. Rev. Microbiol.* **49**:777–805.

Yayanos, A. A., A. S. Dietz, and R. Van Boxtel. 1979. Isolation of a deep-sea barophilic bacterium and some of its growth characteristics. *Science* **205**:808–810.

Yayanos, A. A., A. S. Dietz, and R. Van Boxtel. 1981. Obligately barophilic bacterium from the Mariana trench. *Proc. Natl. Acad. Sci. USA* **78**:5212–5215.

Yayanos, A. A., and E. C. Pollard. 1969. A study of the effects of hydrostatic pressure on macromolecular synthesis in *Escherichia coli. Biophys. J.* **9**:1464–1482.

Yoneta, M., T. Sahara, K. Nitta, and Y. Takada. 2004. Characterization of chimeric isocitrate dehydrogenases of a mesophilic nitrogen-fixing bacterium, *Azotobacter vinelandii*, and a psychrophilic bacterium, *Colwellia maris. Curr. Microbiol.* **48**:383–388.

Yumoto, I., K. Kawasaki, H. Iwata, H. Matsuyama, and H. Okuyama. 1998. Assignment of *Vibrio* sp. strain ABE-1 to *Colwellia maris* sp. nov., a new psychrophilic bacterium. *Int. J. Syst. Bacteriol.* **48**:1357–1362.

ZoBell, C. E. 1970. *Pressure Effects on Morphology and Life Processes of Bacteria.* Academic Press, Inc., London, England.

ZoBell, C. E., and A. B. Cobet. 1950. Some effects of hydrostatic pressure on the multiplication and morphology of marine bacteria. *J. Bacteriol.* **60**:771–781.

ZoBell, C. E., and A. B. Cobet. 1962. Growth, reproduction and death rates of *Escherichia coli* at increased hydrostatic pressures. *J. Bacteriol.* **84**:1228–1236.

ZoBell, C. E., and A. B. Cobet. 1963. Filament formation by *Escherichia coli* at increased hydrostatic pressures. *J. Bacteriol.* **87**:710–719.

VI. HABITAT AND ECOLOGY

The Biology of Vibrios
Edited by F. L. Thompson et al.
© 2006 ASM Press, Washington, D.C.

Chapter 12

Aquatic Environment

HIDETOSHI URAKAWA AND IRMA NELLY G. RIVERA

THE WHEREABOUTS OF VIBRIOS; THE RANGE OF HABITATS AND NICHES OF *VIBRIO* SPECIES

Bacteria belonging to the family *Vibrionaceae* are ubiquitous and are widely distributed in aquatic environments from brackish to deep-sea water, worldwide (Fig. 1). They are commonly associated with marine living species and include many important pathogens for farmed animals and humans who consume contaminated seafood or polluted drinking water. There is no doubt that *Vibrio* spp. have evolved in marine environments, because one specific nature of the group is the sodium requirement for growth. Indeed, most vibrios are solely isolated from the marine and estuarine environments. Such evolutionary and habitat patterns of *Vibrio* and *Photobacterium* species are similar to those of other genera belonging to the *Gammaproteobacteria*, such as *Pseudoalteromonas*, *Moritella*, *Colwellia*, and *Halomonas*, which contain only marine species. The situation is different in *Shewanella* and *Pseudomonas*, which contain both marine and terrestrial species. In this chapter, the nature of vibrios—their habitats, ecology, physiological traits, and evolution—will be considered.

Coastal Water

Vibrio species is a major group of culturable, heterotrophic bacteria, especially in coastal waters. The density of *Vibrionaceae* and its proportion to total plate counts vary widely according to sampling methods, geographical areas, and seasonality (Austin et al., 1979). The sizes of vibrio populations found in aquatic environments have been included in Table 1. In general, *Vibrio* species are frequently detectable in summer, but during winter months they are less common, possibly because of the occurrence of a viable but nonculturable stage. However, in tropical and subtropical waters, the variation in vibrio populations is low. For example, in the coastal waters of the southeastern United States, the composition of vibrios averaged 34.6% of the total bacterial population during the summer (Oliver et al., 1982). In another example, the population of vibrios in the subtropical coastal waters of Hong Kong ranged from 10^1 to 10^3 cells/ml, which was equivalent to 0.4 to 40% of the total bacterial populations in summer months (May to October) (Chan et al., 1986). It appears that the abundance of vibrios in the water column reflects the water temperature. Thus, throughout the study of *Vibrio parahaemolyticus*, a temperature of 14 to 19°C was found to be critical for the detection of culturable cells in the water column during late spring or early summer (Kaneko and Colwell, 1973). In the Chesapeake Bay, *V. parahaemolyticus* could not be detected in the water column during the winter months, whereas cells were detected from sediment throughout the year. A similar conclusion resulted from the study of *Vibrio vulnificus* in the Chesapeake Bay (Wright et al., 1996).

Maeda et al. (2003) investigated the seasonal population change of vibrios in Yoshimi Bay, Japan, and reported that the numbers ranged from 8×10^0 to 7×10^2 CFU/ml of seawater and from 3×10^3 to 1×10^6 CFU/g of sediment. Overall, the vibrio populations increased with temperature, which influenced the CFU and the taxonomic genetic composition. PCR-restriction fragment length polymorphism analysis revealed that the *Vibrio* group comprised *V. parahaemolyticus* and related species predominantly when the water temperature was >20°C, whereas *Vibrio splendidus* and *Vibrio lentus* were most abundant when the water temperature was ≤20°C in seawater and sediment.

Hidetoshi Urakawa • Center for Advanced Marine Research, Ocean Research Institute, The University of Tokyo, 1-15-1 Minamidai, Nakano, Tokyo 164-8639, Japan. **Irma Nelly G. Rivera** • Department of Microbiology, Biomedical Science Institute, University of São Paulo, São Paulo, CEP 05508-900, Brazil.

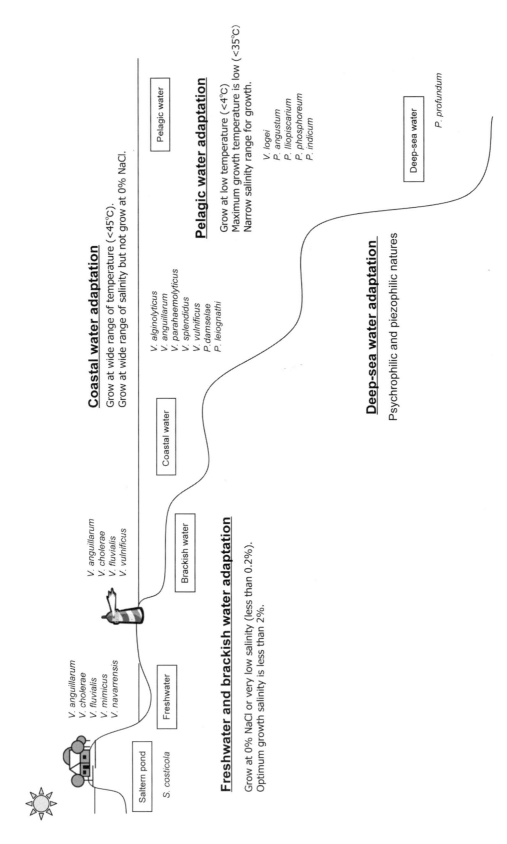

Figure 1. Specific distribution of *Vibrionaceae* representatives in aquatic environments.

Table 1. Abundance of *Vibrio* spp. in water and sediment

Sampling site[a]	Season or month	Source	Bacteria in water or sediment (per ml or g)		% of vibrios[b]	Medium for total bacteria[c]	Medium for vibrios	Reference
			Total bacterial	No. of vibrios				
Southeastern U.S. (average)	Summer	Seawater	1.1×10^3	1.7×10^2	34.6	EA	TCBS	Oliver et al. (1982)
Wilmington, N.C.	Aug.	Seawater	2.8×10^2–3.8×10^2	3.5×10^1–5.5×10^1		EA	TCBS	Oliver et al. (1982)
Ft. Fisher, N.C.	Jun.	Seawater	2.8×10^2–6.5×10^2	4.0×10^1–3.5×10^2		EA	TCBS	Oliver et al. (1982)
	Sep.	Seawater	3.5×10^2–6.8×10^3	1.0×10^2–2.0×10^3		EA	TCBS	Oliver et al. (1982)
North Myrtle Beach, S.C.	Aug.	Seawater	3.8×10^2–2.9×10^3	1.5×10^2–1.8×10^2		EA	TCBS	Oliver et al. (1982)
Charleston, S.C.	Sep.	Seawater	1.8×10^3–1.4×10^4	5.0×10^1–1.8×10^4		EA	TCBS	Oliver et al. (1982)
Savannah, Ga.	Oct.	Seawater	5.4×10^2–1.0×10^4	1.8×10^1–4.0×10^2		EA	TCBS	Oliver et al. (1982)
East Coast U.S. (average of 17 sites)	Summer	Seawater	4.7×10^3	7.1×10^2	31.2	MSWYE	TCBS	Oliver et al. (1983)
		Sediment	2.2×10^5	6.3×10^4	35	MSWYE	TCBS	Oliver et al. (1983)
Chesapeake Bay (Cape Charles)	Jun.	Seawater	2.0×10^4	6.3×10^2		2216	TCBS	Austin et al. (1979)
Japanese coast								
Tokyo Bay	Jul.	Seawater	9.1×10^4	1.8×10^1	0.02	2216	TCBS	Austin et al. (1979)
Tokyo Bay	Aug.	Seawater	1.7×10^2–2.0×10^3	1.7×10^2–2.2×10^3		PPES-II	Vibrio	Simidu and Tsukamoto (1980)
Otsuchi Bay	Nov.	Seawater	1.0×10^1–1.3×10^3	1.1×10^1–7.7×10^2		PPES-II	Vibrio	Simidu and Tsukamoto (1980)
Sagami Bay	Aug.	Seawater	2.2×10^1	1.5×10^1	68	PPES-II	Vibrio	Simidu and Tsukamoto (1980)
Yoshimi Bay	Annual	Seawater	3×10^3–3×10^5	8×10^0–7×10^2		PPES-II	Vibrio	Maeda et al. (2003)
	Annual	Sediment	4×10^5–6×10^6	3×10^3–1×10^6		PPES-II	Vibrio	Maeda et al. (2003)
Hong Kong (average of 6 sites)	Summer	Seawater	4.5×10^4	2.1×10^3	4.6	2216	TCBS	Chan et al. (1986)
Repulse Bay		Seawater	1.0×10^3	9.0×10^1	9	2216	TCBS	Chan et al. (1986)
Mid Bay		Seawater	8.1×10^3	3.0×10^1	0.41	2216	TCBS	Chan et al. (1986)
Aberdeen shelter		Seawater	2.0×10^5	5.5×10^3	2.7	2216	TCBS	Chan et al. (1986)
Shek O		Seawater	2.0×10^3	2.0×10^2	10	2216	TCBS	Chan et al. (1986)
Stanley Bay		Seawater	7.2×10^2	2.9×10^2	40	2216	TCBS	Chan et al. (1986)
Jordan ferry pier		Seawater	5.5×10^4	6.7×10^3	12	2216	TCBS	Chan et al. (1986)

[a] Data are the average of samples if samples are collected from several sampling points.
[b] The percent of vibrios is determined as the number of vibrios per number of the total bacteria.
[c] EA, estuarine agar.

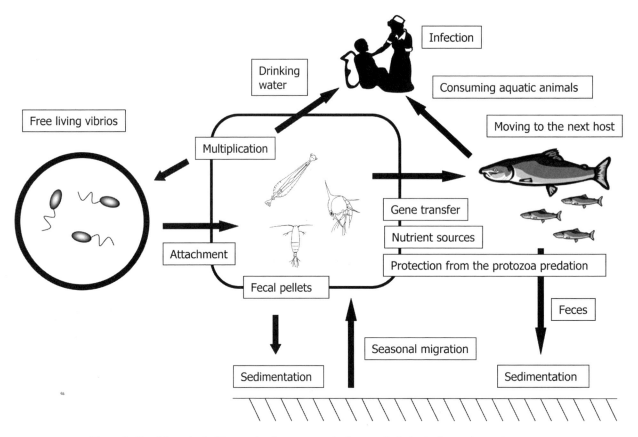

Figure 2. Possible ecological interaction between zooplankton and vibrios in the marine environment.

Some studies have focused on the relationship between water pollution and the natural microbiota, although the association between water pollution and other marine organisms, e.g., phytoplankton, has been investigated as well. For example, Simidu et al. (1977) compared the generic composition of a heterotrophic bacterial population in Tokyo Bay, which is a highly polluted and eutrophic bay, with adjacent deep oceanic bays, i.e., Sagami Bay and Suruga Bay, which are influenced by the Kuroshio current. Members of *Vibrionaceae* predominated in the natural microbiota of seawater samples from Sagami Bay, Suruga Bay, and the mouth of Tokyo Bay. However, *Vibrio* species formed only a small fraction of the bacterial population in water samples from inner Tokyo Bay. In this area, abundant growth of phytoplankton occurs throughout the year, and the *Acinetobactor–Moraxella* group predominated. Thus, the decrease of *Vibrio* species might be a good indicator of the deterioration of the ecosystem in coastal marine environments.

Zooplankton

Vibrio species show a strong association with marine plankton, particularly with zooplankton (Fig. 2)

(Kaneko and Colwell, 1975; Simidu et al., 1971; Sochard et al., 1979). In general, zooplankton harbor more vibrios than phytoplankton (Simidu et al., 1977). *Vibrio* species are the most abundant bacterial species associated with copepod (Simidu et al., 1971; Sochard et al., 1979). *Vibrio* species are also members of bacterial flora for chaetognaths (Nair et al., 1988). The seasonal association of bacteria and zooplankton has been reported previously (Kaneko and Colwell, 1973). In particular, the number of bacteria located inside copepods is generally constant, whereas external bacterial counts vary with water temperature. The colony count of *Vibrio* species sometimes reaches a maximum of 10^9 cells/g of plankton wet weight. This is significantly higher than the number detected from the surrounding water column (10^2 cells/ml). Moreover, during the summer months in Chesapeake Bay, nearly the entire total viable count of plankton consists of *Vibrio* species (Kaneko and Colwell, 1973).

Like *V. parahaemolyticus*, the strong association of *Vibrio cholerae* with zooplankton (i.e., copepods) has been reported (Huq and Colwell, 1995). Attachment of bacterial cells on the surface of copepods is site specific; a dense population of cells is usually seen on the egg case and oral region of the copepod (Huq et al., 1983). Once free-swimming cells of *V. cholerae* attach to copepods and their eggs, the bacteria rapidly increase such that there is enough for an infectious dose of *V. cholerae* (i.e., 10^4 to 10^6 cells/copepod). Therefore, infection possibly occurs via copepods during bathing, swimming, or drinking untreated raw water collected from ponds, rivers, and lakes in countries where cholera is epidemic (Huq and Colwell, 1995). It is interesting to note that that a simple filtration procedure using a sari cloth reduced the incidence of cholera cases by 48% in Bangladeshi villages (Colwell et al., 2003).

Pelagic Water

It is not difficult to culture *Vibrio* and *Photobacterium* strains from deep-sea waters. Currently, the commercially available medium, marine agar 2216E, is used preferentially by many researchers to isolate *Vibrio* species and other marine heterotrophic microorganisms. It is apparent from culturing experiments that *Vibrionaceae* representatives are distributed throughout the water column in the open ocean (Radjasa et al., 2001). However, the species composition may well change with depth (Ruby et al., 1980). It has been reported that *Vibrio* species constitute 21.3 to 39.2% of the microbial communities in the water column of Bengal Bay and the South China Sea, suggesting a wide distribution of *Vibrio* species in the tropical and subtropical open oceanic waters (Simidu et al., 1982). *Vibrio* species were particularly abundant in the water layers from 10 to 400 m, but not so in surface and depths of >400 m. In addition, Ruby et al. (1980) reported that the CFU of luminous bacteria was most abundant in two layers: between the surface and 100 m, and between 250 and 1,000 m. Water samples from depths of >4,000 m contained <0.1 CFU/100 ml. *Photobacterium phosphoreum* was found in great abundance in deep water, peaking in cell density between depths of 200 and 1,000 m.

Although *Vibrio* and *Photobacterium* species have been found from the deep sea, they are not abundant among isolates from Antarctica. Bowman et al. (1997) could not find any *Vibrio* species among many natural isolates from Antarctica, based on 16S rRNA gene sequences. The major components of the isolates identified as gram-negative facultative anaerobes were *Colwellia* species. So far, there is no evidence of *Photobacterium* species from Antarctica. It is not clear whether this phenomenon shows a natural distribution of *Vibrio* and *Photobacterium* species, or whether the absence of *Photobacterium* reflects less effort taken to recover these organisms from Antarctica.

Host-Microbe Interactions

Vibrio species occur frequently in the digestive tract and on the skin surface of marine animals. Although some *Vibrio* species are known as pathogens of marine animals in aquaculture, many researchers reveal that vibrios are members of the normal bacterial biota for aquatic animals such as flatfish (Liston, 1957), jackmackerel (*Trachurus japonicus*) (Aiso et al., 1968), salmonids (Yoshimizu and Kimura, 1976), larval and juvenile sea bream (Muroga et al., 1987), blue crabs (Davis and Sizemore, 1982), shrimps (*Penaeus merguiensis, Litopenaeus vannamei*) (Oxley et al., 2002; Vandenberghe et al., 1999), and oysters (*Crassostrea gigas*) (Olafsen et al., 1993). These findings have gradually fostered a notion that some *Vibrio* species are beneficial and potentially useful for probiotics of commercially important aquaculture animals (Verschuere et al., 2000). Thus, the study of microbial biota in healthy aquatic animals is required as background data for the development of effective and safe probiotics for commercial application in aquaculture.

The composition of the bacterial biota in the digestive tracts of marine animals clearly differs from that in the surrounding seawater (Simidu and Aiso, 1962; Muroga et al., 1987, Yoshimizu and Kimura, 1976). The availability of organic matter for vibrios in the gut of marine animals is several orders of magnitude higher than that in the surrounding seawater. However, the gut environment is represented by low pH, secretion of bile acids, and micro- or anaerobic conditions that are restrictive to the components of the microbial biota. Yet, these environments are suitable for vibrios. Yoshimizu and Kimura (1976) confirmed that *Aeromonas, Enterobacteriaceae,* and *Vibrio* isolates were resistant to low pH and the presence of bile, whereas most coryneforms—*Micrococcus, Achromobacter,* and *Flavobacterium/Cytophaga*—were sensitive.

In crustacea, the hepatopancreas, hemolymph, and digestive tracts are the normal habitats of vibrios. For example, in the case of shrimp *(L. vannamei),* a dense population of vibrios (10^4 to 10^5 CFU/g of tissue) has been detected in the hepatopancreas and the hemolymph of healthy individuals (Gomez-Gil et al., 1998). In vertebrates, only sharks provide exceptional habitats for vibrios: the tissues and organs of healthy sharks containing viable, culturable bacteria (Grimes et al., 1985, 1993).

Several *Vibrio* species are isolated frequently from the same host species, suggesting mutual relationships between host and bacteria (host–microbe interactions) (Sawabe et al., 2003). Indeed, several *Vibrio* species have established and maintained symbiotic relationships with marine animals. Of relevance, in the gut of *Haliotis* abalone, 40% of the total bacterial population were identified as *Vibrio* species on the basis of fluorescence in situ hybridization (Tanaka et al., 2004). Thus, *Vibrio halioticoli*, *Vibrio superstes*, *Vibrio gallicus*, *Vibrio neonatus*, and *Vibrio ezurae* are commonly isolated from the guts of *Haliotis* abalones. Sawabe et al. (2003) reported that *V. halioticoli* was the dominant culturable bacterium from the gut microflora of four *Haliotis* abalone species, representing 40.6 to 64.3% of the total heterotrophic bacterial community that varied in the range of 10^3 to 10^7 CFU/g of gut. The organism was also found in the gut of a turban shell (*Turbo cornutus*); however, only 16% of the population was identified as *V. halioticoli*, indicating a strong association between *V. halioticoli* and *Haliotis* abalones.

To colonize an animal host, the microorganism must overcome or adapt to the defense mechanisms of the host, specifically those that prevent the invasion and growth of unfavorable bacteria. Therefore, genetic traits controlling the animal-host colonization in pathogenic and benign species may have arisen from a common ancestral origin. In the *Vibrionaceae*, it is believed that symbiosis has evolved numerous times in the evolutionary history (Nishiguchi and Nair, 2003).

Sediment

Little is known about the distribution, abundance, survival, and ecological roles of *Vibrio* species in marine sediments. As mentioned above, *Vibrio* species are distributed widely and often predominate in the aquatic environment. However, these bacteria do not usually dominate in marine sediment. Hood and Ness (1982) used water-sediment chambers to investigate in vitro survival of *V. cholerae* and determined that the organism survived equally well in sterile water, nonsterile water, and sterile sediment but not as well in nonsterile sediment. Rapid die-off occurred after 5 h of incubation, which suggested intraspecific competitions among the microbial populations in the nonsterile sediment environment. This result indicates that the sediment environment is more competitive for *Vibrio* species to compete with other microorganisms than other habitats in the aquatic environment.

The seasonal migration of *V. parahaemolyticus* between zooplankton in summer and sediment in winter was reported previously (Kaneko and Colwell, 1973). Although *Vibrio* species constituted only a small portion of sediment bacterial flora, *V. parahaemolyticus* was detectable in marine sediment sam-

ples throughout the year. Thus, sediment could be a potential reservoir of *Vibrio* species under unfavorable conditions.

STRATEGIES TO PERSIST IN CHANGING ENVIRONMENTAL CONDITIONS AND GRAZING PRESSURES

Starvation

Since low bioavailability of nutrients is a common feature of most aquatic environments, it is reasonable to suggest that aquatic microorganisms have developed a starvation response mechanism. The ability to survive in a kind of spore-like or dormant stage is required by microbes as a strategy to persist in starved and/or changing environmental conditions. In general, under starvation or stressed conditions, the bacterium loses its flagellum and rounds up, i.e., becomes spherical/coccoid. The cells are not dead, because they continue to respire and have enzymatic activities. In addition, these viable but nonculturable cells have a slightly thickened periplasmic space and demonstrate resistance to hot and cold and to desiccation. In this dormant state they can also withstand a variety of environmental extremes. Moreover, it is firmly believed that the organism, when in the viable but nonculturable stage, can cause disease (Magariños et al., 1994; Colwell, 2000; Huq et al. 2000). Certainly, *Vibrio* species have been used repeatedly as models in starvation experiments (Rice et al., 2000).

Grazing Pressures

In natural aquatic ecosystems, the distribution and abundance of bacterial species are influenced not only by abiotic factors such as temperature, salinity, and nutrient availability, but also by biotic factors. For example, competition with other bacterial species, grazing by protozoans, and virus infection are considered the representative biotic factors controlling bacterial abundance and genetic structures. In these biotic factors, grazing of bacteria by protozoans has been studied intensively in association with the microbial loop (Sherr and Sherr, 1984). Free-living heterotrophic nanoflagellates (HNF) are major and ubiquitous bacterial grazers in various aquatic environments. They distribute approximately 10^5 to 10^6 cells/liter in seawater. The phenotypic prey selectivity of aquatic HNF has been reported to be mainly determined by size (Jurgens et al., 2000). Cells affiliated with frequently isolated genera such as *Vibrio*, *Alteromonas*, and *Pseudoalteromonas* are significantly larger than the community average. The relative abundance of these bacterial groups is negatively correlated with HNF densities under experimental conditions. Thus, it is concluded that HNF controls the number of specific bacterial populations and affects the genotypic composition of microbial communities (Beardsley et al., 2003). Based on culture-independent detection techniques, it was considered that bacteria affiliated with frequently isolated genera might be rare in coastal picoplankton because of their rapid growth response to changing environmental conditions being counterbalanced by higher grazing mortality rates. This is the reason why a free-living strategy is potentially unfavorable for *Vibrio* and other opportunistically growing *Gammaproteobacteria*, which might be preferred targets of selective predation for HNF.

Vibrio species may use marine animals as their vehicle for survival, where they may escape the grazing pressure of protozoans (Fig. 2). Certainly, many protozoans graze bacteria as a main food source. Many marine microorganisms are capable of surviving in the digestive tract of fish and marine invertebrates and are voided with excrement. Indeed, Sochard et al. (1979) emphasized that the release of copepod gut bacteria via fecal pellets provides a mechanism for the widespread distribution of these microorganisms in the water column and sediment of estuaries and the oceans.

Although some species, such as *V. gallicus* and *V. halioticoli*, are nonmotille, most of the other *Vibrio* species are motile, a phenomenon that is potentially important for chemotaxis in the natural environment. In the case of *Vibrio fischeri*, motility is important for host association. Thus, after colonization of the light organ of the squid, *V. fischeri* loses its flagellae (Millikan and Ruby, 2002). Therefore, it is apparent that *V. fischeri* can regulate gene expression and control its ability to be motile as a reflection of changes in the surrounding environment. The importance of swimming speed for the attachment to solid substrates has been clearly demonstrated in the case of *V. alginolyticus* (Kogure et al., 1998). It is apparent that surface attachment is essential for colonization of microorganisms in natural environments. As mentioned above, *Vibrio* species attach preferentially to substrates, such as marine animals and plants. *V. alginolyticus* carries chitin-binding proteins by means of which the bacteria irreversibly adhere to chitin surfaces. This may explain the ability of *V. alginolyticus* to sometimes dominate in colonizing copepod surfaces. Thus, motility and chitin-binding strategy may contribute prosperity and survival ability to *Vibrio* species in aquatic environments.

SALINITY AND TEMPERATURE AS DETERMINATIVE ENVIRONMENTAL FACTORS FOR THE DISTRIBUTION AND HABITAT SEGREGATION OF *VIBRIO* SPECIES IN THE AQUATIC ENVIRONMENT

Salinity

For the identification of *Vibrio* species, a routine test is for their Na^+ requirement. Growth in 0% (wt/vol) NaCl has been used as an important trait that differentiates *V. cholerae* and *Vibrio mimicus* from the halophilic *Vibrio* species (Farmer and Hickman-Brenner, 1992). Sodium ion is required by Na^+-proton antiports in the energy-transducing cytoplasmic membrane, and inorganic ions are required to maintain the integrity of the cell wall and membrane. The growth requirement for Na^+ is widespread among species of the *Vibrionaceae*. Thus, the distribution of *Vibrio* species in freshwater has been limited but is considered important because their presence increases the risk of encounter between humans and pathogenic vibrios. Of the *Vibrio* and *Photobacterium* species, *V. cholerae*, *V. mimicus*, *Vibrio fluvialis*, and *Vibrio furnissii* are characterized as vibrios that can grow in 0% (wt/vol) NaCl; all are clinical species (Farmer and Hickman-Brenner, 1992). Some *Vibrio* species can survive under low-salinity conditions and occur naturally in a freshwater environment (Fig. 1). However, the presence of high concentrations of organic nutrients or divalent cations can compensate to a degree for the lack of Na^+. For example, *V. cholerae* and *Vibrio anguillarum* are representative species frequently isolated from freshwater (Miyamoto and Eguchi, 1997). Indeed, pathogenic *V. cholerae* can grow in water with low salinity if the water temperature is relatively high and if organic nutrients are present in high concentrations. However, oceanic *Vibrio* isolates have narrow salinity tolerances (Simidu and Tsukamoto, 1985). These bacteria require Na^+ for both growth and starvation survival, whereas *V. anguillarum* requires Na^+ only for starvation survival (Fujiwara-Nagata and Eguchi, 2004). Interestingly, *Photobacterium* species have not been found in brackish and freshwater environments. Indeed, most *Photobacterium* isolates are only found in the sea or in seafood (Dalgaard et al., 1997).

Moderately halophilic vibrios

The family *Vibrionaceae* contains a group that includes moderately halophilic bacteria. This group has been found in salted meats, brines, saltern pond waters, and saline soils. In 1996, *Vibrio costicola*, which has been used as a model halophilic bacterium for the study of osmoregulatory and other physiological mechanisms, was reclassified as *Salinivibrio costicola*, i.e., as a third genus in the family *Vibrionaceae* (Mellado et al., 1996). Therefore, it could be argued that salinity extremes have contributed to bacterial evolution within the *Vibrionaceae*.

Temperature

In general, the ocean is the largest cold environment on Earth. It is likely that ice-cold seawater has fostered many *Vibrio* species in evolutionary history. Despite the fact that all human pathogenic vibrios are mesophilic, an ecological study of isolates belonging to the family *Vibrionaceae* from the seasonally cold coastal waters of Newfoundland indicated that most of the isolates (from cold waters) were different from previously described species (Martin-Kearley and Gow, 1994). This finding was confirmed by other researchers (Benediktsdóttir et al., 1998; Urakawa et al., 1999a,b). For example, Benediktsdóttir et al. (1998) investigated *Vibrio* isolates from salmonid fish reared at water temperatures of <10°C and determined that isolates could be distinguished biochemically and physiologically from other fish pathogens. These bacteria showed similarity with psychrophilic vibrios, although the major phena could not be equated with any recognized taxa in the *Vibrionaceae*. A difference in species composition was also found between the two different cold marine environments. Thus, a large genetic difference in species composition among vibrios was found between seasonally cold and permanently cold environments (Urakawa et al., 1999a). The current classification of the family *Vibrionaceae* precludes psychrophilic species that meet the definition of psychrophile; i.e., the optimal growth temperature is ≤15°C without any growth occurring at ≥20°C) (Morita, 1975). The one exception is *Photobacterium profundum* (Nogi et al., 1998).

Psychrophilic vibrios

P. profundum is definitely the most cold-adapted species in the family *Vibrionaceae*; it is known as a psychrophilic and piezophilic (high-pressure-loving) deep-sea bacterium. *P. profundum* strain SS9 has been intensively studied as a model of typical piezophilic bacteria with regard to the molecular mechanisms of pressure regulation (Bartlett et al., 1989; Bartlett, 1999). Previously, strain SS9 was thought to belong to the genus *Vibrio* but was reclassified as *Photobacterium* as a result of 16S rRNA gene analysis (Nogi et al., 1998). *P. profundum* is the only photobacterium to display piezophily and the only taxon known to produce the long-chain polyunsaturated fatty acid eicosapentaenoic acid. Like strain SS9, some psychrophilic and/or piezophilic isolates were tentatively identified as *Vibrio* species, including *V. psychroerythrus* and *V. marinus*. However, *V. psychroerythrus* has been reclassified as a type species of the genus *Colwellia* (Deming et al., 1988), and *V. marinus* is currently classified as *Moritella marina*, the type species of the genus *Moritella* (Urakawa et al., 1998). Hamamoto et al. (1995) reported docosahexaenoic and eicosapentaenoic acid-producing vibrios in the deep-sea environment. However, these species are likely to belong in the genera *Moritella*, *Colwellia*, *Shewanella*, and possibly *Photobacterium*, as suggested by Russell and Nichols (1999). To date, >10 piezophilic bacterial species have been described, and half are facultative anaerobes. Interestingly, to date, *Vibrio* species have not been included in the list of piezophilic bacteria. However, it is not clear if *Vibrio* species have adapted to cold environments but not to the high-pressure environment, as less effort has been made to recover piezophilic vibrios from the deep sea.

Cold adaptation in the evolutionary history of the family *Vibrionaceae*

A few *Vibrio* species, such as *V. logei*, *V. wodanis*, and *V. salmonicida*, seem to adapt to cold environments better than other species, although they are in the category of "psychrotrophs"; i.e., optimal growth temperature is >20°C, and they grow at ≤4°C. Despite the capability of some vibrios to grow at 4°C, only a few species have the inability to grow at 30°C. This trait might well be an important feature for cold adaptation, because the occurrence and abundance of cold-adapted vibrios that cannot grow at 30°C are restricted in cold marine environments (Urakawa et al., 1999a,b). On the phylogenetic tree based on the 16S rRNA gene, three facultatively psychrophilic groups have been detected among *Vibrio* and *Photobacterium* species (Table 2).

The first adaptation occurred in the *Photobacterium* lineage; i.e., *P. profundum* is a representative psychrophilic species adapted to deep-sea environments. Moreover, *P. iliopiscarium* and *P. indicum* do not grow at 30°C (Onarheim et al., 1994; Xie and Yokota, 2004). The second adaptation took place in the *V. logei* lineage. This group contains three psychrophilic species: *V. wodanis* and *V. salmonicida* were isolated from Atlantic salmon and other cold-water fish (Benediktsdóttir et al., 1998), whereas *V. logei* has been recovered from cold waters and marine animals (Baumann and Baumann, 1992).

Table 2. Psychrophilic vibrios

Lineage	Species[a]	Growth occurs between	Optimum growth temp. (°C)	Isolated from	Reference(s)
V. tapetis lineage	V. tapetis	4 and 22 but not at 30°C	18	Cultured clam	Borrego et al. (1996)
V. logei lineage	V. logei	4 and 15 but not at 30°C	15[b]	Gut of arctic scallop	Bang et al. (1978), Baumann et al. (1980)
	V. salmonicida	1 and 22 but not at 30°C	15	Diseased salmon	Egidius et al. (1986)
	V. wodanis	4 and 25 but not at 30°C	15[b]	Salmon with winter ulcer	Lunder et al. (2000)
Photobacterium lineage	P. iliopiscarium	4 and 22 but not at 30°C	18–22	Intestine of cold-water fish	Onarheim et al. (1994), Urakawa et al. (1999c)
	P. indicum	4 and 25 but not at 30°C	24[b]	Sea mud at a depth of 400 m	Johnson and Weisrock (1969), Xie and Yokota (2004)
	P. profundum	4 and 18 but not at 20°C	10	Ryukyu Trench, at a depth of 5,110 m	Nogi et al. (1998)

[a] Different strains of P. phosphoreum and V. splendidus show variability of growth at 30°C. Thus, these species are omitted from the list.
[b] These temperatures were obtained from the literature, and were used for phenotypic tests and the maintenance of cultures. Thus, they are not exact optimum temperatures.

Cold adaptation does not develop widely in the core species of the genus *Vibrio*. Numerous species, i.e., *V. lentus*, *V. fortis*, *V. hepatarius*, *V. pectenicida*, and *V. rumoiensis*, have a psychrotrophic nature, namely, that growth may occur at 4 to 30°C and possibly higher. *V. tapetis*, which is a causative agent of the brown ring disease of cultured clams, is the sole cold-adapted *Vibrio* that can grow at 4°C but not at 30°C (Borrego et al., 1996). Generally, isolates tentatively identified as *V. splendidus* demonstrate a variability to grow at 4 and 30°C. Urakawa et al. (1999b) reported that *V. splendidus*-like organisms are distributed widely on Japanese coasts and may form a major portion of the culturable vibrio population in low-temperature environments within geographically separated areas. These strains have been recovered from cold waters in Canada, Spain, and Japan (Martin-Kearley and Gow, 1994; Ortigosa et al., 1994, Urakawa et al., 1999b). In general, most isolates from cold marine environments have not been matched with recognized species. Considering the vast extent of the psychrosphere, many psychrophilic vibrios remain unclassified.

Maximum growth temperature for vibrios

In general, clinical species grow at ≥37°C; for example, *V. cholerae* grows at 42°C. *V. diabolicus* is especially heat tolerant, which is not surprising insofar as isolations were made from a deep-sea hydrothermal vent polychaete annelid, *Alvinella pompejana* (Raguénès et al., 1997). Another example is *S. costicola*, which grows at 45°C (Mellado et al., 1996). However, there is no evidence to suggest that thermophilic vibrios exist in nature.

FUNCTION AND ECOLOGICAL ROLES OF VIBRIOS IN THE AQUATIC ENVIRONMENT

Facultatively Aerobic Heterotrophs

Generally, the ecological role of *Vibrio* species is unclear. However, it is important to emphasize that they are facultative anaerobes. *Vibrionaceae* representatives have two different fermentation patterns: mixed-acid fermentation, and 2,3-butanediol fermentation. In general, these two fermentation types are distinguished by the Voges-Proskauer (VP) reaction and the methyl red (MR) test. The microorganisms that have mixed-acid fermentation patterns are positive for the MR test and negative for the VP reaction. In contrast, the microorganisms that have a butanediol fermentation pattern are negative for the MR test but positive for the VP reaction. Although these two fermentation patterns are key differential taxonomic characteristics for enteric bacteria, such as *Escherichia* and *Serratia*, *Vibrio* representatives contain both fermentation patterns. In the case of *Vibrionaceae*, these two fermentation types can be key features to distinguish some species, such as *V. cholerae* and *V. mimicus* (Farmer and Hickman-Brenner, 1992). In mixed-acid fermentation, three major acids, i.e., acetic, lactic, and succinic, are produced, whereas ethanol, CO_2, and H_2 are also formed. In butanediol fermentation, smaller amounts of acids are formed, and butanediol, ethanol, CO_2, and H_2 are the main products. *Vibrio* species are ubiquitous in marine sediments according to both culture-dependent and -independent techniques. Therefore, the bacteria must be important in the decomposition of organic matter via fermentative pathways, leading to the formation of small organic molecules, notably lactate, butyrate, propionate, acetate, formate, H_2, and CO_2. Indeed, these compounds are the main substrates for sulfate reduction and used partly for methane formation.

The Nitrogen Cycle

The primary role of heterotrophic bacteria is considered to be the decomposition and mineralization of dissolved and particulate organic nitrogen. Bacterial NO_3^- assimilation is not a pathway currently considered in pelagic carbon and nitrogen cycle models. Thus, the utilization of inorganic N by bacteria has historically received less attention. A recent review of freshwater and marine studies, however, reported that bacteria may rely on both NH_4^+ and NO_3^- for growth and biomass synthesis, and overall they may be significant consumers of inorganic N; mean consumption values of 30 and 40% have been reported for NH_4^+ and NO_3^-, respectively.

A nested PCR approach was developed to amplify the *nasA* gene, which is a part of the structural gene for nitrate assimilation, to detect the genetic potential for heterotrophic bacterial nitrate utilization in marine environments (Allen et al., 2001). Using marine isolates and DNA extracts obtained from microbial communities, it was confirmed that several groups of heterotrophic bacterial *nasA* genes, especially from the *Gammaproteobacteria*, are common and widely distributed in oceanic environments. In particular, members of *Vibrio* and *Pseudoalteromonas* accounted for 70% of the *nasA*-positive strains isolated. Therefore, *Vibrio* species are potentially important for heterotrophic bacterial nitrate utilization in the marine environment.

Vibrio diazotrophicus is a nitrogen-fixing species isolated from sea urchins (Guerinot and Patriquin, 1981a,b; Guerinot et al., 1982). The ability to fix molecular nitrogen seems to be a unique property for this vibrio. Studies with $^{15}N_2$ have shown incorporation of microbially fixed ^{15}N into sea urchin tissues, demonstrating that this microorganism provides nitrogen by N_2 fixation as the protein nutrition for sea urchins.

Chitin Degradation

Chitin, a $(1 \rightarrow 4)$-β-linked homopolymer of *N*-acetyl-D-glucosamine, is an abundant structural polysaccharide produced by various marine organisms. In particular, chitin is an important element of the exoskeletons of crustacea, such as copepods and invertebrate larvae. The first step of chitin degradation is primarily carried out by microorganisms, and this trait is widespread among many taxonomic groups of prokaryotes. The capability of chitin degradation

would seem to be an important attribute of marine microorganisms, given the presumed high input of detrital chitin into the ocean. Chitinolytic bacteria are typically detected by the production of zones of clearing on agar containing chitin, and it has been suggested that ~10% of culturable bacteria degrade chitin. *Vibrio* and *Photobacterium* species are potentially important chitinolytic bacteria in the ocean because most *Vibrio* species can degrade chitin and have strong association with chitinaceous zooplankton and crustacean (Kaneko and Colwell, 1973). However, chitin degradation is a complex process, including sensing, attaching, transport, and catabolism of the natural chitin surface in the aquatic ecosystem (Meibom et al., 2004).

Cottrell et al. (2000) reported that the chitinase gene sequences of *Vibrio* species were more similar to those of *Alphaproteobacteria* than other gammaproteobacterial chitinases, such as those in some alteromonads. The phylogeny deduced from chitinase genes only partially followed the 16S rRNA gene phylogeny. This deviation in chitinase gene and 16S rRNA gene phylogeny may be explained by lateral gene transfer.

Mucinase production

V. cholerae produces mucinase, a metalloprotease, which allows the bacteria to overcome the mucus barrier that covers the gastrointestinal epithelium (Silva et al., 2003; Colwell, 2004).

Tetrodotoxin (TTX) Production

TTX is a potent neurotoxin, which is the cause of puffer fish poisoning. However, TTX is not restricted to puffer fish, and is widely distributed among various animals inhabiting the marine and terrestrial environments (Mosher, 1964; Noguchi, 1973). The origin of TTX in marine animals has been the subject of numerous studies. Production of TTX by bacteria has been considered to support the food chain origin of TTX in puffer fish and other TTX-containing animals. The food chain origin of TTX was first hypothesized from the observation that the toxin was undetectable from cultured puffer fish, which suggested that puffer fish do not have any ability to produce TTX. These data indicate that TTX-containing animals may have accumulated the toxin and its derivatives from TTX-producing marine microorganisms (Yasumoto et al., 1986). Many bacterial taxa, including *Vibrio* spp. and in particular *V. alginolyticus* (Noguchi et al., 1987), have been associated with TTX production (Noguchi et al., 1986; Lee et al., 2000). However, in contrast to previous work, it has been demonstrated recently that cultured puffer fish do have detectable levels of TTX and that production of the toxin does indeed occur in the fish (Matsumura, 1996, 1998). The chemical structures of TTXs from puffer fish, newt, and blue-ringed octopus have been determined. However, there has not been any description of the structure of TTX derived from bacteria (Matsumura, 1995).

The ability of bacteria to produce TTXs was discussed by Matsumura in a letter to the editor of *Applied and Environmental Microbiology* (**67**:2393–2394). In it, he addressed a previous publication by Lee et al. (2000). Matsumura highlighted experimental problems with high-pressure liquid chromatography and commented that there was no structural determination of TTX in the manuscript of Lee et al. (2000). Lee et al. had demonstrated that TTX and its derivatives purified from bacterial culture cells were strongly toxic in a mouse bioassay. Indeed, one of three *Vibrio* cultures examined produced TTX and its derivatives. Therefore, it is possible that *Vibrio* species and other marine bacteria could be the major source of TTX in nature. It is relevant to heed the opinion of Lee et al. (2000), who concluded that a final resolution of this contradiction over the precise source of TTX in marine animals would be obtained only after elucidation of the genes for TTX synthesis, at the molecular level.

Degradation of Polycyclic Aromatic Hydrocarbons (PAH)

Phenanthrene, a PAH, is present in coal tar and petroleum and is a by-product of petroleum refining. Yet it is relevant to note that a marine *Vibrio* sp. may well be able to degrade PAH (Geiselbrecht et al., 1996). Thus, it is apparent that *Vibrio* species contribute to biodegradation in aquatic environments.

Impact of Vibrios on Marine Aquatic Animal Populations

Members of the *Vibrionaceae* family are pathogenic for aquatic animals. Although this trait is critical to the aquaculture industry, certain pathogenic features of vibrios may well contribute to the control of the number of aquatic animals within natural populations. For example, the blue crab, *Callinectes sapidus*, the basis of a large seafood industry, is commonly found in the coastal waters of the United States, from the Gulf of Mexico to the Atlantic Ocean. The blue crab, whether diseased or healthy, harbors bacteria in the hemolymph. In the case of blue crab collected from Galveston Bay, >75% of the animals were infected with bacteria, whereas only 12% were

bacteria-free (Davis and Sizemore, 1982). Therefore, unlike the normally sterile mammalian circulatory systems, it appears that healthy, marketable blue crabs can tolerate comparatively high numbers of bacteria, especially *Vibrio* spp., in the hemolymph of males rather than females. Thus, there is a possibility that stress, including shortage of food leading to malnutrition, sudden changes in the water temperature, spawning, or injury could lead to the onset of disease, notably septicemia. Also, *Vibrio* species have been found in the hemolymph of healthy shrimps (Gomez-Gil et al., 1998). Therefore, *Vibrio* species may well contribute to mortality among populations of animals in marine ecosystems.

THE ECOLOGICAL RELATIONSHIP BETWEEN CLIMATE AND THE EPIDEMIC PATTERNS OF *VIBRIO* SPECIES, AND THE USE OF REMOTE SENSING TO PREDICT BLOOMS OF *VIBRIO* SPECIES IN THE MARINE ENVIRONMENT

In the last few decades, there have been numerous reports describing an increase in emerging diseases of marine animals and plants as a function of climate and human activities (Harvell et al., 1999). Recently, the ecological relationship between climate and the epidemic patterns of *Vibrio* species has been disputed. However, it has been argued that climate change affects the health and productivity of marine ecosystems. For example, coral bleaching is induced by a variety of environmental stimuli, including increased seawater temperature. Large-scale bleaching events have been linked to global warming (Ben-Haim et al., 2003b). Yet two *Vibrio* species, *V. shiloi* and *V. coralliilyticus* (pathogenic for *Pocillopora damicornis*), are recognized to cause coral disease (Ben-Haim et al., 2003a; Kushmaro et al., 1996, 1997). For example, it was reported that coral bleaching of the Mediterranean coral *Oculina patagonica* is triggered by an infection with *V. shiloi*, with water temperature as a contributing factor (Kushmaro et al., 1997). In particular, it has been suggested that an increase in seawater temperature may well influence the outcome of bacterial infection by lowering the resistance of the coral to disease and/or increasing the virulence of the bacterium. It is apparent that further detailed studies are needed to investigate the ecological relationship between climate and the distribution and epidemic patterns of *Vibrio* species.

The ecology of *V. cholerae* and the epidemiology of cholera give some insights into the disease. Cholera was proposed as a model disease in relation to its interaction with climate (Colwell, 1996; Lipp et al., 2002). Cholera is an endemic disease, especially in India, Bangladesh, and Africa, and there is a well-recognized seasonal variation in India and Bengal (Pollitzer, 1959). However, the risk of cholera outbreaks depends mainly on public health systems and sanitary conditions of the region, mainly the consumption of untreated water.

The ecological relationship between climate and the epidemic patterns of *V. cholerae* has been intensively studied in Southeast Asia and Latin America. For example, there is a relationship between sea surface temperature and cholera cases in Bangladesh (Colwell, 1996). The incidence of the disease along the Peruvian coast was correlated to elevated sea surface temperature from October 1997 to June 2000, which coincides with the 1997 to 1998 El Niño event (Colwell and Huq, 1999; Gil et al., 2004). To reiterate, the association of *V. cholerae* with chitinaceous zooplankton, notably copepods, has been reported previously (Huq et al., 1983). This association provides evidence for the environmental source of cholera, as well as reasons to explain why cholera epidemics are sporadic and erratic. It is relevant to mention a study that modeled the occurrence of *V. cholerae* in time and space. Thus, a variation in populations was recorded over a 3-year period in Chesapeake Bay (Louis et al., 2003). The sporadic and erratic nature of cholera epidemics can now be related to regional climate patterns and global climate events, such as El Niño, insofar as zooplankton have been found to harbor the pathogen, and zooplankton blooms quickly exceed those of phytoplankton. Remote sensing has been employed successfully to determine the relationship of cases of cholera with chlorophyll levels, sea surface temperature, sea surface height, and turbidity (Lobitz et al., 2000). The phytoplankton concentration may be inferred from chlorophyll concentrations estimated from ocean color imagery, and the height of the ocean measures the inland movement of *Vibrio*-contaminated tidal water.

Cholera has been identified as a candidate disease for early warning of evidence of climate sensitivity and interannual variability. The key variables include zooplankton abundance, sea surface temperature, the El Niño-Southern Oscillation, human factors, and socioeconomic variables (World Health Organization, 2004).

In general, the availability of environmental parameters, such as water temperature, nutrient, and chlorophyll concentrations, is limited, although the use of these data is straightforward in the monitoring of pathogens in aquatic environments. By remote sensing, the sea surface temperature, turbidity, chlorophyll, and sea surface height can be monitored; it has been possible to determine which environmental parameters are strongly linked with epidemics. In the near future, the use of remote sensing might be incorporated into monitoring programs for fish and shellfish diseases.

REFERENCES

Aiso, K., U. Simidu, and K. Hasuo. 1968. Microflora in the digestive tract of inshore fish in Japan. *J. Gen. Microbiol.* **52:**361–364.

Allen, A. E., M. G. Booth, M. E. Frischer, P. G. Verity, J. P. Zehr, and S. Zani. 2001. Diversity and detection of nitrate assimilation genes in marine bacteria. *Appl. Environ. Microbiol.* **67:**5343–5348.

Austin, B., S. Garges, B. Conrad, E. E. Harding, R. R. Colwell, U. Simidu, and N. Taga. 1979. Comparative study of the aerobic, heterotrophic bacterial flora of Chesapeake Bay and Tokyo Bay. *Appl. Environ. Microbiol.* **37:**704–714.

Bang, S. S., P. Baumann, and K. H. Nealson. 1978. Phenotypic characterization of *Photobacterium logei* (sp. nov.), a species related to *P. fischeri. Curr. Microbiol.* **1:**285–288.

Bartlett, D., M. Wright, A. A. Yayanos, and M. Silverman. 1989. Isolation of a gene regulated by hydrostatic pressure in a deep-sea bacterium. *Nature* **342:**572–574.

Bartlett, D. H. 1999. Microbial adaptations to the psychrosphere/piezosphere. *J. Mol. Microbiol. Biotechnol.* **1:**93–100.

Baumann, P., L. Baumann, S. S. Bang, and M. J. Woolkalis. 1980. Reevaluation of the taxonomy of *Vibrio, Beneckea,* and *Photobacterium*: abolition of the genus *Beneckea. Curr. Microbiol.* **4:**127–132.

Baumann, P., and L. Baumann. 1992. The marine gram-negative eubacteria: genera *Photobacterium, Beneckea, Alteromonas, Pseudomonas,* and *Alcaligenes,* p. 1302–1331. *In* A. Balows, H. G. Trüper, M. Dworkin, W. Harder, and K.-H. Schleifer (ed.), *The Prokaryotes, a Handbook on the Biology of Bacteria: Ecophysiology, Isolation, Identification, Applications,* 2nd ed., vol. III. Springer-Verlag, New York, N.Y.

Beardsley, C., J. Pernthaler, W. Wosniok, and R. Amann. 2003. Are readily culturable bacteria in coastal North Sea waters suppressed by selective grazing mortality? *Appl. Environ. Microbiol.* **69:**2624–2630.

Benediktsdóttir, E., S. Helgason, and H. Sigurjonsdottir. 1998. *Vibrio* spp. isolated from salmonids with shallow skin lesions and reared at low temperature. *J. Fish Dis.* **21:**19–28.

Ben-Haim, Y., F. L. Thompson, C. C. Thompson, M. C. Cnockaert, B. Hoste, J. Swings, and E. Rosenberg. 2003a. *Vibrio coralliilyticus* sp. nov., a temperature-dependent pathogen of the coral *Pocillopora damicornis. Int. J. Syst. Evol. Microbiol.* **53:**309–315.

Ben-Haim, Y., M. Zicherman-Keren, and E. Rosenberg. 2003b. Temperature-regulated bleaching and lysis of the coral *Pocillopora damicornis* by the novel pathogen *Vibrio coralliilyticus. Appl. Environ. Microbiol.* **69:**4236–4242.

Borrego, J. J., D. Castro, A. Luque, C. Paillard, P. Maes, M. Garcia, and A. Ventosa. 1996. *Vibrio tapetis* sp. nov., the causative agent of the brown ring disease affecting cultured clams. *Int. J. Syst. Bacteriol.* **46:**480–484.

Bowman, J. H., S. A. McCammon, M. V. Brown, D. S. Nichols, and T. A. McMeekin. 1997. Diversity and association of psychrophilic bacteria in Antarctic sea ice. *Appl. Environ. Microbiol.* **63:**3068–3078.

Chan, K.-Y., M. L. Woo, K. W. Lo, and G. L. French. 1986. Occurrence and distribution of halophilic vibrios in subtropical coastal waters of Hong Kong. *Appl. Environ. Microbiol.* **52:**1407–1411.

Colwell, R. R. 1996. Global climate and infectious disease: the cholera paradigm. *Science* 274:2025–2031.

Colwell, R. R. 2000. Viable but nonculturable bacteria: a survival strategy. *J. Infect. Chemother.* 6:121–125.

Colwell, R. R. 2004. Infectious disease and environment: cholera as a paradigm for waterborne disease. *Perspectives* 7:285–289.

Colwell, R. R., and A. Huq. 1999. Global microbial ecology: biogeography and diversity of *Vibrios* as a model. *J. Appl. Microbiol.* 85:134S–137S.

Colwell, R. R., A. Huq, M. S. Islam, K. M. A. Ariz, M. Yunus, N. H. Khan, A. Mahmud, R. B. Sack, G. B. Nair, J. Chakraborty, D. A. Sack, and E. Russek-Cohen. 2003. Reduction of cholera in Bangladeshi villages by simple filtration. *Proc. Natl. Acad. Sci. USA* 100:1051–1055.

Cottrell, M. T., D. N. Wood, L. Yu, and D. L. Kirchman. 2000. Selected chitinase genes in cultured and uncultured marine bacteria in the alpha- and gamma-subclasses of the Proteobacteria. *Appl. Environ. Microbiol.* 66:1195–1201.

Dalgaard, P., O. Mejholm, T. J. Christiansen, and H. H. Huss. 1997. Importance of *Photobacterium phosphoreum* in relation to spoilage of modified atmosphere-packed fish products. *Lett. Appl. Microbiol.* 24:373–378.

Davis, J. W., and R. K. Sizemore. 1982. Incidence of *Vibrio* species associated with blue crabs *(Callinectes sapidus)* collected from Galveston Bay, Texas. *Appl. Environ. Microbiol.* 43:1092–1097.

Deming, J. W., L. K. Somers, W. L. Straube, D. G. Swartz, and M. T. MacDonell. 1988. Isolation of an obligately barophilic bacterium and description of a new genus, *Colwellia* gen. nov. *Syst. Appl. Microbiol.* 10:152–160.

Egidius, E., R. Wiik, K. A. Hoff, and B. Hjeltnes. 1986. *Vibrio salmonicida* sp. nov., a new fish pathogen. *Int. J. Syst. Bacteriol.* 36:518–520.

Farmer, J. J., III, and F. W. Hickman-Brenner. 1992. The genera *Vibrio* and *Photobacterium*, p. 2952–3011. *In* A. Balows, H. G. Trüper, M. Dworkin, W. Harder, and K.-H. Schleifer (ed.), *The Prokaryotes, a Handbook on the Biology of Bacteria: Ecophysiology, Isolation, Identification, Applications*, 2nd ed., vol. III. Springer-Verlag, New York, N.Y.

Fujiwara-Nagata, E., and M. Eguchi. 2004. Significance of Na$^+$ in the fish pathogen, *Vibrio anguillarum*, under energy depleted condition. *FEMS Microbiol. Lett.* 234:163–167.

Geiselbrecht, A. D., R. P. Herwig, J. W. Deming, and J. T. Staley. 1996. Enumeration and phylogenetic analysis of polycyclic aromatic hydrocarbon-degrading marine bacteria from Puget Sound sediments. *Appl. Environ. Microbiol.* 62:3344–3349.

Gil, A. I., V. R. Louis, I. N. G. Rivera, E. Lipp, A. Huq, C. F. Lanata, D. N. Taylor, E. Russek-Cohen, N. Choopun, R. B. Sack, and R. R. Colwell. 2004. Occurrence and distribution of *Vibrio cholerae* in the coastal environment of Peru. *Environ. Microbiol.* 6:699–706.

Gomez-Gil, B., L. Tron-Mayen, A. Rogue, J. F. Turnbull, V. Inglis, and A. L. Guerra-Flores. 1998. Species of *Vibrio* isolated from hepatopancreas, haemolymph and digestive tract of a population of healthy juvenile *Penaeus vannamei. Aquaculture* 163:1–9.

Grimes, D. J., P. Brayton, R. R. Colwell, and S. H. Gruber. 1985. Vibrios as autochthonous flora of neritic sharks. *Syst. Appl. Microbiol.* 6:221–226.

Grimes, D. J., D. Jacobs, D. G. Swartz, P. R. Brayton, and R. R. Colwell. 1993. Numerical taxonomy of gram-negative, oxidase-positive rods from carcharhinid sharks. *Int. J. Syst. Bacteriol.* 43:88–98.

Guerinot, M. L., and D. G. Patriquin. 1981a. N$_2$-fixing vibrios isolated from the gastrointestinal-tract of sea urchins. *Can. J. Microbiol.* 27:311–317.

Guerinot, M. L., and D. G. Patriquin. 1981b. The association of N$_2$-fixing bacteria with sea urchins. *Mar. Biol.* 62:197–207.

Guerinot, M. L., P. A. West, J. V. Lee, and R. R. Colwell. 1982. *Vibrio diazotrophicus* sp. nov, a marine nitrogen-fixing bacterium. *Int. J. Syst. Bacteriol.* 32:350–357.

Hamamoto, T., N. Takata, T. Kudo, and K. Horikoshi. 1995. Characteristic presence of polyunsaturated fatty-acids in marine psychrophilic vibrios. *FEMS Microbiol. Lett.* 129:51–56.

Harvell, C. D., K. Kim, J. M. Burkholder, R. R. Colwell, P. R. Epstein, D. J. Grimes, E. E. Hofmann, E. K. Lipp, A. Osterhaus, R. M. Overstreet, J. W. Porter, G. W. Smith, and G. R. Vasta. 1999. Emerging marine diseases: climate links and anthropogenic factors. *Science* 285:1505–1510.

Hood, M. A., and G. E. Ness. 1982. Survival of *Vibrio cholerae* and *Escherichia coli* in estuarine waters and sediments. *Appl. Environ. Microbiol.* 43:578–584.

Huq, A., and R. R. Colwell. 1995. Vibrios in the marine and estuarine environments. *J. Mar. Biotechnol.* 3:60–63.

Huq, A., I. N. G. Rivera, and R. R. Colwell. 2000. Epidemiological significance of viable but nonculturable microorganisms, p. 301–323. *In* R. R. Colwell and D. J. Grimes (ed.), *Nonculturable Microorganisms in the Environment.* ASM Press, Washington, D.C.

Huq, A., E. B. Small, P. A. West, M. I. Huq, R. Rahman, and R. R. Colwell. 1983. Ecological relationships between *Vibrio cholerae* and planktonic crustacean copepods. *Appl. Environ. Microbiol.* 45:275–283.

Johnson, R. M., and W. P. Weisrock. 1969. *Hyphomicrobium indicum* sp. nov. *Hyphomicrobiaceae* Douglas. *Int. J. Syst. Bacteriol.* 19:295--307.

Jurgens, K., J. M. Gasol, and D. Vaque. 2000. Bacteria-flagellate coupling in microcosm experiments in the Central Atlantic Ocean. *J. Exp. Mar. Biol. Ecol.* 245:127–147.

Kaneko, T., and R. R. Colwell. 1973. Ecology of *Vibrio parahaemolyticus* in Chesapeake Bay. *J. Bacteriol.* 113:24–32.

Kaneko, T., and R. R. Colwell. 1975. Adsorption of *Vibrio parahaemolyticus* onto chitin and copepods. *Appl. Microbiol.* 29:269–274.

Kogure, K., E. Ikemoto, and H. Morisaki. 1998. Attachment of *Vibrio alginolyticus* to glass surfaces is dependent on swimming speed. *J. Bacteriol.* 180:932–937.

Kushmaro, A., Y. Loya, M. Fine, and E. Rosenberg. 1996. Bacterial infection and coral bleaching. *Nature* 380:396.

Kushmaro, A., E. Rosenberg, M. Fine, and Y. Loya. 1997. Bleaching of the coral *Oculina patagonica* by *Vibrio* AK-1. *Mar. Ecol. Prog. Ser.* 147:159–165.

Lee, M. J., D. Y. Jeong, W. S. Kim, H. D. Kim, C. H. Kim, W. W. Park, Y. H. Park, K. S. Kim, H. M. Kim, and D. S. Kim. 2000. A tetrodotoxin-producing *Vibrio* strain, LM-1, from the puffer fish *Fugu vermicularis radiatus. Appl. Environ. Microbiol.* 66:1698–1701.

Lipp, E. K., A. Huq, and R. R. Colwell. 2002. Effects of global climate on infectious disease: the cholera model. *Clin. Microbiol. Rev.* 15:757–770.

Liston, J. 1957. The occurrence and distribution of bacterial types on flatfish. *J. Gen. Microbiol.* 16:205–216.

Lobitz, B., L. Beck, A. Huq, B. Wood, G. Fuchs, A. S. G. Faruque, and R. Colwell. 2000. Climate and infectious disease: use of remote sensing for detection of *Vibrio cholerae* by indirect measurement. *Proc. Natl. Acad. Sci. USA* 97:1438–1443.

Louis, V. R., E. Russek-Cohen, N. Choopun, I. N. G. Rivera, B. Gangle, S. C. Jiang, A. Rubin, J. A. Patz, A. Huq, and R. R. Colwell. 2003. Predictability of *Vibrio cholerae* in Chesapeake Bay. *Appl. Environ. Microbiol.* 69:2773–2785.

Lunder, T., H. Sorum, G. Holstad, A. G. Steigerwalt, P. Mowinckel, and D. J. Brenner. 2000. Phenotypic and genotypic characterization of *Vibrio viscosus* sp. nov. and *Vibrio wodanis* sp. nov. isolated from Atlantic salmon (*Salmo salar*) with "winter ulcer." *Int. J. Syst. Evol. Microbiol.* 50:427–450.

Maeda, T., Y. Matsuo, M. Furushita, and T. Shiba. 2003. Seasonal dynamics in a coastal *Vibrio* community examined by a rapid clustering method based on 16S rDNA. *Fish. Sci.* **69**:385–394.

Magariños, B., J. L. Romalde, J. L. Barja, and A. E. Toranzo. 1994. Evidence of a dormant but infective state of the fish pathogen *Pasteurella piscicida* in seawater and sediment. *Appl. Environ. Microbiol.* **60**:180–186.

Martin-Kearley, J., and J. A. Gow. 1994. Numerical taxonomy of *Vibrionaceae* from Newfoundland coastal waters. *Can. J. Microbiol.* **40**:355–361.

Matsumura, K. 1995. Reexamination of tetrodotoxin production by bacteria. *Appl. Environ. Microbiol.* **61**:3468–3470.

Matsumura, K. 1996. Tetrodotoxin concentrations in cultured puffer fish, *Fugu rubripes. J. Agric. Food Chem.* **44**:1–2.

Matsumura, K. 1998. Production of tetrodotoxin in puffer fish embryos. *Environ. Toxicol. Pharmacol.* **6**:217–219.

Meibom, K. L., X. B. Li, A. T. Nielsen, C.-Y. Wu, S. Roseman, and G. Sckoolnik. 2004. The *Vibrio cholerae* chitin utilization program. *Proc. Natl. Acad. Sci. USA* **101**:2524–2539.

Mellado, E., E. R. B. Moore, J. J. Nieto, and A. Ventosa. 1996. Analysis of 16S rRNA gene sequences of *Vibrio costicola* strains: description of *Salinivibrio costicola* gen. nov., comb. nov. *Int. J. Syst. Bacteriol.* **46**:817–821.

Millikan, D. S., and E. G. Ruby. 2002. Alterations in *Vibrio fischeri* motility correlate with a delay in symbiosis initiation and are associated with additional symbiotic colonization defects. *Appl. Environ. Microbiol.* **68**:2519–2528.

Miyamoto, N., and M. Eguchi. 1997. Direct detection of a fish pathogen, *Vibrio anguillarum* serotype J-0-1, in freshwater by fluorescent antibody technique. *Fish. Sci.* **63**:253–257.

Morita, R. Y. 1975. Psychrophilic bacteria. *Bacteriol. Rev.* **39**:144–167.

Mosher, H. 1964. Tarichatoxin-tetrodotoxin: a potent neurotoxin. *Science* **144**:1100–1110.

Muroga, K., M. Higashi, and H. Keitoku. 1987. The isolation of intestinal microflora of farmed red seabream (*Pagrus major*) and black seabream (*Acanthopagrus schlegeli*) at larval and juvenile stages. *Aquaculture* **65**:79–88.

Nair, S., K. Kita-Tsukamoto, and U. Simidu. 1988. Bacterial flora of healthy and abnormal chaetognaths. *Nipp. Suis. Gakk.* **54**:491–496.

Nishiguchi, M. K., and V. S. Nair. 2003. Evolution of symbiosis in the *Vibrionaceae*: a combined approach using molecules and physiology. *Int. J. Syst. Evol. Microbiol.* **53**:2019–2026.

Noguchi, T. 1973. Isolation of tetrodotoxin from a goby *Gobius criniger. Toxin* **11**:305–310.

Noguchi, T., J. K. Jeon, O. Arakawa, H. Sugita, Y. Deguchi, Y. Shida, and K. Hashimoto. 1986. Occurrence of tetrodotoxin and anhydrotetrodotoxin in *Vibrio* sp. isolated from the intestines of a xanthid crab, *Atergatis floridus. J. Biochem.* (Tokyo) **99**:311–314.

Noguchi, T., D. F. Hwang, O. Arakawa, H. Sugita, Y. Deguchi, Y. Shida, and K. Hashimoto. 1987. *Vibrio alginolyticus*, a tetrodotoxin-producing bacterium, in the intestines of the fish *Fugu vermicularis vermicularis. Mar. Biol.* **94**:625–630.

Nogi, Y., N. Masui, and C. Kato. 1998. *Photobacterium profundum* sp. nov., a new, moderately barophilic bacterial species isolated from a deep-sea sediment. *Extremophiles* **2**:1–7.

Olafsen, J. A., H. V. Mikkelsen, H. M. Glever, and G. H. Hansen. 1993. Indigenous bacteria in hemolymph and tissues of marine bivalves at low temperatures. *Appl. Environ. Microbiol.* **59**:1848–1854.

Oliver, J. D., R. A. Warner, and D. R. Cleland. 1982. Distribution and ecology of *Vibrio vulnificus* and other lactose-fermenting marine vibrios in coastal waters of the southeastern United States. *Appl. Environ. Microbiol.* **44**:1404–1414.

Oliver, J. D., R. A. Warner, and D. R. Cleland. 1983. Distribution of *Vibrio vulnificus* and other lactose-fermenting vibrios in the marine environment. *Appl. Environ. Microbiol.* **45**:985–998.

Onarheim, A. M., R. Wiik, J. Burghardt, and E. Stackebrandt. 1994. Characterization and identification of two *Vibrio* species indigenous to the intestine of fish in cold sea water; description of *Vibrio iliopiscarius* sp. nov. *Syst. Appl. Microbiol.* **17**:370–379.

Ortigosa, M., E. Garay, and M. Pujalte. 1994. Numerical taxonomy of *Vibrionaceae* isolated from oysters and seawater along an annual cycle. *Syst. Appl. Microbiol.* **17**:216–225.

Oxley, A. P. A., W. Shipton, L. Owens, and D. McKay. 2002. Bacterial flora from the gut of the wild and cultured banana prawn, *Penaeus merguiensis. J. Appl. Microbiol.* **93**:214–223.

Pollitzer, R. 1959. *Cholera*, p. 51–96. World Health Organization monograph no. 43, World Health Organization, Geneva, Switzerland.

Radjasa, O. K., H. Urakawa, K. Kita-Tsukamoto, and K. Ohwada. 2001. Characterization of psychrotrophic bacteria in the surface and deep-sea waters from the north-western Pacific Ocean based on 16S rDNA analysis. *Mar. Biotech.* **2**:454–462.

Raguénès, G., R. Christen, J. Guezennec, P. Pignet, and G. Barbier. 1997. *Vibrio diabolicus* sp. nov., a new polysaccharide-secreting organism isolated from a deep-sea hydrothermal vent polychaete annelid, *Alvinella pompejana. Int. J. Syst. Bacteriol.* **47**:989–995.

Rice, S. A., D. McDougald, and S. Kjelleberg. 2000. *Vibrio vulnificus*: a physiological and genetic approach to the viable but nonculturable response. *J. Infect. Chemother.* **6**:115–120.

Ruby, E. G., E. P. Greenberg, and J. W. Hastings. 1980. Planktonic marine luminous bacteria: species distribution in the water column. *Appl. Environ. Microbiol.* **39**:302–306.

Russell, N. J., and D. S. Nichols. 1999. Polyunsaturated fatty acids in marine bacteria: a dogma rewritten. *Microbiology* **145**:767–779.

Sawabe, T., N. Setoguchi, S. Inoue, R. Tanaka, M. Ootsubo, M. Yoshimizu, and Y. Ezura. 2003. Acetic acid production of *Vibrio halioticoli* from alginate: a possible role for establishment of abalone V. *halioticoli* association. *Aquaculture* **219**:671–679.

Sherr, E. B., and B. F. Sherr. 1984. Role of heterotrophic protozoa in carbon and energy flow in aquatic ecosystems, p. 412–423. *In* M. J. Klug and C. A. Reddy (ed.), *Current Perspectives in Microbial Ecology*. American Society for Microbiology, Washington, D.C.

Silva, A. J., K. Pham, and J. A. Benitez. 2003. Haemaglutinin/protease expression and mucin gel penetration in El Tor biotype *Vibrio cholerae. Microbiology* **149**:1883–1891.

Simidu, U., and K. Aiso. 1962. Occurrence and distribution of heterotrophic bacteria in sea-water from the Kamogawa Bay. *Bull. Jpn. Soc. Sci. Fish.* **28**:1133–1144.

Simidu, U., K. Ashino, and E. Kaneko. 1971. Bacterial flora of phyto- and zoo-plankton in the inshore water of Japan. *Can. J. Microbiol.* **17**:1157–1160.

Simidu, U., E. Kaneko, and N. Taga. 1977. Microbial studies of Tokyo Bay. *Microb. Ecol.* **3**:173–191.

Simidu, U., and K. Tsukamoto. 1980. A method of the selective isolation and enumeration of marine *Vibrionaceae. Microb. Ecol.* **6**:181–184.

Simidu, U., and K. Tsukamoto. 1985. Habitat segregation and biochemical activities of marine members of the family *Vibrionaceae. Appl. Environ. Microbiol.* **50**:781–790.

Simidu, U., K. Tsukamoto, and Y. Akagi. 1982. Heterotrophic bacterial population in Bengal Bay and the South China Sea. *Jpn. Soc. Sci. Fish.* **48**:425–431.

Sochard, M. R., D. F. Wilson, B. Austin, and R. R. Colwell. 1979. Bacteria associated with the surface and gut of marine copepods. *Appl. Environ. Microbiol.* **37:**750–759.

Tanaka, R., M. Ootsubo, T. Sawabe, Y. Ezura, and K. Tajima. 2004. Biodiversity and *in situ* hybridization of gut microflora of abalone (*Haliotis discus hannai*) determined by culture-independent techniques. *Aquaculture* **241:**453–463.

Urakawa, H., K. Kita-Tsukamoto, S. E. Steven, K. Ohwada, and R. R. Colwell. 1998. A proposal to transfer *Vibrio marinus* (Russell 1891) to a new genus *Moritella* gen. nov. as *Moritella marina* comb. nov. *FEMS Microbiol. Lett.* **165:**373–378.

Urakawa, H., K. Kita-Tsukamoto, and K. Ohwada. 1999a. 16S rDNA restriction fragment length polymorphism analysis of psychrotrophic vibrios from Japanese coastal water. *Can. J. Microbiol.* **45:**1001–1007.

Urakawa, H., K. Kita-Tsukamoto, and K. Ohwada. 1999b. Restriction fragment length polymorphism analysis of psychrophilic and psychrotrophic *Vibrio* and *Photobacterium* from the northwestern Pacific Ocean and Otsuchi Bay, Japan. *Can. J. Microbiol.* **45:**67–76.

Urakawa, H., K. Kita-Tsukamoto, and K. Ohwada. 1999c. Reassessment of the taxonomic position of *Vibrio iliopiscarius* (Onarheim et al. 1994) and proposal for *Photobacterium iliopiscarium* comb. nov. *Int. J. Syst. Bacteriol.* **49:**257–260.

Vandenberghe, J., L. Verdonck, R. Robles-Arozarena, G. Rivera, A. Bolland, M. Balladares, B. Gomez-Gil, J. Calderon, P. Sorgeloos, and J. Swings. 1999. Vibrios associated with *Litopernaeus vannamei* larvae, postlarvae, broodstock, and hatchery probionts. *Appl. Environ. Microbiol.* **65:**2592–2597.

Verschuere, L., G. Rombaut, P. Sorgeloos, and W. Verstraete. 2000. Probiotic bacteria as biological control agents in aquaculture. *Microbiol. Mol. Biol. Rev.* **64:**655–671.

World Health Organization. 2004. *Using Climate to Predict Infectious Disease Outbreaks: a Review*. WHO/SDE/OEH/04.01. World Health Organization, Geneva, Switzerland.

Wright, A. C., R. T. Hill, J. A. Johnson, M. C. Roghman, R. R. Colwell, and J. G. Morris, Jr. 1996. Distribution of *Vibrio vulnificus* in the Chesapeake Bay. *Appl. Environ. Microbiol.* **62:**717–724.

Xie, C. H., and A. Yokota. 2004. Transfer of *Hyphomicrobium indicum* to the genus *Photobacterium* as *Photobacterium indicum* comb. nov. *Int. J. Syst. Evol. Microbiol.* **54:**2113–2116.

Yasumoto, T., D. Yasumura, M. Yotsu, T. Michishita, A. Endo, and Y. Kotaki. 1986. Bacterial production of tetrodotoxin and anhydrotetrodotoxin. *Agric. Biol. Chem.* **50:**793–795.

Yoshimizu, M., and T. Kimura. 1976. Study on the intestinal microflora of salmonids. *Fish Pathol.* **10:**243–259.

The Biology of Vibrios
Edited by F. L. Thompson et al.
© 2006 ASM Press, Washington, D.C.

Chapter 13

Dynamics of *Vibrio* Populations and Their Role in Environmental Nutrient Cycling

Janelle R. Thompson and Martin F. Polz

INTRODUCTION

The bacterial *Vibrionaceae* family encompasses a diverse group of heterotrophic marine bacteria, collectively referred to as vibrios. These include human pathogens (e.g., *Vibrio cholerae*, *Vibrio parahaemolyticus*, and *Vibrio vulnificus*) and benign planktonic and animal-associated organisms of the genera *Vibrio*, *Salinivibrio*, *Photobacterium*, and *Enterovibrio*. All surveys have confirmed the ubiquity of vibrios, but have, with the exception of one study (Rehnstam et al., 1993), also suggested that these populations are generally <1% of total bacterioplankton. This finding is in contrast to culture-based studies, in which vibrios typically comprise >10% of the easily culturable marine bacteria (Eilers et al., 2000a,b). Rapid growth under nutrient enrichment and the ability to consume a wide array of carbon substrates (Farmer and Hickman-Brenner, 2001) indicate that the biogeochemical significance of vibrios may vary with the nutrient status of the environment.

DYNAMICS AND DISTRIBUTION

Members of the *Vibrionaceae* are ubiquitous in the marine environment and have been found in coastal and open ocean environments, surface and deep waters, as free-living populations, and in association with marine animals, algae, and detritus. Although a number of *Vibrio* species have been described in the context of their association with marine animals (e.g., as pathogens), the extent to which such vibrios are also (active) components of the bacterioplankton in many cases remains to be determined (e.g., Lunder et al., 2000; Gomez-Gil et al., 2003; Faury et al., 2004). The distribution and dynamics of

planktonic *Vibrionaceae* populations are influenced by their occurrence along environmental gradients (e.g., temperature and salinity) as well as ecosystem-level interactions including resource availability (nutrients), predation by protozoans and viruses, or the abundance of host organisms.

Coastal Waters and Open Oceans

Vibrio populations in coastal systems have been studied extensively because of the significance of these environments as a resource for fishing, shellfish harvesting, recreation, and transportation. *Vibrio* strains characterized by similar or identical 16S rRNA types (ribotypes) have been obtained from geographically distant environments, suggesting a cosmopolitan coastal flora (e.g., Barbieri et al., 1999; Urakawa et al., 1999a). The coastal abundance of vibrios has been reported as 10^2 to 10^5 cells/ml (Eilers et al., 2000b; Heidelberg et al., 2002a; J. R. Thompson et al., 2004, 2005b), and the distribution of *Vibrio* populations has been correlated to environmental factors including salinity (Kaspar and Tamplin, 1993; Motes et al., 1998; Randa et al., 2004), temperature (Kaneko and Colwell, 1973, 1978; Wright et al., 1996; Nishiguchi et al., 1998; Jiang and Fu, 2001; Randa et al., 2004; J. R. Thompson et al., 2004), and, in some cases, the abundance of host organisms (Lee and Ruby, 1994).

While an extensive body of literature exists on the genetics and ecology of some *Vibrio* species, the diversity and dynamics of co-occurring *Vibrio* populations have been addressed to a more limited extent (e.g., Rehnstam et al., 1993; Caldini et al., 1997; Barbieri et al., 1999; Heidelberg et al., 2002a,b; Maeda et al., 2003; J. R. Thompson et al., 2004, 2005b), and their quantitative dynamics have been addressed even

Janelle R. Thompson and Martin F. Polz • Department of Civil and Environmental Engineering, Massachusetts Institute of Technology, 77 Massachusetts Avenue, Cambridge, MA 02139.

more rarely (Rehnstam et al., 1993; Heidelberg et al., 2002a,b; J. R. Thompson et al., 2004, 2005b). Most studies have focused on temperate environments, where both strains with mesophilic and psychrophilic growth optima can occur, and mesophilic strains are isolated at water temperatures above ~20°C (Maeda et al., 2003). This was recently confirmed by analysis of *Vibrio* sequence diversity by 16S rRNA gene-targeted quantitative PCR and cloning, which revealed a similar seasonal shift in community structure toward mesophilic populations in a temperate North Atlantic bay (Barnegat Bay, N.J.) (J. R. Thompson et al., 2004). It has been suggested that the seasonal occurrence of mesophilic *Vibrio* species, such as *V. parahaemolyticus* and *V. coralliilyticus*, may reflect "overwintering" within sediments or in association with marine fauna (e.g., Kaneko and Colwell, 1973; Ben-Haim et al., 2003) and, indeed, many associations of vibrios with sediments and zooplankton have been observed (e.g., Kaneko and Colwell, 1973, 1978; Urakawa et al., 2000; Heidelberg et al., 2002b).

Vibrios are also readily isolated from open ocean environments (e.g., Orndorff and Colwell, 1980; Simidu and Tsukamoto, 1985; Urakawa et al., 1999a; Radjasa et al., 2001), but it remains unknown whether open ocean and coastal *Vibrio* populations are distinct. Molecular surveys of bacterioplankton communities in coastal regions and open oceans have yielded similar 16S rRNA sequences (Giovannoni and Rappé, 2000), although coastal sites can differ significantly from the open ocean with respect to primary production rates and terrestrial influence. Indeed, 16S rRNA sequences from vibrios recovered from open ocean environments are phylogenetically similar to sequences from strains detected in coastal environments (Urakawa et al., 1999a,b; Radjasa et al., 2001). Many open ocean environments are nutrient-limited systems with low standing stocks of biomass (e.g., the oligotrophic gyres or iron-limited high-nitrogen/low-chlorophyll regions). Such conditions are in contrast to the general perception of vibrios as "high nutrient" adapted, and it remains to be determined to what extent vibrios detected in open ocean environments are active or passive members of the bacterioplankton. Vibrios have been shown to persist for a month or longer under conditions of nutrient limitation (Hood et al., 1986; Kramer and Singleton, 1992; Eilers et al., 2000a; Armada et al., 2003; Larsen et al., 2004), decreasing in cell volume in response to carbon starvation (Hood et al., 1986; Jiang and Chai, 1996; Denner et al., 2002). One strain (*Vibrio calviensis*) has been isolated from the Mediterranean Sea as a facultative "ultramicrocell" (<0.2 μm in diameter) whereby cell volume expands under nutrient-enriched culture conditions (Denner et al., 2002). While obligate "ultramicrobacteria" have been described from oligotrophic open ocean environments and hypothesized to substantially contribute to environmental nutrient cycling (Cho and Azam, 1988, 1990), the extent to which facultative "ultramicro" *Vibrio* cells contribute to microbial diversity and nutrient cycling in oligotrophic environments has not been addressed; this may reflect the limitation of DNA-based studies that are based on a collection of planktonic biomass on a 0.2-μm-pore-size filter.

Estuarine and Freshwater Habitats

A number of vibrios have growth optima at brackish salinities and are routinely detected in coastal estuaries (e.g., *V. cholerae*, *V. mimicus*, *V. vulnificus*, *V. cincinnatiensis*, *V. fluvialis*) (Grimes, 1991; Farmer and Hickman-Brenner, 2001; Heidelberg et al., 2002a). Relationships between salinity, temperature, and abundance of *V. vulnificus* and *V. parahaemolyticus* have been used to predict abundance in oysters under different environmental conditions (Wright et al., 1996; Motes et al., 1998; Food and Drug Administration, 2000). Estuarine vibrios can also be isolated from marine environments, where they may represent active components of the plankton or be passively advected between estuarine habitats. Certain vibrios, most notably *V. cholerae*/*V. mimicus*, are also found in association with freshwater systems (Bockemuhl et al., 1986; Chowdhury et al., 1992; Halpern et al., 2004), and such environments may mediate the spread of cholera in inland human populations.

Deep-Sea Habitats

Molecular surveys have revealed that microbial populations are highly stratified in the water column, with the biggest shift in community structure near the photic to aphotic zone transition, 50 to 200 m (Giovannoni and Rappé, 2000). Vertical clines in the composition of dissolved organic matter and adaptations to environmental gradients associated with depth are likely to influence microbial community composition from the surface to the deep sea. One study has suggested that vibrios from different depths are specialized with respect to carbon utilization. While surface vibrios have diverse metabolic activities, including the ability to degrade labile polymers (e.g., starch, esculin, casein), deep-water vibrios appear more metabolically restricted but, in contrast to surface isolates, can utilize more refractory compounds (e.g., agar, xylan, mannan) (Simidu and Tsukamoto, 1985). However, such work has not been followed up with molecular methods to determine whether phylogenetic differences exist between surface and deep-water populations.

A wide spectrum of *Vibrionaceae* species have been isolated from the deep sea (>1,000 m depth), including strains related to *Vibrio fischeri*, *Vibrio harveyi*, *Vibrio splendidus*, and *Photobacterium* (Ruby et al., 1980; Urakawa et al., 1999a; Radjasa et al., 2001); such strains may belong to ecologically differentiated deep-sea populations or may be representatives of populations distributed throughout the oceans. For example, *Vibrio diabolicus*, a polysaccharide-secreting mesophile isolated from a hydrothermal vent tube worm (*Alvinella pompejana*), grows between 20 and 45°C despite isolation from the deep sea (temperature 0 to 2°C) (Raguenes et al., 1997); it remains to be determined whether this organism is specialized to a hydrothermal vent habitat or whether it may have an ocean-wide distribution enabling colonization of such deep-sea niches. Certain vibrios maintain specific adaptations to conditions in the deep sea, supporting the theory of vertical habitat segregation proposed by Simidu and Tsukamoto and coworkers (Simidu and Tsukamoto, 1985). *Photobacterium profundum* strains have growth optima exceeding 2,000 atmospheres (DeLong et al., 1997; Nogi et al., 1998) and pressure-regulated genetic systems (e.g., Bartlett et al., 1989), and their production of polyunsaturated fatty acids (PUFAs) is hypothesized to help maintain membrane fluidity under high-pressure, low-temperature conditions (Allen et al., 1999).

Association with Marine Organisms

Vibrios have frequently been identified in association with animals or algal cells (e.g., WardRainey et al., 1996; Gomez-Gil et al., 1998; Gil et al., 2004). When attached to zooplankton and algal cells, vibrios may mediate degradation of chitin or other polymeric surface structures and thus contribute to recycling of particulate matter. Furthermore, the facultative anaerobic growth of vibrios is similar to that of the closely related gammaproteobacterial enteric bacteria (e.g., *Escherichia coli*), and the occurrence of vibrios in the guts of marine fauna suggests a possible commensal role for vibrios in mediating organic matter decomposition as marine enteric bacteria. In addition, more specific symbiotic associations between vibrios and animal hosts have been described, such as that between the bioluminescent vibrios and squid or fish (Haygood and Distel, 1993; Ruby and Lee, 1998), whereby the growth of bioluminescent vibrios is stimulated by organic metabolites supplied from the host (Graf and Ruby, 1998).

Association with larger host organisms may mediate the environmental dynamics of symbiotic or commensal *Vibrio* populations. A daily cycle of *V. fischeri* density has been observed in a tropical bay where its squid host is abundant. This is due to periodic enrichment of bay waters by daily expulsion from the squid light organ (Lee and Ruby, 1994). Similarly, associations with deep-sea or meso-pelagic hosts may play a role in the vertical distribution of the *Vibrionaceae* (Ruby et al., 1980). It has been suggested that the dynamics of *V. cholerae* in coastal waters may also be forced by association with planktonic eukaryotes where cholera outbreaks have been correlated to algal and zooplankton blooms (Colwell, 1996; Lobitz et al., 2000; Gil et al., 2004).

Although vibrios are known to exist as benign commensals of many cultured marine fauna, a considerable number of pathogenic strains have been described (J. R. Thompson et al., 2005a) that differ from their harmless counterparts by relatively few pathogenicity determinant genes. These may arise via horizontal transfer of virulence genes among closely related strains (Waldor and Mekalanos, 1996; Boyd and Waldor, 1999; Boyd et al., 2000) and may thus lead to the emergence and existence of pathogens among benign populations. For example, while oysters generally harbor a diverse array of *V. vulnificus* strains, the occurrence of pathogenic variants appears to be relatively rare (Jackson et al., 1997). Nonetheless, infection by *Vibrio* species is one of many significant problems for the commercial culture of fish and marine invertebrates (e.g., Gomez-Leon et al., 2005). The high-density, high-nutrient conditions characteristic of some aquaculture systems may facilitate the rapid spread of virulent strains. It has been hypothesized that the artificial conditions in aquaculture environments may serve as reservoirs for pathogenic *Vibrio* strains when environmental conditions become incompatible for *Vibrio* growth (Ben-Haim et al., 2003). Indeed, sediments underlying farmed mussels have been observed to support an enriched presence of vibrios relative to surrounding environments (La Rosa et al., 2001).

SYSTEM-LEVEL SIGNIFICANCE

In marine food webs, dissolved organic matter released from primary production is recycled by the activity of heterotrophic bacteria and protists to supply regenerated nutrients (nitrogen and phosphorus) while acting as a net sink for carbon due to respiratory loss as CO_2. Organic matter uptake by bacteria followed by regeneration of nutrients through mineralization or grazing mortality is termed the microbial loop (Fig. 1) (Azam et al., 1983; Ducklow, 1983). However, under conditions of inorganic nutrient lim-

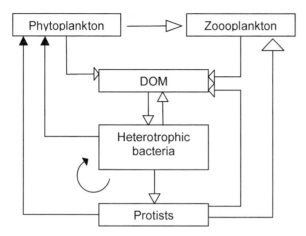

Figure 1. Idealized heterotrophic microbial loop whereby dissolved organic matter is recycled to inorganic nutrients available for primary production by the activities of heterotrophic bacteria and protists. Open arrowheads reflect the flow of organic carbon, and closed arrowheads reflect the flow of inorganic nutrients (N and P). Vibrios mediate biogeochemical cycling through activities such as organic matter uptake and release or competition for inorganic nutrients and by release of cellular materials as a by-product of grazing or viral lysis. (Contributions to nutrient cycling by autotrophic cyanobacteria and protozoan uptake of high-molecular-weight dissolved organic matter [DOM] are not depicted.)

itation, heterotrophic bacteria may compete with primary producers (phytoplankton) for dissolved inorganic nitrogen and phosphorus. The proportion of primary production that flows through a multistep microbial food web versus a shorter phytoplankton-zooplankton food chain has implications for the capacity of marine ecosystems to sequester organic carbon or efficiently produce fish biomass. A number of investigators have proposed scenarios in which pelagic systems characterized by active microbial food webs export less organic carbon (Sherr and Sherr, 2002).

Through heterotrophic growth on organic substrates, the vibrios can contribute to nutrient recycling within the diverse habitats they occupy. Within the plankton, their importance to ecosystem processes may be greater than their relatively low abundance, i.e., <0.1 to 4% (Eilers et al., 2000b; Heidelberg et al., 2002a; J. R. Thompson et al., 2004), would suggest. Proliferation in the plankton is determined by a combination of top-down controls, such as grazing mortality and viral lysis, and bottom-up controls, including resource supply and physical factors (e.g., salinity and temperature). Indeed, observations of explosive growth in response to nutrient enrichment (e.g., Eilers et al., 2000a; Massana et al., 2001; Overas et al., 2003; Pinhassi and Berman, 2003) and selective grazing by protists (Hahn and Hofle, 1998; Beardsley et al., 2003) suggest that vibrios may have high pop-

ulation turnover and thus disproportionately contribute to ecosystem nutrient cycling. The extent to which vibrios mediate environmental processes is thus a product of both their abundance and their activity. Understanding how these variables change with changing environmental conditions is essential to assess their significance in ecosystem nutrient cycling.

Activity in the Plankton

Although some *Vibrio* populations may only grow in association with animal hosts, accumulating data suggest that vibrios are also able to grow in the bacterioplankton. Proliferation of *V. cholerae* (strain N16961) within a bacterioplankton assemblage at rates of up to 2.6 ± 0.3 day^{-1} has been demonstrated in seawater mesocosms containing natural concentrations of phytoplankton bloom-derived, dissolved organic matter (Mourino-Perez et al., 2003). Similar growth rates were observed among naturally occurring vibrios in seawater dilution cultures after 40 μM glucose amendment (2.3 to 3.3 day^{-1} [Pinhassi and Berman, 2003]). Several lines of evidence suggest that vibrios are physiologically adapted to exploit pulses of nutrients in the environment: (i) Respiratory activity under low-nutrient conditions in seawater mesocosms indicates long-term survival in substrate-limiting environments (Ramaiah et al., 2002; Armada et al., 2003). (ii) Maintenance of high ribosome content after a shift to starvation conditions enables rapid growth in response to substrate pulses (Hood et al., 1986; Flardh et al., 1992; Kramer and Singleton, 1992; Eilers et al., 2000a; Pernthaler et al., 2001). (iii) Chemotaxis toward ecologically relevant compounds, including chitin and sugar monomers (Bassler et al., 1991a; Yu et al., 1993), amino acids, and in response to carbon limitation (Gosink et al., 2002; Larsen et al., 2004) indicates an ability to exploit nutrient-rich microenvironments. In addition, filtration of seawater mesocosms to remove protozoan grazers has been shown to allow *V. cholerae* proliferation where growth rates of up to 2.9 day^{-1} could be observed (Worden et al., 2005). Such data suggest that *Vibrio* proliferation in seawater environments can be stimulated by both substrate supply and the release from top-down control (e.g., grazing), supporting the hypothesis that active growth under predation pressure may result in a high turnover of natural *Vibrio* populations.

Extracellular Enzymatic Activity

Most dissolved and particulate organic matter in the marine environment is in the form of complex polymers that must be hydrolyzed prior to cellular uptake. Vibrios, and other gram-negative bacteria, degrade complex organic matter through a defined sequence of steps. Partial hydrolysis of complex polymers must occur extracellularly prior to transport into the periplasmic space, where additional enzymes act to create monomers that can be transported into the cell cytoplasm. Chitinases, proteases, and lipases are among the cell surface or exuded hydrolases that have been described in *Vibrio* species (Riemann and Azam, 2002; Gomez-Leon et al., 2005), whereas enzymatic activities, such as alkaline phosphatase and amino-peptidase, appear to be concentrated in the periplasmic space (Martinez and Azam, 1993). Chitinase activity may reflect one of the most important extracellular enzymatic processes in the marine environment. It has been estimated that 10^{11} tons of chitin are produced annually in marine systems, primarily in the form of zooplankton exoskeletons, and this polymer must be continually remineralized to support sustained primary production in the oceans (Li and Roseman, 2004). While chitinase activity is observed within a subset of marine bacteria (Cottrell et al., 2000), it is prevalent within the *Vibrionaceae* (Suginta et al., 2000; Riemann and Azam, 2002; F. L. Thompson et al., 2004). Thus, vibrios capable of hydrolyzing polymers such as chitin may create important trophic links within bacterioplankton communities.

Extracellular hydrolysis of complex polymers has been suggested as an important cross-feeding mechanism in microbial communities (Martinez et al., 1996; Riemann and Azam, 2002). Diffusion of cell surface products or leakage of products from the periplasmic space may generate a surrounding microenvironment enriched in labile dissolved organic matter, which can be exploited by other (planktonic) bacterial populations; however, the extent to which products of transformations occurring within the periplasm may diffuse into the bulk environment is unknown. Competition for space (i.e., by "chemical warfare") may increase the efficiency by which the products from extracellular enzymatic reactions are utilized. Vibrios have been identified as significant mediators of antagonistic interactions among marine bacteria (Long and Azam, 2001).

Marine isolates have been shown to vary significantly with respect to their cell surface and periplasmic enzymatic profiles and activities (Martinez et al., 1996). Shifts in dominant and active forms of bacteria may strongly influence the pattern of polymer hydrolysis and cycling of dissolved organic matter in aquatic systems. In addition, the extracellular products of *Vibrio* species may be active as virulence factors. Products such as proteases, iron-binding compounds, and toxins have been implicated in the mediation of marine and human diseases, and environmental selection of such activities may contribute to the role of vibrios as opportunistic pathogens (Gomez-Leon et al., 2005).

Food-Web Interactions

Analysis of single-cell activity in a coastal bacterioplankton community has demonstrated that abundance is not correlated to in situ growth rates, pointing to the importance of other factors, such as mortality, in determining community structure (Cottrell and Kirchman, 2004). Larger, actively growing cells appear to be selectively grazed by marine bacterivorous protists (see references in Sherr and Sherr, 2002), and preferential grazing of vibrios by flagellates has been observed in experimental systems and may explain the low abundance of vibrios in marine environments (Hahn and Hofle, 1998; Beardsley et al., 2003; Worden et al., 2005). Indeed, in situ observation of vibrios with fluorescent hybridization probes reveals a characteristic large-cell morphology under coastal conditions (Eilers et al., 2000b). Ovreas et al. (2003) identified a dominant large bacterium in glucose enrichments of seawater as *V. splendidus*; however, such relatively large vibrios (typically 0.5 to 0.8 μm wide and 1.5 to 2 μm long) have been observed to achieve a smaller "microcell" shape under carbon limitation (Hood et al., 1986; Jiang and Chai, 1996; Denner et al., 2002) that may be more resistant to grazing (Sherr and Sherr, 2002); similar coccoid cell morphologies have been described for a number of *Vibrio* species during the onset of the viable but nonculturable form.

Mortality due to viral lysis may also play a significant role in *Vibrio* population dynamics and nutrient cycling by controlling the abundance of specific *Vibrio* populations and pools of growth substrates through cell lysis. Viral lysis can proceed via infection with phage with broad host range or highly specific for individual strains. Strain-specific phage abundance has been inversely correlated to the incidence of phage-sensitive strains of *V. cholerae* and to cholera cases in Bangladesh, suggesting that cholera phages may influence the seasonality of the bacterium and of the disease (Faruque et al., 2005). Organic carbon released by viral lysis of a *Photobacterium* sp. has been shown to provide nutrition for a competing, phage-resistant strain of bacteria, demonstrating that trophic links mediated by strain-specific viral lysis may play an important role in ecosystem nutrient cycling (Middelboe et al., 2003).

MACRONUTRIENT CYCLING

Heterotrophic bacteria, such as the *Vibrionaceae*, are involved in both uptake and remineralization of key elements such as carbon (C), nitrogen (N), and phosphorus (P). When and where bacteria take up or release nutrients are important ecological questions. Because vibrios appear to be selectively grazed by flagellates, it has been suggested that they may have enhanced significance in the cycling of organic matter in aquatic settings. Their contributions to macronutrient cycles are discussed below.

Carbon

Carbon substrates

All currently described members of the family *Vibrionaceae* are obligate heterotrophs and, as such, rely upon organic matter for carbon. Vibrios consume a wide array of carbon substrates (Farmer and Hickman-Brenner, 2001), and this may facilitate their isolation from marine systems. Complex organic macromolecules are degraded through extracellular digestion and subsequent monomer uptake. The high substrate affinity of marine vibrios suggests adaptation to growth under high-nutrient conditions such as would occur in animal guts or in planktonic microenvironments (e.g., microzones of dissolved organic matter enrichment around algal cells or suspended detritus). Indeed, the half-saturation constant for glucose is 29 μM and 500 μM for strains of *Vibrio (Beneckea) natriegens* and *V. parahaemolyticus*, respectively (Linton et al., 1977; Sarker et al., 1994), compared to typical bulk seawater glucose concentrations ranging from <14 nM to 2 μM (Kirchman et al., 2001; Meon and Kirchman, 2001).

Carbon storage

Vibrio species can survive carbon starvation for a month or longer (Jiang and Chai, 1996; Eilers et al., 2000a; Ramaiah et al., 2002; Armada et al., 2003). In *V. cholerae*, carbon inclusion bodies [e.g., glycerol or poly(3-hydroxybutyrate)] that are formed in the presence of excess carbon are consumed within the first week of carbon starvation (Hood et al., 1986). Carbon storage may enable *Vibrio* growth in a fluctuating environment where individual resources may be limiting at different times. Carbon storage may also provide an advantage during competition for limiting nutrients by increasing the diffusional surface area of the cell by "blowing up the balloon" with solid material (Ovreas et al., 2003). Access of *Vibrio* cells to internal carbon pools has a regulatory role in expression of cell surface properties that influence nutrient acquisition and mediate virulence during infection. For example, carbon limitation (and high cyclic AMP levels) has been shown to stimulate protease activity, mediating both detachment from surfaces and penetration into mucosal layers during tissue colonization by *V. cholerae* (Benitez et al., 2001).

Carbon products

Vibrios are able to engage in both respiratory and fermentative metabolism and transform organic carbon into cell material and waste products of energy metabolism. Depending on the energetics of the metabolic reactions, the efficiency of biomass formation per unit substrate can vary a great deal. During aerobic or anaerobic respiration, 30 to 50% of organic carbon is used for biomass formation, while during fermentation, copious amounts of metabolic end products are excreted. These include organic acids, alcohols, and, in some species, H_2, which can stimulate anaerobic food chains (e.g., by interspecies H_2 transfer or growth on exuded products). Characteristic of fermentative growth under anaerobic chemostat conditions, *V. natriegens* converts ~90% of carbon substrates to fermentation end products, while such products make up 10 to 15% of the carbon budget under aerobic conditions (Linton et al., 1975) consistent with observations of carbon "leakage" from algal and microbial cells. *Vibrio halioticoli*, isolated from the gut of a *Haliotis* abalone, fermented algal polysaccharides, producing up to 68 mM acetic and formic acids, which were hypothesized to contribute to host nutrition (Sawabe et al., 2003) and may also represent a trophic link within the gut microbial community. Vibrios have been shown to produce volatile organic compounds, such as acetone, during metabolism of the amino acid leucine (Nemecek-Marshall et al., 1999); whether marine bacteria are a significant source of atmospheric acetone and other volatile organic compounds through such processes remains to be determined. In addition, *P. profundum* produces PUFAs that are essential nutrients for many marine organisms (Nichols, 2003) and are hypothesized to help maintain bacterial membrane fluidity in the deep sea (Allen et al., 1999). Several genera within the gammaproteobacteria and *Cytophaga-Flavobacterium-Bacterioides* grouping are recognized to contain PUFA-producing strains, although most marine PUFAs have been characterized from microalgae (Nichols, 2003); the relative importance of bacterially produced PUFA in marine food webs is an open question.

Nitrogen

The cycling of marine nitrogen involves a series of primarily microbial transformations, including (i) fixation of dinitrogen N_2 to organic N; (ii) dissimilatory reduction of nitrate (NO_3^-) to produce nitrite (NO_2^-) or ammonia (NH_4^+), where nitrite is in many bacteria denitrified to nitrous oxide (N_2O) and dinitrogen (N_2), or assimilatory reduction of nitrate to nitrite and organic N; (iii) nitrification of NH_4^+ to NO_2^-, N_2O, or NO_3^-; and (iv) ammonification of organic N to NH_4^+ (Herbert, 1999). Vibrios are known to participate in many of the reductive pathways (1, 2, and 4) but not in nitrification (3) (Fig. 2).

Nitrogen fixation

Marine systems have traditionally been viewed as nitrogen-limited habitats, while the Earth's atmosphere (79% N_2) represents a major reservoir of nitrogen. Nitrogen-fixing bacteria can access this pool of atmospheric nitrogen and as such have profound effects on net community production by the input of "new" nitrogen to nutrient-limited ecosystems. Owing to high metabolic costs of nitrogen fixation, such capacity is limited by the availability of cellular energy, and in marine environments it has been described primarily in phototrophs that acquire energy directly from sunlight. However, contribution of heterotrophic nitrogen fixation to new nitrogen production in marine ecosystems is increasingly recognized, although its relative significance has not been determined (Karl et al., 2002). In oligotrophic ocean gyres, it has been estimated that up to 50% of the bioavailable nitrogen stems from cyanobacterial nitrogen fixation (Karl et al., 2002), and, in microbial mats, shifts in dominance between cyanobacterial and heterotrophic nitrogen fixation with seasonal environmental changes have been documented (Zehr and Ward, 1995).

Four species of vibrios are known to fix molecular nitrogen: *V. diazotrophicus, V. natriegens, V. (Listonella) pelagius*, and *V. cincinnatiensis* (Guerinot et al., 1982; Urdaci et al., 1988); these organisms have

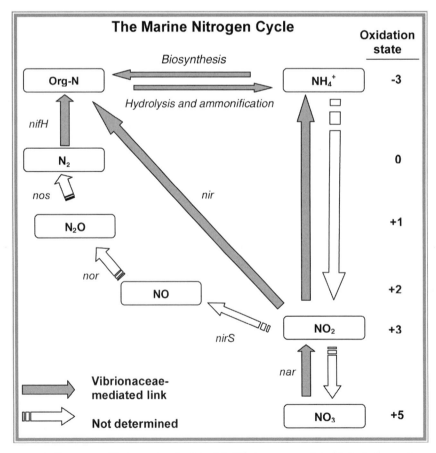

Figure 2. Nitrogen cycles between oxidation states of -3 to $+5$. *Vibrionaceae*-mediated links in the marine environment are shown in gray. Genes encode proteins implicated in mediating nitrogen. Modified from Capone (2000).

been described from animal associations, microbial mats, estuaries, and salt-marsh environments. The diversity of potential nitrogen-fixing organisms in the environment can be surveyed with molecular techniques targeting the *nifH* gene, which encodes the nitrogen-fixing enzyme dinitrogenase reductase (nitrogenase). *nifH* sequences, similar to those found in cultured vibrios, have been detected in picoplankton (Zehr et al., 1998), marine microbial mats (Olson et al., 1999), salt-marsh rhizosphere (Lovell et al., 2000; Bagwell et al., 2002; Brown et al., 2003), and tropical seagrass beds (Bagwell et al., 2002).

Nitrogen fixation in *V. natriegens* is mediated by a cytoplasmic nitrogenase enzyme complex (Coyer et al., 1996). It requires low oxygen concentrations to function and high environmental C:N ratios to stimulate activity and provide energy. The nitrogenase activity has a sharp pH optima at 7 with zero activity at pH 8 (Coyer et al., 1996). Since vibrios have aerobic growth optima at alkaline pH (consistent with the pH of oxic seawater at 7.8 to 8.2), it has been suggested that the neutral pH optima of nitrogen fixation in vibrios is consistent with the pH of anoxic saline environments, rich in organic energy sources, such as sediments and animal guts that should be expected to be neutral to slightly acidic (Urdaci et al., 1988). However, *Vibrio* nitrogenase activity can also be detected in seawater mesocosms and is stimulated by phytoplankton addition, leading to speculation that heterotrophic nitrogen fixation may be coupled to primary production (Guerinot and Colwell, 1985). Nitrogen fixation by vibrios may also contribute substantially to host nutrition. *V. diazotrophicus* was originally described from sea urchin gastrointestinal tracts, and studies with the isotopic tracer ^{15}N have shown incorporation of microbially fixed N into sea urchin tissue (Guerinot and Patriquin, 1981; Guerinot and Colwell, 1985). Thus, heterotrophic nitrogen fixation by symbiotic populations, or in concert with carbon fixation by photosynthetic organisms, may support productivity in low-nutrient environments.

Dissimilatory and assimilatory nitrate reduction

Two pathways of nitrate metabolism have been identified in vibrios: (i) assimilatory reduction of nitrate to biological material and (ii) dissimilatory (respiratory) reduction of nitrate to nitrite or ammonia. To date, no known vibrios possess the capacity or genetic systems for denitrification (reduction of nitrate to gaseous N_2O or N_2, resulting in a net loss of N from an ecosystem). A facultatively anaerobic bacterium originally described as a denitrifying *Vibrio* was recently classified as an alphaproteobacterium based upon DNA sequence data (Shieh et al., 2004).

Assimilation of the charged nitrate ion requires active transport across the cell membrane, while ammonia uptake can occur passively by diffusion, since at seawater alkalinities, ~10% of the ammonia-N occurs as deprotonated NH_3. Active transport of the NH_4^+ ion has also been detected in a marine vibrio (Chou et al., 1999). Elevated temperatures may facilitate nitrate uptake by altering membrane properties (Reay et al., 1999). The occurrence of nitrate assimilation genes (*nasA*) has been shown to correlate with ability to grow on nitrate as a sole N source (Allen et al., 2001), and the diversity of *Vibrio*-related *nasA* sequences in marine systems suggests that vibrios consume nitrate in the marine environment (Allen et al., 2001).

Many facultatively anaerobic bacteria can replace oxygen with nitrate and as a terminal electron acceptor via dissimilatory nitrate reduction. Proctor and Gunsalus (2000) showed that several alternative electron acceptors, including nitrate, fumarate, and trimethylamine N-oxide, but not nitrite, could support anaerobic respiratory growth of vibrios (Proctor and Gunsalus, 2000). While many vibrios are reported to produce nitrite as a by-product of nitrate reduction, the dissimilatory reduction of nitrate to ammonia has been demonstrated in a marine *Vibrio* isolate (Bonin, 1996). It has been estimated that dissimilatory reduction of nitrate to ammonia may govern as much as 80% of overall nitrate consumption in marine sediments (Bonin, 1996).

Nitrification

Nitrification is the process by which ammonium is oxidized to nitrite and then to nitrate (coupled to oxygen or nitrate as a terminal electron acceptor). Although there are nitrifiers within the gammaproteobacteria (e.g., *Nitrococcus*), no vibrios are known to participate in this process.

Ammonification of organic N

The remineralization of nitrogenous compounds, such as nucleic acids, proteins, and polyamino-sugars, to simple carbon compounds and ammonia (ammonification) is a critical link for nutrient recycling via the microbial loop. The nutrient status and C:N ratio of the environment may determine whether the produced ammonia is incorporated into microbial biomass or excreted to the ecosystem. Mechanisms for microbial consumption of polymeric nitrogenous compounds as a source of both nitrogen and carbon involve the extracellular hydrolysis of nitrogenous polymers to simpler subunits followed by cellular uptake of monomers. Many vibrios produce a suite of chitinases and proteases that allow degradation of nitrogenous polymers, including chitin and proteins, as sole carbon and nitrogen sources (Bassler et al., 1991b; Suginta et al., 2000; Riemann and Azam, 2002; Li and Roseman, 2004). Chitin, produced in marine systems at an estimated rate of 10^{11} tons/year, must be continually remineralized by microbial activity to support sustained primary production in the oceans (Li and Roseman, 2004). While chitinase activity is observed within a subset of marine bacteria (Cottrell et al., 2000), it is prevalent within the *Vibrionaceae* (Suginta et al., 2000; Riemann and Azam, 2002; F. L. Thompson et al., 2004). Molecular studies have confirmed the occurrence of *Vibrio*-related chitinase genes (*chiA*) in coastal Pacific and Atlantic waters (Cottrell et al., 2000). Thus, vibrios capable of hydrolyzing nitrogenous polymers such as chitin may be part of an important trophic link within bacterioplankton communities.

Phosphorus

Assimilation of inorganic and organic phosphorus

Phosphorus (P) is required by organisms for fundamental biological processes, including nucleic acid and membrane phospholipid synthesis, signaling pathways, and energy metabolism. While marine environments are typically not considered P-limited, phosphorus bioavailability may influence ecosystem dynamics. In contrast to nitrogen cycling, the metabolism of phosphorus does not involve changes in oxidation state; thus, marine phosphorus cycles are controlled by the partitioning of phosphorus into bioavailable inorganic forms and more refractory organic pools. Vibrios express a number of extracellular enzymes that participate in the degradation of phosphorus-containing macromolecules; these compounds play a prominent role in the recycling of organic P to inorganic forms available for primary production. Dissolved marine phosphorus pools are grouped by their relative bioavailability into the soluble reactive phosphorus pool, which includes inorganic phosphate and polyphosphate ions that can be utilized directly by microbes and phytoplankton, and the soluble nonreactive phosphorus pool, which contains the less bioavailable macromolecular fractions that must be degraded extracellularly before utilization (e.g., monophosphate esters, nucleotides, nucleic acids, and phosphonates [Benitez-Nelson, 2000]). Phosphate-generating exoenzymes important for recycling of marine organic phosphorus pools include alkaline phosphatases, phosphodiesterases, and 5′ nucleotidases (Hoppe, 2003). Alkaline phosphatase has been detected in a number of *Vibrio* species (Woolkalis and Baumann,

1981; Roy et al., 1982; Kaznowski and Wlodarczak, 1991; Hauksson et al., 2000) and is localized to the periplasmic space in *V. cholerae* (Roy et al., 1982). Alkaline phosphatase cleaves inorganic phosphate off phosphorylated compounds under the neutral to alkaline conditions characteristic of the marine environment. Alkaline phosphatase activity is inhibited with increasing levels of free phosphatase (Roy et al., 1982), and activity is low or absent in regions with high soluble reactive phosphorus (Hoppe, 2003), suggesting that this activity is central to supplying phosphate pools when P is limiting.

In contrast, the phosphate-liberating enzymes 5′ nucleotidases are not subject to inhibition by high levels of soluble reactive phosphorus and thus may mediate both the availability of phosphate and the availability of phosphorylated carbon substrates for growth (Ammerman and Azam, 1985). 5′ Nucleotidases degrade 5′ nucleotides to inorganic phosphate and a base prior to their transport into the cytoplasm and subsequent metabolism. 5′ Nucleotidases have been described as periplasmic membrane-bound enzymes in *V. parahaemolyticus* (Sakai et al., 1987) and in *Salinivibrio (Vibrio) costicola* (Bengis-Garber and Kushner, 1981), and genes are present in all sequenced *Vibrio* genomes. Ammerman and Azam (1985) hypothesized that hydrolysis of soluble nonreactive phosphorus by 5′ nucleotidases could supply as much as half the phosphate required by plankton in coastal California waters (Benitez-Nelson, 2000).

A third class of organic phosphate-degrading enzymes widely found in the marine environment is the 3′5′ cyclic nucleotide phosphodiesterases that enable metabolism of extracellular cyclic nucleotides (e.g., cyclic AMP). In *V. fischeri*, a transmembrane cyclic nucleotide phosphodiesterase with an active site in the periplasmic space allows use of extracellular cyclic nucleotides as a sole source of C, N, and P (Callahan et al., 1995). Such periplasmic enzymatic activity contributes to the role of vibrios in remineralizing organic phosphorus compounds to inorganic compounds and carbon substrates for growth and may serve to enrich their local environments with dissolved pools of nutrients that can be utilized by primary producers and other heterotrophs in the community.

GENOMIC PERSPECTIVES

Currently, six *Vibrionaceae* genome sequences have been published in the GenBank database (i.e., *V. cholerae* N16961, *V. parahaemolyticus* RIMD 2210633, *V. vulnificus* YJ016 and CMCP6, *V. fischeri* ES114, and *P. profundum* SS9) (Heidelberg et al., 2000; Chen et al., 2003; Makino et al., 2003;

Ruby et al., 2005). This number is expected to increase as new sequence data become available, providing many opportunities to explore the ecological diversification and evolution of pathogenicity within this group. Analysis and identification of "core" *Vibrionaceae* features such as motility, morphological plasticity, and organic matter cycling will provide a genomic foundation for describing properties of a group that includes plankton active in nutrient cycling, animal commensals, and human pathogens. Comparative genomic approaches between nonpathogens and pathogenic strains will help explain the unifying themes underlying bacterial-host interactions and mechanisms by which pathogenic interactions may emerge (Heidelberg et al., 2000; Chen et al., 2003; Makino et al., 2003; Ruby et al., 2005). Evolutionary genomic and population genetic studies will address the extent to which homologous recombination and horizontal gene transfer (e.g., by phage or plasmids) mediate pathogen emergence from benign strains and influence the evolution of ecologically differentiated *Vibrio* populations (J. R. Thompson et al., 2005b). In addition, environmental genomic approaches to explore the metabolic diversity associated with phylogenetic clades may shed light on how widespread certain features, such as N_2 fixation, bioluminescence, and cell signaling, are among the *Vibrionaceae* and whether vibrios are capable of as-yet-undiscovered metabolic transformations (e.g., denitrification, phototrophy, chemoautotropy).

CONCLUSIONS

Although not numerically dominant in the bacterioplankton, vibrios respond rapidly to carbon and nutrient enrichments and are selectively grazed, suggesting that such populations may disproportionately contribute to environmental nutrient cycling. The dynamics and distribution of bacterioplanktonic *Vibrio* populations are determined by adaptations to environmental gradients, including temperature, salinity, and nutrient concentration. Bottom-up controls of substrate availability and top-down controls such as selective grazing and viral mortality are additional factors influencing *Vibrio* population dynamics. Vibrios may participate in marine macronutrient cycles through processes including remineralization of nutrients through the microbial loop, degradation of complex polymers, and nitrogen fixation. The facultative anaerobic lifestyle and frequent isolation from marine guts suggest that vibrios may also mediate commensal nutrient cycling similar to terrestrial enteric bacteria and may stimulate the activities of anaerobic food chains in anoxic environments.

REFERENCES

Allen, A. E., M. G. Booth, M. E. Frischer, P. G. Verity, J. P. Zehr, and S. Zani. 2001. Diversity and detection of nitrate assimilation genes in marine bacteria. *Appl. Environ. Microbiol.* 67: 5343–5348.

Allen, E. E., D. Facciotti, and D. H. Bartlett. 1999. Monounsaturated but not polyunsaturated fatty acids are required for growth of the deep-sea bacterium *Photobacterium profundum* SS9 at high pressure and low temperature. *Appl. Environ. Microbiol.* 65:1710–1720.

Ammerman, J. W., and F. Azam. 1985. Bacterial 5′-nucleotidase in aquatic ecosystems: a novel mechanism of phosphorus regeneration. *Science* 227:1338–1340.

Armada, S. P., R. Farto, M. J. Perez, and T. P. Nieto. 2003. Effect of temperature, salinity and nutrient content on the survival responses of *Vibrio splendidus* biotype I. *Microbiology* 149:369–375.

Azam, F., T. Fenchel, J. G. Field, J. S. Gray, L. A. Meyerreil, and F. Thingstad. 1983. The ecological role of water-column microbes in the sea. *Mar. Ecol. Prog. Ser.* 10:257–263.

Bagwell, C. E., J. R. LaRocque, G. W. Smith, S. W. Polson, M. J. Friez, J. W. Longshore, and C. R. Lovell. 2002. Molecular diversity of diazotrophs in oligotrophic tropical seagrass bed communities. *FEMS Microbiol. Ecol.* 39:113–119.

Barbieri, E., L. Falzano, and C. Fiorentini. 1999. Occurrence, diversity, and pathogenicity of halophilic *Vibrio* spp. and non-O1 *Vibrio cholerae* from estuarine waters along the Italian Adriatic Coast. *Appl. Environ. Microbiol.* 65:2748–2753.

Bartlett, D., M. Wright, A. A. Yayanos, and M. Silverman. 1989. Isolation of a gene regulated by hydrostatic pressure in a deep-sea bacterium. *Nature* 342:572–574.

Bassler, B. L., P. J. Gibbons, C. Yu, and S. Roseman. 1991a. Chitin utilization by marine bacteria. Chemotaxis to chitin oligosaccharides by *Vibrio furnissii*. *J. Biol. Chem.* 266:24268–24275.

Bassler, B. L., C. Yu, Y. C. Lee, and S. Roseman. 1991b. Chitin utilization by marine-bacteria—degradation and catabolism of chitin oligosaccharides by *Vibrio furnissii*. *J. Biol. Chem.* 266: 24276–24286.

Beardsley, C., J. Pernthaler, W. Wosniok, and R. Amann. 2003. Are readily culturable bacteria in coastal North Sea waters suppressed by selective grazing mortality? *Appl. Environ. Microbiol.* 69:2624–2630.

Bengis-Garber, C., and D. J. Kushner. 1981. Purification and properties of 5′-nucleotidase from the membrane of *Vibrio costicola*, a moderately halophilic bacterium. *J. Bacteriol.* 146: 24–32.

Ben-Haim, Y., F. L. Thompson, C. C. Thompson, M. C. Cnockaert, B. Hoste, J. Swings, and E. Rosenberg. 2003. *Vibrio coralliilyticus* sp. nov., a temperature-dependent pathogen of the coral *Pocillopora damicornis*. *Int. J. Syst. Evol. Microbiol.* 53: 309–315.

Benitez, J. A., A. J. Silva, and R. A. Finkelstein. 2001. Environmental signals controlling production of hemagglutinin/protease in *Vibrio cholerae*. *Infect. Immun.* 69:6549–6553.

Benitez-Nelson, C. R. 2000. The biogeochemical cycling of phosphorus in marine systems. *Earth Sci. Rev.* 51:109–135.

Bockemuhl, J., K. Roch, B. Wohlers, V. Aleksic, S. Aleksic, and R. Wokatsch. 1986. Seasonal distribution of facultatively enteropathogenic vibrios (*Vibrio cholerae*, *Vibrio mimicus*, *Vibrio parahaemolyticus*) in the freshwater of the Elbe River at Hamburg. *J. Appl. Bacteriol.* 60:435–442.

Bonin, P. 1996. Anaerobic nitrate reduction to ammonium in two strains isolated from a coastal marine sediment: a dissimilatory pathway. *FEMS Microbiol. Ecol.* 19:27–38.

Boyd, E., K. Moyer, and L. Shi. 2000. Infectious CTX phi, and the vibrio pathogenicity island prophage in *Vibrio mimicus*: evidence for recent horizontal transfer between *V. mimicus* and *V. cholerae*. *Infect. Immun.* **68**:1507–1513.

Boyd, E., and M. Waldor. 1999. Alternative mechanism of cholera toxin acquisition by *Vibrio cholerae*: generalized transduction of CTX phi by bacteriophage CP-T1. *Infect. Immun.* **67**:5898–5905.

Brown, M. M., M. J. Friez, and C. R. Lovell. 2003. Expression of *nifH* genes by diazotrophic bacteria in the rhizosphere of short form *Spartina alterniflora*. *FEMS Microbiol. Ecol.* **43**:411–417.

Caldini, G., A. Neri, and S. Cresti. 1997. High prevalence of *Vibrio cholerae* non-O1 carrying heat-stable-enterotoxin-encoding genes among *Vibrio* isolates from a temperate-climate river basin of central Italy. *Appl. Environ. Microbiol.* **63**:2934–2939.

Callahan, S. M., N. W. Cornell, and P. V. Dunlap. 1995. Purification and properties of periplasmic 3′:5′-cyclic nucleotide phosphodiesterase. A novel zinc-containing enzyme from the marine symbiotic bacterium *Vibrio fischeri*. *J. Biol. Chem.* **270**:17627–17632.

Capone, D. G. 2000. The marine microbial nitrogen cycle, p. 455–493. *In* D. L. Kirchman (ed.), *Microbial Ecology of the Oceans.* Wiley-Liss, Inc., New York, N.Y.

Chen, C. Y., K. M. Wu, Y. C. Chang, C. H. Chang, H. C. Tsai, T. L. Liao, Y. M. Liu, H. J. Chen, A. B. Shen, J. C. Li, T. L. Su, C. P. Shao, C. T. Lee, L. I. Hor, and S. F. Tsai. 2003. Comparative genome analysis of *Vibrio vulnificus*, a marine pathogen. *Genome Res.* **13**:2577–2587.

Cho, B. C., and F. Azam. 1990. Biogeochemical significance of bacterial biomass in the oceans euphotic zone. *Mar. Ecol. Prog. Ser.* **63**:253–259.

Cho, B. C., and F. Azam. 1988. Major role of bacteria in biogeochemical fluxes in the oceans interior. *Nature* **332**:441–443.

Chou, M., T. Matsunaga, Y. Takada, and N. Fukunaga. 1999. NH$_4$+ transport system of a psychrophilic marine bacterium, *Vibrio* sp. strain ABE-1. *Extremophiles* **3**:89–95.

Chowdhury, M. A., S. Miyoshi, H. Yamanaka, and S. Shinoda. 1992. Ecology and distribution of toxigenic *Vibrio cholerae* in aquatic environments of a temperate region. *Microbios* **72**:203–213.

Colwell, R. R. 1996. Global climate and infectious disease: the cholera paradigm. *Science* **274**:2025–2031.

Cottrell, M. T., and D. L. Kirchman. 2004. Single-cell analysis of bacterial growth, cell size, and community structure in the Delaware estuary. *Aquat. Microb. Ecol.* **34**:139–149.

Cottrell, M. T., D. N. Wood, L. Yu, and D. L. Kirchman. 2000. Selected chitinase genes in cultured and uncultured marine bacteria in the alpha- and gamma-subclasses of the proteobacteria. *Appl. Environ. Microbiol.* **66**:1195–1201.

Coyer, J. A., A. Cabello-Pasini, H. Swift, and R. S. Alberte. 1996. N$_2$ fixation in marine heterotrophic bacteria: dynamics of environmental and molecular regulation. *Proc. Natl. Acad. Sci. USA* **93**:3575–3580.

DeLong, E. F., D. G. Franks, and A. A. Yayanos. 1997. Evolutionary relationships of cultivated psychrophilic and barophilic deep-sea bacteria. *Appl. Environ. Microbiol.* **63**:2105–2108.

Denner, E. B., D. Vybiral, U. R. Fischer, B. Velimirov, and H. J. Busse. 2002. *Vibrio calviensis* sp. nov., a halophilic, facultatively oligotrophic 0.2 micron-filterable marine bacterium. *Int. J. Syst. Evol. Microbiol.* **52**:549–553.

Ducklow, H. W. 1983. Production and fate of bacteria in the oceans. *Bioscience* **33**:494–501.

Eilers, H., J. Pernthaler, and R. Amann. 2000a. Succession of pelagic marine bacteria during enrichment: a close look at cultivation-induced shifts. *Appl. Environ. Microbiol.* **66**:4634–4640.

Eilers, H., J. Pernthaler, F. O. Glockner, and R. Amann. 2000b. Culturability and in situ abundance of pelagic bacteria from the North Sea. *Appl. Environ. Microbiol.* **66**:3044–3051.

Farmer, J. J., and F. W. Hickman-Brenner. 2001. The genera *Vibrio* and *Photobacterium*. *In* M. Dworkin et al. (ed.), *The Prokaryotes: an Evolving Electronic Resource for the Microbiological Community*, 3rd ed., release 3.7. Springer-Verlag, New York, N.Y.

Faruque, S. M., I. B. Naser, M. J. Islam, A. S. Faruque, A. N. Ghosh, G. B. Nair, D. A. Sack, and J. J. Mekalanos. 2005. Seasonal epidemics of cholera inversely correlate with the prevalence of environmental cholera phages. *Proc. Natl. Acad. Sci. USA* **102**:1702–1707.

Faury, N., D. Saulnier, F. L. Thompson, M. Gay, J. Swings, and F. L. Roux. 2004. *Vibrio crassostreae* sp. nov., isolated from the haemolymph of oysters (Crassostrea gigas). *Int. J. Syst. Evol. Microbiol.* **54**:2137–2140.

Food and Drug Administration. 2000. *Draft Risk Assessment on the Public Health Impact of* Vibrio parahaemolyticus *in Raw Molluscan Shellfish*. Center for Food Safety and Applied Nutrition, Food and Drug Administration, Washington, D.C.

Flardh, K., P. S. Cohen, and S. Kjelleberg. 1992. Ribosomes exist in large excess over the apparent demand for protein synthesis during carbon starvation in marine *Vibrio* sp. strain CCUG 15956. *J. Bacteriol.* **174**:6780–6788.

Gil, A. I., V. R. Louis, I. N. Rivera, E. Lipp, A. Huq, C. F. Lanata, D. N. Taylor, E. Russek-Cohen, N. Choopun, R. B. Sack, and R. R. Colwell. 2004. Occurrence and distribution of *Vibrio cholerae* in the coastal environment of Peru. *Environ. Microbiol.* **6**:699–706.

Giovannoni, S., and M. Rappé. 2000. Evolution, diversity, and molecular ecology of marine prokaryotes, p. 47–84. *In* D. L. Kirchman (ed.), *Microbial Ecology of the Oceans.* Wiley-Liss, Inc., New York, N.Y.

Gomez-Gil, B., F. L. Thompson, C. C. Thompson, and J. Swings. 2003. *Vibrio rotiferianus* sp. nov., isolated from cultures of the rotifer *Brachionus plicatilis*. *Int. J. Syst. Evol. Microbiol.* **53**:239–243.

Gomez-Gil, B., L. Tron-Mayen, A. Rogue, J. F. Turnbull, V. Inglis, and A. L. Guerra-Flores. 1998. Species of *Vibrio* isolated from hepatopancreas, haemolymph and digestive tract of a population of healthy juvenile *Penaeus vannamei*. *Aquaculture* **163**:1–9.

Gomez-Leon, J., L. Villamil, M. L. Lemos, B. Novoa, and A. Figueras. 2005. Isolation of *Vibrio alginolyticus* and *Vibrio splendidus* from aquacultured carpet shell clam (*Ruditapes decussatus*) larvae associated with mass mortalities. *Appl. Environ. Microbiol.* **71**:98–104.

Gosink, K. K., R. Kobayashi, I. Kawagishi, and C. C. Hase. 2002. Analyses of the roles of the three cheA homologs in chemotaxis of *Vibrio cholerae*. *J. Bacteriol.* **184**:1767–1771.

Graf, J., and E. G. Ruby. 1998. Host-derived amino acids support the proliferation of symbiotic bacteria. *Proc. Natl. Acad. Sci. USA* **95**:1818–1822.

Grimes, D. J. 1991. Ecology of estuarine bacteria capable of causing human disease: a review. *Estuaries* **14**:345–360.

Guerinot, M. L., and R. R. Colwell. 1985. Enumeration, isolation and characterization of N$_2$-fixing bacteria from seawater. *Appl. Environ. Microbiol.* **50**:350–355.

Guerinot, M. L., and D. G. Patriquin. 1981. The association of N$_2$-fixing bacteria with sea urchins. *Mar. Biol.* **62**:197–207.

Guerinot, M. L., P. A. West, J. V. Lee, and R. R. Colwell. 1982. *Vibrio diazotrophicus* sp. nov., a marine nitrogen-fixing bacteria. *Int. J. Syst. Bacteriol.* **32**:350–357.

Hahn, M. W., and M. G. Hofle. 1998. Grazing pressure by a bacterivorous flagellate reverses the relative abundance of *Coma-*

monas acidovorans PX54 and *Vibrio* strain CB5 in chemostat cocultures. *Appl. Environ. Microbiol.* **64:**1910–1918.

Halpern, M., Y. B. Broza, S. Mittler, E. Arakawa, and M. Broza. 2004. Chironomid egg masses as a natural reservoir of *Vibrio cholerae* non-O1 and non-O139 in freshwater habitats. *Microb. Ecol.* **47:**341–349.

Hauksson, J. B., O. S. Andresson, and B. Asgeirsson. 2000. Heat-labile bacterial alkaline phosphatase from a marine *Vibrio* sp. *Enzyme Microb. Technol.* **2:**66–73.

Haygood, M. G., and D. L. Distel. 1993. Bioluminescent symbionts of flashlight fishes and deep-sea anglerfishes form unique lineages related to the genus *Vibrio. Nature* **363:**154–156.

Heidelberg, J. F., J. A. Eisen, W. C. Nelson, R. A. Clayton, M. L. Gwinn, R. J. Dodson, D. H. Haft, E. K. Hickey, J. D. Peterson, L. Umayam, S. R. Gill, K. E. Nelson, T. D. Read, H. Tettelin, D. Richardson, M. D. Ermolaeva, J. Vamathevan, S. Bass, H. Qin, I. Dragoi, P. Sellers, L. McDonald, T. Utterback, R. D. Fleishmann, W. C. Nierman, O. White, S. L. Salzberg, H. O. Smith, R. R. Colwell, J. J. Mekalanos, J. C. Venter, and C. M. Fraser. 2000. DNA sequence of both chromosomes of the cholera pathogen *Vibrio cholerae. Nature* **406:**477–483.

Heidelberg, J. F., K. B. Heidelberg, and R. R. Colwell. 2002a. Seasonality of Chesapeake Bay bacterioplankton species. *Appl. Environ. Microbiol.* **68:**5488–5497.

Heidelberg, J. F., K. B. Heidelberg, and R. R. Colwell. 2002b. Bacteria of the gamma-subclass Proteobacteria associated with zooplankton in Chesapeake Bay. *Appl. Environ. Microbiol.* **68:**5498–5507.

Herbert, R. A. 1999. Nitrogen cycling in coastal marine ecosystems. *FEMS Microbiol. Rev.* **23:**563–590.

Hood, M. A., J. B. Guckert, D. C. White, and F. Deck. 1986. Effect of nutrient deprivation on lipid, carbohydrate, DNA, RNA, and protein levels in *Vibrio cholerae. Appl. Environ. Microbiol.* **52:**788–793.

Hoppe, H. 2003. Phosphatase activity in the sea. *Hydrobiologia* **493:**187–200.

Jackson, J. K., R. L. Murphree, and M. L. Tamplin. 1997. Evidence that mortality from *Vibrio vulnificus* infection results from single strains among heterogeneous populations in shellfish. *J. Clin. Microbiol.* **35:**2098–2101.

Jiang, S. C., and W. Fu. 2001. Seasonal abundance and distribution of *Vibrio cholerae* in coastal waters quantified by a 16S-23S intergenic spacer probe. *Microb. Ecol.* **42:**540–548.

Jiang, X., and T. J. Chai. 1996. Survival of *Vibrio parahaemolyticus* at low temperatures under starvation conditions and subsequent resuscitation of viable, nonculturable cells. *Appl. Environ. Microbiol.* **62:**1300–1305.

Kaneko, T., and R. R. Colwell. 1973. Ecology of *Vibrio parahaemolyticus* in Chesapeake Bay. *J. Bacteriol.* **113:**24–32.

Kaneko, T., and R. R. Colwell. 1978. Annual cycle of *Vibrio parahaemolyticus* in Chesapeake Bay. *Microb. Ecol.* **4:**135–155.

Karl, D., A. Michaels, B. Bergman, D. G. Capone, E. J. Carpenter, R. Letelier, F. Lipschultz, H. W. Paerl, D. Sigman, and L. J. Stal. 2002. Dinitrogen fixation in the world's oceans. *Biogeochemistry* **57/58:**47–98.

Kaspar, C. W., and M. L. Tamplin. 1993. Effects of temperature and salinity on the survival of *Vibrio vulnificus* in seawater and shellfish. *Appl. Environ. Microbiol.* **59:**2425–2429.

Kaznowski, A., and K. Wlodarczak. 1991. Enzymatic characterization of *Vibrionaceae* strains isolated from environment and cold-blooded animals. *Acta Microbiol. Pol.* **40:**71–76.

Kirchman, D. L., B. Meon, H. W. Ducklow, C. A. Carlson, D. A. Hansell, and G. F. Steward. 2001. Glucose fluxes and concentrations of dissolved combined neutral sugars (polysaccharides)

in the Ross Sea and Polar Front Zone, Antarctica. *Deep Sea Res. Part II-Topical Studies in Oceanography* **48:**4179–4197.

Kramer, J. G., and F. L. Singleton. 1992. Variations in rRNA content of marine *Vibrio* spp. during starvation-survival and recovery. *Appl. Environ. Microbiol.* **58:**201–207.

La Rosa, T., S. Mirto, A. Marino, V. Alonzo, T. L. Maugeri, and A. Mazzola. 2001. Heterotrophic bacteria community and pollution indicators of mussel-farm impact in the Gulf of Gaeta (Tyrrhenian Sea). *Mar. Environ. Res.* **52:**301–321.

Larsen, M. H., N. Blackburn, J. L. Larsen, and J. E. Olsen. 2004. Influences of temperature, salinity and starvation on the motility and chemotactic response of *Vibrio anguillarum. Microbiology* **150:**1283–1290.

Lee, K.-H., and E. G. Ruby. 1994. Effect of the squid host on the abundance and distribution of symbiotic *Vibrio fischeri* in nature. *Appl. Environ. Microbiol.* **60:**1565–1571.

Li, X., and S. Roseman. 2004. The chitinolytic cascade in vibrios is regulated by chitin oligosaccharides and a two-component chitin catabolic sensor/kinase. *Proc. Natl. Acad. Sci. USA* **101:**627–631.

Linton, J. D., D. E. F. Harrison, and A. T. Bull. 1975. Molar growth yields, respiration and cytochrome patterns of *Beneckea natriegens* when grown at different medium dissolved-oxygen tensions. *J. Gen. Microbiol.* **90:**237–246.

Linton, J. D., D. E. Harrison, and A. T. Bull. 1977. Molar growth yields, respiration and cytochrome profiles of *Beneckea natriegens* when grown under carbon limitation in a chemostat. *Arch. Microbiol.* **115:**135–142.

Lobitz, B., L. Beck, A. Huq, B. Wood, G. Fuchs, A. S. G. Faruque, and R. Colwell. 2000. Climate and infectious disease: use of remote sensing for detection of *Vibrio cholerae* by indirect measurement. *Proc. Natl. Acad. Sci. USA* **97:**1438–1443.

Long, R. A., and F. Azam. 2001. Antagonistic interactions among marine pelagic bacteria. *Appl. Environ. Microbiol.* **67:**4975–4983.

Lovell, C. R., Y. M. Piceno, J. M. Quattro, and C. E. Bagwell. 2000. Molecular analysis of diazotroph diversity in the rhizosphere of the smooth cordgrass, *Spartina alterniflora. Appl. Environ. Microbiol.* **66:**3814–3822.

Lunder, T., H. Sorum, G. Holstad, A. G. Steigerwalt, P. Mowinckel, and D. J. Brenner. 2000. Phenotypic and genotypic characterization of *Vibrio viscosus* sp. nov. and *Vibrio wodanis* sp. nov. isolated from Atlantic salmon (*Salmo salar*) with "winter ulcer." *Int. J. Syst. Evol. Microbiol.* **50**(Pt. 2):427–450.

Maeda, T., Y. Matsuo, M. Furushita, and T. Shiba. 2003. Seasonal dynamics in a coastal *Vibrio* community examined by a rapid clustering method based on 16S rDNA. *Fish. Sci.* **69:**385–394.

Makino, K., K. Oshima, K. Kurokawa, K. Yokoyama, T. Uda, K. Tagomori, Y. Iijima, M. Najima, M. Nakano, A. Yamashita, Y. Kubota, S. Kimura, T. Yasunaga, T. Honda, H. Shinagawa, M. Hattori, and T. Iida. 2003. Genome sequence of *Vibrio parahaemolyticus*: a pathogenic mechanism distinct from that of *V. cholerae. Lancet* **361:**743–749.

Martinez, J., and F. Azam. 1993. Periplasmic aminopeptidase and alkaline-phosphatase activities in a marine bacterium—implications for substrate processing in the sea. *Mar. Ecol. Prog. Ser.* **92:**89–97.

Martinez, J., D. C. Smith, G. F. Steward, and F. Azam. 1996. Variability in ectohydrolytic enzyme activities of pelagic marine bacteria and its significance for substrate processing in the sea. *Aquat. Microb. Ecol.* **10:**223–230.

Massana, R., C. Pedros-Alio, E. O. Casamayor, and J. M. Gasol. 2001. Changes in marine bacterioplankton phylogenetic composition during incubations designed to measure biogeochemically significant parameters. *Limnol. Oceanogr.* **46:**1181–1188.

Meon, B., and D. L. Kirchman. 2001. Dynamics and molecular composition of dissolved organic material during experimental phytoplankton blooms. *Mar. Chem.* **75:**185–199.

Middelboe, M., L. Riemann, G. F. Steward, V. Hansen, and O. Nybroe. 2003. Virus-induced transfer of organic carbon between marine bacteria in a model community. *Aquat. Microb. Ecol.* **33:**1–10.

Motes, M. L., A. DePaola, D. W. Cook, J. E. Veazey, J. C. Hunsucker, W. E. Garthright, R. J. Blodgett, and S. J. Chirtel. 1998. Influence of water temperature and salinity on *Vibrio vulnificus* in northern Gulf and Atlantic Coast oysters *(Crassostrea virginica). Appl. Environ. Microbiol.* **64:**1459–1465.

Mourino-Perez, R. R., A. Z. Worden, and F. Azam. 2003. Growth of *Vibrio cholerae* O1 in red tide waters off California. *Appl. Environ. Microbiol.* **69:**6923–6931.

Nemecek-Marshall, M., C. Wojciechowski, W. P. Wagner, and R. Fall. 1999. Acetone formation in the *Vibrio* family: a new pathway for bacterial leucine catabolism. *J. Bacteriol.* **181:** 7493–7499.

Nichols, D. S. 2003. Prokaryotes and the input of polyunsaturated fatty acids to the marine food web. *FEMS Microbiol. Lett.* **219:** 1–7.

Nishiguchi, M., E. Ruby, and M. McFall-Ngai. 1998. Competitive dominance among strains of luminous bacteria provides an unusual form of evidence for parallel evolution in sepiolid squid-vibrio symbioses. *Appl. Environ. Microbiol.* **64:**3209–3213.

Nogi, Y., N. Masui, and C. Kato. 1998. *Photobacterium profundum* sp. nov., a new, moderately barophilic bacterial species isolated from a deep-sea sediment. *Extremophiles* **2:**1–7.

Olson, J. B., R. W. Litaker, and H. W. Paerl. 1999. Ubiquity of heterotrophic diazotrophs in marine microbial mats. *Aquat. Microb. Ecol.* **19:**29–36.

Orndorff, S. A., and R. R. Colwell. 1980. Distribution and identification of luminous bacteria from the Sargasso Sea. *Appl. Environ. Microbiol.* **39:**983–987.

Ovreas, L., D. Bourne, R. A. Sandaa, E. O. Casamayor, S. Benlloch, V. Goddard, G. Smerdon, M. Heldal, and T. F. Thingstad. 2003. Response of bacterial and viral communities to nutrient manipulations in seawater mesocosms. *Aquat. Microb. Ecol.* **31:** 109–121.

Pernthaler, A., J. Pernthaler, H. Eilers, and R. Amann. 2001. Growth patterns of two marine isolates: adaptations to substrate patchiness? *Appl. Environ. Microbiol.* **67:**4077–4083.

Pinhassi, J., and T. Berman. 2003. Differential growth response of colony-forming alpha- and gamma-proteobacteria in dilution culture and nutrient addition experiments from Lake Kinneret (Israel), the eastern Mediterranean Sea, and the Gulf of Eilat. *Appl. Environ. Microbiol.* **69:**199–211.

Proctor, L. M., and R. P. Gunsalus. 2000. Anaerobic respiratory growth of *Vibrio harveyi, Vibrio fischeri* and *Photobacterium leiognathi* with trimethylamine N-oxide, nitrate and fumarate: ecological implications. *Environ. Microbiol.* **2:**399–406.

Radjasa, O. K., H. Urakawa, K. Kita-Tsukamoto, and K. Ohwada. 2001. Characterization of psychrotrophic bacteria in the surface and deep-sea waters from the northwestern Pacific Ocean based on 16S ribosomal DNA analysis. *Mar. Biotechnol.* (NY) **3:**454–462.

Raguenes, G., R. Christen, J. Guezennec, P. Pignet, and G. Barbier. 1997. *Vibrio diabolicus* sp. nov., a new polysaccharide-secreting organism isolated from a deep-sea hydrothermal vent polychaete annelid, *Alvinella pompejana. Int. J. Syst. Bacteriol.* **47:**989–995.

Ramaiah, N., J. Ravel, W. L. Straube, R. T. Hill, and R. R. Colwell. 2002. Entry of *Vibrio harveyi* and *Vibrio fischeri* into the viable but nonculturable state. *J. Appl. Microbiol.* **93:**108–116.

Randa, M. A., M. F. Polz, and E. Lim. 2004. Effects of temperature and salinity on *Vibrio vulnificus* population dynamics as assessed by quantitative PCR. *Appl. Environ. Microbiol.* **70:** 5469–5476.

Reay, D. S., D. B. Nedwell, J. Priddle, and J. C. Ellis-Evans. 1999. Temperature dependence of inorganic nitrogen uptake: reduced affinity for nitrate at suboptimal temperatures in both algae and bacteria. *Appl. Environ. Microbiol.* **65:**2577–2584.

Rehnstam, A.-S., S. Bäckman, D. C. Smith, F. Azam, and Å. Hagström. 1993. Blooms of sequence-specific culturable bacteria in the sea. *FEMS Microbiol. Ecol.* **102:**161–166.

Riemann, L., and F. Azam. 2002. Widespread N-acetyl-D-glucosamine uptake among pelagic marine bacteria and its ecological implications. *Appl. Environ. Microbiol.* **68:**5554–5562.

Roy, N. K., R. K. Ghosh, and J. Das. 1982. Repression of the alkaline phosphatase of *Vibrio cholerae. J. Gen. Microbiol.* **128:** 349–353.

Ruby, E. G., E. P. Greenberg, and J. W. Hastings. 1980. Planktonic marine luminous bacteria—species distribution in the water column. *Appl. Environ. Microbiol.* **39:**302–306.

Ruby, E. G., and K. H. Lee. 1998. The *Vibrio fischeri Euprymna scolopes* light organ association: current ecological paradigms. *Appl. Environ. Microbiol.* **64:**805–812.

Ruby, E. G., M. Urbanowski, J. Campbell, A. Dunn, M. Faini, R. Gunsalus, P. Lostroh, C. Lupp, J. McCann, D. Millikan, A. Schaefer, E. Stabb, A. Stevens, K. Visick, C. Whistler, and E. P. Greenberg. 2005. Complete genome sequence of *Vibrio fischeri*: a symbiotic bacterium with pathogenic congeners. *Proc. Natl. Acad. Sci. USA* **102:**3004–3009.

Sakai, Y., K. Toda, Y. Mitani, M. Tsuda, S. Shinoda, and T. Tsuchiya. 1987. Properties of the membrane-bound 5′-nucleotidase and utilization of extracellular ATP in *Vibrio parahaemolyticus. J. Gen. Microbiol.* **133:**2751–2757.

Sarker, R. I., W. Ogawa, M. Tsuda, S. Tanaka, and T. Tsuchiya. 1994. Characterization of a glucose transport system in *Vibrio parahaemolyticus. J. Bacteriol.* **176:**7378–7382.

Sawabe, T., N. Setoguchi, S. Inoue, R. Tanaka, M. Ootsubo, M. Yoshimizu, and Y. Ezura. 2003. Acetic acid production of *Vibrio halioticoli* from alginate: a possible role for establishment of abalone-*V. halioticoli* association. *Aquaculture* **219:**671–679.

Sherr, E. B., and B. F. Sherr. 2002. Significance of predation by protists in aquatic microbial food webs. *Antonie Leeuwenhoek* **81:**293–308.

Shieh, W. Y., Y. T. Lin, and W. D. Jean. 2004. *Pseudovibrio denitrificans* gen. nov., sp. nov., a marine, facultatively anaerobic, fermentative bacterium capable of denitrification. *Int. J. Syst. Evol. Microbiol.* **54:**2307–2312.

Simidu, U., and K. Tsukamoto. 1985. Habitat segregation and biochemical activities of marine members of the family *Vibrionaceae. Appl. Environ. Microbiol.* **50:**781–790.

Suginta, W., P. A. W. Robertson, B. Austin, S. C. Fry, and L. A. Fothergill-Gilmore. 2000. Chitinases from *Vibrio*: activity screening and purification of chiA from *Vibrio carchariae. J. Appl. Microbiol.* **89:**76–84.

Thompson, F. L., T. Iida, and J. Swings. 2004. Biodiversity of vibrios. *Microbiol. Mol. Biol. Rev.* **68:**403–431.

Thompson, J. R., L. A. Marcelino, and M. F. Polz. 2005a. Diversity, sources, and detection of human bacterial pathogens in the marine environment, p. 29–68. *In* R. Colwell and S. Belkin (ed.), *Oceans and Health: Pathogens in the Marine Environment.* ASM Press, Washington, D.C.

Thompson, J. R., S. Pacocha, C. Pharino, V. Klepac-Ceraj, D. E. Hunt, J. Benoit, R. Sarma-Rupavtarm, D. L. Distel, and M. F. Polz. 2005b. Genotypic diversity within a natural coastal bacterioplankton population. *Science* **307:**1311–1313.

Thompson, J. R., M. A. Randa, L. A. Marcelino, A. Tomita, E. L. Lim, and M. F. Polz. 2004. Diversity and dynamics of a North Atlantic coastal vibrio community. *Appl. Environ. Microbiol.* 70:4103–4110.

Urakawa, H., K. Kita-Tsukamoto, and K. Ohwada. 1999a. 16S rDNA restriction fragment length polymorphism analysis of psychrotrophic vibrios from Japanese coastal water. *Can. J. Microbiol.* 45:1001–1007.

Urakawa, H., K. Kita-Tsukamoto, and K. Ohwada. 1999b. Restriction fragment length polymorphism analysis of psychrophilic and psychrotrophic *Vibrio* and *Photobacterium* from the northwestern Pacific Ocean and Otsuchi Bay, Japan. *Can. J. Microbiol.* 45:67–76.

Urakawa, H., T. Yoshida, M. Nishimura, and K. Ohwada. 2000. Characterization of depth-related population variation in microbial communities of a coastal marine sediment using 16S rDNA-based approaches and quinone profiling. *Environ. Microbiol.* 2:542–554.

Urdaci, M. C., L. J. Stal, and M. Marchand. 1988. Occurrence of nitrogen fixation among *Vibrio* spp. *Arch. Microbiol.* 150:224–229.

Waldor, M. K., and J. J. Mekalanos. 1996. Lysogenic conversion by a filamentous phage encoding cholera toxin. *Science* 272:1910–1914.

WardRainey, N., F. A. Rainey, and E. Stackebrandt. 1996. A study of the bacterial flora associated with *Holothuria atra. J. Exp. Mar. Biol. Ecol.* 203:11–26.

Woolkalis, M. J., and P. Baumann. 1981. Evolution of alkaline phosphatase in marine species of *Vibrio. J. Bacteriol.* 147:36–45.

Worden, A. Z., M. Seidel, S. Smriga, A. Wick, F. Malfatti, D. Bartlett, and F. Azam. 2006. Trophic regulation of *Vibrio cholerae* in coastal marine waters. *Environ. Microbiol.* 8:21–29.

Wright, A. C., R. T. Hill, J. A. Johnson, M. C. Roghman, R. R. Colwell, and J. G. Morris. 1996. Distribution of *Vibrio vulnificus* in the Chesapeake Bay. *Appl. Environ. Microbiol.* 62:717–724.

Yu, C., B. L. Bassler, and S. Roseman. 1993. Chemotaxis of the marine bacterium *Vibrio furnissii* to sugars—a potential mechanism for initiating the chitin catabolic cascade. *J. Biol. Chem.* 268:9405–9409.

Zehr, J. P., M. T. Mellon, and S. Zani. 1998. New nitrogen-fixing microorganisms detected in oligotrophic oceans by amplification of nitrogenase (nifH) genes. *Appl. Environ. Microbiol.* 64:3444–3450.

Zehr, J. P., and B. B. Ward. 1995. Diversity of heterotrophic nitrogen fixation genes in a marine cyanobacterial mat. *Appl. Environ. Microbiol.* 61:2527–2532.

The Biology of Vibrios
Edited by F. L. Thompson et al.
© 2006 ASM Press, Washington, D.C.

Chapter 14

The *Vibrio fischeri–Euprymna scolopes* Light Organ Symbiosis

Eric V. Stabb

INTRODUCTION

The light organ symbiosis between the bioluminescent bacterium *Vibrio fischeri* and the Hawaiian bobtail squid *Euprymna scolopes* has received increasing interest from researchers representing a range of disciplines, including microbiology, zoology, oceanography, immunology, and genetics. Most of the scientists who study this symbiosis are lured to it by a fascination for these two remarkable organisms and their intriguing partnership. This allure is further reflected in the numerous reviews, popularized accounts, and even poetry dedicated to this symbiosis. Such "gee whiz" appeal has engendered great enthusiasm and scientific curiosity, which are powerful motivators, but, ultimately, progress in the field has stemmed from a rare combination of characteristics. Specifically, like all good model systems, this symbiosis is representative of important widespread phenomena while also possessing a unique combination of features that make it experimentally tractable.

WHAT DOES THIS MODEL SYSTEM MODEL?

All animals have associated microorganisms that can contribute to the host's health and normal development, and this native microbiota has recently been recognized as an important underrepresented research focus (Darveau et al., 2003). An improved understanding of the factors underlying these natural symbioses will help face such problems as the (re)emergence of diseases with environmental hosts, the effective application of probiotics, overreactions to commensal bacteria by the immune system, and the negative effects of broad-spectrum antibiotics.

The *V. fischeri–E. scolopes* symbiosis serves to model many animal–bacteria interactions. Like innumerable bacteria–animal interactions, this symbiosis is composed of an extracellular gram-negative bacterium colonizing the microvillous surface of a polarized epithelium. *V. fischeri* also persistently infects the host without causing disease, which is typical of most animal–bacteria associations. Furthermore, like the microbiota of many animals, *V. fischeri* triggers developmental changes in the host. Therefore, insight into the *V. fischeri–E. scolopes* symbiosis may shed light on other host-associated microbiota, including our own.

V. fischeri is also representative of a bacterial family that encompasses important host-associated bacteria, the *Vibrionaceae*. Although often benign colonists of marine animals, some species, such as *Vibrio cholerae*, *Vibrio parahaemolyticus*, and *Vibrio vulnificus*, are also opportunistic or emerging pathogens, and the spread of these pathogens through natural aquatic hosts poses a human health threat. Other species, e.g., *Vibrio harveyi*, *Vibrio anguillarum*, and *Vibrio campbellii*, can cause disease in farmed marine animals. Thus, the *V. fischeri–E. scolopes* symbiosis may also serve as a model for vibrio-host interactions that are important for disease epidemiology and aquaculture.

EXPERIMENTAL TRACTABILITY

As described above, the *V. fischeri–E. scolopes* symbiosis has many features typical of important host-bacteria interactions. However, what sets this interaction apart is that a unique set of features renders it experimentally tractable, allowing approaches that would be difficult to conduct or interpret in other systems.

One requisite feature for tractability is that both partners can be maintained in the laboratory. Typi-

Eric V. Stabb • Department of Microbiology, 828 Biological Sciences, University of Georgia, Athens, GA 30602.

cal of many vibrios, *V. fischeri* is readily cultured. *E. scolopes* requires more husbandry, notably, a need for live prey. Nonetheless, adults, which are generally <5 cm long, can be kept in small aquaria, and inducing the animals to mate requires only that a male and a female are placed in the same tank. A typical female will lay 500 to 1,000 eggs during a 2- to 6-month reproductive life in the laboratory (Fig. 1). *E. scolopes* can even be raised from hatchlings to reproductive adults in captivity (Hanlon et al., 1997), although this requires live prey of varying sizes. In practice, most researchers replenish their adult breeding stock with wild-caught animals. The animals are common in shallow sandy reef areas less than a meter deep, so catching them does not require expensive equipment.

Ideally, maintaining the symbiotic partners separately is a first step toward experimentally reconstituting the symbiosis. Unfortunately, many bacterial symbionts are transmitted through the host's eggs, making it difficult or impossible to generate symbiont-free animals. However, *V. fischeri* is not present in *E. scolopes* eggs, and hatchlings must acquire *V. fischeri* by horizontal transmission from the environment. This allows controlled establishment of the symbiosis in a laboratory setting. Moreover, wild *V. fischeri* populations and the kinetics of natural infections are known, so experimental infections can be performed to mimic natural parameters.

An unusual and useful characteristic of this symbiosis is that colonization results in bacterial biolu-

minescence. As described below, this bioluminescence is apparently used as an antipredatory mechanism. Thus, in contrast to many (e.g., nutritional) symbioses, lab-reared *E. scolopes* are not compromised if they lack *V. fischeri* symbionts. Furthermore, in hatchling squid, bioluminescence intensity is correlated with the degree of bacterial colonization, allowing researchers to monitor the progress of infection noninvasively and nondestructively. As in many other systems, the animals can also be anesthetized, homogenized, and plated to determine the number of symbiont CFU. Luminescence data provide details of infection kinetics and help indicate key time points for plating.

Another unusual feature that makes this system tractable is that only *V. fischeri* colonizes the host's light organ. Many beneficial bacteria, notably, gut symbionts in animals, live in mixed consortia, and the complexity of these communities makes it difficult to reconstitute a natural system or to tease apart the respective role(s) of different symbionts. The specificity of the *V. fischeri–E. scolopes* symbiosis not only obviates these problems, but it also presents a fascinating platform from which to study how hosts can distinguish "good" bacteria from "bad."

Another unusual property of *E. scolopes* is an anatomy that renders symbiotic tissues readily accessible. The light organ is essentially external and can be viewed simply by pulling back the mantle and funnel. Moreover, the light organ tissue is transparent, so fluorescence-based microscopy can be used to view bacterial colonization inside an intact organ. The light organ is also accessible to experimentally added solutes. Pores connect the outside environment with the site of infection in the light organ, so reagents can be added to cure an infection or monitor biochemical processes at the site of infection. Finally, the light organ can be readily plucked away from other tissues. This sort of accessibility stands in contrast to systems that require delicate surgeries to reach symbiotic tissues of interest.

Other features of the *V. fischeri–E. scolopes* symbiosis are less unique but also contribute to its utility as a model system. *V. fischeri* is genetically manipulable, and the introduction and mutation of genes are now common practice. These approaches are being augmented by the genome sequence of *V. fischeri* (Ruby et al., 2005). Similar progress is being made in nascent genomic and proteomic initiatives in the host (Doino and McFall-Ngai, 2000; Kimbell and McFall-Ngai, 2004). Finally, while the lack of a backbone is hardly a unique trait of *E. scolopes*, there are both ethical and regulatory advantages to studying an invertebrate host.

Figure 1. *Euprymna scolopes*. An adult female Hawaiian bobtail squid sits on coralline sand near eggs (indicated by white arrows) that she laid on a PVC half-pipe and on the side of her tank. While housed in a 50-liter tank at the University of Georgia, this female produced 1,616 juveniles over a 4.5-month span.

EVOLUTION AND BASIC BIOLOGY

The first recorded observations of sepiolid squid with bioluminescent bacterial symbionts date to the early 20th century. These squid, now classified in the genera *Sepiola* and *Euprymna*, are found in the Mediterranean Sea and the Pacific Ocean. Although the original reports are hard to find today, most of the seminal works are reviewed in the wonderfully complete book *Bioluminescence* (Harvey, 1952). In each species, a two-lobed organ was found closely associated with the ink gland, in the mantle cavity, near the animal's ventral surface. Of particular note, in 1928 Kishitani made detailed drawings of sections through a *Euprymna* light organ, revealing epithelia-lined pockets filled with bioluminescent bacterial symbionts. Kishitani also described pores connecting these pockets to the light organ surface, and reported that the light organs had reflective tissue on the dorsal surface and a lens on the ventral surface.

Selective Advantages

The light organ architecture suggests that it functions in the camouflaging behavior known as counterillumination (Fig. 2), wherein marine animals emit light downward, matching the light from above and obscuring their silhouette from predators beneath them (Harper and Case, 1999). This is consistent with the light organ morphology of adult *E. scolopes*, wherein reflector and lens tissue direct bioluminescence downward and the ink sac can be used as a shutter to control the amount of light emitted. Furthermore, the amount of light emitted by adult *E. scolopes* is correlated with the intensity of down-welling light

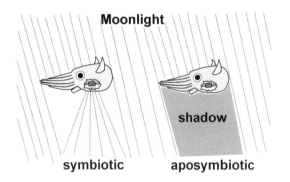

Figure 2. Proposed counterillumination behavior of *E. scolopes*. Based on light organ architecture, it is thought that an adult *E. scolopes* with bioluminescent symbiotic *V. fischeri* can obscure its silhouette with a controllable, ventrally directed luminescence, whereas an aposymbiotic (lacking symbionts) squid would cast a shadow.

(Jones and Nishiguchi, 2004). Camouflaging also seems to be a general strategy for *E. scolopes*, which routinely coats itself with sand (Shears, 1988), changes color to hide, and emits distracting squid-sized ink blobs when threatened (Anderson and Mather, 1996). Conversely, reports of animals emitting brief flashes of light (Moynihan, 1982) or even a persistent bright glow (A. Wier, personal communication) suggest that luminescence may have functions other than camouflaging. More detailed behavioral studies will help determine the benefit(s) of the symbiosis to the host.

The selective advantage of the symbiosis to *V. fischeri* is clear. Each morning, the nocturnal squid bury themselves with sand, and the light of dawn triggers the expulsion of most symbionts into the environment (Boettcher et al., 1996). The remaining *V. fischeri* regrow throughout the day, through two to four generations, so that the light organ is repopulated by the time the animals resume their nocturnal behaviors. As a result of this diurnal "venting," *V. fischeri* populations are relatively large in habitats occupied by *E. scolopes* (Lee and Ruby, 1994b). Further ecological studies and strain typing (Lee and Ruby, 1994a) leave little doubt that in near-shore Hawaiian waters the ability to colonize *E. scolopes* is a strong selective force on *V. fischeri*.

Coevolution

The *V. fischeri*–squid symbiosis appears to be an ancient association in which the host and bacteria have fine-tuned the specifics of their partnership. Many species in the *Sepiola* and *Euprymna* genera have light organs, and their bioluminescent symbionts are either *V. fischeri* or its psychrophilic relative *Vibrio logei* (Fidopiastis et al., 1998; Nishiguchi, 2000). In general, the molecular phylogenies of the hosts and their respective symbionts parallel each other (Nishiguchi et al., 1998; Nishiguchi, 2002), consistent with a model whereby a light organ symbiosis arose in an ancient squid, which coevolved with its symbionts as it speciated. Both early histological and recent molecular studies suggest that this original light organ may have arisen from another tissue colonized by bacteria, the accessory nidamental gland (Harvey, 1952; Nishiguchi et al., 2004).

This model of a coevolved association also fits the observation that *V. fischeri* isolates from the light organs of monocentrid fishes constitute a distinct lineage from the squid-associated *V. fischeri* (Lee, 1994). Coevolution is further intimated by apparent functional coadaptations between hosts and their respective symbionts. For example, *V. fischeri* isolated from the light organs of the fishes *Monocentris japonica* and *Cleidopus gloramaris* can colonize *E. scolopes*, but the populations attained by these foreign strains in the squid light organ are 10- to 100-fold lower than those achieved by native *V. fischeri* isolates (Ruby and Lee, 1998). Similarly, when *V. fischeri* strains compete for host colonization in mixed inocula, the native strain, or the strain closest to the native lineage, tends to dominate (Lee and Ruby, 1994a,b; Nishiguchi et al., 1998; Nishiguchi, 2002).

Taken together, these data indicate that each host has a closely evolved relationship with its respective (native) *V. fischeri* symbionts. This has at least two practical implications for current research. First, it underscores the importance of using native *V. fischeri* in symbiotic studies. Thus, *V. fischeri* MJ1, a type strain used in many pioneering studies of bioluminescence and isolated from the light organ of *M. japonica* (Ruby and Nealson, 1976; Engebrecht et al., 1983), is not an appropriate strain for studies of the *V. fischeri–E. scolopes* symbiosis. Second, although it has become increasingly clear that some bacteria have shifted host range or tissue tropism rapidly through the acquisition of discrete "islands" of genes, this does not seem to be the case for the *V. fischeri* light organ symbionts. Rather, the *V. fischeri*–squid symbiosis probably predates the speciation of *E. scolopes*, and it seems likely that the composition and regulation of many genes will have adapted to maximize the benefits of this partnership.

Basics of the *V. fischeri–E. scolopes* Association

E. scolopes lives in shallow, sandy reef flats in the Hawaiian archipelago. These solitary predators feed mainly on shrimp and polychaetes as they grow from hatchlings just a few millimeters long to thumb-sized adults (Moynihan, 1982; Shears, 1988). The adults may live for several months but probably not more than a year (Singley, 1983; Hanlon et al., 1997). Females lay clutches of tens or hundreds of eggs, and, as mentioned above, each new generation of hatchlings must acquire *V. fischeri* from the surrounding environment (Wei and Young, 1989).

Most studies of this symbiosis focus on the establishment of *V. fischeri* in the light organ through the first 2 to 4 days after hatching. Hatchlings derive energy from a yolk sac; however, after 4 to 5 days, the animals have apparently depleted these reserves and begin to show signs of stress. Interestingly, such stressed animals can clear out established *V. fischeri* symbionts. Researchers generally terminate experiments before this point, although it is possible to extend studies by feeding the juveniles (Hanlon et al., 1997; Claes and Dunlap, 2000).

The light organs of juvenile *E. scolopes* are shown in Color Plate 6. The light organ is loosely covered by the mantle on the animal's ventral surface (Color Plate 6A), and it consists of relatively transparent tissue and a reflector closely associated with the ink sac (Color Plate 6A,B). Symbionts labeled with green fluorescent protein can be visualized inside this transparent tissue (Color Plate 6C), and electron microscopy reveals their extracellular association with epithelial microvilli (Color Plate 6D). In hatchlings, each side of the light organ has a field of ciliated cells on its surface, and each ciliated field includes two appendages that often approximate a discontinuous ring (Color Plate 6E). At the base of these ciliated appendages are three pores, each 10 to 15 μm in diameter, that lead through ciliated ducts into distinct "crypts" lined by microvillous polarized epithelial cells (Color Plate 6F). These crypts span a 10-fold range in size from crypt 1, the largest, to crypt 3, the smallest and also developmentally delayed relative to crypts 1 and 2 (Montgomery and McFall-Ngai, 1994). The crypts are topologically complex, encompassing enlarged "antechambers" between the duct and the largest space as well as deep veins extending medially from crypt 1 (Visick and McFall-Ngai, 2000; Millikan and Ruby, 2004). These extensions of crypt 1 have been termed "medial diverticula" (Visick and McFall-Ngai, 2000), although it is uncertain whether these are actually dead-end diverticular projections or an interconnecting network (as depicted in Color Plate 6F). Regardless, this medial region of crypt 1 is densely packed with bacteria once animals become colonized (Color Plate 6C).

EARLY EVENTS IN SYMBIONT ESTABLISHMENT

In the waters inhabited by *E. scolopes*, aposymbiotic hatchlings become infected very quickly (Wei and Young, 1989), and even the smallest animals that have been caught in the wild and examined are already luminescent (Ruby and Asato, 1993). When fresh hatchlings were placed in near-shore Hawaiian seawater, the onset of luminescence was detectable within 8 h (Ruby and McFall-Ngai, 1991). Initial colonization by small numbers of bacteria, which have not yet begun to generate light, occurs even earlier (Ruby and Asato, 1993).

Obstacles To Overcome

The speed and efficiency of symbiont establishment are astounding considering the obstacles to infection. *V. fischeri* is present in Hawaiian waters at densities of only 100 to 1,000 cells/ml (Lee and Ruby,

1995), representing <0.1% of the total bacterial population. The light organ is located in a ~1.3-μl mantle cavity, which one can estimate contains between zero and two *V. fischeri* cells at any given time, and the squid flush the mantle water every half second (Nyholm et al., 2000). Numerically, the odds seem stacked against infection occurring at all.

The prospects for infection appear even bleaker when taking into account the physiological state of *V. fischeri*. The population of 100 to 1,000 *V. fischeri* cells/ml noted above was determined by quantitative hybridization to environmental DNA; however, the *V. fischeri* cells that form colonies on solid media represent only about 1% of this total (i.e., 1 to 10 CFU/ml) (Lee and Ruby, 1992, 1995). The nonculturable majority appear able to initiate the symbiosis, because dilutions that contained <1 CFU/ml but 10 to 100 unculturable cells were still infective (Lee and Ruby, 1995). Thus, these cells resemble certain other viable but nonculturable (VBNC) vibrios. Little is known about the VBNC state of *V. fischeri*, but based on studies of VBNC *V. cholerae* cells (Kondo et al., 1994), VBNC *V. fischeri* may be relatively dormant and nonmotile.

Thus, one could conclude that between zero and two semidormant *V. fischeri* cells are in the mantle cavity of a hatchling squid at any time, and that these cells have about a second to revive, induce motility, and swim into one of the six small pores on the surface of the light organ before being pushed back out into the surrounding water. Yet juveniles are efficiently infected within hours of hatching! This riddle was solved in an elegant series of microscopic experiments using fluorescent labels and stains to observe the bacteria, abiotic bacteria-sized particles, and host components through the early infection process (Nyholm et al., 2000, 2002; Nyholm and McFall-Ngai, 2003). The results of these observations are described in stages below.

Stages in Symbiont Establishment

Stage 0: hatching

Events leading to symbiont infection begin shortly after *E. scolopes* juveniles emerge from the egg. This early time frame is categorized here as "stage 0," to connote processes that occur before contact with the *V. fischeri* that will ultimately establish an infection. Within 10 to 20 s of hatching, the light organ cilia begin beating, increasing the flow of water across the light organ surface (Nyholm et al., 2000, 2002). For the next hour, the light organ enters a mysterious "permissive" phase, during which particles ~1 μm in diameter or smaller are drawn into the crypts (Nyholm

et al., 2002). Such particles can include small bacterial cells, even nonmotile cells. However, any cells that are brought into the crypts, including *V. fischeri*, are cleared out again and apparently do not establish an infection. For just a few particles to appear inside the crypts, they must be added to water at a high density (millions per milliliter), which is similar to total bacterial populations in seawater but roughly 1,000-fold higher than native *V. fischeri* populations. Furthermore, the absence of *V. fischeri* during the permissive phase does not seem to attenuate the eventual infection process. Taken together, these data indicate that the light organ somehow samples the environment upon hatching; however, the significance of this permissiveness and whether uptake of *V. fischeri* is relevant remain unclear.

An hour after hatching, the animals cease their permissive, nonspecific uptake of external particles. At about the same time, but in an independent event, epithelial cells in the ciliated epithelial appendages begin shedding mucus (Nyholm et al., 2002). Both sialomucin and neutral mucin are released from these epithelial cells, with the former dominating. This mucus secretion is stimulated by peptidoglycan from either gram-negative or gram-positive bacteria, and entry into the crypts is not required for perception of this signal. With the cilia still beating, mucus is moved onto the light organ surface, and seawater is continuously drawn across this mucus matrix.

Stage 1: aggregation

Water carried into the mantle cavity by ventilation and stirred by the ciliated epithelial appendages brings environmental bacteria into contact with the mucus on the light organ surface. Through this process, initial contact is made between the host and its eventual symbiont(s). Bacteria aggregate in the mucus, forming defined balls of cells visible by fluorescence microscopy (Nyholm et al., 2000). At natural densities of *V. fischeri*, these aggregates may be composed of <20 cells (Millikan and Ruby, 2002). Much larger and easily observable aggregates form when denser inocula are used.

Although other gram-negative bacteria will also aggregate on the light organ surface (Nyholm et al., 2000), *V. fischeri* outcompetes them (Nyholm and McFall-Ngai, 2003). The mechanism(s) that favors *V. fischeri* in the aggregate remains to be determined; however, the presence of nitric oxide (NO) in the aggregates and the observation that inhibitors of NO synthesis enable a nonsymbiotic bacterium to form unusually large aggregates suggest a role for this antimicrobial agent in enriching *V. fischeri* (Davidson et al., 2004). Cell surface functions associated with attachment or motility may also contribute to formation of robust aggregates by *V. fischeri* (Millikan and Ruby, 2002), although motility is not required for aggregate formation.

Thus, in the aggregation stage, the squid overcome the challenge of obtaining symbionts in a dilute environment by concentrating bacteria on the light organ surface. An enrichment process in the aggregates further helps overcome the fact that *V. fischeri* represents a small minority of the total bacteria present.

Stage 2: migration

In the next stage of infection, bacteria that have sat essentially motionless in aggregates move toward the light organ pores. Usually, this migration occurs 4 to 6 h after hatching (Nyholm et al., 2000). Based on this timing, it appears most bacteria will have been in an aggregate for a few hours before migrating, allowing ample time for the bacteria to sense their new location and change gene expression accordingly. This could explain how, in nature, semidormant VBNC cells are able to revive and establish an infection.

It is not known what triggers the symbionts to proceed from the aggregation stage to the migration stage. It is tempting to speculate that gene expression in aggregates is controlled via "quorum-sensing" density-dependent regulation, and that a "quorum" of bacteria must be reached in an aggregate before migration proceeds. However, the quorum-sensing LuxIR regulon does not appear to be activated even in large aggregates (Nyholm and McFall-Ngai, 2003).

Migration from the aggregates to the pores requires flagellar motility (Nyholm et al., 2000). Moreover, fast-swimming hyperflagellated mutants move prematurely and do not aggregate or migrate normally (Millikan and Ruby, 2002). Thus, both motility and its coordinated regulation are important in this process. Chemotaxis may play a role in directing migration toward the pores, and *V. fischeri* has the unusual ability to sense and swim toward nucleosides and *N*-acetylneuraminic acid, the latter of which is a component of squid mucus (DeLoney-Marino et al., 2003). If one of these compounds, or some other attractant, is more abundant near the pores, this could help the bacteria determine the correct direction in which to migrate.

However, even if *V. fischeri* responds to an attractant from the host, this does not determine symbiont specificity. Other *Vibrio* species will, like *V. fischeri*, migrate through strands of mucus toward the light organ pores (Nyholm et al., 2000). However, once bacteria proceed through a pore into a duct, the *V. fischeri* cells appear uniquely well adapted. Nonsymbiotic bacteria that proceed beyond the ducts are extremely rare, even when they are present at very high concentrations. The mechanism(s) for this specificity is not known but may be mediated in part by the antimicrobial activity of NO (Davidson et al., 2004).

Stage 3: early colonization

Several *V. fischeri* cells may aggregate and migrate, but relatively few actually establish an infection (McCann et al., 2003). Other *V. fischeri* cells may fall victim to the presumed antimicrobial forces preventing infection by unwanted bacteria. When hatchlings were exposed to three isogenic *V. fischeri* strains, even at the relatively high inoculum dose of 5,000 CFU/ml, most animals were colonized by only one strain 1 day later. Although additional strains may continue to make the journey to the light organ crypts over time, initial colonization triggers a series of events (described below) that minimize the chances of such superinfection.

THE LIGHT ORGAN ENVIRONMENT

Once *V. fischeri* has successfully initiated colonization, it must not only persist but must actually grow through three to four generations each day following the diurnal expulsion of cells. Thus, the crypt environment must be hospitable enough to support bacterial growth. On the other hand, the host must prevent spread of this infection to other tissues and must restrict its hospitality to *V. fischeri*, preventing infection by other bacteria. Accordingly, there is great interest in understanding the environment the bacteria are exposed to in the light organ.

Neatly defining the light organ environment is probably impossible, given its heterogeneous and ever-changing nature. For example, bacteria proximal to the ducts or in antechambers may experience conditions different from those of bacteria deeper in the crypts. Moreover, as the squids grow from juveniles to adults, the light organ itself undergoes many changes (Montgomery and McFall-Ngai, 1995; Foster and McFall-Ngai, 1998; Lamarcq and McFall-Ngai, 1998; Claes and Dunlap, 2000; Doino and McFall-Ngai, 2000; Kimbell and McFall-Ngai, 2004). There is also a diurnal rhythm of changes in the light organ environment, illustrated by a relatively disorganized epithelial layer deep in the crypts immediately after diurnal venting, followed by progressive restructuring of the epithelial boundary (M. McFall-Ngai, personal communication). Taken together, these various considerations make the question "what is the light organ environment?" a complex one to answer in a detailed way. Nonetheless, many studies have suggested how the light organ environment both supports and limits bacterial growth.

Support of *V. fischeri* Growth

A light organ crypt must be nutrient-rich, considering the rapid growth of *V. fischeri* in this environment. Unless unnaturally large inocula are used, colonization starts from a few cells (McCann et al., 2003), but populations reach $>10^5$ cells within hours (Ruby and Asato, 1993). In fact, early in infection, *V. fischeri* may achieve generation times of 20 to 30 min, rivaling the maximum growth rate typically observed in culture (Ruby and Asato, 1993). However, the specific nutrients supporting this growth are not entirely defined.

Numerous free amino acids and peptides are available in the crypts of adult squids (Graf and Ruby, 1998). Glycine is especially abundant in crypt peptides relative to its use in *V. fischeri*, and a glycine auxotroph was unaffected in colonization proficiency (Graf and Ruby, 1998). Arginine, proline, methionine, leucine, cysteine, threonine, serine, and lysine are also available to the symbionts; however, de novo synthesis of these (and possibly other) amino acids is required for full colonization (Graf and Ruby, 1998).

The availability and use of metabolites other than amino acids have gone largely untested, but there is good evidence that other nutrient sources are available. For example, the epithelial cells lining the crypts are glycosylated with mannose (McFall-Ngai et al., 1998). Also, sialomucin is secreted into the light organ crypts, and this could serve as a source of both protein and carbohydrate (Nyholm et al., 2002). Furthermore, dead host cells may be sloughed off the epithelium (Nyholm and McFall-Ngai, 1998), potentially spilling many diverse nutrients to the bacteria. *V. fischeri* grows on a variety of carbon sources, including both common compounds like chitin and unusual growth substrates like cyclic AMP (Dunlap and Callahan, 1993), and the possible roles of these nutrient sources merit further investigation.

How *V. fischeri* generates energy is also unknown. Because bioluminescence requires oxygen, we know that the light organ is not anaerobic. It is possible that aerobic respiration is functional, although its importance is far from certain. Oxygen levels may be relatively low (Boettcher et al., 1996); however, *V. fischeri* apparently possesses cytochrome *bd*- and cytochrome *cbb*$_3$-type oxidases, both of which have high affinity for O_2 in other organisms. Indeed, the latter is related to rhizobial FixN, which has very high O_2 affinity (K_m = 7 nM) and is required for symbiotic bacteroid development and N_2 fixation in legume hosts. In light of this, it is tempting to speculate that *V. fischeri* is well adapted for microaerobic respiration. *V. fischeri* may

also generate energy via anaerobic respiratory pathways, which are induced in the light organ (Proctor and Gunsalus, 2000). Of particular interest, *V. fischeri* can use trimethylamine-*N*-oxide as a terminal electron acceptor, and trimethylamine-*N*-oxide levels are high (1 to 100 mM) in the tissues of many cephalopods (Yancey et al., 1982; Lin and Hurng, 1985, 1989).

Control of *V. fischeri* Populations

The squid must also keep *V. fischeri* in check and prevent growth of other bacteria. Interestingly, this seems to be accomplished largely by barriers and antimicrobial mechanisms similar to those used by animals to restrict pathogens. For example, tight junctions between epithelial cells help form a barrier, and macrophage-like cells monitor the crypts (Nyholm and McFall-Ngai, 1998). Two additional conserved antimicrobial responses elaborated by *E. scolopes* are production of reactive oxygen species (ROS) and antimicrobial peptides (Nyholm and McFall-Ngai, 2004). The squid possess a halide peroxidase, which uses H_2O_2 as a substrate to generate the antimicrobial ROS hypochlorous acid in the light organ (Weis et al., 1996; Small and McFall-Ngai, 1999), and the host may produce the H_2O_2 substrate via a respiratory burst. Consistent with the model that symbionts face extracellular H_2O_2, *V. fischeri* possesses a highly active periplasmic catalase that contributes to colonization competitiveness (Visick and Ruby, 1998). The hosts also secrete NO synthase to produce antimicrobial NO (Davidson et al., 2004).

Restricting *V. fischeri* infection may also be accomplished in part by limiting the availability of key nutrients. For example, iron availability limits symbiont growth in the light organ, much as it does in other animal–bacteria associations (Graf and Ruby, 2000). Furthermore, because *V. fischeri* motility requires Mg^{2+} (O'Shea et al., 2005), the host might prevent spread of the symbiont to other tissues by limiting Mg^{2+} availability. The squid may also control *V. fischeri* growth, while enhancing the bioluminescence of the symbiont, by maintaining high osmolarity (Stabb et al., 2004). The ionic content and composition of the light organ are not known, but based on ionic control in related cephalopods, these hypotheses are not unreasonable (Robertson, 1965; Yancey et al., 1982).

COLONIZATION FACTORS

Discovering the factors that enable *V. fischeri* to colonize *E. scolopes* has been a major research focus, with many studies using mutational approaches to delineate the genes, and thereby the processes, required to colonize the host. As reviewed elsewhere (Ruby, 1996), mutants that are symbiotically attenuated can have a range of specific phenotypes. Some mutants are attenuated in initiating colonization but are normal thereafter, whereas others are just the opposite. Still other mutants may appear to be essentially wild type unless they are forced to compete in a mixed inoculum. Most mutants characterized to date can be categorized as defective in motility, regulation, nutrition, or surface molecules.

Motility Mutants

Nonmotile mutants unable to swim are the only class of mutants that are absolutely unable to colonize the host (Graf et al., 1994). Motility mutants do form aggregates in the mucus on the light organ surface (Nyholm et al., 2000), but they cannot make the journey from the aggregates through the pores and ducts into the light organ crypts. *V. fischeri* may also elaborate surface-mediated "twitching" motility, but this cannot compensate for a defect in swimming.

Regulation of motility may be critical during the infection process and may be tied to the regulation of other colonization factors. Analyses of spontaneous "hyperswimmer" mutants revealed that having too many flagella may interfere with attachment and aggregation (Millikan and Ruby, 2002). Certain hyperswimmer mutants also displayed pleiotropic phenotypes related to colonization, suggesting that motility may be coregulated with other colonization factors. A subsequent study found that the master regulator of motility, FlrA, also regulates several functions unrelated to motility but potentially tied to colonization.

Motility is down-regulated in the light organ (Ruby and Asato, 1993), raising the question of whether this system is expendable once the crypts are colonized. Contrary to this hypothesis, mutants lacking the flagellin subunit FlaA are preferentially expelled from the light organ during the daily venting process, suggesting a role for flagella in competitive retention and possibly even attachment to the light organ epithelium (Millikan and Ruby, 2004). Furthermore, the flagellar basal body might act as a secretion apparatus for extracellular colonization factors (Young et al., 1999). Future work with conditional motility mutants or microinjection of motility mutants into the crypts should yield insight into what role swimming plays in an established infection.

Regulatory Mutants

Certain regulatory mutants are also impaired in colonizing *E. scolopes*. Unlike amotile mutants, regulatory mutants are still able to colonize the host to some degree; however, their colonization deficiency is generally more severe than that displayed by the other mutant classes. The implication is that certain key regulators control the expression of multiple genes that contribute to symbiotic competence.

Perhaps the most intriguing regulator is RscS, which appears to be the sensor part of a classic two-component regulatory system (Visick and Skoufos, 2001). The name RscS derives from "*regulator of symbiotic colonization-sensor*," and *rscS* mutants are severely impaired in initiating infection. *rscS* mutants show no obvious phenotypic deviation from its parent in culture, suggesting that RscS recognizes a host-specific signal and prompts a regulatory response involving symbiosis-specific genes. RscS has a conserved signal transduction domain, but its putative sensor domain is novel, and the signal it recognizes is unknown.

The lack of an obvious phenotype to score in culture has made characterization of Rsc-mediated regulation difficult. Furthermore, *rscS* is not genetically linked to a response regulator, so the putative regulatory RscR component has remained elusive. However, overexpression of RscS elicits changes in expression of a gene cluster critical for symbiotic competence, and this cluster includes a response regulator (K. Visick, personal communication). This exciting development promises to unlock the important symbiotic functions downstream of RscS. A major challenge then will be identifying the environmental cue sensed by RscS.

Mutations in other regulatory systems also affect symbiotic competence, and some but not all of the resultant phenotypes relevant to symbiosis can be identified in culture. As mentioned above, FlrA regulates motility, which is required for symbiosis, but FlrA also regulates other factors that may contribute to colonization. Regulation by FlrA depends on the RNA polymerase sigma factor σ^{54}, and mutants lacking σ^{54} are decreased in motility and therefore do not infect the host (Wolfe et al., 2004). However, σ^{54} also modulates other phenotypes (e.g., biofilm formation) which may be relevant to the symbiosis. Similarly, the two-component global regulator GacS/GacA affects expression of motility, but the symbiotic phenotypes of this mutant suggest that it regulates additional symbiotic factors as well (Whistler and Ruby, 2003; C. A. Whistler, T. A. Koropatnick, M. McFall-Ngai, and E. G. Ruby, Abstr. 104th Gen. Meet. Am. Soc. Microbiol., abstr. N-303, 2004).

Mutant analyses also indicate that the regulators LuxR, AinS, and, to a lesser extent, LuxS can contribute to symbiotic competence (Visick et al., 2000; Lupp et al., 2003; Lupp and Ruby, 2004). Each of these regulates bioluminescence, a known colonization factor described in greater detail below; however, the symbiotic attenuation of *luxR*, *ainS*, and *luxS* mutants may not be fully explained by this phenotype. For example, LuxR also stimulates expression of QsrP, a colonization factor that is periplasmic but otherwise uncharacterized (Callahan and Dunlap, 2000). Another regulator of bioluminescence in culture is LitR; surprisingly, a *litR* mutant actually outcompetes the parent strain for squid colonization in mixed inocula (Fidopiastis et al., 2002). LitR has little or no effect on luminescence levels in the symbiosis, and the enhanced competitiveness of a *litR* mutant is likely due to another property of this pleiotropic mutant.

Thus, several regulators may each control multiple genes involved in the symbiosis. The relatively severe colonization attenuation of these regulatory mutants facilitated the identification of their symbiotic importance, but the challenge ahead lies in delineating how effects on these regulons add up to their net symbiotic phenotype. Defining the suites of genes regulated by σ^{54}, FlrA, GacA, LuxR, AinS, LitR, and LuxS, combined with mutational analyses of those genes, will help tease apart the contributions of various colonization factors in these regulons.

Mutants Affected in Surface and Attachment Functions

Mutant analyses have also indicated that certain cell surface attributes are symbiotically important, although none of these features are essential, and the mutants examined tend to have relatively minor colonization defects. For example, *pilA* encodes a putative type IV-A pilin protein; *pilA* mutants only show a defect in competition with the parent, and even this competition deficiency is quite moderate (Stabb and Ruby, 2003). Similarly, transposon insertions in the *had* locus, which also encodes a putative type IV pilin apparatus, were identified on the basis of a hemagglutination deficiency and were impaired in squid colonization (Feliciano, 2000). Of the two *had* mutants examined, one colonized to about 40% of wild-type levels, and the other displayed diminished symbiotic proficiency only when competed against the parent.

In addition to functioning in attachment, the cell surface acts as a barrier, and colonization defects have been observed for mutants altered in barrier functions. For example, a *pgm* mutant has shorter lipopolysaccharide (LPS) and increased sensitivity to detergents and cationic agents and does not colonize as well as its parent (DeLoney et al., 2002). Similarly, an *ompU* mutant lacks an outer membrane protein that may act as a porin, has increased sensitivity to

detergents and the antimicrobial peptide protamine sulfate, and displays a slight decrease in colonization efficiency at low inoculum doses (Aeckersberg et al., 2001).

These data suggest that surface molecules associated with permeability and attachment may play a role in the symbiosis. However, multiple and somewhat redundant adhesins may be involved in attachment, because colonization phenotypes of strains with single-gene mutations have been relatively subtle so far. Genomic analyses, discussed below, appear to support the notion of semiredundant attachment mechanisms, especially pili (Ruby et al., 2005).

Nutritional Mutants

V. fischeri mutants with specific nutritional requirements are also attenuated in colonization of *E. scolopes*. Although some scientists are reluctant to label metabolic pathways "colonization factors," analyses of these mutants have unquestionably helped elucidate the nutritional environment of the light organ crypts and the requirements for symbiosis. Most notably, mutant analyses have revealed that siderophore-mediated iron acquisition and the ability to synthesize certain amino acids contribute to light organ colonization proficiency (Graf and Ruby, 1998, 2000).

THE IMPORTANCE OF BIOLUMINESCENCE

One of the most intriguing colonization factors is bioluminescence. Long considered an energetic drag on cultured cells, luminescence is critical for full colonization of the *E. scolopes* light organ by *V. fischeri* (Visick et al., 2000). Much is known about the genetics and biochemistry of bioluminescence, and several hypotheses have been proposed to explain why it may be useful to the bacteria during colonization of a host. These hypotheses and approaches for testing them have been reviewed elsewhere (Stabb, 2005) and will be described below.

In bacteria, light is generated by a heterodimeric luciferase composed of the *luxA* and *luxB* gene products. LuxAB sequentially binds reduced riboflavin 5′-phosphate ($FMNH_2$), O_2, and an aliphatic aldehyde, which it converts to flavin mononucleotide (FMN), water, and an aliphatic acid, generating light in the process (Hastings and Nealson, 1977; Tu and Mager, 1995). The *luxC*, *luxD*, and *luxE* genes flank *luxA* and *luxB* and are responsible for (re)generating the aldehyde substrate (Boylan et al., 1989; Meighen, 1994). Regeneration of $FMNH_2$ and the aldehyde substrate relies on reduction by NAD(P)H, and, in the latter case, ATP hydrolysis. *luxG*, located downstream from *luxE*, encodes a protein that shuttles reducing power from NAD(P)H to FMN.

Despite this detailed knowledge, the selective advantage(s) of bioluminescence to bacteria remains a controversial topic (Nealson and Hastings, 1979; Timmins et al., 2001). The apparent costs of bioluminescence have especially motivated curiosity in its utility. These costs include sizable biosynthetic input (LuxAB can make up 5% of the protein in bright cells [Hastings et al., 1965]), ATP hydrolysis, and the consumption of reducing power and oxygen, which could otherwise theoretically be used to generate energy from aerobic respiration. Although an advantage for the host (e.g., antipredation) could confer a fitness advantage on the bacteria, thereby rationalizing such energetic input, this cannot explain why dark *lux* mutants are decreased in fitness in the absence of a selective pressure for the host (Visick et al., 2000).

Several hypotheses have been proposed to explain how bioluminescence aids bacteria directly. Two of these propose that light itself is the key. One hypothesis is that the squid detect luminescence in the light organ and impose sanctions on dark infections. This would allow the host to ensure that it received bioluminescence in exchange for the nutrients it provides the symbionts, much as legumes limit rhizobial symbionts that do not fix nitrogen (Kiers et al., 2003). Another hypothesis states that bioluminescence stimulates DNA repair mediated by photolyase (Czyz et al., 2003). Experimental data indicate that this hypothesis is possibly correct (Czyz et al., 2000), although it seems unlikely that the conditions used mimic the light organ environment.

Other hypotheses posit that the important function of bioluminescence is to burn oxygen. In consuming oxygen, the symbionts may create a hypoxic stress for the nearby squid epithelium, thereby attenuating ROS production or simply generating an environment that the facultative symbionts are better suited to live in than are the obligately aerobic host epithelial cells (Visick et al., 2000). Bioluminescence could also depress intracellular oxygen concentration in the bacteria, either to increase resistance to oxidative stress in general (Timmins et al., 2001) or to protect specific oxygen-sensitive enzymes. The high affinity of luciferase for oxygen is consistent with its having a role in decreasing either ambient or intracellular oxygen levels.

Another hypothesis is that the relevant reactant is reducing power, which is supplied indirectly from NADH. Luciferase could help recycle NAD$^+$ cofactor when it becomes limiting as its reduced form, NADH, builds up (Bourgois et al., 2001). In this model, luminescence acts as a valve to release excess reductant

and becomes important for symbiotic bacteria when biomass production, another electron sink, becomes limited by spatial constraints in the light organ.

At present, none of these hypotheses can be ruled out. Indeed, it is possible that more than one is correct. For example, the host may recognize light production and impose a sanction of ROS on dark infections, while bioluminescence may also prevent or minimize the effects of ROS by lowering the ambient oxygen. Conversely, the models proposing that bioluminescence acts as an electron sink or as an antioxidant may be mutually exclusive, as they imply a reducing or oxidizing environment, respectively. Predictions can be drawn from each of the above hypotheses, and experimental analyses are under way to test them (Stabb, 2005).

INTERSPECIES RECOGNITION AND SIGNALING

Only *V. fischeri* colonizes the *E. scolopes* light organ, and this infection triggers specific developmental responses in the host. If kept in *V. fischeri*-free seawater, *E. scolopes* can be raised to adulthood without the light organ's becoming infected or developing bioluminescence (Hanlon et al., 1997). In most respects, such "aposymbiotic" animals mature normally (Claes and Dunlap, 2000); however, several distinct developmental changes do not occur without *V. fischeri*.

Specific developmental events triggered by *V. fischeri* symbionts include (i) regression of the ciliated fields accompanied by increased trafficking of hemocytes to the ciliated appendages and apoptotic cell death (Montgomery and McFall-Ngai, 1994; Foster and McFall-Ngai, 1998; Koropatnick et al., 2004), (ii) cessation of mucus shedding by the ciliated fields (Nyholm et al., 2002), (iii) constriction of the ducts (Kimbell and McFall-Ngai, 2004), (iv) downregulation of NO synthase (Davidson et al., 2004), (v) swelling of the crypt epithelial cells (Montgomery and McFall-Ngai, 1994), (vi) proliferation of the epithelial microvilli (Lamarcq and McFall-Ngai, 1998), (vii) increased mucus secretion in the crypts (Nyholm et al., 2002), and (viii) many molecular changes that have yet to be linked to distinct processes (Doino and McFall-Ngai, 2000; Kimbell and McFall-Ngai, 2003). Nyholm and McFall-Ngai (2004) reviewed these symbiont-triggered events and their timing.

Many of these developmental events are easily rationalized. Once the squid are colonized by *V. fischeri*, continued harvesting of bacteria from the environment is unnecessary and may encourage undesirable infections, so removal of the mucus-shedding bacteria-collecting ciliated fields and constriction of the ducts are understandable. Meanwhile, inside the crypts, cell swelling and microvillar proliferation may help increase the surface area of contact between symbiont and host, allowing more efficient exchange of metabolites between the partners.

Some of the events triggered by *V. fischeri* are reversible by curing *E. scolopes* of its symbionts with antibiotics (Lamarcq and McFall-Ngai 1998; Nyholm et al., 2002). Although there are many explanations for this observation, one possibility is that some developmental changes are linked to an ongoing metabolic exchange between the animal and its symbionts. Interestingly, the swelling of the crypt epithelial cells is both reversible upon curing the symbionts and requires that the *V. fischeri* cells are bioluminescent (Visick et al., 2000). This swelling is consistent with the epithelium's experiencing hypoxic stress and could be tied to ongoing oxygen consumption by bioluminescence. Cell swelling could also represent a developmental response of the animal to light or to metabolic by-products related to the physiology of bioluminescence.

Other developmental programs cannot be stopped once they are set in motion. This includes the regression of the ciliated epithelial fields, which becomes irreversible 12 h after hatchlings are exposed to *V. fischeri* (Doino and McFall-Ngai, 1995). The regression process is triggered remotely, in that morphological changes occur several cell layers away from the bacterial infection. LPS from the bacteria triggers apoptosis in the ciliated field but not regression of this structure (Foster et al., 2000). A peptidoglycan monomer (PGM) can stimulate regression and acts synergistically with LPS to elicit a response closely resembling that induced by symbiotic infection (Koropatnick et al., 2004). Although LPS and PGM purified from other bacteria can also stimulate these responses in *E. scolopes*, it is likely that the squid only experience high levels of these two signaling molecules in infected light organs.

The observation that LPS and PGM stimulate developmental processes in a mutualistic animal–bacteria association is notable for at least two reasons.

First, these molecules are also recognized by the innate immune systems of animals and often trigger antimicrobial responses. This suggests conserved mechanisms for mutualist and pathogen detection by animals, with host responses being context dependent. Second, the particular PGM molecule shed by *V. fischeri* is identical to "cytotoxins" of pathogens *Bordetella pertussis* and *Neisseria gonorrhoeae* (Koropatnick et al., 2004), revealing an interesting and unanticipated similarity between mutualistic and pathogenic bacteria–animal associations (Table 1).

INSIGHTS FROM GENOMICS

Rapid advances in genomics are transforming research on the *V. fischeri*–*E. scolopes* symbiosis. Given the relative genetic simplicity and manipulability of *V. fischeri*, it is not surprising that genomics is further along in the bacterial symbiont; however, genomic approaches are being applied to both partners and promise to revolutionize investigations of this symbiosis.

The *V. fischeri* Genome

The complete annotated genome of *V. fischeri* strain ES114 was recently published (Ruby et al., 2005). ES114 was picked from among various *E. scolopes* isolates in part because it contains a large (45.8-kb) plasmid, pES100, which represents a group of plasmids often found in *E. scolopes* symbionts. ES114 also lacks small multicopy plasmids, which would be overrepresented in DNA pools, thereby confounding sequencing efforts. Many *E. scolopes* symbionts contain small plasmids, however, so a representative of this group, pES213, was recently sequenced and characterized to complement the genome project (Dunn et al., 2005).

V. fischeri ES114 became the fourth *Vibrio* species to have its genome published, joining pathogens *V. cholerae*, *V. vulnificus*, and *V. parahaemolyticus*. As the lone mutualistic *Vibrio* species sequenced,

Table 1. Varied roles for peptidoglycan monomer

Bacterium	Host	Result of infection	Activity of shed N-acetylglucosaminyl-1,6-anhydro-N-acetyl-muramylalanyl-γ-glutamyldiaminopimelylalanine
Bordetella pertussis	Human	Whooping cough	Causes death in ciliated airway epithelial cells
Neisseria gonorrhoeae	Human	Gonorrhea	Causes death in ciliated fallopian tube epithelial cells
Vibrio fischeri	Hawaiian bobtail squid	Bioluminescent light organ mutualism	Triggers regression of ciliated epithelial fields on light organ surface

these data are enriching studies of genome evolution. Interestingly, each of these *Vibrio* species has two chromosomes, a relatively large chromosome harboring many conserved "housekeeping" genes, and a second, smaller chromosome that is more variable between species. The smaller "chromosome 2" may therefore carry more lifestyle-specific genes and could be rich in genes that function in the symbiosis. Also, *V. fischeri* DNA is more A+T rich than the other *Vibrio* species, which parallels a trend of high A+T content in other bacteria that are especially dependent on associations with an animal host (Rocha and Danchin, 2002).

It would be difficult to overstate the importance of the *V. fischeri* genome database to current and future research. The sequence, which encompasses 4.28 Mbp and 3,802 putative open reading frames, has provided intriguing insight into old questions as well as unanticipated fodder for future research directions. For example, as described above, the subtle symbiotic phenotypes of putative attachment (e.g., pilus) mutants intimated possible functional redundancy. The appearance of no fewer than eight putative type IV pilus loci along with other putative adhesins supports that conjecture. Furthermore, the unanticipated discovery of a homolog to toxin-coregulated pili of *V. cholerae* provides an intriguing new research target (Ruby et al., 2005).

In this and other examples, genomic data provide a rapid way to identify and mutate specific genes to test hypotheses about what is important in the symbiosis. Although targeted mutational approaches were possible before genomics, they involved uncertain and lengthy cloning strategies, negative cloning results were uninterpretable, and possible redundancy was difficult to address. Genomics has streamlined the process of target identification and allows far more rapid cloning and mutant generation, and the resulting data can be interpreted in a broad context.

Analysis of the *V. fischeri* genome has also opened up a qualitatively new approach by allowing a broad assessment of regulatory and metabolic pathways. For example, as described above, RscS, the sensor component of a putative two-component regulatory system, was identified as an important regulator of colonization factors but was not genetically linked to a response regulator (Visick and Skoufos, 2001). A genomic search for putative response regulators unpaired with a sensor revealed several potential targets. This sort of approach would not have been possible without the genome sequence. Genomic analyses have also made it possible to view how the biochemistry of bioluminescence integrates into the overall physiology of *V. fischeri*, which should help define its role in symbiosis (see above and Stabb, 2005).

Microarray-based approaches will also shed light on the symbiosis, for example, by defining the suites of genes controlled by regulators that are important in the symbiosis (e.g., RscS/RscR). Ideally, microarrays will also elucidate the changes in global gene expression associated with entering the symbiosis. One hurdle to this approach is that obtaining sufficient mRNA from the bacterial symbionts currently requires the use of adult animals, which are not necessarily infected by strain ES114. This approach could also miss important regulatory events specific to establishing the symbiosis. If these do present a problem for interpreting microarray results, this should be overcome by technologies that require less input mRNA or by improved methods for large-scale recovery of symbionts from juvenile squid infected with ES114.

Comparative genomics will also be a valuable research tool. Comparison of *V. fischeri* with other sequenced *Vibrio* species already sheds light on the evolution of *V. fischeri* as a symbiont. In the future, sequencing of other *V. fischeri* strains, both those that infect squid efficiently and those that do not, will further define what enables a strain to partner with *E. scolopes*.

E. scolopes Genomics

Proteomic and subtractive cDNA experiments have revealed that the animal undergoes many molecular changes during the initiation of the symbiosis (Doino and McFall-Ngai, 2000; Kimbell and McFall-Ngai, 2003), and this is now being augmented by genomics. Fourteen thousand expressed *E. scolopes* genes have been sequenced and placed on a microarray (Nyholm and McFall-Ngai, 2004), and differences in gene expression in aposymbiotic animals and infected animals are being examined. These data will funnel into existing technologies for quantitating transcript abundance using real-time PCR and visualizing expression patterns in situ with fluorescent probes (Kimbell and McFall-Ngai, 2004). Mutational approaches are not yet possible with *E. scolopes*, but RNA interference technology may ultimately allow researchers to test the symbiotic importance of specific genes by disrupting their expression. The expressed sequence tag database is also being used to identify gene products that may be involved in LPS- and peptidoglycan-mediated signaling, including homologs of the receptors and signal transducers known to function in the innate immune systems of animals. In this way, genomic approaches in *E. scolopes* are helping connect the gross morphological changes associated with the symbiosis to changes in gene expression and underlying molecular mechanisms.

NEW PERSPECTIVES

Studies of the *V. fischeri–E. scolopes* symbiosis have yielded fresh insights into host-animal interactions. The remarkable specificity of this symbiosis has helped dispel the notion that invertebrate immune systems are "primitive" or indiscriminate, as these squid clearly possess sophisticated mechanisms for recognizing and controlling their microbiota. Another new insight has been the revelation that the same molecules that function in pathogenesis and pathogen detection can also function as signaling molecules in an animal–bacteria mutualism.

The *V. fischeri–E. scolopes* symbiosis is also yielding new perspectives as an archetype *Vibrio*-host interaction. Many *Vibrio* species associate with animals, but often the evolutionary and ecological relevance of the laboratory models studied are uncertain (Klose, 2000). The *V. fischeri–E. scolopes* symbiosis ties experimental tractability in the laboratory to an animal model that is relevant to the life history of the bacteria in the real world. Accordingly, studies of regulation, signaling, adherence, evolution, and physiology in the *V. fischeri–E. scolopes* system promise to shed new light on many associations between marine *Vibrio* species and animal hosts.

Acknowledgments. I thank Deborah Millikan for comments on the manuscript and Jeffrey Bose for assistance in preparing figures. I was supported in this work by a CAREER award from the National Science Foundation (MCB-0347317).

REFERENCES

Aeckersberg, F., C. Lupp, B. Feliciano, and E. G. Ruby. 2001. *Vibrio fischeri* outer membrane protein OmpU plays a role in normal symbiotic colonization. *J. Bacteriol.* **183:**6590–6597.

Anderson, R. C., and J. A. Mather. 1996. Escape responses of *Euprymna scolopes* Berry, 1911 (Cephalopoda: Sepiolidae). *J. Moll. Stud.* **62:**543–545.

Boettcher, K. J., E. G. Ruby, and M. J. McFall-Ngai. 1996. Bioluminescence in the symbiotic squid *Euprymna scolopes* is controlled by a daily biological rhythm. *J. Comp. Physiol.* **179:**65–73.

Bourgois, J.-J., F. E. Sluse, F. Baguet, and J. Mallefet. 2001. Kinetics of light emission and oxygen consumption by bioluminescent bacteria. *J. Bioenerg. Biomembr.* **33:**353–363.

Boylan, M., C. Miyamoto, L. Wall, A. Grahm, and E. Meighen. 1989. Lux C, D, and E genes of the *Vibrio fischeri* luminescence operon code for the reductase, transferase, and synthetase enzymes involved in aldehyde biosyntheses. *Photochem. Photobiol.* **49:**681–688.

Callahan, S. M., and P. V. Dunlap. 2000. LuxR- and acyl-homoserine-lactone-controlled non-*lux* genes define a quorum sensing regulon in *Vibrio fischeri. J. Bacteriol.* **182:**2811–2822.

Claes, M. F., and P. V. Dunlap. 2000. Aposymbiotic culture of the sepiolid squid *Euprymna scolopes*: role of the symbiotic bacterium *Vibrio fischeri* in host animal growth, development, and light organ morphogenesis. *J. Exp. Biol.* **286:**280–296.

Czyz, A., B. Wrobel, and G. Wegrzyn. 2000. *Vibrio harveyi* bioluminescence plays a role in stimulation of DNA repair. *Microbiology* **146:**283–288.

Czyz, A., K. Plata, and G. Wegrzyn. 2003. Stimulation of DNA repair as an evolutionary drive for bacterial bioluminescence. *Luminescence* **18:**140–144.

Darveau, R. P., M. McFall-Ngai, E. G. Ruby, S. Miller, and D. F. Mangan. 2003. Host tissues may actively respond to beneficial microbes. *ASM News* **69:**186–191.

Davidson, S. K., T. A. Koropatnick, R. Kossmehl, L. Sycuro, and M. J. McFall-Ngai. 2004. No means "yes" in the squid-vibrio symbiosis: nitric oxide (NO) during the initial stages of a beneficial association. *Cell. Microbiol.* **6:**1139–1151.

DeLoney, C. R., T. M. Bartley, and K. L. Visick. 2002. Role for phosphoglucomutase in *Vibrio fischeri-Euprymna scolopes* symbiosis. *J. Bacteriol.* **184:**5121–5129.

DeLoney-Marino, C. R., A. J. Wolfe, and K. L. Visick. 2003. Chemoattraction of *Vibrio fischeri* to serine, nucleosides, and N-acetylneuraminic acid, a component of squid light-organ mucus. *Appl. Environ. Microbiol.* **69:**7527–7530.

Doino, J. A., and M. J. McFall-Ngai. 1995. A transient exposure to symbiosis-competent bacteria induces light organ morphogenesis in the host squid. *Biol. Bull.* **189:**347–355.

Doino, J., and M. McFall-Ngai. 2000. Alterations in the proteome of the *Euyprymna scolopes* light organ in response to symbiotic *Vibrio fischeri. Appl. Environ. Microbiol.* **66:**4091–4097.

Dunlap, P. V., and S. M. Callahan. 1993. Characterization of a periplasmic 3':5'-cyclic nucleotide phosphodiesterase gene, *cpdP*, from the marine symbiotic bacterium *Vibrio fischeri. J. Bacteriol.* **175:**4615–4624.

Dunn, A. K., M. O. Martin, and E. V. Stabb. 2005. Characterization of pES213, a small mobilizable plasmid from *Vibrio fischeri. Plasmid* **54:**114–134.

Engebrecht, J., K. Nealson, and M. Silverman. 1983. Bacterial bioluminescence: isolation and genetic analysis of functions from *Vibrio fischeri. Cell* **32:**773–781.

Feliciano, B. A. 2000. The identification and characterization of a *Vibrio fischeri* hemagglutination factor. M.S. thesis. University of Hawaii, Honolulu.

Fidopiastis, P. M., C. Miyamoto, M. G. Jobling, E. A. Meighen, and E. G. Ruby. 2002. LitR, a new transcriptional activator in *Vibrio fischeri*, regulates luminescence and symbiotic light organ colonization. *Mol. Microbiol.* **45:**131–143.

Fidopiastis, P. M., S. von Boletzky, and E. G. Ruby. 1998. A new niche for *Vibrio logei*, the predominant light organ symbiont of squids in the genus *Sepiola. J. Bacteriol.* **180:**59–64.

Foster, J. S., M. A. Apicella, and M. J. McFall-Ngai. 2000. Vibrio fischeri lipopolysaccharide induces developmental apoptosis, but not complete morphogenesis, of the *Euprymna scolopes* symbiotic light organ. *Dev. Biol.* **226:**242–254.

Foster, J. S., and M. J. McFall-Ngai. 1998. Induction of apoptosis by cooperative bacteria in the morphogenesis of host epithelial tissues. *Dev. Genes Evol.* **208:**295–303.

Graf, J., P. V. Dunlap, and E. G. Ruby. 1994. Effect of transposon-induced motility mutations on colonization of the host light organ by *Vibrio fischeri. J. Bacteriol.* **176:**6986–6991.

Graf, J., and E. G. Ruby. 1998. Host-derived amino acids support the proliferation of symbiotic bacteria. *Proc. Natl. Acad. Sci. USA* **95:**1818–1822.

Graf, J., and E. G. Ruby. 2000. Novel effects of a transposon insertion in the *Vibrio fischeri glnD* gene: defects in iron uptake and symbiotic persistence in addition to nitrogen utilization. *Mol. Microbiol.* **37:**168–179.

Hanlon, R. T., M. F. Claes, S. E. Ashcraft, and P. V. Dunlap. 1997. Laboratory culture of the sepiolid squid *Euprymna scolopes*: a model system for bacteria-animal symbiosis. *Biol. Bull.* **192:**364–374.

Harper, R. D., and J. F. Case. 1999. Counterillumination and its antipredatory value in the plainfin midshipman fish *Porichthys notatus*. *Mar. Biol.* **134:**529–540.

Harvey, E. N. 1952. *Bioluminescence*. Academic Press, Inc., New York, N.Y.

Hastings, J. W., and K. H. Nealson. 1977. Bacterial bioluminescence. *Annu. Rev. Microbiol.* **31:**549–595.

Hastings, J. W., W. H. Riley, and J. Massa. 1965. The purification, properties, and chemiluminescent quantum yield of bacterial luciferase. *J. Biol. Chem.* **240:**1473–1481.

Jones, B. W., and M. K. Nishiguchi. 2004. Counterillumination in the Hawaiian bobtail squid, *Euprymna scolopes* Berry (Mollusca: Cephalopoda). *Mar. Biol.* **144:**1151–1155.

Kiers, E. T., R. A. Rousseau, S. A. West, and R. F. Denison. 2003. Host sanctions and the legume-rhizobium mutualism. *Nature* **425:**78–81.

Kimbell, J. R., and M. J. McFall-Ngai. 2003. The squid-*Vibrio* symbioses: from demes to genes. *Integr. Comp. Biol.* **43:**254–260.

Kimbell, J. R., and M. J. McFall-Ngai. 2004. Symbiont-induced changes in host actin during the onset of a beneficial animal-bacterial association. *Appl. Environ. Microbiol.* **70:**1434–1441.

Klose, K. E. 2000. The suckling mouse model of cholera. *Trends Microbiol.* **8:**189–191.

Kondo, K., A. Takade, and K. Amako. 1994. Morphology of the viable but nonculturable *Vibrio cholerae* as determined by the freeze fixation technique. *FEMS Microbiol. Ecol.* **123:**179–184.

Koropatnick, T. A., J. T. Engle, M. A. Apicella, E. V. Stabb, W. E. Goldman, and M. J. McFall-Ngai. 2004. Microbial factor-mediated development in a host-bacterial-mutualism. *Science* **306:**1186–1188.

Lamarcq, L. H., and M. J. McFall-Ngai. 1998. Induction of a gradual, reversible morphogenesis of its host's epithelial brush border by *Vibrio fischeri*. *Infect. Immun.* **66:**777–785.

Lee, K.-H. 1994. Ecology of *Vibrio fischeri*, the light organ symbiont of the Hawaiian sepiolid squid *Euprymna scolopes*. Ph.D. thesis. University of Southern California, Los Angeles.

Lee, K.-H., and E. G. Ruby. 1992. Detection of the light organ symbiont, *Vibrio fischeri*, in Hawaiian seawater using *lux* gene probes. *Appl. Environ. Microbiol.* **58:**942–947.

Lee, K.-H., and E. G. Ruby. 1994a. Competition between *Vibrio fischeri* strains during initiation and maintenance of a light organ symbiosis. *J. Bacteriol.* **176:**1985–1991.

Lee, K.-H., and E. G. Ruby. 1994b. Effect of the squid host on the abundance and distribution of symbiotic *Vibrio fischeri* in nature. *Appl. Environ. Microbiol.* **60:**1565–1571.

Lee, K.-H., and E. G. Ruby. 1995. Symbiotic role of the viable but nonculturable state of *Vibrio fischeri* in Hawaiian coastal water. *Appl. Environ. Microbiol.* **61:**278–283.

Lin, J. K., and D. C. Hurng. 1985. Thermal conversion of trimethylamine-N-oxide to trimethylamine and dimethylamine in squids. *Food Chem. Toxicol.* **23:**579–583.

Lin, J. K., and D. C. Hurng. 1989. Potential of ferrous sulphate and ascorbate on the microbial transformation of endogenous trimethylamine N-oxide to trimethylamine and dimethylamine in squid extracts. *Food Chem. Toxicol.* **27:**613–618.

Lupp, C., and E. G. Ruby. 2004. *Vibrio fischeri* LuxS and AinS: comparative study of two signal synthases. *J. Bacteriol.* **186:**3873–3881.

Lupp, C., M. Urbanowski, E. P. Greenberg, and E. G. Ruby. 2003. The *Vibrio fischeri* quorum-sensing systems *ain* and *lux* sequentially induce luminescence gene expression and are important for persistence in the squid host. *Mol. Microbiol.* **50:**319–331.

McCann, J., E. V. Stabb, D. S. Millikan, and E. G. Ruby. 2003. Population dynamics of *Vibrio fischeri* during infection of *Euprymna scolopes*. *Appl. Environ. Microbiol.* **69:**5928–5934.

McFall-Ngai, M., C. Brennan, V. Weis, and L. Lamarcq. 1998. Mannose adhesin-glycan interactions in the *Euprymna scolopes-Vibrio fischeri* symbiosis, p. 272–276. *In* Y. Le Gal and H. O. Halvorson (ed.), *New Developments in Marine Biotechnology*. Plenum Press, New York, N.Y.

Meighen, E. A. 1994. Genetics of bacterial bioluminescence. *Annu. Rev. Genet.* **28:**117–139.

Millikan, D. S., and E. G. Ruby. 2002. Alterations in *Vibrio fischeri* motility correlate with a delay in symbiosis initiation and are associated with additional symbiotic colonization defects. *Appl. Environ. Microbiol.* **68:**2519–2528.

Millikan, D. S., and E. G. Ruby. 2004. *Vibrio fischeri* flagellin A is essential for normal motility and for symbiotic competence during initial squid light organ colonization. *J. Bacteriol.* **186:**4315–4325.

Montgomery, M., and M. McFall-Ngai. 1994. Bacterial symbionts induce host organ morphogenesis during early post embryonic development of the squid *Euprymna scolopes*. *Development* **120:**1719–1729.

Montgomery, M. K., and M. J. McFall-Ngai. 1995. The inductive role of bacterial symbionts in the morphogenesis of a squid light organ. *Am. Zool.* **35:**372–380.

Moynihan, M. 1982. Notes on the behavior of *Euprymna scolopes* (Cephalopoda: Sepiolidae). *Behavior* **85:**25–41.

Nealson, K. H., and J. W. Hastings. 1979. Bacterial bioluminescence: its control and ecological significance. *Microbiol. Rev.* **43:**496–518.

Nishiguchi, M. K. 2000. Temperature affects species distribution in symbiotic populations of *Vibrio* spp. *Appl. Environ. Microbiol.* **66:**3550–3555.

Nishiguchi, M. K. 2002. Host-symbiont recognition in the environmentally transmitted sepiolid squid-*Vibrio* mutualism. *Microb. Ecol.* **44:**10–18.

Nishiguchi, M. K., J. E. Lopez, and S. von Boletzky. 2004. Enlightenment of old ideas from new investigations: more questions regarding the evolution of bacteriogenic light organs in squids. *Evol. Dev.* **6:**41–49.

Nishiguchi, M. K., E. G. Ruby, and M. J. McFall-Ngai. 1998. Competitive dominance among strains of luminous bacteria provides an unusual form of evidence for parallel evolution in Sepiolid squid-*Vibrio* symbioses. *Appl. Environ. Microbiol.* **64:**3209–3213.

Nyholm, S. V., B. Deplancke, H. R. Gaskins, M. A. Apicella, and M. J. McFall-Ngai. 2002. Roles of *Vibrio fischeri* and nonsymbiotic bacteria in the dynamics of mucus secretion during symbiont colonization of the *Euprymna scolopes* light organ. *Appl. Environ. Microbiol.* **68:**5113–5122.

Nyholm, S. V., and M. J. McFall-Ngai. 1998. Sampling the light organ microenvironment of *Euprymna scolopes*: description of a population of host cells in association with the bacterial symbiont. *Biol. Bull.* **195:**89–97.

Nyholm, S. V., and M. J. McFall-Ngai. 2003. Dominance of *Vibrio fischeri* in secreted mucus outside the light organ of *Euprymna scolopes*: the first site of symbiont specificity. *Appl. Environ. Microbiol.* **69:**3932–3937.

Nyholm, S. V., and M. J. McFall-Ngai. 2004. The winnowing: establishing the squid-*Vibrio* symbiosis. *Nat. Rev. Microbiol.* **2:**632–642.

Nyholm, S. V., E. V. Stabb, E. G. Ruby, and M. J. McFall-Ngai. 2000. Establishment of an animal-bacterial association: recruiting symbiotic vibrios from the environment. *Proc. Natl. Acad. Sci. USA* **97:**10231–10235.

O'Shea, T. M., C. R. DeLoney-Marino, S. Shibata, S.-I. Aizawa, A. J. Wolfe, and K. L. Visick. 2005. Magnesium promotes flagellation of *Vibrio fischeri*. *J. Bacteriol.* **187:**2058–2065.

Proctor, L. M., and R. P. Gunsalus. 2000. Anaerobic respiratory growth of *Vibrio harveyi*, *Vibrio fischeri*, and *Photobacterium leiognathi* with trimethylamine N-oxide, nitrate, and fumarate: ecological implications. *Environ. Microbiol.* **2:**399–406.

Robertson, J. D. 1965. Studies on the chemical composition of muscle tissue. III. The mantle muscle of Cephalopod molluscs. *J. Exp. Biol.* **42:**153–175.

Rocha, E. P. C., and A. Danchin. 2002. Base composition might result from competition for metabolic resources. *Trends Genet.* **18:**291–294.

Ruby, E. G. 1996. Lessons from a cooperative bacterial-animal association: the *Vibrio fischeri-Euprymna scolopes* light organ symbiosis. *Annu. Rev. Microbiol.* **50:**591–624.

Ruby, E. G., and L. M. Asato. 1993. Growth and flagellation of *Vibrio fischeri* during initiation of the sepiolid squid light organ symbiosis. *Arch. Microbiol.* **159:**160–167.

Ruby, E. G., and K.-H. Lee. 1998. The *Vibrio fischeri-Euprymna scolopes* light organ association: current ecological paradigms. *Appl. Environ. Microbiol.* **64:**805–812.

Ruby, E. G., and M. J. McFall-Ngai. 1991. Symbiont recognition and subsequent morphogenesis as early events in an animal-bacterial mutualism. *Science* **254:**1491–1494.

Ruby, E. G., and K. H. Nealson. 1976. Symbiotic association of *Photobacterium fischeri* with the marine luminous fish *Monocentris japonica*: a model of symbiosis based on bacterial studies. *Biol. Bull.* **151:**574–586.

Ruby, E. G., M. Urbanowski, J. Campbell, A. Dunn, M. Faini, R. Gunsalus, P. Lohstroh, C. Lupp, J. McCann, D. Millikan, A. Schaefer, E. Stabb, A. Stevens, K. Visick, C. Whistler, and E. P. Greenberg. 2005. Complete genome sequence of *Vibrio fischeri*: a symbiotic bacterium with pathogenic congeners. *Proc. Natl. Acad. Sci. USA* **102:**3004–3009.

Shears, J. S. 1988. The use of a sand-coat in relation to feeding and diel activity in the sepiolid squid *Euprymna scolopes*. *Malacologia* **29:**121–133.

Singley, C. T. 1983. *Euprymna scolopes*. *In* P. R. Boyle (ed.), *Cephalopod Life Cycles*. Academic Press, Inc., London, England.

Small, A. L., and M. J. McFall-Ngai. 1999. Halide peroxidase in tissues that interact with bacteria in the host squid *Euprymna scolopes*. *J. Cell. Biochem.* **72:**445–457.

Stabb, E. V. 2005. Shedding light on the bioluminescence "paradox." *ASM News* **71:**223–229.

Stabb, E. V., M. S. Butler, and D. M. Adin. 2004. Correlation between osmolarity and luminescence of symbiotic *Vibrio fischeri* strain ES114. *J. Bacteriol.* **186:**2906–2908.

Stabb, E. V., and E. G. Ruby. 2003. Contribution of *pilA* to competitive colonization of *Euprymna scolopes* by *Vibrio fischeri*. *Appl. Environ. Microbiol.* **69:**820–826.

Timmins, G. S., S. K. Jackson, and H. M. Swartz. 2001. The evolution of bioluminescent oxygen consumption as an ancient oxygen detoxification mechanism. *J. Mol. Evol.* **52:**321–332.

Tu, S.-C., and H. I. X. Mager. 1995. Biochemistry of bacterial bioluminescence. *Photochem. Photobiol.* **62:**615–624.

Visick, K. L., J. Foster, J. Doino, M. McFall-Ngai, and E. G. Ruby. 2000. *Vibrio fischeri lux* genes play an important role in colonization and development of the host light organ. *J. Bacteriol.* **182:**4578–4586.

Visick, K. L., and E. G. Ruby. 1998. The periplasmic, group III catalase of *Vibrio fischeri* is required for normal symbiotic competence and is induced both by oxidative stress and approach to stationary phase. *J. Bacteriol.* **180:**2087–2092.

Visick, K. L., and M. J. McFall-Ngai. 2000. An exclusive contract: specificity in the *Vibrio fischeri-Euprymna scolopes* partnership. *J. Bacteriol.* **182:**1779–1787.

Visick, K. L., and L. M. Skoufos. 2001. A two-component sensor required for normal symbiotic colonization of *Euprymna scolopes* by *Vibrio fischeri*. *J. Bacteriol.* **183:**835–842.

Wei, S. L., and R. E. Young. 1989. Development of symbiotic bacterial bioluminescence in a nearshore cephalopod, *Euprymna scolopes*. *Mar. Biol.* **103:**541–546.

Weis, V. M., A. L. Small, and M. J. McFall-Ngai. 1996. A peroxidase related to the mammalian antimicrobial protein myeloperoxidase in the *Euprymna-Vibrio* mutualism. *Proc. Natl. Acad. Sci. USA* **93:**13683–13688.

Whistler, C. A., and E. G. Ruby. 2003. GacA regulates symbiotic colonization traits of *Vibrio fischeri* and facilitates a beneficial association with an animal host. *J. Bacteriol.* **185:**7202–7212.

Wolfe, A. J., D. S. Millikan, J. M. Campbell, and K. L. Visick. 2004. *Vibrio fischeri* σ^{54} controls motility, biofilm formation, luminescence, and colonization. *Appl. Environ. Microbiol.* **70:**2520–2524.

Yancey, P. H., M. E. Clark, S. C. Hand, R. D. Bowlus, and G. N. Somero. 1982. Living with water stress: evolution of osmolyte systems. *Science* **217:**1214–1222.

Young, G. M., D. H. Schmiel, and V. L. Miller. 1999. A new pathway for the secretion of virulence factors by bacteria: the flagellar export apparatus functions as a protein-secretion system. *Proc. Natl. Acad. Sci. USA* **96:**6456–6461.

The Biology of Vibrios
Edited by F. L. Thompson et al.
© 2006 ASM Press, Washington, D.C.

Chapter 15

The Mutual Partnership between *Vibrio halioticoli* and Abalones

TOMOO SAWABE

INTRODUCTION

Intensive investigations of the microflora associated with zooplankton, shrimp, mollusks, and other marine invertebrates have been conducted for the last 50 years. The pathogenic potential of vibrios is commonly recognized in connection with serious outbreaks of food- or waterborne infections, especially those resulting from the consumption of seafood (Colwell and Liston, 1960, 1962; Kaneko and Colwell, 1973, 1974, 1975a,b; Colwell, 1984; Colwell et al., 1977). Consequently, there has been increasing attention given to understanding the roles of the gut microflora of aquatic animals (Harris, 1993; Sawabe et al., 1995). In particular, vibrios are among the most commonly reported groups of gut bacteria in marine vertebrates and invertebrates (Colwell and Liston, 1960, 1962; Harris, 1993; Thompson et al., 2004). However, there has been little conclusive evidence about positive (beneficial) relationships between gut microbes and marine animals (Harris, 1993). Even the taxonomy of the nonpathogenic vibrios has yet to be fully resolved; often, they are given informal names, such as "gut group *Vibrio*" (Colwell and Liston, 1960, 1962).

Most of the gut microbial ecosystems are improperly understood. Commonly, gut microbial ecosystems comprise hundreds of microbial species in dense populations (Hungate, 1966; Breznak, 1982; Stewart and Bryant, 1988; Russell and Rychlik, 2001; Hooper et al., 2002; Xu et al., 2003, 2004; Hooper, 2004; Bäckhed et al., 2004). The variety and complexity of the indigenous microflora have been determined for each animal species and population (Breznak, 1982; Harris, 1993). However, it is clear that the vast and complex consortium of gut microorganisms forms dynamic ecosystems coexisting with the animal hosts from birth to death and is affected by postnatal development of the digestive tract (Hooper and Gordon, 2001). Eukaryotic microbes are also major components of gut microbial communities, especially of ruminants (Hungate, 1966; Stewart and Bryant, 1988) and termites (Breznak, 1982; Dolan, 2001), being involved in the supply of nutrients for the host or as symbionts with prokaryotic partners. However, recent progress in the study of gut microbe–host interactions of herbivorous animals, xylophagous insects, and omnivorous humans has revealed the presence of so-called uncultured microbes (Ohkuma and Kudo, 1996; Whitford et al., 1998; Suau et al., 1999). These have been studied with gnotobiotic animal models (Falk et al., 1998). It is noteworthy that *Bacterioides thetaiotaomicron*, which has been considered a human commensal bacterium, has a number of important physiological roles in mammalian development, including glycan production, angiogenesis, and nutrient uptake (Stappenbeck et al., 2002; Xu et al., 2003, 2004; Hooper, 2004; Bäckhed et al., 2004). The findings on human symbionts are the second paradigm of microbe-induced morphogenesis of host animals, followed by *Vibrio fischeri*-induced squid light organ morphogenesis (MacFall-Ngai and Ruby, 1991; Hooper, 2004). Production of volatile short-chained fatty acids is also recognized as being of major importance in *B. thetaiotaomicron* and in rumen and termite gut microbes (Russell and Rychlik, 2001; Hooper et al., 2002).

Gut microbes make continuous and intimate contact through mucus cells in the epithelial tissues of animal digestive tracts from birth (Breznak, 1982; Stewart and Bryant, 1988; Russell and Rychlik, 2001; Hooper et al., 2002, 2004). Vibrios have important functions in balancing host physiology, specifically involving the gut ecosystem of marine vertebrates and invertebrates (Harris, 1993). Ruminants, termites, and herbivorous marine animals consume plant materials,

Tomoo Sawabe • Laboratory of Microbiology, Graduate School of Fisheries Sciences, Hokkaido University, 3-1-1 Minato-cho, Hakodate 041-8611, Japan.

which are rich in carbohydrates. However, the majority of seaweed carbohydrates are not digestible by terrestrial herbivorous animals. Therefore, it is believed that the gut microbial ecosystem of marine herbivorous animals must be a good model for studying the symbiotic association of gut vibrios. Among the huge number of marine herbivorous animals, abalone has been selected as a model because of (i) the availability of specimens from cultured and wild-caught populations, (ii) the possession of a ruminant-like digestive tract, (iii) the long life span, and (iv) its economic importance. In fact, *Vibrio* cells occupy 40% of the total gut bacterial community of abalone, as determined by whole-cell hybridization techniques (Tanaka et al., 2004). The abundance and diversity of *Vibrio halioticoli* and related species in the gut of abalones, and the possible mutual partnership of vibrios in the abalone gut microbial ecosystem, have now been clearly demonstrated.

AN OVERVIEW OF ABALONE BIOLOGY

Abalone is the common name for a member of the family *Haliotidae*, which is grouped in the class *Gastropoda* and the phylum *Mollusca*, and is a well-known marine invertebrate of considerable economic value (Oakes and Ponte, 1996). More than 100 abalone species are recognized, and they inhabit rocky shores along the Red Sea, Indian Ocean, Madagascar, South Africa, West Africa, Mediterranean, Northeastern Atlantic, Caribbean Sea, South America, Panamaic Province, Northeastern Pacific, temperate Northwestern Pacific, Indo-Malayan Archipelago, Central Pacific Australia, and New Zealand (ABMAP project by D. L. Gieger; http://www.vetigastropoda.com/ABMAP/text/index.html). Because of the hard shell of abalones, the oldest fossils of California and Caribbean species have been found from the Maastrichian age of the Upper Cretaceous period (65 to 73 million years ago) (Ino, 1952; Geiger and Groves, 1999).

Abalone taxonomy is based on "shell" morphology. Fossils and molecular phylogenetic techniques provide important clues to determining the evolutionary history (Lee and Vacquier, 1995; Coleman and Vacquier, 2002). The Cretaceous fossils found in California and the Caribbean share similarities to modern species of *Haliotis iris* (a modern New Zealand species) and *Haliotis cyclobates* (a modern Australian species), respectively (Geiger and Groves, 1999).

The major habitat of the abalones is the rocky shore with kelp forests >50 m depth. The feeding behavior of abalones is herbivorous; and the animals bite off bits of algae using a tongue-like buccal mass and radula, which is the major feeding apparatus of

the *Gastropoda* (Ino, 1952; Kohn, 1983). Wild abalones show selective preferences for brown algae, including a variety of *Laminariales* (*Laminaria*, *Undaria*, *Eisenia*, *Ecklonia*, *Alaria*, *Egregia*, *Macrocystis*, and *Nereocystis*) and *Fucales* (*Desmarestia*), and red algae, including *Chondrus*, *Pterocladis*, *Gigaritina*, and *Asparagopsis* (Ino, 1952; Leighton and Boolootian, 1963). The choice of seaweed is restricted to the indigenous fauna for each species and/or each population of abalones. For example, Japanese abalones *Haliotis discus hannai* prefer *Laminaria*; the South African abalones *Haliotis midae* like another brown alga, *Ecklonia*; but Australian abalones seek out and grow better on indigenous red algae (Fleming and Hone, 1996).

Development and Aquaculture

Abalones spawn at different times in different habitats. The variation in this process is largely dependent on water temperature (Ino, 1952; Bevelander, 1988). Along the Japanese coastline, abalones spawn during late August to December, when the seawater temperature falls to ~20°C (Ino, 1952). Sperms and eggs are released from respiratory pores on the shell through the right renal organ of matured abalones. The tiny (0.2 mm in diameter) green or green-brownish eggs are fertilized in seawater. Within a day, these fertilized eggs develop into veliger larvae. The veliger has a swimming organ, a thick shell, and a lobed velum, mainly consisting of an upper and a lower epithelium (Kohn, 1983). After the short swimming stage, the veligers metamorphose and settle on the rocky sea floor, where they start eating diatom on substrata. The metamorphosed snail-like juveniles develop the feeding apparatus (buccal mass and radula) and behave as adult abalones (Kohn, 1983; Bevelander, 1988). Commonly, juveniles of >5 mm in size start ingesting seaweed in accordance with the development of the digestive system. Sensory cells, called taste buds, are observed in the tissue, which means that the animals are capable of selective feeding for "tasty" food (Ino, 1952). The most notable feature of the anatomy of abalone is the bending (V-shaped) stomach, which consists of four histologically differentiated parts (Ino, 1952; Bevelander, 1988) (Fig. 1). The first and second parts of the bending stomach are apparent in all abalone species (Ino, 1952; Bevelander, 1988; Erasmus et al., 1997). These are referred to as the crop and stomach, or simply as the first and second stomach, respectively (Fig. 1). Mucus cells are not observed inside the crop, which may be involved with food storage. Several ducts pass from the hepatopancreas in the upper part of the stomach. The inside of the (second) stomach is cov-

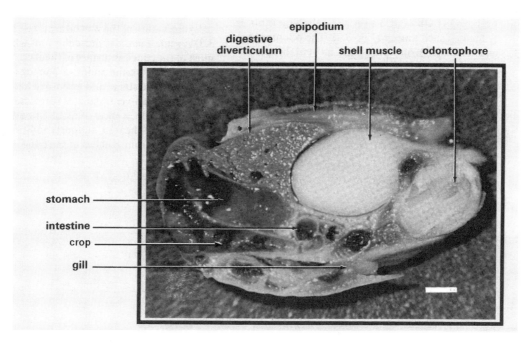

Figure 1. Sectioned view of *Haliotis* abalone. The bending stomach consists of the crop and stomach. Bar, 3 mm. The picture is reproduced from Bevelander (1988) with permission of the publisher.

ered by epithelium (Ino, 1952). In the Japanese abalone *H. discus*, third and fourth stomachs have also been observed. Of interest, the third stomach is covered with long epithelial cells (Ino, 1952). The digestive system seems to be a miniaturized version of that which exists in ruminant animals (Erasmus et al., 1997). The recorded levels of <0.38 mg of dissolved oxygen per liter in the digestive tract of greenlip abalone *Haliotis laevigata* suggest that the environment is microaerobic or anaerobic. Furthermore, the pH of the crop and the stomach has been recorded as pH 5.3 and 5.5, respectively (Harris et al., 1998).

Intensive aquaculture of abalone was developed during the 1950s in Japan to satisfy the need for conservation and human consumption (Ino, 1952). The success of artificial fertilization and (artificial) feed has gone a long way toward establishing the successful culture of abalone. Currently, many countries, including China, the United States, and Australia, have adopted abalone aquaculture (Oakes and Ponte, 1996).

DISTRIBUTION, ABUNDANCE, AND DIVERSITY OF *V. HALIOTICOLI* AND RELATED SPECIES

Currently, five *Vibrio* species have been found in the gut ecosystem of the ruminant-like abalone (Sawabe et al., 1995, 1998, 2003, 2004a,b; Hayashi

et al., 2003). The distribution, abundance, and diversity of these vibrios are discussed in this section.

Distribution and Abundance of *V. halioticoli* and Related Species in the Abalone Gut

V. halioticoli was originally described by Sawabe et al. (1998) as an alginolytic, nonmotile *Vibrio* (Fig. 2). *V. halioticoli*-like strains are commonly found in the gut of *Haliotis* spp. in Japan (Sawabe et al., 1995, 1998, 2002, 2004b), South Africa (Sawabe et al., 2003), Australia (Hayashi et al., 2003), and France (Sawabe et al., 2004a). *V. halioticoli* has also been isolated from water in aquaculture facilities raising abalones (Tanaka et al., 2002b, 2003). It is noteworthy that *V. halioticoli* has never been isolated from any other mollusc or echinoderm (Sawabe et al., 2003). Furthermore, protozoans, fungi, and archaea have never been detected in the abalone gut (Tanaka et al., 2004).

V. halioticoli was isolated as an alginolytic, nonmotile, unflagellated facultative anaerobic bacterium from the gut of Japanese abalone *H. discus hannai* (Sawabe et al., 1995). Of interest, there have not been any reports of unflagellated *Vibrio* species before 1998. At the time, it was unclear whether these unflagellated organisms should have been included in *Vibrio* or *Photobacterium*. Later, phylogenetic analysis of the 16S rRNA gene sequence of four *V. halioticoli* cultures clearly demonstrated that the organism should

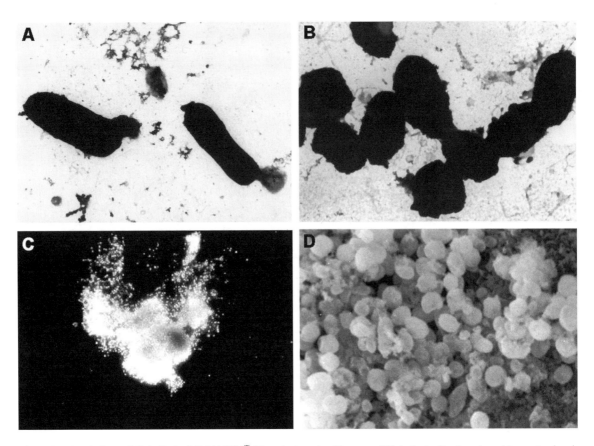

Figure 2. Morphology of *V. halioticoli* IAM14596[T]. Negatively stained images of *V. halioticoli* cells cultured in marine broth (A) and broth containing 0.5% (wt/vol) sodium alginate (B). (C) The DAPI (4′,6′-diamidino-2-phenylindole)-stained cells are attached to a seawater-derived alginate gel matrix; the scanning electron microscope image is also shown (D). Parts of panels A, B, and D are reproduced from Sawabe et al. (1998) with permission of the publisher.

be classified into the genus *Vibrio* (Sawabe et al., 1998). Furthermore, polyphasic taxonomy concluded that these isolates belong in a new species, for which the name *Vibrio halioticoli* (the species name is derived from the Latin "from the gut of *Haliotis* abalone") was proposed (Sawabe et al., 1998). The organism utilizes a narrow range of carbohydrates, i.e., glucose, fructose, maltose, mannitol, D-glucosamine, N-acetyl-D-glucosamine, laminarin, and alginate. It is noteworthy that the organism grows well on marine agar supplemented with sodium alginate. In terms of population sizes, the number of *V. halioticoli* cells in *H. discus hannai* has been reported to be in the range of 2.6 × 10[6] to 9.9 × 10[8] CFU/g of fresh gut, and that they constitute ~70% of the viable bacterial population (Sawabe et al., 1995). It has been noted that *V. halioticoli* produces multiple alginate-degrading enzymes, which are specific for the polyguluronate block in the alginate molecule, and may be involved in the cooperative degradation of seaweed polysaccharide with the host (Sawabe et al., 1995).

In the years immediately after the discovery of *V. halioticoli* (Sawabe et al., 1998), the cells in a variety of abalone species were enumerated with a genomic DNA probe derived from the type strain (Tanaka et al., 2002a; Sawabe et al., 2003). Positive reactions were recorded in *Haliotis discus discus*, *Haliotis diversicolor aquatilis*, and *Haliotis diversicolor diversicolor*, which are all modern Japanese species of abalones. As before, it was noted that all of the *V. halioticoli*-like strains are nonmotile and are capable of alginate degradation. The *V. halioticoli*-like strains have also been isolated from turbo shells, but have never been detected from other seaweed-consuming invertebrates, for example, sea hare, sea urchin, and *Trochidae* and *Littoridae* shells (Sawabe et al., 2003). Subsequently, polyphasic taxonomy, including amplified fragment length polymorphism fingerprinting, and sequencing of the housekeeping gene concluded that isolates regarded as comprising *V. halioticoli*-like taxa could be better elevated into separate species. Thus, *Vibrio ezurae* was established for the *Haliotis diversicolor* isolates, and *Vibrio neonatus* was named for the *H. discus discus* cultures (Sawabe et al., 2002, 2004b). The proportion of *V. halioticoli*-related species to total microflora in the gut is 42.4, 64.3, 40.6, and 19.2% for *H. discus discus*, *H. diversicolor aquatilis*, *H. diversicolor diversicolor*, and *Turbo cornatus*, respectively (Sawabe et al., 2003).

Nonmotile alginolytic vibrios have been isolated from wild populations of Australian greenlip and blacklip abalones, *H. laevigata* and *Haliotis rubra*, collected at Clifton Springs, Victoria (Hayashi et al., 2003). The bacterial strains were classified as *Vibrio* spp. and made up 96.4% of the 9.4×10^6 CFU/g of the fresh gut material. However, nonmotile alginolytic vibrios consisted of only 10% of these populations. Polyphasic taxonomy of these nonmotile vibrios led to the description of *Vibrio superstes*, in which "superstes" is derived from the Latin meaning "a survivor" (Hayashi et al. 2003). The remainder of the 90% of culturable bacteria were tentatively identified as *Vibrio* spp. (Hayashi et al., 2003). It has been speculated that a small population of *V. superstes* might be correlated with the feeding behavior of the Australian abalones, which prefer red algae—these are common along the Australian coast. The gut microflora, which is dominated by "gut group *Vibrio*," might well be adapted to the host feeding behavior. It is noted that *V. superstes* demonstrates an ability to utilize a wide range of carbohydrate sources, more so than other *V. halioticoli*-related species (Hayashi et al., 2003).

V. halioticoli has been found in the gut of the South African abalone *H. midae*, which was collected at Robin Island, Cape Town (Sawabe et al., 2003). The major food source for these South African abalones is brown algae, i.e., *Ecklonia maxima*. Experiments have revealed that the proportion of *V. halioticoli* in *H. midae* amounts to 67% of the total bacterial population of 3.7×10^6 CFU/g of fresh gut.

Vibrio gallicus, the species name of which is derived from the Latin "from France," is the most common vibrio in the gut of the French abalone *Haliotis tuberculata*, specimens of which have been collected from the coast of Brest, Brittany (Sawabe et al., 2004a). The species is regarded as being divergent from *V. halioticoli* on the basis of 16S rRNA gene phylogeny, which revealed a similarity of 97%. Furthermore, isolates clustered with other psychrophilic vibrios (Sawabe et al. 2004a). Experiments revealed that *V. gallicus* constitutes up to 55% of the total bacterial population of 3.0×10^6 CFU/g of fresh gut.

Vibrios from the guts of abalones from California, New Zealand, and Taiwan have not been examined, to date. Japanese *H. discus hannai* has been transplanted to well-controlled sites in several coun-

tries outside of Japan. The gut microbial ecosystem of these transplanted abalones could provide good experimental material for studying microbial evolution as affected by man-made environmental conditions.

Diversity of *V. halioticoli* and Related Species

Different phenotypic traits among *V. halioticoli*, *V. ezurae*, *V. neonatus*, *V. gallicus*, and *V. superstes* have been observed in 20 out of 78 features examined (Sawabe et al., 2004b); most differences are in utilization patterns of carbohydrates and organic acids. It has been noted that *V. ezurae*, *V. neonatus*, and *V. gallicus* do not utilize as wide a range of carbon compounds as *V. halioticoli* does (Sawabe et al., 1998). In contrast, *V. superstes* is more active metabolically and is capable of utilizing many carbon compounds (14 have been documented) (Hayashi et al., 2003; Sawabe et al., 2004b). Intraspecific phenotypic variations have been regarded as insignificant in *V. halioticoli*, *V. ezurae*, *V. neonatus*, and *V. gallicus* (Sawabe et al., 2004b).

V. halioticoli, *V. neonatus*, *V. ezurae*, *V. gallicus*, and *V. superstes* are included in a single robust clade that does not feature any other validly described species, as determined from small subunit rRNA gene sequences (Sawabe et al., 2004b). These 16S rRNA gene sequences are most similar for *V. halioticoli*, *V. ezurae*, and *V. neonatus*. In contrast, the 16S rRNA gene sequence of *V. gallicus* is more distant (<98%) from the other four *V. halioticoli*-related species (Hayashi et al., 2003). A robust clade has also been observed by *gap* (glyceraldehyde-3-phosphate dehydrogenase) phylogeny (Sawabe et al., 2004b).

V. halioticoli, *V. gallicus*, and *V. superstes* are most definitely separate species, insofar as the DNA: DNA similarities are <70% (Hayashi et al., 2003; Sawabe et al., 2004a). Conversely, both pairwise and reciprocal DNA:DNA similarity values among *V. halioticoli*, *V. ezurae*, and *V. neonatus* are at the boundary of a species definition (Sawabe et al., 2002, 2004b). However, the genomic diversities of the species are different, as determined by amplified fragment length polymorphism and repetitive extragenic palindrome-PCR fingerprinting (Sawabe et al., 2002) and the sequencing of housekeeping genes (Sawabe et al., 2004b).

Motility of *V. halioticoli* and Sister Species

The genus name *Vibrio* is derived from the Latin for "vibrate" (Farmer and Hickman-Brenner, 1999). All 63 currently recognized species of *Vibrio* possess sheathed flagella (Baumann and Schubert, 1984; Thompson et al., 2004), with the exception of *V. halio-*

ticoli-related species (Sawabe et al., 1995, 1998, 2003, 2004b; Hayashi et al., 2003) and *V. rumoiensis* (Yumoto et al., 1999). It is interesting that most of the nonflagellated vibrios are found in the gut of abalones. In addition to morphological observations on the lack of flagella among the *V. halioticoli*-related species (Sawabe et al., 1998, 2004a,b; Hayashi et al., 2003), PCR amplifications of genes responsible for *Escherichia coli* flagellin (*fliC*) and *Vibrio parahaemolyticus* flagellins (*flaA* and *flaC*) have not been detected in *V. halioticoli*, *V. neonatus*, or *V. ezurae* (unpublished data).

A possible explanation for the absence of flagella in *V. halioticoli*-related species is being sought by comparing the organism with the lifestyles of *V. fischeri* (Visick and McFall-Ngai, 2000) and *Bordetella* spp. (Parkhill et al., 2003; Nierman and Fraser, 2004). *V. fischeri* is a well-known symbiont colonizing the light organ of the Hawaiian bobtail squid *Euprymna scolopes*. It has been observed that *V. fischeri* loses its flagella after colonization of the light organ crypt (Visick and McFall-Ngai, 2000). Another possibility resides with whole-genome sequence analysis of *Bordetella pertussis* and *Bordetella parapertussis*, which are respiratory tract colonizers. The analyses have revealed that there are large-scale inactivations by frame-shift mutation or transposon insertion in the flagellar operons (Parkhill et al., 2003; Nierman and Fraser, 2004). Downregulation of flagellar expression by long-term symbiotic association or parasitic colonization might well lead to loss of flagellation or gene disruption of flagellar operons.

Detection Methods

Species-specific detection methods are available for *V. halioticoli*, *V. neonatus*, and *V. ezurae* (Sugimura et al., 2000b; Tanaka et al., 2001, 2002a; Sawabe et al., 2004b). In situ PCR specific to an alginate lyase gene of *V. halioticoli* was capable of discriminating between *V. halioticoli* and related vibrios (Sugimura et al., 2000b). The method was considered to be possibly effective in detection of single cells of *V. halioticoli*. Other detection methods include 16S rDNA PCR-restriction fragment length polymorphism and colony hybridization with the *V. halioticoli* genome as probe (Tanaka et al., 2001, 2002a). Both methods are reliable and rapid and should be useful for studying the ecology of *V. halioticoli*. However, it should be emphasized that cultures of *V. halioticoli* could not be discriminated from *V. ezurae* and *V. neonatus* (Sawabe et al., 2003). In fact, sequencing of the *gap* gene is the only effective method for differentiating *V. halioticoli* from *V. neonatus* and *V. ezurae* (Sawabe et al., 2004b).

ECOPHYSIOLOGICAL ROLES OF *V. HALIOTICOLI*

Abalones contain dense populations of *V. halioticoli*. The questions to be answered concern the nature of any contributions that *V. halioticoli* may make to the well-being of the host abalone. This aspect is considered in the following sections.

Cooperative Degradation of Carbohydrates

Algal polysaccharide, cellulose, alginate, and laminarin may well be important energy sources for abalones (Leighton and Boolootian, 1963; Takami et al. 1998). In particular, alginate degradation has been intensively studied as a key biochemical process in the digestive processes of abalone. Alginate is a linear heteropolymeric polysaccharide consisting of a uronic acid backbone (Gacesa, 1988) and is a major component of the cell wall matrix of brown algae (Kloareg and Quatrano, 1988). There are three heterogeneous block structures: polymannuronate, polyguluronate, and mannuronate- and guluronate-mixed blocks. The first characterization of abalone alginate degradation enzymes occurred during the 1960s and involved Japanese and Californian abalones (Tsujino and Saito, 1961; Tujino, 1962; Nakada and Sweeny, 1967). At least two alginases with different substrate preferences have been characterized (Nakada and Sweeny, 1967). As there are technical difficulties with the complete separation of the digestive tract, especially the crop and stomach, cross-contamination of enzymes from gut microbes and host abalone could occur. So the overriding concern is the origin of any enzyme studied. Quite simply, is the enzyme microbial or abalone?

It is apparent that *V. halioticoli* and related species produce alginate lyase (Sawabe et al., 1995, 1998, 2004a,b; Hayashi et al., 2003). In addition, *V. halioticoli* has been recognized to produce polyguluronate-specific alginases (Sawabe et al., 1995). The alginate-loving bacterium also shows unique behavior during alginate degradation in vitro (Fig. 2B-D). Thus, the bacterial cells attach and grow onto a calcium-induced alginate gel matrix and make dense clusters (Fig. 2C,D). The insoluble gel matrix is gradually shrunk by vigorous degradation by *V. halioticoli*. On gels, the organism forms rounded shapes and is chained (Fig. 2B). These cells are larger than those recovered from marine broth without carbohydrate (Fig. 2A). Attempts to purify the alginate-degrading enzyme(s) have not been successful owing to an unsatisfactory separation of the enzyme(s) on size filtration chromatography. However, three kinds of genes responsible for polyguluronate-specific alginases have been cloned (Sugimura et al., 2000a). In short, *V. halioticoli* is a polyguluronate-specific-alginase producer.

It is clear that abalone secretes alginase (Shimizu et al., 2003), and a polymannuronate-specific alginase (HdAly) has been purified from the hepatopancreas of *H. discus hannai* by Professor T. Ojima (Hokkaido University). Moreover, the gene responsible for the purified enzyme has been cloned from an abalone hepatopancreas eukaryotic mRNA gene library. HdAly has an active pH range of 6.5 to 9 and a pH optimum of 8.0. The major end product is tri-uronide (Shimizu et al., 2003). It is noteworthy that abalones produce their own polymannuronate-specific alginase, which is active at neutral pH (Nakada and Sweeny, 1967; Heyraud et al., 1996; Shimizu et al., 2003). Therefore, it is speculated that there may well be cooperative (symbiotic) degradation between *V. halioticoli* and the abalone in the sharing of energy-rich alginate.

Entry of Vibrios into the Gut Ecosystem

The entry of key symbionts into the environment of the aquatic host gut or symbiotic organs is an uncertain procedure (Lee and Ruby, 1994; Ruby and Lee, 1998; Gros et al., 1996; Millikan et al., 1999). The spatial distribution of symbionts outside of the gut environment and the timing to colonize the gut need to be understood in terms of the development of the host.

Postlarval development of polysaccharide degradation activity in the gut of *H. discus discus* was measured by Takami et al. (1998) using the diatom *Cocconeis scutellum* in well-controlled feeding experiments. Enzyme activities of *H. discus discus* for cellulose, alginate, and laminarin larvae were detected after 17 days of settlement in the case of juveniles, which were fed with diatoms. The enzyme activities increased gradually until 37 days after settlement at 20°C and rapidly increased after 37 days. Chrysolaminarin, which is a β1, 3-glucan homologous to laminarin, is one of the major polysaccharides in *C. scutellum*. Although laminarinase activity might be induced by the diatom laminarin, other polysaccharide-degrading enzymes, i.e., cellulase and alginase, could be constitutively expressed without substrate induction. This means that abalones are ready to eat macroalgae within 45 days in standard rearing conditions (Takami et al., 1998).

Takami et al. (1997) reported on the induction of settlement and the effect of growth on the trail mucus. Commonly, wild abalone larvae swim on trail mucus; these workers designed experiments involving use of (i) trail mucus only, (ii) diatoms only, and (iii) trail mucus and diatoms, in an attempt to better understand the behavior of abalone settlement. The best

system for settlement and growth involved use of trail mucus plus diatoms. This led to 97.3% settlement and 70% survival during the 4 weeks of the experiment. Finally, it was apparent that the abalone larvae normally grew up to 1.4 mm in length.

Changes in the gut microflora have been reported during different stages of abalone development (Tanaka et al., 2003). It has been revealed that the gut microflora of 80-day-old juveniles of *H. discus hannai* is affected by microflora of the surrounding seawater, which is composed largely of strictly aerobes, notably *Pseudomonas* and *Alteromonas* spp. The first appearance of *V. halioticoli* in the gut environment is at 80 days. Then by 110 days, the gut becomes populated predominantly by *Vibrio* spp., with *V. halioticoli* accounting for >50% of the total culturable bacterial numbers. *V. halioticoli* cells have also been recovered from water (3 CFU/liter), diatom beds used to culture juveniles (8×10^2 CFU/g), and feces (up to 10^5 CFU/g) (Tanaka et al., 2001, 2002a,b).

So far, scientists have only snapshots of the microecology of *V. halioticoli* in terms of abalone development (Fig. 3). It is interesting to speculate upon the likely timing of *V. halioticoli* entry into the host gut ecosystem. It may well be that the best time is when juveniles start feeding on diatoms, which are covered with the trail mucus of the mother abalones (Fig. 3). This timing is supported by the fact that the survival of juvenile abalones is extended when feeding on diatoms covered with trail mucus (Takami et al., 1997). Furthermore, dense populations of *V. halioticoli* have been observed on diatom beds (Tanaka et al., 2002b). Host abalones start ingesting seaweed, preferring brown algae, and this is accompanied by development of the host digestive system. It is reasoned that sustainable nutrient supplies for *V. halioticoli* and/or other gut microbes lead to stable microbial ecosystem in the gut of abalones (Fig. 3). Increased populations of *V. halioticoli* could lead to distribution into the aquatic environment, via feces, and thus to the seafloor. The nonmotile vibrios could then be available for future entry into and colonization of the abalone gut (Fig. 3).

Fermentation: an Extending Concept for Vibrio-Abalone Symbiosis

Volatile short-chained fatty acids (VSCFAs) are available to the host animal as fermentation products converted from energy-rich carbohydrates. VSCFAs are recognized as important energy sources not only in herbivorous ruminant animals (Hungate, 1966; Stewart and Bryant, 1988; Russell and Rychlik, 2001) and termites (Breznak, 1982) but also in omnivorous human beings (Hooper et al., 2002). Vibrios ferment

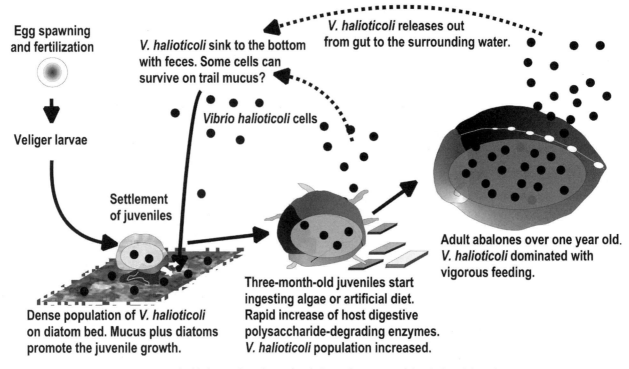

Figure 3. The likely ecophysiology of *V. halioticoli* in terms of the abalone life cycle.

a wide variety of carbohydrates (Baumann and Schubert, 1984). In particular, *V. halioticoli*, *V. neonatus*, and *V. ezurae* are capable of fermentation not only of simple hexoses but also of seaweed carbohydrates through the acetic acid/formic acid pathway (Sawabe et al., 2003). Acetic acid is probably involved as an available energy source or metabolic precursor for abalones (Sawabe et al., 2003). In the case of gut microbial ecosystems, formic acid should be eliminated by methanogens to reduce the possibility of toxicity. However, there is no evidence for the presence of methanogens in the abalone gut ecosystem (Tanaka et al., 2004). Moreover, in situ fermentation experiments of alginate by mixtures of abalone gut microbes with or without the presence of the vibriostatic agent O/129 revealed a reduction in acetic acid and formic acid production in cultures when O/129 was present (unpublished data). It is likely that vibrios may be responsible for alginate fermentation. Amounts of acetic acid and formic acid were also detected from gut homogenates of *H. discus hannai*, which had been fed with artificial diets and *Laminaria* (unpublished data).

The oxygen level inside the gut of *Haliotis* spp. is <0.38 mg dissolved oxygen per liter, as determined by microelectrode analysis (Harris et al., 1998). The gut environment, which could be anaerobic or microaerobic, has a pH of 5.3 to 5.6 in the crop and pH 5.5 to 6.5 in the stomach (Erasmus et al., 1997; Harris et al., 1998). The pH of the intestine is >pH 6.0. Thus, it is possible that *V. halioticoli* and related species are capable of fermenting carbohydrates inside the crop and/or stomach for production of VSC-FAs, which may then be absorbed in the posterior part of the stomach and intestine, leading to small pH increases (Ino, 1952).

Contribution of Gut Microbes to the Nitrogen Cycle of Abalone

The carbon and nitrogen balance is an important aspect of animal physiology. In ruminants, it is believed that a major amount of nitrogen is provided by whole cell fractions of rumen microbes (Hungate, 1966; Stewart and Bryant, 1988). Proteolytic enzyme activities have been reported in five species of *Haliotis* (García-Carreño et al., 2003). The Californian green abalone, *Haliotis fulgens*, when fed with the brown alga *Macrocystis purifera*, demonstrated the highest growth rate but the lowest protease activity. Conversely, the lowest growth and highest protease activity occurred as a result of feeding with the red alga *Gelidium robstum* (García-Carreño et al., 2003). Moreover, the total protein content is higher in *G. robstum* than in *M. purifera*. Thus, it seems likely that the protease activities of abalones depend directly on the protein content of the food. However, nitrogen metabolism is only just beginning to be considered.

COEVOLUTION OF VIBRIO–ABALONE SYMBIOSIS

Host-parasite systems are intrinsically interesting to evolutionary biologists because they potentially signal a long and intimate association between two or more groups of organisms that are often distantly related and quite dissimilar biologically (Page and Holmes, 1998). This long history of association often leads to reciprocal adaptation in the host and parasites (classical coevolution or coadaptation) as well as contemporaneous cladogenetic events in two lineages (cospeciation) (Page and Holmes, 1998). Reconstructing the history of host–parasite systems by comparing their gene trees could test hypotheses of evolutionary events in which host and parasite have cospeciated, the parasites have switched to the host, and a population of parasites has become extinct (Hafner and Nadler, 1988; Page and Holmes, 1998).

The host–parasite theory has also been applied by many biologists to reconstruct marine host–microbe symbioses. Reconstructing the history of host–microbe symbioses has been tried in (i) squid and the light organ symbiotic *Vibrio* (Nishiguchi et al., 1998), (ii) deep-sea clam and the gill symbiotic sulfur oxi-

dizers (Peek et al., 1998), and (iii) algal and gill ectosymbionts (Ashen and Goff, 2000). Parallel evolution (coevolution) between host and symbiont has been observed in *V. fischeri* and squid (Nishiguchi et al., 1998) and the sulfur oxidizer and clam (Peek et al., 1998).

Abalones have a long life history, as supported by fossil records (Geiger and Groves, 1999). It is interesting to estimate that the observed diversity among *V. halioticoli* and related species might be attributed to long symbiotic associations. I have attempted to reconstruct the phylogenetic history of the *Vibrio*–abalone association using available internal transcribed spacer (ITS) sequences for host abalones (Coleman and Vacquier, 2002) and the *gap* gene sequences for the symbiotic vibrios (Sawabe et al., 2004b) (Fig. 4). Both maximum likelihood trees were incongruent, but potentially ancient *V. halioticoli* populations could be divided into at least four populations, with one population cospeciated to French and Australian lineage. The other three populations could be inherited through Japanese and South African lineages with a number of host-sorting and extinction events (Fig. 4). These events might be supported by the fact that mixed populations of *V. halioticoli* and *V. neonatus* are observed in the Japanese abalone *H. discus discus* (Sawabe et al., 2002).

It is not known whether the reconciled tree is real or not because the phylogenies of both vibrio and

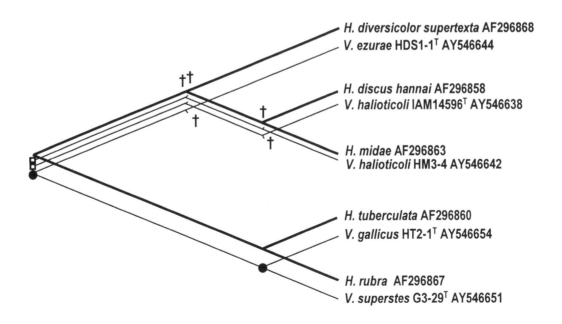

Figure 4. Reconciled tree of *Vibrio*–abalone symbiosis. Each phylogenetic tree was constructed by the maximum likelihood method based on *gap* and ITS genes for symbiotic *Vibrio* and host, respectively. Maximum likelihood trees were compared with the TreeMap program (Page, 1995). The reconciled tree estimated cospeciation (closed circle), duplication (open square), and a sorting event (short branch plus sword mark).

host are not yet fixed. Coleman and Vacquier (2002) simply grouped (i) the North Pacific species, (ii) the European species, and (iii) the Australian species; ungrouped *H. midae* and *H. diversicolor* organisms both diverged below that of the European and Australian species, as determined by ITS phylogeny. The age of *Vibrio–Escherichia* (or *Enterobacteriaceae*) radiation cannot yet be estimated. Global comparative genome analysis would provide good information on resolving the evolutional events of vibrio–animal symbiosis.

THE FUTURE

Out of the chaos of the abalone gut microbial ecosystem, symbiotic harmony has been found between *V. halioticoli* and its host. The central part of this harmony is the cooperative metabolism of algal carbohydrates, specifically alginate. Abalones are clearly capable of breaking down part of the alginate molecule (polymannuronates) to series of di- and trisaccharides as sources of energy. The symbiotic gut vibrios utilize another part of the alginate molecule, polyguluronate, which the abalone is incapable of using, and convert it to host-available waste, i.e., acetic acid. It has been determined that the vibrios are unwilling to use the waste product (Sawabe et al., 1998). Thus, the vibrios appear to be effective partners of the abalones. This vibrio–abalone symbiotic harmony is a good example of vibrio–aquatic animal interactions.

However, many aspects of the vibrio–abalone symbiosis have been determined from snapshots obtained from field material. The absence of sophisticated experimental models has been an obstacle for studying these interactions. Certainly, there is an excellent well-established model for *V. fischeri*–squid symbiosis (McFall-Ngai and Ruby, 1991; Nyholm et al., 2000), and this should be useful for the future. The goal of further work will be to determine whether these vibrios contribute to host morphogenesis, and whether the abalone transfers the vibrio to the next generation. This knowledge may be directly applicable to aquaculture and the conservation of wild stocks.

Acknowledgments. I thank Dr. Yoshio Ezura (Professor Emeritus, Hokkaido University) and Dr. E. P. Ivanova (Swinburne University of Technology) for critically reading the manuscript.

REFERENCES

Ashen, J. B., and L. J. Goff. 2000. Molecular and ecological evidence for species specificity and coevolution in a group of marine algal-bacterial symbioses. *Appl. Environ. Microbiol.* **66:**3024–3030.

Bäckhed, F., H. Ding, T. Wang, L. V. Hooper, G. Y. Koh, A. Nagy, C. F. Semenkovich, and J. I. Gordon. 2004. The gut microbiota as an environmental factor that regulates fat storage. *Proc. Natl. Acad. Sci. USA* **101:**15718–15723.

Baumann, P., and R. H. W. Schubert. 1984. *Vibrionaceae,* p. 516–550. *In* N. R. Krieg and J. G. Holt (ed.), *Bergey's Manual of Systematic Bacteriology,* vol. 1. Lippincott Williams & Wilkins, Baltimore, Md.

Bevelander, G. 1988. *Abalone, Gross and Fine Structure.* Boxwood Press, Pacific Grove, Calif.

Breznak, J. A. 1982. Intestinal microbiota of termites and other xylophagous insects. *Annu. Rev. Microbiol.* **36:**323–343.

Coleman, A. W., and V. D. Vacquier. 2002. Exploring the phylogenetic utility of ITS sequences for animals: a test case for abalone (*haliotis*). *J. Mol. Evol.* **54:**246–257.

Colwell, R. R., and J. Liston. 1960. Microbiology of shellfish. Bacteriological study of the natural flora of Pacific oyster (*Crassostrea gigas*). *Appl. Microbiol.* **8:**104–109.

Colwell, R. R., and J. Liston. 1962. The natural bacterial flora of certain marine invertebrate. *J. Insect Pathol.* **4:**23–33.

Colwell, R. R., J. Kaper, and S. W. Joseph. 1977. *Vibrio cholerae, Vibrio parahaemolyticus,* and other vibrios: occurrence and distribution in Chesapeake Bay. *Science* **198:**394–396.

Colwell, R. R. 1984. Vibrios in the environment, p. 1–12. *In* R. R. Colwell (ed.), *Vibrios in the Environment.* John Wiley & Sons, New York, N.Y.

Dolan, M. F. 2001. Speciation of termite gut protests: the role of bacterial symbionts. *Int. Microbiol.* **4:**203–208.

Erasmus, J. H., P. A. Cook, and V. E. Coyne. 1997. The role of bacteria in the digestion of seaweed by the abalone *Haliotis midae. Aquaculture* **155:**377–386.

Falk, P. G., L. V. Hooper, T. Midtvedt, and J. I. Gordon. 1998. Creating and maintaining the gastrointestinal ecology: what we know and need to know from gnotobiology. *Microbiol. Mol. Biol. Rev.* **62:**1157–1170.

Farmer, J. J., III, and F. W. Hickman-Brenner. 1999. The genera *Vibrio* and *Photobacterium. In* M. Dworkin et al. (ed.), *The Prokaryotes: an Evolving Electronic Resource for the Microbiological Community,* 3rd ed., release 3.0. Springer-Verlag, New York, N.Y. http://link.springer-ny.com/link/service/books/10125/.

Fleming, A. E., and P. W. Hone. 1996. Abalone aquaculture. *Aquaculture* **140:**1–4.

Gacesa, P. 1988. Alginates. *Carbohydr. Polymer* **8:**161–182.

García-Carreño, F. L., M. A. Navarrete del Toro, E. Serviere-Zaragoza. 2003. Digestive enzymes in juvenile green abalone, *Haliotis fulgens,* fed natural food. *Comp. Biochem. Physiol.* **134:**143–150.

Geiger, D. L., and L. T. Groves. 1999. Review of fossil abalone (Gastropoda: Vetigastropoda: Haliotidae) with comparison to recent species. *J. Paleont.* **73:**872–885.

Gros, O., A. Darrasse, P. Durand, L. Frenkiel, and M. Mouza. 1996. Environmental transmission of a sulfur-oxidizing bacterial gill endosymbiont in the tropical Lucinid bivalve *Codakia orbicularis. Appl. Environ. Microbiol.* **62:**2324–2330.

Hafner, M. S., and S. A. Nadler. 1988. Phylogenetic trees support the coevolution of parasites and their hosts. *Nature* **332:**258–259.

Harris, J. M. 1993. The presence, nature, and role of gut microflora in aquatic invertebrates: a synthesis. *Microb. Ecol.* **25:**195–231.

Harris, J. O., C. M. Burke, and G. B. Maguire. 1998. Characterization of the digestive tract of greenlip abalone, *Haliotis laevigata* Donovan. II. Microenvironment and bacterial flora. *J. Shellfish Res.* **17:**989–994.

Hayashi, K., J. Moriwaki, T. Sawabe, F. L. Thompson, J. Swings, N. Gudkovs, R. Christen, and Y. Ezura. 2003. *Vibrio superstes* sp. nov., isolated from the gut of Australian abalones *Haliotis laevigata* and *Haliotis rubra*. *Int. J. Syst. Evol. Microbiol.* **53:** 1813–1817.

Heyraud, A., R. Colin-Morel, S. Girond, C. Richard, and B. Kloareg. 1996. HPLC analysis of saturated or unsaturated oligoguluronates and oligomannuronates. Application to the determination of the action pattern of *Haliotis tuberculata* alginate lyase. *Carbohydr. Res.* **291:**115–126.

Hooper, L. V. 2004. Bacterial contributions to mammalian gut development. *Trends Microbiol.* **12:**129–134.

Hooper, L. V., and J. I. Gordon. 2001. Commensal host-bacterial relationships in the gut. *Science* **292:**1115–1118.

Hooper, L. V., T. Midtvedt, and J. I. Gordon. 2002. How host-microbial interactions share the nutrient environment of the mammalian intestine. *Annu. Rev. Nutr.* **22:**283–307.

Hungate, R. E. 1966. *The Rumen and Its Microbes*. Academic Press, Inc., New York, N.Y.

Ino, T. 1952. Biological studies on the propagation of the Japanese abalone (genus *Haliotis*). *Bull. Tokai Reg. Fish Res. Lab.* **5:**1–101.

Kaneko, T., and R. R. Colwell. 1973. Ecology of *Vibrio parahaemolyticus* in Chesapeake Bay. *J. Bacteriol.* **113:**24–32.

Kaneko, T., and R. R. Colwell. 1974. Distribution of *Vibrio parahaemolyticus* and related organisms in the Atlantic Ocean off South Carolina and Georgia. *Appl. Microbiol.* **28:**1009–1017.

Kaneko, T., and R. R. Colwell. 1975a. Adsorption of *Vibrio parahaemolyticus* onto chitin and copepods. *Appl. Microbiol.* **29:** 296–274.

Kaneko, T., and R. R. Colwell. 1975b. Incidence of *Vibrio parahaemolyticus* in Chesapeake Bay. *Appl. Microbiol.* **30:**251–257.

Kloareg, B., and R. S. Quatrano. 1988. Structure of the cell walls of marine algae and ecophysiological functions of the matrix polysaccharides. *Oceanogr. Mar. Biol. Annu. Rev.* **26:**259–315.

Kohn, A. J. 1983. Feeding biology of gastropods, p. 1–63. *In* A. S. M. Saleuddin and K. M. Wilbur (ed.), *The Mollusca*, vol. 5. Academic Press, Inc., New York, N.Y.

Lee, K.-H., and E. G. Ruby. 1994. Effect of the squid host on the abundance and distribution of symbiotic *Vibrio fischeri* in nature. *Appl. Environ. Microbiol.* **60:**1565–1571.

Lee, Y.-H., and V. D. Vacquier. 1995. Evolution and systematics in Haliotidae (Mollusca: Gastropoda): inferences from DNA sequences of sperm lysine. *Mar. Biol.* **124:**267–278.

Leighton, D. L., and R. A. Boolootian. 1963. Diet and growth in the black abalone *Haliotis cracherodii*. *Ecology* **44:**227–238.

MacFall-Ngai, M. J., and E. G. Ruby. 1991. Symbiont recognition and subsequent morphogenesis as early events in an animal-bacterial mutualism. *Science* **254:**1491–1494.

Millikan, D. S., H. Felbeck, and J. L. Stein. 1999. Identification and characterization of a flagellin gene from the endosymbiont of the hydrothermal vent tubeworm *Riftia pachyptila*. *Appl. Environ. Microbiol.* **65:**3129–3133.

Nakada, H. I., and P. C. Sweeny. 1967. Alginic acid degradation by eliminases from abalone hepatopancreas. *J. Biol. Chem.* **242:** 845–851.

Nierman, W. C., and C. M. Fraser. 2004. The power in comparisons. *Trends Microbiol.* **12:**62–63.

Nishiguchi, M. K., E. G. Ruby, and M. J. McFall-Ngai. 1998. Competitive dominance among strains of luminous bacteria provides an unusual form of evidence for parallel evolution in Sepiolid squid-*Vibrio* symbioses. *Appl. Environ. Microbiol.* **64:**3209–3213.

Nyholm, S. V., E. V. Stabb, E. G. Ruby, and M. J. McFall-Ngai. 2000. Establishment of an animal-bacterial association: recruiting symbiotic vibrios from the environment. *Proc. Natl. Acad. Sci. USA* **97:**10231–10235.

Oakes, F. R., and R. D. Ponte. 1996. The abalone market: opportunities for cultured abalone. *Aquaculture* **140:**187–195.

Ohkuma, M., and T. Kudo. 1996. Phylogenetic diversity of the intestinal bacterial community in the termite *Reticulitermes speratus*. *Appl. Environ. Microbiol.* **62:**461–468.

Page, R. D. M. 1995. Parallel phylogenies: reconstructing the history of host-parasite assemblages. *Cladistics* **10:**155–173.

Page, R. D. M., and E. C. Holmes. 1998. *Molecular Evolution*. Blackwell Scientific Publications, Oxford, England.

Parkhill, J., M. Sebaihia, A. Preston, et al. 2003. Comparative analysis of the genome sequences of *Bordetella pertussis*, *Bordetella parapertussis*, and *Bordetella bronchiseptica*. *Nat. Genet.* **35:**32–40.

Peek, A. S., R. A. Feldman, R. A. Lutz, and R. C. Vruenhoex. 1998. Cospeciation of chemoautotrophic bacteria and deep sea clam. *Proc. Natl. Acad. Sci. USA* **95:**9962–9966.

Ruby, E. G., and K.-H. Lee. 1998. The *Vibrio-fischeri-Euprymna scolopes* light organ association: current ecological paradigms. *Appl. Environ. Microbiol.* **64:**805–812.

Russell, J. B., and J. L. Rychlik. 2001. Factors that alter rumen microbial ecology. *Science* **292:**1119–1122.

Sawabe, T., Y. Oda, Y. Shiomi, and Y. Ezura. 1995. Alginate degradation by bacteria isolated from the gut of sea urchins and abalones. *Microb. Ecol.* **30:**192–202.

Sawabe, T., I. Sugimura, M. Ohtsuka, K. Nakano, K. Tajima, Y. Ezura, and R. Christen. 1998. *Vibrio halioticoli* sp. nov., a non-motile alginolytic marine bacterium isolated from the gut of abalone *Haliotis discus hannai*. *Int. J. Syst. Bacteriol.* **48:** 573–580.

Sawabe, T., F. L. Thompson, J. Heyrman, M. Cnockaert, K. Hayashi, R. Tanaka, M. Yoshimizu, B. Hoste, J. Swings, and Y. Ezura. 2002. Fluorescent amplified fragment length polymorphism (FAFLP) and repetitive extragenic palindrome (rep)-PCR fingerprinting reveal host specific-genetic diversity of *Vibrio halioticoli*-like strains isolated from the gut of Japanese abalone. *Appl. Environ. Microbiol.* **68:**4140–4144.

Sawabe, T., N. Setoguchi, S. Inoue, R. Tanaka, M. Ootsubo, M. Yoshimizu, and Y. Ezura. 2003. Acetic acid production of *Vibrio halioticoli* from alginate: a possible role for establishment of abalone-*Vibrio halioticoli* association. *Aquaculture* **219:** 671–679.

Sawabe, T., K. Hayashi, J. Moriwaki, F. L. Thompson, J. Swings, P. Potin, R. Christen, and Y. Ezura. 2004a. *Vibrio gallicus* sp. nov., isolated from the gut of French abalone *Haliotis tuberculata*. *Int. J. Syst. Evol. Microbiol.* **54:**843–846.

Sawabe, T., K. Hayashi, J. Moriwaki, Y. Fukui, F. L. Thompson, J. Swings, and R. Christen. 2004b. *Vibrio neonatus* sp. nov. and *Vibrio ezurae* sp. nov. isolated from the gut of Japanese abalones. *Syst. Appl. Microbiol.* **27:**527–534.

Shimizu, E., T. Ojima, and K. Nishita. 2003. cDNA cloning of an alginate lyase from abalone, *Haliotis discus hannai*. *Carbohydr. Res.* **338:**2841–2852.

Stappenbeck, T. S., L. V. Hooper, and J. I. Gordon. 2002. Developmental regulation of intestinal angiogenesis by indigenous microbes via Paneth cells. *Proc. Natl. Acad. Sci. USA* **99:**15451–15455.

Stewart, C. S., and M. P. Bryant. 1988. The rumen bacteria, p. 21–75. *In* P. N. Hodoson (ed.), *The Rumen Microbial Ecosystem*. Elsevier Applied Science, London, England.

Suau, A., R. Bonnet, M. Sutren, J.-J. Godon, G. R. Gibson, M. D. Collins, and J. Dore. 1999. Direct analysis of genes encoding 16S rRNA from complex communities reveals many novel molecular species within the human gut. *Appl. Environ. Microbiol.* **65:**4799–4807.

Sugimura, I., T. Sawabe, and Y. Ezura. 2000a. Cloning and sequence analysis of *Vibrio halioticoli* genes encoding three types of polyguluronate lyase. *Mar. Biotechnol.* **2:**65–73.

Sugimura, I., T. Sawabe, and Y. Ezura. 2000b. In situ PCR visualization of *Vibrio halioticoli* using alginate lyase gene alyVG2. *Mar. Biotechnol.* **2**:74–79.

Takami, H., T. Kawamura, and Y. Yamashita. 1997. Survival and growth rates of post-larval abalone *Haliotis discus hannai* fed conspecific trail mucus and/or benthic diatom *Cocconeis scutellum* var. *parva. Aquaculture* **152**:129–138.

Takami, H., T. Kawamura, and Y. Yamashita. 1998. Development of polysaccharide degradation activity in postlarval abalone *Haliotis discus hannai. J. Shellfish Res.* **17**:723–727.

Tanaka, R., T. Sawabe, K. Tajima, J. Vandenberghe, and Y. Ezura. 2001. Identification of *Vibrio halioticoli* using 16S rDNA PCR/RFLP (restriction fragment length polymorphism) analysis. *Fisheries Sci.* **67**:185–187.

Tanaka, R., M. Ootsubo, T. Sawabe, K. Tajima, J. Vandenberghe, and Y. Ezura. 2002a. Identification of *Vibrio halioticoli* by colony hybridization with non-radioisotope labeled genomic DNA probe. *Fisheries Sci.* **68**:227–229.

Tanaka, R., T. Sawabe, M. Yoshimizu, and Y. Ezura. 2002b. Distribution of *Vibrio halioticoli* around an abalone farming center in Japan. *Microb. Environ.* **17**:6–9.

Tanaka, R., I. Sugimura, T. Sawabe, M. Yoshimizu, and Y. Ezura. 2003. Gut microflora of abalone *Haliotis discuss hannai* in culture change coincident with a change in diet. *Fisheries Sci.* **69**:949–956.

Tanaka, R., M. Ootsubo, T. Sawabe, Y. Ezura, and K. Tajima. 2004. Biodiversity and *in situ* abundance of gut microflora of abalone (*Haliotis discus hannai*) determined by culture-independent techniques. *Aquaculture* **241**:453–463.

Thompson, F. L., T. Iida, and J. Swings. 2004. Biodiversity of vibrios. *Microbiol. Mol. Biol. Rev.* **68**:403–431.

Tsujino, I., and T. Saito. 1961. Studies on alginase. Part II. A new unsaturated uronide isolated from alginase hydrolysate. *Agric. Biol. Chem.* **26**:115–118.

Tujino, I. 1962. A new unsaturated uronide isolated from alginase hydrolysate. *Nature* **192**:970–971.

Visick, K. L., and M. J. McFall-Ngai. 2000. An exclusive contact: specificity in the *Vibrio fischeri-Euprymna scolopes* partnership. *J. Bacteriol.* **182**:1779–1787.

Whitford, M. F., R. J. Foster, C. E. Beard, J. Gong, and R. M. Teather. 1998. Phylogenetic analysis of rumen bacteria by comparative sequence analysis of cloned 16S rRNA genes. *Anaerobe* **4**:153–163.

Xu, J., M. K. Bjursell, J. Himord, S. Deng, L. K. Carmichael, H. C. Chiang, L. V. Hooper, and J. I. Gordon. 2003. A genomic view of the human-*Bacteroides thetaiotaomicron* symbiosis. *Science* **299**:2074–2076.

Xu, J., H. C. Chiang, M. K. Bjursell, and J. I. Gordon. 2004. Message from a human gut symbiont: sensitivity is a prerequisite for sharing. *Trends Microbiol.* **12**:21–28.

Yumoto, I., H. Iwata, T. Sawabe, K. Ueno, N. Ichise, H. Matsuyama, H. Okuyama, and K. Kawasaki. 1999. Characterization of a facultative psychrophilic bacterium, *Vibrio rumoiensis* sp. nov., that exhibits high catalase activity. *Appl. Environ. Microbiol.* **65**:67–72.

The Biology of Vibrios
Edited by F. L. Thompson et al.
© 2006 ASM Press, Washington, D.C.

Chapter 16

Vibrios in Coral Health and Disease

EUGENE ROSENBERG AND OMRY KOREN

INTRODUCTION

Coral reefs are the largest and most spectacular structures produced by living organisms. They maintain high levels of biodiversity, comparable to rain forests (Porter et al., 2001). Although corals are present in oligotrophic waters, they are highly productive owing to the successful symbiosis between the coral animal and endosymbiotic algae, referred to as zooxanthellae. Much of the photosynthetic carbon compounds produced by the zooxanthellae provide nutrients for the coral (Muscatine, 1990) as well as many other coral reef organisms (Hatcher, 1988). Coral reefs protect coastlines from erosion (Goreau et al., 1979). For many countries, coral reefs provide a major source of income from fishing and tourism. However, during the last few years, coral reefs have become an important source of new drugs for the pharmaceutical industry. In short, coral reefs are a major resource that should be protected.

Unfortunately, during the last 3 decades, coral reefs around the world have undergone mass destruction as a result of diseases and man-made factors. It has been estimated that 30% of the corals have been destroyed during the last 30 years (Hughes et al., 2003). On a global scale, coral bleaching has been the most devastating disease. Hoegh-Guldberg (2004) predicted that only remnants (<5%) will remain by 2050. In the first stage (1990 to 2010), the coral population on the reefs will change, and only the more resistant corals will remain, leading to a loss of biodiversity. This change has already been well documented in a study by Loya et al. (2001) in the Sea of Japan, in what the authors refer to as "the winners and the losers." In the second stage (2010–2050), even the more resistant corals will be destroyed. It should be emphasized that this prediction is based on projections of continuing increases in the rapid rise of seawater temperature and the inability of corals to adapt rapidly enough to these changes.

What is the cause of the ongoing coral decline? To begin with, there is more than one cause, and even the major cause, coral bleaching, probably has more than one etiology. The most obvious symptom of coral bleaching is loss of color, either as a result of loss of the zooxanthellae or the loss of pigments associated with the algae. The environmental factors known to cause bleaching are high seawater temperature (Glynn, 1993), low seawater temperature (Coles and Fadlallah, 1991), too much or too little sunlight (Gleason and Wellington, 1993; Lesser et al., 1990), reduced salinity (van Woesik et al., 1995), bacterial infection (Kushmaro et al., 1997; Ben-Haim et al., 2003b), or a combination (Glynn, 1993; Kushmaro et al., 1997).

The hypothesis accepted by most coral biologists who study coral bleaching is mass bleaching, which is the result of photobleaching of the endosymbiotic zooxanthellae (Warner et al., 1999). Basically, this hypothesis states that the photosynthetic apparatus of the algae is constantly undergoing photodamage in the light. The algae have a mechanism to repair this damage. However, when the temperature is too high, the proteins that repair the damage are denatured and cannot repair the damage. By an unknown mechanism, the coral expels the damaged algae. The algal photoinhibition hypothesis has led to a second hypothesis, the adaptive hypothesis of bleaching (Buddemeier et al., 2004). This hypothesis is based on the fact that there are several zooxanthellae clades, some of which are more resistant to high temperature than others. Thus, removal of temperature-sensitive algae allows the coral to enter into a new symbiotic relationship with temperature-resistant algae: in effect,

Eugene Rosenberg and Omry Koren • Department of Molecular Microbiology and Biotechnology, George S. Wise Faculty of Life Sciences, Tel Aviv University, Ramat Aviv 69978, Israel.

forming a new "ecospecies." The advantage of this mechanism is that it allows for adaptation over shorter time scales than ordinary evolution via mutation and selection.

Table 1 lists the reported coral diseases that have been associated with different *Vibrio* species. In three of the diseases—bleaching of *Oculina patagonica* and *Pocillopora damicornis* and the yellow blotch disease of *Monastraea*—the causative agent(s) has been demonstrated by satisfying Koch's postulates. In the case of skeletal tumors, only a correlation between the disease and an increase in the population of *Vibrio* species has been established.

INCREASES IN *VIBRIO* POPULATIONS DURING BLEACHING OF *MONASTRAEA*

Ritchie and Smith (1995) and Ritchie et al. (1994) were the first to demonstrate shifts in the bacterial population during bleaching of corals. For example, before bleaching, the surface mucopolysaccharide layer of *Monastraea annularis* contained ~20% *Vibrio* and 20% *Pseudomonas* cells. However, at the peak of bleaching, the population changed to >60% *Vibrio* and 2% *Pseudomonas* cells. When the corals recovered from the bleaching, the population returned to its initial values. The authors suggested that the population change was the result of the availability of different carbon compounds during the bleaching event. Also, they suggested that during bleaching there are large changes in the physical environment (i.e., pH and oxygen tension). Although these pioneering studies clearly pointed out a relationship between an increase in *Vibrio* populations and bleaching, the authors were careful to avoid concluding that *Vibrio* species were responsible for the bleaching. In light of the subsequent demonstrations that specific *Vibrio* species were responsible for bleaching other corals, it would be interesting to test if any of the *Vibrio* species that increased during the bleaching were causative agents of the bleaching, by performing infection experiments at different temperatures.

BLEACHING OF *OCULINA PATAGONICA* BY *VIBRIO SHILOI*

The zooxanthellae scleractinian (hard) coral *Oculina patagonica* was first observed in the Mediterranean in 1966 in Savona Harbor, Gulf of Gonoa (Zibrowius, 1974). This species is the only scleractinian coral reported to have invaded a new region. Since *O. patagonica* is a known fouling organism, it was suggested that it crossed the Atlantic Ocean by adhering to the hull of a ship (Fine et al., 2002). At present it is known to inhabit many natural sites and harbors along hundreds of kilometers of the Mediterranean coastline. Highest abundances were recorded along the Spanish coast, reaching 30 colonies/10 m transect in shallow water, and the Israeli coast, 10 colonies/transect (Fine et al., 2001). The coral is present in tidal pools, shallow sandstone reefs, and at depths up to 8 m. In addition, the coral can be found in dark caves in the azooxanthellae form (lacking the algae). The ease of obtaining genetically identical coral fragments from the same colony and maintaining them in aquaria makes *O. patagonica* an excellent experimental coral model system.

Bleaching of *O. patagonica* was first recorded along the Israeli shoreline in the summer of 1993 (Fine and Loya, 1995). A photograph of a partially bleached *O. patagonica* colony is shown in Color Plate 7. Since 1995, the extent of bleaching and seawater temperature have been monitored continuously. Bleaching begins to increase late in spring, rises in the summer, reaching ~80% in September, and then the corals slowly recover in the autumn and winter. By early spring most of the corals have completely recovered. Maximum bleaching always occurs about 1 month after the maximum seawater temperature.

V. shiloi Is the Causative Agent of Bleaching of *O. patagonica*

Koch's postulates were applied to demonstrate that *V. shiloi* is the causative agent of the bleaching disease of *O. patagonica* (Kushmaro et al., 1996, 1997).

Table 1. Role of *Vibrio* spp. in coral diseases

Disease	Coral host	Location	*Vibrio* spp.	Reference
Bleaching	*Monastraea annularis*	Caribbean	Unidentified	Ritchie et al., 1994
Bleaching	*Oculina patagonica*	Mediterranean	*V. shiloi*	Kushmaro et al., 1996
Bleaching	*Pocillopora damicornis*	Indian Ocean, Red Sea	*V. coralliilyticus*	Ben-Haim et al., 2003b
Yellow blotch	*Monastraea*	Caribbean	*V. alginolyticus* and 3 other species	Cervino et al., 2004
Skeletal tumors	*Porites compressa*	Hawaii	Unidentified	Breitbart et al., 2005
Bleaching	Several	Australia	*P. eurosenbergii; E. coralii*	Thompson et al., 2005

1. The bacterium was isolated from all 28 samples taken from bleached corals and was absent from all 24 samples taken from unbleached corals.

2. The bacterium was obtained in pure culture, classified as a new species of *Vibrio* (see below), and given the name *V. shiloi* (after the late microbiologist Moshe Shilo).

3. In controlled aquaria experiments, it was demonstrated that *V. shiloi* caused bleaching of healthy corals. At 29°C, as few as 120 bacteria/ml induced 60% bleaching of *O. patagonica* in 10 days and 100% bleaching in 20 days. No bleaching occurred for several months under the same conditions if *V. shiloi* was not added or if antibiotics were administered at the same time as the inoculum.

4. *V. shiloi* was reisolated from the tissues of the infected bleached corals.

Classification of *V. shiloi*

Exponentially growing *V. shiloi* is a gram-negative, motile, rod-shaped bacterium (2.4 × 1.6 μm) that has a single, polar, sheathed flagellum. When growing intracellularly in coral tissue, the cells lack the flagellum, and are smaller (2.0 × 1.0 μm). These properties, together with its ability to form yellow colonies on TCBS (thiosulfate citrate bile salt sucrose) agar and its sensitivity to the *Vibrio* static agent O129 (2,4-diamino-6,7-diisopropylpteridine), indicated that it is a species of the genus *Vibrio* (Farmer and Hickmann-Brenner, 1992). Based on its genotypic and phenotypic characteristics, it was classified as a new species and given the name *V. shiloi* (Kushmaro et al., 2001). The type strain is *V. shiloi* AK1 (ATCC BAA-91; DSM 13774).

The 16S rDNA sequence of strain AK1 was determined and compared with data for known *Vibrio* species. The sequence of *V. shiloi* AK1 is related most closely to *Vibrio mediterranei* (99.4%). DNA:DNA hybridization data showed that AK1 and *V. mediterranei* are 72% related at the total DNA level. However, *V. shiloi* AK1 contains several plasmids, including two major high-copy-number ones (2.5 and 50 kb) that are not seen in *V. mediterranei* (unpublished data). *V. shiloi* and *V. mediterranei* also show large differences in their cellular fatty acid compositions, salt tolerances, and ability to utilize saccharides. Nevertheless, Thompson et al. (2001) argued that AK1 should not be classified as a new species, but rather as a subspecies of *V. mediterranei*. Nevertheless, it should be emphasized that *V. mediterranei* does not infect corals, nor is it known to grow intracellularly.

Temperature-Regulated Mechanisms of Infection

Bleaching of *O. patagonica* in the sea occurs only at elevated seawater temperatures. High temperature also plays a key role in bleaching by *V. shiloi* in laboratory aquaria experiments (Kushmaro et al., 1998). At 29°C, bleaching is rapid and complete; at 25°C, it is slower and incomplete; and below 20°C, no bleaching occurs even with a very high inoculum size of *V. shiloi*. In principle, the inability of *V. shiloi* to bleach *O. patagonica* at the winter temperatures could be due to increased resistance of the coral or decreased virulence of the pathogen. The data summarized in this chapter indicate that several critical *V. shiloi* virulence factors are only produced at the elevated summer water temperatures, suggesting that the primary effect of temperature is on the pathogen, not the host.

The first step in the infectious process is the adhesion of *V. shiloi* to the coral surface. Using the capillary tube assay, it was shown that *V. shiloi* is attracted to the mucus obtained from *O. patagonica* (Banin et al., 2001a). Motility and chemotactic behavior were present when the bacteria were grown at either 16 or 25°C. The bacteria then adhere to a β-galactoside-containing receptor on the coral surface (Toren et al., 1998). This was demonstrated by the inhibition of binding to corals by methyl-β-D-galactopyranoside and the specific binding of *V. shiloi* to Sephadex beads containing covalently bound β-galactoside. Spontaneous mutants, which did not adhere to the beads, did not bind to the coral and were avirulent. The temperature of bacterial growth was critical for the adhesion of *V. shiloi* to the coral. When the bacteria were grown at the low winter temperature (16 to 20°C), there was no adhesion to the coral, regardless of the temperature at which the coral had been maintained. However, bacteria grown at elevated summer temperatures (25 to 30°C) adhered avidly to corals maintained either at low or high seawater temperatures. The important ecological significance of these findings is that the environmental stress condition (high temperature) is necessary for the coral bleaching pathogen to initiate the infection and become virulent.

The β-galactoside-containing receptor that *V. shiloi* recognizes on the *O. patagonica* surface is present in the coral mucus. This was demonstrated by binding the coral mucus to enzyme-linked immunosorbent assay plates: the bacteria adhered avidly to the mucus-coated plates (Banin et al., 2001a). Adhesion to the coated wells was inhibited by methyl-β-D-galactopyranoside. Interestingly, the receptor is present only in corals that contain photosynthetically active zooxanthellae. *V. shiloi* does not adhere to bleached corals or to white (azooxanthellate) *O. patagonica* cave corals. When mucus was removed from healthy corals, adhesion of *V. shiloi* to the coral was inhibited. Furthermore, addition of 3-(3,4-dichlorophenyl)-1,1-dimethyl urea, an inhibitor of algal photosynthesis, to mucus-depleted corals prevented adhesion, suggesting that synthesis of the receptor or its transport to the mucus was inhibited. Thus, *V. shiloi* is able to distinguish healthy corals from those which are photosynthetically impaired, and it only adheres to healthy zooxanthellate *O. patagonica*.

Electron micrographs of thin sections of *O. patagonica* following infection with *V. shiloi* demonstrated large numbers of bacteria in the epidermal layer of the coral (Banin et al., 2000). Moreover, using monoclonal antibodies specific to *V. shiloi*, it was shown that the observed intracellular bacteria were, in fact, *V. shiloi*. The gentamicin invasion assay was used to measure the kinetics of *V. shiloi* penetration into the epidermal cells. This assay relies on the fact that the antibiotic gentamicin does not penetrate into eukaryotic cells (Isberg and Falkow, 1985). Thus, only *V. shiloi* cells that have penetrated into the coral cells escape the killing action of gentamicin. After adhesion was complete (ca. 12 h), the bacteria began to penetrate into the coral, as determined by both total counts and CFU. By 24 h, 40 to 50% of the inoculated *V. shiloi* had penetrated into coral cells. From 24 to 72 h, the intracellular bacteria multiplied (based on total counts), reaching 3×10^8 bacteria/1 cm^3 coral fragment. When the infected corals were maintained at the high summer temperatures, the bacteria remained at 10^8 to 10^9 cells/cm^3 for at least 2 weeks.

At the same time that *V. shiloi* multiplies inside the coral tissue (24 to 48 h after infection), the number of CFU decreases more than 1,000-fold. Entry of bacteria into a viable but nonculturable (VBNC) state has been reported repeatedly with a large number of bacterial species, including several *Vibrio* species (Colwell et al., 1985; Lee and Ruby, 1995; Oliver et al., 1991). A bacterium in the VBNC state has been defined as "a cell which can be demonstrated to be metabolically active, while being incapable of undergoing the sustained cellular division required for growth on a medium normally supporting growth of that cell" (Oliver, 1993). Intracellular *V. shiloi* cells fit that definition, but unlike most cases of VBNC that have been studied, this is not brought about by starvation or low temperature. Rather, the entry of *V. shiloi* into the VBNC state occurs inside the coral epidermis, where nutrients are abundant. The fact that intracellular *V. shiloi* is viable, i.e., metabolically active, is apparent from the observation that the cells multiply intracellularly and score as viable with the Live/Dead Baclight Bacterial Viability Kit. It is possible that the differentiated intracellular VBNC *V. shiloi* becomes dependent on one or more nutrients present in the coral cell. In this regard, it has been shown that VBNC *V. shiloi* is infectious (Israely et al., 2001).

V. shiloi produces extracellular toxins that block photosynthesis and bleach and lyse zooxanthellae (Ben-Haim et al., 1999). The toxin responsible for inhibition of photosynthesis, referred to as toxin P, is the following proline-rich peptide: PYPVYAPPPVVP (Banin et al., 2001b). In the presence of NH$_4$Cl, the toxin causes a rapid decrease in the photosynthetic quantum yield of zooxanthellae. Evidence has been presented that the toxin binds irreversibly to algal membranes, forming a channel that allows NH$_3$, but not NH$_4^+$, to rapidly pass, thereby destroying the pH gradient across the thylakoid membrane and blocking photosynthesis. This mode of action of toxin P can help explain the mechanism of coral bleaching. Toxin P is produced at >10-fold higher levels at 29°C than at 16°C. Also, *V. shiloi* produces heat-sensitive, high-molecular-weight toxins that bleach and lyse isolated zooxanthellae. These toxins have not yet been isolated and characterized.

The demonstration that the bacterial adhesin is not produced at temperatures below 20°C provides an explanation for the failure of *V. shiloi* to infect *O. patagonica* in the field during the winter and in the laboratory at low temperatures. However, it does not explain why corals recover during the winter. When corals are infected with *V. shiloi* at the permissive temperature of 28°C, the bacteria adhere, penetrate, and begin to multiply intracellularly. If the infected corals are then shifted slowly (0.5°C per day) to lower temperatures, the bacteria die, and the infection is aborted (Israely et al., 2001). The failure of *V. shiloi* to survive inside coral tissue at <20°C is because it does not produce superoxide dismutase (SOD) at these low temperatures (Banin et al. 2003). This hypothesis was supported by constructing an SOD-minus mutant. At 28°C, the mutant adhered to the coral, penetrated into the tissue, and then died. However, death occurred only when the coral was exposed to light. The most reasonable explanation for these data is that the high concentration of oxygen and resulting oxygen radicals produced by the zooxanthellae during photosynthesis is highly toxic to bacteria and is one of the mechanisms by which corals resist infection. At high temperatures, *V. shiloi* produces a potent SOD that helps it to survive in the coral tissue.

In a more general sense, the data on the importance of SOD for the survival of *V. shiloi* inside coral tissue provides strong evidence for the role of zooxanthellae in coral resistance to bacterial infection. Thus, zooxanthellae not only provide nutrients for coral growth, but also protect the coral against infection by potential pathogens.

TRANSMISSION OF THE DISEASE

The observations that *V. shiloi* could not be found inside *O. patagonica* during the winter and that the bacterium could not survive in the coral below 20°C (Banin et al., 2003) indicate that bleaching of *O. patagonica* requires a fresh infection each spring, rather than the activation of dormant intracellular bacteria. This posed a question: where does *V. shiloi* reside during the winter?

Using fluorescence in situ hybridization with a *V. shiloi*-specific deoxyoligonucleotide probe, Sussman et al. (2003) found that the marine fireworm *Hermodice carunculata* is a winter reservoir for *V. shiloi*. Worms taken directly from the sea during the winter contained 0.6×10^8 to 2.9×10^8 *V. shiloi* cells per worm. When worms were infected with *V. shiloi* in the laboratory, most of the bacteria adhered to the worm within 24 h and then penetrated into the epidermal cells. By 48 h, $<10^{-4}$ of the intact *V. shiloi* cells in the worm gave rise to colonies, suggesting that they differentiated inside the worm into the VBNC state.

To test if worms carrying *V. shiloi* could serve as vectors for transmitting the pathogen to *O. patagonica*, specimens infected with *V. shiloi* were placed in aquaria containing *O. patagonica*. Corals that came into contact with the infected worms showed small patches of bleached tissue in 7 to 10 days and total bleaching in 17 days. Uninfected worms did not cause bleaching. Thus, *H. carunculata* is not only a winter reservoir for *V. shiloi*, but also a potential vector for transmitting the bleaching disease to *O. patagonica* (Color Plate 8). It should be emphasized that knowledge about reservoirs and modes of transmission has proven useful in the past for developing technologies for controlling the spread of disease.

BLEACHING OF *P. DAMICORNIS* BY *VIBRIO CORALLIILYTICUS*

Vibrio coralliilyticus is an etiological agent of bleaching of the coral *P. damicornis* on coral reefs in the Indian Ocean and Red Sea (Ben-Haim and Rosenberg, 2002; Ben-Haim et al., 2003b). Based on phenotypic and genotypic characteristics, *V. coralliilyticus* was classified as a new species (Ben-Haim et al., 2003a). Strains of *V. coralliilyticus* have been isolated from diseased corals on the Eilat coral reef in the Red Sea and in the Indian Ocean near Zanzibar, from diseased oyster larvae near Kent, United Kingdom, and from bivalve larvae near Brazil. All of these *V. coralliilyticus* strains bleached *P. damicornis* in controlled aquaria experiments. The *V. coralliilyticus* strains form a tight cluster, with more than 99.5% 16S rDNA similarity. Their closest phylogenetic neighbors are *Vibrio tubiashii* (97.2%), *Vibrio nereis* (96.8%), and *V. shiloi* (96.6%).

The infection of *P. damicornis* by *V. coralliilyticus* shows strong temperature dependence (Ben-Haim and Rosenberg, 2002; Ben-Haim et al., 2003b). Thus, at <22°C no infection occurred, but at 24 to 26°C the infection resulted in bleaching (Color Plate 9), whereas at 27 to 29°C the infection caused rapid tissue lysis and death of the coral. These data are similar to the report of Jokiel and Coles (1990), which stated that seawater temperatures slightly above the normal maximum caused bleaching of *P. damicornis*, while temperatures a few degrees higher caused lysis and death. These investigators did not perform microbiological studies, so it is not known if the corals were infected before the temperature was elevated.

V. coralliilyticus produces a potent metalloproteinase at temperatures >26°C (Ben-Haim et al., 2003b). This enzyme shows high levels of amino acid sequence homology to a range of proteases found in members of the family *Vibrionaceae*, including the well-studied hemagglutinin protease of *Vibrio cholerae* (Finkelstein and Hanne, 1982). Since the purified protease caused tissue lysis of corals, it was suggested that at the elevated seawater temperature, where the protease is produced, the bacterium attacks the coral tissues, whereas at the lower temperatures the intracellular algae are the target, and the outcome of the infection is bleaching.

VIBRIO AND YELLOW BLOTCH/BAND DISEASE

Yellow blotch (also called yellow band) disease of the major reef-building coral of the Caribbean Sea, *Monastraea* spp., is well documented (Brown et al., 1995). During the early stages of the disease, pale yellow blotches appear on the coral tissue. The blotches expand and form bands that can spread slowly across the colony at rates of 0.5 to 1.0 cm/month (Santavy et al., 1999). As recently demonstrated by Cervino et al. (2004), the paleness is a result of lower densities of the intracellular zooxanthellae and large decreases in chlorophyll and C_2 pigments.

In an attempt to isolate the causative agent of yellow blotch disease, Cervino et al. (2004) isolated bacteria that were present on diseased corals but absent from healthy ones. After a series of infection/

Table 2. Causative agents of yellow blotch disease[a]

Vibrio strain	GenBank accession no.	Closest relative
FLG2A	AY770830	V. alginolyticus (AY373027)
YBFL3122	AY770831	Vibrio sp. strain CJ11052 (AF500207)
YB36	AY770833	Vibrio sp. strain NAP-4 (AF064637)
YBM23	AY770832	Vibrio sp. strain NAP-4 (AF064637)

[a]Data obtained from Cervino et al., 2004.

reisolation experiments, they found four *Vibrio* spp. that induced the disease in controlled aquaria experiments (Table 2). Colony morphology and carbon compound utilization pattern were used to further characterize the four strains. Although only slight paling was observed when each of the four strains was inoculated separately onto *Monastraea* spp. and incubated at 30°C for 48 h, inoculation of a mixture of the four strains resulted in signs resembling those of yellow blotch disease. The rate of spread of the disease induced by the mixed *Vibrio* culture was proportional to temperature from 25 to 33°C. The authors concluded that the mixed *Vibrio* infection targets the zooxanthellae rather than the host tissue of *Monastraea* spp. and that the temperature plays a key role in spread of the disease. These two conclusions suggest a close similarity to the previously described bleaching of corals by *V. shiloi* and *V. coralliilyticus*. The close similarity between the *Vibrio* spp. responsible for the yellow blotch disease and *Vibrio alginolyticus* and *Vibrio parahaemolyticus* suggests that toxins may be involved in the disease (i.e., hemolysins).

VIBRIO AND SKELETAL TUMORS ON *PORITES*

The presence of a tumor on a coral was first described 40 years ago (Squires, 1965). Since then, hard tissue tumors in scleractinian corals have been reported from >10 coral families on reefs worldwide (Peters et al., 1986; Cheney, 1975). The tumors are characterized by enlarged skeletal elements, atypical and rapid growth, a fewer number of polyps per surface area, and a large reduction in intracellular zooxanthellae (Yamashiro et al., 2000; Gateno et al., 2003; Cheney, 1975). Coral tumors are not transmitted between colonies, even after fusion of healthy and tumor coral fragments (Gateno et al., 2003). Studies have indicated that healthy parts of a coral that contains a tumor eventually deteriorate (Yamashiro et al., 2001). Rapid growth of tumors is accompanied by a lower growth rate in healthy parts of the colony. Apparently, there is translocation of limiting nutrients

from healthy tissue to the tumors via the gastrovascular system that connects polyps.

Recently, the microbial communities associated with skeletal tumors of the reef-building coral *Porites compressa* have been analyzed (Breitbart et al., 2005). Although the numbers of microorganisms per surface area of the tumor and healthy corals were similar, the growth rates of the microbes associated with tumors were significantly higher. Interestingly, the occurrence of *Vibrio* spp. (measured by culturing on TCBS agar) was >10 times higher on tumors than on healthy colonies. Yet the significance of the increased *Vibrio* spp. population on the tumors is presently unknown. It is also not clear whether the increase in population is the result of multiplication or conversion from a VBNC state to a culturable form.

BLEACHING ON THE GREAT BARRIER REEF

Recently, Thompson et al. (2005) characterized two new vibrio-like isolates from bleached corals in the Magnetic Islands (Australia). The two new taxa, named *Photobacterium eurosenbergii* sp. nov. and *Enterovibrio coralii* sp. nov., share the main phenotypic features of *Vibrio*, but 16S rRNA, recA, and rpoA gene sequences clearly suggest that they represent two new species. *P. eurosenbergii* was isolated from tissue extracts of the bleached coral *Pachyseris speciosa*, while *E. coralii* was isolated from water extracts of the bleached coral *Merulina ampliata*. The possible role of these new species in coral bleaching remains to be determined.

Acknowledgments. This work was supported by the Israel Center for the Study of Emerging Diseases and the Pasha Gol Chair for Applied Microbiology.

REFERENCES

Banin, E., T. Israely, A. Kushmaro, Y. Loya, E. Orr, and E. Rosenberg. 2000. Penetration of the coral-bleaching bacterium *Vibrio shiloi* into *Oculina patagonica*. *Appl. Environ. Microbiol.* **66:** 3031–3036.

Banin, E., T. Israely, M. Fine, Y. Loya, and E. Rosenberg. 2001a. Role of endosymbiotic zooxanthellae and coral mucus in the adhesion of the coral-bleaching pathogen *Vibrio shiloi* to its host. *FEMS Microbiol. Lett.* **199:**33–37.

Banin, F., K. H. Sanjay, F. Naider, and E. Rosenberg. 2001b. A proline-rich peptide from the coral pathogen *Vibrio shiloi* that inhibits photosynthesis of zooxanthellae. *Appl. Environ. Microbiol.* **67:**1536–1541.

Banin E., D. Vassilakos, E. Orr, R. J. Martinez, and E. Rosenberg. 2003. SOD is a virulence factor produced by the coral bleaching pathogen *Vibrio shiloi*. *Curr. Microbiol.* **46:**418–422.

Ben-Haim, Y., E. Banin, A. Kushmaro, Y. Loya, and E. Rosenberg. 1999. Inhibition of photosynthesis and bleaching of zooxanthellae by the coral pathogen *Vibrio shiloi*. *Environ. Microbiol.* **1:**223–229.

Ben-Haim, Y., and E. Rosenberg. 2002. A novel *Vibrio* sp. pathogen of the coral *Pocillopora damicornis*. *Mar. Biol.* **141:**47–55.

Ben-Haim, Y., F. L. Thompson, C. C. Thompson, M. C. Cnockaert, B. Hoste, J. Swings, and E. Rosenberg. 2003a. *Vibrio coralliilyticus* sp nov., a temperature-dependent pathogen of the coral *Pocillopora damicornis*. *Int. J. Syst. Evol. Microbiol.* **53:**309–315.

Ben-Haim, Y., M. Zicherman-Keren, and E. Rosenberg. 2003b. Temperature-regulated bleaching and lysis of the coral *Pocillopora damicornis* by the novel pathogen *Vibrio coralliilyticus*. *Appl. Environ. Microbiol.* **69:**4236–4242.

Breitbart, M., R. Bhagooli, S. Griffin, and F. Rohwer. 2005. Microbial communities associated with skeletal tumors on *Porites compressa*. *FEMS Microbiol. Lett.* **243:**431–436.

Brown, B. E., M. D. A. Le Tissier, and J. C. Bythell. 1995. Mechanisms of bleaching deduced from histological studies of reef corals sampled during a natural bleaching event. *Mar. Biol.* **122:**655–663.

Buddemeier, W., A. C. Baker, D. G. Fautin, and J. R. Jacobs. 2004. The adaptive hypothesis of bleaching, p. 427–440. *In* E. Rosenberg and Y. Loya (ed.), *Coral Health and Disease.* Springer-Verlag, New York, N.Y.

Cervino, J. M., R. L. Hayes, S. W. Polson, S. C. Polson, T. J. Goreau, R. J. Martinez, and G. W. Smith. 2004. Relationship of *Vibrio* species infection and elevated temperatures to yellow blotch/band disease in Caribbean corals. *Appl. Environ. Microbiol.* **70:**6855–6864.

Cheney, D. 1975. Hard tissue tumors in scleractinian corals. *Adv. Exp. Med. Biol.* **64:**77–87.

Coles, S. L., and Y. H. Fadlallah. 1991. Reef coral survival and mortality at low temperatures in the Arabian Gulf: new species-specific lower temperature limits. *Coral Reefs* **9:**231–237.

Colwell, R. R., P. R. Brayton, D. J. Grimes, D. B. Roszak, S. A. Huq, and L. M. Palmer. 1985. Viable but non-culturable *Vibrio cholerae* and related pathogens in the environment: implications for release of genetically engineered microorganisms. *Bio/Technology* **3:**817–820.

Farmer, J. J., III, and F. W. Hickman-Brenner. 1992. The genera *Vibrio* and *Photobacterium*, p. 2952–3011. *In* A. Barlows, H. G. Truper, M. Dworkin, W. Harder, and H. K. Schleifer (ed.), *The Prokaryotes*, 2nd ed. Springer-Verlag, New York, N.Y.

Fine, M., and Y. Loya. 1995. The coral *Oculina patagonica*: a new immigrant to the Mediterranean coast of Israel. *Isr. J. Zool.* **41:**81.

Fine, M., H. Zibrowius, and Y. Loya. 2001. *Oculina patagonica*, a non lessepsian scleractinian coral invading the Mediterranean Sea. *Mar. Biol.* **138:**1195–1203.

Fine, M., U. Oren, and Y. Loya. 2002. Bleaching effect on regeneration and resource translocation in the coral *Oculina patagonica*. *Mar. Ecol. Prog. Ser.* **234:**119–125.

Finkelstein, R. A., and L. F. Hanne. 1982. Purification and characterization of the soluble hemagglutinin (cholera lectin) produced by *Vibrio cholerae*. *Infect. Immun.* **36:**1199–1208.

Gateno, D., A. Leon, Y. Barki, J. Cortes, and B. Rinkevich. 2003. Skeletal tumor formation in the massive coral *Pavona clavus*. *Mar. Ecol. Prog. Ser.* **258:**97–108.

Gleason, D. F., and G. M. Wellington. 1993. Ultraviolet radiation and coral bleaching. *Nature* **365:**836–838.

Glynn, P. W. 1993. Coral-reef bleaching—ecological perspectives. *Coral Reefs* **12:**1–17.

Goreau, T. F., N. I. Goreau, and T. J. Goreau. 1979. Corals and coral reefs. *Sci. Am.* **241:**124–135.

Hatcher, B. G. 1988. Coral reef primary productivity: a beggar's banquet. *Trends Ecol. Evol.* **3:**106–111.

Hoegh-Guldberg, O. 2004. Coral reefs and projections of future change, p. 463–484. *In* E. Rosenberg and Y. Loya (ed.), *Coral Health and Disease.* Springer-Verlag, New York, N.Y.

Hughes, T. P., A. H. Baird, D. R. Bellwood, M. Cad, S. R. Connolly, C. Folke, R. Grosberg, O. Hoegh-Guldberg, J. B. C. Jackson, J. Kleypas, J. M. Lough, P. Marshall, M. Nystrom, S. R. Palumbi, J. M. Pandolfi, B. Rosen, and J. Roughgarden. 2003. Climate change, human impacts, and the resilience of coral reefs. *Science* **301:**929–933.

Isberg, R. R., and S. Falkow. 1985. A single genetic locus encoded by *Yersinia pseudotuberculosis* permits invasion of cultured animal cells by *Escherichia coli* K-12. *Nature* **317:**262–264.

Israely, T., E. Banin, and E. Rosenberg. 2001. Growth, differentiation and death of *Vibrio shiloi* in coral tissue as a function of seawater temperature. *Aquat. Microb. Ecol.* **24:**1–8.

Jokiel, P. L., and S. L. Coles. 1990. Response of Hawaiian and other Indo Pacific reef corals to elevated temperatures. *Coral Reefs* **8:**155–162.

Kushmaro, A., Y. Loya, M. Fine, and E. Rosenberg. 1996. Bacterial infection and coral bleaching. *Nature* **380:**396–396.

Kushmaro, A., E. Rosenberg, M. Fine, and Y. Loya. 1997. Bleaching of the coral *Oculina patagonica* by *Vibrio* AK-1. *Mar. Ecol. Prog. Ser.* **147:**159–165.

Kushmaro, A., E. Rosenberg, M. Fine, Y. Ben-Haim, and Y. Loya. 1998. Effect of temperature on bleaching of the coral *Oculina patagonica* by *Vibrio shiloi* AK-1. *Mar. Ecol. Prog. Ser.* **171:**131–137.

Kushmaro, A., E. Banin, E. Stackebrandt, and E. Rosenberg. 2001. *Vibrio shiloi* sp. nov: the causative agent of bleaching of the coral *Oculina patagonica*. *Int. J. Syst. Evol. Microbiol.* **51:**1383–1388.

Lee, K. H., and E. G. Ruby. 1995. Symbiotic role of the viable but nonculturable state of *Vibrio fischeri* in Hawaiian coastal seawater. *Appl. Environ. Microbiol.* **61:**278–283.

Lesser, M. P., W. R. Stochaj, D. W. Tapley, and J. M. Shick. 1990. Bleaching in coral reef anthozoans: effects of irradiance, ultraviolet radiation, and temperature on the activities of protective enzymes against active oxygen. *Coral Reefs* **8:**225–232.

Loya, Y., K. Sakai, K. Yamazato, Y. Nakano, H. Sembali, and R. van Woesik. 2001. Coral bleaching: the winners and the losers. *Ecol. Lett.* **4:**122–131.

Muscatine, L. 1990. The role of symbiotic algae in carbon and energy flux in reef corals. *Coral Reefs* **25:**1–29.

Oliver, J. D., L. Nilsson, and S. Kjelleberg. 1991. Formation of nonculturable *Vibrio vulnificus* cells and its relationship to the starvation state. *Appl. Environ. Microbiol.* **57:**2640–2644.

Oliver, J. D. 1993. Formation of viable but nonculturable cells, p. 239–272. *In* S. Kjelleberg (ed.), *Starvation in Bacteria.* Plenum Press, New York, N.Y.

Peters, E., J. Halas, and H. McCarty. 1986. Calicoblastic neoplasms in *Acropora palmata*, with a review of reports on anomalies of growth and form in corals. *J. Natl. Cancer. Inst.* **76:**895–912.

Porter, J. W., P. W. Dustan, W. Jaap, K. Patterson, V. Kosmyin, M. Patterson, and M. Parsons. 2001. Patterns of spread of coral disease in the Florida Keys. *Hydrobiology* **460:**1–24.

Ritchie, K. B., J. H. Dennis, T. McGrath, and G. W. Smith. 1994. Bacteria associated with bleached and non-bleached areas of *Monastraea annularis*. *Proc. Symp. Natl. Hist. Bahamas* **5**: 75–80.

Ritchie, K. B., and G. W. Smith. 1995. Preferential carbon utilization by surface bacterial communities from water mass, normal, and white-band diseased *Acropora cervicornis*. *Mol. Mar. Biol. Biotech.* **4**:345–354.

Santavy, D. L., E. C. Peters, C. Quirolo, J. W. Porter, and C. N. Bianchi. 1999. Yellow blotch disease outbreak on reefs of the San Blas Islands, Panama. *Coral Reefs* **18**:97.

Squires, D. F. 1965. Neoplasia in a coral? *Science* **148**:503–505.

Sussman, M., Y. Loya, M. Fine, and E. Rosenberg. 2003. The marine fireworm *Hermodice carunculata* is a winter reservoir and spring-summer vector for the coral-bleaching pathogen *Vibrio shiloi*. *Environ. Microbiol.* **5**:250–255.

Thompson, F. L., B. Hoste, C. C. Thompson, G. Huys, and J. Swings. 2001. The coral bleaching *V. shiloi* Kushmaro et al. 2001 is a later synonym of *Vibrio mediterranei* Pujalte and Garay 1986. *Syst. Appl. Microbiol.* **24**:516–519.

Thompson, F. L., C. C. Thompson, S. Naser, B. Hoste, K. Vandemeulebroecke, C. Munn, D. Bourne, and J. Swings. 2005. *Photobacterium rosenbergii* sp. nov. and *Enterovibrio coralii* sp. nov., vibrios associated with coral bleaching. *Int. J. Syst. Evol. Microbiol.* **55**:913–917.

Toren, A., L. Landau, A. Kushmaro, Y. Loya, and E. Rosenberg. 1998. Effect of temperature on adhesion of *Vibrio* strain AK-1 to *Oculina patagonica* and on coral bleaching. *Appl. Environ. Microbiol.* **64**:1379–1384.

van Woesik, R., L. M. De Vantier, and J. S. Glazebrook. 1995. Effects of cyclone Joy on nearshore coral communities of the Great Barrier Reef. *Mar. Ecol. Prog. Ser.* **128**:261–270.

Warner, M. E., W. K. Fitt, and G. W. Schmidt. 1999. Damage to photosystem II in symbiotic dinoflagellates: a determinant of coral bleaching. *Proc. Natl. Acad. Sci. USA* **96**:8007–8012.

Yamashiro, H., M. Yamamoto, and R. van Woesik. 2000. Tumor formation on the coral *Montipora informis*. *Dis. Aquat. Org.* **41**:211–217.

Yamashiro, H., H. Oku, K. Onaga, H. Iwasaki, and K. Takara. 2001. Coral tumors store reduced level of lipids. *J. Exp. Mar. Biol. Ecol.* **265**:171–179.

Zibrowius, H. 1974. *Oculina patagonica*, a hermatypic scleractinian coral introduced to the Mediterranean Sea. *Helgoland. Wiss. Meer.* **26**:153–173.

The Biology of Vibrios
Edited by F. L. Thompson et al.
© 2006 ASM Press, Washington, D.C.

Chapter 17

Vibrio cholerae Populations and Their Role in South America

ANA CAROLINA P. VICENTE, IRMA N. G. RIVERA, MICHELLE D. VIEIRA, AND ANA COELHO

CHOLERA IN SOUTH AMERICA: HISTORICAL AND EPIDEMIOLOGICAL ASPECTS

The etiological agent of Asiatic cholera, *Vibrio cholerae*, was present in ancient India. Since 1817, seven pandemics have been described, reporting the spread of cholera out of India. South America was affected by epidemic cholera at various times before the recent epidemic in the 1990s. Some cases occurred in 1829 in Chile and Peru during the epidemic that devastated several countries in Europe (Pollitzer, 1959; Barua, 1992). Venezuela and Brazil were affected by the pandemic cholera in 1855. The first area affected in Brazil was Grão-Pará Province (Pará State). The Deffensor ship arrived from Portugal, with 37 members (12.8%) of the crew dead as a consequence of acute and severe diarrhea (Pollitzer, 1959; Beltrão, 2004). The disease spread rapidly, causing outbreaks in several local villages. Two months later, after disseminating to Salvador (Bahia State), the disease reached Rio de Janeiro and subsequently spread southward to the region of the De La Plata River (Argentina). During that decade, incursions of the disease in its fourth pandemic occurred in all regions of Brazil, killing more people in the southern region than the concomitant Brazil-Paraguay war (1864 to 1870). Indeed, it is estimated that ~200,000 cholera deaths occurred during this war (Cooper, 1986). In 1866, Paraguay, Brazil, and Argentina had cholera cases among troops involved in the war, and it spread for the first time to Uruguay, Bolivia, and Peru. Cholera subsided in South America, with the last reports of outbreaks dated around 1895. The disease disappeared from South America by 1900 (Pollitzer, 1959; Barua, 1992).

The spread of pandemic cholera to South America in the 19th century was probably the consequence of the uncontrolled transport of infected people from regions in Europe where the disease was epidemic. Cholera reached Brazil by the Atlantic Ocean trade routes (Color Plate 10). The seventh cholera pandemic began in 1961 in Indonesia, spreading to other countries in Asia and the Middle East, and, after a decade, it reached Africa and southern Europe. In 1974, Brazil implemented a surveillance program because of the possibility of a cholera epidemic imported from Africa through refugees from the Portuguese colonies of Mozambique and Angola, who came in high numbers to Brazil. However, there was no cholera outbreak at that time (Martins, 1988; Tauxe et al., 1994).

THE 1990s CHOLERA EPIDEMIC

Cholera struck South and Central America again in 1991. The first cases occurred in the Peruvian port city of Chancay, which is located 60 km north of the capital, Lima. Cholera outbreaks quickly occurred along the entire Peruvian coast. Later, the epidemic crossed the Andean mountains, spreading to neighboring countries such as Brazil, essentially following the travelers' routes down the Amazon and other rivers (Color Plate 10) (Tauxe et al., 1994; Gangarosa and Tauxe, 1992). More than 1 million cholera cases were reported in South America between 1991 and 2002, ca. 70% of them in Peru. Starting in 1991 with 390,000 cases, the total number of yearly cases decreased steadily, but in 1998 there was an increase, from 13,000 cases in 1997 to 50,000 in 1998, decreasing again to 7,000 cases in 1999 (PAHO, 2003;

Ana Carolina P. Vicente • Department of Genetics, Instituto Oswaldo Cruz, Rio de Janeiro, CEP 21045-900, Brazil. **Irma Nelly G. Rivera** • Department of Microbiology, Biomedical Science Institute, University of São Paulo, São Paulo, CEP 05508-900, Brazil. **Michelle D. Vieira and Ana Coelho** • Department of Genetics, Instituto de Biologia, Universidade Federal do Rio de Janeiro, Rio de Janeiro, CEP 21944-970, Brazil.

http://www.paho.org/english/ad/dpc/cd/cholera-1991-2002.htm).

Isolates from the early cholera cases (1991 to 1993) were characterized using a number of phenotypic and molecular typing methods, including multilocus enzyme electrophoresis (MLEE) and ribotyping. These isolates could be distinguished from the seventh pandemic clone by MLEE at just a single locus, leucine aminopeptidase (Evins et al., 1995). Ribotyping was applied to a few isolates obtained from the beginning of the epidemic (Popovic et al., 1993; Evins et al., 1995); the results also revealed the close relationship of Latin American strains with the seventh pandemic clone.

Epidemic cholera normally occurs in coastal areas and is associated with travelers' movement worldwide (Colwell, 1996). Several theories were proposed to explain the reemergence of cholera in South America. It was theorized that the bacteria were released with the discharge of sewage from ships arriving from Asia (Tauxe et al., 1994) or the discharge of ballast water (coastal water contained inside ballast tanks to maintain ship stability). The major problem with ballast water is the transportation of a wide variety of organisms from port to port. Indeed, scientists estimate that >3,000 alien species are transported in ships around the world every day (National Research Council, 1995; Ruiz et al., 2000). There are many examples of biological invasions around the world; however, there is little information about the spread of exotic microorganisms. Yet McCarthy and Khambaty (1994) recovered toxigenic *V. cholerae* O1 El Tor from ballast water, bilge water, and sewage from five ships arriving at the U.S. Gulf of Mexico in 1991 and 1992. These isolates were indistinguishable from the Latin America epidemic strain. The National Agency of Sanitary Surveillance of the Brazilian Ministry of Health, in 2001 to 2002, evaluated the microbiological quality of 105 ballast water samples collected in nine Brazilian ports. The data revealed that toxigenic *V. cholerae* O1 was isolated in two samples collected from ships arriving at Belém and Recife ports, confirming the transport of pathogenic strains by ballast water and the possible risk to public health. Another theory proposed that the introduction of cholera stems from the new understanding that nontoxigenic strains can be infected by bacteriophage and thus become toxigenic (Faruque et al., 1998). However, a more likely possibility is that *V. cholerae* O1 is present at all times, albeit as viable but nonculturable forms. The organism presumably multiplies and spreads during plankton blooms (Colwell, 1996). In this way, a major epidemic could start at several points along the coast, as happened in Peru. Seas et al. (2000) investigated the presence of severe diarrhea cases (cholera) in Peru

before the epidemics occurred and found a few cases in various cities 1,000 km apart. This indicates that *V. cholerae* O1 was already present in the area some months before the epidemic appeared. The El Niño phenomenon probably had a role in plankton blooms, leading to an increase in seawater temperature and other environmental factors. As a result of the presence of high numbers of an epidemic strain of *V. cholerae*, there was undoubtedly contamination of seafood and drinking water. The increase in the number of cases in 1998 was also related to the 1997 to 1998 El Niño event (Chavez et al., 1999). In particular, Gil et al. (2004) demonstrated that cholera cases that occurred in Peru during the summer of 1998 correlated with the peak in sea surface temperatures.

An unexpectedly large outbreak of cholera took place in the Port Region of Paranaguá (Parana State, southern Brazil) in 1999. At least 467 cholera cases and 3 deaths were reported in this outbreak. However, only 6 cholera cases had occurred there in 1993. Epidemiological data suggested that the main source of *V. cholerae* O1 infection was the consumption of the raw bivalve *Modiolus brasiliensis* (also called "sururu") (Passos, 1999).

MOLECULAR ANALYSES OF *VIBRIO CHOLERAE* POPULATIONS IN SOUTH AMERICA

V. cholerae is a free-living bacterium found in the water column and in association with plankton, fish, and shellfish. *V. cholerae* non-O1/non-O139 strains are frequently isolated from aquatic ecosystems. Several studies in South America, analyzing both clinical and environmental isolates, were conducted during the epidemic of the 1990s.

Argentina

Two populations of *V. cholerae* O1 were defined among isolates from 1992 to 1998 in Argentina, based on analysis by randomly amplified polymorphic DNA (RAPD) and pulsed-field gel electrophoresis (PFGE). Isolates carrying the major virulence genes from the CTX phage and *tcpA* were linked to the Latin American epidemic clone, whereas isolates lacking these pathogenicity-associated factors and showing distinct RAPD and PFGE profiles characterized the other clone named Tucumán (Pichel et al., 2003). Interestingly, even sharing various phenotypic traits with a variant from Brazil called Amazonia (see below), Tucumán and Amazonia strains were not related genetically. In fact, the designation "variant" is not appropriate for them because they are not genetically related to any *V. cholerae* O1 previously characterized. These strains, which were isolated from patients with diarrhea, caused restricted outbreaks and were probably autochthonous from these countries (Color Plate 10).

Colombia

The cholera epidemic occurred between 1991 and 1996 in Colombia. Ribotyping analysis of 173 *V. cholerae* O1 strains isolated between 1991 and 1994, mainly from cholera patients' stools, revealed that 165 isolates belonged to the ribotype B5a, the predominant ribotype of the seventh cholera pandemic in Africa and Asia (Tamayo et al., 1997), as is the case for other isolates from South America.

Peru

PFGE and ribotyping of Peruvian *V. cholerae* O1 strains, which were recovered between 1991 and 1995 from patients, indicated the clonality of the South American strains (Dalsgaard et al., 1997). The occurrence of *V. cholerae* on the Peruvian coast was analyzed between 1997 and 2000 (Gil et al., 2004). Water and plankton samples ($n = 178$) were collected and screened by both plating and a direct fluorescent antibody assay. *V. cholerae* O1 was present along the coast, but on several occasions it was as viable but nonculturable cells.

Brazil

The standard of laboratory diagnosis of *V. cholerae* is the isolation of sucrose-fermenting colonies on TCBS. In the South American pandemic, isolates with a sucrose late fermenting phenotype emerged in French Guyana and spread quickly over some regions of the Brazilian Amazon, displacing the regular *V. cholerae* O1 strain during 1994 and 1995 (Ramos et al., 1997) (Color Plate 10). The reexamination, by us, of isolates from this population, using MLEE and PFGE, revealed homology with the major epidemic South American clone. The sequence analysis of genes from the two major virulence gene clusters, the *ctx*B from CTXφ and the *tcp*A from toxin-coregulated pilus–vibrio pathogenicity island, also showed their homology with those from the El Tor biotype.

Enterobacterial repetitive intergenic consensus (ERIC)-PCR was applied to a Brazilian *V. cholerae* collection, including *V. cholerae* O1 and non-O1, toxigenic and nontoxigenic, environmental and clinical strains, isolated before and after the seventh cholera pandemic (1977 to 1994). The methodology allowed strain differentiation and could be used in surveillance programs (Rivera et al., 1995). The same methodology was applied to *V. cholerae* O1 isolates from areas where the disease was endemic (Bangladesh), supporting the hypothesis that there are spatial and temporal fluctuations in the composition of toxigenic *V. cholerae* populations in the aquatic environment that can cause shifts in the dynamics of the disease (Zo et al., 2002). A total of 63 different ERIC-PCR patterns were revealed in an analysis of 256 environmental isolates of *V. cholerae* obtained between 1994 and 1997 in mangrove areas of Rio de Janeiro at Coroa Grande, Barra de Guara-tiba, Catalão, and Praia da Luz (Barros, 2000). The cultures were recovered from mangrove water and shellfish: *Anomalocardia brasiliana*, *Mytella guyanensis*, *Tagelus plebeius*, *Codakia orbiculares*, *Uca uruguayensis*, *Goniopsis cruentata*, and *Eurytium limosum*. All the isolates were non-O1 and had the main phenotypic features of *V. cholerae* (typical reactions on API-20E strips). The various ERIC-PCR patterns consisted of 8 to 17 bands each. A few *V. cholerae* strains with the same ERIC pattern were isolated from different organisms (even from different species) and places. For example, identical isolates were obtained in *Anomalocardia brasiliana*, *Mytella guyanensis*, and *Tagelus plebeius* at Coroa Grande, with a 1-year interval. Identical isolates were obtained, with a 5-month interval, from *Anomalocardia brasiliana* and *Mytella guyanensis* at two distant sites, Praia da Luz (Rio de Janeiro) and Coroa Grande, located on opposite sides of the city. Overall, the data indicated a high level of intraspecific genomic diversity, but with a few strains restricted to a particular geographical site, and even fewer widespread strains.

In another study, the presence of eight virulence-associated genes was verified in a *V. cholerae* population from clinical and environmental sources (1977 to 1994), and the study concluded that for human health risk assessment, *ctxA*, *stn/sto*, and *tcpA* genes should be monitored (Rivera et al., 2001).

Using MLEE, Campos et al. (2004) analyzed 92 clinical O1 and 56 environmental *V. cholerae* O1 and non-O1 isolates (1991 to 1999) from Brazil. They showed that all clinical and environmental O1 isolates belong to a unique MLEE type, whereas the non-O1 isolates are included in 15 types. Also using MLEE, Farfán et al. (2000) analyzed 10 environmental *V. cholerae* O1 and 20 environmental non-O1 strains from Brazil. They indicated that each isolate (except for 2 non-O1 isolates belonging to the same type) had a different MLEE type. Similar results were found among the 5 Peruvian *V. cholerae* clinical isolates analyzed by the same authors. These results are hard to reconcile with many previous analyses by MLEE, particularly because 9 of the O1 strains from Brazil and the 5 strains from Peru were labeled as El Tor, which many other researchers had found to be much more uniform in South America (Freitas et al., 2002). In addition, the date of isolation of the strains analyzed by Farfán et al. (2000) is not given, making it difficult to interpret their data.

Amplified fragment length polymorphism analysis has been a most valuable tool for investigating the diversity of V. cholerae worldwide (Jiang et al., 2000). In particular, clinical and environmental strains, both O1 and O139, share high genomic similarity. In some cases, indistinguishable O1 or O139 clones were isolated several decades apart from one another and from different continents (Jiang et al., 2000). It was suggested that a single clone from Asia spread to other continents, resulting in the current seventh pandemic (Jiang et al., 2000). V. cholerae O1 and non-O1/non-O139 Brazilian strains ($n = 106$) isolated from the environment and clinical specimens between 1991 and 2001 were characterized using amplified fragment length polymorphism (Thompson et al., 2003). Seven main groups of genomes were found, all of which originated from a variety of places in different years, suggesting the presence of different V. cholerae populations in Brazil, as is the case for the V. cholerae Amazonia strain (see below). Most O1 and non-O1/non-O139 strains formed distinct groups. This study also demonstrated the persistence of some strains of highly related genomes during several years in distinct geographical regions, suggesting that these populations are successful in adapting to changing environmental conditions. Six representative V. cholerae O1 strains, i.e., R-18306, R-18325, R-18333, R-18337, R-18335, and R-18355, and the non-O1 strain R-18246 isolated in Brazil were further characterized by sequencing (480 to 522 bp) five housekeeping genes, i.e., mdh, idh, lap, asd, and epd (Thompson, 2003). The results were compared with those of Farfán et al. (2002), who analyzed 31 V. cholerae O139 strains by multilocus sequence typing. The sequence variation of the V. cholerae strains for the loci asd, lap, idh, and mdh was between 1.5 and 3%, whereas the variation in epd was 6%. Strains R-18306, R-18308, R-18325 and BO1, CO402, MDO90, and NT 329 had identical sequences in all loci and formed the most abundant group. Conversely, the Amazonia isolates R-18333, R-18337, R-18335, and R-18355 had different alleles at all loci compared to the group above, except for lap and mdh, which were identical to those of strains CO391, CO407, and 653/36 (Farfán et al., 2002).

To check the possibility of recombination in the five loci examined in these strains, the index of asso-ciation (I_A) was applied (Smith et al., 1993; Feil and Spratt, 2001) and resulted in a value (3.08) that suggested that V. cholerae is a clonal species. Populations in linkage equilibrium (free recombination) have values equal to zero. However, linkage disequilibrium alone does not warrant clonality, as sampling of some dominant clones of the population may bias the statistics (Smith et al., 2000; Feil et al., 2003). An analysis by SplitsTree decomposition (Huson, 1998) revealed a network-like tree for asd, lap, and epd genes, suggesting that recombination may have indeed occurred within this collection of V. cholerae strains (Thompson, 2003).

Genetic Characterization of V. cholerae Amazonia

A molecular survey of clinical isolates from Northern Brazil began in the early 1990s. It revealed that, along with the epidemic strain, there was a distinct V. cholerae O1 nontoxigenic strain, the Amazonia strain, which is probably autochthonous from the Amazon region (Coelho et al., 1995). This strain, responsible for a localized outbreak, showed distinct PFGE, ribotyping, and MLEE patterns. The Amazonia strain has been shown to carry toxR but not toxT, ctx, or tcpA genes. The major virulence factors, cholera toxin genes, and toxin-coregulated pilus cluster were not present, but various virulence-associated systems were characterized (Pellegrini, 2003). The RTX toxin clusters of Amazonia and El Tor are similar and are distinct from the classical biotype. A milder form of cholera disease can be elicited by export of enterotoxins other than cholera toxin, including the V. cholerae RTX (VcRtxA) toxin encoded by the rtxA gene. The ubiquitous presence of VcRtxA suggests that this toxin is an important virulence factor of V. cholerae (Sheahan et al., 2004). Another accessory toxin produced by V. cholerae of both biotypes is the metalloprotease hemagglutinin/protease (Hap), encoded by the hapA gene (Finkelstein et al., 1992; Silva et al., 2003). Hap has been shown to have mucinolytic and cytotoxic activity. Other studies have suggested that Hap plays an important role in cholera pathogenesis by promoting mucin gel penetration, detachment, and spreading of infection along the gastrointestinal tract. The hapA gene in the Amazonia strain shares 98% sequence similarity with the same gene of the El Tor strain (Pellegrini, 2003). Hemolysin is a potential virulence factor, and the Amazonia strain produces a large amount of it. A 30-fold increase in the 50% lethal dose in infant mice has been reported from an El Tor hlyA mutant (Williams et al., 1993). It is possible that in the case of the Amazonia strains, hemolysin could play a more important role in the disease, perhaps as a result of higher expression or as a consequence of the six amino acid differences in the carboxy half, leading to increased activity (Coelho et al., 2000).

PROTEOMIC ANALYSIS OF THE AMAZONIA STRAIN

Proteomic technology has been used previously in a comparison of a Latin American El Tor epidemic strain to the sequenced El Tor N16961 strain. This comparison of bidimensional protein maps helped to establish a protein map for *V. cholerae* El Tor, taking into consideration abundant proteins present in both strains (Coelho et al., 2004). However, the gels showed marked differences between the two strains, with many spots (proteins) in different positions, indicating the need for further characterization of these strain-specific proteins. A similar comparison was performed between the Amazonia strain and El Tor N16961. Experiments were run under standardized conditions, with the protein prepared and focused at the same time. Cells were grown at 30°C for 16 h in M9 medium with casamino acids and glucose before centrifugation. The resulting supernatant was concentrated ~1,000-fold. Isoelectric focusing with 400 μg of protein was done on 7-cm Immobiline strips, in the pI range 4 to 7. Medium-sized 15% polyacrylamide gels were prepared for the second dimension, and the gels were stained with Coomassie brilliant blue. A comparison of a fraction enriched in supernatant proteins from both strains is shown in Color Plate 11. Panel A shows the two-dimensional (2D) gels, and panel B shows the identified spots on each gel. Each different protein was identified by a unique number, and repeats, in the same gel or in the other, were given the same protein number. A striking difference is seen in the gel patterns, with many spots unique to each strain. Fifty-six spots from the Amazonia strain and 40 from the El Tor strains were identified by mass spectrometry (matrix-assisted laser desorption ionization–time of flight [MALDI-TOF] or MALDI TOF/TOF) (Table 1). One important result from these experiments was the identification of at least three spots as the secreted Amazonia hemolysin (*V. cholerae* cytotoxin; Coelho et al., 2000). Both 65-kDa and 50-kDa forms were found, in agreement with the published literature for the El Tor strain (Yamamoto et al., 1990; Ikigai et al., 1999). Two 65-kDa isoforms were identified: a new result probably indicating posttranslational modification. The average pI of the three isoforms was 6.09, higher than the theoretical expected pI of 5.73. Another important identification was Hcp, the hemolysin-coregulated protein (Williams et al., 1996). Growth conditions were optimized for hemolysin production, and the fact that Hcp was also found in this study corroborates the hypothesis of coregulation of both proteins and is evidence of a similar mechanism for the Amazonia strain. In short, this strain seems to be a good system for analysis of hemolysin and its regulation.

Other secreted proteins were detected in this system. For the Amazonia strain, two spots for chitinase were found. The outer membrane protein OmpV, an extracellular nuclease-related protein, and flagellins were also identified. For the El Tor strain, OMPs OmpU, OmpA, and OmpT were detected, and the well-studied hemagglutinin protease HapA protein (Finkelstein et al., 1992; Silva and Benitez, 2004). HapA is an abundant protein in the conditions used. Moreover, the spot for HapA is within a "train" of spots. If all these spots turn out to be HapA, its total amount will be a significant fraction of the total proteins of this fraction of the El Tor cells. It is noteworthy that several of the identifications presented by Coelho et al. (2004) were confirmed. Thirty-five of the spots identified correspond to proteins in similar positions in the map presented in that paper. Undoubtedly, additional proteomic studies will help in the description of the metabolic pathways, regulatory pathways, and molecular complexes of *V. cholerae* and the difference among these pathways and complexes in different strains.

PERSPECTIVES

The emergence of a new cholera epidemic caused by *V. cholerae* El Tor biotype strains in South America in the early 1990s, and the occurrence of outbreaks thereafter, highlights the impact of this pathogen in the health of South American inhabitants. Apart from the obvious predisposing factors associated with poor potable water supply and sanitation in many areas of this region, other environmental and genetic parameters might well have led to the emergence of epidemic cholera in South America.

Research on cholera and *V. cholerae* is at a very exciting point. New tools are at hand for a more prompt global genetic analysis of strains, allowing a better description of strains and populations of vibrios. Whole-genome sequences of vibrios have underpinned novel total gene expression analyses, with the use of transcriptomes and proteomes. Clearly, the whole-genome sequencing of other *Vibrio* strains will allow the study of a more varied set of genes for these global studies. In particular, the Amazonia and Tucumán pathogenic strains are good local candidates for sequencing.

Table 1. Mass spectrometry identification of proteins present in 2D gel spots[a]

Spot[b]	Strain description	El Tor ID[c]
Amazonia strains		
1	Extracellular nuclease-related ptn	VC2621
2	Aconitate hydratase 2	VC0604
3	Isocitrate dehydrogenase, NADP-dependent, monomeric type	VC1141
4	Enterobactin receptor	VC0475
5	DnaK protein	VC0855
6	Chitinase, putative	VCA0811
7	Hemolysin	VCA0219
7	Hemolysin	VCA0219
8	Phosphoenolpyruvate carboxykinase	VC2738
8A	Chaperonin 60-kDa subunit	VC2664
9	Glutamate-ammonia ligase	VC2746
10	Phosphoribosylaminoimidazolecarboxamide	VC0276
11	Oligopeptide ABC transporter, periplasmic oligopeptide-binding	VC1091
6	Chitinase, putative	VCA0811
12	Phosphoglycerate mutase, 2,3-bisphosphoglycerate-	VC0336
13	Peptide ABC transporter, periplasmic peptide-binding ptn	VC0171
13	Peptide ABC transporter, periplasmic peptide-binding ptn	VC0171
13	Peptide ABC transporter, periplasmic peptide-binding ptn	VC0171
14	Glycerol kinase	VCA0744
15	Pyruvate kinase I	VC0485
7	Hemolysin	VCA0219
16	Pyruvate dehydrogenase, E3 component, lipoamide	VC2412
17	Enolase	VC2447
18	Phosphoglucomutase/phosphomannomutase	VC0611
19	Isocitrate lyase	VC0736
20	Elongation factor Tu	VC0321
21	Flagellin FlaC	VC2187
22	Alanine dehydrogenase	VC1905
23	Flagellin FlaB	VC2142
24	Phosphoglycerate kinase	VC0477
25	Elongation factor TU	VC0362
26	Fructose-bisphosphate aldolase, class II	VC0478
27	Iron (III) ABC transporter, periplasmic iron-compound-binding	VC0608
28	Glyceraldehyde 3-phosphate dehydrogenase	VC2000
29	Amino acid ABC transporter, periplasmic amino acid-binding	VC1362
30	Spermidine/putrescine ABC transporter, periplasmic	VCA1113
31	Malate dehydrogenase	VC0432
29	Amino acid ABC transporter, periplasmic amino acid-binding	VC1362
32	Thiosulfate ABC transporter, periplasmic thiosulfate-binding ptn	VC0538
33	Elongation factor Ts	VC2259
27	Iron (III) ABC transporter, periplasmic iron-compound-binding	VCA0227
34	Immunogenic ptn	VC0430
35	Hypothetical protein	AAL59733
36	Ribose ABC transporter, periplasmic D-ribose-binding ptn	VCA0130
37	Uridine phosphorylase	VC1034
38	Triosephosphate isomerase	VC2670
39	Tryptophan synthase, alpha subunit	VC1169
40	Outer membrane OmpV	VC1318
40	Outer membrane OmpV	VC1318
41	Hcp protein	VC1415
42	Amino acid ABC transporter, periplasmic amino acid-binding	VC0010
25	Elongation factor TU	VC0362
43	Arginine ABC transporter, periplasmic arginine-binding protein	VCA0759
44	Antioxidant, AhpC/Tsa family	VC0731
44	Antioxidant, AhpC/Tsa family	VC0731
45	Superoxide dismutase, Fe	VC2045
El Tor strains		
3	Isocitrate dehydrogenase, NADP-dependent, monomeric type	VC1141
3	Isocitrate dehydrogenase, NADP-dependent, monomeric type	VC1141

Continued on following page

Table 1. *Continued*

Spot[b]	Strain description	El Tor ID[c]
46	2',3'-cyclic-nucleotide 2'-phosphodiesterase	VC2562
8	Phosphoenolpyruvate carboxykinase	VC2738
11	Oligopeptide ABC transporter, periplasmic oligopeptide-binding	VC1091
11	Oligopeptide ABC transporter, periplasmic oligopeptide-binding	VC1091
47	Hypothetical protein	VC1485
12	Phosphoglycerate mutase, 2,3-bisphosphoglycerate-	VC0336
13	Peptide ABC transporter, periplasmic peptide-binding ptn	VC0171
13	Peptide ABC transporter, periplasmic peptide-binding ptn	VC0171
16	Pyruvate dehydrogenase, E3 component, lipoamide	VC2412
48	Pyruvate kinase I	VC0485
49	DnaK protein	VC0855
22	Alanine dehydrogenase	VC1905
50	Hemagglutinin/protease	VCA0865
24	Phosphoglycerate kinase	VC0477
51	Outer membrane protein OmpU	VC0633
51	Outer membrane protein OmpU	VC0633
51	Outer membrane protein OmpU	VC0633
29	Amino acid ABC transporter, periplasmic amino acid-bdg ptn	VC1362
52	Outer membrane protein OmpA	VC2213
30	Spermidine/putrescine ABC transporter, periplasmic	VC1425
53	Hydrolase, putative	VCA0877
54	Galactosidase ABC transporter, periplasmic D-galactose/D-	VC1325
55	Enolase	VC2447
38	Triosephosphate isomerase	VC2670
38	Triosephosphate isomerase	VC2670
56	Probable amino acid ABC transporter, periplasmic amino acid-	VCA0978
57	Hypothetical protein	VCA0981
43	Arginine ABC transporter, periplasmic arginine-binding ptn	VCA0759
58	Inorganic pyrophosphatase	VC2545
59	OmpT protein	VC1854
44	Antioxidant, AhpC/Tsa family	VC0731
60	Fructose-bisphosphate aldolase, class II	VC0478
60	Fructose-bisphosphate aldolase, class II	VC0478
60	Fructose-bisphosphate aldolase, class II	VC0478
44	Antioxidant, putative	VC1350
61	Riboflavin synthase, beta subunit	VC2268
62	Hypothetical protein	VCA0881
63	Elongation factor TU	VC0362

[a]The Proteomic Network of Rio de Janeiro MALDI-TOF was used for the mass spectrometry analysis, and the MALDI-TOF/TOF identifications were done at LNLS, Campinas, Brazil. Criteria for identification were the same as in Coelho et al. (2004).
[b]Spot numbers in this table correspond to numbers in Color Plate 11. For the Amazonia strain, the spots were numbered in increasing order. For the El Tor strain, the same was done, but those with corresponding spots in the Amazonia gel were given the corresponding numbers.
[c]El Tor ID, the accession code for the corresponding gene in The Institute for Genomic Research–Comprehensive Microbiol Resource for *V. cholerae* El Tor N16961.

The description of fundamental regulatory networks within the cells in the various environmental conditions encountered by pathogenic *V. cholerae* strains in South America, as well as the characterization and monitoring of these strains in the environment, needs to be further carried out. Work is under way in our laboratories to shed light on the biology of South American *V. cholerae* strains, with environmental surveys of both free-living and plankton-associated vibrios in different ecosystems, including the Amazonian rain forest. Also, analyses are needed of vibrios associated with disease in humans, probing at the interaction of pathogen and host. Themes such as quorum sensing, biofilm development, motility,

and virulence intertwine, and several regulatory chains are yet to be discovered in South American *V. cholerae* strains. Transition to an infective state during human infection has been described (Merrell et al., 2002), and this process needs to be evaluated in a broad collection of *V. cholerae*.

Models for horizontal gene transfer describing detailed interactions of bacteria and phage that mimic transfer of new genes to the superintegrons or other parts of the chromosomes of *Vibrio* are not yet described. Vibrio phages are in evidence, with studies of lysogen induction and proposals of dynamic interactions between bacteria and phage leading to the end of epidemics (Espeland et al., 2004; Faruque et al., 2005). Much will be learned through the study of phage biodiversity, evolution, and distribution in South America. The field of modes of gene transfer and recombination is open. Little is known about the extent and effects of conjugation and transformation in *V. cholerae* genomes. These issues clearly call for deeper investigation. The interplay of all these interactions—building new genomes capable of survival in different conditions through the fine-tuning of various regulatory networks, virulent in the human host, and persistent in the environment through interactions with other organisms and abiotic substrates—is probably the source of diversity leading to new pathogenic strains.

REFERENCES

Barros, C. 2000. Isolamento e análise intra-específica de *Vibrio cholerae* por ERIC-PCR e RAPD. M.Sc. thesis, p. 136. Ciências Biológicas (Genética), Universidade Federal do Rio de Janeiro, Brasil.

Barua, D. 1992. History of cholera, p. 1–36. *In* D. Barua and W. B. Greenough III (ed.), *Cholera*. Plenum Medical Book Company, New York, N.Y.

Beltrão, J. F. 2004. Cólera, o flagelo da Belém do Grão-Pará. Ed. Universitária UFPA, Brasil.

Campos, L. C., V. Zahner, K. E. Avelar, R. M. Alves, D. S. Pereira, B. J. Vital, F. S. Freitas, C. A. Salles, and D. K. Karaolis. 2004. Genetic diversity and antibiotic resistance of clinical and environmental *Vibrio cholerae* suggests that many serogroups are reservoirs of resistance. *Epidemiol. Infect.* **132:**985–992.

Chavez, F., P. Strutton, G. Friederich, R. Feely, G. Feldman, D. Foley, and M. McPhaden. 1999. Biological and chemical response of the equatorial Pacific ocean to the 1997–1998 El Niño. *Science* **286:**2126–2131.

Coelho, A., J. R. C. Andrade, A. C. P. Vicente, and C. A. Salles. 1995. A new variant of *Vibrio cholerae* O1 from clinical isolates in Amazonia. *J. Clin. Microbiol.* **33:**114–118.

Coelho, A., J. R. C. Andrade, A. C. P. Vicente, and V. J. Dirita. 2000. Cytotoxic cell vacuolating activity from *Vibrio cholerae* hemolysin. *Infect. Immun.* **68:**1700–1705.

Coelho, A., E. O. Santos, M. L. Faria, D. P. de Carvalho, M. R. Soares, W. M. von Kruger, and P. M. Bisch. 2004. A proteome reference map for *Vibrio cholerae* El Tor. *Proteomics* **4:**1491–1504.

Colwell, R. R. 1996. Global climate and infectious disease: the cholera paradigm. *Science* **274:**2025–2031.

Cooper, D. B. 1986. The new "Black Death": cholera in Brazil 1855–1856. *Social Science History* **10:**467–488.

Dalsgaard, A., M. N. Skov, S. O. Serichantalerg, P. Echeverria, R. Meza, and D. N. Taylor. 1997. Molecular evolution of *Vibrio cholerae* O1 strains isolated in Lima, Peru, from 1991–1995. *J. Clin. Microbiol.* **35:**1151–1156.

Espeland, E. M., E. K. Lipp, A. Huq, and R. R. Colwell. 2004. Polylysogeny and prophage induction by secondary infection in *Vibrio cholerae*. *Environ. Microbiol.* **6:**760–763.

Evins, G. M., D. N. Cameron, J. G. Wells, K. D. Greene, T. Popovic, S. Giono-Cerezo, K. Wachsmuth, and R. V. Tauxe. 1995. The emerging diversity of the electrophoretic types of *Vibrio cholerae* in the Western Hemisphere. *J. Infect. Dis.* **172:**173–179.

Farfán, M., D. Minana, M. C. Fuste, and J. G. Loren. 2000. Genetic relationships between clinical and environmental *Vibrio cholerae* isolates based on multilocus enzyme electrophoresis. *Microbiology* **146:**2613–2626.

Farfán, M., D. Minana-Galbis, M. C. Fuste, and J. G. Loren. 2002. Allelic diversity and population structure in *Vibrio cholerae* O139 Bengal based on nucleotide sequence analysis. *J. Bacteriol.* **184:**1304–1313.

Faruque, S. M., M. J. Albert, and J. J. Mekalanos. 1998. Epidemiology, genetics, and ecology of toxigenic *Vibrio cholerae*. *Microbiol. Mol. Biol. Rev.* **62:**1301–1314.

Faruque, S. M., I. B. Naser, M. J. Islam, A. S. Faruque, A. N. Ghosh, G. B. Nair, D. A. Sack, and J. J. Mekalanos. 2005. Seasonal epidemics of cholera inversely correlate with the prevalence of environmental cholera phages. *Proc. Natl. Acad. Sci. USA* **102:**1702–1707.

Feil, E. J., J. E. Cooper, H. Grundmann, D. A. Robinson, M. C. Enright, T. Berendt, S. J. Peacock, J. M. Smith, M. Murphy, B. G. Spratt, C. E. Moore, and N. P. J. Day. 2003. How clonal is *Staphylococcus aureus*? *J. Bacteriol.* **185:**3307–3316.

Feil, E. J., and B. G. Spratt. 2001. Recombination and the population structures of bacterial pathogens. *Annu. Rev. Microbiol.* **55:**561–590.

Finkelstein, R. A., M. Boesman-Finkelstein, Y. Chang, and C. C. Hase. 1992. *Vibrio cholerae* hemagglutinin/lectin/protease hydrolyzes fibronectin and ovomucin: F.M. Burnet revisited. *Proc. Natl. Acad. Sci. USA* **80:**1092–1095.

Freitas, F. S., H. Momen, and C. A. Salles. 2002. The zymovars of *Vibrio cholerae*: multilocus enzyme electrophoresis of *Vibrio cholerae*. *Mem. Inst. Oswaldo Cruz* **97:**511–516.

Gangarosa, E. J., and R. V. Tauxe. 1992. Epilogue: the Latin American cholera epidemic, p. 351–358. *In* D. Barua and W. B. Greenough III (ed.), *Cholera*. Plenum Medical Book Company, New York, N.Y.

Gil, A. I., V. R. Louis, I. N. G. Rivera, E. Lipp, A. Huq, C. F. Lanata, D. N. Taylor, E. Russek-Cohen, N. Choopun, R. B. Sack, and R. R. Colwell. 2004. Occurrence and distribution of *V. cholerae* in the coastal environment of Peru. *Environ. Microbiol.* **6:**699–706.

Huson, D. H. 1998. SplitsTree: analyzing and visualizing evolutionary data. *Bioinformatics* **14:**68–73.

Ikigai, H., T. Ono, T. Nakae, H. Otsuru, and T. Shimamura. 1999. Two forms of *Vibrio cholerae* O1 El Tor hemolysin derived from identical precursor protein. *Biochim. Biophys. Acta* **1415:**297–305.

Jiang, S. C., M. Matte, G. Matte, A. Huq, and R. R. Colwell. 2000. Genetic diversity of clinical and environmental isolates of *Vibrio cholerae* determined by amplified fragment length polymorphism fingerprinting. *Appl. Environ. Microbiol.* **66:**148–153.

Martins, M. T. 1988. Ecologia de *Vibrio cholerae* no ecossistema aquático. Tese Livre Docência ICB-USP, Universidade de São Paulo, S. Paulo. Brazil.

McCarthy, S. A., and F. Khambaty. 1994. International dissemination of epidemic *Vibrio cholerae* by cargo ship ballast and other nonpotable waters. *Appl. Environ. Microbiol.* **60:**2597–2601.

Merrell, D. S., S. M. Butler, F. Qadri, N. A. Dolganov, A. Alam, M. B. Cohen, S. B. Calderwood, G. K. Schoolnik, and A. Camilli. 2002. Host-induced epidemic spread of the cholera bacterium. *Nature* **417:**642–645.

National Research Council. 1995. *Understanding Marine Biodiversity: a Research Agenda for the Nation*. Ocean Studies Board, National Research Council. National Academy Press, Washington, D.C.

Passos, A. D. C. 1999. [Cholera epidemiology in Southern Brazil.] *Lett. Cad. Saúde Pública* **15:**426–427. (In Portuguese.)

Pellegrini, M. P. 2003. [Genetic characterization of the *Vibrio cholerae* Amazonia strain.] M.Sc. thesis. Instituto Oswaldo Cruz, Rio de Janeiro, Brazil. (In Portuguese.)

Pichel, M., M. Rivas, I. Chinen, F. Martín, C. Ibarra, and N. Binsztein. 2003. Genetic diversity of *V. cholerae* O1 in Argentina and emergence of a new variant. *J. Clin. Microbiol.* **41:**124–134.

Pollitzer, R. 1959. *Cholera,* p. 11–50. World Health Organization monograph 43. World Health Organization, Geneva, Switzerland.

Popovic, T., C. Bopp, Ø. Olsvik, and K. Wachsmuth. 1993. Epidemiologic application of a standardized ribotype scheme for *V. cholerae* O1. *J. Clin. Microbiol.* **31:**2474–2482.

Ramos, F. L., Z. C. Lins-Lainson, E. L. da Silva, A. A. Proietti, Jr., M. L. Mareco, and M. L. Lamarao. 1997. Cholera in north Brazil: on the occurrence of strains of *Vibrio cholerae* O1 which fail to ferment sucrose during routine plating on thiosulphate-citrate-bile salt-sucrose agar (TCBS). A new problem in diagnosis and control? *Rev. Latinoam. Microbiol.* **39:**141–144.

Rivera, I. N. G., M. A. R. Chowdhury, A. Huq, D. Jacobs, M. T. Martins, and R. R. Colwell. 1995. Enterobacterial repetitive intergenic consensus sequences and the PCR generate fingerprints of genomic DNAs from *Vibrio cholerae* O1, O139 and non-O1 strains. *Appl. Environ. Microbiol.* **61:**2898–2904.

Rivera, I. N. G., J. Chun, A. Huq, R. B. Sack, and R. R. Colwell. 2001. Genotypes associated with virulence in environmental isolates of *Vibrio cholerae.* *Appl. Environ. Microbiol.* **67:**2421–2429.

Ruiz, G. M., T. K. Rawlings, F. C. Dobbs, L. A. Drake, T. Mullady, A. Huq, and R. R. Colwell. 2000. Global spread of microorganisms by ships. *Nature* **408:**49–50.

Seas, C., J. Miranda, A. I. Gil, R. Leon-Barua, J. Patz, A. Huq, R. R. Colwell, and R. B. Sack. 2000. New insights on the emergence of cholera in Latin America during 1991: the Peruvian experience. *Am. J. Trop. Med. Hyg.* **62:**513–517.

Sheahan, K. L., C. L. Cordero, and K. J. Satchell. 2004. Identification of a domain within the multifunctional *Vibrio cholerae* RTX toxin that covalently cross-links actin. *Proc. Natl. Acad. Sci. USA* **101:**9798–9803.

Silva, A. J., and J. A. Benitez. 2004. Transcriptional regulation of *Vibrio cholerae* hemagglutinin/protease by the cyclic AMP receptor protein and RpoS. *J. Bacteriol.* **186:**6374–6382.

Silva, A. J., K. Pham, and J. A. Benitez. 2003. Haemagglutinin/protease expression and mucin gel penetration in El Tor biotype *Vibrio cholerae.* *Microbiology* **149:**1883–1891.

Smith, J. M., N. H. Smith, M. O'Rourke, and B. G. Spratt. 1993. How clonal are bacteria? *Proc. Natl. Acad. Sci. USA* **90:**4384–4388.

Smith, J. M., E. J. Feil, and N. H. Smith. 2000. Population structure and evolutionary dynamics of pathogenic bacteria. *Bioessays* **22:**1115–1122.

Tamayo, M., S. Koblavi, F. Grimont, E. Castañeda, and P. A. D. Grimont. 1997. Molecular epidemiology of *V. cholerae* O1 isolates from Colombia. *J. Med. Microbiol.* **46:**611–616.

Tauxe, R., L. Seminario, R. Tapia, and M. Libel. 1994. The Latin America epidemic, p. 321–344. *In* I. K. Waschmuth, P. A. Black, and Ø. Olsvik (ed.), Vibrio cholerae *and Cholera: Molecular to Global Perspectives.* ASM Press, Washington, D.C.

Thompson, C. C. 2003. Application of multilocus sequence typing in the taxonomy of the family *Vibrionaceae.* M.Sc. thesis. Ghent University, Belgium.

Thompson, F. L., C. C. Thompson, A. C. P. Vicente, G. N. D. Theophilo, E. Hofer, and J. Swings. 2003. Genomic diversity of clinical and environmental *Vibrio cholerae* strains isolated in Brazil between 1991 and 2001 as revealed by fluorescent amplified fragment length polymorphism analysis. *J. Clin. Microbiol.* **41:**1946–1950.

Williams, S. G., S. R. Attridge, and P. A. Manning. 1993. The transcriptional activator HlyU of *Vibrio cholerae*: nucleotide sequence and role in virulence gene expression. *Mol. Microbiol.* **9:**751–760.

Williams, S. G., L. T. Varcoe, S. R. Attridge, and P. A. Manning. 1996. *Vibrio cholerae* Hcp, a secreted protein coregulated with HlyA. *Infect. Immun.* **64:**283–289.

Yamamoto, K., Y. Ichinose, H. Shinagawa, K. Makino, A. Nakata, M. Iwanaga, T. Honda, and T. Miwatani. 1990. Two-step processing for activation of the cytolysin/hemolysin of *Vibrio cholerae* O1 biotype El Tor: nucleotide sequence of the structural gene (*hlyA*) and characterization of the processed products. *Infect. Immun.* **58:**4106–4116.

Zo, Y., I. N. G. Rivera, E. Russek-Cohen, M. S. Islam, A. K. Siddique, M. Yunus, R. B. Sack, A. Huq, and R. R. Colwell. 2002. Genomic profiles of clinical and environmental isolates of *Vibrio cholerae* O1 in cholera-endemic areas of Bangladesh. *Proc. Natl. Acad. Sci. USA* **99:**12409–12414.

VII. ANIMAL PATHOGENS

The Biology of Vibrios
Edited by F. L. Thompson et al.
© 2006 ASM Press, Washington, D.C.

Chapter 18

The Biology and Pathogenicity
of *Vibrio anguillarum* and *Vibrio ordalii*

JORGE H. CROSA, LUIS A. ACTIS, AND MARCELO E. TOLMASKY

INTRODUCTION

Vibriosis is one of the most prevalent fish diseases caused by bacteria belonging to the genus *Vibrio*. Vibriosis caused by *Vibrio anguillarum* has been particularly devastating in the marine culture of salmonid fish. This pathogen was first described by Bergman in 1909 as the etiological agent of "Red Pest of eels" in the Baltic Sea. Before this report, in 1893 Canestrini described epizootics in migrating eels (*Anguilla vulgaris*) dating back to 1817 that implicated the bacterium named *Bacillus anguillarum*. The pathology of the disease and the characteristic of the bacterium in these two reports suggested that the etiological agents were the same. Vibriosis was first reported in North America in 1953, when *V. anguillarum* was isolated from chum salmon (*Oncorhynchus keta*). Outbreaks affecting close to 50 species of fresh- and saltwater fishes have been reported in several countries in the Pacific and the Atlantic coasts. The losses are so devastating that vibriosis caused by *V. anguillarum* has been recognized as a major obstacle for salmonid marine culture. Some time ago, it was demonstrated that isolates of *V. anguillarum*, the most important etiological agent of vibriosis, exhibited a marked heterogeneity, which led to the division of these vibrios into two separate biotypes, 1 and 2. Later, a new species for the *V. anguillarum* biotype 2 was proposed on the basis of cultural and biochemical characteristics as well as on DNA homology with the biotype 1. This new species was named *Vibrio ordalii*.

V. ANGUILLARUM

V. anguillarum, which has been designated *Beneckea anguillara* biotype 1 (Baumann et al., 1978) and, more recently, *Listonella anguillarum* (MacDonell and Colwell, 1985), is a polar flagellated gramnegative curved rod. This bacterium is actually very similar to *Vibrio vulnificus*, *Vibrio cholerae*, and *Vibrio parahaemolyticus* (>90% identity in many important gene clusters), and most authors continue to use the species name *V. anguillarum*. It is a facultative anaerobe with a G+C content of 43 to 46%. It grows rapidly at 25 to 30°C in rich media such as brain heart infusion, brain heart infusion agar, trypticase soy broth, or trypticase soy agar containing 1.5% (wt/vol) sodium chloride. On solid medium, it produces circular, cream-colored colonies. *V. anguillarum* belongs to one of the halophilic groups of vibrios and survives at different salinities. Hoff (1989) has shown that it is able to survive in seawater for >50 months. A list of the phenotypic properties, serotypes, as well as the salt and temperature range for *V. anguillarum* has been published (Schiewe et al., 1981; Pedersen et al., 1999).

Identification and Classification

Epizootics of *V. anguillarum* may lead to a rapid loss of farmed fish. Consequently, a prompt diagnosis is essential, and includes a culture medium for presumptive identification, filter assays to test sensitivity to the vibriostatic agent O/129, nitrate reduction, the presence of oxidase, catalase, and arginine decarboxylase, reaction with monoclonal antibodies and antiflagellar antiserum, and hybridization with specific 16S rRNA oligonucleotides. Other identification and diagnostic methods have also been described, including 16S- and 23S-based PCR (Shewan et al., 1954; Tassin et al., 1983; Larsen and Mellergaard, 1984; Rehnstam et al., 1989; Alsina et al., 1994; Martinez-

Jorge H. Crosa • Department of Molecular Microbiology and Immunology, School of Medicine, Oregon Health and Science University, Portland, OR 97201-3098. Luis A. Actis • Department of Microbiology, Miami University, Oxford, OH 45056. Marcelo E. Tolmasky • Department of Biology, College of Natural Sciences and Mathematics, California State University—Fullerton, Fullerton, CA 92834-6850.

Picardo et al., 1994; Sparagano et al., 2002) and multiplex PCR in combination with DNA microarrays (Gonzalez et al., 2004). Enzyme-linked immunosorbent assay and fluorescein isothiocyanate immunofluorescence tests showed that genus-specific monoclonal antibodies were very useful for identifying vibrios, whereas the species-specific monoclonal antibodies were useful for completing the diagnosis (Chen et al., 1992). The problem with these biochemical and immunological methods is that they require the culture and isolation of the infecting bacteria from the fish. Conversely, newer hybridization techniques do not require a pure culture. More recently, synthetic oligonucleotides have been used as specific probes for the identification of *V. anguillarum* (Rehnstam et al., 1989; Martinez-Picardo et al., 1994). These oligonucleotides were designed using information from sequencing the 16S rRNA from several *V. anguillarum* strains. The radiolabeled probes were used in DNA hybridization assays carried out using purified DNA and homogenized fish tissues, such as kidney, with no cross-hybridization against other bacterial species, permitting the identification of *V. anguillarum* within 24 h. A pitfall with this molecular technique is that pathogenic, environmental, and reference strains are similar (Larsen, 1983). However, because there are no cross-reacting antigens present in these strains, the classification of *V. anguillarum* by serological methods has proved to be convenient (Pacha and Kiehn, 1969; Johnsen, 1977; Kitao et al., 1983; Sorensen and Larsen, 1986).

A serotyping scheme has been proposed based on the detection by slide agglutination of *V. anguillarum* O antigens (Sorensen and Larsen, 1986). With this test, it is possible to detect the presence of this pathogen in infected fish (Toranzo and Barja, 1990). Ten serotypes (O1 to O10) have been described in *V. anguillarum*; however, most vibriosis outbreaks involving cultured salmonids and feral marine fish have been shown to be caused by strains belonging to serotypes O1, O2, and O3 (Toranzo and Barja, 1990; Toranzo et al., 1997). Strains of serotypes O3 to O10 have been mainly isolated from marine environmental samples, including water, sediment, and phyto- and zooplankton. Bolinches et al. (1990) established, using a combination of immunological methods, that the O2 group can be subdivided into two subgroups, O2a and O2b. This finding agreed with that of Rasmussen (1987a), who demonstrated that two lipopolysaccharide (LPS) species, designated O2a and O2b, were present in O2 strains. Bacteria belonging to the O2a group were isolated from salmonids and nonsalmonid fish; strains belonging to the O2b group were isolated only from nonsalmonids. The presence of capsular antigens (K antigens) has also been reported for *V. anguillarum* of the O1, O4, O5, and O6 groups (Rasmussen, 1987a,b; Tajima et al., 1987a,b). However, the role of these capsular antigens in virulence remains to be determined. Later on, new serotypes were defined, bringing the total of O serogroups to 23, with the possibility that new serogroups could be defined and added to the current list as new *V. anguillarum* isolates are classified with the current tools (Pedersen et al., 1999).

The correlation between serotype and virulence may reflect the ability of the bacterial surface antigens to interact with the host tissues. Furthermore, studies of other surface components, such as outer membrane proteins (OMPs) and LPS, of *V. anguillarum* strains demonstrated that these bacterial components are related to the serotypes of these pathogens (Aoki et al., 1981; Nomura and Aoki, 1985).

Agglutination has also been used to type environmental and fish isolates (Larsen and Mellergaard, 1984). All *V. anguillarum* isolates exerted either mannose-sensitive hemagglutination or were non-agglutinating. The *V. anguillarum* agglutinins showing specificity against human, poultry, guinea pig, and trout erythrocytes, as well as yeast cells and eight different agglutination types (A through H), were defined by this method.

Control of *V. anguillarum* includes the use of antibiotics, vaccination, and the culture of other bacteria that inhibit the growth of the pathogen. Antibiotics and other chemotherapeutic agents are included as feed additives or added directly to the water to prevent and treat vibriosis. Ampicillin, chloramphenicol, nalidixic acid derivatives, nitrofuran derivatives, sulfonamides, and trimethoprim have been used routinely in Japan to treat vibriosis (Aoki et al., 1984). However, the overuse of these compounds has resulted in the emergence of drug-resistant strains (Watanabe et al., 1971; Aoki et al., 1977, 1980; Hayashi et al., 1982; Toranzo et al., 1984; Zhao and Aoki, 1992a,b; Pedersen et al., 1995). For example, tetracycline-resistant isolates were recovered routinely from cultured ayu (*Plecoglossus altivelis*) up until 1977 when the use of this antibiotic was discontinued. Analysis of the resistance profile of *V. anguillarum* recovered since 1978 has shown that only one isolate was resistant to tetracycline (Aoki et al., 1984), demonstrating a correlation between the use of the compound and the appearance of antibiotic resistance. Molecular and genetic analyses show that the genes encoding antibiotic resistance are often found in plasmids. Furthermore, in some cases these plasmids are shown to be conjugative (Aoki et al., 1984; Toranzo et al., 1984). Further studies are needed to demonstrate whether these genes are also present in transposable elements (Kleckner, 1981; Bennett and Hawkey, 1991) or integrons (Stokes and Hall, 1989).

There are several licensed vaccines, which have proved successful in the field (http://www.vetcare.gr/vibriosis.htm). For example, Vibrogen-2 bacterin consists of an aqueous suspension of formalin-inactivated cultures of 3 strains of *V. anguillarum*, without adjuvants or preservatives. As is the case with this vaccine, traditional vaccines consist of formalin-killed *V. anguillarum* or bacterial membrane components (Agius et al., 1983). While some authors claim that the vaccines give better protection when administered by intraperitoneal (i.p.) injection than by immersion or oral administration (Kawano et al., 1984; Ward et al., 1985; Bowden et al., 2002), the former method has limitations, and for practical reasons the other approaches are currently used. Several studies have recently been performed using bacterins or other formulations (Bergh et al., 2001; Acosta et al., 2004). Attempts to develop a live vaccine have been made using avirulent derivatives generated by transposition mutagenesis. Norqvist et al. (1989) constructed avirulent mutants capable of inducing protective immunity after bath vaccination. Avirulent strains were also constructed by curing the virulence plasmid pJM1 (Crosa et al., 1980), by performing other genetic manipulations of this plasmid such as deletion of the iron-uptake region (Walter et al., 1983), and by mutagenesis by either transposition (Tolmasky et al., 1988a) or in vitro insertion (Singer et al., 1991). In the latter case, the avirulent derivative persisted in inoculated fish. The mutants could be isolated 9 days postinoculation, suggesting that these avirulent *V. anguillarum* derivatives could be good candidates for a live attenuated vaccine.

The addition of absorption enhancers such as sodium salicylate, sodium caprate, and vitamin E to *V. anguillarum* O2 antigen increased the serum antigen levels and specific antibody titers in the systemic circulation and increased antibody levels in the skin mucus after oral vaccination (Vervarcke et al., 2004). Xia et al. (2005) demonstrated that the anti-idiotype monoclonal antibody 1E10 can mimic the protective epitope of *V. anguillarum* and be used as vaccine to prevent fish infection. These authors cloned the variable heavy domain and the variable light domain of mAb1E10 and linked them to each other, introducing a disulfide bond. The derivative thus generated was inserted into a phagemid that upon cotransfection with the helper phage M13KO7 induced the host *Escherichia coli* to secrete disulfide-stabilized Fv fragment that displayed on the surface of the phage. Protection experiments using Japanese flounder showed that this phage-displayed disulfide-stabilized Fv may be used as vaccine against *V. anguillarum*.

Another strategy to combat vibriosis has been proposed on the basis of observations that bacteria of the normal gut flora produce inhibitory substances (Lemos et al., 1985; Onarheim and Raa, 1990; Westerdahl et al., 1991). After studying more than 400 intestinal isolates from turbot (*Scophthalmus maximus*, L.), Westerdahl et al. (1991) found that 28% of those isolates exhibited inhibitory effects against *V. anguillarum*. Thus, it is at least theoretically possible that strains with enhanced inhibitory action could provide protection after oral administration and might be generated by genetic and molecular biology methods.

The Disease

Vibriosis occurs in cultured and in wild marine fish in salt or brackish water, particularly in shallow waters during late summer. There is evidence that *V. anguillarum* is normally present in the food of cultured and wild healthy fish (Roberts, 1989). The temperature and quality of the water, the virulence of the *V. anguillarum* strain, and stress on the fish are important elements influencing the onset of disease outbreaks. The characteristic clinical signs of vibriosis include red spots on the ventral and lateral areas of the fish, swollen and dark skin lesions that ulcerate, releasing a blood exudate. There are also corneal lesions characterized by an initial opacity, followed by ulceration and evulsion of the orbital contents. However, in acute and severe epizootics, the course of the infection is rapid, and most of the infected fish die without showing any clinical signs.

Ransom et al. (1984) studied and compared the histopathology of vibriosis caused by *V. anguillarum* and *V. ordalii* in rainbow trout and salmon and found significant differences between the diseases caused by these bacteria. This disease is characterized by a hemorrhagic septicemia. The number of leukocytes is reduced (Ransom et al., 1984), and darkening of the diseased fish is noted with petechiae at the base of fins and skin. Ulcers can also be observed. Cisar and Fryer (1969) reported that the intestine becomes distended and fills with a clear, viscous liquid. *V. anguillarum* was found in large numbers in the blood and hematopoietic tissues. The pathology is more severe in the descending gastrointestinal tract and rectum than in the anterior region, owing to a pH gradient that is alkaline in the rectum and becomes acidic toward the anterior gastrointestinal tract. Ransom et al. (1984) demonstrated that *V. anguillarum* cannot grow in an acidic medium. Histological examination of infected rainbow trout tissues can demonstrate the location of the *V. anguillarum* during infection (Nelson et al., 1985a,b). The organism was initially found in the spleen, but as the number of cells in this organ increased, bacteria appeared in the kidney. At the time of death, most tissues were septic, and no phagocytosis by macrophages was detected (Nelson et al., 1985a). Severe cardiac myopathy, renal and splenic necrosis, and periorbital edema were also described in pre-acute vibriosis cases (Roberts, 1989).

V. anguillarum infection has recently been visualized at the whole fish and single bacterium levels using microscopy, with zebrafish and green fluorescent protein-labeled V. anguillarum cells as a model. These experiments showed that the V. anguillarum cells were first detected in the gastrointestinal tract. While the enteric localization occurred independently of the flagellum or motility, chemotactic motility was essential for the association of the bacterial cells with the fish surface. These studies present evidence that the intestine and skin are sites of V. anguillarum infection and suggest a host site where chemotaxis may function in virulence (Actis et al., 1998; O'Toole et al., 2004).

Virulence Factors

The pathogenesis of vibriosis is due to the ability of V. anguillarum to produce several virulence factors that allow this organism to colonize and persist in the host. The flagella of V. anguillarum were also suggested to be a virulence factor. The flagella consist of flagellin A, which is essential for virulence (Milton et al., 1996), and three additional flagellin proteins, FlaB, -C, and -D, as well as another flagellin, FlaE (McGee et al., 1996). It was of interest that mutants of the RNA polymerase gene rpoN resulted in the abrogation of expression of the flagellin subunits and a concomitant loss of motility. The mutations in the rpoN gene did not result in a change in the virulence of V. anguillarum when using the i.p. route. However, virulence was reduced significantly when fish were infected by immersion in water containing bacteria. Complementation of this rpoN mutant with the wild-type rpoN gene of V. anguillarum or Pseudomonas restored flagella production, motility, and virulence in the immersion model. Therefore, it appears that RpoN must be important in the infection of fish by waterborne bacteria (O'Toole et al., 1997). The expression of the virB and virC genes, which are responsible for the production of a major surface antigen located on the outer sheath of the flagellum, is important for the virulence of V. anguillarum (Norqvist and Wolf-Watz, 1993). The product of virC, which is a 51.4-kDa protein with no significant homology to other proteins in GenBank, is another bacterial factor that plays a role in the virulence of V. anguillarum (Milton et al., 1995). However, the mechanisms by which it participates in the pathogenesis of the infections caused in fish remain to be elucidated. The role

of the flagellum and chemotactic motility of V. anguillarum for phagocytosis by and intracellular survival in fish macrophages was recently studied. The results of these studies indicated that the presence of a flagellum did not influence the uptake by the macrophages, but the smooth swimming phenotype of a nonchemotactic mutant increased its intracellular presence (Larsen and Boesen, 2001).

V. anguillarum also produces a 38-kDa major OMP, identical to OmpU from V. cholerae, which may be involved in environmental adaptation (Wang et al., 2002, 2003). Expression of this V. anguillarum ompU gene increases in the presence of bile salts and decreases 50- to 100-fold in a toxR mutant when compared to that in the wild type, demonstrating that ompU expression is positively regulated by ToxR and induced by bile salts. Similar to a toxR mutant, an ompU mutant was characterized by an increased sensitivity to bile salts and by formation of a thicker biofilm with increased surface area coverage as compared to that resulting from the wild-type strain (Wang et al., 2002, 2003). The toxR and toxS genes of V. anguillarum were recently cloned, and the deduced protein sequences were 59 and 67% identical to the V. cholerae ToxR and ToxS proteins, respectively. After performing deletion mutations, it was determined that the toxR mutant was slightly decreased in virulence, suggesting that ToxR may not be a regulator of major virulence factors. One interesting fact is that this toxR mutant was 20% less motile than the wild type and that both the toxR and toxS genes positively regulate the 38-kDa OMP homologue of OmpU (Wang et al., 2002). It is noteworthy that ToxR and to a lesser extent ToxS enhanced resistance to bile, and that bile in the growth medium increased expression of the OmpU homologue but did not affect expression of ToxR. Furthermore, a toxR mutant also forms a better biofilm on a glass surface as compared to the wild type, indicating that ToxR might operate in response to environmental stimuli.

Quorum sensing occurs in V. anguillarum, and this bacterium possesses at least two N-acyl-L-homoserine lactone (AHL) quorum-sensing circuits. One of them is a LuxRI-type quorum-sensing system (VanRI) (Milton et al., 1997) that produces N-(3-oxodecanoyl) homoserine lactone. Despite this, it was of interest that a vanI null mutant still activated AHL biosensors, suggesting the existence of an additional quorum-sensing circuit in this bacterium. Two additional AHLs (N-hexanoylhomoserine lactone and N-[3-hydroxyhexanoyl]homoserine lactone) were identified (Milton et al., 2001). Downstream of vanM is vanN, a homologue of luxN encoding a hybrid sensor kinase that is part of a phosphorelay cascade involved in the regulation of bioluminescence in V. harveyi. Mutations in *vanM* or *vanN* did not affect the production of either proteases or virulence in a fish infection model. These data indicate that *V. anguillarum* possesses a hierarchy of sensing systems with the VanRI homologues of *V. fischeri* LuxRI and the VanMN homologues of *V. harveyi* LuxMN. Recently, Milton's team cloned an additional gene, *vanT*, which codes for a LuxR-like transcriptional regulator (Croxatto et al., 2002). Mutations in *vanT* affect total protease activity as a consequence of a loss of expression of the metalloprotease EmpA, but there are no changes in either AHL production or virulence (Croxatto et al., 2002). Furthermore, use of *lacZ* fusions demonstrated that three additional genes are also affected: *serA*, encoding a 3-phosphoglycerate dehydrogenase that catalyzes the first step in the serine-glycine biosynthesis pathway; *hgdA*, encoding a protein with identity with homogentisate dioxygenases; and *hpdA*, encoding a homologue of 4-hydroxyphenylpyruvate dioxygenases involved in pigment production. It is interesting that certain *V. anguillarum* strains require an active VanT to produce high levels of an L-tyrosine-induced brown color via the 4-hydroxyphenylpyruvate dioxygenase-mediated mechanism, suggesting that VanT must regulate pigment production (Croxatto et al., 2002). A *V. anguillarum vanT* mutant and a mutant carrying a polar mutation in the *sat-vps73* DNA locus (Vps73 and Sat are related to *V. cholerae* proteins encoded within a DNA locus required for biofilm formation) were shown to produce defective biofilms (Croxatto et al., 2002).

V. anguillarum produces and secretes proteases that could play a role in the virulence of this fish pathogen (Farrell and Crosa, 1991; Milton et al., 1992). An elastolytic metalloprotease of 36 kDa, requiring Zn^{2+} for its activity and Ca^{2+} for its stability, was found to be associated with the invasion properties of the *V. anguillarum* NB10 strain. Virulence studies using proteolytic activity mutants showed that they have a 1,000-fold-higher 50% lethal dose (LD_{50}) when assayed by immersion and a 10-fold-higher LD_{50} when assayed by i.p. injection (Nelson et al., 1985a and b). In addition, Farrell and Crosa (1991) purified a metalloprotease from *V. anguillarum* strain 514 that seems to be either related to or identical to that described in strain NB10. The molecular mass of this purified protease was calculated to be 38 kDa. In this case, the protease activity was EDTA sensitive and could be restored by the addition of both Ca^{2+} and Zn^{2+}. Biochemical analysis as well as determination of the N-terminal amino acid sequence showed that the *V. anguillarum* protease(s) is highly related to proteases already described in other *Vibrio* species, such as *V. cholerae* and *V. vulnificus*, as well as to the elastase of *Pseudomonas aeruginosa* and the

protease of *Legionella pneumophila* (Norqvist et al., 1990; Farrell and Crosa, 1991).

The zinc metalloprotease EmpA has also been implicated as a virulence factor in the fish pathogen *V. anguillarum* (Milton et al., 1992; Denkin and Nelson, 2004). EmpA is secreted as a 48-kDa proenzyme that is activated extracellularly by the removal of a 10-kDa peptide (Staroscik et al., 2005). Apparently, expression of *empA* is differentially regulated in certain strains of *V. anguillarum* 93Sm and requires in some of the cases RpoS, quorum-sensing components, and gastrointestinal mucus (Denkin and Nelson, 1999).

V. anguillarum can utilize hemin and hemoglobin as sole iron sources in transport that can be mediated by either the *tonB1* or the *tonB2* systems (Stork et al., 2004). HuvA, the *V. anguillarum* outer membrane heme receptor, was identified by complementation of a heme utilization mutant with a cosmid clone (pML1). This cosmid encodes nine potential heme uptake and utilization proteins: HuvA, the heme receptor; HuvZ and HuvX; TonB1, ExbB1, and ExbD1; and the homologues of HuvB, a periplasmic binding protein; HuvC, an inner membrane permease; and HuvD, an ABC transporter ATPase (Mazoy et al., 2003; Mourino et al., 2004). The gene arrangement of this heme uptake cluster shows some differences in gene arrangement as compared to the homologue clusters described for other *Vibrio* species.

One proven virulence factor for many pathogenic serotype O1 *V. anguillarum* strains, including the prototype strain 775, is a very efficient iron-sequestering system encoded by the 65-kb pJM1 plasmid (Crosa et al., 1977; Crosa, 1984, 1987, 1989; Tolmasky and Crosa, 1990, 1991). This virulence plasmid (Color Plate 12) provides the organism with an iron-sequestering system consisting of the siderophore anguibactin (Actis et al., 1986) and a specific transport complex for the ferric-siderophore (Actis et al., 1988). Anguibactin is synthesized from 2,3-dihydroxybenzoic acid (DHBA), L-cysteine, and N-hydroxyhistamine. The ω-N-hydroxy-ω-N-[2′-(2″,3″-dihydroxyphenyl) thiazolin-4′-yl]carboxy] histamine anguibactin molecule contains, besides the catechol moiety of DHBA, the N-hydroxyl group of hydroxyhistamine. As a consequence, anguibactin can be classified as both a catechol and a hydroxamate siderophore. From the structure of the siderophore and from genetic, biochemical, and in silico studies of biosynthetic genes identified on the pJM1 plasmid from *V. anguillarum* strain 775 (Di Lorenzo et al., 2003), a pathway for anguibactin biosynthesis has been proposed (Fig. 1). Most of the enzymes involved in anguibactin biosynthesis show similarities with domains of nonribosomal peptide synthetases.

The anguibactin precursors DHBA and L-cysteine are activated by the adenylation (A) domains of the

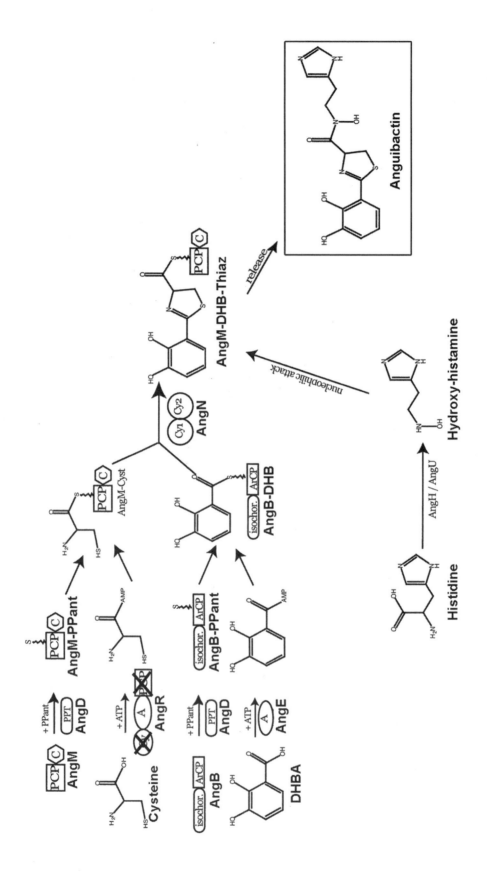

Figure 1. Model of the anguibactin biosynthesis pathway.

AngE protein and the AngR protein, respectively. We have reported that there are two homologues of AngE, one encoded on the pJM1 plasmid and the other encoded on the chromosome (Alice et al., 2005). This is also true for other genes, such as *angC*, *angA*, and the isochorismate lyase domain of *angB*, all of which are essential for DHBA biosynthesis. The next step requires the phosphopantetheinylation of both the ArCP of the AngB/G protein and the peptidyl carrier protein (PCP) domain of the AngM protein (Di Lorenzo et al., 2004). The phosphopantetheinylation enzyme AngD is also encoded as plasmid and chromosomal homologues. Another laboratory has demonstrated that a pJM1-like plasmid also encodes homologues of *angE* and *angD* genes that are functional in vitro, although, contrary to our findings, this plasmidic *angE* is not repressed at high iron concentrations (Liu et al., 2004; Wu et al., 2004). The AngE-activated DHBA is tethered on the phosphopantetheinylate arm of the ArCP of the AngB/G protein, while activated cysteine is tethered to the phosphopantetheinylate arm of the PCP domain of the AngM protein. The AngR protein also possesses a PCP domain on which cysteine could be tethered but the PCP domain of AngR is not functional because an essential serine is replaced by an alanine. The C domain of AngM catalyzes the formation of the peptide bond between DHBA and cysteine. In this or later steps of anguibactin biosynthesis, cysteine is cyclized to a thiazoline ring by AngN, which possesses two Cy groups. Another Cy domain is found in AngR, but this domain is also not likely to be functional since an essential aspartic acid is replaced by asparagine. Next, the dihydroxyphenylthiazoline dipeptide is released from the PCP domain of AngM by nucleofilic attack of N-hydroxyhistamine, resulting in the free anguibactin molecule. N-hydroxyhistamine is obtained by modification of histidine catalyzed by two tailoring enzymes, the mono-oxygenase AngU and the histidine decarboxylase AngH.

Ferric-anguibactin is transported to the cells' cytosol by the FatABCD complex. The genes encoding these proteins are located on the virulence plasmid and transcribed as one polycistronic message with the biosynthesis genes *angR* and *angT* (iron transport biosynthesis [ITB] operon) (Color Plate 12). The ferric-anguibactin receptor FatA is an 86-kDa protein that is essential for anguibactin transport. The FatA amino acid sequence is similar to other receptors involved in iron transport, e.g., FhuA and FepA of *E. coli*, and a TonB box can be identified at its amino-terminal end. This TonB box is one of the interaction sites between the outer membrane receptor and the energy transducer TonB. The chromosomes of *V. anguillarum*

harbor two *tonB* systems: *tonB1* and *tonB2* (Stork et al., 2004). A mutant in *tonB2* is unable to transport this siderophore, indicating that uptake of anguibactin requires only the TonB2 protein. Once in the periplasm, the ferric-anguibactin is bound by the periplasmic binding protein FatB, a 35-kDa lipoprotein that is anchored in the inner membrane. FatB shuttles ferric-anguibactin to the permeases FatC and FatD in the inner membrane that internalize the ferric-anguibactin complex to the cytoplasm, using the energy generated by ATP hydrolysis. The ABC transporters necessary for ATP hydrolysis are not yet identified in *V. anguillarum*.

In *V. anguillarum* the siderophore-mediated iron uptake system is negatively regulated by the chromosomally encoded Fur. It has been shown that, upon binding at the Fur box of the ITB operon promoter, Fur bends the DNA-blocking RNA polymerase from binding. Two positive regulators, the AngR protein and products encoded in the transacting factor region, were shown to act at the same promoter. AngR is a bifunctional protein that besides its role in regulation is also involved in the biosynthesis of anguibactin. The transacting factor product is encoded in a region of the virulence plasmid noncontiguous to the ITB operon. Besides the positive regulators AngR and TAF, the ITB operon promoter is also positively regulated by the siderophore anguibactin itself. Curiously, within this operon, the ratio of the expression between the *fat* and the *ang* genes is ~25:1. Evidence exists that the reduced relative expression of the *ang* genes within this operon is due to regulation by an antisense RNA, RNAβ.

Many *V. anguillarum* strains isolated from various geographical regions carry plasmids highly related, albeit not identical, to pJM1 (Tolmasky et al., 1985, 1988b). Some of those pJM1-like plasmids, chiefly those present in strains isolated from the Atlantic Ocean, specified an increased anguibactin production as compared to the pJM1-mediated anguibactin levels found with *V. anguillarum* strains isolated from the Pacific Ocean (Tolmasky et al., 1988b). The *angR* gene of one of these pJM1-like plasmids has recently been identified as the element responsible for the enhanced biosynthesis of anguibactin in the strain 531A (Salinas et al., 1989; Tolmasky et al., 1988b, 1993).

More recently, polymorphisms in pJM1-like plasmids were also observed in several other *V. anguillarum* strains belonging to serotype O1 (Wiik et al., 1989; Olsen and Larsen, 1990). Several pathogenic plasmidless *V. anguillarum* strains belonging to serotype O1 and O2 have recently been isolated. All these strains were shown to possess a chromosomally encoded iron-uptake system (Lemos et al., 1988; Conchas et al., 1991). DNA hybridization experiments showed that this system is not related to that encoded by pJM1. Biochemical characterization of the chromosomally encoded siderophore demonstrated that it belongs to the family of the catechols.

The opportunistic human pathogen *Acinetobacter baumannii* 19606, which causes severe infections in compromised patients (Yamamoto et al., 1994; Villegas and Hartstein, 2003), produces acinetobactin (Yamamoto et al., 1994). Interestingly, this catechol siderophore is highly related to anguibactin. The only difference between these two iron-scavenging compounds is that anguibactin contains cysteine while threonine is the amino acid found in acinetobactin as a functional group. Further analysis using siderophore utilization bioassays showed that acinetobactin enhances the growth of *V. anguillarum* under iron-stress conditions (Dorsey et al., 2004). Random insertion mutagenesis of *A. baumannii* 19606 led to the identification of a gene encoding a predicted OMP named BauA, highly similar to FatA, the receptor for ferric anguibactin. Western blot analysis and siderophore utilization assays proved that BauA and FatA are immunologically and functionally related, the expression of which is required for both pathogens to use either acinetobactin or anguibactin under iron-limiting conditions (Dorsey et al., 2004). Further analysis showed that *bauA* is part of a polycistronic locus that includes the *bauDCEB* coding regions, which has the same arrangement of the *fatD-CBA* locus with the exception of the presence of *bauE*. The product of the latter gene is related to the ATPase component of gram-positive ATP-binding cassette (ABC) transport systems. This entire locus is flanked by genes encoding predicted proteins related to AngU and AngN, which are *V. anguillarum* proteins required for the biosynthesis of anguibactin. Furthermore, these genes are part of an *A. baumannii* 19606 chromosomal region containing acinetobactin biosynthetic and transport genes, which code for proteins related to the anguibactin utilization system (Mihara et al., 2004). Taken together, these results demonstrate that these two pathogens, which cause serious infections in unrelated hosts, express very similar siderophore-mediated iron-acquisition systems. Transformation, conjugation, and transposition occur in *Acinetobacter* (Vivian, 1991), and the pJM1 iron-uptake genes are flanked by insertion sequences in a composite transposon-like structure (Tolmasky and Crosa, 1995). Furthermore, *Acinetobacter* is a component of the bacterial flora of salmonid fishes (Cahill, 1990). Thus, plasmid conjugation, transformation, and/or transposition might have played a role in the transmission of these essential genes between these two unrelated bacterial strains present in the microbial flora of salmon and trout.

V. ORDALII

V. ordalii causes vibriosis in wild and cultured marine salmonids in the Pacific Northwest of the United States and Japan. This gram-negative curved rod with a polar flagellum has been reported as *Vibrio* sp. 1669 (Harrell et al., 1976), *Vibrio* sp. RT (Ohnishi and Muroga, 1976), *V. anguillarum* biotype 2 (Schiewe et al., 1977), *B. anguillara* biotype II (Baumann et al., 1978) but later amended to *V. anguillarum* biotype II and *V. anguillarum* phenon II, until its taxonomical status was clarified by Schiewe et al. (1981). DNA analysis of a number of strains showed that the mol% G+C content of this organism is 43 to 44% and demonstrated that isolates of *V. ordalii* have more than 83% homology within the group, while showing only 58 to 69% homology with *V. anguillarum* (Schiewe et al., 1981). There is little or no homology with DNA isolated from *V. parahaemolyticus*, *V. alginolyticus*, and other unclassified vibrios. *V. ordalii* is also known as *Listonella ordalii* owing to the similarity of its 5S rRNA nucleotide sequence to that of other members of *Listonella* (Pillidge and Colwell, 1988). More recently, the fish-pathogenic vibrios were reclassified using ribotyping and restriction analysis. These more current approaches showed that North American strains of *V. ordalii* share restriction profiles with isolates obtained in Australia and Japan, although each of the geographical groups has its own distinct profile (Pedersen et al., 1996). Comparative 16S rRNA sequence analysis (Wiik et al., 1995) proved to be a suitable tool for the classification of fish-pathogenic vibrios. Overall, the evolutionary relationships described with this approach and the previous taxonomy of this bacterial family are in agreement. However, the observation that two *V. ordalii* strains were almost indistinguishable from the *V. anguillarum* NCMB 6 type strain shows the need to validate the classification of *V. ordalii* as it was defined initially by Schiewe et al. (1981).

Identification and Classification

V. ordalii has been isolated from natural infections in coho salmon (*Oncorhynchus kisutch*) as well as from experimental infections in chum and spring chinook salmon (*Oncorhynchus tshawytscha*) (Ransom et al., 1984). It grows at 30°C in trypticase soy broth or brain heart infusion with the addition of 1% (wt/vol) NaCl. The biochemical characteristics of *V. ordalii* have been described before (Schiewe, 1983); the biochemical tests frequently used to distinguish between *V. ordalii* and *V. anguillarum* are listed in Table 1. These two fish pathogens are also distinguished by the antigenic differences of their

LPS, which can be detected with specific monoclonal or polyclonal antibodies that can be used as diagnostic and serotyping reagents (Mutharia et al., 1993; Mutharia and Amor, 1994). The structure of the LPS O antigen and the capsular polysaccharide of *V. ordalii* explains the antigenic similarities as well as the differences between this pathogen and *V. anguillarum* (Sadovskaya et al., 1998).

The analysis of a collection of isolates showed the presence of a 30-kb cryptic plasmid in all strains examined (Schiewe and Crosa, 1981). This extrachromosomal element, which was named pMJ101, is unrelated to the pJM1 virulence plasmid present in *V. anguillarum* 775 and does not share significant homology with plasmids belonging to already defined incompatibility groups. These initial observations were confirmed by more recent studies (Tiainen et al., 1995; Pedersen et al., 1996), one of which showed that, although highly conserved, the pMJ101-like plasmids isolated from different strains have variations in their restriction patterns (Pedersen et al., 1996). It has been suggested that these variations together with ribotyping, which showed three *V. ordalii* ribotypes clearly different from those of *V. anguillarum*, could be useful in epidemiological studies of *V. ordalii* (Pedersen et al., 1996).

Restriction mapping and cloning and transformation experiments in combination with Tn5 transposition mutagenesis localized the essential pMJ101 replication functions within a 2.4-kb EcoRV-HindIII restriction fragment (Bidinost et al., 1994). Recombinant clones carrying this fragment replicate in *E. coli* cells deficient in DNA polymerase I or integration host factors. However, the replication of plasmid derivatives containing the pMJ101 *ori* region were highly dependent on the presence of a functional dam methylase, suggesting that methylation is a requirement for the replication of this plasmid. The ability of pMJ101 derivatives to replicate in the absence of polymerase I suggested that this plasmid encodes its own replication protein. Electrophoretic analysis of radiolabeled plasmid-encoded proteins showed that a 36-kDa protein is encoded within the replication region and is essential for the replication of pMJ101. The expression of this protein was correlated with the ability of different recombinant plasmids harboring this pMJ101 DNA region to replicate in an *E. coli* PolA-deficient strain. Replication typing showed that pMJ101 is not related to any of the plasmid incompatibility groups contained in the bank of *rep* probes. Furthermore, a recombinant derivative harboring the pMJ101 replication region proved to be compatible with the *V. anguillarum* 775 pJM1 plasmid.

Sequence analysis of a 1.56-kb fragment harboring the pMJ101 replication region (Bidinost et al., 1999) revealed the presence of an AT-rich region, 11 *dam*-methylation sites, of which 5 are within the putative *ori* region, and 5 copies of the 9-bp consensus sequence for DnaA binding (Fig. 2). Gel retardation experiments showed that the latter replication element indeed binds DnaA purified from *E. coli*. A potential open reading frame (ORF) encoding a hydrophilic protein with a predicted pI of 10.3 and an M_r of 33,826 Da was found adjacent to the *ori* region. Although these properties are typical of DNA-binding proteins, no significant homology was found between this predicted protein, named RepM, and other previously characterized proteins. Reverse transcription-PCR analysis of total RNA demonstrated the presence of *repM* mRNA in *V. ordalii*. The major initiation site of this mRNA was located 187 nucleotides upstream of the GTG initiation codon. This transcription initiation site is preceded by putative -10 and -35 promoter sequences that control the

Table 1. Phenotypic properties used to differentiate *V. anguillarum* from *V. ordalii*[a]

Biochemical tests and growth temp	Result for:	
	V. anguillarum	*V. ordalii*
Arginine-alkaline reaction	+	−
Citrate, Christensen	+	−
Citrate, Simmons	+	−
Lipase	+	−
o-Nitrophenyl-β-D-galactopyranoside	+	−
Starch hydrolysis	+	−
Voges-Proskauer reaction	+	−
Acid production from:		
Cellobiose	+	−
Glycerol	+	−
Sorbitol	+	−
Trehalose	+	−
Growth at 37°C	+	−

[a]Data taken from Schiewe et al. (1981).

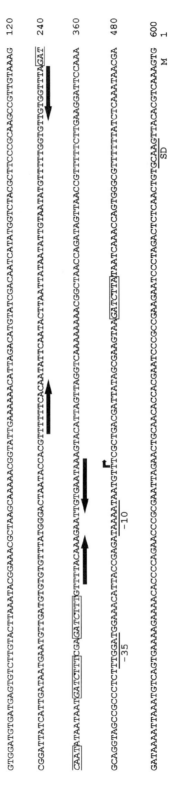

Figure 2. Nucleotide sequence of the pMJ101 origin of replication. The locations of the predicted Shine-Dalgarno (SD) and −10 and −35 promoter elements are indicated by underlined nucleotides. The bent arrow identifies the main transcription initiation site. Predicted methylation sites (GATC) are indicated in italics, and the conserved 7-bp repeats containing them are shown within the open boxes. The position and orientation of putative DnaA boxes are indicated by the horizontal arrows. The first amino acid of RepM is shown below the last nucleotide triplet.

expression of the *repM* replication gene (Fig. 2). Southern blot analysis of denatured and nondenatured DNA indicated that pMJ101 replicates via a theta replication mechanism as is normally found in most gram-negative bacteria. Despite these analyses, the role of pMJ101 in the *V. ordalii* life cycle and/or virulence remains unknown. Therefore, the isolation of a plasmidless derivative will be an invaluable tool to assess the function of this universal plasmid present in *V. ordalii* isolates. Alternatively, the determination and analysis of the complete nucleotide sequence could provide information about the role of pMJ101 in the pathobiology of this fish pathogen.

The Disease

Studies in naturally acquired vibriosis in chum salmon showed that *V. ordalii* localizes more frequently in the muscle and skin, with bacterial colonies or aggregations that can replace large areas of host tissues—a tissue tropism different from that displayed by *V. anguillarum* (Schiewe, 1983; Ransom et al., 1984). This bacterium also causes the necrosis and hemorrhaging of the surrounding tissues in some of the infected places. These findings indicate that *V. ordalii* can enter the host by invasion of the salmonid integument. Bacterial colonies are commonly found in loose connective tissues, the gills, throughout the digestive tract, and in the pyloric caeca, suggesting that the infection could also begin at these sites. Occasionally, *V. ordalii* can be observed as microcolonies in spleen and liver, and low counts in blood may be observed during the initial stages of the infection. A similar pathology was observed when either chum, coho, or chinook salmons were exposed to a large number of bacterial cells in experimental waterborne infections. Juvenile salmon exposed to *V. ordalii* by parenteral challenge developed a systemic infection, and the bacterium was recovered from liver, kidneys, spleen, and blood immediately after the infection (Schiewe, 1983). However, the number of bacteria in the liver declined after 1 h and then increased 22 h after infection; bacterial numbers were high in all the organs, and 100% mortality occurred 6 days after infection. The latter observations demonstrate that artificial infection of juvenile salmon is a valid experimental model to study the mechanisms and the bacterial virulence factors involved in the pathogenesis of the infections caused by *V. ordalii*.

Virulence Factors

There is a correlation between the presence of *V. ordalii* in the host blood and a marked decrease in white blood cell counts in moribund fish during the progression of the infection (Harbell et al., 1979; Ransom et al., 1984). This observation suggests the production of a leukocytolytic factor, which may play an important role in the pathogenesis of the infection. Other factors can also play an important role in pathogenesis, such as the ability of this bacterium to resist the bactericidal activity of normal nonimmune rainbow trout serum, which has been correlated with the virulence of this bacterium (Trust et al., 1981). *V. ordalii* strains agglutinate trout and human erythrocytes as well as yeast cells, although some discrepancies were reported, probably due to technical artifacts (Trust et al., 1981; Larsen and Mellergaard, 1984). The ability of *V. ordalii* to agglutinate different eukaryotic cell types suggests that it could attach to and interact with the host cells. However, a correlation between the agglutination phenotype and the virulence properties of the strains studied could not be obtained. The potential role of these agglutinins during the onset and progress of infections in salmonids remains to be tested.

REFERENCES

Acosta, F., K. Lockhart, S. K. Gahlawat, F. Real, and A. E. Ellis. 2004. Mx expression in Atlantic salmon (*Salmo salar* L.) parr in response to *Listonella anguillarum* bacterin, lipopolysaccharide and chromosomal DNA. *Fish Shellfish Immunol.* 17:255–263.

Actis, L. A., M. E. Tolmasky, and J. H. Crosa. 1998. Vibriosis, p. 523–557. *In* P. T. K. Woo and D. W. Bruno (ed.), *Fish Diseases and Disorders, vol. 3: Viral, Bacterial, and Fungal Infections.* CAB International, Wallingford, United Kingdom.

Actis, L. A., W. Fish, J. H. Crosa, K. Kellerman, S. R. Ellenberger, F. M. Hauser, and J. Sanders-Loehr. 1986. Characterization of anguibactin, a novel siderophore from *Vibrio anguillarum* 775(pJM1). *J. Bacteriol.* 167:57–65.

Actis, L. A., M. E. Tolmasky, D. H. Farrell, and J. H. Crosa. 1988. Genetic and molecular characterization of essential components of the *Vibrio anguillarum* plasmid-mediated iron-transport system. *J. Biol. Chem.* 263:2853–2860.

Agius, C., M. Horne, and P. Ward. 1983. Immunization of rainbow trout, *Salmo gairdneri* Richardson, against vibriosis: comparison of an extract antigen with whole cell bacterins by oral and intraperitoneal routes. *J. Fish Dis.* 6:129–134.

Alice, A. F., C. S. Lopez, and J. H. Crosa. 2005. Plasmid- and chromosome-encoded redundant and specific functions are involved in biosynthesis of the siderophore anguibactin in *Vibrio anguillarum* 775: a case of chance and necessity? *J. Bacteriol.* 187:2209–2214.

Alsina, M., J. Martinez-Picado, J. Jofre, and A. R. Blanch. 1994. A medium for presumptive identification of *Vibrio anguillarum. Appl. Environ. Microbiol.* 60:1681–1683.

Aoki, T., Y. Jo, and S. Egusa. 1980. Frequent occurrence of drug resistance bacteria in ayu (*Plecoglossus altivelis*). *Fish Pathol.* 15:1–6.

Aoki, T., T. Kitao, and T. Arai. 1977. R plasmids in fish pathogens, p. 39–45. *In* S. Mitsuhashi, L. Rosival, and W. Kremery (ed.), *Plasmids: Medical and Theoretical Aspects.* Czechoslovak Medical Press, Prague.

Aoki, T., T. Kitao, T. Itabashi, Y. Wada, and M. Sakai. 1981. Proteins and lipopolysaccharides in the membrane of *Vibrio anguillarum. Dev. Biol. Stand.* 49:226–232.

Aoki, T., T. Kitao, S. Watanabe, and S. Takeshita. 1984. Drug resistance and R plasmids in *Vibrio anguillarum* isolated in cultured ayu (*Plecoglossus altivelis*). *Microbiol. Immunol.* **28**:1–9.

Baumann, P., S. Bang, and L. Baumann. 1978. Phenotypic characterization of *Beneckea anguillara* biotypes I and II. *Curr. Microbiol.* **1**:85–88.

Bennett, P. M., and P. M. Hawkey. 1991. The future contribution of transposition to antimicrobial resistance. *J. Hosp. Infect.* **18** (Suppl. A):211–221.

Bergh, O., L. Vikanes, P. Makridis, J. Skjermo, D. Knappskog, and O. M. Rodseth. 2001. Uptake and processing of a *Vibrio anguillarum* bacterin in *Artemia franciscana* measured by ELISA and immunohistochemistry. *Fish Shellfish Immunol.* **11**:15–22.

Bidinost, C., J. H. Crosa, and L. A. Actis. 1994. Localization of the replication region of the pMJ101 plasmid from *Vibrio ordalii*. *Plasmid* **31**:242–250.

Bidinost, C., P. J. Wilderman, C. W. Dorsey, and L. A. Actis. 1999. Analysis of the replication elements of the pMJ101 plasmid from the fish pathogen *Vibrio ordalii*. *Plasmid* **42**:20–30.

Bolinches, J., M. Lemos, B. Fouz, A. Toranzo, M. Cambra, and J. L. Larsen. 1990. Serological relationships within *Vibrio anguillarum* strains. *J. Aquat. Anim. Health* **2**:12–20.

Bowden, T. J., D. Menoyo-Luque, I. R. Bricknell, and H. Wergeland. 2002. Efficacy of different administration routes for vaccination against *Vibrio anguillarum* in Atlantic halibut (*Hippoglossus hippoglossus* L.). *Fish Shellfish Immunol.* **12**:283–285.

Cahill, M. M. 1990. Bacterial flora of fishes: a review. *Microb. Ecol.* **19**:21–41.

Chen, D., P. J. Hanna, K. Altmann, A. Smith, P. Moon, and L. S. Hammond. 1992. Development of monoclonal antibodies that identify *Vibrio* species commonly isolated from infections of humans, fish, and shellfish. *Appl. Environ. Microbiol.* **58**:3694–3700.

Cisar, J. O., and J. L. Fryer. 1969. An epizootic of vibriosis in Chinook salmon. *Bull. Wildlife Dis. Assoc.* **5**:73–75.

Conchas, R. F., M. L. Lemos, J. L. Barja, and A. E. Toranzo. 1991. Distribution of plasmid- and chromosome-mediated iron uptake systems in *Vibrio anguillarum* strains of different origins. *Appl. Environ. Microbiol.* **57**:2956–2962.

Crosa, J. H. 1987. Bacterial iron metabolism, plasmids and other virulence factors, p. 139–170. *In* J. Bullen and E. Griffiths (ed.), *Iron and Infection*. John Wiley and Sons, Chichester, England.

Crosa, J. H. 1989. Genetics and molecular biology of siderophore-mediated iron transport in bacteria. *Microbiol. Rev.* **53**:517–530.

Crosa, J. H. 1984. The relationship of plasmid-mediated iron transport and bacterial virulence. *Annu. Rev. Microbiol.* **38**:69–89.

Crosa, J. H., L. Hodges, and M. Schiewe. 1980. Curing of a plasmid is correlated with an attenuation of virulence in the marine fish pathogen *Vibrio anguillarum*. *Infect. Immun.* **27**:897–902.

Crosa, J. H., M. Schiewe, and S. Falkow. 1977. Evidence for plasmid contribution to the virulence of the fish pathogen *Vibrio anguillarum*. *Infect. Immun.* **18**:509–513.

Croxatto, A., V. J. Chalker, J. Lauritz, J. Jass, A. Hardman, P. Williams, M. Camara, and D. L. Milton. 2002. VanT, a homologue of *Vibrio harveyi* LuxR, regulates serine, metalloprotease, pigment, and biofilm production in *Vibrio anguillarum*. *J. Bacteriol.* **184**:1617–1629.

Denkin, S. M., and D. R. Nelson. 1999. Induction of protease activity in *Vibrio anguillarum* by gastrointestinal mucus. *Appl. Environ. Microbiol.* **65**:3555–3560.

Denkin, S. M., and D. R. Nelson. 2004. Regulation of *Vibrio anguillarum empA* metalloprotease expression and its role in virulence. *Appl. Environ. Microbiol.* **70**:4193–4204.

Di Lorenzo, M., M. Stork, M. E. Tolmasky, L. A. Actis, D. Farrell, T. J. Welch, L. M. Crosa, A. M. Wertheimer, Q. Chen, P. Salinas, L. Waldbeser, and J. H. Crosa. 2003. Complete sequence of

virulence plasmid pJM1 from the marine fish pathogen *Vibrio anguillarum* strain 775. *J. Bacteriol.* **185**:5822–5830.

Di Lorenzo, M., S. Poppelaars, M. Stork, M. Nagasawa, M. E. Tolmasky, and J. H. Crosa. 2004. A nonribosomal peptide synthetase with a novel domain organization is essential for siderophore biosynthesis in *Vibrio anguillarum*. *J. Bacteriol.* **186**:7327–7336.

Dorsey, C. W., A. P. Tomaras, P. L. Connerly, M. E. Tolmasky, J. H. Crosa, and L. A. Actis. 2004. The siderophore-mediated iron acquisition systems of *Acinetobacter baumannii* ATCC 19606 and *Vibrio anguillarum* 775 are structurally and functionally related. *Microbiology* **150**:3657–3667.

Farrell, D. H., and J. H. Crosa. 1991. Purification and characterization of a secreted protease from the pathogenic marine bacterium *Vibrio anguillarum*. *Biochemistry* **30**:3432–3436.

Gonzalez, S. F., M. J. Krug, M. E. Nielsen, Y. Santos, and D. R. Call. 2004. Simultaneous detection of marine fish pathogens by using multiplex PCR and a DNA microarray. *J. Clin. Microbiol.* **42**:1414–1419.

Harbell, S., O. Hodgins, and M. Schiewe. 1979. Studies on the pathogenesis of vibriosis in coho salmon *Oncorhynchus kisutch* (Walbaum). *J. Fish Dis.* **2**:391–404.

Harrell, L., A. J. Novotny, M. H. Schiewe, and H. Hodgins. 1976. Isolation and description of two vibrios pathogenic to Pacific salmon in Puget Sound, Washington. *Fish. Bull.* **74**:447–449.

Hayashi, F., B. Araki, K. Harada, M. Inove, and S. Mitsuhashi. 1982. Epidemiological studies of drug resistant strains in cultured in cultured fish and water. *Bull. Jpn. Soc. Sci. Fish.* **48**:1121–1127.

Hoff, K. A. 1989. Survival of *Vibrio anguillarum* and *Vibrio salmonicida* at different salinities. *Appl. Environ. Microbiol.* **55**:1775–1786.

Johnsen, G. S. 1977. Immunological studies on *Vibrio anguillarum*. *Aquaculture* **10**:212–230.

Kawano, K., T. Aoki, and T. Kitao. 1984. Duration of protection against vibriosis in ayu, *Plecoglossus altivelis*, vaccinated by immersion and oral administration with *Vibrio anguillarum*. *Bull. Jpn. Soc. Sci. Fish.* **50**:771–774.

Kitao, T., T. Aoki, K. Fukudome, K. Kawano, Y. Wada, and Y. Mizuno. 1983. Serotyping of *Vibrio anguillarum* isolated from diseased freshwater fish in Japan. *J. Fish Dis.* **6**:175–181.

Kleckner, N. 1981. Transposable elements in prokaryotes. *Annu. Rev. Genet.* **15**:341–404.

Larsen, J., and S. Mellegaard. 1984. Agglutination typing of *Vibrio anguillarum* isolates from diseased fish and from the environment. *Appl. Environ. Microbiol.* **47**:1261–1265.

Larsen, J. L. 1983. *Vibrio anguillarum*: a comparative study of fish pathogenic, environmental, and reference strains. *Acta Vet. Scand.* **24**:456–479.

Larsen, M. H., and H. T. Boesen. 2001. Role of flagellum and chemotactic motility of *Vibrio anguillarum* for phagocytosis by and intracellular survival in fish macrophages. *FEMS Microbiol. Lett.* **203**:149–152.

Lemos, M., A. Toranzo, and J. Barja. 1985. Antibiotic activity of epiphytic bacteria isolated from intertidal seaweeds. *Microbiol. Ecol.* **11**:149–163.

Lemos, M. L., P. Salinas, A. E. Toranzo, J. L. Barja, and J. H. Crosa. 1988. Chromosome-mediated iron uptake system in pathogenic strains of *Vibrio anguillarum*. *J. Bacteriol.* **170**:1920–1925.

Liu, Q., Y. Ma, H. Wu, M. Shao, H. Liu, and Y. Zhang. 2004. Cloning, identification and expression of an *entE* homologue *angE* from *Vibrio anguillarum* serotype O1. *Arch. Microbiol.* **181**:287–293.

MacDonell, M., and R. R. Colwell. 1985. Phylogeny of the Vibrionaceae, and recommendation for two new genera, *Listonella* and *Shewanella*. *Syst. Appl. Microbiol.* **6**:171–182.

Martinez-Picado, J., A. R. Blanch, and J. Jofre. 1994. Rapid detection and identification of *Vibrio anguillarum* by using a specific oligonucleotide probe complementary to 16S rRNA. *Appl. Environ. Microbiol.* **60:**732–737.

Mazoy, R., C. R. Osorio, A. E. Toranzo, and M. L. Lemos. 2003. Isolation of mutants of *Vibrio anguillarum* defective in haeme utilisation and cloning of *huvA*, a gene coding for an outer membrane protein involved in the use of haeme as iron source. *Arch. Microbiol.* **179:**329–338.

McGee, K., P. Horstedt, and D. L. Milton. 1996. Identification and characterization of additional flagellin genes from *Vibrio anguillarum*. *J. Bacteriol.* **178:**5188–5198.

Mihara, K., T. Tanabe, Y. Yamakawa, T. Funahashi, H. Nakao, S. Narimatsu, and S. Yamamoto. 2004. Identification and transcriptional organization of a gene cluster involved in biosynthesis and transport of acinetobactin, a siderophore produced by *Acinetobacter baumannii* ATCC 19606T. *Microbiology* **150:**2587–2597.

Milton, D. L., V. J. Chalker, D. Kirke, A. Hardman, M. Camara, and P. Williams. 2001. The LuxM homologue VanM from *Vibrio anguillarum* directs the synthesis of N-(3-hydroxyhexanoyl) homoserine lactone and N-hexanoylhomoserine lactone. *J. Bacteriol.* **183:**3537–3547.

Milton, D. L., A. Hardman, M. Camara, S. R. Chhabra, B. W. Bycroft, G. S. Stewart, and P. Williams. 1997. Quorum sensing in *Vibrio anguillarum*: characterization of the *vanI/vanR* locus and identification of the autoinducer N-(3-oxodecanoyl)-L-homoserine lactone. *J. Bacteriol.* **179:**3004–3012.

Milton, D. L., A. Norqvist, and H. Wolf-Watz. 1992. Cloning of a metalloprotease gene involved in the virulence mechanism of *Vibrio anguillarum*. *J. Bacteriol.* **174:**7235–7244.

Milton, D. L., A. Norqvist, and H. Wolf-Watz. 1995. Sequence of a novel virulence-mediating gene, *virC*, from *Vibrio anguillarum*. *Gene* **164:**95–100.

Milton, D. L., R. O'Toole, P. Horstedt, and H. Wolf-Watz. 1996. Flagellin A is essential for the virulence of *Vibrio anguillarum*. *J. Bacteriol.* **178:**1310–1319.

Mourino, S., C. R. Osorio, and M. L. Lemos. 2004. Characterization of heme uptake cluster genes in the fish pathogen *Vibrio anguillarum*. *J. Bacteriol.* **186:**6159–6167.

Mutharia, L. M., and P. A. Amor. 1994. Monoclonal antibodies against *Vibrio anguillarum* O2 and *Vibrio ordalii* identify antigenic differences in lipopolysaccharide O-antigens. *FEMS Microbiol. Lett.* **123:**289–298.

Mutharia, L. W., B. T. Raymond, T. R. Dekievit, and R. M. Stevenson. 1993. Antibody specificities of polyclonal rabbit and rainbow trout antisera against *Vibrio ordalii* and serotype 0:2 strains of *Vibrio anguillarum*. *Can. J. Microbiol.* **39:**492–499.

Nelson, J., J. S. Rohovec, and J. L. Fryer. 1985a. Location of *Vibrio anguillarum* in tissues of infected rainbow trout (*Salmo gairdneri*) using the fluorescent antibody technique. *Fish Pathol.* **20:**229–235.

Nelson, J., J. S. Rohovec, and J. L. Fryer. 1985b. Tissue location of *Vibrio* bacterin delivered by intraperitoneal injection, immersion and oral routes to *Salmo gairdneri*. *Fish Pathol.* **19:**263–269.

Nomura, J., and T. Aoki. 1985. Morphological analysis of lipopolysaccharide from gram-negative fish pathogenic bacteria. *Fish Pathol.* **20:**193–197.

Norqvist, A., A. Hagstrom, and H. Wolf-Watz. 1989. Protection of rainbow trout against vibriosis and furunculosis by the use of attenuated strains of *Vibrio anguillarum*. *Appl. Environ. Microbiol.* **55:**1400–1405.

Norqvist, A., B. Norrman, and H. Wolf-Watz. 1990. Identification and characterization of a zinc metalloprotease associated with invasion by the fish pathogen *Vibrio anguillarum*. *Infect. Immun.* **58:**3731–3736.

Norqvist, A., and H. Wolf-Watz. 1993. Characterization of a novel chromosomal virulence locus involved in expression of a major surface flagellar sheath antigen of the fish pathogen *Vibrio anguillarum*. *Infect. Immun.* **61:**2434–2444.

Ohnishi, K., and K. Muroga. 1976. *Vibrio* sp. as a cause of disease in rainbow trout cultured in Japan. I. Biochemical characteristics. *Fish Pathol.* **11:**159–165.

Olsen, J. E., and J. L. Larsen. 1990. Restriction fragment length polymorphism of the *Vibrio anguillarum* serovar O1 virulence plasmid. *Appl. Environ. Microbiol.* **56:**3130–3132.

Onarheim, A., and J. Raa. 1990. Characteristics and possible biological significance of an autochthonous flora in the intestinal mucosa of sea water fish, p. 197–201. *In* R. Lesel (ed.), *Microbiology in Poecilotherms*. Elsevier Science Publishers, Amsterdam, The Netherlands.

O'Toole, R., D. L. Milton, P. Horstedt, and H. Wolf-Watz. 1997. RpoN of the fish pathogen *Vibrio* (*Listonella*) *anguillarum* is essential for flagellum production and virulence by the waterborne but not intraperitoneal route of inoculation. *Microbiology* **143:**3849–3859.

O'Toole, R., J. Von Hofsten, R. Rosqvist, P. E. Olsson, and H. Wolf-Watz. 2004. Visualisation of zebrafish infection by GFP-labelled *Vibrio anguillarum*. *Microb. Pathog.* **37:**41–46.

Pacha, R. E., and E. D. Kiehn. 1969. Characterization and relatedness of marine vibrios pathogenic to fish: physiology, serology, and epidemiology. *J. Bacteriol.* **100:**1242–1247.

Pedersen, K., L. Grisez, R. van Houdt, T. Tiainen, F. Ollevier, and J. L. Larsen. 1999. Extended serotyping scheme for *Vibrio anguillarum* with the definition and characterization of seven provisional O-serogroups. *Curr. Microbiol.* **38:**183–189.

Pedersen, K., S. Koblavi, T. Tiainen, and P. A. Grimont. 1996. Restriction fragment length polymorphism of the pMJ101-like plasmid and ribotyping in the fish pathogen *Vibrio ordalii*. *Epidemiol. Infect.* **117:**385–391.

Pedersen, K., T. Tiainen, and J. L. Larsen. 1995. Antibiotic resistance of *Vibrio anguillarum*, in relation to serovar and plasmid contents. *Acta. Vet. Scand.* **36:**55–64.

Pillidge, C. J., and R. R. Colwell. 1988. Nucleotide sequence of the 5S rRNA from *Listonella* (*Vibrio*) *ordalii* ATCC 33509 and *Listonella* (*Vibrio*) *tubiashii* ATCC 19105. *Nucleic Acids Res.* **16:**3111.

Ransom, D. P., C. N. Lannan, J. S. Rohovec, and J. L. Fryer. 1984. Comparison of histopathology caused by *Vibrio anguillarum* and *Vibrio ordalii* in three species of Pacific salmon. *J. Fish Dis.* **7:**107–115.

Rasmussen, H. B. 1987a. Subgrouping of lipopolysaccharide O antigens from *Vibrio anguillarum* serogroup O2 by immunoelectrophoretic analyses. *Curr. Microbiol.* **16:**39–42.

Rasmussen, H. B. 1987b. Evidence for two new *Vibrio anguillarum* K antigens. *Curr. Microbiol.* **16:**105–107.

Rehnstam, A. S., A. Norqvist, H. Wolf-Watz, and A. Hagstrom. 1989. Identification of *Vibrio anguillarum* in fish by using partial 16S rRNA sequences and a specific 16S rRNA oligonucleotide probe. *Appl. Environ. Microbiol.* **55:**1907–1910.

Roberts, R. 1989. The bacteriology of teleosts, p. 289–319. *In* R. Roberts (ed.), *Fish Pathology*, 2nd ed. Bailliere Tindall, London, England.

Sadovskaya, I., J. R. Brisson, N. H. Khieu, L. M. Mutharia, and E. Altman. 1998. Structural characterization of the lipopolysaccharide O-antigen and capsular polysaccharide of *Vibrio ordalii* serotype O:2. *Eur. J. Biochem.* **253:**319–327.

Salinas, P. C., M. E. Tolmasky, and J. H. Crosa. 1989. Regulation of the iron uptake system in *Vibrio anguillarum*: evidence for a cooperative effect between two transcriptional activators. *Proc. Natl. Acad. Sci. USA* **86:**3529–3533.

Schiewe, M. H. 1983. *Vibrio ordalii* as a cause of vibriosis in salmonid fish, p. 31–40. *In* J. H. Crosa (ed.), *Bacterial and Viral Diseases of Fish*. Washington Sea Grant, Seattle, Wash.

Schiewe, M. H., and J. H. Crosa. 1981. Molecular characterization of *Vibrio anguillarum* biotype 2. *Can. J. Microbiol.* **27:**1011–1018.

Schiewe, M. H., J. H. Crosa, and E. Ordal. 1977. Deoxyribonucleic acid relationships among marine vibrios pathogenic to fish. *Can. J. Microbiol.* **23:**954–958.

Schiewe, M. H., T. Trust, and J. H. Crosa. 1981. *Vibrio ordalii* sp. nov.: a causative agent of vibriosis in fish. *Curr. Microbiol.* **6:**343–348.

Shewan, J., W. Hodgkiss, and J. Liston. 1954. A method for the rapid differentiation of certain non-pathogenic asporogenous bacilli. *Nature* **173:**208–209.

Singer, J., W. Choe, and K. A. Schmidt. 1991. Use of a restriction-defective variant for the construction of stable attenuated strains of the marine fish pathogen *Vibrio anguillarum*. *J. Microbiol. Methods* **13:**49–60.

Sorensen, U. B., and J. L. Larsen. 1986. Serotyping of *Vibrio anguillarum*. *Appl. Environ. Microbiol.* **51:**593–597.

Sparagano, O. A., P. A. Robertson, I. Purdom, J. McInnes, Y. Li, D. H. Yu, Z. J. Du, H. S. Xu, and B. Austin. 2002. PCR and molecular detection for differentiating *Vibrio* species. *Ann. N. Y. Acad. Sci.* **969:**60–65.

Staroscik, A. M., S. M. Denkin, and D. R. Nelson. 2005. Regulation of the *Vibrio anguillarum* metalloprotease EmpA by post-translational modification. *J. Bacteriol.* **187:**2257–2260.

Stokes, H. W., and R. M. Hall. 1989. A novel family of potentially mobile DNA elements encoding site-specific gene-integration functions: integrons. *Mol. Microbiol.* **3:**1669–1683.

Stork, M., M. Di Lorenzo, S. Mourino, C. R. Osorio, M. L. Lemos, and J. H. Crosa. 2004. Two tonB systems function in iron transport in *Vibrio anguillarum*, but only one is essential for virulence. *Infect. Immun.* **72:**7326–7329.

Tajima, K., Y. Ezura, and T. Kimura. 1987a. The possibility of use of a thermolabile antigen in detection of *Vibrio anguillarum*. *Fish Pathol.* **22:**237–242.

Tajima, K., Y. Ezura, and T. Kimura. 1987b. Serological analysis of the thermolabile antigens of *Vibrio anguillarum*. *Fish Pathol.* **22:**221–226.

Tassin, M. G., R. J. Siebeling, N. C. Roberts, and A. D. Larson. 1983. Presumptive identification of *Vibrio* species with H antiserum. *J. Clin. Microbiol.* **18:**400–407.

Tiainen, T., K. Pedersen, and J. L. Larsen. 1995. Ribotyping and plasmid profiling of *Vibrio anguillarum* serovar O2 and *Vibrio ordalii*. *J. Appl. Bacteriol.* **79:**384–392.

Tolmasky, M. E., L. A. Actis, and J. H. Crosa. 1988a. Genetic analysis of the iron uptake region of the *Vibrio anguillarum* plasmid pJM1: molecular cloning of genetic determinants encoding a novel trans activator of siderophore biosynthesis. *J. Bacteriol.* **170:**1913–1919.

Tolmasky, M. E., P. C. Salinas, L. A. Actis, and J. H. Crosa. 1988b. Increased production of the siderophore anguibactin mediated by pJM1-like plasmids in *Vibrio anguillarum*. *Infect. Immun.* **56:**1608–1614.

Tolmasky, M. E., L. A. Actis, and J. H. Crosa. 1993. A single amino acid change in AngR, a protein encoded by pJM1-like virulence plasmids, results in hyperproduction of anguibactin. *Infect. Immun.* **61:**3228–3233.

Tolmasky, M. E., L. A. Actis, A. E. Toranzo, J. L. Barja, and J. H. Crosa. 1985. Plasmids mediating iron uptake in *Vibrio anguillarum* strains isolated from turbot in Spain. *J. Gen. Microbiol.* **131:**1989–1997.

Tolmasky, M. E., and J. H. Crosa. 1995. Iron transport genes of the pJM1-mediated iron uptake system of *Vibrio anguillarum* are included in a transposonlike structure. *Plasmid* **33:**180–190.

Tolmasky, M. E., and J. H. Crosa. 1990. Plasmid-mediated iron transport and virulence in the fish pathogen *Vibrio anguillarum*, p. 49–54. *In* O. Olsvik and G. Bukholm (ed.), *Application of Molecular Biology in Diagnosis of Infectious Diseases*. Norwegian College of Veterinary Medicine, Oslo, Norway.

Tolmasky, M. E., and J. H. Crosa. 1991. Regulation of plasmid-mediated iron transport and virulence in *Vibrio anguillarum*. *Biol. Methods* **4:**33–35.

Toranzo, A. E., and J. A. Barja. 1990. A review of the taxonomy and seroepizootiology of *Vibrio anguillarum*, with special reference to aquaculture in the northwest of Spain. *Dis. Aquat. Organ.* **9:**73–82.

Toranzo, A. E., P. Combarro, M. L. Lemos, and J. L. Barja. 1984. Plasmid coding for transferable drug resistance in bacteria isolated from cultured rainbow trout. *Appl. Environ. Microbiol.* **48:**872–877.

Toranzo, A. E., Y. Santos, and J. L. Barja. 1997. Immunization with bacterial antigens: *Vibrio* infections. *Dev. Biol. Stand.* **90:**93–105.

Trust, T., I. Courtice, A. G. Khouri, J. H. Crosa, and M. H. Schiewe. 1981. Serum resistance and hemagglutination ability of marine vibrios pathogenic for fish. *Infect. Immun.* **34:**702–707.

Vervarcke, S., F. Ollevier, R. Kinget, and A. Michael. 2004. Oral vaccination of African catfish with *Vibrio anguillarum* O2: effect on antigen uptake and immune response by absorption enhancers in lag time coated pellets. *Fish Shellfish Immunol.* **16:**407–414.

Villegas, M. V., and A. I. Hartstein. 2003. *Acinetobacter* outbreaks, 1977-2000. *Infect. Control Hosp. Epidemiol.* **24:**284–295.

Vivian, A. 1991. Genetic organization of *Acinetobacter*, p. 191–200. *In* K. J. Towner, E. Bergogne-Berenzin, and C. A. Fewson (ed.), *The Biology of* Acinetobacter. Plenum Press, New York, N.Y.

Walter, M. A., S. A. Potter, and J. H. Crosa. 1983. Iron uptake system medicated by *Vibrio anguillarum* plasmid pJM1. *J. Bacteriol.* **156:**880–887.

Wang, S. Y., J. Lauritz, J. Jass, and D. L. Milton. 2003. Role for the major outer-membrane protein from *Vibrio anguillarum* in bile resistance and biofilm formation. *Microbiology* **149:**1061–1071.

Wang, S. Y., J. Lauritz, J. Jass, and D. L. Milton. 2002. A ToxR homolog from *Vibrio anguillarum* serotype O1 regulates its own production, bile resistance, and biofilm formation. *J. Bacteriol.* **184:**1630–1639.

Ward, P., M. Tatner, and M. Horne. 1985. Factors influencing the efficacy of vaccines against vibriosis caused by Vibrio anguillarum, p. 221–229. *In* M. Manning and M. Tatner (ed.), *Fish Immunology*. Academic Press, Inc., London, England.

Watanabe, T., Y. Ogata, and S. Egusa. 1971. R factors related to fish culturing. *Ann. N. Y. Acad. Sci.* **182:**383–410.

Westerdahl, A., J. C. Olsson, S. Kjelleberg, and P. L. Conway. 1991. Isolation and characterization of turbot (*Scophtalmus maximus*)-associated bacteria with inhibitory effects against *Vibrio anguillarum*. *Appl. Environ. Microbiol.* **57:**2223–2228.

Wiik, R., K. A. Hoff, K. Andersen, and F. L. Daae. 1989. Relationships between plasmids and phenotypes of presumptive strains of *Vibrio anguillarum* isolated from different fish species. *Appl. Environ. Microbiol.* **55:**826–831.

Wiik, R., E. Stackebrandt, O. Valle, F. L. Daae, O. M. Rodseth, and K. Andersen. 1995. Classification of fish-pathogenic vibrios based on comparative 16S rRNA analysis. *Int. J. Syst. Bacteriol.* **45:**421–428.

Wu, H., Y. Ma, Y. Zhang, and H. Zhang. 2004. Complete sequence of virulence plasmid pEIB1 from the marine fish pathogen *Vibrio anguillarum* strain MVM425 and location of its replication region. *J. Appl. Microbiol.* **97:**1021–1028.

Xia, Y. J., W. H. Wen, W. Q. Huang, and B. C. Huang. 2005. Development of a phage displayed disulfide-stabilized Fv fragment vaccine against *Vibrio anguillarum*. *Vaccine* **23:**3174–3180.

Yamamoto, S., N. Okujo, and Y. Sakakibara. 1994. Isolation and structure elucidation of acinetobactin, a novel siderophore from *Acinetobacter baumannii*. *Arch. Microbiol.* **162:**249–254.

Zhao, J., and T. Aoki. 1992a. Cloning and nucleotide sequence analysis of a chloramphenicol acetyltransferase gene from *Vibrio anguillarum*. *Microbiol. Immunol.* **36:**695–705.

Zhao, J., and T. Aoki. 1992b. Nucleotide sequence analysis of the class G tetracycline resistance determinant from *Vibrio anguillarum*. *Microbiol. Immunol.* **36:**1051–1060.

The Biology of Vibrios
Edited by F. L. Thompson et al.
© 2006 ASM Press, Washington, D.C.

Chapter 19

Vibrio harveyi: Pretty Problems in Paradise

LEIGH OWENS AND NANCY BUSICO-SALCEDO

INTRODUCTION

Despite being known for almost half a century, *Vibrio harveyi* was considered only a pretty, bioluminescent, environmental bacterium (Color Plate 13) that made balmy nights more magical when it contributed to water bioluminescence as bathers splashed around tropical lagoons. With the advent of large-scale prawn aquaculture, *V. harveyi* started to get attention as a pathogen, particularly in Australia and the Philippines. From the mid-1980s, more and more animals, particularly invertebrates in the tropics and fish in the more temperate areas, were recorded as being infected by *V. harveyi*. Recently, literature on *V. harveyi* has gained a wider audience as control of its bioluminescent pathway is elucidated and shown to be substantially different from and more complicated than that of *Vibrio fischeri*. Similarly, the bacteriophage-mediated virulence in *V. harveyi* has also made scientists more interested in this bacterium.

CLASSIFICATION AND IDENTIFICATION OF *VIBRIO HARVEYI*

History of Classification

V. harveyi was first described as a species of *Acromonobacter* by Johnson and Shunk (1936). Later studies in classification of luminous bacteria reported three major groups. The first group contains *Photobacterium fischeri*, the second group consists of *Photobacterium leiognathi* and *Photobacterium phosphoreum*, and the third group contains *Beneckea harveyi*, *Beneckea splendida*, and *V. cholerae* biotype *albensis* (Hendrie et al., 1970; Reichelt and Baumann, 1973; Baumann and Baumann, 1977). Later, this bacterium was included in the *Approved Lists of Bacterial Names* under the denominations of *Beneckea*

harveyi and *Lucibacterium harveyi*. In 1980, Baumann et al. abolished *Beneckea* and *Lucibacterium* and transferred the various species into *Vibrio*. The phenotypic characteristics of *V. harveyi* are very close to those of *Vibrio carchariae*. Indeed, Pedersen et al. (1998) concluded that *V. carchariae* is a junior synonym of *V. harveyi*, which was confirmed by Gauger and Gomez-Chiarri (2002). Thompson et al. (2002) also reported *Vibrio trachuri* as a junior synonym of *V. harveyi* on the basis of phylogenetic analysis and a polyphasic approach using 16S rDNA sequencing, fluorescent amplified fragment length polymorphism, DNA-DNA hybridization experiments, and G+C content of DNA.

The general morphology, characteristics, and biochemical traits of *V. harveyi* are listed in Table 1.

Identification of *Vibrio harveyi*

The great diversity of *V. harveyi* poses certain difficulties in the biochemical determination and identification of environmental *Vibrio* species. It was concluded that biochemical criteria are not always sufficient to distinguish between some *Vibrio* species because of their variable characters (West et al., 1986). In the biochemical characterization tests, *V. harveyi* and *V. campbellii* exhibited very similar characteristics because of their close phenotypic and genotypic relationship (Bryant et al., 1986). The work of Gomez-Gil et al. (2004) has created much debate about the importance of *V. harveyi*, as their work suggested that *V. campbellii* was much more important as a disease-causing organism. However, this confusion should not have occurred, as Alsina and Blanch (1994) showed *V. harveyi* as arginine and lysine negative and ornithine positive, whereas *V. campbellii* is always ornithine negative. Indeed, the ornithine-positive nature of *V. harveyi* was used to create

Leigh Owens • Microbiology and Immunology, School of Veterinary and Biomedical Sciences, James Cook University, Townsville 4811, Australia. **Nancy Busico-Salcedo** • College of Veterinary Medicine, University of Southern Mindanao, Kabacan, 9407 Cotabato, The Philippines.

Table 1. Morphological, physical, and nutritional characteristics of *V. harveyi*[a]

Parameter	Result
General	Gram negative, facultatively anaerobic, oxidase positive, straight rods, fermentative and respiratory metabolism competent, chemo-organotrophic, sodium required for growth, motile via sheathed polar flagella
Growing temp.	No growth at 4°C; optimum growth at 30–35°C
Mol% G+C content of DNA	46–48
Flagella	Synthesizing lateral flagella on solid medium
Able to utilize:	D-Mannose, trehalose, cellobiose, D-gluconate, heptanoate, α-ketoglutarate, L-serine, L-glutamate, L-tyrosine
Unable to utilize:	D-Xylose, melibiose, D-galacturonate, β-hydroxybutyrate, D-sorbitol, ethanol, L-leucine, γ-aminobutyrate, putrescine
Positive for:	Production of amylase, gelatinase, lipase, chitinase
Negative for:	Production of arginine dihydrolase and acetoin and/or diacetyl

[a] Modified from Baumann et al. (1984) as cited by Munro (2001).

V. harveyi agar (Harris et al., 1996). Furthermore, 85% of *V. harveyi* cells are bioluminescent, whereas *V. campbellii* cells are not.

To identify and differentiate environmental marine bacteria, it was suggested that API 20NE should be coupled with additional tests beyond those provided in the API kit (Analytab Products, Plainview, N.Y.) to produce a more conclusive and realistic result (Breschel and Singleton, 1992). The API style of biochemical test kit comprises only 20 reactions and profiles only the medically important bacteria, such as *Vibrio alginolyticus*, *Vibrio parahaemolyticus*, and *Vibrio vulnificus*. The profiles of isolates of *V. harveyi* from disease outbreaks in the Philippines matched those of *V. alginolyticus* and *V. parahaemolyticus* obtained by using the API 20NE and 20NFT systems (Lightner et al., 1992). Of 16 *V. harveyi* strains, 5 isolates were positively identified as *V. alginolyticus* and *V. parahaemolyticus* using API 20NE kits. The other 11 isolates returned less definitive matches to either *V. alginolyticus*, *V. parahaemolyticus*, or *V. vulnificus*. The use of the API-NFT system was probably responsible for misidentification of *V. harveyi* isolates as luminescent strains of a range of bacteria (e.g., *V. parahaemolyticus*, *V. alginolyticus*, or *V. vulnificus*) that were not characterized as luminescent in the investigation of "sindroma gaviota" in Ecuador (Mohney et al., 1994). Identification of *Vibrio* isolates using the Biolog GN technique was not accurate when subjected to amplified fragment length polymorphism genomic fingerprinting (Vandenberghe et al., 1998). The patterns of these strains showed only 30% similarity with the cluster containing the type strain and reference strains of *V. harveyi*.

Thiosulfate citrate bile salt sucrose agar is a widely used medium for isolating and enumerating *Vibrio* species. However, it is not a differential medium

for *V. harveyi* owing to its variability to utilize sucrose and cannot be distinguished from other sucrose-positive or sucrose-negative species (Harris et al., 1996). A selective medium was developed for the culture of *V. harveyi*. This medium, called VHA (*Vibrio harveyi* agar), has been shown to have definitive potential as a *Vibrio*-specific, primary differential medium on which it is possible to differentiate *V. harveyi* colonies from colonies formed by other *Vibrio* species (Harris et al., 1996). It is characterized by a pH of 9, a strong NaCl concentration (30 g/liter), and the presence of ornithine and cellobiose (Harris et al., 1996). The decarboxylation of ornithine alkalizes the medium, which contains indicators of pH, whereas the acidification of cellobiose lowers the pH. After 48 h of incubation at 28°C, colonies of *V. harveyi* have a diameter of 2 to 5 mm; they generally have regular contours, and their color is blue or slightly green with a dark-green center that is sometimes surrounded by yellow halation.

Identification of *Vibrio* species using the classical numerical taxonomy methods has been used by a number of authors (Baumann and Schubert, 1984; Bryant, 1991; Austin and Lee, 1992). These still represent the best means for proper identification and differentiation of *V. harveyi*. Once an isolate has been characterized as a straight or curved rod, gram-negative, oxidase-positive, and facultatively anaerobic, taxonomic methods of identifying and differentiating *V. harveyi* from other *Vibrio* species usually rely on ornithine and lysine decarboxylase and arginine dihydrolase tests as a major means of grouping bacterial isolates. Alsina and Blanch (1994) reported a fast and presumptive identification for environmental isolates of *Vibrio* spp. This set of biochemical keys can be used for strains that are gram negative and oxidase positive, grow on thiosulfate citrate bile salt sucrose agar, and are facultative anaerobes. This method can deliver presumptive identification of *Vibrio* species using a maximum of 10 tests with 90% confidence. This system was designed for routine environmental screening and cannot resolve some of the more complex issues of the taxonomy of *Vibrionaceae*. However, it is still an excellent method for preliminary identification of environmental isolates of *V. harveyi*.

To screen for bacteria pathogenic to cultured tiger prawns (*Penaeus monodon*), such as *V. harveyi*, through their hemolytic activity, Chang et al. (2000) studied a newly developed prawn blood agar consisting of 1 ml of tiger prawn hemolymph in medium containing 200 mg/liter Rose Bengal. It was noted that this blood agar was faster and more accurate for determining prawn hemolytic activity of bacterial isolates than sheep blood agar, which is the conventional method of hemocytolytic assay. For detection of subclinical vibriosis, an indirect immunodot blot assay using avidin-biotin complex proved to be the best method (Song et al., 1992). This method can detect 10^3 organisms per 100 μl of hepatopancreatic homogenates. The monoclonal antibodies in enzyme immunoassays were shown to react with 24 *V. harveyi* strains, including the type strain ATCC 14126. In addition, this method would be suitable for both confirmation of the identification of *V. harveyi* isolates and field monitoring of *V. harveyi* levels in aquaculture farms (Song et al., 1992). The monoclonal antibodies developed by Hanna et al. (1991) showed adequate specificity against pure bacterial suspensions. However, only one strain of *V. harveyi* was used for developing antibodies against this species, and there has been no further report of these antibodies being screened against a greater number of isolates.

Identification and typing of *Vibrio* strains using genomic approaches such as DNA-DNA hybridization, PCR, and ribotyping are useful for taxonomic studies and identification to the subspecies level (Austin et al., 1995). Protein profile analysis and M13 DNA fingerprinting in *V. harveyi* demonstrated genetic diversity (Pizzutto and Hirst, 1995), and studies using randomly amplified polymorphic DNA similarly demonstrated the genetic heterogeneity of this species (Karunasagar et al., 1998). On the basis of DNA-DNA hybridization experiments, Pedersen et al. (1998) concluded that *V. carchariae* is a junior synonym of *V. harveyi*, which showed 88% DNA binding between these type strains. On the other hand, arbitrarily primed PCR proved to be a reliable tool for strain differentiation since it generates fingerprints that can be used to compare microorganisms at the species level and within species with high discriminating power (Welsh and McClelland, 1990). It also

allows identification of isolates at the genospecies level and studies of infraspecific population structures of epidemiological interest (Goarant et al., 1999). Recently, a PCR using 16S ribosomal DNA (16S rDNA) sequences was used to confirm the identification of *V. harveyi* in conjunction with no swarming on sheep's blood agar and negative Voges Proskauer reaction to separate it from *V. alginolyticus* (Oakey et al., 2003). This technique was more rapid and economical than the conventional biochemical and morphological testing, thus making diagnosis faster. However, its use was limited to confirmation of identification of a bacterial isolate and therefore not applicable as a single tool for the detection of *V. harveyi* in a mixed population.

BACTERIAL LUMINESCENT DISEASE

Epizootics

Outbreaks of vibriosis have been reported worldwide. A minority of *V. harveyi* strains have been shown to cause disease in a variety of aquatic animal hosts, including marine fish, bivalves, and crustaceans (Table 2). Studies have shown that *V. harveyi* has been isolated from diseased but not healthy larvae and often from the water in the rearing facilities (Jiravanichpaisal et al., 1994; Karunasagar et al., 1994).

In general, the infections associated with *V. harveyi* in fish are more opportunistic in nature than reports of disease in invertebrates. It was reported as an opportunistic pathogen of common snook, *Centropomus undecimalis* (Bloch) (Kraxberger-Beatty et al., 1990): isolated from diseased seahorse, *Hippocampus* sp. (Alcaide et al., 2001); from pearl oyster, *Pinctada maxima* (Pass et al., 1987); from rock lobster, *Jasus verreauxi* (Diggles et al., 2000); from diseased marine fish such as cultured silvery black porgy, *Acanthopagrus cuvieri* (Saaed, 1995), brown-spotted grouper, *Epinephelus tauvina* (Rasheed, 1989), sea bream, *Sparus aurata* (Balebona et al., 1995), and dentex, *Dentex dentex* (Company et al., 1999); from sunfish, *Mola mola* (Hispano et al., 1997); and from cage-cultured sea bass, *Lates calcarifer* Bloch (Tendencia, 2002).

Some infections were only reported in stressed or experimental situations. For example, skin infections in barramundi (*L. calcarifer*) associated with *V. harveyi* isolates have only been reported in fish thought to be stressed in a captive environment or already infected with other agents (Anderson and Norton, 1991). Tendencia (2002) reported *V. harveyi* infection in sea bass, *L. calcarifer*, in the Philippines, where the disease usually infects fish when they are stressed.

Table 2. Hosts in which *V. harveyi* has been implicated as a pathogen and the associated disease

Host	Nature of disease
Gorgonian corals	Possible role in surface fouling (*V. harveyi* used as surface foulant to test antimicrobial activity of corals)[a]
Pinctada maxima (pearl oyster)	Mortality, tissue lesions (disease reproduced by injection of *V. harveyi*)[a]
Larvae and postlarvae of *Penaeus monodon* (black tiger prawn) and *Penaeus merguiensis* (banana prawn)	Luminous vibriosis (disease reproduced by bath inoculation and *V. harveyi* reisolated from larvae)[a]
Adult *Penaeus japonicus* (kuruma prawn)	Mortality, septicemia (*V. harveyi* isolated from moribund prawns in pure culture)[a]
Penaeus esculentus spawners (brown tiger prawn)	Septicemia, mortality (mortality reproduced by injection of *V. harveyi*)[a]
Postlarvae of *Macrobrachium rosenbergii* (freshwater prawn)	Luminous vibriosis[a]
Panulirus homarus (spiny lobster)	Septicemia, lesions on exoskeleton (*V. harveyi*-like bacteria isolated from hemolymph and lesions, mortality produced by i.m. injection)[b]
Jasus verreauxi (rock lobster)	Luminous vibriosis[c]
Centropomus undecimalis (common snook)	Infection of cornea causing blindness (*V. harveyi* most common and numerous isolate)[a]
Lates calcarifer (barramundi)	Skin infection, septicemia (*V. harveyi* most common and numerous isolate)[a]
Salmo salar (Atlantic salmon), *Oncorhynchus kisutch* (Coho salmon)	Bacteria closely related to *V. fischeri*-*V. harveyi* group isolated as secondary invaders[a]
Scophthalmus maximus (turbot)	Ulcerative lesions, hemorrhagia (*V. harveyi* commonly isolated in diseased fish, often in pure culture in liver and kidney)[a]
Hippocampus sp. (seahorse)	Bacterial luminescent disease[d]
Chanos chanos (milkfish)	Eye lesions, disease recreated by bath challenge[e]

[a] Harris (1998).
[b] Vandenberghe et al. (1998).
[c] Diggles et al. (2000).
[d] Alcaide et al. (2001).
[e] Ishimaru and Muroga (1997).

The signs exhibited included anorexia and darkening of the whole fish. There were also local hemorrhagic ulcers on the mouth or skin surface and focal necrotic lesions in the muscle or eye opacity. In a study by Owens et al. (1996), *V. harveyi* strains were pathogenic by intraperitoneal (i.p.) inoculation with 10^3 to 10^5 cells in Atlantic salmon (*Salmo salar*); however, this was an artificial route of infection, and there have been no published reports of disease outbreaks caused by *V. harveyi* in this fish species.

An outbreak of disease in *P. maxima* (pearl oyster) with 80% mortality following resettlement on leases after collection from wild sites was shown to be caused by *V. harveyi* (Pass et al., 1987). These oysters were thought to be predisposed to infection by cold temperatures, overcrowding, and inadequate water circulation during transport from wild sites. Lavilla-Pitogo and de la Pena (1998) also reported an epizootic of luminescent, non-sucrose-fermenting *V. harveyi* in larvae of *P. monodon* in the Philippines. The exposure of the larvae to 10^2 *V. harveyi* cells/ml resulted in significant mortality within 48 h. It is noteworthy to include the contribution made by environmental conditions or variation in the susceptibility of the host animals, particularly when raised under intensive culture when outbreak of disease takes place.

LUMINOUS VIBRIOSIS

Luminous vibriosis is the term describing the disease of penaeid prawns caused by luminescent *V. harveyi*. Dead and dying animals visibly luminesce during a serious outbreak and have been observed to collect in a luminescent mat at the bottom of larval rearing tanks (Lavilla-Pitogo et al., 1992). In larval prawns, it was probable that virulent luminous bacteria initiated infection by entering through the mouth and feeding apparatus. Scanning electron micrographs revealed colonization of the bacteria specifically on the feeding apparatus and oral cavity of the larvae (Lavilla-Pitogo et al., 1990). External signs in diseased prawns were brittle shells, brown or black spots on the shell, darkened or red body surface, pink or brown gills, murky whitish muscle, a lack of food in the midgut, and a folded base of the tail (Lavilla-Pitogo et al., 1990). Affected larvae of penaeid shrimp developed luminescence, reduced feeding, and had poor development, sluggish swimming, reduced escape mechanisms, degeneration of hepatopancreatic tissue with resultant formation of necrotic bundles, and increased mortality (Robertson et al., 1998). These infections were usually septicemic, with pure cultures of the causative strain isolated from the hemolymph and hepatopancreas (Jiravanichpaisal et al., 1994; Liu et al., 1996).

It has also been suggested that exotoxins may be involved in the disease process of luminous vibriosis. Owens and Hall-Mendelin (1990) and Muir (1991) reported that the histopathology of *Penaeus monodon* larvae experimentally infected with *V. harveyi* showed changes suggestive of toxicity. The ganglionic neurophiles of infected larvae showed extensive vacuolation and granulation compared to uninoculated controls. Owens et al. (1992) reported extensive colonization of the connective tissue of experimentally infected *Penaeus esculentus*. Connective tissue infection was also characteristic of experimentally infected pearl oyster in the study of Pass et al. (1987). Histopathology of moribund shrimp showed extensive hepatopancreatic tubular necrosis and replacement by bacterial hemocytic nodules, often melanized and marked hemocytic enteritis (Nithimathachoke et al., 1995) (Color Plate 14).

Factors Contributing to Luminous Vibriosis

Many factors have been associated with luminous vibriosis. The increased amounts of organic matter in the pond water, particularly in the nursery reservoir pond, tanks, and contaminated equipment used between ponds, were the probable factors (Nithimathachoke et al., 1995). Cultures of marine algae are prone to contamination with *Vibrio* spp. due to either aerosol transmission or from cross contamination from the hands of workers or equipment (Harris and Owens, 1997). *Artemia* cysts are known to carry a residual flora consisting primarily of gram-positive bacteria, such as *Bacillus*, *Micrococcus*, and coryneforms (Igarashi et al., 1989). However, upon hatching in seawater of 35‰ salinity, the microflora associated with *Artemia nauplii* rapidly becomes dominated by members of the autochthonous microflora of seawater such as *Vibrio* and *Pseudomonas* species (Igarashi et al., 1989).

The seasonality of outbreaks of luminous vibriosis has been reported in some countries (Sunaryanto and Mariam, 1986; Ramaiah et al., 2000). The level of *V. harveyi* in sites affected by run-off from heavy rainfall can rise as much as 100-fold as a consequence of nutrient runoff (Ramesh et al., 1989; Harris and Owens, 1997).

Sources of *V. harveyi*

Studies on the sources of *V. harveyi* by Lavilla-Pitogo et al. (1992) reported that the midgut contents of spawners of *P. monodon*, which are shed into the water almost simultaneously with the eggs during spawning, are the main source of the luminescent bacteria. It was also found as a minor component of the microflora on the exoskeleton of female black tiger prawns, *P. monodon*, in Thailand (Jiravanichpaisal et al., 1994). Moreover, this species of *Vibrio* seems to dominate the luminous bacterial population in nearshore seawater. Lavilla-Pitogo et al. (1990) found 5 to 7 CFU/ml of *V. harveyi* in nearshore water in the Philippines. Similarly, Makemson et al. (1992) reported 26 to 58 CFU/ml of *V. harveyi* from surface seawater samples in the Arabian Gulf. Ramesh et al. (1989) reported levels of 10^3 CFU/g, of which 70% were *V. harveyi* in estuary sediments from southern India. It was elucidated that the ability of this species to utilize a wide variety of organic compounds as carbon and energy sources contributed to the survival of this species when competing for the scarce nutrients present in seawater (Ramesh et al., 1989).

Control of Luminous Vibriosis

Reducing the bacterial load

Harris and Owens (1997) identified seven critical control points in a commercial *P. monodon* hatchery in continuous or semicontinuous production, where bacterial levels should be monitored and controlled to avoid contamination of the hatchery with undesirable levels of virulent *V. harveyi* strains. The critical points were as follows:

1. incoming waters,
2. spawners,
3. algal cultures,
4. *Artemia* cultures,
5. larval and postlarval rearing water,
6. tank and pipe surfaces, and
7. hatchery workers.

The first four of these critical control points are the major sources of importance for the initial introduction of virulent bacterial strains into the hatchery environment (Lavilla-Pitogo et al., 1990, 1992; Harris and Owens, 1997). The last three critical control points are of more importance for avoiding cross contamination or persistence of virulent strains within the hatchery (Karunasagar et al., 1996; Harris and Owens, 1997). Sterilization of incoming seawater, the reduction of cross contamination of tanks within the hatchery and particularly the separation and sterilization of eggs as soon as possible from spawners and the spawners' feces were points of hygiene addressed by Lavilla-Pitogo et al. (1992). The persistence of virulent strains of *V. harveyi* on tank surfaces was shown to be an important point of contamination in an Indian hatchery (Karunasagar et al., 1996). In that study, *V. harveyi* living in a biofilm on high-density polyethylene or concrete slabs survived treatment with 100 mg/liter chlorine and 50 mg/liter chloramphenicol and tetracycline. It was suggested that scrubbing the tank surfaces with strong cleansing agents or dryout was the best measure to remove biofilms on these surfaces in order to decrease the population of pathogenic *V. harveyi* strains.

Abdel-Aziz (2001) recommended some measures to prevent the occurrence of *V. harveyi*-associated mortalities.

1. maintenance of good husbandry practices and proper nutrition to reduce stress,

2. quarantine practices for live and newly acquired shrimp,

3. strict sanitation procedures prior to and during rearing of shrimp larvae,

4. control of pathogenic vibrios using UV-irradiated water as well as employing a series of good-quality filtration systems,

5. the use of previously chlorinated water during spawning and rearing,

6. periodic siphoning of sediments and debris from the bottom of rearing tanks,

7. disinfection of infected stock prior to discarding, followed by cleaning and disinfection of hatchery after each larval rearing period, and

8. daily water exchange of 80 to 90%.

Antibiotics

The use of antibiotics as prophylactic agents to prevent bacterial infection of penaeid larvae has been employed in many shrimp hatcheries (Baticados and Paclibare, 1992). However, the development of antibiotic resistance is one of the major consequences resulting from prophylactic antibiotic use. Indeed, there have been several reports of antibiotic-resistant pathogens isolated from both Indonesia (Supriyadi and Rukyani, 1992) and the Philippines (Baticados and Paclibare, 1992). Baticados et al. (1990) determined the effect of 24 commonly used chemotherapeutic agents on strains of *V. harveyi* and *V. splendidus* and on *P. monodon* larvae. Of the used chemotherapeutics, only 6 of these antibiotics showed MIC of <25 µg/ml, and even at these concentrations these antibiotics caused deformities in the carapace, rostrum, and setae of larvae. The conclusions of this study were that chemical treatment of luminous vibriosis would be limited and ineffectual as a long-term application to hatcheries for *P. monodon*. This was based on the relatively high dosage necessary to exert an effect on the target bacteria, the corresponding cost of treatment,

the risk of development of resistant strains, and the damaging effects of antibiotic exposure to both larval and human health. The biofilm formation of *V. harveyi* on surfaces like high-density polyethylene plastics causes resistance to sanitizers (Karunasagar et al., 1996). Also, the use of antibiotics as prophylactics to control bacteria led to the persistence of antibiotic-resistant *V. harveyi* in hatchery tanks (Karunasagar et al., 1994). These results suggest the need for physical removal of biofilm on tank surfaces and periodic drying of tanks to reduce the chance of infection by organisms such as *V. harveyi* (Karunasagar et al., 1996).

In addition, many of the most commonly used antibiotics, such as erythromycin, oxytetracycline, and chloramphenicol, are dangerous to human health. The use of chloramphenicol as antibiotic for food animals is banned in certain countries because of the risk to human health. Therefore, alternative methods of disease control are urgently needed, and an understanding of virulence mechanisms and environmental factors controlling pathogens is of primary importance (Prayitno and Latchford, 1995).

Probiotics

The use of probiotics is another approach to bacterial control to maintain a beneficial balance of bacteria and other microorganisms in the culture systems (Maeda, 1988). Probiotics are microorganisms or their products which, when applied to humans or animals beneficially affect the host by improving its intestinal balance (Havenaar et al., 1992). Fuller (1989) recommended some properties of an ideal probiotic: probiotics should provide actual benefit to the host, be able to survive in the digestive tract, be capable of commercialization, and be stable and viable for prolonged storage conditions and in the field. Effective probiotic treatments may provide broader-spectrum and greater nonspecific disease protection than vaccination or immunostimulation (Rengpipat et al., 2000), and they are more desirable and more environmentally benign than antibiotics and chemicals.

Chythanya et al. (2002) reported a probiotic, *Pseudomonas* I-2, that inhibited the growth of shrimp pathogenic vibrios, including *V. harveyi*, *V. fluvialis*, *V. parahaemolyticus*, *Photobacterium damselae*, and *V. vulnificus*. *V. alginolyticus* has been shown to have antagonistic effects against pathogenic bacteria such as *V. harveyi*. However, the properties of this probiotic need further testing (Vandenberghe et al., 1999). Beneficial microorganisms found to be effective for growth inhibition of *V. harveyi* on a laboratory scale include *V. alginolyticus* (Ruangpan, 1998), *Chlorella* sp. (Direkbusarakom et al., 1997), and *Skeletonema costatum* (Panichsuke et al., 1997). Challenge of *V. harveyi* with *Bacillus* sp. as probiotic feed achieved 74% relative survival (Phianpark et al., 1997). Moreover, the use of *Bacillus* S11 increased the survival and growth of *P. monodon* exposed to *V. harveyi* (Rengpipat et al., 2000). This probiotic provided disease protection by activating both cellular and humoral immune defenses.

The use of bacteriophage in the control of vibriosis seems very promising. An experiment with bacteriophage from diseased ayu, *Plecoglossus altivelis*, provided protection against infection by *Pseudomonas plecoglossicida*, a pathogen of ayu. These bacteriophages, representatives of *Myoviridae* and *Podoviridae*, reduced the number of bacterial cells in the kidneys of affected ayu and the underlying water environment (Park et al., 2000). A follow-up study in ayu by Park and Nakai (2003) showed that phage-impregnated feeds administered to an ayu pond resulted in a decreased mortality rate and the appearance of phage in the kidneys. The successful effect of phage in phage therapy can be seen in the increased presence of phage in survivors and the death of host bacterial cells.

VIRULENCE FACTORS OF *VIBRIO HARVEYI*

Most *V. harveyi* strains are not harmful to larvae of *P. monodon*; however, some strains are extremely pathogenic. Highly virulent strains of *V. harveyi* cause acute, devastating disease outbreaks with 50 to 100% mortality from inocula as low as 100 cells/ml (Harris, 1998). Despite extensive studies, the virulence mechanisms of this species are not well understood.

Environmental Enhancement

During the rainy season, the salinity of seawater falls to as low as 10% (Prayitno and Latchford, 1995), which triggers *V. harveyi* strains to increase to a high level, thus increasing their virulence. However, growth of these strains in media of pH 5.5 prior to challenge of *P. monodon* has an opposite effect, significantly decreasing the virulence of these strains (Prayitno and Latchford, 1995). These effects probably indicate the expression of virulence genes being affected in response to bacterial processing of environmental signals.

Adhesion

It is generally known that the ability to attach to external and mucosal surfaces is an initial step and often a virulence determinant for bacteria. In a study

of the attachment of *Vibrio* spp. to the mucous layer of sea bream, *S. aurata*, 40% of 30 strains tested adhered, with only one of 12 tested strains of *V. harveyi* having that ability.

Exotoxins and Other Exoenzymes

Virulence in *V. harveyi* is reported to be associated with the presence of extracellular proteins (ECPs). ECPs are produced during the mid-exponential phase of growth and have amino acid sequence similarity to virulence-associated proteins in *Salmonella*, *Shigella*, and *Bacillus* species (Manefield et al., 2000). Some *Vibrio* species produce exotoxins and exoenzymes that play a role in their pathogenicity. For example, TDH, a major virulence factor of *V. parahaemolyticus* (Raimondi et al., 2000), has an enterotoxigenic effect on rabbit small intestine and could be responsible for inducing a watery diarrhea (Nishibuchi et al., 1992). An extracellular cytolytic toxin has been reported in *V. vulnificus* that possesses cytolytic, lethal, and vascular permeability factor activities in the guinea pig and the mouse (Gray and Kreger, 1985). Fouz et al. (1993) reported ECPs of *P. damselae* that exhibit phospholipase, hemolytic, and cytotoxic activities in diseased fish. The production of proteinase by *V. alginolyticus* is involved in the pathogenesis of bivalve vibriosis (Nottage and Birkbeck, 1987). The ECPs (exotoxins) in *V. penaeicida*, *V. alginolyticus*, and *Vibrio nigripulchritudo* demonstrated toxic effects in shrimp but not on a fish cell line (Goarant et al., 2000). A shellfish-pathogenic *Vibrio* sp. isolated from moribund American oyster larvae produced a heat-labile extracellular toxin that caused malformations and mortality in developing oyster embryonic culture (Brown and Roland, 1984).

V. harveyi was shown to produce proteases, phospholipase, hemolysins, or exotoxins that are important for pathogenicity (Liu et al., 1996). Liu and Lee (1999) also reported that a cysteine protease is the major exotoxin lethal to the tiger prawn, *P. monodon*. This cysteine protease may markedly interfere with hemostasis, leading to the occurrence of unclottable hemolymph, thus significantly contributing to the pathogenicity of *V. harveyi* (Lee et al., 1999). A thermostable exotoxin in *V. harveyi* that was recovered from diseased postlarval *Penaeus vannamei* was also reported to be lethal to Dublin Bay prawns (Montero and Austin, 1999). These exotoxins were shown to have proteolytic, hemolytic, and cytotoxic activities. In a study by Zhang and Austin (2000), *V. harveyi* strain VIB 645 produced ECPs with a maximal effect on salmonids. These ECPs contain caseinase, gelatinase, phospholipase, lipase, and hemolysins with the highest titer of hemolytic activity toward *Salmo salar* (Atlantic salmon) and *Oncorhynchus mykiss* (rainbow trout) erythrocytes. In addition, the possession of double hemolysin genes has been associated with the virulence of *V. harveyi* (Zhang et al., 2001). Marine toxins, such as tetrodotoxin and anhydrotetrodotoxins, were produced by *V. harveyi* (Simidu et al., 1987). Toxins of this nature would be extremely capable of harming host animals. Two luminous strains of *V. harveyi* (642 and 47666-1) were shown to produce proteinaceous exotoxins that caused mortality in CBA mice and juvenile *P. monodon* at low doses (Harris and Owens, 1999). Toxin T1, produced by *V. harveyi* strain 47666-1, had a 50% lethal dose (LD_{50}) of 2.1 μg/g by i.p. injection in CBA mice and 1.8 μg/g by intramuscular (i.m.) injection in juvenile *P. monodon*. Toxin T2, produced by *V. harveyi* strain 642, had an LD_{50} of 3.1 μg/g by i.p. injection in CBA mice and 2.2 μg/g by i.m. injection in juvenile *P. monodon* (Harris and Owens, 1999).

Chitinases

Chitin is one of the most abundant biopolymers in nature and is produced in the marine environments by many marine organisms, such as zooplankton and phytoplankton (Gooday, 1990). The great diversity of chitin structures present in the environment necessitates the bacterial production of different chitinases to efficiently hydrolyze the different forms of chitin (Svitil et al., 1997). The presence of chitinases presumably aids the invasion of the pathogen and provides nutrients directly in the form of amino acids or indirectly by exposing other host tissues to enzymatic degradation (Gooday, 1990). Chitinases consist of a group of hydrolytic enzymes that are able to break down polymeric chitin to chitin oligosaccharides, diacetylchitobiose, and *N*-acetylglucosamine (Thompson et al., 2001). There are two genes encoding two enzymes involved in chitin degradation. The first gene is chitiobiase, which cleaves the bond joining the two *N*-acetylglucosamine units in chitobiose (Jannatipour et al., 1987); the other gene is *chiA*, which encodes a chitinase tentatively identified as the main chitinase of *V. harveyi* (Soto-Gil, 1988).

V. harveyi excretes ~10 chitinases when grown on chitin, and the composition of the excreted chitinases varies when cells are exposed to different chitins; some are always present, whereas others are excreted only with particular chitins (Svitil et al., 1997). Based on restriction analysis, immunological data, and enzymatic properties, it has also been shown that six genes are responsible for the coding of a total of 10 excreted chitinases in this species (Svitil et al., 1997). Montgomery and Kirchman (1993) showed that the attachment of *V. harveyi* to chitin

was specific and was mediated by at least two peptides associated with the outer membrane of the cell. One of these peptides that appeared to mediate initial attachment to chitin was a 53-kDa peptide, and the other peptide, which was induced by chitin and presumably involved in time-dependent attachment, was a 150-kDa chitin-binding peptide.

In natural marine systems, most bacteria attached to chitinaceous particles are vibrios (Nagasawa et al., 1987). Baumann et al. (1980) reported that all pathogenic *Vibrio* species elaborate an extracellular chitinase, as shown by Baumann and Schubert (1984) in a study in which chitinase-positive *Vibrio* isolates were pathogenic to animals. Jiravanichpaisal et al. (1994) erroneously concluded that their chitinase-positive strain of *V. harveyi* was less virulent when injected into *P. monodon* than their chitinase-negative strain. Their experiment was flawed because, by injecting the isolates, they completely bypassed the chitinous cuticle and bypassed the selective advantage of the virulence determinant they were assessing.

Siderophores

The possession of a siderophore-mediated iron transport system is related to the increased virulence of some bacterial pathogens (Crosa, 1984; Griffiths, 1987). Iron acquisition in microbial pathogens is limited due to strong binding capacity of this element to the high-affinity iron-binding proteins of animal body fluids (Neilands, 1981). Many bacteria have complex systems to transport iron into the cell in the form of siderophores (Aznar et al., 1989) coupled with iron-repressible outer membrane receptors for the siderophore/iron complex. A siderophore is a low-molecular-weight, Fe (III)-specific ligand (Neilands, 1981). Siderophores act as growth or germination factors and as potent antibiotics. They have a role in receptor-dependent iron transport and function as virulence factors in animal and plant diseases (Neilands, 1981). They are molecules designed to trap traces of iron under the form of very stable complexes. Considerable evidence exists for the influence of iron on microbial growth. Indeed, a soluble iron fed or injected into an infected animal may greatly increase the virulence of some pathogens (Madigan et al., 2000).

Siderophores were reported as a major virulence determinant for pathogens such as *Aeromonas salmonicida* (Hirst et al., 1994), *V. anguillarum* (Sigel and Payne, 1982), *Shigella flexneri* (Crosa, 1989), *V. cholerae* non-O1 (Amaro et al., 1990), *Aeromonas hydrophila* (Barghouti et al., 1989), and *Salmonella enterica* serovar Typhimurium (Payne, 1983). Siderophore activity in *V. harveyi* appeared to be linked to pathogenicity in vertebrates but not in invertebrates

(Owens et al., 1996). This may be due to the tight binding of iron by high-affinity iron-binding proteins such as transferrin and lactoferrin in serum and secretions (Crosa, 1989). However, invertebrates seem to lack iron-binding compounds such as lactoferrin and transferrin (Owens et al., 1996). It has been suggested that the lack of competition for iron between invertebrates and invertebrate pathogens may have resulted in a decreased necessity for these bacteria to produce siderophores (Harris, 1993; Owens et al., 1996). In *V. harveyi*, a bacterium that seems to infect fish opportunistically and invertebrates, particularly larvae, as a primary pathogen, the distinction between iron chelation in vertebrate and invertebrate physiology seems to be reflected in a relationship between strain origin and siderophore production (Owens et al., 1996). These authors concluded that siderophores must be considered a virulence factor in piscine-infecting *V. harveyi* but not in strains infecting invertebrates.

Antibiotic Resistance Plasmids

Bacterial plasmids are molecules of double-stranded DNA and are responsible for the virulence of bacteria causing plague, dysentery, anthrax, and tetanus as well as many other diseases of humans, animals, fish, and plants (Hardy, 1984). Bacteria that contain antibiotic resistance plasmids have been shown to exhibit higher rates of survival in aquatic environments (Baya et al., 1986). Plasmids may act as a vector to quickly spread genetic information throughout a bacterial population (Richmond, 1973). The genes that encode resistance are the resistance determinants on an R factor. These are products that inactivate the antibiotics or prevent the antibacterial drug from contacting its target within the cell (Richmond, 1973). The presence of R factor in virulent strains of bacteria enables them to become dominant within natural flora of the aquatic environment and thus increase virulence. The increase in resistance can be attributed to the selection of resistant strains that have the ability to exchange plasmids encoding resistance (Baya et al., 1986).

A higher prevalence of plasmid-bearing strains in marine vibrios of the Gulf of Mexico was found in a polluted rather than an unpolluted site (Hada and Sizemore, 1981). Bacteria containing plasmid DNA that demonstrate antibiotic resistance are more frequently encountered in isolates acquired from toxic chemical waste than from sewage-impacted waters or from uncontaminated open ocean sites (Baya et al., 1986).

A conjugative R factor plasmid in a strain of *V. harveyi* virulent to *P. monodon* was reported by Harris (1993). This R factor conferred resistance to erythromycin, streptomycin, kanamycin, sulfafurazole, and cotrimoxazole. Karunasagar et al. (1994) also reported a highly virulent antibiotic-resistant strain of *V. harveyi* that colonized larval tanks in a hatchery in India. That strain showed resistance against cotrimoxazole, chloramphenicol, streptomycin, and the vibriostatic agent O/129.

Bacteriocins

Bacteriocins are plasmid-derived proteins produced by bacteria that exhibit an antimicrobial mode of action against sensitive, usually closely related bacterial species (Tagg et al., 1976). McCall and Sizemore (1979) reported a bacteriocin-like substance in *V. harveyi* (formerly classified in the genus *Beneckea*) mediated by a plasmid. This lethal substance, termed harveyicin, was the first well-documented bacteriocin in a marine bacterium. The likely role for *V. harveyi* bacteriocin is to confer a competitive advantage on bacteriocin-producing strains over closely related, nonbacteriocinogenic strains in the enteric environment (Hoyt and Sizemore, 1982).

N-(β-Hydroxybutyryl) Homoserine Lactone: a Possible Role for Bioluminescence

Expression of the luminescence system in many bacteria is controlled by a cell density-dependent induction called autoinduction (Sun et al., 1994). An autoinducer serves as a pheromone which signals cells in a light organ symbiosis to luminesce (Greenberg et al., 1979). N-(β-Hydroxybutyryl) homoserine lactone is the autoinducer molecule regulating bioluminescence in *V. harveyi* (Cao and Meighen, 1989). This autoinducer enables the bacterium to monitor its own population (quorum sensing) and regulate virulence gene expression (Milton et al., 1997). Quorum sensing probably enhances bioluminescence, virulence factor expression, antibiotic production, and biofilm development (Chen et al., 2002). The *lux* genes encode the enzyme for bioluminescence, luciferase, and they are expressed only when the autoinducer signal accumulates to a critical external concentration or when the cells reach a high population density (Von Bodman and Farrand, 1995). The enzyme luciferase, consisting of two subunits (α and β), catalyzes the oxidation of a reduced flavin and a long-chain aldehyde (Ziegler and Baldwin, 1981).

Pathogenic bacteria, e.g., *Erwinia stewartii* (Von Bodman and Farrand, 1995), *Erwinia carotovora* (Pirhonen et al., 1993), and *Pseudomonas aeruginosa* (Jones et al., 1993), use acyl-homoserine lactone-mediated autoinduction to control the expression of virulence functions. The marine bacterium *V. fischeri* also possesses autoinducer to signal cells in a light organ symbiosis to luminescence (Eberhard, 1972; Greenberg et al., 1979). This autoinducer is similar in structure to the *V. harveyi* autoinducer, which suggests that the regulation of luminescence induction in these bacteria may be related, despite their differences in *lux* gene organization. *V. anguillarum* also uses acyl-homoserine lactone to regulate virulence gene expression (Milton et al., 1997).

A recently discovered autoinducer, AI-2, is found to be produced by a large number of bacterial species, including *V. harveyi*; hence, it was proposed to serve as a "universal" signal for interspecies communication (Chen et al., 2002). The autoinducer structure and the development of light emission in *V. harveyi* is similar to that in *V. fischeri*. Thus, it was suggested that both have a common mechanism in the regulation of autoinduction of the *lux* system (Cao and Meighen, 1989). However, the regulatory system that operates in *V. harveyi* is very different and more complex than that in *V. fischeri* (Bassler et al., 1997). Luminescence in *V. harveyi* is thought to be affected by two autoinducer molecules. Only one of these molecules, N-(3-hydroxybutanoyl) homoserine lactone, has been characterized (Cao and Meighen, 1989). The production of these molecules is encoded by genes, which are not contiguous with the luminescence operon, composed of *lux*CDABEGH. The *V. harveyi* autoinducers interact with separate sensor proteins produced by the *lux* N and *lux* PQ loci, respectively, to alter the activity by phosphorylation of the regulatory protein *lux* O (Bassler et al., 1997).

It has been shown that halogenated furanones produced by marine microalga *Delisea pulchra* interfere with N-acyl-L-homoserine lactone-regulated gene expression. Manefield et al. (2000) investigated the ability of a halogenated furanone to inhibit the quorum sensing-regulated luminescence phenotype of the pathogenic *V. harveyi* strain 47666-1 and found that this inhibited luminescence as well as extracellular toxin production.

Bacteriophages

Bacteriophages are infectious agents made up of DNA or RNA and a protein capsid which protects their genetic material and displays structural components necessary for phage to infect its bacterial host (Lindqvist, 1998). Phages thrive in bacterial populations, where they constantly transfer their genetic material in a horizontal fashion. A phage infection may lead to death of the cell. If the cell survives, the phage genome may establish itself as a prophage in the form of an integrated copy in the host chromosome, or be maintained as a plasmid, or may simply abort, lead-

ing to cell survival (Lindqvist, 1998). Phages can multiply by two alternative mechanisms: the lytic cycle or the lysogenic cycle. The former ends with the lysis and death of the host cell, whereas the latter results in survival of the host cell (Tortora et al., 2000). The lysogenic cycle can confer virulence to *V. harveyi* (Oakey and Owens, 2000).

Ruangpan et al. (1999) found a bacteriophage and bacterial cells of *Vibrio* morphology which have lethal toxicity to cultivated *P. monodon*. They suggested that the presence of bacteriophage may sometimes mediate toxicity of *V. harveyi* by the transfer of a toxin gene or a gene controlling toxin production. A temperate phage in *V. harveyi* VH1039 isolated from tea brown gill syndrome in black tiger prawn, *P. monodon*, is a lysogenic bacteriophage and morphologically identified as a siphovirus (Pasharawipas et al., 1998). Lysogenic VH1039 caused no symptoms in the shrimp but, when combined with *V. harveyi* VHN1, led to shrimp death several hours after i.m. injection. At least two bacteriophages have been isolated from two different strains of *V. harveyi* that have been shown to produce toxin (Harris and Owens, 1999). The first isolated phage was from strain 47666-1 and classified in the family *Podoviridae* (Oakey, 2000); the second, isolated from strain 642, was classified in the family *Myoviridae* (Oakey and Owens, 2000).

Oakey and Owens (2000) isolated a bacteriophage from a toxin-producing strain of *V. harveyi* (strain 642) in tropical Australia. This bacteriophage is classified in the family *Myoviridae* on the basis of morphological characteristics such as an icosahedral head, a neck/collar region, and a sheathed rigid tail and nucleic acid characteristics (double-stranded linear DNA) and was termed bacteriophage VHML (*V. harveyi* myovirus-like). Munro et al. (2003) studied this bacteriophage and confirmed that infection of bacteriophage to previously naive *V. harveyi* strains 12, 20, 45, and 645 (Oakey and Owens, 2000) confers virulence to *V. harveyi* strain 642. Results from sodium dodecyl sulfate-polyacrylamide electrophoretic gels showed that the bacteriophage VHML caused an upregulation of certain extracellular proteins from *V. harveyi* strains 12, 20, 45, and 645. With regard to hemolysin production, the halo size and colony size from strains of *V. harveyi* that were bacteriophage infected were significantly larger as a group ($P < 0.001$) than those from strains without the bacteriophage VHML. Moreover, the bioassay experiment using the bath challenge method demonstrated that strains of *V. harveyi* infected with the bacteriophage VHML caused significantly ($P < 0.001$) higher mortality (LD_{50}) in larval prawns than did the same strains without the bacteriophage. Monoclonal antibodies confirmed that the infection of the bacteriophage VHML resulted in proteins' being synthesized and excreted by the previously naive bacteria, as shown by the production of bands that were recognized by the monoclonal antibodies, whereas no bands were detected in the strains not infected with the bacteriophage (Munro et al., 2003).

The creation of phenotypic profiles using more than 50 biochemical, antibiotic resistance, and carbon utilization tests for strains of *V. harveyi* infected and uninfected with VHML demonstrated changes to phenotypic profiles when infected with VHML (Vidgen, 2004). VHML-infected *V. harveyi* strain 645 demonstrated an inability to use gluconate as a sole carbohydrate source, when uninfected strain 645 was able to do so. Additionally, VHML-infected *V. harveyi* 20 was not able to hydrolyze L-glutamic acid 5-(4-nitroanilide), indicating a lack of operational glutamyltranspeptidase, whereas uninfected strain 20 was able to hydrolyze this compound. The variation in the phenotypic profiles indicates that VHML does integrate into the host's genome and causes changes to the profiles that are used to identify these bacteria. While these phenotypic changes would not cause a misidentification of *V. harveyi*, they decreased the surety of the identification (Vidgen, 2004). This raises the intriguing possibility that integration of bacteriophage into other hosts' genomes actually changes the phenotypic identification of bacteria.

A study (Busico-Salcedo, 2004) of a podovirus-like bacteriophage (VHPL) isolated from a virulent strain of *V. harveyi* (47666-1) was conducted to see if the pattern of changes in phage VHML from strain 642 would occur. Sodium dodecyl sulfate-polyacrylamide gel electrophoresis analysis showed an upregulation and production of extracellular proteins in previously naive *V. harveyi* strains receiving this phage. Noticeably, these were not all the same strains that took up VHML: strains 12 and 20 were identical, but different strains, 13 and 643, also took up VHPL. Hemolysin assays indicated a significant increase ($P < 0.001$) in both the halo of clearing and colony diameter in VHPL-infected strains of *V. harveyi*. Chitin degradation resulted in significantly greater zones of clearing ($P < 0.001$) in strains infected with bacteriophage from *V. harveyi* 47666-1. In bath challenge assays of *P. monodon* larvae, the results indicated that, as a group, there was a significant difference in mortality rate among strains infected with bacteriophage from *V. harveyi* 47666-1 ($F = 82.824$, df = 9, 40, $P < 0.001$) than strains of *V. harveyi* without the bacteriophage. However, siderophore production was not significant, as all of the inducible strains did not respond positively on chrome azurol-S agar.

CONCLUSIONS

The marine bacterium *V. harveyi* occurs naturally in tropical marine waters. It can be found as a free-living organism or as part of the gut flora of marine animals, as an opportunistic pathogen, and as both a primary and an opportunistic pathogen of marine invertebrates. Many virulence factors have been implicated in the severity of outbreaks caused by luminous vibriosis. The possession of toxic extracellular proteins such as proteases, phospholipases, hemolysins, and cysteine proteases can increase the virulence of this marine bacterium. In addition, virulence factors such as siderophores (in vertebrates), chitinases, bacteriocins, resistance plasmids, and bacteriophages (Oakey et al., 2002) also contribute to the pathogenicity of *V. harveyi*.

It was suggested that virulent isolates within *V. harveyi* are rare and that virulence is likely to be due to a mobile genetic element such as plasmids or bacteriophages (Pizzutto and Hirst, 1995). The involvement of bacteriophage in the virulence of *V. harveyi* has been documented by several authors (Ruangpan et al., 1999; Oakey and Owens, 2000; Munro et al., 2003; Busico-Salcedo, 2004). At least two strains of *V. harveyi* (strains 47666-1 and 642) were shown to cause devastating disease in prawns in northern Australia and to carry bacteriophage associated with their virulence.

REFERENCES

Abdel-Aziz, E. 2001. Association of asymptomatic mortalities in cultured white shrimp, *Penaeus indicus* with dominance of non-luminescent *Vibrio harveyi* biotypes in mariculture environment. *Vet. Med. J. Giza* 50:247–260.

Alcaide, E., C. Gil-Sanz, E. Sanjuan, D. Esteve, and C. Amaro. 2001. *Vibrio harveyi* causes disease in seahorse, *Hippocampus* sp. *J. Fish Dis.* 24:311–313.

Alsina, M., and A. Blanch. 1994. A set of keys for biochemical identification of environmental *Vibrio* species. *J. Appl. Bacteriol.* 76:79–85.

Amaro, C., R. Aznar, E. Alcaide, and M. Lemos. 1990. Iron-binding compounds and related outer membrane proteins in *Vibrio cholerae* non-O1 strains from aquatic environments. *Appl. Environ. Microbiol.* 56:2410–2416.

Anderson, I., and J. Norton. 1991. Diseases of barramundi in aquaculture. *Austasia Aquaculture* 5:21–24.

Austin, B., M. Alsina, D. Austin, A. Blanch, F. Grimont, P. Grimont, J. Joffre, S. Koblavi, J. Larsen, K. Pedersen, T. Tiainen, L. Verdonck, and J. Swings. 1995. Identification and typing of *Vibrio anguillarum*: a comparison of different methods. *Syst. Appl. Microbiol.* 18:285–302.

Austin, B., and J. Lee. 1992. *Aeromonadaceae* and *Vibrionaceae*, p. 163–182. *In* R. Board, D. Jones, and F. Skinner (ed.), *Identification Methods in Applied and Environmental Microbiology*. Society for Applied Bacteriology, Oxford, England.

Aznar, R., C. Amaro, E. Alcaide, and M. Lemos. 1989. Siderophore production by environmental strains of *Salmonella* species. *FEMS Microbiol. Lett.* 57:7–12.

Balebona, M., M. Moriñigo, A. Faris, K. Krovacek, I. Mansson, M. Bordas, and J. Borrego. 1995. Influence of salinity and pH on the adhesion of pathogenic *Vibrio* strains to *Sparus aurata* skin mucus. *Aquaculture* **132**:113–120.

Barghouti, S., R. Young, M. Olson, J. Arceaux, L. Clem, and B. Byers. 1989. Amonabactin, a novel tryptophan- or phenylalanine-containing phenolate siderophore in *Aeromonas hydrophila*. *J. Bacteriol.* **171**:1811–1816.

Bassler, B. L., E. P. Greenberg, and A. M. Stevens. 1997. Cross-species induction of luminescence in the quorum-sensing bacterium *Vibrio harveyi*. *J. Bacteriol.* **179**:4043–4045.

Baticados, M., C. Lavilla-Pitogo, E. Cruz-Lacierda, L.D. Peña, and N. Sunaz. 1990. Studies on the chemical control of luminous bacteria *Vibrio harveyi* and *V. splendidus* isolated from diseased *Penaeus monodon* larvae and rearing water. *Dis. Aquat. Org.* **9**:133–139.

Baticados, M., and J. Paclibare. 1992. The use of chemotherapeutic agents in aquaculture in the Philippines, p. 531–546. *In* M. Shariff, R. Subasinghe, and A. Arthur (ed.), *Diseases in Asian Aquaculture I.* Asian Fisheries Society, Manila, Philippines.

Baumann, P., and L. Baumann. 1977. Biology of the marine enterobacteria: genera *Beneckea* and *Photobacterium*. *Annu. Rev. Microbiol.* **31**:39–61.

Baumann, P., L. Baumann, S. Bang, and M. Woolkalis. 1980. Reevaluation of the taxonomy of *Vibrio*, *Beneckea*, and *Photobacterium*: abolition of the genus *Beneckea*. *Curr. Microbiol.* **4**:127–132.

Baumann, P., A. Furniss, and J. Lee. 1984. Genus I: *Vibrio*, p. 516–575. *In* N. R. Krieg and J. G. Holt (ed.), *Bergey's Manual of Systematic Bacteriology.* Lippincott Williams & Wilkins, Philadelphia, Pa.

Baumann, P., and R. Schubert. 1984. *Vibrionaceae*, p. 516–575. *In* N. R. Krieg and J. G. Holt (ed.), *Bergey's Manual of Systematic Bacteriology.* Lippincott Williams & Wilkins, Philadelphia, Pa.

Baya, A., P. Brayton, V. Brown, D. Grimes, E. Russek-Cohen, and R. Colwell. 1986. Coincident plasmids and antimicrobial resistance in marine bacteria isolated from polluted and unpolluted Atlantic Ocean samples. *Appl. Environ. Microbiol.* **51**:1285–1292.

Breschel, T. S., and F. L. Singleton. 1992. Use of the API rapid NFT system for identifying nonfermentative and fermentative marine bacteria. *Appl. Environ. Microbiol.* **58**:21–26.

Brown, C., and G. Roland. 1984. Characterisation of exotoxin produced by a shellfish-pathogenic *Vibrio* sp. *J. Fish Dis.* **7**:117–126.

Bryant, T. 1991. Bacterial identifier: *A Utility for Probabilistic Identification of Bacteria.* Blackwell Scientific Publications, Oxford, England.

Bryant, T., J. Lee, P. West, and R. Colwell. 1986. A probability matrix for the identification of species of *Vibrio* and related genera. *J. Appl. Bacteriol.* **61**:1183–1189.

Busico-Salcedo, N. 2004. Podovirus bacteriophage mediated virulence in *Vibrio harveyi* strain 47-666-1. M.V.Sc. thesis. James Cook University of North Queensland, Australia.

Cao, J. G., and E. A. Meighen. 1989. Purification and structural identification of an autoinducer for the luminescence system of *Vibrio harveyi*. *J. Biol. Chem.* **264**:21670–21676.

Chang, C., W. Liu, and C. Shyu. 2000. Use of prawn blood agar hemolysis to screen for bacteria pathogenic to cultured tiger prawns *Penaeus monodon*. *Dis. Aquat. Org.* **43**:153–157.

Chen, X., S. Schauder, N. Potier, A. Van Dorsselaer, I. Pelczer, B. L. Bassler, and F. Hughson. 2002. Structural identification of a bacterial quorum-sensing signal containing boron. *Nature* **415**:545–549.

Chythanya, R., I. Karunasagar, and I. Karunasagar. 2002. Inhibition of shrimp pathogenic vibrios by a marine Pseudomonas I-2 strain. *Aquaculture* **208**:1–10.

Company, R., A. Sitja-Bobadilla, M. Pujalte, E. Garay, P. Alvarez-Pellitero, and J. Perez-Sanchez. 1999. Bacterial and parasitic pathogens in cultured common dentex, *Dentex dentex* L. *J. Fish Dis.* **22**:299–309.

Crosa, J. 1984. The relationship of plasmid mediated iron transport and bacterial virulence. *Annu. Rev. Microbiol.* **38**:69–89.

Crosa, J. 1989. Genetics and molecular biology of siderophore-mediated iron transport in bacteria. *Microbiol. Rev.* **53**:517–530.

Diggles, B. K., G. A. Moss, J. Carson, and C. D. Anderson. 2000. Luminous vibriosis in rock lobster *Jasus verreauxi* (Decapoda: Palinuridae) phyllosoma larvae associated with infection by *Vibrio harveyi*. *Dis. Aquat. Org.* **43**:127–137.

Direkbusarakom, S., T. Pechmanee, M. Assavaaree, and Y. Danayadol. 1997. Effect of *Chlorella* on the growth of *Vibrio* isolated from diseased shrimp, p. 355–358. *In* T. Flegel and I. MacRae (ed.), *Disease in Asian Aquaculture III.* Fish Health Section, Asian Fisheries Society, Manila, Philippines.

Eberhard, A. 1972. Inhibition and activation of bacterial luciferase synthesis. *J. Bacteriol.* **109**:1101–1105.

Fouz, B., J. Barja, C. Amaro, C. Rivas, and A. Toranzo. 1993. Toxicity of the extracellular products of *Vibrio damsela* isolated from diseased fish. *Curr. Microbiol.* **27**:341–347.

Fuller, R. 1989. A review: probiotics in man and animals. *J. Appl. Bacteriol.* **66**:365–378.

Gauger, E. J., and M. Gomez-Chiarri. 2002. 16S ribosomal DNA sequencing confirms the synonymy of *Vibrio harveyi* and *V. carchariae*. *Dis. Aquat. Org.* **52**:39–46.

Goarant, C., J. Herlin, R. Brizard, A. Marteau, C. Martin, and B. Martin. 2000. Toxic factors of *Vibrio* strains pathogenic for shrimp. *Appl. Environ. Microbiol.* **40**:101–107.

Goarant, C., F. Merien, F. Berthe, I. Mermoud, and P. Perolat. 1999. Arbitrarily primed PCR to type *Vibrio* spp. pathogenic for shrimp. *Appl. Environ. Microbiol.* **65**:1145–1151.

Gomez-Gil, B., S. Soto-Rodriguez, A. Garcia-Gasca, A. Roque, R. Vazquez-Juarez, F. L. Thompson, and J. Swings. 2004. Molecular identification of *Vibrio harveyi*-related isolates associated with diseased aquatic organisms. *Microbiology* **150**:1769–1777.

Gooday, G. 1990. The ecology of chitin degradation. *Adv. Microb. Ecol.* **11**:387–430.

Gray, L., and A. Kreger. 1985. Purification and characterisation of an extracellular cytolysin produced by *Vibrio vulnificus*. *Infect. Immun.* **48**:62–72.

Greenberg, E. P., J. Hastings, and S. Ulitzur. 1979. Induction of luciferase synthesis in *Beneckea harveyi* by other marine bacteria. *Arch. Microbiol.* **120**:87–91.

Griffiths, E. 1987. The iron-uptake systems of pathogenic bacteria. *In* J. Bullen and E. Griffiths (ed.), *Iron and Infection: Molecular, Biological and Clinical Aspects.* John Wiley and Sons, Chichester, England.

Hada, H., and R. Sizemore. 1981. Incidence of plasmids in marine *Vibrio* spp. isolated from an oil field in the northwestern Gulf of Mexico. *Appl. Environ. Microbiol.* **41**:199–202.

Hanna, P., K. Altmann, D. Chen, A. Smith, and S. Cosic. 1991. Development of monoclonal antibodies for the identification of epizootic *Vibrio* species. *J. Fish Dis.* **15**:63–69.

Hardy, K. 1984. Introduction: other plasmids, p. 82–87. *In* K. Hardy (ed.), *Bacterial Plasmids*, 2nd ed. Van Nostrand Reinhold, Berkshire, United Kingdom.

Harris, L. J. 1993. An investigation into the virulence of strains of *Vibrio harveyi* pathogenic to larvae of the tiger prawn, *Penaeus monodon*. Honours thesis. James Cook University of North Queensland, Australia.

Harris, L. J. 1998. An investigation into the virulence and the control of strains of *Vibrio harveyi* pathogenic to larvae of the black tiger prawn, *Penaeus monodon*. Ph.D. thesis. James Cook University of North Queensland, Australia.

Harris, L. J., and L. Owens. 1997. Guidelines for controlling luminous vibriosis in *Penaeus monodon*. *Austasia Aquaculture* **11**:59–62.

Harris, L. J., and L. Owens. 1999. Production of exotoxins by two luminous *Vibrio harveyi* strains known to be primary pathogens of *Penaeus monodon* larvae. *Dis. Aquat. Org.* **38**:11–22.

Harris, L. J., L. Owens, and S. Smith. 1996. A selective and differential medium for *Vibrio harveyi*. *Appl. Environ. Microbiol.* **62**:3548–3550.

Havenaar, R., B. T. Brink, and J. H. Veld. 1992. Selection of strain for probiotic use, p. 209–224. *In* R. Fuller (ed.), *Probiotics. The Scientific Basis.* Chapman and Hall, London, England.

Hendrie, M., W. Hodgkiss, and J. Shewan. 1970. The identification, taxonomy, and classification of luminous bacteria. *J. Gen. Microbiol.* **64**:151–169.

Hirst, I., T. Hastings, and A. Ellis. 1994. Utilisation of haem compounds by *Aeromonas salmonicida*. *J. Fish Dis.* **17**:365–373.

Hispano, C., Y. Nebra, and A. Blanch. 1997. Isolation of *Vibrio harveyi* from an ocular lesion in the short sunfish (*Mola mola*). *Bull. Eur. Assoc. Fish Pathol.* **17**:104–107.

Hoyt, R., and R. Sizemore. 1982. Competitive dominance by a bacteriocin-producing *Vibrio harveyi* strain. *Appl. Environ. Microbiol.* **44**:653–658.

Igarashi, M., H. Sugita, and Y. Deguchi. 1989. Microflora associated with eggs and nauplii of *Artemia salina*. *Nipp. Suis. Gakk.* **55**:2045.

Ishimaru, K., and K. Muroga. 1997. Taxonomic re-examination of two pathogenic *Vibrio* species isolated from milkfish and swimming crab. *Fish Pathol.* **32**:59–64.

Jannatipour, M., R. W. Soto-Gil, L. C. Childers, and J. W. Zyskind. 1987. Translocation of *Vibrio harveyi* N,N′-diacetylchitobiase to the outer membrane of *Escherichia coli*. *J. Bacteriol.* **169**:3785–3791.

Jiravanichpaisal, P., T. Miyazaki, and C. Limsuwan. 1994. Histopathology, biochemistry and pathogenicity of *Vibrio harveyi* infecting black tiger prawn *Penaeus monodon*. *J. Aquat. Anim. Health* **6**:27–35.

Johnson, F., and I. Shunk. 1936. An interesting new species of luminous bacteria. *J. Bacteriol.* **31**:585–592.

Jones, S., B. Yu, N. J. Bainton, M. Birdsall, B. W. Bycroft, S. R. Chhabra, A. J. Cox, P. Golby, P. J. Reeves, S. Stephens, et al. 1993. The lux autoinducer regulates the production of exoenzyme virulence determinants in *Erwinia carotovora* and *Pseudomonas aeruginosa*. *EMBO J.* **12**:2477–2482.

Karunasagar, I., S. Otta, and I. Karunasagar. 1996. Biofilm formation by *Vibrio harveyi* on surfaces. *Aquaculture* **140**:241–245.

Karunasagar, I., S. Otta, and I. Karunasagar. 1998. Disease problems affecting cultured Penaeid shrimp in India. *Fish Pathol.* **33**:413–419.

Karunasagar, I., R. Pai, G. Malathi, and I. Karunasagar. 1994. Mass mortality of *Penaeus monodon* larvae due to antibiotic resistant *Vibrio harveyi* infection. *Aquaculture* **128**:203–209.

Kraxberger-Beatty, T., D. McGarey, H. Grier, and D. Lim. 1990. *Vibrio harveyi*, an opportunistic pathogen of common snook, *Centropomus undecimalis* (Bloch), held in captivity. *J. Fish Dis.* **13**:557–560.

Lavilla-Pitogo, C., L. Albright, M. Paner, and N. Sunaz. 1992. Studies on the sources of luminescent *Vibrio harveyi* in *Penaeus monodon* hatcheries, p. 157–164. *In* M. Shariff, R. Subasinghe, and J. Arthur (ed.), *Diseases in Asian Aquaculture*. Fish Health Section, Asian Fisheries Society, Manila, Philippines.

Lavilla-Pitogo, C., M. Baticados, E. Cruz-Lacierda, and L. dela Peña. 1990. Occurrence of luminous bacterial disease of *Penaeus monodon* larvae in the Philippines. *Aquaculture* **91**:1–13.

Lavilla-Pitogo, C., and L. dela Peña. 1998. Bacterial disease in shrimp (*Penaeus monodon*) cultured in the Philippines. *Fish Pathol.* **33**:405–411.

Lee, K.-K., Y. L. Chen, and P. C. Liu. 1999. Hemostasis of tiger prawn *Penaeus monodon* affected by *Vibrio harveyi*, extracellular products, and a toxic cysteine protease. *Blood Cells Mol. Dis.* **25**:180–192.

Lightner, D., T. Bell, R. Redman, L. Mohney, J. Natividad, A. Rukyani, and A. Poernomo. 1992. A review of some major diseases of economic significance in penaeid prawns/shrimp of the Americas and Indopacific, p. 58–75. *In* S. Subasinghe and J. Arthur (ed.), *Diseases in Asian Aquaculture 1*. Asian Fisheries Society, Manila, Philippines.

Lindqvist, B. 1998. Bacteriophage and gene transfer. *APMIS* (Suppl.)**84**:15–18.

Liu, P. C., and K. K. Lee. 1999. Cysteine protease is a major exotoxin of pathogenic luminous *Vibrio harveyi* in the tiger prawn, *Penaeus monodon*. *Lett. Appl. Microbiol.* **28**:428–430.

Liu, P. C., K. K. Lee, K. C. Yii, G. H. Kou, and S. N. Chen. 1996. Isolation of *Vibrio harveyi* from diseased kuruma prawns *Penaeus japonicus*. *Curr. Microbiol.* **33**:129–132.

Madigan, M., J. Martinko, and J. Parker (ed.). 2000. *Brock Biology of Microorganisms*, 9th ed., p. 784–786. Prentice Hall International, Englewood Cliffs, N.J.

Maeda, M. 1988. Microorganisms and protozoea as feed in mariculture. *Progr. Ocean.* **21**:201–206.

Makemson, J. C., N. Fulayfil, and P. Basson. 1992. Association of luminous bacteria with artificial and natural surfaces in Arabian Gulf seawater. *Appl. Environ. Microbiol.* **58**:2341–2343.

Manefield, M., L. Harris, S. A. Rice, R. de Nys, and S. Kjelleberg. 2000. Inhibition of luminescence and virulence in the black tiger prawn (*Penaeus monodon*) pathogen *Vibrio harveyi* by intercellular signal antagonists. *Appl. Environ. Microbiol.* **66**:2079–2084.

McCall, J. O., and R. K. Sizemore. 1979. Description of a bacteriocinogenic plasmid in *Beneckea harveyi*. *Appl. Environ. Microbiol.* **38**:974–979.

Milton, D. L., A. Hardman, M. Camara, S. R. Chhabra, B. W. Bycroft, G. S. Stewart, and P. Williams. 1997. Quorum sensing in *Vibrio anguillarum*: characterisation of the *vanI/vanR* locus and identification of the autoinducer N-(3-oxodecanoyl)-L-homoserine lactone. *J. Bacteriol.* **179**:3004–3012.

Mohney, L. L., D. V. Lightner, and T. A. Bell. 1994. An epizootic of vibriosis in Ecuadorian pond-reared *Penaeus vannamei* (Crustacea: Decapoda). *J. World Aquacult. Soc.* **25**:116–125.

Montero, A., and B. Austin. 1999. Characterisation of extracellular products from an isolate of *Vibrio harveyi* recovered from diseased post-larval *Penaeus vanname* (Bonne). *J. Fish Dis.* **22**:377–386.

Montgomery, M., and D. L. Kirchman. 1993. Role of chitin-binding proteins in the specific attachment of the marine bacterium *Vibrio harveyi* to chitin. *Appl. Environ. Microbiol.* **59**:373–379.

Muir, P. 1991. Factors affecting the survival of penaeid prawns in culture with particular reference to the larval stages. Ph.D. thesis. James Cook University of North Queensland, Australia.

Munro, J. 2001. Bacteriophage-mediated exotoxin production in naive bacterial hosts. Honours thesis. James Cook University of North Queensland, Australia.

Munro, J., H. J. Oakey, E. Bromage, and L. Owens. 2003. Experimental bacteriophage-mediated virulence in strains of *Vibrio harveyi*. *Dis. Aquat. Org.* **54**:187–194.

Nagasawa, S., M. Terazaki, and T. Nemoto. 1987. Bacterial attachment to the epipelagic copepod *Acartia* and the bathypelagic copepod *Calanus*. *Proc. Jpn. Acad. Serv. B* **63**:33–35.

Neilands, J. 1981. Microbial iron compounds. *Annu. Rev. Biochem.* **50**:715–731.

Nishibuchi, M., A. Fasano, R. Russel, and J. B. Kaper. 1992. Enterotoxigenicity of *Vibrio parahaemolyticus* with and without genes encoding thermostable direct haemolysin. *Infect. Immun.* **60**:3539–3545.

Nithimathachoke, C., P. Pratanpipat, K. Thongdaeng, and G. Nash. 1995. Luminous bacterial infection in pond-reared *Penaeus monodon*. *Asian Shrimp News* 23:1–4.

Nottage, A., and T. H. Birkbeck. 1987. Purification of a proteinase produced by the bivalve pathogen *Vibrio alginolyticus* NCMB 1339. *J. Fish Dis.* 10:211–220.

Oakey, H. J. 2000. The mode of action and risk assessment of bacteriophage associated with virulent strains of the marine pathogen, *Vibrio harveyi*. *Annual Report, Fish Bites Newsletter*, vol. 2002. Aquafin CRC Education and Training Industry Programs, South Australia.

Oakey, H. J., B. R. Cullen, and L. Owens. 2002. The complete nucleotide sequence of the *Vibrio harveyi* bacteriophage VHML. *J. Appl. Microbiol.* 93:1089–1098.

Oakey, H. J., N. Levy, D. G. Bourne, B. Cullen, and A. Thomas. 2003. The use of PCR to aid in the rapid identification of *Vibrio harveyi* isolates. *J. Appl. Microbiol.* 95:1293–1303.

Oakey, H. J., and L. Owens. 2000. A new bacteriophage, VHML, isolated from a toxin-producing strain of *Vibrio harveyi* in tropical Australia. *J. Appl. Microbiol.* 89:702–709.

Owens, L., and S. Hall-Mendelin. 1990. Recent advances in Australian penaeid diseases and pathology. Advances in Tropical Aquaculture, Tahiti (1989). *Actes Colloques* 9:103–112.

Owens, L., D. Austin, and B. Austin. 1996. Effect of strain origin on siderophore production in *Vibrio harveyi* isolates. *Dis. Aquat. Org.* 27:157–160.

Owens, L., P. Muir, D. Sutton, and M. Wingfields. 1992. The pathology of microbial disease in tropical Australian crustacea, p. 165–172. *In* M. Shariff, R. Subasinghe, and A. Arthur (ed.), *Diseases in Asian Aquaculture*. Fish Health Section, Asian Fisheries Society, Manila, Philippines.

Panichsuke, P., L. Ruangpan, and T. Pechmanee. 1997. Inhibitory effect of *Skeletonema costatum* on the growth of *Vibrio harveyi*, p. 122. *In Proc. 2nd Asian-Pacific Marine Biotechnology Conference, 7–10 May 1997, Phuket, Thailand*.

Park, S., I. Shimamura, M. Fukunaga, K, Mori, and T. Nakai. 2000. Isolation of bacteriophage specific to a fish pathogen, *Pseudomonas plecoglossida*, as a candidate for disease control. *Appl. Environ. Microbiol.* 66:1416–1422.

Park, S. C., and T. Nakai. 2003. Bacteriophage control of *Pseudomonas plecoglossicida* infection in ayu *Plecoglossus altivelis*. *Dis. Aquat. Org.* 53:33–39.

Pasharawipas, T., S. Sriurairatana, S. Direkbusarakom, Y. Danayadol, S. Thaikua, L. Ruangpan, and T. Flegel. 1998. Luminous *Vibrio harveyi* associated with tea brown gill syndrome in black tiger shrimp. *In* T. Flegel (ed.), *Advances in Shrimp Biotechnology*. National Centre for Genetic Engineering and Biotechnology, Bangkok, Thailand.

Pass, D., R. Dybdahl, and M. Mannion. 1987. Investigation into the causes of mortality of the pearl oyster, *Pinctada maxima* (Jamson), in Western Australia. *Aquaculture* 65:149–169.

Payne, S. 1983. Siderophores and acquisition of iron by gram-negative pathogens, p. 346–349. *In* D. Schlessinger (ed.), *Microbiology*. American Society for Microbiology, Washington, D.C.

Pedersen, K., L. Verdonck, D. Austin, A. Blanch, P. A. D. Grimont, J. Jofre, S. Koblavi, J. Larsen, T. Tiainen, M. Vigneullle, and J. Swings. 1998. Taxonomic evidence that *Vibrio carchariae* Grimes et al., 1985 is a junior synonym of *Vibrio harveyi* (Johnson and Shunk 1936, Baumann et al., 1981). *Int. J. Syst. Bacteriol.* 48:749–758.

Phianpark, W., S. Piyatiratitivorakul, P. Menasvita, and S. Rengpipat. 1997. Use of probiotics in *Penaeus monodon*, p. 18. *In Proc. 2nd Asian-Pacific Marine Biotechnology Conference, 7–10 May 1997, Phuket, Thailand*.

Pirhonen, M., D. Flego, R. Heikinheimo, and E. Palva. 1993. A small diffusible molecule is responsible for the global control of virulence and exoenzyme production in the plant pathogen *Erwina carotovora*. *EMBO J.* 12:2467–2476.

Pizzutto, M., and R. Hirst. 1995. Classification of isolates of *Vibrio harveyi* virulent to *Penaeus monodon* larvae by protein profile analysis and M13 DNA fingerprinting. *Dis. Aquat. Org.* 21:61–68.

Prayitno, S., and J. Latchford. 1995. Experimental infections of crustaceans with luminous bacteria related to *Photobacterium* and *Vibrio*: effect of salinity and pH on infectiosity. *Aquaculture* 132:105–112.

Raimondi, F., J. Kao, C. Fiorentini, A. Fabbri, G. Donelli, N. Gasparini, A. Rubino, and A. Fasano. 2000. Enterotoxicity and cytotoxicity of *Vibrio parahaemolyticus* thermostable direct haemolysin in *in vitro* systems. *Infect. Immun.* 68:3180–3185.

Ramaiah, N., J. Chun, J. Ravel, W. L. Straube, R. T. Hill, and R. R. Colwell. 2000. Detection of luciferase gene sequences in nonluminescent bacteria from the Chesapeake Bay. *FEMS Microbiol. Ecol.* 33:27–34.

Ramesh, A., B. Loganathan, and V. Venugopalan. 1989. Seasonal distribution of luminous bacteria in the sediments of a tropical estuary. *J. Gen. Appl. Microbiol.* 35:363–368.

Rasheed, V. 1989. Diseases of cultured brown-spotted grouper *Epinephelus tauvina* and silvery black porgy *Acanthopagrus cuviere* in Kuwait. *J. Aquat. Anim. Health* 1:102–107.

Reichelt, J. L., and P. Baumann. 1973. Taxonomy of the marine luminous bacteria. *Arch. Microbiol.* 94:283–330.

Rengpipat, S., S. Rukpratanporn, S. Piyatiratitivorakul, and P. Menasvita. 2000. Immunity enhancement in black tiger shrimp (*Penaeus monodon*) by a probiont bacterium (*Bacillus* S11). *Aquaculture* 191:271–288.

Richmond, M. 1973. Resistance factors and their ecological importance to bacteria and to man, p. 191–248. *In* J. Davidson and W. Cohn (ed.), *Progress in Nucleic Acid Research and Molecular Biology*. Academic Press, Inc., New York, N.Y.

Robertson, P. A., J. Calderon, L. Carrera, J. Stark, M. Zherdmant, and B. Austin. 1998. Experimental *Vibrio harveyi* infections in *Penaeus vannamei* larvae. *Dis. Aquat. Org.* 32:151–155.

Ruangpan, L. 1998. Luminous bacteria associated with shrimp mortality, p. 205–211. *In* T. Flegel (ed.), *Advances in Shrimp Biotechnology*. National Centre for Genetic Engineering and Biotechnology, Bangkok, Thailand.

Ruangpan, L., Y. Danayadol, S. Direkbusarakom, S. Siurairatana, and T. Flegel. 1999. Lethal toxicity of *Vibrio harveyi* to cultivated *Penaeus monodon* induced by a bacteriophage. *Dis. Aquat. Org.* 35:195–201.

Saaed, M. 1995. Association of *Vibrio harveyi* with mortalities in cultured marine fish in Kuwait. *Aquaculture* 136:21–29.

Sigel, S., and S. Payne. 1982. Effect of iron limitation on growth, siderophore production and expression of outer membrane proteins of *Vibrio cholerae*. *J. Bacteriol.* 150:148–155.

Simidu, U., T. Noguchi, D. Hwang, Y. Shida, and K. Hashimoto. 1987. Marine bacteria which produce tetrodotoxin. *Appl. Environ. Microbiol.* 53:1714–1715.

Song, Y., S. Lee, Y. Lin, and C. Chen. 1992. Enzyme immunoassay for shrimp vibriosis. *Dis. Aquat. Org.* 14:43–50.

Soto-Gil, R. W. 1988. Chitobiase and chitinase from *Vibrio harveyi*. Ph.D. dissertation. University of California, San Diego.

Sun, W., J. G. Cao, K. Teng, and E. A. Meighen. 1994. Biosynthesis of poly-3-hydroxybutyrate in the luminescent bacterium, *Vibrio harveyi*, and regulation by the lux autoinducer, *N*-(3-hydroxybutanoyl)homoserine lactone. *J. Biol. Chem.* 269:20785–20790.

Sunaryanto, A., and A. Mariam. 1986. Occurrence of a pathogenic bacteria causing luminescence in penaeid larvae in Indonesian hatcheries. *Bull. Brackishwater Aquacult. Dev. Centre* 8:105–112.

Supriyadi, H., and A. Rukyani. 1992. The use of chemotherapeutic agents for the treatment of bacterial disease of fish and shrimp in

Indonesia, p. 515–517. *In* M. Shariff, R. Subasinghe, and A. Arthur (ed.), *Diseases in Asian Aquaculture 1.* Asian Fisheries Society, Manila, Philippines.

Svitil, A. L., S. Chadhain, J. Moore, and D. L. Kirchman. 1997. Chitin degradation proteins produced by the marine bacterium *Vibrio harveyi* growing on different forms of chitin. *Appl. Environ. Microbiol.* **63:**408–413.

Tagg, J., A. Dajani, and L. Wannamaker. 1976. Bacteriocins of gram-positive bacteria. *Bacteriol. Rev.* **40:**722–756.

Tendencia, E. A. 2002. *Vibrio harveyi* isolated from cage-cultured seabass *Latex calcarifer* Bloch in the Philippines. *Aquacult. Res.* **33:**455–458.

Thompson, F. L., B. Hoste, K. Vandemeulebroecke, K. Engelbeen, R. Denys, and J. Swings. 2002. *Vibrio trachuri* Iwamoto et al. 1995 is a junior synonym of *Vibrio harveyi* (Johnson and Shunk 1936) Baumann et al. 1981. *Int. J. Syst. Evol. Microbiol.* **52:**973–976.

Thompson, S., M. Smith, M. Wilkinson, and K. Peek. 2001. Identification and characterisation of a chitinase antigen from *Pseudomonas aeruginosa* strain 385. *Appl. Environ. Microbiol.* **67:**4001–4008.

Tortora, G. J., B. R. Funke, and C. L. Case. 2000. *Microbiology: an Introduction,* 7th ed. Benjamin Cummings, San Francisco, Calif.

Vandenberghe, J., Y. Li, L. Verdonck, J. Li, P. Sorgeloos, H.-S. Xu, and J. Swings. 1998. Vibrios associated with *Penaeus chinensis* larvae in Chinese shrimp hatcheries. *Aquaculture* **169:**121–132.

Vandenberghe, J., L. Verdonck, R. Robles-Arozarena, G. Rivera, A. Bolland, M. Balladares, B. Gomez-Gil, J. Calderon, P. Sorgeloos, and J. Swings. 1999. Vibrios associated with *Litopenaeus vannamei* larvae, postlarvae, broodstock, and hatchery probionts. *Appl. Environ. Microbiol.* **65:**2592–2597.

Vidgen, M. 2004. Investigation of the integration site of the bacteriophage *Vibrio harveyi* myovirus-like (VHML) in *Vibrio harveyi.* Honours thesis. James Cook University of North Queensland, Australia.

Von Bodman, S., and S. Farrand. 1995. Capsular polysaccharide biosynthesis and pathogenicity in *Erwinia stewartii* require induction by an *N*-acylhomoserine lactone autoinducer. *J. Bacteriol.* **177:**5000–5008.

Welsh, J., and M. McClelland. 1990. Fingerprinting genomes using PCR with arbitrary primers. *Nucleic Acids Res.* **18:**7213–7218.

West, P., P. Brayton, T. Bryant, and R. Colwell. 1986. Numerical taxonomy of vibrios isolated from aquatic environments. *Int. J. Syst. Bacteriol.* **36:**531–543.

Zhang, X.-H., and B. Austin. 2000. Pathogenicity of *Vibrio harveyi* to salmonids. *J. Fish Dis.* **23:**93–102.

Zhang, X.-H., P. G. Meaden, and B. Austin. 2001. Duplication of hemolysin genes in a virulent isolate of *Vibrio harveyi. Appl. Environ. Microbiol.* **67:**3161–3167.

Ziegler, M. M., and T. O. Baldwin. 1981. Biochemistry of bacterial luminescence. *Curr. Top. Bioenerg.* **12:**65–113.

The Biology of Vibrios
Edited by F. L. Thompson et al.
© 2006 ASM Press, Washington, D.C.

Chapter 20

Vibrio salmonicida

BRIAN AUSTIN

INTRODUCTION

Norway has become a dominant producer of farmed Atlantic salmon (*Salmo salar*), and with the intensity of production, disease is a regular occurrence even in the most well managed facilities. In 1979, an apparently "new" disease was recognized in Atlantic salmon farms around the island of Hitra, which is located south of Trondheim, Norway. A few years later (during 1983), this disease was diagnosed in Stavanger, particularly in a large number of salmon farms in the region of Bergen. With its recognition as a likely new condition, the disease was named cold-water vibriosis (because of the association with cold water, i.e., not normally >10°C) (Colquhoun and Sørum, 2001; Colquhoun et al., 2002) or Hitra disease (after the area from which it was originally recognized) (Egidius et al., 1981). The disease was noted to occur mainly during the colder months, from late autumn to early spring. After its original recognition in the vicinity of Hitra, the disease spread throughout Norway and moved to other Atlantic salmon-producing areas, including Scotland and Shetland (Bruno et al., 1985), the Atlantic seaboard of Canada and possibly Chile.

CHARACTERISTICS OF THE DISEASE

Resembling a generalized hemorrhagic septicemia, coldwater vibriosis is characterized by hemorrhaging around the abdomen (Holm et al., 1985). Internally, anemia and hemorrhaging of the organs, swim bladder, abdominal wall, and posterior gastrointestinal tract may be seen (Poppe et al., 1985; Holm et al., 1985; Egidius et al., 1986). Microscopic examination has revealed that the bacteria are spread throughout infected fish, being especially plentiful in the blood and kidney of moribund and newly dead animals. Using isolated macrophages from Atlantic salmon and rainbow trout coupled with immunofluorescence techniques, the pathogen was observed to be internalized (Brattgjerd et al., 1995).

ISOLATION OF THE PATHOGEN

Cultures have been recovered from blood and kidney on tryptic soy agar supplemented with 1.5% (wt/vol) NaCl and incubation at 15°C for up to 5 days (Holm et al., 1985; Egidius et al., 1986). Dense pure culture growth of small, i.e., ≤1 mm in diameter, round, raised, entire, and translucent colonies developed.

CHARACTERISTICS OF THE ORGANISM

Cultures, which were subsequently named as a new species, *Vibrio salmonicida*, were recovered from diseased salmon (Egidius et al., 1986). Yet, unlike most previously recognized vibrios, this pathogen was comparatively slow growing, did not produce a great deal of biomass on conventional media, was prone to losing viability without regular subbing or by incubation at supraoptimum temperatures, and was fairly unreactive (Austin et al., 1997). Cultures were observed to contain motile (with cells possessing ~9 polarly arranged flagella), fermentative, gram-negative, curved pleomorphic rods of 2×0.5 to 3×0.5 μm in size. Luminescence has been reported, with *V. salmonicida* serving as an autoinducer of *Vibrio fischeri* and conversely responding to (*V. fischeri*) autoinducer (Fidopiastis et al., 1999). Catalase and oxidase were produced, but not arginine dihydrolase, β-galactosidase, H_2S, or indole. Nitrates were not reduced, nor was the Voges Proskauer reaction positive.

Brian Austin • School of Life Sciences, John Muir Building, Heriot-Watt University, Riccarton, Edinburgh EH14 4AS, Scotland, United Kingdom.

Sodium citrate was not utilized. Neither blood, chitin, gelatin, lipids, nor urea was degraded. *N*-acetylglucosamine, glucose, glycerol (slowly), maltose, ribose, sodium gluconate, and trehalose were utilized, but not adonitol, amygdalin, D- or L-arabinose, D- or L-arabitol, arbutin, D-cellobiose, dulcitol, erythritol, D- or L-fucose, inositol, β-gentobiose, lactose, D-lycose, D-mannose, melezitose, melibiose, D-raffinose, rhamnose, salicin, sorbitol, sucrose, L-sorbose, D-tagatose, D-turanose, or D- or L-xylose. Growth was determined to occur at 1 to 22°C, optimally at 15°C, but not at 37°C, and in 0 to 4% but not 7% (wt/vol) NaCl. Sensitivity was displayed to the vibriostatic agent, O/129, but not to novobiocin. The G+C ratio of the DNA was reported as 44 mol% (Holm et al., 1985; Egidius et al., 1986).

Strains have been grouped into four categories on the basis of plasmid composition (Wiik et al., 1989). These plasmids were of 2.6, 3.4, and 24 MDa in size, with the largest being common to all four groups. However, there was not any apparent difference in biochemical traits among the four plasmid groups. Separately, Sørum et al. (1990) described plasmids of 2.8, 3.4, 21, and 61 MDa from isolates recovered from Atlantic salmon and cod. These workers reported 11 plasmid profiles for *V. salmonicida*. Indeed, a similar plasmid composition has been reported for isolates from the Faroe Islands (Nielsen and Dalsgaard, 1991). When comparing isolates from Canada, Faroe Islands, Norway, and Shetland, Sørum et al. (1993) noted a similarity in plasmid profile, with three plasmids of 2.8, 3.4, and 21 MDa revealed. Furthermore, a conclusion has been reached that all strains carried plasmids (Valla et al., 1992).

DNA hybridization of four cultures confirmed homogeneity (DNA homology was 82 to 100%) but with low relatedness to *Vibrio anguillarum* (30%), *Vibrio ordalii* (34%), and *Vibrio parahaemolyticus* (40%) (Wiik and Egidius, 1986). These data were used to substantiate the unique position of *V. salmonicida* and confirm the wisdom of naming a new species to accommodate the isolates. The validity of *V. salmonicida* has now been attested by other studies (e.g., Austin et al., 1997).

ECOLOGY

V. salmonicida has been determined to survive for >14 months in laboratory-based experiments with seawater, when seeded at a comparatively high dose of ~10^6 cells/ml (Hoff, 1989). This indicates that the pathogen has the capability of surviving for comparatively long times in the aquatic environment, such as in the direct vicinity of fish farms (Husevåg

et al., 1991). Further work revealed that the pathogen could be detected in the sediment (12 to 43 cells/ml) below fish farms several months after a disease outbreak. However, it is more worrying to note that *V. salmonicida* has been found in sediments even from fish farms that were not experiencing clinical disease (Enger et al., 1989, 1991).

PATHOGENICITY

Experimental challenge by means of intraperitoneal injection, immersion, and cohabitation of Atlantic salmon with *V. salmonicida* has led to the development of clinical disease with pathogenicity related to water temperature (Nordmo and Ramstad, 1999). The data indicated that Atlantic salmon was more susceptible than rainbow trout (Egidius et al., 1986; Hjeltnes et al., 1987), with a 50% lethal dose (for Atlantic salmon) of 4×10^6 to 1×10^8 cells/fish (Wiik et al., 1989). However, the susceptibility of smolts in seawater may reflect the overall quality of the fish (Eggset et al., 1997). Also, the nature of the feeding regimen has been determined to influence susceptibility to infection, with optimal feeding having less effect than restricted feeding regimens (Damsgard et al., 2004). Susceptibility to *V. salmonicida* was influenced by prior infection. Thus, in fish displaying acute infectious pancreatic necrosis with concomitant high titers to the virus, higher mortalities resulted from subsequent infection with *V. salmonicida* than when the bacterium was the only pathogen present (Johansen and Sommer, 2001).

Incidentally, plasmids do not appear to exert any direct effect on virulence (Valla et al., 1992). Yet, a siderophore and iron-regulated outer membrane protein system has been documented (Colquhoun and Sørum, 2001) and may well have relevance in terms of scavenging for iron. The dihydroxamate siderophore bisucaberin (Winkelmann et al., 2002) was produced significantly only at ≤10°C (Colquhoun and Sørum, 2001), which is when the disease is most troublesome to Atlantic salmon. Furthermore, expression of high-molecular-weight iron-regulated outer membrane protein was apparently suppressed at 15°C but not at 10°C. A proposal was made that temperature-sensitive iron sequestration could well constitute an important virulence mechanism for *V. salmonicida* (Colquhoun and Sørum, 2001).

In addition to the interest in salmonids, one report has pointed to the susceptibility of cod to *V. salmonicida* (Jørgensen et al., 1989). In this species, an elimination mechanism of lipopolysaccharide (LPS), a potent endotoxin of gram-negative bacteria, from blood was described (Seternes et al., 2001). Data revealed that the endocardial endothelial cells (possibly via the endosomal/lysosomal vesicles) of the heart took up LPS by endocytosis when administered to cod by injection, thus removing the toxin from circulation (Seternes et al., 2001).

DISEASE CONTROL

There has been success with the development of whole-cell formalized vaccines, leading to the commercialization of more than one polyvalent product. Immersion of Atlantic salmon in these formulations resulted in protection for at least 6 months (Holm and Jørgensen, 1987), with humoral antibodies being developed to the LPS component, particularly the O-side chain (Steine et al., 2001). Furthermore, *V. salmonicida* vaccines have adjuvant activities on T-dependent and T-independent antigens in salmonids, with vaccine preparations enhancing the antibody response. Thus, the inclusion of inactivated *V. salmonicida* antigens in vaccine preparations may well have an overall beneficial effect on the recipient fish (Hoel et al., 1998).

As an alternative strategy, there may be benefit in investigating natural disease resistance among stocks of Atlantic salmon, with previous work suggesting a significant variation in resistance patterns (Gjedrem and Aulstad, 1974).

REFERENCES

Austin, B., D. A. Austin, A. R. Blanch, M. Cerdà, F. Grimont, P. A. D. Grimont, J. Jofre, S. Koblavi, J. L. Larsen, K. Pedersen, T. Tiainen, L. Verdonck, and J. Swings. 1997. A comparison of methods for the typing of fish-pathogenic *Vibrio* spp. *Syst. Appl. Microbiol.* 20:89–101.

Brattgjerd, S., Ø. Evensen, L. Speilberg, and A. Lauve. 1995. Internalization of *Vibrio salmonicida* in isolated macrophages from Atlantic salmon *(Salmo salar)* and rainbow trout *(Oncorhynchus mykiss)* evaluated by a paired immunofluorescence technique. *Fish Shellfish Immunol.* 5:121–135.

Bruno, D. W., T. S. Hastings, A. E. Ellis, and R. Wootten. 1985. Outbreak of a cold-water vibriosis in Atlantic salmon in Scotland. *Bull. Eur. Assoc. Fish Pathol.* 5:62–63.

Colquhoun, D. J., and H. Sørum. 2001. Temperature dependent siderophore production in *Vibrio salmonicida*. *Microb. Pathog.* 31:213–219.

Colquhoun, D. J., K. Alvheim, K. Dommarsnes, C. Syvertsen, and H. Sørum. 2002. Relevance of incubation temperature for *Vibrio salmonicida* vaccine production. *J. Appl. Microbiol.* 92:1087–1096.

Damsgard, B., U. Sørum, I. Ugelstad, R. A Eliassen, and A. Mortensen. 2004. Effects of feeding regime on susceptibility of Atlantic salmon *(Salmo salar)* to cold water vibriosis. *Aquaculture* 239:37–46.

Eggset, G., A. Mortensen, L.-H. Johansen, and A.-I. Sommer. 1997. Susceptibility to furunculosis, cold water vibriosis, and infectious pancreatic necrosis (IPN) in post-smolt Atlantic salmon

(Salmo salar L.) as a function of smolt status by seawater transfer. *Aquaculture* 158:179–191.

Egidius, E., K. Andersen, E. Causen, and J. Raa. 1981. Cold water vibriosis or "Hitra disease" in Norwegian salmonid farming. *J. Fish Dis.* 4:353–354.

Egidius, E., R. Wiik, K. Andersen, K. A. Hoff, and B. Hjeltnes. 1986. *Vibrio salmonicida* sp. nov., a new fish pathogen. *Int. J. Syst. Bacteriol.* 36:518–520.

Enger, Ø., B. Husevåg, and J. Goksøyr. 1989. Presence of the fish pathogen *Vibrio salmonicida* in fish farm sediments. *Appl. Environ. Microbiol.* 55:2815–2818.

Enger, Ø., B. Husevåg, and J. Goksøyr. 1991. Seasonal variation in presence of *Vibrio salmonicida* and total bacterial counts in Norwegian fish farm water. *Can. J. Microbiol.* 37:618–623.

Fidopiastis, P. M., H. Sørum, and E. G. Ruby. 1999. Cryptic luminescence in the cold-water fish pathogen *Vibrio salmonicida*. *Arch. Microbiol.* 171:205–209.

Gjedrem, T., and D. Aulstad. 1974. Selection experiments with salmonids. 1. Differences in resistance to vibrio disease of salmon parr *(Salmo salar)*. *Aquaculture* 3:51–59.

Hjeltnes, B., K. Andersen, H.-M. Ellingsen, and E. Egidius. 1987. Experimental studies on the pathogenicity of a *Vibrio* sp. isolated from Atlantic salmon, *Salmo salar* L., suffering from Hitra disease. *J. Fish Dis.* 10:21–27.

Hoel, K., G. Holstad, and A. Lillehaug. 1998. Adjuvant activities of a *Vibrio salmonicida* bacterin on T-dependent and T-independent antigens in rainbow trout *(Oncorhynchus mykiss)*. *Fish Shellfish Immunol.* 8:287–293.

Hoff, K. A. 1989. Survival of *Vibrio anguillarum* and *Vibrio salmonicida* at different salinities. *Appl. Environ. Microbiol.* 55: 1775–1786.

Holm, K. O., E. Strøm, K. Stemsvåg, J. Raa, and T. Jørgensen. 1985. Characteristics of a *Vibrio* sp. associated with the "Hitra disease" of Atlantic salmon in Norwegian fish farms. *Fish Pathol.* 20:125–129.

Holm, K. O., and T. Jørgensen. 1987. A successful vaccination of Atlantic salmon, *Salmo salar* L., against "Hitra disease" or cold-water vibriosis. *J. Fish Dis.* 10:85–90.

Husevåg, B., B. T. Lunestad, P. J. Johannessen, Ø. Enger, and O. B. Samuelsen. 1991. Simultaneous occurrence of *Vibrio salmonicida* and antibiotic-resistant bacteria in sediments at abandoned aquaculture sites. *J. Fish Dis.* 14:631–640.

Johansen, L. H., and A. I. Sommer. 2001. Infectious pancreatic necrosis virus infection in Atlantic salmon *Salmo salar* postsmolts affects the outcome of secondary infections with infectious salmon anaemia virus or *Vibrio salmonicida*. *Dis. Aquat. Org.* 47:109–117.

Jørgensen, T., K. Midling, S. Espelid, R. Nilsen, and K. Stensvåg. 1989. *Vibrio salmonicida*, a pathogen in salmonids, also causes mortality in net-pen captured cod *(Gadus morhua)*. *Bull. Eur. Assoc. Fish Pathol.* 9:42–44.

Nielsen, B., and I. Dalsgaard. 1991. Plasmids in *Vibrio salmonicida* isolates from the Faroe Islands. *Bull. Eur. Assoc. Fish Pathol.* 11:206–207.

Nordmo, R., and A. Ramstad. 1999. Variables affecting the challenge pressure of *Aeromonas salmonicida* and *Vibrio salmonicida* in Atlantic salmon *(Salmo salar* L.). *Aquaculture* 171: 1–12.

Poppe, T. T., T. Håstein, and R. Salte. 1985. Hitra disease (haemorrhagic syndrome) in Norwegian salmon farming: present status, p. 223–229. *In* A. E. Ellis (ed.), *Fish and Shellfish Pathology*. Academic Press, Inc., New York, N.Y.

Seternes, T., R. A. Dalmo, J. Hoffman, J. Bogwald, S. Zykova, and B. Smedsrod. 2001. Scavenger-receptor-mediated endocytosis of lipopolysaccharide in Atlantic cod *(Gadus morhua* L.). *J. Exp. Biol.* 204:4055–4064.

Sørum, H., A. B. Hvaal, M. Heum, F. L. Daae, and R. Wiik. 1990. Plasmid profiling of *Vibrio salmonicida* for epidemiological studies of cold-water vibriosis in Atlantic salmon (*Salmo salar*) and cod (*Gadus morhua*). *Appl. Environ. Microbiol.* **56:**1033–1037.

Sørum, H., E. Myhr, B. M. Zwicker, and A. Lillehaug. 1993. Comparison by plasmid profiling of *Vibrio salmonicida* strains isolated from diseased fish from different North European and Canadian coastal areas of the Atlantic Ocean. *Can. J. Fish. Aquat. Sci.* **50:**247–250.

Steine, N. O., G. O. Melingen, and H. I. Wergeland. 2001. Antibodies against *Vibrio salmonicida* lipopolysaccharide (LPS) and whole bacteria in sera from Atlantic salmon (*Salmo salar* L.) vaccinated during the smolting and early post-smolt period. *Fish Shellfish Immunol.* **1:**39–52.

Valla, S., K. Frydenlund, D. H. Coucheron, K. Haugan, B. Johansen, T. Jørgensen, G. Knudsen, and A. Strøm. 1992. Development of a gene transfer system for curing plasmids in the marine fish pathogen *Vibrio salmonicida*. *Appl. Environ. Microbiol.* **58:**1980–1985.

Wiik, R., and E. Egidius. 1986. Genetic relationships of *Vibrio salmonicida* sp. nov. to other fish-pathogenic vibrios. *Int. J. Syst. Bacteriol.* **36:**521–523.

Wiik, R., K. Andersen, F. L. Daae, and K. A. Hoff. 1989. Virulence studies based on plasmid profiles of the fish pathogen *Vibrio salmonicida*. *Appl. Environ. Microbiol.* **55:**819–825.

Winkelmann, G., D. G. Schmid, G. Nicholson, G. Jung, and D. J. Colquhoun. 2002. Bisucaberin: a dihydroxamate siderophore isolated from *Vibrio salmonicida*, an important pathogen of farmed Atlantic salmon (*Salmo salar*). *Biometals* **15:**153–160.

The Biology of Vibrios
Edited by F. L. Thompson et al.
© 2006 ASM Press, Washington, D.C.

Chapter 21

Vibrio splendidus

Frédérique Le Roux and Brian Austin

INTRODUCTION

Vibrio splendidus is a dominant culturable vibrio in coastal marine sediments (Sobecky et al., 1998), seawater (Urakawa et al., 1999), and bivalves, namely, oysters (Farto et al., 1999). The organism has long been considered an environmental organism without any pathogenic significance. However, for several years, strains phenotypically related to *V. splendidus* have been associated with mortalities in a wide range of marine animals. The present controversial status of *V. splendidus*, i.e., mutualistic, opportunistic, or pathogenic, seems to be attributed to a lack of pertinent diagnostic tools for identification and for the evaluation of its potential pathogenic capacity.

Epidemiological studies of *V. splendidus* strains associated with mortalities in mollusks have demonstrated the genetic diversity within this group and suggested its polyphyletic nature (Le Roux et al., 2002, 2004). To date, seven species phenotypically related to *V. splendidus* have been described by means of genotyping methods, although biochemical methods are not available to clearly discriminate between species within this group (Macian et al., 2001; Hedlund and Staley, 2001; Thompson et al., 2003a,b; Faury et al., 2004). Therefore, it is important to use a polyphasic approach (phenotypic and molecular techniques) to the identification of *V. splendidus*-related strains (Thompson et al., 2001; Stackebrandt et al., 2002; Le Roux et al., 2004).

Virulence has been commonly observed among strains related to *V. splendidus* following experimental infections involving different animal models (Sugumar et al., 1998; Farto et al., 1999; Gatesoupe et al., 1999; Le Roux et al., 2002; Gay et al., 2004a). Studies of ribosomal and *gyrB* gene polymorphisms or randomly amplified polymorphic DNA have not led to the identification of markers associated with path-

ogenicity. Thus, so far, there are not any phenotypic or genotypic features to distinguish pathogenic from nonpathogenic strains, and the only way to determine the virulence of a given culture remains experimental infection. However, intra- and interlaboratory comparisons suggest problems with reproducibility of data. The development of cellular and/or molecular tests to evaluate the potential virulence of strains is necessary. It is apparent that the development of such bioassays requires the preliminary description of host alterations to define the virulence mechanisms implicated in pathogenesis.

Bacterial pathogenicity is known to be associated with structural components of the cells or active secretions of substances that either damage host tissues or protect the bacteria against host defenses. Recently, pathologists have begun to describe genes in order to understand the regulation of virulence factors. Nevertheless, little is known about marine bacterial pathogens, and until now there has not been any information concerning the *V. splendidus* group. Therefore, much research is sorely needed to improve the basic knowledge of *V. splendidus*-related strains and their interactions with farmed aquatic animals. Clarification of the taxonomy and elucidation of molecular pathogenicity mechanisms are an appropriate target for the development of new prophylactic approaches to fight against this pathogen.

GENETIC STRUCTURE OF *V. SPLENDIDUS*-RELATED BACTERIA

As aquaculture production is highly vulnerable to the impact of infectious diseases, prophylaxis is of central importance. Given this, it is necessary to develop new diagnostic tests to detect specific pathogens in fish, shellfish, and the environment, and to monitor

Frédérique Le Roux • Laboratoire de Génétique et Pathologie, Ifremer, BP 133, Ronce les bains, 17390 La Tremblade, France. **Brian Austin** • School of Life Sciences, John Muir Building, Heriot-Watt University, Riccarton, Edinburgh EH14 4AS, Scotland, United Kingdom.

the spread and evolution of diseases. In this context, the definition and identification of bacterial species and strains raise taxonomy to its pivotal role. Since its first description by Reichelt et al. (1976), the identification of *V. splendidus* and neighboring species has evolved as a result of developments in taxonomy.

V. splendidus was defined initially as a luminous marine species with two biotypes distinguishable by phenotypic features (Baumann et al., 1980; Baumann and Schubert, 1984; Alsina and Blanch, 1994). However, these authors suspected that the biotypes could be separate species, because biotype I and II showed a maximum of only 61% DNA:DNA homogeneity. Nevertheless, several researchers are still using the biotype designation (e.g., Sugumar et al., 1998; Gomez-Leon et al., 2005). Other authors recorded a heterogeneity among *V. splendidus* isolates based on phenotypic data, suggesting the existence of more than two biotypes/species (e.g., Austin et al., 1997). Genotyping by ribotyping, amplified fragment length polymorphism (AFLP), or PCR-restriction fragment length polymorphism has revealed a remarkably high genetic diversity within this group and suggested its polyphyletic nature (Farto et al., 1999; Urakawa et al., 1999; Macian et al., 2000; Thompson et al., 2001; Le Roux et al., 2002). Some groups have been validated as distinct species by DNA:DNA hybridization, which is still the definitive method for bacterial systematics (Stackebrandt et al., 2002). Currently, seven species, i.e., *V. lentus*, *V. cyclitrophicus*, *V. kanaloae*, *V. pomeroyi*, *V. tasmaniensis*, *V. chagasii*, and *V. crassostreae*, have been described within the *V. splendidus* group (Macian et al., 2001; Hedlund and Staley, 2001; Thompson et al., 2003a,b; Faury et al., 2004). Unfortunately, the current range of biochemical tests is not sufficiently discriminating to distinguish between species related to *V. splendidus* (C. C. Thompson et al., 2004). Even though the description of a new species based on a relatively low number of cultures may provide discriminatory phenotypic tests, they are highly variable when applied to a broader collection of strains belonging to the same species. This could be explained by metabolic versatility among strains exposed to the vagaries of the marine environment. Of course, bacterial identification procedures should take this variation in characteristics into consideration by incorporating DNA-based methods. A polyphasic approach using molecular techniques is now widely recognized and recommended for species definition (Thompson et al., 2001; Stackebrandt et al., 2002). This should provide reliable identification tools suitable for use in diagnostic laboratories.

Although small subunit ribosomal DNA (SSU rDNA) sequence analysis is a standard method to or-

der prokaryotic taxa hierarchically among the ranks of genera and kingdoms (Garrity and Holt, 2001; Ludwig and Klenk, 2001; Stackebrandt et al., 2002), results of SSU rDNA-based analyses often do not correlate with DNA:DNA hybridization (Fox et al., 1992; Stackebrandt and Goebel, 1994). In fact, SSU rDNA sequences of *V. splendidus*-related strains are very similar (Macian et al., 2001; Hedlund and Staley, 2001; Thompson et al., 2003a,b; Faury et al., 2004). Only one cluster, containing the type strain of *V. chagasii*, is supported by a high bootstrap value (Le Roux et al., 2004). This confirms the current opinion that SSU rDNA sequences essentially permit phylogenetic analysis of distantly related strains (Gupta, 1998) but could be less relevant for closely related strains or even species. Furthermore, a dimorphism of sequences was reported for several strains related to *V. splendidus* and could result from microheterogeneity between cistrons (Fox et al., 1992; Le Roux et al., 2004). In this case, divergence between cistrons is equivalent to divergence between strains and species. Therefore, it is impossible from the data to conclude anything about the identity of the strains. Bearing in mind the limitations cited above, use of protein-coding genes as a source of information should be considered. These genes should provide suitable sequence data for bacterial phylogenies, i.e., possessing essential attributes such as limited horizontal transmission and presence in all bacterial groups.

RecA is a multifunctional protein contributing to homologous recombination, DNA repair, and the SOS response (Cox, 2003). The gene coding for RecA has been suggested as a potential marker to unravel phylogenetic relationships among higher taxonomic ranks, such as families, classes, and phyla (Lloyd and Sharp, 1993). The usefulness of RecA gene sequences as an alternative phylogenetic marker for vibrios has been demonstrated by F. L. Thompson et al. (2004). These authors proposed that strains of the same species have >94% RecA sequence similarity. The *V. splendidus*-related group was clearly separate from other species. However, inside the group, even some species shared similarities ranging from 91 to 94%, i.e., *V. chagasii* and *V. pomeroyi*; *V. pomeroyi*, *V. splendidus*, and *V. cyclitrophicus*; others shared higher similarities, namely, 97% for *V. lentus* and *V. splendidus*. Furthermore, the majority of species were represented by only one strain, and, as a consequence, low bootstrap values were obtained.

DNA gyrase subunit, gyrB, is a type II topoisomerase found in bacteria. Since 1995, when universal primers for the gene became available, several publications have suggested that *gyrB* provides suitable sequence data for bacterial phylogenies, especially for

16S homogenous group of species (Yamamoto and Harayama, 1995; Watanabe et al., 2001).

The phylogenetic grouping of the different species phenotypically related to *V. splendidus* based on *gyrB* sequences is given in Fig. 1. It is supported by strong bootstrap values and is almost congruent with data based on quantitative DNA:DNA hybridization (Le Roux et al., 2004). Strains belonging to *V. splendidus*, *V. lentus*, and *V. chagasii* were clearly separated, and sequence similarities ranged between 98% and 85%. In the case of *V. pomeroyi*, *V. kanaloae*, and *V. tasmaniensis*, the *gyrB*-based analysis appeared less discriminative than fluorescent AFLP fingerprintings when applied to this group (Thompson et al., 2003a,b). Further work incorporating more strains belonging to these species should be conducted to compare both approaches. These results led the authors to propose the use of *gyrB*-based phylogenetic structure as an interim measure to cluster strains before validation of species affiliation by DNA:DNA hybridization and

the description of phenotypic features (Le Roux et al., 2004; Faury et al., 2004). *gyrB* gene sequencing has been used to characterize a collection of vibrios originating from diseased oysters (Gay et al., 2004a,b). The outcome was that most strains were related to *V. splendidus*, although some isolates could not be assigned clearly to any species within this group. For example, strain LGP 32 was separated from a cluster containing strain LGP 31, *V. kanaloae*, *V. tasmaniensis*, and *V. pomeroyi* type strains (Fig. 1). DNA:DNA hybridization was performed with strains LGP 31 or LGP 32 as probes (Table 1), and the rates of hybridization ranged between 40% (LGP 31, *V. cyclitrophicus*) and 92% (LGP 31, *V. kanaloae*). Both strain LGP 31 and strain LGP 32 probes hybridized at >70% with *V. kanaloae* (92 and 87%, respectively) and *V. pomeroyi* (72 and 76%, respectively). Other housekeeping genes (*rpoD*, *rctB*, and *toxR*) failed to clarify the taxonomic position of LGP 31 and LGP 32 (Fig. 2) (F. Le Roux et al., unpublished data).

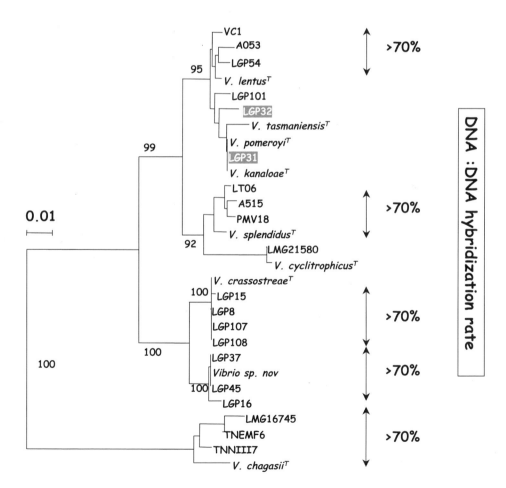

Figure 1. Phylogenetic tree of *V. splendidus*-related strains based on partial *gyrB* sequences; 1,064 gap-free sites were compared. Horizontal branch lengths are proportional to evolutionary divergence. Bootstrap values in percent of 1,000 replicates appear next to the corresponding branch when >50%.

Table 1. Intraspecific DNA-DNA homology among strains LGP 31 and LGP 32 and type strain belonging to the *V. splendidus* polyphyletic group

Strain or species	% DNA homology to LGP strain:	
	31	32
31	100	81
32	79	100
V. kanaloae[T]	92	87
V. tasmaniensis[T]	—[a]	80
V. pomeroyi[T]	72	76
V. lentus[T]	64	63
V. splendidus[T]	56	62
V. cyclitrophicus[T]	40	53

[a]—, not determined.

Data suggested that *V. splendidus*-related organisms formed a group that appeared to be homogeneous on the basis of phenotype, 16S and interspecific DNA:DNA hybridization rate ranging between 50 and 70% (Macian et al., 2001; Thompson et al., 2003a,b). The different species are only discriminable by genotyping techniques, such as ribotyping or AFLP. However, by increasing the number of strains introduced in such studies, the delineation between species appeared less clear.

A new concept, the ecotype, has been proposed (Cohan, 2002). This theory is based on the principle of multilocus sequence typing (MLST)-based population genetics. An ecotype is a cluster of strains that has at least five or six identical 500-bp gene fragments out of seven housekeeping genes. Authors argue that most of the currently recognized species harbor several ecotypes, each of which occupies a different ecological niche. Cohan (2002) concluded that actual species could be regarded as genera. The question that we want to resolve is: does *V. splendidus* correspond to a continuum of strains or to a group of numerous ecotypes? In other words, if the apparent intermediate position of LGP 31, LGP 32, and others is due to (i) the lack of isolates permitting the grouping of LGP 31/32 in distinct clades (sampling bias), or (ii) the nonexistence of delineation between proposed species. MLST analysis appears as an emergency method to evaluate linkage between alleles (congruence in phylogenetic tree) or, as in the case of *Neisseria gonorrhoeae*, random associations between loci (Smith et al., 1993). The conclusion about genomic plasticity and in consequence the evolution of *V. splendidus*-related species will be, of course, very different: numerous ecotypes strongly suggest the existence of numerous microniches to select them.

V. splendidus-related strains are frequently isolated from mollusks. In oysters there may be 10^5 CFU/g compared to 10^2 CFU/ml in seawater (Pujalte et al., 1999). Other authors reported 10^5 bacteriophages/g in oysters (DePaola et al., 1998). Therefore, oysters could be a suitable place for horizontal gene transfer by transformation, conjugation, or transduction. A high diversity of microniches could be suspected from the observation that *V. splendidus* may be detected in different anatomical compartments, e.g., gonad, hemolymph, and digestive tract, the remarkable genetic polymorphism between animals of the same species, and the numerous species from which these vibrios have been isolated.

In studies analyzing the genetic diversity of natural populations of *V. splendidus* recovered from oysters and their temporal variation, ribotyping revealed an alternate predominance of some *V. splendidus* ribotypes instead of a constant replacement of different genotypes that would be only transiently associated with oysters (Macian et al., 2000). These findings together with the constant presence of strains related to *V. splendidus* in oysters (Pujalte et al., 1999) are suggestive of close relationships between these two organisms.

Why do these bacteria develop such evolutionary strategies when a close relationship with their host is suspected? In vertebrates, genomic plasticity can confer a selective advantage to infectious agents by bypassing specific immunity. In marine invertebrates, no such specific immunity has been demonstrated. Genomic plasticity could be at the origin of variable adaptative functions, leading to the capacity to colonize different ecological niches. This must be linked to the extraordinary machinery of natural cloning, the superintegron found in many *Vibrio* species, and the relationship between bacteriophage and bacteria as a key to virulence in different pathogenic species.

VIBRIO AND HOST INTERACTIONS

V. splendidus has long been considered an environmental organism without any pathogenic significance (Baticados et al., 1990; Paillard and Maes, 1990; Myhr et al., 1991; Castro et al. 1992). However, for several years, different strains phenotypically related to this species have been associated with deaths, mainly in mollusks (Nicolas et al., 1996; Austin et al., 1997; Sugumar et al., 1998; Lacoste et al., 2001a,b; Waechter et al., 2002; Farto et al., 2003; Gay et al., 2004a; Gomez-Leon et al., 2005), but also fish (Angulo et al., 1994; Austin et al., 1997; Santos et al., 1997; Balebona et al., 1998; Jensen et al., 2003), shrimps (Baticados et al., 1990; Leano et al., 1998), and gorgonians (Martin et al., 2002)

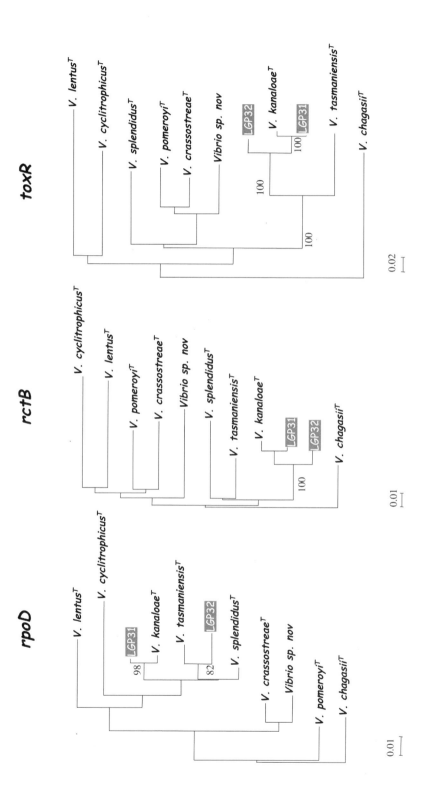

Figure 2. Phylogenetic tree of *V. splendidus*-related strains based on partial *rpoD*, *rctB*, and *toxR* sequences; 500 gap-free sites were compared. Other features are as described in the legend to Fig. 1.

(Table 2). In part, the present uncertain pathogenic status of *V. splendidus* is due to

1. a lack of pertinent diagnostic tools for identification,

2. the absence of in vivo and ex vivo models for the evaluation of its potential pathogenicity, and

3. an opportunistic status that considers the host immune capacity and environment as cofactors in disease development.

Most *V. splendidus*-related strains isolated from diseased marine animals and suspected to be the etiological agent of disease have been identified by means of classical phenotypic schemes (e.g., Alsina and Blanch, 1994). However, these methods have led to many misidentifications due to the variability/versatility of results for certain tests and the existence of many taxa not yet described. For example, a strain pathogenic for oysters (*Crassostrea gigas)* and initially identified as *V. splendidus* biovar II was later assigned to a new species, *V. chagasii*, by genomic analysis (Waechter et al., 2002; Le Roux et al., 2004). Therefore, it is essential to reevaluate the taxonomic position of virulent isolates previously described as *V. splendidus* biovar I or II or *V. lentus* by polyphasic approaches that should include new genotyping and phenotyping methods. Certainly, MLST analysis is now recommended by many authors to investigate the biodiversity of bacterial isolates. Based on a common set of genes, sequence data could lead to the development of diagnostic tools to identify species or subspecies of interest (virulence, ecotype). Phenotypic tests should also be improved and take into account virulence factors. Since the interactions between hosts and pathogens are mediated by surface structure, phenotypic characterization should include serology and surface properties of bacterial cells (agglutination, hydrophobicity, lipopolysaccharide [LPS], outer membrane proteins). The antigenic characteristics of *V. splendidus*-related strains isolated from diseased turbot or cod have been reported, with authors observing that LPS and outer membrane protein profiles were correlated with the origin of strain and the virulence in experimental challenge (Santos et al., 1997). However, we still do not have any phenotypic or genotypic feature to distinguish pathogenic from nonpathogenic strains belonging to single species, and as a consequence the only way to determine the virulence of a given strain remains experimental infection.

Experimental challenge of larvae seems to give reproducible results, and as a consequence the pathogenic status of strains is less controversial than that in adult animals (Nicolas et al., 1996; Sugumar et al., 1998; Gatesoupe et al., 1999). Generally, bacterial isolates are incubated directly with larvae cultures, and mortalities appear rapidly, i.e., from 1 to 10 days after infection. These tests can be miniaturized, and therefore permit the screening of large numbers of isolates. However, because several strains are pathogenic only for certain stages in development, experimental challenge of larvae cannot be used as a sole test for virulence determination. Additionally, the virulence of bacteria may vary with the host. For example, a *V. splendidus*-related strain was pathogenic for turbot larvae but not for halibut and cod larvae (Gatesoupe et al., 1999).

Putative pathogens have been selected by injection of bacterial suspensions in the adductor muscle of mollusks (Lacoste et al., 2001a; Le Roux et al., 2002; Gay et al., 2004a) or intraperitoneally into fish (Angulo et al., 1994; Santos et al., 1997; Balebona et al., 1998). The results are obtained quickly because mortality appears often within 24 h of injection in the case of mollusks. However, there are drawbacks, such as disruption of tissues and variability in the results. Furthermore, injection techniques do not mimic the natural route of infection.

The injection of isolates into the palleal cavity is an alternative measure insofar as it allows the transmission of the disease without disrupting the tissues. Yet, this method does not exactly reflect the natural infection due to the artificial crossing of the mechanical barrier (shell and mantle) and the stress induced by a long period of time under static conditions (up to 1 month). However, this method has permitted the demonstration of virulence in several *V. splendidus*-related strains in oysters (*C. gigas*) and clams (*Ruditapes decussates*) (Gay et al., 2004b; Gomez-Leon et al., 2005).

Experimental infections by bathing have succeeded in demonstrating the virulence of certain strains against oysters and clam spat (Waechter et al., 2002; Gomez-Leon et al., 2005), gorgonians (Martin et al., 2002), and octopus (Farto et al., 2003). However, all too frequently, the method does not result in any mortalities at all. Since bath exposure of *C. gigas* with *V. splendidus* did not lead to any mortality, Lacoste et al. (2001a) suggested that the organism cannot be transmitted horizontally. Furthermore, because *V. splendidus* was shown to accumulate notably in the gonads of oysters, Sugumar et al. (1998) suggested that brood stock could be the source and route of transmission, as already pointed out for *Ostrea edulis* (Lodeiros et al., 1987) and *Argopecten purpuratus* (Riquelme et al., 1995). However, another explanation may well be that these negative results reflect the variability often observed in the different inoculation methods. First, culture media or environmental conditions may modulate the expres-

Table 2. *V. splendidus*-related strains isolated from animals suffering abnormal deaths and demonstrated to be pathogenic by experimental challenge

Host	Stage	Strains	Virulence assays	Host alteration	Phenotyping	Genotyping	Reference(s)
Crassostrea gigas (Japanese oysters)	Larvae	Unnamed	+/*C. gigas* and *O. edulis* larvae	Bacillary necrosis	Few phenotypic tests	None	Jeffries, 1982
Scophthalmus maximus (turbot)	Juvenile	16N, 43N	+/rainbow trout, +/turbot	Hemorrhagic areas in the mouth	Numerical taxonomy (>90 tests)	Ribotyping	Angulo et al., 1994; Farto et al., 1999
Pecten maximus	Larvae	A515, A053	+/*P. maximus*, larvae	ND[a]	Numerical taxonomy (>90 tests)	G+C content, 16S	Nicolas et al., 1996
Gadus morhua (cod)	Adult	95-3-72, 89, 90, 92	+/rainbow trout	ND	Few phenotypic tests; LPS serotyping	None	Santos et al., 1997
Crassostrea gigas	Larvae	58, 59, 60, Q3, 102	+/*C. gigas*, larvae	Bacillary necrosis	Alsina and Blanch (1994) key of identification (35 tests)	None	Sugumar et al., 1998
Scophthalmus maximus	Larvae	VS5, VS6, VS9, VS11	+/*S. maximus*, larvae	ND	Numerical taxonomy (>90 tests)	16S, DNA-DNA hybridization, RAPD	Gatesoupe et al., 1999
Sparus aurata (sea bream)	Adult	A058, 25/900 DC1 OR3	+/*S. aurata*, juvenile	ND	Numerical taxonomy (74 tests)	None	Balebona et al., 1998
Crassostrea gigas	Juvenile	Unnamed	+/*C. gigas*, juvenile; −/*C. gigas*, adult	ND	API20E	16S	Lacoste et al., 2001a,b
Crassostrea gigas	Juvenile	TNEMF6	+/*C. gigas*, juvenile; +/*R. phillippinarum*, adult	ND	Alsina and Blanch (1994) key of identification (35 tests)	16S, *gyrB*, DNA-DNA hybridization, PCR-RFLP	Waechter et al., 2002; Le Roux et al., 2002, 2004
E. cavolinii (gorgons)	Adult	A, B, C	+/*E. cavolinii*	Necrosis	Alsina and Blanch (1994) key of identification (35 tests)	None	Martin et al., 2002
Octopus vulgaris	Adult	P58	+/octopus adult; −/sea bream and turbot	Lesion in the arms or head mantle, loss of skin	Numerical taxonomy (85 tests)	16S, ribotyping	Farto et al., 2003
Crassostrea gigas	Juvenile	LGP31, 32, 7, 8, 15, 35	+/*C. gigas*, juvenile; +/*R. phillippinarum*, adult	Muscle alterations	Numerical taxonomy (80 tests)	16S, *gyrB*, DNA-DNA hybridization	Gay et al. 2004a,b; Faury et al., 2004
Ruditapes phillippinarum (clams)	Larvae	TA2	+/clam spat	Bacillary necrosis	Alsina and Blanch (1994) key of identification (35 tests)	16S (312 bp)	Gomez-Leon et al., 2005

[a] ND, not determined.

sion of virulence factors or subcultures and lead to a loss of virulence. Second, a particular status of the host (genetic and/or physiological) may be necessary for initiation of pathogenesis and/or susceptibility of the host.

In *C. gigas*, because of a low reproducibility of results yielded by experimental infection trials, authors have emphasized physiological and environmental factors as necessary stressors triggering vibriosis (Lacoste et al., 2001b; Le Roux et al., 2002). Initially, Lacoste et al. (2001b) demonstrated that *V. splendidus* strains induced mortalities in a temperature-dependent manner: 7°C > 16°C > 26°C (Lacoste et al., 2001a). In a second study, these authors showed that both mortality and *V. splendidus* loads increased in stressed oysters, whereas their numbers remained low in unstressed animals. Injection of noradrenaline or adrenocorticotropic hormone, which are two key components of the oyster's neuroendocrine stress response system, caused higher mortalities and increased accumulation of *V. splendidus* in challenged oysters. These results suggest that physiological changes imposed by stress or stress hormones influenced host-pathogen interactions in oysters and increased juvenile *C. gigas* vulnerability to *V. splendidus*.

In oysters, maturation induces an increased susceptibility to infection. This is supported by the knowledge that gametogenesis is a period of negative energy budget (Soletchnik et al., 1997) wherein most of the acquired energy is used for the production of gametes to the detriment of the defense mechanisms (Perdue et al., 1981; Myrand et al., 2000). A correlation between maturation stage and sensitivity to vibrios has been demonstrated by experimental challenge with different *V. splendidus*-related strains (Gay, 2004). Also, it was demonstrated that host genetic origin is an important factor in this opportunistic disease. In another study, a collection of *Vibrio* strains, isolated from *C. gigas*, genotyped by *gyrB*-based phylogenetic analysis, and screened for their virulence by experimental infection, was established (Gay et al., 2004a). Overall, few strains displayed individual pathogenicity. However, some cultures displayed enhanced virulence when inoculated concomitantly, suggesting an agonistic action. In particular, this collaboration has been shown to be statistically significant for the strains LGP 31 and LGP 32. These strains are closely related but may each possess a specific virulence feature, which, when simultaneously expressed, leads to increased pathogenicity. It is unknown whether the observed phenomenon reveals synergy-based effects on the interdependence of certain virulence mechanisms or if additive effects stemming from reciprocal benefits facilitate counteraction of the host immune response.

In facing the difficulties encountered by experimental infection in nondomesticated animals, the development of cellular and/or molecular tests to evaluate the potential virulence of strains is necessary. The development of such bioassays could be oriented by description of host alterations to define the virulence mechanisms implicated in the pathogenesis. Though *V. splendidus*-related strains have been associated with outbreaks of mortality among several aquatic animals, pathogenesis has not yet been well defined. In larvae, bacillary necrosis caused by various *Vibrio* spp., including *V. splendidus*, corresponds to lesions that affect mainly the mantle, the velum, and the connective tissues (Jeffries, 1982; Gomez-Leon et al., 2005). Martin et al. (2002) described some *V. splendidus* strains that induced tissue necrosis and mass mortality in gorgons. Furthermore, a *V. lentus* culture, which was isolated from diseased octopus, was able to induce round, hard lesions in the arm or head mantle (Farto et al., 2003). Moreover, *C. gigas* alterations induced by injection of strains LGP 31 and/or LGP 32 have been described (Gay et al., 2004b). In this case, bacteria became localized at the periphery of the muscle and induced extensive lesions in the translucent part of the adductor muscle. No major difference was observed when the strains were inoculated individually or together. Thus, strain-specific virulence apparently does not induce differential structural changes in altered cells.

According to our experience, it is difficult to devise a specific clinical table of disease characteristics for three main reasons:

1. It is necessary to analyze a high number of samples to investigate a specific observation.
2. Postmortem changes may be confused with actual disease signs.
3. Mollusks are normally already colonized by commensal bacteria.

Specific molecular tools (in situ hybridization and green fluorescent protein-expressing bacteria) should be developed to describe the infection route and the specific localization and migration patterns of vibrios. In the absence of reproducible animal models, it is necessary to reduce external influences by using animals maintained in standardized conditions in hatcheries. For example, by modulating temperature and feeding regimes, it is possible to control maturation in oysters (S. Pouvreau, personal communication).

Highly standardized in vitro studies are necessary to screen for overall virulence activities: adherence, cytolytic effect, apoptosis factors, hemolysin, extracellular enzymes, and stress proteins. It is noteworthy that the development of a clam primary cell culture has been used successfully to study the interaction between cells and pathogenic *Vibrio* (Lopez-Cortes et al., 1999; Choquet et al., 2003). In vitro tests performed on fish cell lines can be very relevant to explaining pathogenic processes (Ormonde et al., 2000).

Bacterial pathogenicity is associated with structural components of the cells (e.g., capsules, fimbriae, LPS, and endotoxins) or active secretions of substances that either damage host tissues or protect the bacteria against host defenses (invasin, enzyme, hemolysin, coagulase, toxins, i.e., diphtheria, botulism, tetanus, or cholera). Recently, pathologists have begun to describe genes in order to understand the regulation of virulence factors. Nevertheless, little is known concerning marine bacterial pathogens, although some toxins have been described in *V. anguillarum*, *V. vulnificus*, *V. harveyi*, or *V. penaeicida* isolated from fish and crustacea. In some bivalve pathogenic vibrios, virulence factors have been demonstrated, such as the adherence factors and hemocyte lysin in *V. tapetis*, and toxins in *V. pectenicida*, *V. alginolyticus*, or *Vibrio* sp. However, despite descriptions of invasivity and exotoxicity by several authors (Elston and Leibovitz, 1980; Lodeiros et al., 1987; Nicolas et al., 1996, Sugumar et al., 1998; Gay et al., 2004b; Gomez-Leon et al., 2005), information is not yet available about *V. splendidus*.

COMPLETE GENOME SEQUENCE OF *VIBRIO SPLENDIDUS*

Vibrio species represent an important fraction of the heterotrophic bacterial population of marine environments and estuaries. Vibrios play a significant role in nutrient cycling in these niches; however, a number of species are also severe pathogens for fish, shellfish, and mammals, including humans. If human pathogens, such as *V. cholerae*, are well studied, fish or shellfish pathogens are far less well characterized. In collaboration with Didier Mazel (Unité de plasticité du génome bactérien, Institut Pasteur, Paris, France), we initiated sequencing of the *V. splendidus* genome at the Institut Pasteur. The project has a dual objective: (i) to allow a broader analysis of the *Vibrio* genome plasticity, and (ii) to have complete access to the pathogenicity repertoire of this species.

The *V. splendidus* strain that we have chosen, LGP 32, is an oyster (*C. gigas*) pathogen, associated with "summer mortality syndrome," which has been responsible for high mortality rates in oyster beds in France since 1991. In a previous study (Gay et al., 2004a), LGP 31 and LGP 32 were selected as models because they were related to the *V. splendidus* group previously associated with several mollusk and fish

mortality events. They also illustrated the collaborative effect observed for different strains that will allow us to analyze the shared and/or different mechanisms of virulence implicated in this agonistic action. We have started to study the physiology and pathogenicity of these bacteria (Gay, 2004; Gay et al., 2004b) and, in addition, considered that accessing the complete genome of one of the strains would be highly beneficial as it would probably accelerate the identification of pathogenicity and virulence determinants. Indeed, a considerable amount of data concerning human pathogenic *Vibrio* species is already available. Moreover, the genomes of three species have already been sequenced, i.e., *V. cholerae* (Heidelberg et al., 2000), *V. parahaemolyticus* (Makino et al., 2003), and *V. vulnificus* (Chen et al., 2003). More recently, the complete genome of *V. fischeri*, a symbiotic bacterium, has been sequenced (Ruby et al., 2005). These data make further comparative genomic studies possible.

Bacterial genomes can show a large degree of variability, in both content and organization, not only between related genera and species but even between different strains of the same species. Indeed, different isolates from the same bacterial species have been observed to show differences of as much as 20% in their gene content and arrangement. In addition to internal reshuffling through recombination (be it illegitimate or homologous between repeated sequences such as insertion elements), lateral gene transfer plays an indisputable role in overall genome plasticity. The importance of lateral gene transfer in shaping bacterial genomes has become increasingly obvious over the past 10 years. This is largely, but not entirely, due to the rapid accumulation of the complete sequences of >120 prokaryotic genomes. However, despite the number and diversity of the genetic elements responsible (bacteriophage, genomic islands, transposons, insertion sequences, integrons), studies have largely been confined to assembling a catalogue rather than investigating their impact on genome dynamics and evolution. Indeed, these aspects are poorly understood, and the behavior of only a very limited number of elements composing the horizontal gene pool has been analyzed in any detail. In parallel with the LGP 32 genome sequencing, an approach based on subtractive hybridization enrichment between the sequenced strain and LGP 31 or an avirulent *V. splendidus* isolate has been carried out. The objectives of this project are to explore (i) the specific genes implicated in the virulence of LGP 32 or LGP 31 and (ii) the origin of those genes' acquisition.

In a second stage, our expertise in *Vibrio* genetics will allow us to carry on inverse genetics and knockouts of pathogenicity gene candidates. In the future, macroarrays should be developed to investigate the biodiversity of vibrios isolated from fish and shellfish disease. Gene expression will be investigated in vivo (in situ hybridization) or ex vivo (Northern blot, reverse transcription-PCR, macroarray) if culture conditions permit the expression of an interesting phenotype (exotoxicity, adherence, invasion). Inverse genetics could also permit the introduction of reporter genes (green fluorescent protein, luciferase) in operons containing virulence genes. Such mutant strains could be used to investigate biotic and abiotic factors modulating virulence.

CONCLUSION

The emergence of bacterial diseases in bivalves is an increasing problem for aquaculture. To date, little information has been published about mollusk-pathogenic vibrios. By achieving a basic understanding (bacterial taxonomy, identification of virulence mechanisms, genome evolution) and developing new diagnostic test systems based on pathogenic abilities, we intend to achieve better knowledge of vibriosis of bivalves. This should allow the development of methods of disease prevention suitable for each farmed species, especially for natural and hatchery-reared invertebrate seed stock. A survey of hatchery production is of primary importance, because the production of spat tends to increase and replace the natural spatfall. Indeed, the genetic improvement of bivalves (triploid or selected oyster) will lead to the production of ever-increasing quantities of spat in hatcheries.

By evaluating the real impact of the emergent diseases and following their evolution, science will contribute to maintaining the dynamism of aquacultural development by promoting maximal productivity while minimizing losses caused by infectious diseases.

REFERENCES

Alsina, M., and A. R. Blanch. 1994. A set of keys for biochemical identification of environmental *Vibrio* species. *J. Appl. Bacteriol.* **76**:79–85.

Angulo, L., J. E. Lopez, A. Vicente, and A. M. Saborido. 1994. Haemorrhagic areas in the mouth of farmed turbot, *Scophthalmus maximus* (L.). *J. Fish Dis.* **17**:163–169.

Austin, B., D. A. Austin, A. R. Blanch, et al. 1997. A comparison of methods for the typing of fish-pathogenic *Vibrio* spp. *Syst. Appl. Microbiol.* **20**:89–101.

Balebona, M. C., I. Zorilla, M. A. Morinigo, and J. J. Borrego. 1998. Survey of bacterial pathologies affecting farmed gilt-head sea bream (*Sparus aurata* L.) in southwestern Spain from 1990 to 1996. *Aquaculture* **166**:19–35.

Baticados, M. C., C. R. Lavilla-Pitogo, E. R. Cruz-Lacierda, L. D. de la Pena, and N. A. Sunaz. 1990. Studies on the chemical control of luminous bacteria *Vibrio harveyi* and *V. splendidus* iso-

lated from diseased *Penaeus monodon* larvae and rearing water. *Dis. Aquat. Org.* **9**:133–139.

Baumann, P., L. Baumann, S. S. Bang, and M. J. Woolkalis. 1980. Reevaluation of the taxonomy of *Vibrio*, *Beneckea*, and *Photobacterium*: abolition of the genus *Beneckea*. *Curr. Microbiol.* **4**:127–132.

Baumann, P., and R. H. W. Schubert. 1984. Family II *Vibrionaceae* Véron 1965, 5245[AL], p. 516–517. *In* N. R. Krieg and J. G. Holt (ed.), *Bergey's Manual of Systematic Bacteriology*, vol. 1. Lippincott Williams & Wilkins, Philadelphia, Pa.

Castro, D., E. Martinez-Manzanares, and A. Luque. 1992. Characterization of strains related to brown ring disease outbreaks in southwestern Spain. *Dis. Aqua. Org.* **14**:229–236.

Chen, C. Y., K. M. Wu, Y. C. Chang, C. H. Chang, H. C. Tsai, T. L. Liao, Y. M. Liu, H. J. Chen, A. B. Shen, J. C. Li, T. L. Su, C. P. Shao, C. T. Lee, L. I. Hor, and S. F. Tsai. 2003. Comparative genome analysis of *Vibrio vulnificus*, a marine pathogen. *Genome Res.* **13**:2577–2587.

Choquet, G., P. Soudant, C. Lambert, J. L. Nicolas, and C. Paillard. 2003. Reduction of adhesion properties of *Ruditapes philippinarum* hemocytes exposed to *Vibrio tapetis*. *Dis. Aquat. Org.* **57**:109–116.

Cohan, F. M. 2002. What are bacterial species? *Annu. Rev. Microbiol.* **56**:457–487.

Cox, M. M. 2003. Bacterial RecA protein as a motor protein. *Annu. Rev. Microbiol.* **57**:551–577.

DePaola, A., M. L. Motes, A. M. Chan, and C. A. Suttle. 1998. Phages infecting *Vibrio vulnificus* are abundant and diverse in oysters (*Crassostrea virginica*) collected from the Gulf of Mexico. *Appl. Environ. Microbiol.* **64**:346–351.

Elston, R., and L. Leibovitz. 1980. Detection of vibriosis in hatchery reared larval oysters: correlation between clinical, histological and ultrastructural observations in experimentally induced disease. *Proc. Natl. Shellfish Assoc.* **70**:122–123.

Farto, R., M. Montes, M. J. Perez, T. P. Nieto, J. L. Larsen, and K. Pedersen. 1999. Characterization by numerical taxonomy and ribotyping of *Vibrio splendidus* biovar I and *Vibrio scophthalmi* strains associated with turbot cultures. *J. Appl. Microbiol.* **86**:796–804.

Farto, R., S. P. Armada, M. Montes, J. A. Guisande, M. J. Perez, and T. P. Nieto. 2003. *Vibrio lentus* associated with diseased wild octopus (*Octopus vulgaris*). *J. Invertebr. Pathol.* **83**:149–156.

Faury, N., D. Saulnier, F. L. Thompson, M. Gay, J. Swings, and F. Le Roux. 2004. *Vibrio crassostreae* sp. nov., isolated from the haemolymph of oysters (*Crassostrea gigas*). *Int. J. Syst. Evol. Microbiol.* **54**:2137–2140.

Fox, G. E., J. D. Wisotzkey, and P. J. Jurtshuk. 1992 How close is close: 16S rRNA sequence identity may not be sufficient to guarantee species identity. *Int. J. Syst. Bacteriol.* **42**:166–170.

Garrity, G. M., and J. G. Holt. 2001. The road map to the *Manual*, p. 119–166. *In Bergey's Manual of Systematic Bacteriology*, 2nd ed. Springer-Verlag, New York, N.Y.

Gatesoupe, F. J., C. Lambert, and J. L. Nicolas. 1999. Pathogenicity of *Vibrio splendidus* strains associated with turbot larvae, *Scophthalmus maximus*. *J. Appl. Microbiol.* **87**:757–763.

Gay, M., F. C. Berthe, and F. Le Roux. 2004a. Screening of *Vibrio* isolates to develop an experimental infection model in the Pacific oyster *Crassostrea gigas*. *Dis. Aquat. Org.* **59**:49–56.

Gay, M., T. Renault, A. M. Pons, and F. Le Roux. 2004b. Two *Vibrio splendidus* related strains collaborate to kill *Crassostrea gigas*: taxonomy and host alterations. *Dis. Aquat. Org.* **62**:65–74.

Gay, M. 2004. Ph.D. thesis. Université de la Rochelle, France.

Gomez-Leon, J., L. Villamil, M. L. Lemos, B. Novoa, and A. Figueras. 2005. Isolation of *Vibrio alginolyticus* and *Vibrio splendidus* from aquacultured carpet shell clam (*Ruditapes de-*

cussatus) larvae associated with mass mortalities. *Appl. Environ. Microbiol.* **71:**98–104.

Gupta, R. S. 1998. Protein phylogenies and signature sequences: a reappraisal of evolutionary relationships among Archaebacteria, Eubacteria, and Eukaryotes. *Microbiol. Mol. Biol. Rev.* **62:** 1425–1491.

Hedlund, B. P., and J. T. Staley. 2001. *Vibrio cyclotrophicus* sp. nov., a polycyclic aromatic hydrocarbon (PAH)-degrading marine bacterium. *Int. J. Syst. Evol. Microbiol.* **51:**61–66.

Heidelberg, J. F., J. A. Eisen, W. C. Nelson, R. A. Clayton, M. L. Gwinn, R. J. Dodson, D. H. Haft, E. K. Hickey, J. D. Peterson, L. Umayam, S. R. Gill, K. E. Nelson, T. D. Read, H. Tettelin, D. Richardson, M. D. Ermolaeva, J. Vamathevan, S. Bass, H. Qin, I. Dragoi, P. Sellers, L. McDonald, T. Utterback, R. D. Fleishmann, W. C. Nierman, O. White, S. L. Salzberg, H. O. Smith, R. R. Colwell, J. J. Mekalanos, J. C. Venter, and C. M. Fraser. 2000. DNA sequence of both chromosomes of the cholera pathogen *Vibrio cholerae. Nature* **406:**477–483.

Jeffries, V. E. 1982. Three *Vibrio* strains pathogenic to larvae of *Crassostrea gigas* and *Ostrea edulis. Aquaculture* **29:**201–226.

Jensen, S., O. B. Samuelsen, K. Andersen, L. Torkildsen, C. Lambert, G. Choquet, C. Paillard, and O. Bergh. 2003. Characterization of strains of *Vibrio splendidus* and *V. tapetis* isolated from corkwing wrasse *Symphodus melops* suffering vibriosis. *Dis. Aquat. Org.* **53:**25–31.

Lacoste, A., F. Jalabert, S. Malham, A. Cueff, F. Gélébart, C. Cordevant, M. Lange, and S. A. Poulet. 2001a. A *Vibrio splendidus* strain is associated with summer mortality of juvenile oysters *Crassostrea gigas* in the Bay of Morlaix (North Brittany, France). *Dis. Aquat. Org.* **46:**139–145.

Lacoste, A., F. Jalabert, S. K. Malham, A. Cueff, and S. A. Poulet. 2001b. Stress and stress-induced neuroendocrine changes increase the susceptibility of juvenile oysters (*Crassostrea gigas*) to *Vibrio splendidus. Appl. Environ. Microbiol.* **67:**2304–2309.

Leano, E. M., C. R. Lavilla-Pitogo, and M. G. Paner. 1998. Bacterial flora in the hepatopancreas of pond-reared *Penaeus monodon* juveniles with luminous vibriosis. *Aquaculture* **164:**367–374.

Le Roux, F., M. Gay, C. Lambert, M. Waechter, S. Poubalanne, B. Chollet, J. L. Nicolas, and F. C. Berthe. 2002. Comparative analysis of *Vibrio splendidus* related strains isolated during *Crassostrea gigas* mortality events. *Aquat. Living Resources* **15:** 251–258.

Le Roux, F., M. Gay, C. Lambert, J. L. Nicolas, M. Gouy, and F. Berthe. 2004. Phylogenetic study and identification of *Vibrio splendidus*-related strains based on gyrB gene sequences. *Dis. Aquat. Organ.* **58:**143–150.

Lloyd, A. T., and P. M. Sharp. 1993. Use of the recA gene and the molecular phylogeny of bacteria. *J. Mol. Evol.* **37:**399–407.

Lodeiros, C., J. Bolinches, C. P. Dopazo, and A. E. Toranzo. 1987. Bacillary necrosis in hatcheries of *Ostrea edulis* in Spain. *Aquaculture* **65:**15–29.

Lopez-Cortes, L., A. Luque, E. Martinez-Manzanares, D. Castro, and J. J. Borrego. 1999. Adhesion of *Vibrio tapetis* to clam cells. *J. Shellfish Res.* **18:**91–97.

Ludwig, W., and H. P. Klenk. 2001. Overview: a phylogenetic backbone and taxonomic frame work for prokaryotic systematics, p. 49–65. *In Bergey's Manual of Systematic Bacteriology*, 2nd ed. Springer-Verlag, New York, N.Y.

Macian, M. C., E. Garay, F. Gonzalez-Candelas, M. J. Pujalte, and R. Aznar. 2000. Ribotyping of vibrio populations associated with cultured oysters (*Ostrea edulis*). *Syst. Appl. Microbiol.* **23:** 409–417.

Macian, M. C., W. Ludwig, R. Aznar, P. A. Grimont, K. H. Schleifer, E. Garay, and M. J. Pujalte. 2001. *Vibrio lentus* sp. nov., isolated from Mediterranean oysters. *Int. J. Syst. Evol. Microbiol.* **51:**1449–1456.

Makino, K., K. Oshima, K. Kurokawa, K. Yokoyama, T. Uda, K. Tagomori, Y. Iijima, M. Najima, M. Nakano, A. Yamashita, Y. Kubota, S. Kimura, T. Yasunaga, T. Honda, H. Shinagawa, M. Hattori, and T. Iida. 2003. Genome sequence of *Vibrio parahaemolyticus*: a pathogenic mechanism distinct from that of *V. cholerae. Lancet* **361:**743–749.

Martin, Y., J. L. Bonnefont, and L. Chancerelle. 2002. Gorgonians mass mortality during the 1999 late summer in French Mediterranean coastal waters: the bacterial hypothesis. *Water Res.* **36:** 779–782.

Myhr, E., J. L. Larsen, A. Lillehaug, R. Gudding, M Heum, and T. Hastein. 1991. Characterization of *Vibrio anguillarum* and closely related species isolated from farmed fish in Norway. *Appl. Environ. Microbiol.* **57:**2750–2757.

Myrand, B., H. Guderley, and J. H. Himmelman. 2000. Reproduction and summer mortality of blue mussels *Mytilus edulis* in the Magdalene Islands, Southern Gulf of St. Lawrence. *Mar. Ecol. Prog. Ser.* **197:**193–207.

Nicolas, J. L., S. Corre, G. Gauthier, R. Robert, and D. Ansquer. 1996. Bacterial problems associated with scallop *Pecten maximus* larval culture. *Dis. Aquat. Org.* **27:**67–76.

Ormonde, P., P. Horstedt, R. O'Toole, and D. L. Milton. 2000. Role of motility in adherence to and invasion of a fish cell line by *Vibrio anguillarum. J. Bacteriol.* **182:**2326–2328.

Paillard, C., and P. Maes. 1990. Etiologie de la maladie de l'anneau brun chez *Tapes philippinarum*: pathogénicité d'un *Vibrio* sp. *C. R. Acad. Sci. Paris* **310:**15–20.

Perdue, J. A., J. H. Beattie, and K. K. Chew. 1981. Some relationships between gametogenic cycle and summer mortality phenomenon in the Pacific oyster *C. gigas* in the Washington state. *J. Shellfish Res.* **1:**9–16.

Pujalte, M. J., M. Ortigosa, M. C. Macian, and E. Garay. 1999. Aerobic and facultative anaerobic heterotrophic bacteria associated to Mediterranean oysters and seawater. *Int. Microbiol.* **2:**259–266.

Reichelt, J. L., P. Baumann, and L. Baumann. 1976. Study of genetic relationships among marine species of the genera Beneckea and Photobacterium by means of in vitro DNA/DNA hybridization. *Arch. Microbiol.* **110:**101–120.

Riquelme, C., G. Hayashida, N. Vergara, A. Vasquez, Y. Morales, and P. Chavez. 1995. Bacteriology of the scallop *Argopecten purpuratus* (Lamarck, 1819) cultured in Chile. *Aquaculture* **138:**49–60.

Ruby, E. G., M. Urbanowski, J. Campbell, A. Dunn, M. Faini, R. Gunsalus, P. Lostroh, C. Lupp, J. McCann, D. Millikan, A. Schaefer, E. Stabb, A. Stevens, K. Visick, C. Whistler, and E. P. Greenberg. 2005. Complete genome sequence of *Vibrio fischeri*: a symbiotic bacterium with pathogenic congeners. *Proc. Natl. Acad. Sci. USA* **102:**3004–3009. (First published 9 February 2005; 10.1073/pnas.0409900102.)

Santos, Y., F. Pazos, S. Nunez, and A. E. Toranzo. 1997. Antigenic characterization of *Vibrio anguillarum* related organisms isolated from turbot and cod. *Dis. Aquat. Org.* **28:**45–50.

Smith, J. M., N. H. Smith, M. O'Rourke, and B. G. Spratt. 1993. How clonal are bacteria? *Proc. Natl. Acad. Sci. USA* **90:**4384–4388.

Sobecky, P. A., T. J. Mincer, M. C. Chang, A. Toukdarian, and D. R. Helinski. 1998. Isolation of broad-host-range replicons from marine sediment bacteria. *Appl. Environ. Microbiol.* **64:** 2822–2830.

Soletchnik, P., D. Razet, P. Geairon, N. Faury, and P. Goulletquer. 1997. Ecophysiologie de la maturation sexuelle et de la ponte de l'huître creuse *C. gigas*: réponses métaboliques (respiration) et alimentaires (filtration, absorption) en fonction des différents stades de maturation. *Aquat. Living Resources* **10:**177–185.

Stackebrandt, E., and B. M. Goebel. 1994. A place for DNA-DNA reassociation and 16S ribosomal-RNA sequence-analysis in the present species definition in bacteriology. *Int. J. Syst. Bacteriol.* **44:**846–849.

Stackebrandt, E., W. Frederiksen, G. M. Garrity, P. A. Grimont, P. Kampfer, M. C. Maiden, X. Nesme, R. Rossello-Mora, J. Swings, H. G. Trüper, L. Vauterin, A. C. Ward, and W. B. Whitman. 2002. Report of the ad hoc committee for the re-evaluation of the species definition in bacteriology. *Int. J. Syst. Evol. Microbiol.* **52**:1043–1047.

Sugumar, G., T. Nakai, Y. Hirata, D. Matsubara, and K. Muroga. 1998. *Vibrio splendidus* biovar II as the causative agent of bacillary necrosis of Japanese oyster *Crassostrea gigas* larvae. *Dis. Aquat. Organ.* **33**:111–118.

Thompson, C. C., F. L. Thompson, K. Vandemeulebroecke, B. Hoste, P. Dawyndt, and J. Swings. 2004. Use of recA as an alternative phylogenetic marker in the family *Vibrionaceae. Int. J. Syst. Evol Microbiol.* **54**:919–924.

Thompson, F. L., B. Hoste, K. Vandemeulebroecke, and J. Swings. 2001. Genomic diversity amongst Vibrio isolates from different sources determined by fluorescent amplified fragment length polymorphism. *Syst. Appl. Microbiol.* **24**:520–538.

Thompson, F. L., C. C. Thompson, Y. Li, B. Gomez-Gil, J. Vandenberghe, B. Hoste, and J. Swings. 2003a. *Vibrio kanaloae* sp. nov., *Vibrio pomeroyi* sp. nov. and *Vibrio chagasii* sp. nov.,

from sea water and marine animals. *Int. J. Syst. Evol. Microbiol.* **53**:753–759.

Thompson, F. L., C. C. Thompson, and J. Swings. 2003b. *Vibrio tasmaniensis* sp. nov., isolated from Atlantic salmon (*Salmo salar* L.). *Syst. Appl. Microbiol.* **26**:65–69.

Thompson, F. L., T. Iida, and J. Swings. 2004. Biodiversity of vibrios. *Microbiol. Mol. Biol. Rev.* **68**:403–431.

Urakawa, H., K. Kita-Tsukamoto, and K. Ohwada. 1999. 16S rDNA restriction fragment length polymorphism analysis of psychrotrophic vibrios from Japanese coastal water. *Can. J. Microbiol.* **45**:1001–1007.

Waechter, M., F. Le Roux, J. L. Nicolas, E. Marissal, and F. Berthe. 2002. [Characterization of pathogenic bacteria of the cupped oyster *Crassostrea gigas*.] (In French.) *C. R. Biol.* **325**:231–238.

Watanabe, K., J. Nelson, S. Harayama, and H. Kasai. 2001. ICB database: the gyrB database for identification and classification of bacteria. *Nucleic Acids Res.* **29**:344–345.

Yamamoto, S., and S. Harayama. 1995. PCR amplification and direct sequencing of gyrB genes with universal primers and their application to the detection and taxonomic analysis of *Pseudomonas putida* strains. *Appl. Environ. Microbiol.* **61**:1104–1109.

The Biology of Vibrios
Edited by F. L. Thompson et al.
© 2006 ASM Press, Washington, D.C.

Chapter 22

Miscellaneous Animal Pathogens

BRIAN AUSTIN

INTRODUCTION

Vibrios are the scourge of marine and estuarine vertebrates and invertebrates. Apart from those species detailed elsewhere in this book, many other taxa have been associated with disease. In some cases the evidence for involvement in disease processes is spurious, whereas other bacterial species are recognized as serious pathogens. However, it is apparent that a wide range of vibrios is capable of causing disease in aquatic animals, with the list of putative pathogens being regularly extended. Undoubtedly, many more fish- and shellfish-pathogenic vibrios will be recognized in the future. Emphasis will be placed here on those bacterial taxa for which published evidence supports a role in fish/shellfish pathology.

Moritella marina

A new condition was recognized among farmed Atlantic salmon in Iceland in which the affected fish developed extensive but shallow skin lesions over a substantive part of the flank at low water temperatures, i.e., ~10°C (Benediktsdóttir et al., 1998). In total, 19 Icelandic and 1 Norwegian isolate were recovered on 5% (vol/vol) horse blood agar supplemented with 1.5% (wt/vol) sodium chloride with incubation at 15°C for 7 days, and equated with *Vibrio marinus* as a result of a numerical taxonomy study (Benediktsdóttir et al., 1998). Subsequently, this species was transferred to *Moritella*, as *M. marina*, according to 16S rRNA sequencing data (Urakawa et al., 1998).

Moritella viscosa

Two groups of vibrios were recovered from Atlantic salmon with "winter ulcer" disease, i.e., large, shallow skin lesions that occurred at low water temperatures (Lunder et al., 2000), often in large (2 to

3 kg) fish (Bruno et al., 1998). Internally, petechial hemorrhages, splenomegaly, congestion, and general necrosis were observed (Bruno et al., 1998). On the basis of phenotypic and genotypic data, these vibrios were grouped into two new species, *Vibrio wodanis* and *Vibrio viscosus* (Lunder et al., 2000), the latter of which was subsequently reclassified as *M. viscosa* (Benediktsdóttir et al., 2000). From the initial infection in Norway, the disease has been subsequently recognized in Atlantic salmon in Norway and Scotland (Bruno et al., 1998; Bjornsdottir et al., 2004). Isolates have been recovered from cod, and turbot and rainbow trout have been infected artificially (Greger and Goodrich, 1999; Bjornsdottir et al., 2004). Intraperitoneal (i.p.) injection of oil-adjuvanted vaccines containing *M. viscosa* antigens was protective to rainbow trout and Atlantic salmon (relative percent survival [RPS] = 97%) (Greger and Goodrich, 1999).

Photobacterium damselae subsp. *damselae*

A new organism (*Vibrio damsela*) was recovered initially from ulcers that occurred in the region of the pectoral fin and caudal peduncle during summer and autumn on blacksmith (*Chromis punctipinnis*), a damselfish, found in coastal waters of southern California (Love et al., 1981). The ulcerated fish accounted for 10 to 70% of the population in King Harbor, Redondo Beach, Calif., during August to October and at a second site (Ship Rock, Catalina Island) during June to October. These data suggested a seasonal distribution in the incidence of disease, possibly coinciding with warmer water temperatures and low resistance caused by physiological changes in the host during sexual maturity. Subsequently, the organism has been recovered from sharks (Grimes et al., 1984a; Fujioka et al., 1988; Pedersen et al., 1997), rainbow trout (Pedersen et al., 1997), turbot (Fouz et al., 1991,

Brian Austin • School of Life Sciences, John Muir Building, Heriot-Watt University, Riccarton, Edinburgh EH14 4AS, Scotland, United Kingdom.

1992), and yellowtail (Sakata et al., 1989) and from human wounds (Love et al., 1981). Initially named *V. damsela*, low DNA homology values confirmed its distinctiveness from other vibrios (Grimes et al., 1984b), but the pathogen was subsequently reclassified to *Listonella* (MacDonell and Colwell, 1985) and then to *Photobacterium*, as *P. damsela* (Smith et al., 1991) with the specific epithet corrected to *P. damselae* (Trüper and De'Clari, 1997). Evidence of heterogeneity has been reported as a result of phenotypic analysis (i.e., based on biochemical traits such as urease production, which is useful in separating the two subspecies of fish pathogens) and by amplified fragment length polymorphism (13 clusters were defined among 71 isolates from sea bass and sea bream in Spain; Botella et al., 2002).

Pathogenicity has been confirmed in laboratory-based challenges involving *C. punctipinnis*. Thus, four to six scales were removed from fish, the dermis was scarified, and the wound was swabbed with 10^7–10^8 viable cells of *P. damselae*. The data revealed that at water temperatures of 16.0 to 16.5°C, the infected fish developed large ulcers in 3 days, with death following 24 h later. Interestingly, fish (other than damselfish), i.e., representatives of *Atherinidae*, *Clinidae*, *Cottidae*, *Embiotocidae*, *Girellidae*, and *Gobiidae*, which cohabited the reefs with blacksmith, were more resistant to the experimental challenges (Love et al., 1981). The pathogen was highly cytotoxic, with a neurotoxic acetylcholinesterase being described (Peréz et al., 1998).

Extracellular proteins (ECPs) have been implicated in cytotoxicity, with the 50% lethal dose (LD$_{50}$) ranging from 0.02 to 0.43 μg of protein/g of fish and death occurring between 4 and 72 h after administration (Fouz et al., 1993). The ECP possessed low proteolytic activity, without any evidence of caseinase, elastinase, or gelatinase, but with pronounced phospholipase and hemolytic activity. This was recorded for human, sheep, and turbot erythrocytes. It was considered likely that lipopolysaccharide (LPS) contributed to heat stability of the toxic fractions (Fouz et al., 1993). A siderophore-mediated iron sequestering system has been described and is likely to be involved in pathogenicity (Fouz et al., 1994, 1997).

Evidence suggests that the organism is transmitted through water, with the portal of entry possibly involving the skin mucus, insofar as cultures resisted the bacteriocidal effect of mucus (Fouz et al., 2000). Instead, cultures adhered to mucus (Fouz et al., 2000).

Photobacterium damselae subsp. *piscicida*

A new disease was recognized during the summer of 1963 in white perch and striped bass from the upper region of the Chesapeake Bay. Isolation of pure cultures from kidney or spleen may be readily achieved using marine 2216E agar, nutrient agar, blood agar, brain heart infusion agar, or trypticase soy agar (TSA), each supplemented with 1% (wt/vol) NaCl or TCBS (thiosulfate citrate bile salt sucrose agar) with incubation at 25°C for 48 to 72 h (Fujioka et al., 1988; Sakata et al., 1989; Fouz et al., 1991; Liu et al., 2003). Cultures have been recovered from water in the vicinity of fish 8 days before the outbreak of disease, by using a filtration method involving the filtration of 250-ml volumes of water through 0.45-μm cellulose nitrate filters, before transfer of the filters to marine 2216E agar supplemented with 1% (wt/vol) mannitol and 0.5% (wt/vol) phenol red. On this medium, the nonfermenting *P. damselae* subsp. *piscicida* produced red colonies (Reali et al., 1997).

The initial microbiological examination revealed the presence of a previously unrecognized organism that was considered to resemble *Pasteurella* (Snieszko et al., 1964), and thus the disease was termed pasteurellosis. The possible association with *Pasteurella* was reinforced by cross-precipitin reactions with *Pasteurella (Yersinia) pestis* (Janssen and Surgalla, 1968). Therefore, the name *P. piscicida* was coined. Separately, in Japan, the same condition occurred but was named pseudotuberculosis because of the distinctive pathology, and the causal agent was named *Pasteurella seriola;* however, the synonymy with *P. piscicida* was quickly recognized, and that name was given preference (Kusuda and Yamaoka, 1972).

The disease has been surmised to cause heavy losses in menhaden and striped mullet from Galveston Bay, Texas (Lewis et al., 1970), in farmed yellowtail from Japan (Egusa, 1983), and in farmed cobria from Taiwan (Liu et al., 2003). In addition, pasteurellosis has been diagnosed in a range of other fish species, including gilthead sea bream (Balebona et al., 1998), red sea bream (Yasunaga et al., 1983), black sea bream (Muroga et al., 1977; Ohnishi et al., 1982), hybrid striped bass (Hawke et al., 2003), cobria (Liu et al., 2003), and many farmed and wild fish stocks in France, Italy, and Spain (Magariños et al., 1992). The disease is a septicemia with few readily recognizable pathological signs. Granulomatous-like deposits, consisting of gray-white bacterial colonies of 0.5 to 1.0 mm^2 in size, have been seen on the kidney and spleen, leading to the coining of the descriptive term pseudotuberculosis (Kusuda and Yamaoka, 1972). Also, purulent matter may be present in the abdominal cavity (Lewis et al., 1970).

Cultures have been considered to be phenotypically and serologically homogeneous (Magariños et al., 1992; Hawke et al., 2003). Ribotyping of 29 isolates revealed two major ribotypes for European and Japanese isolates (Magariños et al., 1997). A taxonomic reevaluation based on small-subunit rRNA sequencing and DNA:DNA hybridization revealed a high relatedness (>80% DNA relatedness) to *P. damselae*, which led to reclassification as *P. damselae* subsp. *piscicida* (Gauthier et al., 1995). Effective diagnosis has been achieved by means of 16S rRNA sequencing and amplified fragment length polymorphism (Kvitt et al., 2002).

Infection may occur in seawater at ~25°C (Yasunaga et al., 1983). Yet the pathogen has been regarded to be short-lived in freshwater (≤48 h) and estuarine (salinity = 12%) habitats (4 to 5 days [Toranzo et al., 1982]). However, the pathogen may well exhibit a nonculturable, dormant phase in the aquatic environment (Magariños et al., 1994a). Magariños et al. (1994a) published plate count data that demonstrated that *P. damselae* subsp. *piscicida* survived for 6 to 12 days in seawater and sediment, with metabolism being reduced by 80%. When culturing showed a reduction in bacterial numbers, microscopy using acridine orange suggested that the populations remained unchanged (Magariños et al., 1994a).

Experimental infection has been achieved by intramuscular (i.m.) injection, orally or by immersion. ECPs are lethal (1.26 μg of protein/fish) (Liu et al., 2003). Following i.m. injection, the pathogen was detected initially in the kidney and spleen, before its appearance in the gills, heart, intestine, and pyloric ceca. Yet after oral uptake, the organism was detected in the stomach before it reached the internal organs. In comparison, following immersion, the pathogen was found in the gills, and then the heart, kidney, liver, pyloric ceca, and spleen (Kawahara et al., 1989). Specifically, *P. damselae* subsp. *piscicida* accumulated in the macrophages (Nelson et al., 1989), but data suggested that these intracellular bacteria were killed in 3 to 5 h (Skarmeta et al., 1995); other workers reported that replication occurred in macrophages (Elkamel et al., 2003). In common with many other gram-negative bacteria, pathogenicity involved an initial cell adherence stage (Magariños et al., 1996a,b) involving capsular material (Magariños et al., 1996b). A siderophore-mediated iron-sequestering mechanism has been found with iron-regulated outer membrane proteins of 75, 105, 118, and 145 kDa in size (Magariños et al., 1994b).

Resistance to the pathogen has been documented and may well reflect the size of the fish and the efficiency of the phagocytes (Noya et al., 1995) and components of the (fish) mucus (Magariños et al., 1995). However, the bacterial capsule may well give protection against phagocytosis (Arijo et al., 1998).

Vaccine development programs have centered on passive immunization (Fukuda and Kusuda, 1981a) and bacterins, i.e., formalin-inactivated whole-cell preparations, both monovalent and divalent (with *Vibrio alginolyticus* [Morinigo et al., 2002]) (Kusuda and Fukuda, 1980; Fukuda and Kusuda, 1981b; Acosta et al., 2005), the latter of which was combined with Freund's complete adjuvant and applied by i.p. injection to yellowtail, which developed agglutinating antibody titers of 1:256 to 1:2,048 within 5 weeks (Kusuda and Fukuda, 1980). Toxoid-enriched whole cells used by immersion led to a low antibody response and an RPS of 37 to 41% in sea bream (Magariños et al., 1994c). An improved RPS of >60% after 35 days resulted with an LPS mixed chloroform-killed whole-cell vaccine (Kawakami et al., 1997). A divalent vaccine with formalized cells of *P. damselae* subsp. *piscicida* and *V. alginolyticus* incorporating ECP led to an RPS of >70% in juvenile gilthead sea bream when applied by immersion (Morinigo et al., 2002). Moreover, bacterins with or without Freund's complete adjuvant and application by i.p. injection, a 5- to 7-s spray, hyperosmotic infiltration, and oral uptake led Fukuda and Kusuda (1981b) to report promising results within 21 days following artificial challenge with the pathogen. The best results, conferring 100% protection to the fish, were obtained by use of i.p. injection or by spraying. Also, these authors highlighted the value of subcellular components, especially LPS (Fukuda and Kusuda, 1982). A ribosomal vaccine has shown promise following administration by i.p. injection into yellowtail (Kusuda et al., 1988; Ninomiya et al., 1989). Moreover, a potassium thiocyanate extract and acetic acid-treated "naked cells" obtained from a virulent culture have been examined and gave reasonable protection (RPS = 36.5 [Muraoka et al., 1991]), although the antibody titers were low, suggesting that the humoral immune response did not exert an important role in protection (Muraoka et al., 1991). An auxotroph, namely, an aroA mutant, has been evaluated in hybrid striped bass, and excellent protection developed (Thune et al., 2003). In an interesting development, Nitzan et al. (2004) noted that the salinity of the bacterial growth medium influenced the immune response after vaccination. Specifically, medium containing 2.5% (wt/vol) NaCl used at 25°C produced the most effective vaccine (Nitzan et al., 2004).

An immunostimulant, specifically, glucan, enhanced resistance of gilthead bream to experimental challenge with *P. damselae* subsp. *piscicida* (Couso et al., 2003). This approach is worthy of further consideration to expand the current range of disease control strategies applicable to pasteurellosis.

Shewanella putrefaciens

During spring of 1985, *S. putrefaciens* was recovered from diseased rabbitfish, *Siganus rivulatus*, farmed in sea cages in the Red Sea (Saeed et al., 1987). In addition, a similar organism has been recovered from diseased carp and rainbow trout in freshwater (Kozinska and Pekala, 2004). Cultures may be recovered from the kidney, liver, and spleen following inoculation onto brain heart infusion agar supplemented with 3% (wt/vol) NaCl (Saeed et al., 1987). Experimental challenge resulted from i.p. injection of fish having an average weight of 50 g, which developed clinical disease, with 80% mortality within 48 h (Saeed et al., 1987). A bacterin was fairly effective at controlling mortality when applied twice by i.p. injection (Saeed et al., 1987).

Vibrio alginolyticus

V. alginolyticus has been recognized as pathogenic to both fin- and shellfish. In finfish, the organism has been associated with septicemia in sea bream (*Sparus aurata* [Colorni et al., 1981]), exophthalmia and corneal opaqueness, which have been reported to occur in grouper (*Epinephelus malabaricus* [Lee, 1995]), ascites, lethargy, and melanosis in cobia (*Rachycentron canadum* [Liu et al., 2004]), and ulcers (Akazaka, 1968). In addition, the organism has been believed to cause large-scale mortalities in silver sea bream (*Sparus sarba*) in Hong Kong (Woo et al., 1995; Deane et al., 2004), gilthead sea bream in Spain (Balebona et al., 1998), cultured black sea bream fry in Japan (Kusuda et al., 1986), and cobia (Liu et al., 2004). Additionally, *V. alginolyticus* has been reported as a secondary invader of sea mullet diagnosed with "red spot" (Burke and Rodgers, 1981). Certainly, the lethality of cultures has been confirmed. For example, Liu et al. (2004) calculated the LD_{50} as 3.28×10^4 for cobia, and Lee (1995) described an ECP that was lethal at 0.52 µg/g of fish and contained a 44-kDa toxic protease with a minimum lethal dose at 0.17 µg/g of fish. ECP, with hemolytic and proteolytic activity, led to an effect on hepatic heat shock protein (Deane et al., 2004).

The organism has been readily isolated using TSA prepared with seawater, TCBS, or seawater agar with incubation at 15 to 25°C for 2 to 7 days (e.g., Selvin and Lipton, 2003). Apart from diseased animals, the organism is a common inhabitant of seawater and has been recovered from the water in marine fish tanks (Gilmour, 1977).

In penaeids, *V. alginolyticus* has established a reputation as the cause of significant losses, especially in Asia. The organism has been associated with white spot in *Penaeus monodon* in India (Selvin and Lipton, 2003) and Taiwan (Lee et al., 1996). Here, the diseased shrimp developed an overall reddish color and displayed white spots in the carapace (Selvin and Lipton, 2003). Isolates have been reported to cause mortalities when administered to shrimp, with the LD_{50} calculated variously as 1.13×10^5/g (Lee et al., 1996)

and 5×10^6 CFU/animal (Selvin and Lipton, 2003). These authors reported that the vibrio was more problematic in shrimp, which were infected with the white spot syndrome virus. ECPs have been lethal to *P. monodon*, with the LD_{50} dose calculated as 0.23 µg of protein/g of body weight (Lee et al., 1996). It has been reported that disease is exacerbated in the presence of 1 mg/liter of Cu^{2+}. This led to an effect on the immune response, specifically, an increase in respiratory burst, but a decreased hemocyte count and phenoloxidase and phagocytic activities, and the ability of the shrimp to clear cells of *V. alginolyticus* (Yeh et al., 2004). Also, nitrite (dosed at 5.14 mg/liter) increased susceptibility to the pathogen and led to a reduction in total hemocyte count and phenoloxidase activity, i.e., adversely affected the immune response (Tseng and Chen, 2004).

There has been an indication that the organism may occur as a component of mixed cultures in diseased animals. Thus, *V. alginolyticus* and *V. splendidus* have been recovered together from larval and spat carpet shell clams (*Ruditapes decussatus*) that were experiencing large-scale mortalities in Spain (Gómez-León et al., 2005).

A divalent vaccine containing formalized cells and ECP of *V. alginolyticus* has been developed (Morinigo et al., 2002). Also, the administration of sodium alginate in the diet has improved the resistance of white shrimp (*Litopenaeus vannamei*) to infection by *V. alginolyticus* (Cheng et al., 2005). Here, shrimp, which were fed with diets supplemented with 2.0 g of sodium alginate/kg of feed, cleared the pathogen more rapidly than controls and revealed heightened immune parameters, namely, an enhanced phagocytic index, phenoloxidase activity, respiratory burst, and superoxide dismutase activity, but decreased glutathione peroxidase activity (Cheng et al., 2005).

V. cholerae (non-O1)

V. cholerae became associated with fish pathology in 1977 when an epizootic occurred in a wild population of ayu in the River Amano, Japan, with disease signs including petechial hemorrhages on the body surface and congestion of the organs (Muroga et al., 1979; Kiiyukia et al., 1992). From these fish, cultures were isolated by inoculating swabbed kidney material onto the surface of nutrient agar plates, with incubation at 25°C (Muroga et al., 1979; Kiiyukia et al., 1992). Later, a similar organism was recovered from septicemic goldfish in Australia (Reddacliff et al., 1993).

The phenotypic characteristics published by Muroga et al. (1979), Kiiyukia et al. (1992), and to some extent Reddacliff et al. (1993) agreed largely with the description of *V. cholerae* (Baumann et al., 1984). Two differences displayed by the Japanese cultures were noted: they had the characteristics of the El Tor biotype, namely, production of ornithine decarboxylase and utilization of mannose, which were both negative for the fish pathogens, but there was an 86% DNA:DNA homology with a reference culture of *V. cholerae*. However, the fish isolates did not agglutinate with antisera to *V. cholerae*. Therefore, the conclusion was reached that the fish pathogen was indeed *V. cholerae* non-O1.

The published evidence suggested that *V. cholerae* was highly virulent to ayu and eels following immersion in only 1.26×10^4 cells/ml and 1.26×10^2 cells/ml, respectively. Yamanoi et al. (1980) noted that with ayu, mortalities began in 2 to 7 days at water temperatures of 21 and 26°C, but deaths did not occur if the water temperature was only 16°C. In comparison, eels experienced only 10% mortality within 5 days at a water temperature of 21°C and 30% mortality in 3 to 7 days at 26°C.

V. coralliilyticus

The newly described species *Vibrio coralliilyticus* was regarded as pathogenic for coral, *Pocillopora damicornis*, in the Red Sea. However, an additional culture has been recovered from diseased oyster larvae in England (Ben-Haim et al., 2003).

V. fischeri

During autumn 1988, visceral tumors (neoplasia) and skin papillomas were observed in juvenile turbot, farmed in northwest Spain. Many of the diseased fish displayed whitish nodules on the skin of the dorsal surface, hemorrhagic ulceration, and tumors in the pancreas and bile duct. Losses amounted to 39% of the fish population within 1 year (Lamas et al., 1990). Although viral involvement was suspected, bacteria, approximating the description of *Vibrio fischeri*, were isolated on TSA supplemented with 2% (wt/vol) NaCl, marine 2216E agar, and TCBS (Lamas et al., 1990). Additional isolates have been obtained from gilthead sea bream in Spain (Balebona et al., 1998).

V. furnissii

Vibrio furnissii has been associated, albeit tenuously, with eel disease in Spain (Esteve, 1995).

V. ichthyoenteri

In 1971, opaque intestines and intestinal necrosis accompanied by high mortalities were reported in Japanese flounder hatcheries from which a new species, *Vibrio ichthyoenteri*, was described (Ishimaru et al., 1996).

V. logei

An organism with similarities to *Vibrio logei* was recovered on 5% (vol/vol) horse blood agar supplemented with 1.5% (wt/vol) NaCl with incubation at 15°C for 7 days from shallow skin lesions of Atlantic salmon farmed in Iceland at low temperatures, i.e., ~10°C (Benediktsdóttir et al., 1998).

V. mimicus

Vibrio mimicus has been associated as a secondary invader, causing mortalities in red claw crayfish. Here, the stressors were regarded as overcrowding, poor management, and/or adverse water quality (Eaves and Ketterer, 1994).

V. parahaemolyticus

There is controversy over the role of *Vibrio parahaemolyticus* as a fish pathogen, although its significance as a cause of shellfish disease is well established. Tilapias have been infected with *V. parahaemolyticus*, but the authenticity of the isolates is questionable (Balfry et al., 1997). However, the organism has been associated with mortalities in Iberian toothcarp from Spain, with disease signs including external hemorrhages and tail rot (Alcaide et al., 1999). In addition, organisms with intermediate characteristics between *V. alginolyticus* and *V. parahaemolyticus* have been recovered from diseased milkfish in the Philippines (Muroga et al., 1984).

In contrast, the role of *V. parahaemolyticus* as a pathogen of invertebrates, notably penaeids, is well established, with heavy losses ensuing on an all too regular basis. Pathogenicity has been established in tiger prawns, with the LD_{50} dose calculated for a culture and its ECP as 1×10^5 CFU/penaeid, and 8 μg protein/penaeid, respectively (Sudheesh and Xu, 2001). Mortalities may be exacerbated in the presence of nitrite in the water, which adversely affects the total hemocyte count; phagocytic, phenoloxidase, and respiratory burst activities; and the ability of animals to clear cells of the pathogen (Cheng et al., 2004). Consequently, much effort has been devoted to developing strategies to ameliorate the infections. One interesting approach was the use of butanolic extracts of plants (terrestrial herbs and seaweeds), which conferred positive health benefits to *Penaeus indicus* when applied bioencapsulated in *Artemia* (Immanuel et al., 2004).

V. pectenicida

Isolates from moribund scallop (*Pecten maximus*) in France, demonstrated to be pathogenic to scallop larvae, were described as a new species, *Vibrio pectenicida* (Lambert et al., 1998). It was determined that the pathogen inhibited the chloride activity of hemocytes, leading to their death within a few hours (Lambert et al., 2001). A heat-stable, protease-resistant toxin, i.e., the vibrio hemocyte-killer toxin, of <3 kDa in molecular mass was considered to be responsible (Lambert et al., 2001).

V. pelagius

In winter 1991, an epizootic occurred in juvenile farmed turbot in northwest Spain when the water temperature was 12 to 15°C. The diseased fish displayed eroded dorsal fins and tail, hemorrhages at the base of the finage, hemorrhages on the internal organs, and intestines full of mucoid liquid (Angulo et al., 1992). Bacteria were recovered on TSA supplemented with 2% (wt/vol) NaCl, marine 2216E agar, and TCBS with incubation at 25°C for 48 h, and equated with *Vibrio pelagius* (Angulo et al., 1992). Infectivity experiments with 10 g of rainbow trout and 5 g of turbot confirmed the virulence of the organism and established the LD_{50} as 1.9×10^5 cells/fish and 9.5×10^4 cells/fish, respectively (Angulo et al., 1992).

V. penaeicida

Vibrio penaeicida has been described to accommodate cultures recovered from diseased kuruma prawns (*Penaeus japonicus*) initially in Japan during 1995 (Ishimaru et al., 1995; Takahashi et al., 1998) and has been associated with "syndrome 93" in shrimp (*Litopenaeus stylirostris*) from New Caledonia (Goarant et al., 2000). Here, the syndrome, which occurred predominantly during winter, was linked to temperature drops and was related specifically to the toxic effects of ECPs (Goarant et al., 2000). The disease signs of syndrome 93 included erratic swimming, lethargy, and weakness (Goarant et al., 1998). In Japan, the disease signs included brown spots in the gills and on the lymphoid organs. Also, there was a cloudy appearance in the muscle, specifically in the sixth abdominal segment (Takahashi et al., 1998). The organism was recovered from hemolymph on marine agar and TCBS (Costa et al., 1998).

A reverse transcription-PCR and an arbitrary primed PCR have been developed for diagnosis (Genmoto et al., 1996; Goarant et al., 1999). The sensitivity of PCR was attested to by Saulnier et al. (2000), who reported the detection of only 20 cells in seawater or hemolymph.

Cultures of *V. penaeicida* were used to induce experimental infections in kuruma shrimp (dela Peña et al., 1998) and blue shrimp (Aguirre-Guzman et al., 2003). Mortalities of up to 60% were recorded in blue shrimp within 96 h, and the LD_{50} dose was calculated as 10^4 CFU/ml (Aguirre-Guzman et al., 2003). Yet in kuruma shrimp, only 10% mortality resulted following immersion in 10^8 CFU/ml; in contrast, mortalities did not ensue at 10^7 CFU/ml. Saulnier et al. (2000) reported better success for immersion challenge, with the LD_{50} dose calculated as 1.3×10^4 CFU/ml; i.m. injection was even more successful insofar as the LD_{50} was only 5 CFU/shrimp. Following oral intubation, high levels of mortalities resulted, with the LD_{50} dose estimated as 10^3 to 10^4 CFU/shrimp (de la Peña et al., 1998). These workers concluded that the main portal of entry for the pathogen was the digestive tract, although the surface cuticle and, in particular, wounds were also likely entry points. Avarre et al. (2003) highlighted a link between vitellogenesis and immunity in *P. indicus*. In short, after challenge, mortalities in previtellogenic shrimps peaked earlier than those in their vitellogenic counterparts, although the total number of deaths was similar in both groups (Avarre et al., 2003). The number of hemocytes was significantly less in the vitellogenic animals, 24 h after infection, when compared with the previtellogenic group (Avarre et al., 2003).

Proteinaceous ECPs from some broth cultures incubated at 20°C led to mortalities of up to 98% in blue shrimp within 48 h. In contrast, the supernatants from cultures incubated at 30°C were barely harmful (Aguirre-Guzman et al., 2003).

The disease resistance of kuruma shrimp was strengthened by the oral administration of peptidoglycan from *Bifidobacterium thermophilum* (Itami et al., 1998). Moreover, the phagocytic activity of the shrimp granulocytes, which were fed with peptidoglycan, was higher than that of the controls (Itami et al., 1998). Antibacterial peptides, coined penaeicidins, have been implicated with disease resistance in *L. stylirostris* (Muñoz et al., 2004).

V. proteolyticus

There is some indication that *Vibrio proteolyticus* may be pathogenic for brine shrimp, *Artemia*, affecting the microvilli, (gut) epithelial cells, and tissues surrounding the gut (Verschuere et al., 2000).

V. tapetis

A new bacterial species, *Vibrio tapetis*, was described to accommodate isolates considered to cause brown ring disease in Manila clams (*Tapes philippinarum* [Borrego et al., 1996]) and carpet shell clams (*R. decussatus* [Novoa et al., 1998]). The disease, which was initially recognized in 1987 and has caused severe problems in areas of England, France, Portugal, and Spain, is characterized by the presence of brown deposits on the inner surface of the shells, with resulting high levels of mortality (Castro et al., 1997a; Allam et al., 2000). There is evidence that the organism adheres to and disrupts the production of the periostracal lamina, which leads to deposition of periostracum around the inner shell (Allam et al., 2002). In animals that survived infection, the pathogen was cleared from hemolymph and soft tissues. However, in clams that eventually died, there was evidence that the number of cells of *V. tapetis* increased substantially, leading to tissue damage (Allam et al., 2002). In vitro experiments indicated that in the presence of the pathogen, hemocytes lost filopods and became rounded-up, which indicated cytotoxicity (Choquet et al., 2003). Work has demonstrated that temperature may affect the development of the disease. Thus, at 21°C, there were higher hemocyte counts and leucine aminopeptidase activity and lysozyme activities, less disease, and better recovery than at 14°C (Paillard et al., 2004).

The homogeneity of isolates has been indicated after an examination of 22 isolates, in which most showed the same pulsed-field gel electrophoretic pattern, and all harbored a common 74.5-kb plasmid and belonged in the same ribotype (Castro et al., 1997b). Yet using pulsed-field gel electrophoresis, Romalde et al. (2002) defined two major clusters for 27 isolates and distinguished the isolates of *R. decussatus* from other species of clam.

Experimental infections have been achieved, with losses over 30 days ranging from 6 to 80% (Novoa et al., 1998). Furthermore, work has demonstrated that the most likely transmission route was direct contact with infected clams (Martinez-Manzanares et al., 1998). It has been suggested that the disease may be controlled by means of antimicrobial compounds, e.g., nitrofurantoin, flumequine, or oxolinic acid, administered as bath treatments (Martinez-Manzanares et al., 1998). However, there are concerns regarding the use of pharmaceutical compounds in the natural environment.

V. tubiashii

V. tubiashii is the causal agent of bacillary necrosis in marine bivalves, such as oysters, e.g., the Pacific oyster, *Crassostrea gigas* (Hada et al., 1984; Gibson et al., 1998; Takahashi et al., 2000), with disease signs among larvae including anomalous swimming and detachment/necrosis of cilia and/or vela (Takahashi et al., 2000). Cultures may be highly pathogenic, insofar as 24-h exposure to 10^5 bacterial cells may result in 100% larval mortality (Takahashi et al., 2000). The pathogenicity mechanisms may well include proteases (a 35-kDa zinc-containing metalloprotease has been described [Delston et al., 2003]) and an extracellular heat- and protease-sensitive hemolytic 59-kDa vulnificolysin-like cytolysin (Kothary et al., 2001). Interestingly, mortalities in oyster larvae may be reduced by the use of probiotics, notably *Aeromonas media* (Gibson et al., 1998), and ovoglobulins (a dose of 10 μg/ml was successful) from the whites of chicken eggs (Takahashi et al., 2000).

V. vulnificus

Starting during 1975 in Japan, serious outbreaks of disease characterized by redness (hemorrhages) on the body, notably flank and/or tail, and involvement of the gastrointestinal tract, gills, heart, liver, and spleen (Miyazaki et al., 1977) occurred among cultured eels. The causal agent was isolated on seawater agar and TSA supplemented with NaCl, with incubation for up to 7 days at 20 to 25°C, and identified initially as *Vibrio anguillicida* (the name was resurrected after Bruun and Heiberg, 1935) and then as *Vibrio vulnificus* (Muroga et al., 1976a,b; Nishibuchi and Muroga, 1977, 1980; Nishibuchi et al., 1979). Subsequently, a similar condition was recognized in Spain (Biosca et al., 1991; Amaro et al., 1992) and Denmark (Dalsgaard et al., 1999). Initially, the eel isolates were classified in a new biotype, i.e., biogroup 2 (Tison et al., 1982), although this classification was subsequently questioned (Arias et al., 1997a,b), and the concept of serovars (rather than biotypes) was suggested, with biogroup 2 being regarded as serovar E (Biosca et al., 1997; Marco-Noales et al., 2001). Nevertheless, DNA:DNA hybridization, which showed 90% homology between *V. vulnificus* and the eel pathogens (Tison et al., 1982), confirmed the identification of the isolates.

Experimental infection of eels by i.m. injection established that high doses, i.e., 4.85×10^8, resulted in 80% mortality within 7 days (Dalsgaard et al. [1999] reported an LD_{50} of $<9.4 \times 10^3$ to 2.3×10^5 CFU/eel), at a water temperature of 25°C. ECPs, which are lethal to fish (Lee et al., 1997), included hemolysins, lipases, phospholipases, and proteases (Amaro et al., 1992). In addition, the LPS O side chain is toxic (Amaro et al., 1997).

The organism is common in coastal and estuarine environments (Oliver et al., 1983), which probably constitute the reservoir for infection (Høi et al.,

1998; Marco-Noales et al., 2001) with uptake possibly involving fish skin (Amaro et al., 1995), the digestive tract, and gills (Marco-Noales et al., 2001).

V. vulnificus has been the target for the development of effective disease control strategies, with success in eels achieved after an initial immersion vaccination followed by oral booster in which serum, skin mucus, and bile antibody production was stimulated, and the bacteriocidal action of the bodily fluids increased (Esteve-Gassent et al., 2004a). A vaccine, known as Vulnivaccine, which was administered by immersion on three separate occasions, has shown promise in eels (Esteve-Gassent et al., 2004b). Also, these workers highlighted a new grouping within *V. vulnificus* termed serovar A, and they discussed the development of a bivalent vaccine, which was evaluated by i.p. injection, immersion, and oral and anal intubation, the latter of which led to commendable levels of protection with a significant humoral and mucosal immune response (Esteve-Gassent et al., 2004b). Flounder (*Paralicththys olivaceus*) developed a superior antibody response and better protection after the i.p. administration of an uncoated heat-killed whole-cell vaccine compared with its formalized counterpart (Park et al., 2001). Also, these workers pointed to the benefit of oral vaccination with the uncoated heat-killed vaccine.

Vibrio spp.

Difficult-to-identify *Vibrio* spp. have been regularly recovered from diseased fish (see, e.g., Yasunobu et al., 1988; Masumura et al., 1989; Muroga et al., 1990). Conceivably, these organisms represent new species.

Other Fish-Pathogenic Vibrios

It has been suggested that *Vibrio campbellii* and *Vibrio nereis* may be responsible for disease in gilthead sea bream in Spain (Balebona et al., 1998). However, confirmation is awaited.

REFERENCES

Acosta, F., F. Real, A. E. Ellis, C. Tabraue, D. Padilla, and C. M. R. de Galarreta. 2005. *Photobacterium damselae* subsp. *piscicida* (= *Pasteurella piscicida*) influence of vaccination on the nitric oxide response of gilthead seabream following infection with *Photobacterium damselae* subsp. *piscicida*. *Fish Shellfish Immunol.* 18:31–38.

Aguirre-Guzman, G., Y. Labreuche, D. Ansquer, B. Espiau, P. Levy, F Ascencio, and D. Saulnier. 2003. Proteinaceous exotoxins of shrimp-pathogenic isolates of *Vibrio penaeicida* and *Vibrio nigripulchritudo*. *Ciencias Marinas* 29:77–88.

Akazaka, H. 1968. Bacterial disease of marine fishes. *Bull. Jpn. Soc. Sci. Fish.* 34:271–272.

Alcaide, E., C. Amaro, R. Todoli, and R. Oltra. 1999. Isolation and characterization of *Vibrio parahaemolyticus* causing infection in Iberian toothcarp *Aphanius iberus*. *Dis. Aquat. Org.* 35:77–80.

Allam, B., C. Paillard, A. Howard, and M. Le Pennec. 2000. Isolation of the pathogen *Vibrio tapetis* and defense parameters in brown ring diseased Manila clams *Ruditapes philippinarum* cultivated in England. *Dis. Aquat. Org.* 41:105–113.

Allam, B., C. Paillard, and S. E. Ford. 2002. Pathogenicity of *Vibrio tapetis*, the etiological agent of brown ring disease in clams. *Dis. Aquat. Org.* 48:221–231.

Amaro, C., E. G. Biosca, C. Esteve, B. Fouz, and A. E. Toranzo. 1992. Comparative study of phenotypic and virulence properties in *Vibrio vulnificus* biotypes 1 and 2 obtained from a European eel farm experiencing mortalities. *Dis. Aquat. Org.* 13:29–35.

Amaro, C., E. G. Biosca, B. Fouz, E. Alcaide, and C. Esteve. 1995. Evidence that water transmits *Vibrio vulnificus* biotype 2 infections to eels. *Appl. Environ. Microbiol.* 61:1133–1137.

Amaro, C., B. Fouz, E. G. Biosca, E. Marco-Noales, and R. Collado. 1997. The lipopolysaccharide O side chain of *Vibrio vulnificus* serogroup E is the virulence determinant for eels. *Infect. Immun.* 65:2475–2479.

Angulo, L., J. E. Lopez, C. Lema, and J. A. Vicente. 1992. *Vibrio pelagius* associated with mortalities in farmed turbot, *Scophthalmus maximus*. *Thalassas* 10:129–133.

Arias, C. R., L. Verdonck, J. Swings, E. Garay, and R. Aznar. 1997a. Intraspecific differentiation of *Vibrio vulnificus* biotypes by amplified fragment length polymorphism and ribotyping. *Appl. Environ. Microbiol.* 63:2600–2606.

Arias, C. R., L. Verdonck, J. Swings, R. Aznar, and E. Garay. 1997b. A polyphasic approach to study the intraspecific diversity amongst *Vibrio vulnificus* isolates. *Syst. Appl. Microbiol.* 20:622–633.

Arijo, S., J. J. Borrego, I. Zorilla, M. C. Balebona, and M. A. Moriñigo. 1998. Role of the capsule of *Photobacterium damselae* subsp. *piscicida* in protection against phagocytosis and killing by gilt-head seabream (*Sparus aurata*, L) macrophages. *Fish Shellfish Immunol.* 8:63–72.

Avarre, J. C., D. Saulnier, Y. Labreuche, D. Ansquer, A. Tietz, and E. Lubzens. 2003. Response of *Penaeus indicus* females at two different stages of ovarian development to a lethal infection with *Vibrio penaeicida*. *J. Invert. Pathol.* 82:23–33.

Balebona, M. C., I. Zorrilla, M. A. Moriñigo, and J. J. Borrego. 1998. Survey of bacterial pathogens affecting farmed gilt-head sea bream (*Sparus aurata* L.) in southwestern Spain from 1990 to 1996. *Aquaculture* 166:19–35.

Balfry, S. K., M. Shariff, and G. K. Iwama. 1997. Strain differences in non-specific immunity of tilapia *Oreochromis niloticus* following challenge with *Vibrio parahaemolyticus*. *Dis. Aquat. Org.* 30:77–80.

Baumann, P., A. L. Furniss, and J. V. Lee. 1984. Genus I, *Vibrio* Pacini 1854, 411[AL], p. 518–538. *In* N. R. Krieg and J. G. Holt (ed.), *Bergey's Manual of Systematic Bacteriology*, vol. 1. Lippincott Williams and Wilkins, Philadelphia, Pa.

Benediktsdóttir, E., S. Helgason, and H. Sigurjónsdóttir. 1998. *Vibrio* spp. isolated from salmonids with shallow skin lesions and reared at low temperature. *J. Fish Dis.* 21:19–28.

Benediktsdóttir, E., L. Verdonck, C. Sproer, S. Helgason, and J. Swings. 2000. Characterization of *Vibrio viscosus* and *Vibrio wodanis* isolated at different geographical locations: a proposal for reclassification of *Vibrio viscosus* as *Moritella viscosa* comb. nov. *Int. J. Syst. Evol. Microbiol.* 50:479–488.

Ben-Haim, Y., F. L. Thompson, C. C. Thompson, M. C. Cnockaert, B. Hoste, J. Swings, and E. Rosenberg. 2003. *Vibrio corallilyticus* sp. nov., a temperature-dependent pathogen of the coral *Pocillopora damicornis*. *Int. J. Syst. Evol. Microbiol.* 53:309–315.

Biosca, E. G., C. Amaro, C. Esteve, E. Alcaide, and E. Garay. 1991. First record of *Vibrio vulnificus* biotype 2 from diseased European eel, *Anguilla anguilla* L. *J. Fish Dis.* 14:103–109.

Biosca, E. G., C. Amaro, J. L. Larsen, and K. Pedersen. 1997. Phenotypic and genotypic characterization of *Vibrio vulnificus*: proposal for the substitution of the subspecific taxon biotype for serovar. *Appl. Environ. Microbiol.* 63:1460–1466.

Bjornsdottir, B., S. Gudmundsdottir, S. H. Bambir, B. Magnadottir, and B. K. Gudmundsdottir. 2004. Experimental infection of turbot, *Scophthalmus maximus* (L.), by *Moritella viscosa*, vaccination effort and vaccine-induced side-effects. *J. Fish Dis.* 27:645–655.

Borrego, J. J., D. Castro, A. Luque, C. Paillard, P. Maes, M. T. Garcia, and A. Ventosa. 1996. *Vibrio tapetis* sp nov, the causative agent of the brown ring disease affecting cultured clams. *Int. J. Syst. Bacteriol.* 46:480–484.

Botella, S., M. J. Pujalte, M. C. Macian, M. A. Ferrus, J. Hernandez, and E. Garay. 2002. Amplified fragment length polymorphism (AFLP) and biochemical typing of *Photobacterium damselae* subsp *damselae*. *J. Appl. Microbiol.* 93:681–688.

Bruno, D. W., J. Griffiths, J. Petrie, and T. S. Hastings. 1998. *Vibrio viscosus* in farmed Atlantic salmon *Salmo salar* in Scotland: field and experimental observations. *Dis. Aquat. Org.* 34:161–166.

Bruun, A. F., and B. Heiberg. 1935. Weitere Untersuchungen über die Rotseuche des Aales in den dänischen Gewässern. *Z. Fisch. Hilf.* 33:379–382.

Burke, J., and L. Rodgers. 1981. Identification of pathogenic bacteria associated with the occurrence of "red spot" in sea mullet, *Mugil cephalus* L., in south-eastern Queensland. *J. Fish Dis.* 3:153–159.

Castro, D., J. A. Santamaria, A. Luque, E. Martinez-Manzanares, and J. J. Borrego. 1997a. Determination of the etiological agent of brown ring disease in southwestern Spain. *Dis. Aquat. Org.* 29:181–188.

Castro, D., J. L. Romalde, J. Vila, B. Magariños, A. Luque, and J. J. Borrego. 1997b. Intraspecific characterization of *Vibrio tapetis* strains by use of pulsed-field gel electrophoresis, ribotyping, and plasmid profiling. *Appl. Environ. Microbiol.* 63:1449–1452.

Cheng, W. T., I. S. Hsiao, and J. C. Chen. 2004. Effect of nitrite on immune response of Taiwan abalone *Haliotis diversicolor supertexta* and its susceptibility to *Vibrio parahaemolyticus*. *Dis. Aquat. Org.* 60:157–164.

Cheng, W. T., C. H. Liu, C. M. Kuo, and J. C. Chen. 2005. Dietary administration of sodium alginate enhances the immune ability of white shrimp *Litopenaeus vannamei* and its resistance against *Vibrio alginolyticus*. *Fish Shellfish Immunol.* 18:1–12.

Choquet, G., P. Soudant, C. Lambert, J. L. Nicolas, and C. Paillard. 2003. Reduction of adhesion properties of *Ruditapes philippinarum* hemocytes exposed to *Vibrio tapetis*. *Dis. Aquat. Org.* 57:109–116.

Colorni, A., I. Paperna, and H. Gordin. 1981. Bacterial infections in gilthead sea bream *Sparus aurata* cultured in Elat. *Aquaculture* 23:257–267.

Costa, R., I. Mermoud, S. Koblavi, B. Morlet, P. Haffner, F. Berthe, M. Legroumellec, and P. Grimont. 1998. Isolation and characterization of bacteria associated with a *Penaeus stylirostris* disease (syndrome 93) in New Caledonia. *Aquaculture* 164:297–309.

Couso, N., R. Castro, B. Magariños, A. Obach, and J. Lamas. 2003. Effect of oral administration of glucans on the resistance of gilthead seabream to pasteurellosis. *Aquaculture* 219:99–109.

Dalsgaard, I., L. Hoi, R. J. Siebeling, and A. Dalsgaard. 1999. Indole-positive *Vibrio vulnificus* isolated from disease outbreaks on a Danish eel farm. *Dis. Aquat. Org.* 35:187–194.

Deane, E. E., J. Li, and N. Y. S. Woo. 2004. Modulated heat shock protein expression during pathogenic *Vibrio alginolyticus* stress of sea bream. *Dis. Aquat. Organ.* 62:205–215.

de la Peña, L. D., T. Naka, and K. Muroga. 1998. Experimental infection of kuruma prawn (*Penaeus japonicus*) with *Vibrio penaeicida*. *Israeli J. Aquacult.-Bamidgeh* 50:128–133.

Delston, R. B., M. H. Kothary, K. A. Shangraw, and B. D. Tall. 2003. Isolation and characterization of a zinc-containing metalloprotease expressed by *Vibrio tubiashii*. *Can. J. Microbiol.* 49:525–529.

Eaves, L. E., and P. J. Ketterer. 1994. Mortalities in red claw crayfish *Cherax quadricarinatus* associated with systemic *Vibrio mimicus* infection. *Dis. Aquat. Org.* 19:233–237.

Egusa, S. 1983. Disease problems in Japanese yellowtail, *Seriola quinqueradiata*, culture: a review, p. 10–18. *In* J. E. Stewart (ed.), *Diseases of Commercially Important Marine Fish and Shellfish*. Conseil International pour l'Exploration de la Mer, Copenhagen, Denmark.

Elkamel, A. A., J. P. Hawke, W. G. Henk, and R. L. Thune. 2003. *Photobacterium damselae* subsp *piscicida* is capable of replicating in hybrid striped bass macrophages. *J. Aquat. Anim. Health* 15:175–183.

Esteve, C. 1995. Numerical taxonomy of Aeromonadaceae and Vibrionaceae associated with reared fish and surrounding fresh and brackish water. *Syst. Appl. Microbiol.* 18:391–402.

Esteve-Gassent, M. D., R. Barrera, and C. Amaro. 2004a. Vaccination of market-size eels against vibriosis due to *Vibrio vulnificus* serovar E. *Aquaculture* 241:9–19.

Esteve-Gassent, M. D., B. Fouz, and C. Amaro. 2004b. Efficacy of a bivalent vaccine against eel diseases caused by *Vibrio vulnificus* after its administration by four different routes. *Fish Shellfish Immunol.* 16:93–105.

Fouz, B. I., J. L. Larsen, and A. E. Toranzo. 1991. *Vibrio damsela* as a pathogenic agent causing mortalities in cultured turbot (*Scophthalmus maximus*). *Bull. Eur. Assoc. Fish Pathol.* 11:80–81.

Fouz, B., J. L. Larsen, B. Nielsen, J. L. Barja, and A. E. Toranzo. 1992. Characterization of *Vibrio damsela* strains isolated from turbot, *Scophthalmus maximus* in Spain. *Dis. Aquat. Org.* 12:155–166.

Fouz, B., J. L. Barja, C. Amaro, C. Rivas, and A. E. Toranzo. 1993. Toxicity of the extracellular products of *Vibrio damsela* isolated from diseased fish. *Curr. Microbiol.* 27:341–347.

Fouz, B., A. E. Toranzo, E. G. Biosca, R. Mazoy, and C. Amaro. 1994. Role of iron in the pathogenicity of *Vibrio damsela* for fish and mammals. *FEMS Microbiol. Lett.* 121:181.

Fouz, B., E. G. Biosca, and C. Amaro. 1997. High affinity iron-uptake systems in *Vibrio damsela*: role in the acquisition of iron from transferrin. *J. Appl. Microbiol.* 82:157–167.

Fouz, B., A. E. Toranzo, M. Milan, and C. Amaro. 2000. Evidence that water transmits the disease caused by the fish pathogen *Photobacterium damselae* subsp *damselae*. *J. Appl. Microbiol.* 88:531–535.

Fujioka, R. S., S. B. Greco, M. B. Cates, and J. P. Schroeder. 1988. *Vibrio damsela* from wounds in bottlenose dolphins, *Tursiops truncatus*. *Dis. Aquat. Org.* 4:1–8.

Fukuda, Y., and R. Kusuda. 1981a. Passive immunization of cultured yellowtail against pseudotuberculosis. *Fish Pathol.* 16:85–89.

Fukuda, Y., and R. Kusuda. 1981b. Efficacy of vaccination for pseudotuberculosis in cultured yellowtail by various routes of administration. *Bull. Jpn. Soc. Sci. Fish.* 47:147–150.

Fukuda, Y., and R. Kusuda. 1982. Detection and characterization of precipitating antibody in the serum of immature yellowtail immunized with *Pasteurella piscicida* cells. *Fish Pathol.* 17:125–127.

Gauthier, G., B. Lafay, R. Ruimy, V. Breittmayer, J. L. Nicolas, M. Gauthier, and R. Christen. 1995. Small-subunit rRNA sequences and whole DNA relatedness concur for the re-assignment of *Pasteurella piscicida* (Snieszko et al.) Janssen and Surgalla to the genus *Photobacterium* as *Photobacterium damsela* subsp. *piscicida* comb. nov. *Int. J. Syst. Bacteriol.* **45**:139–144.

Genmoto, K., T. Nishizawa, T. Nakai, and K. Muroga. 1996. 16S rRNA targeted RT-PCR for the detection of *Vibrio penaeicida*, the pathogen of cultured kuruma prawn *Penaeus japonicus*. *Dis. Aquat. Org.* **24**:185–189.

Gibson, L. F., J. Woodworth, and A. M. George. 1998. Probiotic activity of *Aeromonas media* on the Pacific oyster, *Crassostrea gigas*, when challenged with *Vibrio tubiashii*. *Aquaculture* **169**:111–120.

Gilmour, A. 1977. Characteristics of marine vibrios isolated from fish farm tanks. *Aquaculture* **11**:51–62.

Goarant, C., F. Regnier, R. Brizard, and A. L. Marteau. 1998. Acquisition of susceptibility to *Vibrio penaeicida* in *Penaeus stylirostris* postlarvae and juveniles. *Aquaculture* **169**:291–296.

Goarant, C., F. Merien, F. Berthe, I. Mermoud, and P. Perolat. 1999. Arbitrarily primed PCR to type *Vibrio* spp. pathogenic for shrimp. *Appl. Environ. Microbiol.* **65**:1145–1151.

Goarant, C., J. Herlin, R. Brizard, A. L. Marteau, C. Martin, and B. Martin. 2000. Toxic factors of *Vibrio* strains pathogenic to shrimp. *Dis. Aquat. Org.* **40**:101–107.

Gómez-León, J., L. Villamil, M. L. Lemos, B. Novoa, and A. Figueras. 2005. Isolation of *Vibrio alginolyticus* and *Vibrio splendidus* from aquacultured carpet shell clam (*Ruditapes decussates*) larvae associated with mass mortalities. *Appl. Environ. Microbiol.* **71**:98–104.

Greger, E., and T. Goodrich. 1999. Vaccine development for winter ulcer disease, *Vibrio viscosus*, in Atlantic salmon, *Salmo salar* L. *J. Fish Dis.* **22**:193–199.

Grimes, D. J., J. Stemmler, H. Hada, E. B. May, D. Maneval, F. M. Hetrick, R. T. Jones, M. Stoskopf, and R. R. Colwell. 1984a. *Vibrio* species associated with mortality of sharks held in captivity. *Microb. Ecol.* **10**:271–282.

Grimes, D. J., R. R. Colwell, J. Stemmler, H. Hada, D. Maneval, F. M. Hetrick, E. B. May, R. T. Jones, and M. Stoskopf. 1984b. *Vibrio* species as agents of elasmobranch disease. *Helgoländer Meeresunter.* **37**:309–315.

Hada, H. S., P. A. West, J. V. Lee, J. Stemmler, and R. R. Colwell. 1984. *Vibrio tubiashii* sp nov, a pathogen of bivalve mollusks. *Int. J. Syst. Bacteriol.* **34**:1–4.

Hawke, J. P., R. L. Thune, R. K. Cooper, E. Judice, and M. Kelly-Smith. 2003. Molecular and phenotypic characterization of strains of *Photobacterium damselae* subsp *piscicida* isolated from hybrid striped bass cultured in Louisiana, USA. *J. Aquat. Anim. Health* **15**:189–201.

Høi, L., J. L. Larsen, I. Dalsgaard, and A. Dalsgaard. 1998. Occurrence of *Vibrio vulnificus* biotypes in Danish marine environments. *Appl. Environ. Microbiol.* **64**:7–13.

Immanuel, G., V. C. Vincybai, V. Sivaram, A. Palavesam, and M. P. Marian. 2004. Effect of butanolic extracts from terrestrial herbs and seaweeds on the survival, growth and pathogen (*Vibrio parahaemolyticus*) load on shrimp *Penaeus indicus* juveniles. *Aquaculture* **236**:53–65.

Ishimaru, K., M. Akagawa-Matsushita, and K. Muroga. 1995. *Vibrio penaeicida* sp nov, a pathogen of kuruma prawns (*Penaeus japonicus*). *Int. J. Syst. Bacteriol.* **45**:134–138.

Ishimaru, K., M. Akagawa-Matsushita, and K. Muroga. 1996. *Vibrio ichthyoenteri* sp. nov., a pathogen of Japanese flounder (*Paralichthys olivaceus*). *Int. J. Syst. Bacteriol.* **46**:155–159.

Itami, T., M. Asano, K. Tokushige, K. Kubono, A. Nakagawa, N. Takeno, H. Nishimura, M. Maeda, M., Kondo, and Y. Takahashi. 1998. Enhancement of disease resistance of kuruma shrimp, *Penaeus japonicus*, after oral administration of peptido-

glycan derived from *Bifidobacterium thermophilum*. *Aquaculture* **164**:277–288.

Janssen, W. A., and M. J. Surgalla. 1968. Morphology, physiology and serology of a *Pasteurella* species pathogenic for white perch (*Roccus americanus*). *J. Bacteriol.* **96**:1606–1610.

Kawahara, E., K. Kawai, and R. Kusuda. 1989. Invasion of *Pasteurella piscicida* in tissues of experimentally infected yellowtail *Seriola quinqueradiata*. *Nipp. Suis. Gakk.* **55**:499–501.

Kawakami, H., N. Shinohara, Y. Fukuda, H. Yamashita, H. Kihara, and M. Sakai. 1997. The efficacy of lipopolysaccharide mixed chloroform-killed cell (LPS-CKC) bacterin of *Pasteurella piscicida* on yellowtail, *Seriola quinqueradiata*. *Aquaculture* **154**:95–105.

Kiiyukia, C., A. Nakajima, T. Nakai, K. Muroga, H. Kawakami, and H. Hashimoto. 1992. *Vibrio cholerae* non-01 isolated from ayu fish (*Plecoglossus altivelis*) in Japan. *Appl. Environ. Microbiol.* **58**:3078–3082.

Kothary, M. H., R. B. Delston, S. K. Curtis, B. A. McCardell, and B. D. Tall. 2001. Purification and characterization of a vulnificolysin-like cytolysin produced by *Vibrio tubiashii*. *Appl. Environ. Microbiol.* **67**:3707–3711.

Kozinska, A., and A. Pekala. 2004. First isolation of *Shewanella putrefaciens* from freshwater fish—a potential new pathogen of fish. *Bull. Eur. Assoc. Fish Pathol.* **24**:189–193.

Kusuda, R., and Y. Fukuda. 1980. Agglutinating antibody titers and serum protein changes of yellowtail after immunization with *Pasteurella piscicida* cells. *Bull. Jpn. Soc. Sci. Fish.* **46**:801–807.

Kusuda, R., and M. Yamaoka. 1972. Etiological studies on bacterial pseudotuberculosis in cultured yellowtail with *Pasteurella piscicida* as the causative agent. 1. On the morphological and biochemical properties. *Bull. Jpn. Soc. Sci. Fish.* **38**:1325–1332.

Kusuda, R., J. Yokoyama, and K. Kawai. 1986. Bacteriological study on cause of mass mortalities in cultured black sea bream fry. *Bull. Jpn. Soc. Sci. Fish.* **52**:1745–1751.

Kusuda, R., M. Itaoka, and K. Kawai. 1988. Drug sensitivity of *Pasteurella piscicida* strains isolated from cultured yellowtail from 1984 to 1985. *Nipp. Suis. Gakk.* **54**:1521–1526.

Kvitt, H., M. Ucko, A. Colorni, C. Batargias, A. Zlotkin, and W. Knibb. 2002. *Photobacterium damselae* ssp *piscicida*: detection by direct amplification of 16S rRNA gene sequences and genotypic variation as determined by amplified fragment length polymorphism (AFLP). *Dis. Aquat. Org.* **48**:187–195.

Lamas, J., R. Anadon, S. Devesa, and A. E. Toranzo. 1990. Visceral neoplasia and epidermal papillomas in cultured turbot *Scophthalmus maximus*. *Dis. Aquat. Org.* **8**:179–187.

Lambert, C., J. L. Nicolas, V. Cilia, and S. Corre. 1998. *Vibrio pectenicida* sp. nov.: a pathogen of scallop (*Pecten maximus*) larvae. *Int. J. Syst. Bacteriol.* **48**:481–487.

Lambert, C., J. L. Nicolas, and V. Bultel. 2001. Toxicity to bivalve hemocytes of pathogenic *Vibrio* cytoplasmic extract. *J. Invert. Pathol.* **77**:165–172.

Lee, K.-K. 1995. Pathogenesis studies on *Vibrio alginolyticus* in the grouper, *Epinephelus malabaricus*, Bloch et Schneider. *Microb. Pathog.* **19**:39–48.

Lee, K.-K., S. R. Yu, F. R. Chen, T. I. Yang, and P. C. Liu. 1996. Virulence of *Vibrio alginolyticus* isolated from diseased tiger prawn, *Penaeus monodon*. *Curr. Microbiol.* **32**:229–231.

Lee, K.-K., H. T. Chiang, K. C. Yii, W. M. Su, and P. C. Liu. 1997. Effects of extracellular products of *Vibrio vulnificus* on *Acanthopagrus schlegeli* serum components *in vitro* and *in vivo*. *Microbios* **92**:209–217.

Lewis, D. H., L. C. Grumbles, S. McConnell, and A. I. Flowers. 1970. *Pasteurella*-like bacteria from a epizootic in menhaden and mullet in Galveston Bay. *J. Wildl. Dis.* **6**:160–162.

Liu, P. C., J. Y. Liu, and K. K. Lee. 2003. Virulence of *Photobacterium damselae* subsp *piscicida* in cultured cobia *Rachycentron canadum*. *J. Basic Microbiol.* **43**:499–507.

Liu, P. C., J. Y. Liu, P. T. Hsiao, and K. K. Lee. 2004. Isolation and characterization of pathogenic *Vibrio alginolyticus* from diseased, cobia *Rachycentron canadum*. *J. Basic Microbiol.* **44**:23–28.

Love, M., D. Teebken-Fisher, J. E. Hose, J. J. Farmer III, F. W. Hickman, and G. R. Fanning. 1981. *Vibrio damsela*, a marine bacterium, causes skin ulcers on the damselfish *Chromis punctipinnis*. *Science* **214**:1139–1140.

Lunder, T., H. Sorum, G. Holstad, A. G. Steigerwalt, P. Mowinckel, and D. J. Brenner. 2000. Phenotypic and genotypic characterization of *Vibrio viscosus* sp nov and *Vibrio wodanis* sp nov isolated from Atlantic salmon (*Salmo salar*) with "winter ulcer." *Int. J. Syst. Evol. Microbiol.* **50**:427–450.

MacDonell, M. T., and R. R. Colwell. 1985. Phylogeny of the Vibrionaceae and recommendations for two new genera, *Listonella* and *Shewanella*. *Syst. Appl. Microbiol.* **6**:171–182.

Magariños, B., J. L. Romalde, I. Bandín, B. Fouz, and A. E. Toranzo. 1992. Phenotypic, antigenic, and molecular characterization of *Pasteurella piscicida* strains isolated from fish. *Appl. Environ. Microbiol.* **58**:3316–3322.

Magariños, B., J. L. Romalde, J. L. Barja, and A. E. Toranzo. 1994a. Evidence of a dormant but infective state of the fish pathogen *Pasteurella piscicida* in seawater and sediment. *Appl. Environ. Microbiol.* **60**:180–186.

Magariños, B., J. L. Romalde, M. L. Lemos, J. L. Barja, and A. E. Toranzo. 1994b. Iron uptake by *Pasteurella piscicida* and its role in pathogenicity for fish. *Appl. Environ. Microbiol.* **60**:2990–2998.

Magariños, B., J. L. Romalde, Y. Santos, J. F. Casal, J. L. Barja, and A. E. Toranzo. 1994c. Vaccination trials on gilthead sea bream *(Sparus aurata)* against *Pasteurella piscicida*. *Aquaculture* **120**:201–208.

Magariños, B., F. Pazos, Y. Santos, J. L. Romalde, and A. E. Toranzo. 1995. Response of *Pasteurella piscicida* and *Flexibacter maritimus* to skin mucus of marine fish. *Dis. Aquat. Org.* **21**:103–108.

Magariños, B., J. L. Romalde, M. Noya, J. L. Barja, and A. E. Toranzo. 1996a. Adherence and invasive capacities of the fish pathogen *Pasteurella piscicida*. *FEMS Microbiol. Lett.* **138**:29–34.

Magariños, B., R. Bonet, J. L. Romalde, M. L. Martínez, F. Congregado, and A. E. Toranzo. 1996b. Influence of the capsular layer on the virulence of *Pasteurella piscicida* for fish. *Microb. Pathog.* **21**:289–297.

Magariños, B., C. R. Osorio, A. E. Toranzo, and J. L. Romalde. 1997. Applicability of ribotyping for intraspecific classification and epidemiological studies of *Photobacterium damsela* subsp. *piscicida*. *Syst. Appl. Microbiol.* **20**:634–639.

Marco-Noales, E., M. Milan, B. Fouz, E. Sanjuan, and C. Amaro. 2001. Transmission to eels, portals of entry, and putative reservoirs of *Vibrio vulnificus* serovar E (biotype 2). *Appl. Environ. Microbiol.* **67**:4717–4725.

Martinez-Manzanares, E., D. Castro, J. I. Navas, M. L. Lopez-Cortes, and J. J. Borrego. 1998. Transmission routes and treatment of brown ring disease affecting manila clams (Tapes philippinarum). *J. Shellfish Res.* **17**:1051–1056.

Masumura, K., H. Yasunobu, N. Okada, and K. Muroga. 1989. Isolation of a *Vibrio* sp., the causative bacterium of intestinal necrosis of Japanese flounder larvae. *Fish Pathol.* **24**:135–141.

Miyazaki, T., Y. Jo, S. S. Kubota, and S. Egusa. 1977. Histopathological studies on vibriosis of the Japanese eel *Anguilla japonica*. Part 1. Natural infection. *Fish Pathol.* **12**:163–170.

Morinigo, M. A., J. L. Romalde, M. Chabrillon, B. Magariños, S. Arijo, C. Balebona, and A. E. Toranzo. 2002. Effectiveness of a divalent vaccine for gilt-head sea bream (*Sparus aurata*) against

Vibrio alginolyticus and *Photobacterium damselae* subsp *piscicida*. *Bull. Eur. Assoc. Fish Pathol.* **22**:298–303.

Muñoz, M., F. Vandenbulcke, J. Garnier, Y. Gueguen, P. Bulet, D. Saulnier, and E. Bachere. 2004. Involvement of penaeidins in defense reactions of the shrimp *Litopenaeus stylirostris* to a pathogenic vibrio. *Cell. Mol. Life Sci.* **61**:961–972.

Muraoka, A., K. Ogawa, S. Hashimoto, and R. Kusuda. 1991. Protection of yellowtail against pseudotuberculosis by vaccination with a potassium thiocyanate extract of *Pasteurella piscicida* and co-operating protective effect of acid-treated, naked bacteria. *Nipp. Suis. Gakk.* **57**:249–253.

Muroga, K., Y. Jo, and M. Nishibuchi. 1976a. Pathogenic *Vibrio* isolated from cultured eels. I. Characteristics and taxonomic status. *Fish Pathol.* **12**:141–145.

Muroga, K., M. Nishibuchi, and Y. Jo. 1976b. Pathogenic *Vibrio* isolated from cultured eels. II. Physiological characteristics and pathogenicity. *Fish Pathol.* **12**:147–151.

Muroga, K., T. Sugiyama, and N. Ueki. 1977. Pasteurellosis in cultured black sea bream (*Mylio macrocephalus*). *J. Fac. Fish. Anim. Husb., Hiroshima Univ.* **16**:17–21.

Muroga, K., S. Takahashi, and H. Yamanoi. 1979. Non-cholera *Vibrio* isolated from diseased ayu. *Bull. Jpn. Soc. Sci. Fish.* **45**:829–834.

Muroga, K., G. Lio-Po, C. Pitogo, and R. Imada. 1984. *Vibrio* sp. isolated from milkfish (*Chanos chanos*) with opaque eyes. *Fish Pathol.* **19**:81–87.

Muroga, K., H. Yasunobu, N. Okada, and K. Masumura. 1990. Bacterial enteritis of cultured flounder *Paralichthys olivaceus* larvae. *Dis. Aquat. Org.* **9**:121–125.

Nelson, J. S., E. Kawahara, K. Kawai, and R. Kusuda. 1989. Macrophage infiltration in pseudotuberculosis of yellowtail, *Seriola quinqueradiata*. *Bull. Mar. Sci. Fish., Kochi Univ.* **11**:17–22.

Ninomiya, M., A. Muraoka, and R. Kusuda. 1989. Effect of immersion vaccination of cultured yellowtail with a ribosomal vaccine prepared from *Pasteurella piscicida*. *Nipp. Suis. Gakk.* **55**:1773–1776.

Nishibuchi, M., and K. Muroga. 1977. Pathogenic *Vibrio* isolated from cultured eels. III. NaCl tolerance and flagellation. *Fish Pathol.* **12**:87–92.

Nishibuchi, M., K. Muroga, R. J. Seidler, and J. L. Fryer. 1979. Pathogenic *Vibrio* isolated from cultured eels. IV. Deoxyribonucleic acid studies. *Bull. Jpn. Soc. Sci. Fish.* **45**:1469–1473.

Nishibuchi, M., and K. Muroga. 1980. Pathogenic *Vibrio* isolated from cultured eels. V. Serological studies. *Fish Pathol.* **14**:117–124.

Nitzan, S., B. Shwartsburd, and E. D. Heller. 2004. The effect of growth medium salinity on the immune response of *Photobacterium damselae* subsp *piscicida* of hybrid bass (*Morone saxatilis* x M-*chrysops*). *Fish Shellfish Immunol.* **16**:107–116.

Novoa, B., A. Luque, D. Castro, J. J. Borrego, and A. Figueras. 1998. Characterization and infectivity of four bacterial strains isolated from brown ring disease-affected clams. *J. Invert. Pathol.* **71**:34–41.

Noya, M., B. Magariños, and J. Lamas. 1995. Interactions between peritoneal exudate cells (PECs) of gilthead bream (*Sparus aurata*) and *Pasteurella piscicida*. A morphological study. *Aquaculture* **131**:11–21.

Ohnishi, K., K. Watanabe, and Y. Jo. 1982. *Pasteurella* infection in young black seabream. *Fish Pathol.* **16**:207–210.

Oliver, J. D., R. A. Warner, and D. R. Cleland. 1983. Distribution of *Vibrio vulnificus* and other lactose-fermenting vibrios in the marine environment. *Appl. Environ. Microbiol.* **45**:985–998.

Paillard, C., B. Allam, and R. Oubella. 2004. Effect of temperature on defense parameters in Manila clam *Ruditapes philippinarum* challenged with *Vibrio tapetis*. *Dis. Aquat. Org.* **59**:249–262.

Park, J. H., W. J. Park, and H. D. Jeong. 2001. Immunological efficacy of *Vibrio vulnificus* bacterins given as an oral vaccine in the flounder, *Paralichthys olivaceus*. *Aquaculture* **201**:187–197.

Pedersen, K., I. Dalsgaard, and J. L. Larsen. 1997 *Vibrio damsela* associated with diseased fish in Denmark. *Appl. Environ. Microbiol.* **63**:3711–3715.

Peréz, M. J., L. A. Rodríguez, and T. P. Nieto. 1998. The acetylcholinesterase ichthyotoxin is a common component of the extracellular products of Vibrionaceae strains. *J. Appl. Microbiol.* **84**:47–52.

Reali, D., C. Pretti, L. Tavanti, and A. M. Cognetti-Varriale. 1997. *Pasteurella piscicida* (Janssen & Surgalla, 1964): a simple method of isolation and identification from rearing-ponds. *Bull. Eur. Assoc. Fish Pathol.* **17**:51–53.

Reddacliff, G. L., M. Hornitsky, J. Carson, R. Petersen, and R. Zelski. 1993. Mortalities of goldfish, *Carassius auratus* (L.), associated with *Vibrio cholerae* (non-O1). *J. Fish Dis.* **16**:517–520.

Romalde, J. L., D. Castro, B. Magariños, L. Lopez-Cortes, and J. J. Borrego. 2002. Comparison of ribotyping, randomly amplified polymorphic DNA, and pulsed-field gel electrophoresis, for molecular typing of *Vibrio tapetis*. *Syst. Appl. Microbiol.* **25**:544–550.

Saeed, M. O., M. M. Alamoudi, and A. H. Al-Harbi. 1987. A *Pseudomonas* associated with disease in cultured rabbitfish *Siganus rivulatus* in the Red Sea. *Dis. Aquat. Org.* **3**:177–180.

Sakata, T., M. Matsuura, and Y. Shimokawa. 1989. Characteristics of *Vibrio damsela* isolated from diseased yellowtail *Seriola quinqueradiata*. *Nip. Suis. Gakk.* **55**:135–141.

Saulnier, D., J. C. Avarre, G. Le Moullac, D. Ansquer, P. Levy, and V. Vonau. 2000. Rapid and sensitive PCR detection of *Vibrio penaeicida*, the putative etiological agent of syndrome 93 in New Caledonia. *Dis. Aquat. Org.* **40**:109–115.

Selvin, J., and A. P. Lipton. 2003. *Vibrio alginolyticus* associated with white spot disease of *Penaeus monodon*. *Dis. Aquat. Org.* **57**:147–150.

Skarmeta, A. M., I. Bandín, Y. Santos, and A. E. Toranzo. 1995. *In vitro* killing of *Pasteurella piscicida* by fish macrophages. *Dis. Aquat. Org.* **23**:51–57.

Smith, S. K., D. C. Sutton, J. A. Fuerst, and J. L. Reichelt. 1991. Evaluation of the genus *Listonella* and reassignment of *Listonella damsela* (Love et al) MacDonell and Colwell to the genus *Photobacterium* as *Photobacterium damsela* comb. nov. with an emended description. *Int. J. Syst. Bacteriol.* **41**:529–534.

Snieszko, S. F., G. L. Bullock, E. Hollis, and J. G. Boone. 1964. *Pasteurella* sp. from an epizootic of white perch (*Roccus americanus*) in Chesapeake Bay tidewater areas. *J. Bacteriol.* **88**:1814–1815.

Sudheesh, P. S., and H. S. Xu. 2001. Pathogenicity of *Vibrio parahaemolyticus* in tiger prawn *Penaeus monodon* Fabricius: possible role of extracellular proteases. *Aquaculture* **196**:37–46.

Takahashi, K. G., A. Nakamura, and K. Mori. 2000. Inhibitory effects of ovoglobulins on bacillary necrosis in larvae of the Pacific oyster, *Crassostrea gigas*. *J. Invert. Pathol.* **75**:212–217.

Takahashi, Y., T. Itami, M. Maeda, and M. Kondo. 1998. Bacterial and viral diseases of kuruma shrimp (*Penaeus japonicus*) in Japan. *Fish Pathol.* **33**:357–364.

Thune, R. L., D. H. Fernandez, J. P. Hawke, and R. Miller. 2003. Construction of a safe, stable, efficacious vaccine against *Photobacterium damselae* ssp *piscicida*. *Dis. Aquat. Org.* **57**:51–58.

Tison, D. L., M. Nishibuchi, J. D. Greenwood, and R. J. Seidler. 1982. *Vibrio vulnificus* biogroup 2: a new biogroup pathogenic for eels. *Appl. Environ. Microbiol.* **44**:640–646.

Toranzo, A. E., J. L. Barja, and F. H. Hetrick. 1982. Survival of *Vibrio anguillarum* and *Pasteurella piscicida* in estuarine and freshwaters. *Bull. Eur. Assoc. Fish Pathol.* **3**:43–45.

Trüper, H. G., and L. De'Clari. 1997. Taxonomic note: necessary correction of specific epithets formed as substantives (nouns) "in apposition." *Int. J. Syst. Bacteriol.* **47**:908–909.

Tseng, I. T., and J. C. Chen. 2004. The immune response of white shrimp *Litopenaeus vannamei* and its susceptibility to *Vibrio alginolyticus* under nitrite stress. *Fish Shellfish Immunol.* **17**:325–333.

Urakawa, H., K. Kita-Tsukamoto, S. E. Stevens, K. Ohwada, and R. R. Colwell. 1998. A proposal to transfer *Vibrio marinus* (Russell 1891) to a new genus *Moritella* gen. nov. as *Moritella marina* comb. nov. *FEMS Microbiol. Lett.* **165**:373–378.

Verschuere, L., H. Heang, G. Criel, P. Sorgeloos, and W. Verstraete. 2000. Selected bacterial strains protect *Artemia* spp. from the pathogenic effects of *Vibrio proteolyticus* CW8T2. *Appl. Environ. Microbiol.* **66**:1139–1146.

Woo, N. Y. S., J. L. M. Ling, and K. M. Lo. 1995. Pathogenic *Vibrio* spp. in the sea bream, *Sparus sarba*. *J. Sun Yetsen Univ.* **Suppl. 3**:192–193.

Yamanoi, H., K. Muroga, and S. Takahashi. 1980. Physiological characteristics and pathogenicity of NAG vibrio isolated from diseased ayu. *Fish Pathol.* **15**:69–73.

Yasunaga, N., K. Hatai, and J. Tsukahara. 1983. *Pasteurella piscicida* from an epizootic of cultured red seabream. *Fish Pathol.* **18**:107–110.

Yasunobu, H., K. Muroga, and K. Maruyama. 1988. A bacteriological investigation on the mass mortalities of red seabream *Pagrus major* larvae with intestinal swelling. *Suisanzoshoku* **1**:11–20.

Yeh, S. T., C. H. Liu, and J. C. Chen. 2004. Effect of copper sulfate on the immune response and susceptibility to *Vibrio alginolyticus* in the white shrimp *Litopenaeus vannamei*. *Fish Shellfish Immunol.* **17**:437–446.

VIII. THE IMPACT OF GENOMICS AND PROTEOMICS IN THE STUDY OF HUMAN PATHOGENS

The Biology of Vibrios
Edited by F. L. Thompson et al.
© 2006 ASM Press, Washington, D.C.

Chapter 23

Vibrio cholerae: the Genetics of Pathogenesis and Environmental Persistence

MICHAEL G. PROUTY AND KARL E. KLOSE

INTRODUCTION

Historical accounts of cholera have been recorded among ancient civilizations along the Ganges delta (Barua, 1992). Regularly occurring epidemics were greatly feared due to the extremely high mortality associated with the outbreaks. Early beliefs were that cholera was spread via a fog emanating from the rivers (Snow et al., 1936). However, diligent research by men such as John Snow and Robert Koch ushered in a greater understanding of the disease and its cause. The modern era of cholera epidemics began with an outbreak in India beginning in 1817. This outbreak would eventually spread throughout Asia, creating what is referred to as the first pandemic (Rosenberg, 1962). At the time of this writing, there have been six additional recorded pandemics, with the emergence of epidemics caused by the O139 serogroup in 1992 often referred to as the beginning of an eighth pandemic (International Centre for Diarrhoeal Diseases Research, 1993; Ramamurthy et al., 1993). Virtually no inhabited region of the globe has remained untouched by this disease.

Interestingly, the causative agent has changed through these pandemics. After the discovery by Koch of the bacterium *Vibrio cholerae* as the etiological agent of cholera (Kotloff et al., 1992), further investigation was directed toward studying the organisms responsible for the pandemics. The fifth and sixth pandemics are attributed to the serogroup O1 biotype classical strain of *V. cholerae*, whereas the seventh pandemic, which began in 1961, was caused by the serogroup O1 biotype El Tor strain (Barua, 1992). The biotype distinction of O1 strains into "classical" versus "El Tor" was originally based upon several phenotypic differences (e.g., resistance to polymyxin B, expression of HlyA) (Roy and Mukerjee, 1962; Feeley, 1965; Barrett and Blake, 1981), but the differences between these biotypes have more recently been shown to be due to specific differences in the genomes of these strains (Dziejman et al., 2002). O1 El Tor strains were actually first isolated as early as 1905 (Pollitzer, 1959); however, for unknown reasons, they began to spread pandemically some 60 years later, eventually displacing the classical strains (Sack et al., 2004). The epidemic outbreak of cholera in 1992 in Bangladesh and India, which has since spread throughout Southeast Asia, was caused by an O139 serogroup strain, demonstrating the capacity for non-O1 strains to cause pandemic cholera (International Centre for Diarrhoeal Diseases Research, 1993; Ramamurthy et al., 1993). Whereas this epidemic is regarded as the eighth pandemic, it is important to note that O139 strains have not replaced the O1 El Tor strains, which remain in the environment and are responsible for continuing epidemics (Sack et al., 2004). Non-O1/O139 strains can also be responsible for sporadic human diarrheal disease but do not cause large epidemics and generally do not contain the virulence genes associated with epidemic cholera (discussed below). Because the vast majority of research has involved O1/O139 strains, this chapter focuses on our current understanding of O1/O139 virulence and environmental persistence.

Cholera is acquired through the consumption of contaminated water or food. After passing through the acid barrier of the stomach, the bacteria penetrate the mucous lining of the small intestine and adhere to intestinal epithelial cells to establish an infection (Color Plate 15A). The dose required to cause an infection in healthy human volunteers is fairly high: 10^6 to 10^{11} CFU, depending on the inoculating conditions (Hornick et al., 1971; Sack et al., 1998). This relatively high inoculum is most likely due to the acid sensitivity of the bacteria as they pass through the stomach. During colonization of the small intestine,

Michael G. Prouty and Karl E. Klose • South Texas Center for Emerging Infectious Diseases and Department of Biology, The University of Texas at San Antonio, San Antonio, TX 78249.

the bacteria begin to express several key virulence factors, including cholera toxin (CT). The action of CT on intestinal epithelial cells is responsible for the massive diarrhea associated with cholera. As a result of the diarrhea, large numbers of the bacteria are flushed from the intestine and presumably back into the environment.

The symptoms of cholera are quite easy to distinguish. The hallmark symptom is the profuse watery diarrhea. In acute cases, the stools acquire a characteristic "rice water" appearance in which the mucus that lines the intestines becomes suspended in the fluid, resulting in a milky or rice water appearance. Over the course of 3 days, up to 90 liters of diarrhea may be produced. The massive loss of fluid leads to severe dehydration that, if untreated, leads to circulatory failure and ultimately death (Sack et al., 2004). However, survival from an episode of cholera appears to provide protection against a second infection of the same serogroup.

VIRULENCE FACTORS

Key to the ability of *V. cholerae* to cause disease are the virulence factors that it produces. Extensive research efforts involving innovative techniques have been directed at understanding the mechanisms involved in *V. cholerae*'s establishing a successful infection. The two most important virulence factors required for the ability of the organism to cause cholera are the CT and the toxin-coregulated pilus (TCP). As the name implies, these two factors are coordinately regulated during infection.

Cholera Toxin

The most critical virulence factor of *V. cholerae* is CT, which is responsible for the main symptom of the disease. CT is an ADP-ribosylating toxin that is composed of two subunits (Finkelstein and LoSpalluto, 1969; King and van Heyningen, 1973; Gill and King, 1975). The B subunit, which forms a pentamer, binds to GM_1 gangliosides and tethers the A subunit of the toxin to the cell surface (King and van Heyningen, 1973; Gill, 1976). At the cell surface, the binding of CT to GM_1 has been shown to allow the toxin to associate with lipid rafts, which are specialized membrane domains that contain a concentration of cholesterol and glycosphingolipids in the cell's membrane receptor (Simons and Ikonen, 1997; London and Brown, 2000). Following secretion from *V. cholerae*, the A subunit is cleaved at residue 192 into two subunits, A_1 and A_2 (Gill and Rappaport, 1979); this cleavage can be mediated by the *V. cholerae* hemag-glutinin/protease (Booth et al., 1984). The A_2 fragment, attached to the A_1 fragment by a disulfide bond, then serves to tether the A_1 fragment to the B subunits. The A_1 subunit is the enzymatically active portion of the toxin and must translocate the cell membrane and dissociate from the B subunit pentamer for activity (Dickinson and Lencer, 2003). Internalization of the toxin has been shown to occur via both clatherin-dependent and -independent pathways, depending on the cell type (Nichols et al., 2001; Shogomori and Futerman, 2001; Sandvig and van Deurs, 2002). Recently, it was demonstrated that membrane cholesterol is important for CT endocytosis and eventual trafficking to the Golgi apparatus, possibly by allowing CT–GM_1 complexes to associate with the lipid rafts (Wolf et al., 2002).

Following internalization, CT can be found in early and recycling endosomes, the Golgi apparatus, and the endoplasmic reticulum (Majoul et al., 1996, 1998; Richards et al., 2002). After reaching the endoplasmic reticulum, the A_1 subunit separates from the B subunit–GM_1 complex, unfolds, and then translocates across the endoplasmic reticulum membrane into the cytosol (Dickinson and Lencer, 2003). In the cytosol of the cell, the A_1 subunit transfers the ADP-ribose moiety of NAD to the $G_s\alpha$ protein at arginine-201 (Gill and King, 1975; Mekalanos et al., 1979). ADP ribosylation of the G protein results in an increase of adenylate cyclase activity, leading to a significant increase in cyclic AMP (cAMP) levels. The increased cAMP in intoxicated cells leads to a stimulation of intestinal secretion-inducing neurotransmitters within the cell and subsequently an increase in Cl^- secretion (Jodal and Lungren, 1986; de Jonge, 1991). Normally, ions such as Na^+ and Cl^- are absorbed in the intestinal mucosa by Na^+/H^+ and Cl^-/HCO_3^- exchangers; however, these channels are blocked by CT-induced cAMP and serotonin (5-HT) (Sundaram et al., 1991a,b). As a result of the osmotic imbalance, large amounts of water flow into the lumen of the intestine, causing the massive diarrhea and subsequent dehydration (Field et al., 1969).

The genes encoding CT—*ctxA* and *ctxB*—are encoded by the lysogenic filamentous bacteriophage CTXφ (Waldor and Mekalanos, 1996). Since *ctxAB* are carried by the phage, a nontoxigenic strain of *V. cholerae* (i.e., lacking *ctxAB*) can be converted to a toxigenic strain by phage infection, allowing the evolution of toxigenic *V. cholerae* (Boyd et al., 2000).

Recent studies have demonstrated that other toxins produced by *V. cholerae* are responsible for an inflammatory response during infection. These accessory toxins may play a role in "reactogenicity" that is often seen in volunteers administered CT^- vaccine strains (Fullner et al., 2002). The HA protease (HAP), hemolysin (HlyA), and RTX were identified as the candidates for causing the inflammatory response (Fullner et al., 2002).

Toxin Coregulated Pilus (TCP)

A second key virulence factor of *V. cholerae* is the TCP. This is a type IV bundle-forming pilus that is expressed during colonization of the human intestinal tract (Taylor et al., 1987). The gene encoding the pilus subunit (TcpA), a regulatory protein (ToxT), and those required for pilin biosynthesis are encoded in the *tcp* operon, which resides in the vibrio pathogenicity island (VPI) (Ogierman et al., 1993; Everiss et al., 1994; Kovach et al., 1996; Karaolis et al., 1998; Peterson and Mekalanos, 1998; Taylor et al., 1998; Heidelberg et al., 2000). The pilus serves two key roles in the pathogenic potential of *V. cholerae*. The first role involves the colonization of the host. Studies have shown that this pilus is absolutely required for successful colonization of both humans and suckling mice (Taylor et al., 1987; Herrington et al., 1988). Since successful colonization is required for the disease symptoms associated with cholera, expression of TCP is critical for the disease process. The pilus is composed of a homopolymer of repeating subunits of TcpA (Taylor et al., 1988). The TcpA amino acid sequence is divergent between biotypes, with approximately 80% identity between the O1 classical and O1 El Tor biotypes (Jonson et al., 1992). Furthermore, when nucleotide sequence polymorphisms were compared among O1, O139, and non-O1/O139 strains, *tcpA* was found to be 50% different among the strains (Boyd and Waldor, 2002). The divergence found within *tcpA* does not necessarily apply to other genes found within the VPI, as *aldA* demonstrates only a 3% difference (Boyd and Waldor, 2002).

The requirement for TCP in intestinal colonization is still not fully understood. To date, receptors within the intestine for TCP have yet to be identified. Instead, the pilus may mediate bacterial interaction and microcolony formation by direct pilus–pilus contact (Kirn et al., 2000). This interaction can be observed in the classical O1 strain grown under specific laboratory conditions, as expression of TCP results in autoagglutination of the bacteria and the formation of visible "clumps" (Taylor et al., 1987).

A second key role for TCP is to serve as the receptor for CTXφ (Waldor and Mekalanos, 1996). Since the *tcp* locus is required for colonization and contains the gene encoding the transcriptional activator (*toxT*) for both TCP and CT, the use of the pilus as a receptor by CTXφ may serve to "link" CTX production to the virulence regulon (Waldor and Mekalanos, 1996; Faruque et al., 1998; Lee, 1999).

In a study to determine the function of the proteins encoded within the *tcp* operon, the product encoded by *tcpF* was identified as being required for infant mouse colonization (Kirn et al., 2003). An insertion in *tcpF* did not affect in vitro phenotypes such as pilus expression or bacterial autoagglutination; however, the *tcpF* strain was unable to colonize the infant mouse intestine (Kirn et al., 2003). TcpF has been characterized as a novel colonization factor that is secreted by means of the TCP biogenesis machinery (Kirn et al., 2003). How TcpF mediates host colonization by *V. cholerae* remains to be elucidated.

Lipopolysaccharide (LPS)

The LPS of gram-negative bacteria serves as a barrier to protect the bacteria from external stresses (Nikaido and Vaara, 1985; Nikaido, 1988). Strains of *V. cholerae* are classified on the basis of their specific LPS O side chain. Over 200 different *V. cholerae* O serogroups have been identified, yet only the O1 and O139 serogroups are capable of causing large-scale epidemics. As demonstrated by molecular and genetic analysis, O139 strains are genetically very similar to O1 El Tor strains, and horizontal gene transfer has been postulated to be responsible for the emergence of O139 from an O1 El Tor strain (Higa et al., 1993; Behari et al., 2001; Dziejman et al., 2002). The evidence for this serotype conversion is the presence in O139 strains of a 35-kb (*wbf*) locus encoding the O139 polysaccharide that has partially replaced the *rfb* region found in the O1 chromosome that encodes the O1 antigen (Comstock et al., 1996; Stroeher et al., 1998). The *wbf* region also contains the genes that allow the O139 strains to express a capsule (Johnson et al., 1994) that is composed of O139 antigenic material (Fig. 1). The O1 antigen is a homopolymer of perosamine consisting of dideoxyphosphomannose substituted with tetronate (Redmond, 1979; Kenne et al., 1982), whereas the O139 antigen is a complex polysaccharide containing repeats of different carbohydrates (Knirel et al., 1995, 1997). Given the significant structural differences between these two O antigens, a major question is why only these two serogroups are responsible for pandemic disease.

O1 strains lacking the O1 antigen demonstrate significant colonization defects, as do O139 mutants lacking the O139 antigen attached to the LPS, the O139 capsule, or both O antigen and capsule (Waldor and Mekalanos, 1994; Waldor et al., 1994; Iredell et al., 1998; Angelichio et al., 1999; Chiang and Mekalanos, 1999; Nesper et al., 2001). In vitro, strains lacking the O1 LPS side chain or the O139 capsule are more sensitive to complement killing (Waldor

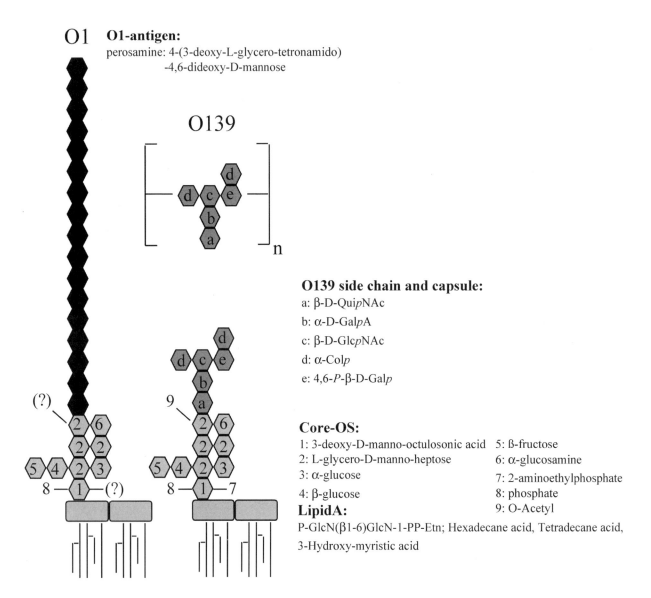

O1   **O1-antigen:**
perosamine: 4-(3-deoxy-L-glycero-tetronamido)
-4,6-dideoxy-D-mannose

O139

O139 side chain and capsule:
a: β-D-QuipNAc
b: α-D-GalpA
c: β-D-GlcpNAc
d: α-Colp
e: 4,6-P-β-D-Galp

Core-OS:
1: 3-deoxy-D-manno-octulosonic acid   5: ß-fructose
2: L-glycero-D-manno-heptose   6: α-glucosamine
3: α-glucose   7: 2-aminoethylphosphate
4: β-glucose   8: phosphate
9: O-Acetyl

LipidA:
P-GlcN(β1-6)GlcN-1-PP-Etn; Hexadecane acid, Tetradecane acid,
3-Hydroxy-myristic acid

Figure 1. LPS structure of O1 and O139 serogroups of *V. cholerae*. Reprinted from Reidl and Klose (2002) with permission of Elsevier Science B.V.

et al., 1994; Chiang and Mekalanos, 1999; Nesper et al., 2002b). In vivo, strains lacking the O1 antigen have been shown to be cleared early from the small intestine (Angelichio et al., 1999). Analysis of the *wav* genes that encode the LPS core oligosaccharide demonstrated that different environmental and epidemic strains have varying genes and organization. One specific *wav* gene cluster is highly associated with pathogenic strains, suggesting that a particular core oligosaccharide structure may also contribute to virulence (Nesper et al., 2002a). Disruption of the biosynthetic pathway for LPS core oligosaccharide also results in a colonization defect (Lee et al., 1999; Nesper et al., 2001).

Motility and Chemotaxis

V. cholerae is motile by means of a single polar sheathed flagellum, and motility has been identified as a virulence factor. In early studies, strains that were nonmotile were shown to have reduced virulence in rabbit models (Freter and O'Brien, 1981a; Richardson, 1991), but not significantly decreased ability for colonization in the infant mouse model (Richardson, 1991; Klose and Mekalanos, 1998). In human vaccine studies, nonmotile mutants caused fewer disease symptoms but were still able to colonize the intestine (Coster et al., 1995; Kenner et al., 1995). Motility and virulence factor expression have been postulated to be inversely regulated, based on the phenotypes of spontaneous motility mutants. Nonmotile strains produced higher levels of CT and TCP than did wild-type strains under noninducing in vitro conditions, whereas "hyperswarming" mutants exhibited minimal virulence factor expression (Gardel and Mekalanos, 1994). It should be noted that all the studies mentioned above were performed with undefined genetic mutations.

V. cholerae flagellar biosynthesis has become better understood, and defined flagellar mutants have made it clear that the influence of flagellar-mediated motility on virulence is complicated, made even more so by strain differences. For example, a structural mutation that prevents formation of a functional flagellum (*flaA*) causes a modest colonization defect (less than twofold) in an O1 classical strain (Klose and Mekalanos, 1998) but a more severe colonization defect (14-fold) in an O1 El Tor strain (Lee et al., 2001). Furthermore, this same mutation causes a rugose phenotype in an O139 strain and results in a significant colonization defect, but this defect is attributed to exopolysaccharide expression and not to a lack of motility (Watnick et al., 2001). Flagellar regulatory factors also influence virulence. Mutant alleles of the σ^{54}-dependent flagellar regulator FlrC caused significant colonization defects, and these studies suggested that "locking" FlrC into either an active or an inactive state is more deleterious to intestinal colonization than removing the protein entirely (Correa et al., 2000). Expression of a hemolytic activity against human O red blood cells is clearly induced in nonmotile mutants (Gardel and Mekalanos, 1996; Correa et al., 2000); however, this hemolysin has not yet been identified, so its role in virulence is currently unknown.

Early data utilizing spontaneous nonchemotactic mutants suggested that chemotaxis was important for *V. cholerae* to penetrate the mucous lining and colonize the intestinal epithelia (Freter et al., 1979, 1981a; Freter and O'Brien, 1981a,b). Recent data from Camilli and colleagues utilizing defined nonchemotactic mutants have expanded on these earlier observations and found an interesting connection between chemotaxis and virulence. Microarray analysis of *V. cholerae* in the stool of human patients demonstrated a repression of chemotaxis-related genes, and this was associated with a "hyperinfectious" state of robust intestinal colonization, suggesting that downregulation of chemotaxis might increase infectivity (Merrell et al., 2002a). A genetic selection for genes important for CT expression within the intestine identified chemotaxis genes, indicating that chemotaxis was important for the induction of CT in vivo (Lee et al., 2001). Nonchemotactic mutants were found to colonize the intestine to a much greater extent than wild-type strains (Lee et al., 2001). Taken together, these data implied that

chemotaxis is important to lead the bacteria to preferred niches within the intestine (presumably within intestinal crypts), where signals important for the induction of CT expression are present. Failure to chemotax increases inappropriate intestinal colonization that does not lead to the expression of CT. Mutants created to display either clockwise (tumble) or counterclockwise (smooth swimming) bias of flagellar rotation allowed these researchers to demonstrate that a counterclockwise rotational bias increases the infectivity of *V. cholerae* (Butler and Camilli, 2004). The *V. cholerae* genome contains a large number of methyl-accepting chemotaxis proteins (Heidelberg et al., 2000), so the exact chemotactic signals being perceived within the intestine may be difficult to elucidate, but several reports have implicated the mucous lining itself as a chemoattractant (Freter et al., 1979; Freter and O'Brien, 1981a).

Porins

Outer membrane (OM) porins in gram-negative bacteria are important channels for the flux and efflux of low-molecular-weight molecules. These porins generally form hydrophilic channels with pore size ranging from 1 to 2 nm (Nikaido, 1996). To date, porin activity has been demonstrated for three *V. cholerae* OM proteins: OmpS, OmpU, and OmpT (Lang and Palva, 1993; Chakrabarti et al., 1996; Benz et al., 1997; Wibbenmeyer et al., 2002; Simonet et al., 2003). The virulence regulator ToxR, in addition to its role in the regulation of *toxT* transcription (see below), also serves to independently regulate the expression of the porins OmpU and OmpT (Miller and Mekalanos, 1988; Champion et al., 1997). Specifically, ToxR (independent of TcpP) activates the transcription of *ompU* and represses the transcription of *ompT* (Miller and Mekalanos, 1988; Champion et al., 1997). A *V. cholerae* strain expressing OmpU in the OM is more resistant to the presence of bile and other anionic detergents than a strain expressing OmpT (Provenzano et al., 2000, 2001), which explains why mutant *toxR V. cholerae* strains are more sensitive to anionic detergents. The ancestral function of ToxR appears to be as a modulator of OM proteins and subsequent resistance to anionic detergents, since *toxR* mutants of a variety of *Vibrio* spp. demonstrate reduced bile resistance and OM protein alterations (Provenzano et al., 2000). In addition to increased bile susceptibility, a *V. cholerae* strain expressing OmpT instead of OmpU in the OM also exhibits decreases in virulence factor (CT and TCP) expression and intestinal colonization (Provenzano and Klose, 2000). The difference in permeability of the porins likely accounts for the phenotypic changes observed.

In the presence of bile salts, OmpU is more restrictive to solute flux, whereas OmpT is more permissive (Wibbenmeyer et al., 2002). Further patch clamp analysis of the porins demonstrated that OmpU and OmpT possess distinctly different electrophysiologic properties (Simonet et al., 2003); for example, OmpT exhibits more frequent closures and for a longer period of time than OmpU. OmpU is more cation selective than OmpT, which probably accounts for the greater bile resistance of OmpU-expressing strains to negatively charged anionic detergents such as bile salts (Simonet et al., 2003). Because bile induces ToxR-dependent *ompU* transcription (Provenzano et al., 2000), this suggests a regulatory loop of OmpT-expressing *V. cholerae* entering the intestine, where bile flux across the OM induces OmpU expression, which then protects cells against further bile movement across the OM. An additional role for OmpU is in the ability of *V. cholerae* to undergo an adaptive stress response known as acid tolerance response (ATR). ToxR-dependent expression of OmpU mediates the organic ATR (Merrell et al., 2001). An additional OM protein (VCA1008) has been identified as being important for intestinal colonization (Osorio et al., 2004), but it has not yet been demonstrated if this OM protein has porin activity.

Other Factors Implicated in Virulence

RTX toxin

A region of the chromosome directly downstream from the CTX element was shown to have similarity to the RTX (repeat in toxin) family of toxins, which are exotoxins produced by a large number of gram-negative bacteria (Lin et al., 1999). The *V. cholerae* RTX gene cluster is composed of four genes, termed *rtxA*, *B*, *C*, and *D*. *rtxA* encodes the toxin, *B* and *D* encode secretory proteins, and *C* encodes the toxin activator (Lin et al., 1999). Interestingly, classical O1 strains of *V. cholerae* contain a 7,869-bp deletion encompassing *rtxC* and portions of *rtxA* and *B*, rendering the RTX toxin system nonfunctional; O1 El Tor and O139 strains, as well as nontoxigenic non-O1 strains, contain functional *rtx* genes (Lin et al., 1999). In general, RTX toxins contain an N-terminal hydrophobic region that is responsible for pore formation, central toxin activation sites, C-terminal calcium-binding GD-rich repeats involved in target cell binding, and a C-terminal secretion signal. Exposure of HEp-2 cells to *V. cholerae* strains with an intact RTX locus results in rounding and detachment of the cells (Lin et al., 1999). RtxA causes the covalent cross-linking of actin fibers, resulting in depolymerization of actin stress fibers and

rounding of HEp-2 cells in culture (Fullner and Mekalanos, 2000; Sheahan et al., 2004). Additionally, RTX toxin has been demonstrated to disrupt the paracellular tight junction of polarized T84 intestinal epithelial cells (Fullner et al., 2001). Within the RtxA protein, a 412-amino-acid region has been identified that is recognized as the actin cross-linking domain (Sheahan et al., 2004). Deletion of this domain from the toxin eliminated actin cross-linking but had no effect on cell rounding (actin cross-linking domain). Export of RTX toxins occurs via a type I secretion system (TISS) which, in general, utilizes a homodimer of an inner membrane transport ATPase, a trimer of a periplasmic linker protein, and a trimer of an OM porin (Andersen, 2003). Boardman and Satchell identified a putative TISS within the RTX gene cluster that appears to be required for toxin secretion (Boardman and Satchell, 2004). Unlike previously described TISS, this TISS requires two transport ATPases that may function as a heterodimer (Boardman and Satchell, 2004).

The role that RTX toxin plays in cholera pathogenesis remains unclear. Expression of RTX toxin in the mouse pulmonary model results in a diffuse pneumonia characterized by a proinflammatory response and tissue damage (Fullner et al., 2002). During vaccine trials it has been observed that patients who ingested CT$^-$ RTX$^+$ strains of *V. cholerae* exhibited a higher level of lactoferrin, a physiological marker for the presence of neutrophils (Silva et al., 1996), suggestive of an inflammatory response possibly mediated by the RTX toxin.

Hemagglutinin/protease

V. cholerae strains of both the O1 and O139 serogroups secrete a Zn-dependent metalloprotease known as hemagglutinin/protease (HA/protease), which is encoded by *hapA* (Hase and Finkelstein, 1991). Analysis of the HA/protease sequence suggests that it undergoes several cleavage steps prior to secretion (Hase and Finkelstein, 1991). Secretion of HA/protease is achieved through the same type II secretion system responsible for CT secretion (Sandkvist et al., 1997).

HA/protease appears to possess a number of different activities, but defining a role for this protein in animal models of virulence has proven elusive (Finkelstein et al., 1992). Early research demonstrated the ability of the enzyme to nick and activate the CT A subunit to form the A$_1$ and A$_2$ subunits (Booth et al., 1984), and it was recently shown that HA/protease can inactivate CTXϕ particles in culture supernatants (Kimsey and Waldor, 1998b). HA/protease also contains a mucinase activity, which facilitates the translocation of *V. cholerae* through mucin gels in vitro (Hase and Finkelstein, 1990). Additionally, HA/protease can perturb the barrier function of epithelial cell lines, specifically the tight junction and the F-actin cytoskeleton. The tight junction protein occludin is targeted by HA/protease for degradation, which results in a loss of tight junction integrity (Wu et al., 1996, 2000). *hapA* mutants show no decrease in intestinal colonization in the infant rabbit model (Finkelstein et al., 1992), and addition of purified HA/protease caused no detectable effect in rabbit ileal loops (Hase and Finkelstein, 1990). Additionally, HA/protease is not significantly involved in the inflammatory response in the murine pulmonary model (Fullner et al., 2002). Expression of HA/protease is regulated by the quorum-sensing response activator HapR (Jobling and Holmes, 1997), described below.

HlyA hemolysin

Hemolytic activity caused by the HlyA hemolysin was initially used as a method to differentiate O1 El Tor (hemolysin-plus) from O1 classical strains (hemolysin-minus) (Pollitzer, 1959)—hence the term "El Tor hemolysin" applied to HlyA. In classical strains, the gene encoding the hemolysin, *blyA*, contains a deletion resulting in the production of a nonfunctional, truncated protein (Rader and Murphy, 1988). HlyA is a secreted cytolytic protein that is expressed by many O1 El Tor and non-O1 strains (Yamamoto et al., 1984; Kaper et al., 1995). HlyA permeabilizes a wide variety of eukaryotic cells, including human and chicken erythrocytes (Yamamoto et al., 1984; Richardson et al., 1986), by forming pentameric pores in the membrane (Zitzer et al., 1999). The toxin binds specifically to cholesterol (Zitzer et al., 1999) and interacts with β1-galactosyl-terminated glycoconjugates (Saha and Banerjee, 1997). Introduction of purified HlyA into rabbit ligated ileal loops results in fluid accumulation, suggesting a possible role in pathogenesis (Ichinose et al., 1987). HlyA can also induce cellular vacuolation in HeLa and Vero mammalian cells (Coelho et al., 2000; Mitra et al., 2000; Figueroa-Arredondo et al., 2001). The exact role HlyA plays in pathogenesis remains unclear; however, it may contribute to the reactogenicity associated with *ctx*⁻ strains (Fullner et al., 2002).

Type II secretion system (EPS)

V. cholerae utilizes a type II secretion pathway to secrete a variety of virulence-associated proteins. Type II secretion in *V. cholerae* requires 14 extracellular protein secretion genes (*epsA–N*) and *vcpD* (*pilD*) (Hirst and Holmgren, 1987a,b; Sandkvist

et al., 1997; Marsh and Taylor, 1998; Fullner and Mekalanos, 1999). The proteins encoded by these genes form a secretion apparatus that secretes cholera toxin, neuriminidase, lipase, chitinase, and HA/protease (Sandkvist et al., 1997; Connell et al., 1998). Curiously, evidence has indicated that OM localization of OmpU is also dependent on the EPS system (Sandkvist et al., 1997), as is the expression of the extracellular polysaccharide VPS (Ali et al., 2000). The EPS proteins were demonstrated to be localized specifically to the old pole of the bacterium, which also contains the flagellum, after cell division (Scott et al., 2001). The advantage of this polarized secretion is not known; however, it may serve to direct putative virulence factors to the cell surface (Scott et al., 2001). Further work has begun to elucidate the crystal structures and protein-protein interactions of the various components of the EPS system (Robien et al., 2003; Abendroth et al., 2004 a,b).

REGULATION OF VIRULENCE FACTOR EXPRESSION

The ToxR Regulon

Successful colonization of the human host requires that expression of the virulence factors be tightly controlled for optimal expression. The virulence regulatory system that evolved in *V. cholerae* is quite elaborate, involving interplay between both ancestral proteins and those acquired through horizontal transfer. This regulatory cascade, referred to often as the ToxR regulon (Fig. 2), is constantly expanding as further studies reveal new regulatory elements.

Expression of the genes required for colonization and disease depends on ToxT, an AraC-family transcriptional activator (DiRita et al., 1991; Higgins et al., 1992). ToxT is encoded within the VPI, downstream of the *tcpA* operon (Higgins et al., 1992). ToxT stimulates the expression of the two major virulence factors, CT and TCP, as well as accessory colonization factors, by directly binding to the promoter regions of the genes encoding these virulence factors and activating transcription (DiRita et al., 1991; Carroll et al., 1997). Additionally, *toxT* is able to autoregulate its own expression by virtue of its location downstream of the ToxT-activated *tcpA* operon (Yu and DiRita, 1999). Thus, under favorable conditions, ToxT upregulates its own production and consequently that of the virulence factors. The initiation of *toxT* transcription occurs specifically within the intestine or under defined laboratory inducing conditions (Angelichio et al., 1999, 2004; Murley et al., 2000) and is controlled by the regulatory cascade

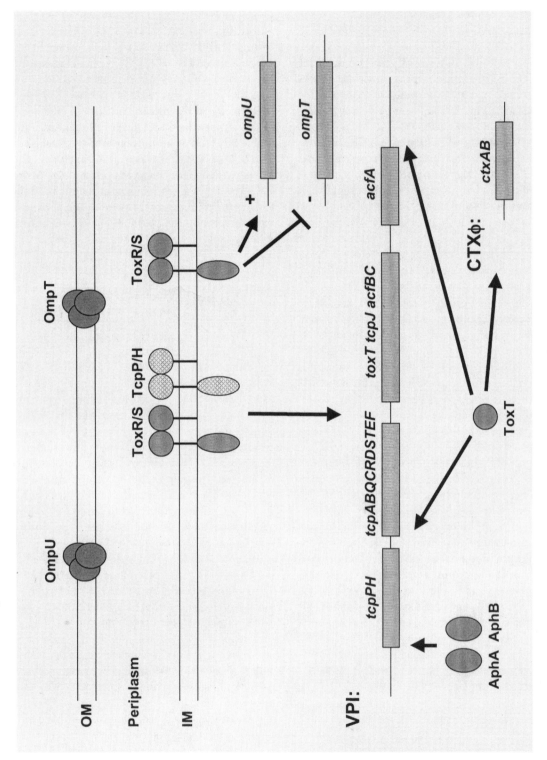

Figure 2. The regulatory cascade that controls virulence factor expression in *V. cholerae.* Reprinted from Reidl and Klose (2002) with permission of Elsevier Science B. V.

described below. However, ToxT itself is able to respond to specific environmental signals, including temperature and the presence of bile, allowing for optimal expression of virulence factors (Gupta and Chowdhury, 1997; Schuhmacher and Klose, 1999). The direct modulation of ToxT by environmental signals provides an additional level of regulation of virulence factor expression, perhaps allowing *V. cholerae* to fine-tune key factors during the course of infection.

The critical event in the induction of virulence factor expression is induction of *toxT* transcription. Two integral cytoplasmic membrane transcriptional activators, ToxR and TcpP, have been shown to be required for maximal *toxT* transcription (Miller and Mekalanos, 1984, 1985; Hase and Mekalanos, 1998; Krukonis and DiRita, 2000). Both ToxR and TcpP are members of the OmpR/PhoB family of winged helix-turn-helix transcriptional activators (Miller and Mekalanos, 1985; Miller et al., 1987; Martinez-Hackert and Stock, 1997; Hase and Mekalanos, 1998). The ToxR protein has been shown to interact with another membrane protein, ToxS, and TcpP also interacts with another membrane protein, TcpH (DiRita and Mekalanos, 1991; Carroll et al., 1997). Evidence suggests that TcpH is critical for TcpP function, as it stabilizes TcpP in the membrane through interactions in the periplasmic domain, preventing degradation (Hase and Mekalanos, 1998; Beck et al., 2004). Overexpression of TcpP alone is capable of activating the *toxT* promoter; however, coexpression with ToxR increases this activation approximately fivefold (Hase and Mekalanos, 1998; Murley et al., 1999; Krukonis et al., 2000). Both ToxR and TcpP bind directly to the *toxT* promoter, with TcpP binding from -54 to -32 and ToxR binding from -104 to -68 with respect to the transcription start site (Krukonis et al., 2000). ToxR binds to the *toxT* promoter with greater affinity than TcpP. Krukonis et al. (2000) demonstrated that ToxR's interaction with TcpP enhances the activity of TcpP at the promoter. The current model indicates that ToxR binds the *toxT* promoter, and then interactions between ToxR and TcpP facilitate TcpP binding and productive interaction with RNA polymerase, which results in transcription activation of *toxT* (Krukonis and DiRita, 2003).

The *toxR* gene appears to be present in all *Vibrio* and *Photobacterium* species and thus can be considered an ancestral gene (Lin et al., 1993; Reich and Schoolnik, 1994; Welch and Bartlett, 1998; Osorio and Klose, 2000). ToxR homologues normally function to regulate the expression of OM proteins, and, in fact, *V. cholerae* ToxR (independent of TcpP) also regulates the expression of two OM porins, OmpU and OmpT (discussed above), as well as a number of

other genes in *V. cholerae* (Champion et al., 1997; Bina et al., 2003). The *tcpPH* genes are located in the VPI and are thus only found in strains containing this element, i.e., predominantly clinical O1/O139 strains (Hase and Mekalanos, 1998; Sarkar et al., 2002). Expression of *toxRS* is believed to be relatively constitutive, at least under laboratory conditions, while transcription of *tcpPH* is known to occur only under in vitro virulence factor-inducing conditions (Carroll et al., 1997; Murley et al., 1999). Artificial expression of TcpPH under noninducing conditions leads to CT and TCP expression (Carroll et al., 1997; Hase and Mekalanos, 1998; Murley et al., 1999; Yu and DiRita, 1999), indicating that the induction of CT and TCP expression, at least within the laboratory, is due to the induction of *tcpPH* transcription.

In an elegant series of experiments, it was found that transcription of *tcpPH* is controlled synergistically by two proteins, AphA and AphB (Kovacikova and Skorupski, 1999; Skorupski and Taylor, 1999). Evidence suggests that these two proteins both bind to the *tcpPH* promoter to stimulate transcription; neither protein is able to activate *tcpPH* transcription independently (Kovacikova and Skorupski, 1999). AphA and AphB are believed to interact directly on the *tcpPH* promoter, as interaction with AphB has been demonstrated to rescue an AphA DNA binding mutant to allow transcription activation (Kovacikova et al., 2004). AphA binds the promoter between -101 and -71 with respect to the start site of transcription, whereas AphB binds a partially overlapping region encompassing -78 to -43 (Kovacikova and Skorupski, 2001).

Regulation of AphA- and AphB-dependent transcription is still not well understood; however, some interesting data have been acquired. The difference between in vitro culture conditions required for classical and El Tor biotypes to express CT and TCP was recently determined to be due to a single base pair change between the classical and El Tor *tcpPH* promoters, which significantly affects the DNA binding ability of AphB in response to environmental signals (Kovacikova and Skorupski, 2000). The exact mechanism for environmental sensing by AphA and AphB remains elusive. The regulatory protein, HapR, directly represses *aphA* expression, which subsequently downregulates the virulence cascade (Kovacikova and Skorupski, 2002). Since HapR expression is regulated by quorum sensing, this mechanism ties population density signaling to virulence factor expression (see below). The *tcpPH* promoter is also subject to negative regulation by cAMP–cAMP receptor protein (CRP), and the binding site for cAMP-CRP overlaps with the binding sites for AphA and AphB (Kovacikova and Skorupski, 2001), raising the possibility that cAMP-CRP regulates virulence factor expression in response to environmental signals.

Spatial Expression of Virulence Factors

Analysis of virulence regulation has led to a central theme of environmental sensing and response by *V. cholerae*. Throughout the virulence cascade, environmental signals are able to provide input to optimally control virulence factor expression. Of particular interest are the expression patterns of the two principal virulence factors, CT and TCP. It has been shown that under laboratory conditions that induce virulence factor expression, CT and TCP are expressed concurrently (Iwanaga et al., 1986; Taylor et al., 1987). A comparison of spatial and temporal expression of CtxA and TcpA between laboratory and in vivo growth yielded interesting results. Induction of *tcpA* transcription was found to occur biphasically within the mouse small intestine (Lee et al., 1999). An early induction period occurred while the bacteria were still in the upper gastrointestinal tract, which preceded the induction of *ctxA* transcription. The second induction of *tcpA* transcription occurred approximately 4 h later when the bacteria colonized the distal end of the small intestine (Lee et al., 1999). Induction of *ctxA* transcription was found to occur monophasically and be dependent upon prior expression of TCP, as transcription of *ctxA* in a strain lacking *tcpA* was approximately 80% less than that of a $tcpA^+$ strain in vivo (Lee et al., 1999). This may indicate that, in vivo, maximal expression of CT requires a TCP-mediated signal, hypothesized by the researchers to be a signal present at the epithelial cell surface.

Prior to colonization of the small intestine, *V. cholerae* must pass through the acid barrier of the stomach. The exposure to the acidic environment of the stomach may provide a signal to the bacterium that facilitates intestinal colonization. Acid adaptation (exposure to acid prior to inoculation) has been demonstrated to increase the ability of *V. cholerae* to colonize the infant mouse (Merrell and Camilli, 1999). Exposure to acid of many bacterial pathogens has been shown to cause an adaptive stress response known as the ATR (Meng and Bennett, 1992; Park et al., 1996; Merrell and Camilli, 1999). Further analysis of the ATR in *V. cholerae* indicates that acid exposure does not increase survival of the bacteria's passage through the stomach nor does it alter colonization behavior or induction of *ctxA* and *tcpA* transcription (Angelichio et al., 2004). Rather, the ATR of *V. cholerae* may allow for an increased multiplication rate within the small intestine, when compared to non-acid-adapted bacteria, that results in a net increase in bacterial numbers within the intestine (Angelichio et al., 2004).

Bile has been shown to be an important signal for gene expression in several enteric bacteria (Gupta and Chowdhury, 1997; Schuhmacher and Klose, 1999; Prouty and Gunn, 2000; Rosenberg et al., 2003; Prouty et al., 2004; Hung and Mekalanos, 2005). During colonization of the small intestine, *V. cholerae* comes into contact with bile, which may serve as a specific signal to facilitate intestinal colonization. Early work demonstrated that CT and TCP expression was decreased in the presence of bile (Gupta and Chowdhury, 1997). Schuhmacher and Klose (1999) confirmed the observed effects and demonstrated that bile inhibited virulence factor expression at the level of ToxT by inhibiting ToxT's ability to activate transcription. Recently, it was demonstrated that the presence of bile can stimulate ToxR-dependent (and ToxT-independent) transcription of the *ctxA* promoter (Hung and Mekalanos, 2005). Additionally, exposure of *V. cholerae* to bile stimulates the ToxR-dependent expression of OmpU, a porin important in bile resistance (Provenzano et al., 2000). How bile is sensed by *V. cholerae* remains unknown, but, at least in the case of ToxR, it may be through interaction with the transmembrane region of ToxR (Hung and Mekalanos, 2005). ToxR in a closely related deep-sea *Photobacterium* sp. was originally identified as a pressure-sensitive transcriptional regulator that responds to membrane perturbation (Welch and Bartlett, 1998), so it seems reasonable to hypothesize that the ToxR response to bile in *V. cholerae* may be a result of membrane disruption by this anionic detergent (Provenzano et al., 2000). Bile has additionally been shown to stimulate the motility and/or chemotaxis of *V. cholerae* (Gupta and Chowdhury, 1997). It is tempting to speculate that the high concentrations of bile within the intestinal lumen coordinate virulence by (i) stimulating OmpU expression, which is necessary for bile resistance and optimal virulence factor expression (Provenzano et al., 2000; Provenzano and Klose, 2000), (ii) stimulating ToxT expression but inhibiting ToxT activity, to avoid premature expression of virulence factors within the lumen, rather than at the epithelial cell surface (Schuhmacher and Klose, 1999), and (iii) stimulating motility and/or chemotaxis to drive the bacteria into the mucus and underlying intestinal crypts (Camilli and Mekalanos, 1995; Gupta and Chowdhury, 1997; Butler and Camilli, 2004).

Expression of virulence factors is subject to a wide variety of other environmental stimuli within the laboratory. The effects of temperature and pH are especially noticeable when comparing O1 classical and El Tor biotypes. The classical biotype expresses maximal CT and TCP in vitro in Luria Bertani broth at pH 6.5 and 30°C, while the El Tor biotype requires biphasic inducing conditions referred to as "AKI," which includes a microaerophilic growth phase followed by vigorous aeration (Iwanaga et al., 1986; Gardel and Mekalanos, 1994). Both classical and El Tor (AKI) laboratory inducing conditions have been shown to induce the expression of TcpP (Murley et al., 2000). Additional in vitro environmental signals that affect CT and TCP expression include amino acids (Miller and Mekalanos, 1988), osmolarity (Gardel and Mekalanos, 1994), and carbon dioxide (Shimamura et al., 1985; Voss and Attridge, 1993). The specific environmental signals within the host have yet to be elucidated; however, it is possible that the observed in vitro signals may all contribute to virulence factor regulation in the host as the bacteria encounter the various environments of the gastrointestinal tract.

Quorum Sensing

Within *V. cholerae* there are three putative quorum-sensing systems that function to regulate genes in response to cell density (Miller et al., 2002). The three systems converge and regulate virulence expression through the response regulator LuxO (Freeman and Bassler, 1999; Miller et al., 2002). By analogy with the quorum-sensing system in *Vibrio harveyi*, it is presumed that quorum-sensing regulation in *V. cholerae* occurs via modulation of the phosphorylation state of LuxO (Lilley and Bassler, 2000; Miller et al., 2002). At low cell density, LuxO (presumably phosphorylated) represses the expression of the LuxR homologue HapR (Jobling and Holmes, 1997), which allows for high-level expression of CT and TCP (Vance et al., 2003; Zhu et al., 2002). The mechanism by which LuxO represses HapR expression was recently elucidated (Lenz et al., 2004). LuxO-P activates the σ^{54}-dependent transcription of four small RNAs, which act to destabilize the *hapR* transcript; the Hfq protein facilitates the interaction of the sRNAs with the *hapR* transcript. At high cell density, LuxO (presumably unphosphorylated) no longer represses HapR expression. HapR then binds to the promoter region of *aphA*, effectively shutting down the virulence cascade (Kovacikova and Skorupski, 2002). HapR is also the transcriptional activator of the HA/protease, so while CT and TCP expression may be low at high cell densities, HA/P expression is high, which is predicted to facilitate exit of the bacteria from the intestine (Miller et al., 2002; Zhu et al., 2002). This same quorum-sensing system has also been shown to regulate biofilm formation in a similar fashion (i.e., promote biofilm formation at low cell density and inhibit biofilm formation at high cell density) (Hammer and Bassler, 2003). It should be noted that certain fully

virulent clinical strains of *V. cholerae* (e.g., El Tor N16961 and classical O395) contain natural frameshift mutations in *hapR* (Heidelberg et al., 2000), and it has been demonstrated that Hfq has no effect on CT/TCP expression in strain N16961 (Ding et al., 2004) (see below), thus confusing quorum sensing's role in pathogenesis.

Other Regulators

Virulence factor expression is further regulated by factors that influence various steps of the ToxR regulon. *cya* and *crp* mutants of O1 El Tor strains, which encode adenylate cyclase and the cAMP–CRP, respectively, exhibit increased transcription of the *ctx* and *tcp* genes even under noninducing conditions (Skorupski and Taylor, 1997). Utilizing purified CRP, it was demonstrated that the protein does not bind to the *toxT*, *tcpA*, or *ctx* promoters; however, CRP binds to the *tcpPH* promoter at nearly identical sites as AphA (Kovacikova and Skorupski, 2001), suggesting that CRP may compete for binding of the *tcpPH* promoter with AphA and downregulate virulence factor expression.

H-NS is an abundant histone-like nucleoid-associated protein found in many members of *Enterobacteriaceae*. In *V. cholerae*, H-NS can also negatively regulate the expression of virulence factors (Nye et al., 2000). In an *hns* mutant strain of *V. cholerae*, a high level of transcription of *toxT*, *ctx*, and *tcpA* occurs even in the absence of ToxR, TcpP, or ToxT, indicating that H-NS serves to "silence" expression of these genes (Nye et al., 2000). Yu and DiRita (2002) further demonstrated that for ToxT-dependent gene expression, ToxT binds to specific AT-rich promoters and displaces H-NS prior to interacting with RNA polymerase for maximal activation.

In a genetic screen to identify genes that repress *toxT* expression, a homologue of the *E. coli pepA* was identified. In *E. coli*, PepA is a hexameric aminopeptidase. Disruption of *V. cholerae pepA* resulted in an increase in the expression of *ctx*, *tcpA*, *toxT*, and *tcpP* under the noninducing condition of high pH (Behari et al., 2001). A genetic screen for in vivo-induced genes identified the *vieSAB* operon (Lee et al., 1998). Sequence analysis of the operon indicates that *vieS* encodes a sensor kinase, while both *vieA* and *vieB* encode response regulators of the two-component signal transduction family (Lee et al., 1998). The VieSAB system was shown to positively regulate the expression of CT both in vitro and in vivo (Tischler et al., 2002). VieA was recently shown to contain an EAL domain, which is involved in the regulation of cyclic diguanylate levels and biofilm formation (Tischler and Camilli, 2004); it is not yet clear if this activity is also involved in regulating CT expression. A mutation in *hfq* has been shown to attenuate *V. cholerae* intestinal colonization by an unknown mechanism, since this mutation had no effect on TCP expression (Ding et al., 2004). Hfq is involved in mediating sRNA-dependent mechanisms of regulation, so this result implicates sRNAs in intestinal colonization. However, as noted above, sRNAs and Hfq are also involved in LuxO-dependent quorum sensing, which directly impacts CT/TCP expression through HapR (Lenz et al., 2004). The seemingly contradictory results of these two studies may be attributed to strain differences, since the N16961 El Tor strain used in the study by Waldor and colleagues contains a natural frameshift mutation in *hapR* (Heidelberg et al., 2000).

MOBILE GENETIC ELEMENTS OF *V. CHOLERAE*

The evolution of the pathogenic potential of *V. cholerae* has clearly occurred through the acquisition of several mobile genetic elements. Arguably, the most important of these is the bacteriophage that carries the genes encoding CT. The VPI was reported to correspond to a lysogenic bacteriophage (Karaolis et al., 1999), but this finding has failed to be replicated by others (Faruque et al., 2003). Thus, although the VPI is clearly a horizontally acquired element, the mechanism of transfer remains controversial.

CTXφ

The *ctx* genes reside on a lysogenic filamentous bacteriophage, CTXφ, illuminating the ability of *V. cholerae* to acquire virulence factors through horizontal transfer (Waldor and Mekalanos, 1996). The CTXφ encodes proteins for replication, integration, and regulation (RstA, B, and R, respectively) as well as proteins for phage packaging and secretion (Psh, Cep, OrfU, Ace and Zot) (Waldor and Mekalanos, 1996; Waldor et al., 1997). The two genes that encode cholera toxin, *ctxA* and *ctxB*, are also encoded within the phage genome; however, these genes are not required for phage integration or replication (Waldor and Mekalanos, 1996), and some CTXφ variants have been found in environmental strains that lack *ctxAB* (Boyd et al., 2000). CTXφ is similar to the Ff filamentous phages such as M13, with notable differences. CTXφ integrates into the chromosome and forms lysogens, unlike M13 (Waldor and Mekalanos, 1996). CTXφ also appears to be more dependent on host proteins for both chromosomal integration and subsequent secretion of the phage (Davis et al., 2000a; Huber and Waldor, 2002).

CTXφ integrates in a site-specific manner into the chromosome. Integration of the phage in O1 El Tor and O139 occurs near the terminus of the large chromosome (chromosome I) (Heidelberg et al., 2000), whereas in O1 classical strains, the phage integrates near the termini of both chromosomes (Davis et al., 2000b). It was recently shown that the host-encoded recombinases XerC and XerD are required for phage integration (Huber and Waldor, 2002). XerC and XerD normally function to convert chromosomal dimers formed during replication into monomers, by catalyzing recombination at *dif* sites. The integration site for CTXφ overlaps the *dif* sites, and these sites are required for successful phage integration (Huber and Waldor, 2002). In O1 El Tor and O139 isolates, the prophage is typically found in tandem or in multiple arrays composed of CTXφ plus a phage-related element, RS1 (Mekalanos, 1983; Waldor and Mekalanos, 1996). RS1 is a "satellite" phage of CTXφ encoding RstA, B, and R plus RstC, an antirepressor of CTXφ gene expression. Since RS1 lacks the coat and secretion proteins required for genome packaging, it relies on CTXφ for propagation (Davis et al., 2002).

Virion production by CTXφ varies from that of other filamentous phages in that an extrachromosomal DNA replicative form is required. CTXφ does not excise itself from the chromosome; rather, the replicative form is generated by a replicative process that requires that tandem elements, either CTXφ-CTXφ or CTXφ-RS1, be present within the chromosome (Davis and Waldor, 2000). The prophages found in classical O1 strains are not present in tandem; consequently, classical biotypes do not produce CTXφ virions (Davis et al., 2000b). This mechanism of replication allows the prophage to remain in the chromosome of the original host, while virion progeny are transferred to a new host. Release of CTXφ from the bacterium occurs via secretion rather than cell lysis. Secretion requires the secretin EpsD, a component of the *V. cholerae* extracellular protein secretion (Eps) type II secretion system, which also is responsible for CT secretion (Davis et al., 2000a).

Infection of *V. cholerae* by CTXφ occurs similarly to filamentous phage infection of other bacteria in that it requires a pilus as a receptor. CTXφ utilizes the TCP as a receptor (Waldor and Mekalanos, 1996) and requires host proteins TolQ, TolR, and TolA for infection (Heilpern and Waldor, 2000). OrfU, recently renamed pIII[CTX], functions as an infection protein, possibly interacting with TCP, although the exact protein–protein interaction has yet to be determined (Heilpern and Waldor, 2003).

V. cholerae produces relatively few virions in comparison to other filamentous phages. This is most likely due to repression of prophage expression by RstR, which binds to the *rstA* promoter and prevents expression of the phage morphogenesis genes (Davis et al., 2002; Kimsey and Waldor, 2004). A second protein, RstC, also modulates CTXφ expression by binding to and sequestering RstR (Davis et al., 2002); thus, when high levels of RstC are present, RstR cannot bind and repress the *rstA* promoter, allowing the production of virions.

SXT Element

Another example of a *V. cholerae* transmissible element is the conjugative-transposon-like element (Ctn) SXT. SXT is a 99.5-kb element that is thought to exist in a nonreplicative circular form prior to transmission (Waldor et al., 1996). There appear to be 87 open reading frames (ORFs) that are composed of genes derived from phages, plasmids, and unknown sources (Beaber et al., 2002b). The first description of SXT was from an O139 strain and it was shown to carry genes encoding multiple antibiotic resistances, including those against sulfamethoxazole, chloramphenicol, and streptomycin (Waldor et al., 1996). SXT elements can currently be found in nearly all clinical *V. cholerae* isolates from the Indian subcontinent, Asia, and Africa, as well as in isolates of *Providencia alcalifaciens* from Bangladesh (Hochhut et al., 2001).

SXT is a self-transmissible element and carries the genes required for its integration, excision, and transmission (Beaber et al., 2002a, 2004; Burrus and Waldor, 2003). Interestingly, the SXT element appears to be transmissible to a large number of gram-negative organisms, at least under laboratory conditions (Burrus and Waldor, 2003). In *V. cholerae* the element integrates into the 5' end of *prfC*, which encodes peptide release factor (Hochhut et al., 2001). Integration and excision of SXT require a tyrosine recombinase (Int) and the Xis protein (Burrus and Waldor, 2003). SetC and SetD, SXT-encoded transcriptional activators, appear to activate the expression of transfer functions for the element (Beaber et al., 2002b; Burrus and Waldor, 2003). SetR represses the activators, and SetR-mediated repression is alleviated by the SOS response to DNA damage in a RecA-dependent manner stimulating genetic transfer (Beaber et al., 2004).

CP-T1

To date, only one generalized transducing phage has been characterized for *V. cholerae* (Ogg et al., 1981). This phage, CP-T1, was isolated by UV induction of a *V. cholerae* lysogen. The genome of CP-T1 is composed of linear double-stranded DNA of approximately 46 kb in size (Guidolin et al., 1984). The sur-

face receptor for CP-T1 is the O antigen component of the lipopolysaccharide of O1 strains (Guidolin and Manning, 1985).

The ability of CP-T1 to transduce large portions of the *V. cholerae* chromosome, and its specificity for O1 strains, suggests that it contributes to horizontal gene transfer among pathogenic strains. Boyd and Waldor (1999) demonstrated the ability of the CP-T1 phage to transduce CTXφ from an El Tor lysogen to several El Tor strains that lack the CTXφ, suggesting a mechanism by which *V. cholerae* strains lacking the CTXφ receptor (i.e., TCP) can undergo toxigenic conversion. CP-T1 can also transduce the 39.5-kb region encompassing the *Vibrio* pathogenicity island (O'Shea and Boyd, 2002). CP-T1 also provides an important genetic tool for *V. cholerae* research. Hava and Camilli (2001) isolated a temperature-sensitive phage CP-T1ts, which can be utilized to transduce genetic elements between O1 strains.

Cryptic Plasmid

A 4.7-kb low-copy-number "small cryptic plasmid" was found to be present in many classical and El Tor biotype *V. cholerae* strains (Cook et al., 1984). This plasmid is designated pTLC for toxin-linked cryptic, since *ctxAB*$^+$ strains all seem to harbor this plasmid. The plasmid can be isolated in its extrachromosomal form as a circular double-stranded DNA molecule (Rubin et al., 1998). Persistence of pTLC is most likely under the control of its own replication functions. The largest ORF of pTLC encodes a protein predicted to be similar to the *E. coli* F-specific filamentous phage replication protein pII/X. Questions as to the origin of pTLC led to a search of the *V. cholerae* genome, where the TLC element was found tandemly duplicated on the chromosome. While not confirmed, the extrachromosomal pTLC may arise through homologous recombination within the TLC element. The TLC element was found to be only 842 bp 5' of the RS1 element, which lies immediately upstream of the CTX prophage, thus linking it to the CTX prophage. To date, the ability of the TLC element to transfer DNA between *V. cholerae* strains has not been demonstrated, so its relevance to virulence is currently unclear (Rubin et al., 1998).

Integron

Integrons are gene expression elements that capture ORFs and convert them into functional genes, generally involving antimicrobial resistance (Recchia and Hall, 1995). These elements, termed multiresistant integrons (MRI), usually contain fewer than five genes and are carried by mobile elements such as plasmids (Rowe-Magnus et al., 2001; Clark et al., 2000). Insertion of the newly acquired gene cassette occurs via site-specific recombination requiring an integrase gene (*intI*) that mediates recombination between an attachment site (*attI*) and a target recombination sequence (*attC*) (Hall and Stokes, 1993; Recchia and Hall, 1995; Stokes et al., 1997). In *V. cholerae*, VCRs that have similar organization to integrons have been identified on the chromosome (Recchia and Hall, 1997). Mazel et al. (1998) identified a VCR cassette not involving antibiotic resistance, which instead involves a heat stabile toxin gene (*sto*) and the mannose-fucose-resistant hemagglutinin gene (*mrhA*). Further study demonstrated that this and other VCRs contain integron-like structures, which led to the conclusion that an integron exists in *Vibrio* for the purpose of gene capture (Mazel et al., 1998). The *Vibrio* integron is extremely large, containing minimally 10 times more gene cassettes than the largest described MRI. Thus, the *Vibrio* integron has been termed superintegron (SI) (Clark et al., 2000; Biskri et al., 2005). In addition to size, the SI differs from MRIs in that it appears to be sedentary in the chromosome and utilizes a highly homologous *attC* site for which the *V. cholerae* SI integrase (VchIntIA) is specific (Biskri et al., 2005).

ENVIRONMENTAL PERSISTENCE OF *V. CHOLERAE*

Biofilms

V. cholerae is a natural inhabitant of aquatic ecosystems, both marine and freshwater (Colwell et al., 1977; Garay et al., 1985; Islam et al., 1993, 1994; Colwell, 1996). While able to survive in both fresh- and saltwater, the bacterium is more stable in a saline environment (Singleton et al., 1982). Within the aquatic environment, *V. cholerae* is often found associated with surfaces, specifically, plants, zooplankton, crustaceans, filamentous green algae, and insects (Huq et al., 1990; Colwell, 1996). A correlation has been made between seasonal algal blooms and cholera outbreaks; however, there is no direct evidence for the enrichment of toxigenic *V. cholerae* during these periods (Epstein, 1993; Islam et al., 1994; Colwell, 1996). Nontoxigenic environmental strains are more commonly isolated from the environment than the toxigenic O1 and O139 strains (Colwell and Spira, 1992; Faruque et al., 2004).

The ability of *V. cholerae* to persist in the environment is critical for its ability to cause pandemic disease. One mode of environmental persistence employed by a wide variety of microorganisms is to form large multicellular structures on surfaces known as biofilms. Biofilms provide protection to the bacteria from such stresses as harsh chemicals or antibiotics, allowing for survival and persistence (Davey and O'Toole, 2000). Biofilm formation by *V. cholerae* has been extensively studied.

VPS Exopolysaccharide

Both O1 El Tor and O139 biotypes are able to form a three-dimensional biofilm under laboratory conditions (Color Plate 15B) (Watnick and Kolter, 1999; Yildiz and Schoolnik, 1999). Formation of this biofilm is dependent on the expression of an exopolysaccharide (VPS), and expression of the VPS during growth on solid media results in a rugose colony phenotype (Wai et al., 1998; Yildiz and Schoolnik, 1999; Watnick et al., 2001). Strains that are unable to express the exopolysaccharide are able to colonize surfaces, but cannot form the large three-dimensional structures indicative of mature biofilms under normal laboratory conditions (Watnick and Kolter, 1999; Watnick et al., 2001).

The genes encoding the VPS (*vps*) are encoded primarily within two operons on the chromosome encompassing a 30.7-kb region (Yildiz and Schoolnik, 1999). Thus far, two positive regulators for VPS biosynthesis have been identified, VpsR and VpsT. VpsR shares homology with members of the σ^{54}-dependent class of transcriptional regulators and contains a response regulatory domain found in two-component systems of bacterial signal transduction (Klose et al., 1998; Yildiz and Schoolnik, 1999; Yildiz et al., 2001). Lauriano et al. (2004) presented indirect evidence, utilizing VpsR proteins with mutations at the putative site of phosphorylation, that indicated that VpsR requires phosphorylation for transcriptional activation; however, phosphorylation has not been proved biochemically because the cognate histidine kinase has not yet been identified. Recently, VpsT was identified as a second positive regulator of the *vps* genes (Casper-Lindley and Yildiz, 2004). VpsT is a transcriptional response regulator of the UphA (FixJ) family and shares homology to CsgD and AgfD of *Escherichia coli* and *Salmonella enterica*, respectively, which are required for extracellular matrix formation (Casper-Lindley and Yildiz, 2004). Strains lacking VpsR and/or VpsT express decreased levels of VPS, but direct regulation of the *vps* gene promoters by these proteins has not yet been demonstrated (Yildiz et al., 2001, 2004; Casper-Lindley and Yildiz, 2004; Lauriano et al., 2004).

Repression of biofilm formation was recently discovered to be modulated by CytR (Haugo and Watnick, 2002). CytR functions as a transcriptional repressor of *vps* gene synthesis by downregulating *vps* gene expression in the absence of nucleosides (Haugo and Watnick, 2002). Additional factors have been shown to be necessary for biofilm formation, including the mannose-sensitive hemagglutinin type IV pilus and flagellar-mediated motility. However, both of these have only been demonstrated to be required for biofilm formation of an O1 El Tor strain and were dispensable for biofilm formation by an O139 strain (Watnick et al., 1999, 2000).

High-level expression of VPS results in a rugose colony morphology, which has been a useful phenotype in identifying signaling components involved in the expression of exopolysaccharide. Spontaneous rugose colonies appear at higher frequencies following nutrient starvation, and there appear to be at least three different signaling mechanisms that stimulate VPS expression in different strains of O1/O139 *V. cholerae*: a quorum-sensing mechanism, a flagellar-dependent mechanism, and a third mechanism independent of the other two (Watnick et al., 2001; Hammer and Bassler, 2003; Lauriano et al., 2004). Regulation of biofilm formation in many bacterial species involves quorum-sensing (density-dependent) signals (Stanley and Lazazzera, 2004). In *V. cholerae*, several quorum-sensing pathways converge on regulating the expression of the transcriptional activator HapR, which is a positive regulator of the HA/protease and in some strains a negative regulator of CT/TCP expression. In many *V. cholerae* strains, HapR also represses the expression of the *vps* genes and biofilm formation (Zhu et al., 2002; Hammer and Bassler, 2003), and thus *hapR* mutations in these strains result in a rugose phenotype (Lauriano et al., 2004; Yildiz et al., 2004). The current model for quorum sensing in *V. cholerae* indicates that under low cell density the quorum-sensing system leads to phosphorylation of LuxO and repression of HapR, allowing for biofilm formation and paradoxically also virulence factor expression (Miller et al., 2002; Zhu et al., 2002). Under conditions of high cell density, the quorum-sensing pathway causes the dephosphorylation of LuxO, resulting in the expression of HapR, and biofilm formation as well as CT/TCP expression are repressed (Miller et al., 2002; Zhu et al., 2002; Hammer and Bassler, 2003).

The flagellum also regulates VPS exopolysaccharide expression in a number of *V. cholerae* strains (Watnick et al., 2001; Lauriano et al., 2004). Specifically, nonflagellated mutants produce copious amounts of VPS and exhibit a rugose phenotype (Watnick et al., 2001; Lauriano et al., 2004). The sodium-driven flagellar motor is necessary for flagellar-dependent VPS expression, suggesting that the motor acts as a mechanosensor, signaling for the induction of VPS expression upon loss of the flagellum

(Lauriano et al., 2004). This may be relevant during the development of mature biofilms, as the cells transition from motile adherent organisms to sessile components of the three-dimensional structure. It has been observed that some *V. cholerae* strains can become rugose in the lab by a HapR- and flagellum-independent mechanism (e.g., N16961), indicating that there are at least three different signaling pathways leading to VPS expression (Miller et al., 2002; Yildiz et al., 2004). The diversity in VPS induction seen in different strains of O1/O139 *V. cholerae* suggests that such diversity may be inherent in natural populations of the organisms and may provide a way for subsets of the population to take advantage of specific environmental conditions to ensure persistence of the species; such a strategy has been proposed for *Pseudomonas aeruginosa* (Boles et al., 2004).

These various signaling systems that induce VPS expression may converge on regulating a common signaling molecule, which in turn induces VpsR- and VpsT-dependent expression of the *vps* genes. One possibility for such a signaling molecule is cyclic diguanylate. Proteins containing GGDEF and EAL domains have been demonstrated to have activities that regulate cyclic diguanylate levels (Ausmees et al., 2001) where the GGDEF domain stimulates cyclic-di-guanylic acid (cyclic-di-GMP) production and the EAL domain stimulates cyclic-di-GMP degradation (Simm et al., 2004). A protein containing GGDEF and EAL domains, MbaA (VC0703), was found to be involved in VPS expression and mature biofilm architecture (Bomchil et al., 2003); a second protein containing GGDEF and EAL domains, RocS (VC0653), was also found to be important for VPS expression and biofilm formation by another laboratory (Rashid et al., 2003). It was recently shown that altering the synthesis or degradation of cyclic diguanylate, by manipulating proteins containing GGDEF and EAL domains, respectively, altered *V. cholerae vps* transcription and biofilm formation (Tischler and Camilli, 2004).

A VPS-independent biofilm has been described for O139 *V. cholerae* grown in seawater (Kierek and Watnick, 2003a). Analysis of the VPS-independent biofilms demonstrated that this biofilm is dependent on the presence of Ca^{2+} and the O antigen (Kierek and Watnick, 2003a,b); removal of Ca^{2+} from the saltwater media results in dissolution of the VPS-independent biofilm (Kierek and Watnick, 2003a). The authors have postulated that the introduction of seawater biofilms into freshwater drinking sources that have lower Ca^{2+} levels may be a trigger for the organisms to leave the biofilm and initiate cholera epidemics (Kierek and Watnick, 2003b).

Extensive microarray analysis of rugose (VPS-expressing) and smooth variant *V. cholerae* strains, as well as defined regulatory mutants in these strains, has provided a wealth of interesting data, confirming that the rugose phenotype is a complex combination of signaling pathways (Yildiz et al., 2004). Rugose strains demonstrate increased transcription of several of the *eps* genes that encode the type II secretion pathway (Yildiz et al., 2004), consistent with previous observations that *eps* genes are required for VPS expression (Ali et al., 2000). Additional genes that are differentially regulated in rugose strains were also identified, including genes involved in regulation, chemotaxis, iron scavenging, and flagellar synthesis (Yildiz et al., 2004).

Mannose-Sensitive Hemagglutinin

O1 El Tor and O139 *V. cholerae* strains are able to agglutinate chicken erythrocytes, whereas O1 classical strains are not (Finkelstein and Mukerjee, 1963; Attridge et al., 1996). The agglutination phenotype is attributed to expression of the mannose-sensitive hemagglutinin type IV pilus (MSHA) (Hase et al., 1994; Jonson et al., 1994). While another type IV pilus, TCP, is required for human colonization, MSHA does not appear to play a role in disease (Attridge et al., 1996; Thelin and Taylor, 1996; Tacket et al., 1998). MSHA does appear to be involved in environmental persistence. The ability of *V. cholerae* O1 El Tor strains to form biofilms on borosilicate tubes is dependent on the expression of MSHA, and the addition of mannose derivatives (which bind the MSHA) abolishes biofilm formation (Watnick et al., 1999). Cholera outbreaks have been associated with blooms of zooplankton and phytoplankton (Epstein, 1993; Colwell, 1996). In studies with the zooplankton *Daphnia pulex*, Chiavelli et al. (2001) demonstrated that O1 El Tor MSHA mutants bound to the surface of zooplankton less efficiently than wild type. Interestingly, the O1 classical biotype, which does not assemble MSHA pili, was able to bind to zooplankton, indicating an alternative mechanism at work (Chiavelli et al., 2001).

Viable but Nonculturable State

V. cholerae can also enter a state of "dormancy" in the environment allowing for persistence. This state is referred to as the viable but nonculturable (VBNC) state (Xu et al., 1982). VBNC bacteria cannot be recovered by conventional culture media; instead, they are identified by other techniques, such as PCR or direct fluorescent antibody (Xu et al., 1982; Huq et al., 1990; Hasan et al., 1994). As *V. cholerae* transits to a VBNC state, the cells become reduced in size and ovoid; however, they still respond to nalidixic acid (a DNA gyrase inhibitor) and will take up radiolabeled substrate (Stevenson, 1978; Roszak et al., 1984). The VBNC state of *V. cholerae* may serve as an environmental reservoir allowing for reinfection of the human population.

GENETIC TECHNIQUES FOR IDENTIFICATION OF NOVEL VIRULENCE FACTORS

A number of different genetic techniques have been utilized to identify *V. cholerae* genes important for virulence. For example, *TnphoA* mutagenesis was the technique that allowed for the identification of TCP (Taylor et al., 1987), as well as additional virulence genes (Peterson and Mekalanos, 1988). The following discussion of genetic analyses of *V. cholerae* virulence is not meant to ignore the pioneering work in this field, but rather focuses on some of the newer technologies that have been applied in the last few years to the study of cholera pathogenesis, as well as biofilm formation.

RIVET

Analysis of host-induced genes during *V. cholerae* infection is predicted to identify putative virulence determinants. In vivo expression technology (IVET) was developed as a "promoter trap" genetic selection to identify in vivo-induced promoters. The original IVET technique utilized a promoterless reporter gene whose product was required during in vivo growth, thus providing a selectable phenotype during infection (Mahan et al., 1993, 1995). One drawback with the original IVET system was that promoters that are transiently active in vivo may be missed, due to the need for continued activation of the reporter gene to survive the in vivo selection.

To identify promoters that are only transiently activated during infection, a modified IVET system that utilizes γδ resolvase as a reporter was developed (Camilli et al., 1994): recombination-based in vivo expression technology (RIVET). In this system, promoters that drive γδ resolvase expression catalyze an irreversible chromosomal recombination that leads to the loss of antibiotic resistance during infection, creating a heritable phenotypic change (Camilli et al., 1994). This clever system has been utilized at great length by Camilli and colleagues to study various aspects of *V. cholerae* gene expression during infection.

In the initial RIVET screen of *V. cholerae* colonization of the infant mouse intestine, 13 new in vivo-activated genes were identified, including genes associated with metabolism, motility, biosynthesis, and unknown functions (Camilli and Mekalanos, 1995). A refined RIVET screen in infant mice that incorporated several changes to enhance detection of genes with some basal level of expression outside the host identified additional genes that are involved in colonization of the infant mouse (Osorio et al., 2005). The RIVET technology has also been used to characterize temporal and spatial patterns of in vivo expression of *ctx* and *tcp* genes (Lee et al., 1999), as well as to perform genetic screens to identify regulators of in vivo induction of *ctx* transcription (Lee et al., 2001).

In Vivo-Induced Antigen Technology

In vivo-induced antigen technology allows for the identification of antigens expressed during a natural (i.e., human) infection, which may be missed in screens utilizing an animal model (Handfield et al., 2000). Briefly, convalescent sera from *V. cholerae* patients were collected and extensively absorbed against in vitro-grown *V. cholerae*, then used to probe an expression library of *V. cholerae* ORFs (Hang et al., 2003). Using this method, a total of 38 positive clones were identified, including several novel antigenic proteins such as LuxP, CheA, and CheR (Hang et al., 2003). Additionally, the in vivo-induced antigen technology screen identified PilA, a structural subunit of a third type IV pilus, as a major antigen in the majority of patient sera tested (Hang et al., 2003).

Signature-Tagged Mutagenesis

Signature-tagged mutagenesis (STM) is a comparative hybridization technique utilizing transposons which contain unique DNA sequence tags (Hensel et al., 1995). STM allows for the identification of genes that are necessary for in vivo survival and growth. STM has been utilized by two different groups to identify *V. cholerae* genes necessary during infant mouse colonization (Chiang et al., 1999; Merrell et al., 2002b). The application of STM to *V. cholerae* identified not only previously known colonization factors (i.e., TCP biogenesis) but also several novel genes required for infant mouse colonization (Chiang and Mekalanos, 1998) as well as factors important for the ATR (Merrell et al., 2002b).

Microarrays

The complete sequence of the two chromosomes of the *V. cholerae* O1 El Tor strain N16961 (Heidelberg et al., 2000) has facilitated genomics-based studies of *V. cholerae*. Microarray technology utilizing the complete *V. cholerae* genome has provided a wealth of information in regard to both comparative genomic analysis and gene expression under various conditions. Comparative genomic analysis has allowed for the comparison of gene content between the sequenced El Tor strain N16961 and other *V. cholerae* isolates (Dziejman et al., 2002). Results from this type of analysis revealed there are genes present in El Tor strains which are absent in classical strains, as well as genes that are found only in pandemic strains or only in seventh pandemic El Tor strains (Dziejman et al., 2002). An interesting outcome of the analysis is that, despite various strains of *V. cholerae* exhibiting differences in virulence and environmental persistence, as a group, *V. cholerae* only contains about a 1% difference in genomic makeup (Dziejman et al., 2002).

Whole-genome RNA expression profile studies utilizing microarrays in *V. cholerae* have been used to study gene expression in regard to both pathogenesis and environmental persistence. In the study of pathogenesis, analysis of the transcriptome has provided some interesting insights into the infection process. In vitro expression studies using mutant strains confirmed much of what is known about gene regulation within the ToxR regulon and also identified numerous additional ToxR- as well as TcpP-dependent genes (Color Plate 16) (Bina et al., 2003).

Several in vivo transcriptome studies have also shed light on aspects of *V. cholerae* pathogenesis. Transcriptome profiling of *V. cholerae* isolated from human stools using microarray analysis (Merrell et al., 2002a; Bina et al., 2003) has provided the closest understanding of gene expression relevant to the natural host. Expression of chemotaxis genes was downregulated in these bacteria that have already exited the host, and this "nonchemotactic" state correlates with a transient "hyperinfectious" state, similar to the enhanced colonization rates of defined nonchemotactic mutants (Merrell et al., 2002a; Butler and Camilli, 2004). One study of human stool *V. cholerae* reported that expression of the ToxR virulence regulon was not significantly upregulated (Merrell et al., 2002a), while another study noted increased levels of CT and TCP transcripts in shed *V. cholerae* (Bina et al., 2003). Overall, however, the results of the two microarray studies of human stool *V. cholerae* suggested that during growth in the intestine, the bacteria experience nutrient deprivation, anaerobiosis, and iron limitation (Merrell et al., 2002a; Bina et al., 2003).

To determine gene expression during colonization of the intestine, Xu et al. analyzed the transcriptome of *V. cholerae* isolated from rabbit ligated ileal loops (Xu et al., 2003). Interestingly, a large portion of genes found on the small chromosome were upregulated in vivo, as were genes involved in motility, colonization, and toxin expression. Among the genes that were most highly upregulated in vivo were genes involved in metabolic processes, as well as transporters and binding proteins, indicating the requirement of these genes for in vivo survival (Xu et al., 2003).

Transcriptome profiling of gene expression associated with environmental persistence has provided a wealth of information on *V. cholerae*'s presumed lifestyle outside the host. In the environment, *V. cholerae* has been hypothesized to associate with zooplankton and to utilize chitin as a carbon source (Keyhani and Roseman, 1999; Chiavelli et al., 2001; Colwell et al., 2003). Microarray analysis of *V. cholerae* grown on chitin allowed for the elucidation of the chitin utilization program (Meibom et al., 2004). Transcriptome analysis of the differences in gene expression between a rugose (i.e., VPS expressing) and a smooth variant of the same strain of *V. cholerae* demonstrated that phase variation, important in environmental persistence, involves a complex regulatory pathway that incorporates contributions from VpsR-, σ^{54}-, and RpoS-dependent mechanisms (Color Plate 16) (Yildiz et al., 2004). Additionally, microarray analysis has confirmed the previously proposed flagellar gene transcription hierarchy (Prouty et al., 2001) and demonstrated regulation of virulence factor expression by flagellar proteins (Color Plate 16) (Klose and Yildiz, unpublished data). The use of microarrays for transcriptome profiling has revolutionized how gene regulation in *V. cholerae* can be studied, and has provided an enormous amount of data for future study.

ANIMAL MODELS

Humans are the only adult mammals that are naturally colonized by *V. cholerae*. To study virulence during infection, a variety of animal models have been employed. Currently, the two most commonly used models are the suckling mouse and the ligated ileal loop of adult rabbits.

Suckling mice (typically 4 to 6 days old) are susceptible to intestinal colonization by *V. cholerae*, whereas adult mice are not colonized, for unknown reasons (Klose, 2000). The suckling mouse has both advantages and disadvantages as a model system for cholera. Many of the known virulence factors, such as CT and TCP, are expressed in the mouse intestine (Taylor et al., 1987; Peterson and Mekalanos, 1988; Lee et al., 1999). Additionally, virulence factors identified in the mouse model have been confirmed to be required for human intestinal colonization (Herring-

ton et al., 1988). In this model, the bacteria are orally inoculated and typically allowed to colonize for 24 h, and then the organisms that successfully colonize the small intestine are recovered and enumerated by plate count. This model measures intestinal colonization but not other aspects of cholera, such as disease symptoms (e.g., fluid accumulation) or dissemination. The largest drawback of this model is the inability to perform long-term vaccine studies. In general, two main assays are utilized: competition assays and 50% lethal dose (LD_{50}) assays. In competition assays, the infant mice are coinfected with both wild-type and mutant strains, such that the colonization behavior of the mutant strain is measured in relation to the coinoculated wild-type strain. This avoids mouse-to-mouse variations that can occur with single-strain inoculations. For LD_{50} assays, a single strain is inoculated into groups of mice, and LD_{50} can be determined at specific time points after infection.

The rabbit ligated ileal loop provides an alternative model to test the enterotoxicity of *V. cholerae* strains. Because the bacteria do not naturally colonize the intestine of adult rabbits, the ileum is tied off in loops, which are then inoculated with a specific strain and allowed to incubate for ~18 h (De and Chatterje, 1953). Following incubation, the loops are measured for fluid accumulation, indicative of cholera toxin expression. The rabbit ligated ileal loop can serve as an excellent indicator of CT production; however, colonization defects cannot be accurately assayed in this system. Still, since the ligated loops induce expression of CT from inoculated strains, this intestinal environment is likely a good mimic of the human intestine and has been exploited to characterize gene expression patterns (Xu et al., 2003), as well as to enrich for environmental strains that contain pathogenic determinants (Faruque et al., 2004). Other animals, such as dogs, rats, mice, and chickens, have also been used in ligated ileal loop models (Sack, 1992), but the rabbit remains the most common form of this model.

Several additional model systems have been used sporadically in the past, such as infant rabbits (Dutta and Habbu, 1955) or dogs (Sack and Carpenter, 1969), the sealed adult mouse (Richardson et al., 1984), and the removable intestinal tie adult rabbit diarrhea (Spira et al., 1981), but none of these models has gained the popularity of the suckling mouse or the rabbit ligated ileal loop among cholera researchers. Additionally, several tissue culture models have been utilized to look at adherence or toxic effects of *V. cholerae*, but the relevance of many of these models to cholera pathogenesis is not yet clear. Adherence studies have used HEp-2 cells (Sperandio et al., 1995; Gardel and Mekalanos, 1996) and HT-29-Rev MTX (Schild et al., 2005), whereas toxicity and trafficking assays have used polarized T84 intestinal epithelial cells (Lencer et al., 1995; Fullner et al., 2001) and Vero mammalian cells (Coelho et al., 2000).

VACCINE DEVELOPMENT

Considering the severe global impact of cholera, much effort has been directed at vaccine development. This topic is only briefly discussed here; more extensive coverage can be found in several excellent reviews (Cryz et al., 1995; Finkelstein, 1995; Kaper et al., 1995; Ryan and Calderwood, 2000; Sack et al., 2004). A killed injectable vaccine was developed shortly after *V. cholerae*'s discovery in the 1880s; however, it resulted in painful injection site reaction, offered limited protection (about 6 months), and was extremely expensive (Sack et al., 2004). Natural infection with a pathogenic serogroup of *V. cholerae* confers significant protection against reinfection by the same serogroup, indicating the development of an infection-derived immunity which may be co-opted for vaccine uses (Levine et al., 1983; Levine and Kaper, 1993). Considering that *V. cholerae* is a noninvasive intestinal pathogen, it is believed that an effective vaccine needs to induce protective mucosal immunity within the intestine; thus, current vaccine strategies primarily center on various oral formulations. The emergence of O139 pandemic cholera has highlighted the need to include this serogroup, in addition to the O1 serogroup, in future vaccine development strategies (Ledon et al., 2003; Sack et al., 2004).

A killed whole-cell oral vaccine that also included purified CT-B was developed and tested extensively for efficacy (Clemens et al., 1986; Svennerholm and Holmgren, 1986). In extensive field trials in Bangladesh, the vaccine protected 85% of adults initially through 6 months; however, by 36 months the protection fell to 50% (Clemens et al., 1986). In children and persons of blood group O (a known risk factor for cholera), the efficacy was significantly lower (Clemens et al., 1986, 1989). This vaccine has been extensively tested and provides a substantial immune response in certain populations; however, its use is hampered by the lack of efficacy in children and the relative expense of preparing purified CT-B for large-scale vaccinations (Kaper et al., 1995).

In general, live attenuated vaccine strains are thought to be capable of inducing more protective mucosal immune responses than killed vaccines, but live attenuated vaccine strains have also been plagued with the induction of mild disease symptoms, referred to as reactogenicity, in volunteer studies (Levine et al.,

1988a,b; Tacket et al., 1993; Taylor et al., 1994; Benitez et al., 1999; Rodriguez et al., 2001; Fullner et al., 2002). The basis of reactogenicity has been elusive, mainly because animal models fail to reproduce all aspects of human cholera; however, some studies suggest that motility and the HA/protease contribute (Benitez et al., 1999; Chiang and Mekalanos, 2000). Another complication of live vaccine strategies is the inherent susceptibility of *V. cholerae* within the intestine (i.e., TCP$^+$) to toxigenic conversion by the CTXϕ (Waldor and Mekalanos, 1996); since TCP is a required component of intestinal colonization, TCP$^-$ bacteria would make ineffective live vaccines. Although a clever strategy to inhibit CTXϕ infection by a phage-mediated immunity mechanism has been proposed (Kimsey and Waldor, 1998a), the presence of alternative CTXϕ types within the environment (Nusrin et al., 2004) suggests that diversity within the natural phage populations may make this strategy relatively ineffective. Mutant TcpA alleles that are resistant to CTXϕ infection and still allow *V. cholerae* to colonize the intestine have not yet been identified (Kirn et al., 2000); such alleles would be useful components of a live attenuated vaccine.

There are currently several promising approaches to live attenuated oral vaccines being developed by researchers at the Center for Vaccine Development at the University of Maryland (Kaper and Levine, 1981; Kaper et al., 1984; Levine et al., 1988a,b) and at Harvard University (Mekalanos et al., 1983; Pearson et al., 1990; Chiang and Mekalanos, 2000). The most advanced vaccine candidate utilizes an avirulent *V. cholerae* strain (CVD 103-HgR) which is administered in a single dose (Tacket et al., 1999). CVD 103-HgR is a derivative of a classical O1 strain in which cholera toxin A subunit (*ctxA*) was deleted; additionally, the strain incorporates a mercury resistance gene in the *hlyA* locus to distinguish it from wild-type O1 strains (Levine et al., 1988a; Kenner et al., 1995). Considerable research has been directed toward determining the safety and efficacy of this vaccine, and it appears to be well tolerated and to provide seroconversion in >90% of healthy adults (Levine et al., 1988a; Cryz et al., 1990; Kotloff et al., 1992; Su-Arehawarantana et al., 1992; Suharyono et al., 1992). Vaccination with CVD 103-HgR has been shown to afford protection for at least 6 months (Tacket et al., 1992).

Additional live and killed vaccines are in development, including vaccines against O1 El Tor and O139 strains (Coster et al., 1995; Ledon et al., 2003). The use of transposon delivery and FLP recombinase-mediated excision has been proposed as a means to allow for the genetic modification of vaccine candidate strains without the introduction of antibiotic resistance (Chiang and Mekalanos, 2000). A key feature of new vaccines will be to make them economically feasible and easily dispensable to populations that are at risk (Sack et al., 2004).

CONCLUDING REMARKS

The scientific study of *V. cholerae* has been exceedingly productive in uncovering pathogenic and environmental survival mechanisms of the bacterium. From the early identification of the causative agent by Robert Koch (Koch, 1894) and the epidemiological studies of John Snow, to the development of extensive genetic screens, *V. cholerae* research has continued to remain on the forefront of scientific discoveries. Contributions from the fields of epidemiology, bacteriology, immunology, biochemistry, and molecular biology have provided a better understanding of and subsequent effective treatments for cholera. However, despite these efforts, cholera continues to be a significant cause of morbidity and mortality in the world.

Further research is still needed to understand the intricacies of this ancient scourge. The recent emergence of epidemic O139 strains indicates that the pathogenic potential of *V. cholerae* is continuously changing. Cholera is not a disease that is ready to be relegated to history; rather, it is a persistent and deadly disease that continues to demand scientific study.

Acknowledgments. We thank M. N. Guentzel, F. Yildiz, J. Bina, J. Mekalanos, and J. Reidl for kindly supplying figures. This work was supported by research grants AI-43486 and AI-51333 from the National Institutes of Health to K. E. Klose.

REFERENCES

Abendroth, J., M. Bagdasarian, M. Sandkvist, and W. G. Hol. 2004a. The structure of the cytoplasmic domain of EpsL, an inner membrane component of the type II secretion system of *Vibrio cholerae*: an unusual member of the actin-like ATPase superfamily. *J. Mol. Biol.* **344:**619–633.

Abendroth, J., A. E. Rice, K. McLuskey, M. Bagdasarian, and W. G. Hol. 2004b. The crystal structure of the periplasmic domain of the type II secretion system protein EpsM from *Vibrio cholerae*: the simplest version of the ferredoxin fold. *J. Mol. Biol.* **338:**585–596.

Ali, A., J. A. Johnson, A. A. Franco, D. J. Metzger, T. D. Connell, J. G. Morris, Jr., and S. Sozhamannan. 2000. Mutations in the extracellular protein secretion pathway genes (eps) interfere with rugose polysaccharide production in and motility of *Vibrio cholerae*. *Infect. Immun.* **68:**1967–1974.

Andersen, C. 2003. Channel-tunnels: outer membrane components of type I secretion systems and multidrug efflux pumps of gram-negative bacteria. *Rev. Physiol. Biochem. Pharmacol.* **147:**122–165.

Angelichio, M. J., J. Spector, M. K. Waldor, and A. Camilli. 1999. *Vibrio cholerae* intestinal population dynamics in the suckling mouse model of infection. *Infect. Immun.* **67:**3733–3739.

Angelichio, M. J., D. S. Merrell, and A. Camilli. 2004. Spatiotemporal analysis of acid adaptation-mediated *Vibrio cholerae* hyperinfectivity. *Infect. Immun.* **72:**2405–2407.

Attridge, S. R., P. A. Manning, J. Holmgren, and G. Jonson. 1996. Relative significance of mannose-sensitive hemagglutinin and toxin-coregulated pili in colonization of infant mice by *Vibrio cholerae* El Tor. *Infect. Immun.* **64:**3369–3373.

Ausmees, N., R. Mayer, H. Weinhouse, G. Volman, D. Amikam, M. Benziman, and M. Lindberg. 2001. Genetic data indicate that proteins containing the GGDEF domain possess diguanylate cyclase activity. *FEMS Microbiol. Lett.* **204:**163–167.

Barrett, T. J., and P. A. Blake. 1981. Epidemiological usefulness of changes in hemolytic activity of *Vibrio cholerae* biotype El Tor during the seventh pandemic. *J. Clin. Microbiol.* **13:**126–1129.

Barua, D. 1992. *History of Cholera*. Plenum Medical Book Company, New York, N.Y.

Beaber, J. W., V. Burrus, B. Hochhut, and M. K. Waldor. 2002a. Comparison of SXT and R391, two conjugative integrating elements: definition of a genetic backbone for the mobilization of resistance determinants. *Cell. Mol. Life Sci.* **59:**2065–2070.

Beaber, J. W., B. Hochhut, and M. K. Waldor. 2002b. Genomic and functional analyses of SXT, an integrating antibiotic resistance gene transfer element derived from *Vibrio cholerae*. *J. Bacteriol.* **184:**4259–4269.

Beaber, J. W., B. Hochhut, and M. K. Waldor. 2004. SOS response promotes horizontal dissemination of antibiotic resistance genes. *Nature* **427:**72–74.

Beck, N. A., E. S. Krukonis, and V. J. DiRita. 2004. TcpH influences virulence gene expression in *Vibrio cholerae* by inhibiting degradation of the transcription activator TcpP. *J. Bacteriol.* **186:**8309–8316.

Behari, J., L. Stagon, and S. B. Calderwood. 2001. pepA, a gene mediating pH regulation of virulence genes in *Vibrio cholerae*. *J. Bacteriol.* **183:**178–188.

Benitez, J. A., L. Garcia, A. Silva, H. Garcia, R. Fando, B. Cedre, A. Perez, J. Campos, B. L. Rodriguez, J. L. Perez, T. Valmaseda, O. Perez, M. Ramirez, T. Ledon, M. D. Jidy, M. Lastre, L. Bravo, and G. Sierra. 1999. Preliminary assessment of the safety and immunogenicity of a new CTXPhi-negative, hemagglutinin/protease-defective El Tor strain as a cholera vaccine candidate. *Infect. Immun.* **67:**539–545.

Benz, R., E. Maier, and T. Chakraborty. 1997. Purification of OmpU from *Vibrio cholerae* classical strain 569B: evidence for the formation of large cation-selective ion-permeable channels by OmpU. *Microbiologia* **13:**321–330.

Bina, J., J. Zhu, M. Dziejman, S. Faruque, S. Calderwood, and J. Mekalanos. 2003. ToxR regulon of *Vibrio cholerae* and its expression in vibrios shed by cholera patients. *Proc. Natl. Acad. Sci. USA* **100:**2801–2806.

Biskri, L., M. Bouvier, A. M. Guerout, S. Boisnard, and D. Mazel. 2005. Comparative study of class 1 integron and *Vibrio cholerae* superintegron integrase activities. *J. Bacteriol.* **187:**1740–1750.

Boardman, B. K., and K. J. Satchell. 2004. *Vibrio cholerae* strains with mutations in an atypical type I secretion system accumulate RTX toxin intracellularly. *J. Bacteriol.* **186:**8137–8143.

Boles, B. R., M. Thoendel, and P. K. Singh. 2004. Self-generated diversity produces "insurance effects" in biofilm communities. *Proc. Natl. Acad. Sc. USA* **101:**16630–16635.

Bomchil, N., P. Watnick, and R. Kolter. 2003. Identification and characterization of a *Vibrio cholerae* gene, mbaA, involved in maintenance of biofilm architecture. *J. Bacteriol.* **185:**1384–1390.

Booth, B. A., M. Boesman-Finkelstein, and R. A. Finkelstein. 1984. *Vibrio cholerae* hemagglutinin/protease nicks cholera enterotoxin. *Infect. Immun.* **45:**558–560.

Boyd, E. F., A. J. Heilpern, and M. K. Waldor. 2000. Molecular analyses of a putative CTXphi precursor and evidence for independent acquisition of distinct CTX(phi)s by toxigenic *Vibrio cholerae*. *J. Bacteriol.* **182**:5530–5538.

Boyd, E. F., and M. K. Waldor. 1999. Alternative mechanism of cholera toxin acquisition by *Vibrio cholerae*: generalized transduction of CTXPhi by bacteriophage CP-T1. *Infect. Immun.* **67**:5898–5905.

Boyd, E. F., and M. K. Waldor. 2002. Evolutionary and functional analyses of variants of the toxin-coregulated pilus protein TcpA from toxigenic *Vibrio cholerae* non-O1/non-O139 serogroup isolates. *Microbiology* **148**:1655–1666.

Burrus, V., and M. K. Waldor. 2003. Control of SXT integration and excision. *J. Bacteriol.* **185**:5045–5054.

Butler, S. M., and A. Camilli. 2004. Both chemotaxis and net motility greatly influence the infectivity of *Vibrio cholerae*. *Proc. Natl. Acad. Sci. USA* **101**:5018–5023.

Camilli, A., D. T. Beattie, and J. J. Mekalanos. 1994. Use of genetic recombination as a reporter of gene expression. *Proc. Natl. Acad. Sci. USA* **91**:2634–2638.

Camilli, A., and J. J. Mekalanos. 1995. Use of recombinase gene fusions to identify *Vibrio cholerae* genes induced during infection. *Mol. Microbiol.* **18**:671–683.

Carroll, P. A., K. T. Tashima, M. B. Rogers, V. J. DiRita, and S. B. Calderwood. 1997. Phase variation in *tcpH* modulates expression of the ToxR regulon in *Vibrio cholerae*. *Mol. Microbiol.* **25**:1099–1111.

Casper-Lindley, C., and F. H. Yildiz. 2004. VpsT is a transcriptional regulator required for expression of vps biosynthesis genes and the development of rugose colonial morphology in *Vibrio cholerae* O1 El Tor. *J. Bacteriol.* **186**:1574–1578.

Chakrabarti, S. R., K. Chaudhuri, K. Sen, and J. Das. 1996. Porins of *Vibrio cholerae*: purification and characterization of OmpU. *J. Bacteriol.* **178**:524–530.

Champion, G. A., M. N. Neely, M. A. Brennan, and V. J. DiRita. 1997. A branch in the ToxR regulatory cascade of *Vibrio cholerae* revealed by characterization of *toxT* mutant strains. *Mol. Microbiol.* **23**:323–331.

Chiang, S. L., and J. J. Mekalanos. 2000. Construction of a *Vibrio cholerae* vaccine candidate using transposon delivery and FLP recombinase-mediated excision. *Infect. Immun.* **68**:6391–6397.

Chiang, S. L., and J. J. Mekalanos. 1999. rfb mutations in *Vibrio cholerae* do not affect surface production of toxin-coregulated pili but still inhibit intestinal colonization. *Infect. Immun.* **67**:976–980.

Chiang, S. L., and J. J. Mekalanos. 1998. Use of signature-tagged transposon mutagenesis to identify *Vibrio cholerae* genes critical for colonization. *Mol. Microbiol.* **27**:797–805.

Chiang, S. L., J. J. Mekalanos, and D. W. Holden. 1999. In vivo genetic analysis of bacterial virulence. *Annu. Rev. Microbiol.* **53**:129–154.

Chiavelli, D. A., J. W. Marsh, and R. K. Taylor. 2001. The mannose-sensitive hemagglutinin of *Vibrio cholerae* promotes adherence to zooplankton. *Appl. Environ. Microbiol.* **67**:3220–3225.

Clark, C. A., L. Purins, P. Kaewrakon, T. Focareta, and P. A. Manning. 2000. The *Vibrio cholerae* O1 chromosomal integron. *Microbiology* **146**:2605–2612.

Clemens, J. D., D. A. Sack, J. R. Harris, J. Chakraborty, M. R. Khan, S. Huda, F. Ahmed, J. Gomes, M. R. Rao, A. M. Svennerholm, et al. 1989. ABO blood groups and cholera: new observations on specificity of risk and modification of vaccine efficacy. *J. Infect. Dis.* **159**:770–773.

Clemens, J. D., D. A. Sack, J. R. Harris, J. Chakraborty, M. R. Khan, B. F. Stanton, B. A. Kay, M. U. Khan, M. Yunus, W. Atkinson, et al. 1986. Field trial of oral cholera vaccines in Bangladesh. *Lancet* **ii**:124–127.

Coelho, A., J. R. Andrade, A. C. Vicente, and V. J. Dirita. 2000. Cytotoxic cell vacuolating activity from *Vibrio cholerae* hemolysin. *Infect. Immun.* **68**:1700–1705.

Colwell, R., and W. M. Spira. 1992. The ecology of *Vibrio cholerae*, p. 107–127. *In* D. Barua and W. B. Greennough III (ed.), *Cholera*. Plenum Press, New York, N.Y.

Colwell, R. R. 1996. Global climate and infectious disease: the cholera paradigm. *Science* **274**:2025–2031.

Colwell, R. R., A. Huq, M. S. Islam, K. M. Aziz, M. Yunus, N. H. Khan, A. Mahmud, R. B. Sack, G. B. Nair, J. Chakraborty, D. A. Sack, and E. Russek-Cohen. 2003. Reduction of cholera in Bangladeshi villages by simple filtration. *Proc. Natl. Acad. Sci. USA* **100**:1051–1055.

Colwell, R. R., J. Kaper, and S. W. Joseph. 1977. *Vibrio cholerae*, *Vibrio parahaemolyticus*, and other vibrios: occurrence and distribution in Chesapeake Bay. *Science* **198**:394–396.

Comstock, L. E., J. A. Johnson, J. M. Michalski, J. G. Morris, Jr., and J. B. Kaper. 1996. Cloning and sequence of a region encoding a surface polysaccharide of *Vibrio cholerae* O139 and characterization of the insertion site in the chromosome of *Vibrio cholerae* O1. *Mol. Microbiol.* **19**:815–826.

Connell, T. D., D. J. Metzger, J. Lynch, and J. P. Folster. 1998. Endochitinase is transported to the extracellular milieu by the eps-encoded general secretory pathway of *Vibrio cholerae*. *J. Bacteriol.* **180**:5591–5600.

Cook, W. L., K. Wachsmuth, S. R. Johnson, K. A. Birkness, and A. R. Samadi. 1984. Persistence of plasmids, cholera toxin genes, and prophage DNA in classical *Vibrio cholerae* O1. *Infect. Immun* **45**:222–226.

Correa, N. E., C. M. Lauriano, R. McGee, and K. E. Klose. 2000. Phosphorylation of the flagellar regulatory protein FlrC is necessary for *Vibrio cholerae* motility and enhanced colonization. *Mol. Microbiol.* **35**:743–755.

Coster, T. S., K. P. Killeen, M. K. Waldor, D. T. Beattie, D. R. Spriggs, J. R. Kenner, A. Trofa, J. C. Sadoff, J. J. Mekalanos, and D. N. Taylor. 1995. Safety, immunogenicity, and efficacy of live attenuated *Vibrio cholerae* O139 vaccine prototype. *Lancet* **345**:949–952.

Cryz, S. J., Jr., J. Kaper, C. Tacket, J. Nataro, and M. M. Levine. 1995. *Vibrio cholerae* CVD103-HgR live oral attenuated vaccine: construction, safety, immunogenicity, excretion and nontarget effects. *Dev. Biol. Stand.* **84**:237–244.

Cryz, S. J., Jr., M. M. Levine, J. B. Kaper, E. Furer, and B. Althaus. 1990. Randomized double-blind placebo controlled trial to evaluate the safety and immunogenicity of the live oral cholera vaccine strain CVD 103-HgR in Swiss adults. *Vaccine* **8**:577–580.

Davey, M. E., and A. O'Toole G. 2000. Microbial biofilms: from ecology to molecular genetics. *Microbiol. Mol. Biol. Rev.* **64**:847–867.

Davis, B. M., H. H. Kimsey, A. V. Kane, and M. K. Waldor. 2002. A satellite phage-encoded antirepressor induces repressor aggregation and cholera toxin gene transfer. *EMBO J.* **21**:4240–4249.

Davis, B. M., E. H. Lawson, M. Sandkvist, A. Ali, S. Sozhamannan, and M. K. Waldor. 2000a. Convergence of the secretory pathways for cholera toxin and the filamentous phage, CTXphi. *Science* **288**:333–335.

Davis, B. M., K. E. Moyer, E. F. Boyd, and M. K. Waldor. 2000b. CTX prophages in classical biotype *Vibrio cholerae*: functional phage genes but dysfunctional phage genomes. *J. Bacteriol.* **182**:6992–6998.

Davis, B. M., and M. K. Waldor. 2000. CTXphi contains a hybrid genome derived from tandemly integrated elements. *Proc. Natl. Acad. Sci. USA* **97**:8572–8577.

De, S. N., and D. N. Chatterje. 1953. An experimental study of the mechanism of action of Vibriod cholerae on the intestinal mucous membrane. *J. Pathol. Bacteriol.* **66**:559–562.

de Jonge, H. R. 1991. Intracellular mechanisms regulating intestinal secretion, p. 107–114. *In* T. M. Wadstrom, P. H. Makela, A. M.

Svennerholm, and H. Wolf-Waltz (ed.), *Molecular Pathogenesis of Gastrointestinal Infections*. Plenum Press, New York, N.Y.

Dickinson, B. L., and W. I. Lencer. 2003. Transcytosis of bacterial toxins across mucosal barriers, p. 173–186. *In* D. L. Burns, J. T. Barbieri, B. H. Iglewski, and R. Rappouli (ed.), *Bacterial Protein Toxins*. ASM Press, Washington, D.C.

Ding, Y., B. M. Davis, and M. K. Waldor. 2004. Hfq is essential for *Vibrio cholerae* virulence and downregulates sigma expression. *Mol. Microbiol.* 53:345–354.

DiRita, V. J., and J. J. Mekalanos. 1991. Periplasmic interaction between two membrane regulatory proteins, ToxR and ToxS, results in signal transduction and transcriptional activation. *Cell* 64:29–37.

DiRita, V. J., C. Parsot, G. Jander, and J. J. Mekalanos. 1991. Regulatory cascade controls virulence in *Vibrio cholerae*. *Proc. Natl. Acad. Sci. USA* 88:5403–5407.

Dutta, N. K., and M. K. Habbu. 1955. Experimental cholera in infant rabbits: a method for chemotherapeutic investigation. *Br. J. Pharmacol.* 10:153–159.

Dziejman, M., E. Balon, D. Boyd, C. M. Fraser, J. F. Heidelberg, and J. J. Mekalanos. 2002. Comparative genomic analysis of *Vibrio cholerae*: genes that correlate with cholera endemic and pandemic disease. *Proc. Natl. Acad. Sci. USA* 99:1556–1561.

Epstein, P. 1993. Algal blooms in the spread and persistence of cholera. *Biosystems* 31:209–221.

Everiss, K. D., K. J. Hughes, and K. M. Peterson. 1994. The accessory colonization factor and toxin-coregulated pilus gene clusters are physically linked on the *Vibrio cholerae* 0395 chromosome. *DNA Seq.* 5:51–55.

Faruque, S. M., M. J. Albert, and J. J. Mekalanos. 1998. Epidemiology, genetics, and ecology of toxigenic *Vibrio cholerae*. *Microbiol. Mol. Biol. Rev.* 62:1301–1314.

Faruque, S. M., N. Chowdhury, M. Kamruzzaman, M. Dziejman, M. H. Rahman, D. A. Sack, G. B. Nair, and J. J. Mekalanos. 2004. Genetic diversity and virulence potential of environmental *Vibrio cholerae* population in a cholera-endemic area. *Proc. Natl. Acad. Sci. USA* 101:2123–2128.

Faruque, S. M., J. Zhu, Asadulghani, M. Kamruzzaman, and J. J. Mekalanos. 2003. Examination of diverse toxin-coregulated pilus-positive *Vibrio cholerae* strains fails to demonstrate evidence for *Vibrio* pathogenicity island phage. *Infect. Immun.* 71:2993–2999.

Feeley, J. C. 1965. Classification of *Vibrio cholerae* (*Vibrio comma*), including El Tor vibrios, by infrasubspecific characteristics. *J. Bacteriol.* 89:665–670.

Field, M., D. Fromm, C. K. Wallace, and W. B. Greenough. 1969. Stimulation of active chloride secretion in small intestine by cholera exotoxin. *J. Clin. Invest.* 48:24a.

Figueroa-Arredondo, P., J. E. Heuser, N. S. Akopyants, J. H. Morisaki, S. Giono-Cerezo, F. Enriquez-Rincon, and D. E. Berg. 2001. Cell vacuolation caused by *Vibrio cholerae* hemolysin. *Infect. Immun.* 69:1613–1624.

Finkelstein, R. A. 1995. Why do we not yet have a suitable vaccine against cholera? *Adv. Exp. Med. Biol.* 371B:1633–1640.

Finkelstein, R. A., M. Boesman-Finkelstein, Y. Chang, and C. C. Hase. 1992. *Vibrio cholerae* hemagglutinin/protease, colonial variation, virulence, and detachment. *Infect. Immun.* 60:472–478.

Finkelstein, R. A., and J. J. LoSpalluto. 1969. Pathogenesis of experimental cholera: preparation and isolation of choleragen and choleragenoid. *J. Exp. Med.* 130:185–202.

Finkelstein, R. A., and S. Mukerjee. 1963. Hemagglutination: a rapid method for differentiating *Vibrio cholerae* and El Tor vibrios. *Proc. Soc. Exp. Biol. Med.* 112:355–359.

Freeman, J. A., and B. L. Bassler. 1999. Sequence and function of LuxU: a two-component phosphorelay protein that regulates quorum sensing in *Vibrio harveyi*. *J. Bacteriol.* 181:899–906.

Freter, R., and P. C. O'Brien. 1981a. Role of chemotaxis in the association of motile bacteria with intestinal mucosa: chemotactic responses of *Vibrio cholerae* and description of motile nonchemotactic mutants. *Infect. Immun.* 34:215–221.

Freter, R., and P. C. O'Brien. 1981b. Role of chemotaxis in the association of motile bacteria with intestinal mucosa: fitness and virulence of nonchemotactic *Vibrio cholerae* mutants in infant mice. *Infect. Immun.* 34:222–233.

Freter, R., P. C. O'Brien, and M. S. Macsai. 1979. Effect of chemotaxis on the interaction of cholera vibrios with intestinal mucosa. *Am. J. Clin. Nutr.* 32:128–132.

Freter, R., P. C. O'Brien, and M. S. Macsai. 1981. Role of chemotaxis in the association of motile bacteria with intestinal mucosa: *in vivo* studies. *Infect. Immun.* 34:234–240.

Fullner, K. J., J. C. Boucher, M. A. Hanes, G. K. Haines, 3rd, B. M. Meehan, C. Walchle, P. J. Sansonetti, and J. J. Mekalanos. 2002. The contribution of accessory toxins of *Vibrio cholerae* O1 El Tor to the proinflammatory response in a murine pulmonary cholera model. *J. Exp. Med.* 195:1455–1462.

Fullner, K. J., W. I. Lencer, and J. J. Mekalanos. 2001. *Vibrio cholerae*-induced cellular responses of polarized T84 intestinal epithelial cells are dependent on production of cholera toxin and the RTX toxin. *Infect. Immun.* 69:6310–6317.

Fullner, K. J., and J. J. Mekalanos. 1999. Genetic characterization of a new type IV-A pilus gene cluster found in both classical and El Tor biotypes of *Vibrio cholerae*. *Infect. Immun.* 67:1393–1404.

Fullner, K. J., and J. J. Mekalanos. 2000. *In vivo* covalent cross-linking of cellular actin by the *Vibrio cholerae* RTX toxin. *EMBO J.* 19:5315–5323.

Garay, E., A. Arnau, and C. Amaro. 1985. Incidence of *Vibrio cholerae* and related vibrios in a coastal lagoon and seawater influenced by lake discharges along an annual cycle. *Appl. Environ. Microbiol.* 50:426–430.

Gardel, C. L., and J. J. Mekalanos. 1996. Alterations in *Vibrio cholerae* motility phenotypes correlate with changes in virulence factor expression. *Infect. Immun.* 64:2246–2255.

Gardel, C. L., and J. J. Mekalanos. 1994. Regulation of cholera toxin by temperature, pH, and osmolarity. *Methods Enzymol.* 235:517–526.

Gill, D. M. 1976. The arrangement of subunits in cholera toxin. *Biochemistry* 15:1242–1248.

Gill, D. M., and C. A. King. 1975. The mechanism of action of cholera toxin in pigeon erythrocyte lysates. *J. Biol. Chem.* 250:6424–6432.

Gill, D. M., and R. S. Rappaport. 1979. Origin of the enzymatically active A1 fragment of cholera toxin. *J. Infect. Dis.* 139:674–680.

Guidolin, A., and P. A. Manning. 1985. Bacteriophage CP-T1 of *Vibrio cholerae*. Identification of the cell surface receptor. *Eur. J. Biochem.* 153:89–94.

Guidolin, A., G. Morelli, M. Kamke, and P. A. Manning. 1984. *Vibrio cholerae* bacteriophage CP-T1: characterization of bacteriophage DNA and restriction analysis. *J. Virol.* 51:163–169.

Gupta, S., and R. Chowdhury. 1997. Bile affects production of virulence factors and motility of *Vibrio cholerae*. *Infect. Immun.* 65:1131–1134.

Hall, R. M., and H. W. Stokes. 1993. Integrons: novel DNA elements which capture genes by site-specific recombination. *Genetica* 90:115–132.

Hammer, B. K., and B. L. Bassler. 2003. Quorum sensing controls biofilm formation in *Vibrio cholerae*. *Mol. Microbiol.* 50:101–104.

Handfield, M., L. J. Brady, A. Progulske-Fox, and J. D. Hillman. 2000. IVIAT: a novel method to identify microbial genes expressed specifically during human infections. *Trends Microbiol.* 8:336–339.

Hang, L., M. John, M. Asaduzzaman, E. A. Bridges, C. Vanderspurt, T. J. Kirn, R. K. Taylor, J. D. Hillman, A. Progulske-Fox, M. Handfield, E. T. Ryan, and S. B. Calderwood. 2003. Use of in vivo-induced antigen technology (IVIAT) to identify genes uniquely expressed during human infection with Vibrio cholerae. *Proc. Natl. Acad. Sci. USA* 100:8508–8513.

Hasan, J. A., D. Bernstein, A. Huq, L. Loomis, M. L. Tamplin, and R. R. Colwell. 1994. Cholera DFA: an improved direct fluorescent monoclonal antibody staining kit for rapid detection and enumeration of *Vibrio cholerae* O1. *FEMS Microbiol. Lett.* 120:143–148.

Hase, C. C., M. E. Bauer, and R. A. Finkelstein. 1994. Genetic characterization of mannose-sensitive hemagglutinin (MSHA)-negative mutants of *Vibrio cholerae* derived by Tn5 mutagenesis. *Gene* 150:17–25.

Hase, C. C., and R. A. Finkelstein. 1991. Cloning and nucleotide sequence of the *Vibrio cholerae* hemagglutinin/protease (HA/protease) gene and construction of an HA/protease-negative strain. *J. Bacteriol.* 173:3311–3317.

Hase, C. C., and R. A. Finkelstein. 1990. Comparison of the *Vibrio cholerae* hemagglutinin/protease and the *Pseudomonas aeruginosa* elastase. *Infect. Immun.* 58:4011–4015.

Hase, C. C., and J. J. Mekalanos. 1998. TcpP protein is a positive regulator of virulence gene expression in *Vibrio cholerae*. *Proc. Natl. Acad. Sci. USA* 95:730–734.

Haugo, A. J., and P. I. Watnick. 2002. *Vibrio cholerae* CytR is a repressor of biofilm development. *Mol. Microbiol.* 45:471–483.

Hava, D. L., and A. Camilli. 2001. Isolation and characterization of a temperature-sensitive generalized transducing bacteriophage for *Vibrio cholerae*. *J. Microbiol. Methods* 46:217–225.

Heidelberg, J. F., J. A. Eisen, W. C. Nelson, R. A. Clayton, M. L. Gwinn, R. J. Dodson, D. H. Haft, E. K. Hickey, J. D. Peterson, L. Umayam, S. R. Gill, K. E. Nelson, T. D. Read, H. Tettelin, D. Richardson, M. D. Ermolaeva, J. Vamathevan, S. Bass, H. Qin, I. Dragoi, P. Sellers, L. McDonald, T. Utterback, R. D. Fleishmann, W. C. Nierman, O. White, S. L. Salzberg, H. O. Smith, R. R. Colwell, J. J. Mekalanos, J. C. Venter, and C. M. Fraser. 2000. DNA sequence of both chromosomes of the cholera pathogen *Vibrio cholerae*. *Nature* 406:477–483.

Heilpern, A. J., and M. K. Waldor. 2000. CTXphi infection of *Vibrio cholerae* requires the tolQRA gene products. *J. Bacteriol.* 182:1739–1747.

Heilpern, A. J., and M. K. Waldor. 2003. pIIICTX, a predicted CTXphi minor coat protein, can expand the host range of coliphage fd to include *Vibrio cholerae*. *J. Bacteriol.* 185:1037–1044.

Hensel, M., J. E. Shea, C. Gleeson, M. D. Jones, E. Dalton, and D. W. Holden. 1995. Simultaneous identification of bacterial virulence genes by negative selection. *Science* 269:400–403.

Herrington, D. A., R. H. Hall, G. Losonsky, J. J. Mekalanos, R. K. Taylor, and M. M. Levine. 1988. Toxin, toxin-coregulated pili, and the *toxR* regulon are essential for *Vibrio cholerae* pathogenesis in humans. *J. Exp. Med.* 168:1487–1492.

Higa, N., Y. Honma, M. J. Albert, and M. Iwanaga. 1993. Characterization of *Vibrio cholerae* O139 synonym Bengal isolated from patients with cholera-like disease in Bangladesh. *Microbiol. Immunol.* 37:971–974.

Higgins, D. E., E. Nazareno, and V. J. DiRita. 1992. The virulence gene activator ToxT from *Vibrio cholerae* is a member of the AraC family of transcriptional activators. *J. Bacteriol.* 174:6974–6980.

Hirst, T. R., and J. Holmgren. 1987a. Conformation of protein secreted across bacterial outer membranes: a study of enterotoxin translocation from *Vibrio cholerae*. *Proc. Natl. Acad. Sci. USA* 84:7418–7422.

Hirst, T. R., and J. Holmgren. 1987b. Transient entry of enterotoxin subunits into the periplasm occurs during their secretion from *Vibrio cholerae*. *J. Bacteriol.* 169:1037–1045.

Hochhut, B., Y. Lotfi, D. Mazel, S. M. Faruque, R. Woodgate, and M. K. Waldor. 2001. Molecular analysis of antibiotic resistance gene clusters in *Vibrio cholerae* O139 and O1 SXT constins. *Antimicrob. Agents Chemother.* 45:2991–3000.

Hornick, R. B., S. I. Music, R. Wenzel, R. Cash, J. P. Libonati, M. J. Snyder, and T. E. Woodward. 1971. The Broad Street pump revisited: response of volunteers to ingested cholera vibrios. *Bull. N. Y. Acad. Med.* 47:1181–1191.

Huber, K. E., and M. K. Waldor. 2002. Filamentous phage integration requires the host recombinases XerC and XerD. *Nature* 417:656–659.

Hung, D. T., and J. J. Mekalanos. 2005. Bile acids induce cholera toxin expression in *Vibrio cholerae* in a ToxT-independent manner. *Proc. Natl. Acad. Sci. USA* 102:3028–3033. (First published 7 February 2005; 10.1073/pnas.0409559102.)

Huq, A., R. R. Colwell, R. Rahman, A. Ali, M. A. Chowdhury, S. Parveen, D. A. Sack, and E. Russek-Cohen. 1990. Detection of *Vibrio cholerae* O1 in the aquatic environment by fluorescent-monoclonal antibody and culture methods. *Appl. Environ. Microbiol.* 56:2370–2373.

Ichinose, Y., K. Yamamoto, N. Nakasone, M. J. Tanabe, T. Takeda, T. Miwatani, and M. Iwanaga. 1987. Enterotoxicity of El Tor-like hemolysin of non-O1 *Vibrio cholerae*. *Infect. Immun.* 55:1090–1093.

International Centre for Diarrhoeal Diseases Research. 1993. Large epidemic of cholera-like disease in Bangladesh caused by *Vibrio cholerae* O139 synonym Bengal. Cholera Working Group, International Centre for Diarrhoeal Diseases Research, Bangladesh. *Lancet* 342:387–390.

Iredell, J. R., U. H. Stroeher, H. M. Ward, and P. A. Manning. 1998. Lipopolysaccharide O-antigen expression and the effect of its absence on virulence in *rfb* mutants of *Vibrio cholerae* O1. *FEMS Immunol. Med. Microbiol.* 20:45–54.

Islam, M. S., B. S. Drasar, and R. B. Sack. 1993. The aquatic environment as a reservoir of *Vibrio cholerae*: a review. *J. Diarrhoeal Dis. Res.* 11:197–206.

Islam, M. S., B. S. Drasar, and R. B. Sack. 1994. The aquatic flora and fauna as reservoirs of *Vibrio cholerae*: a review. *J. Diarrhoeal Dis. Res.* 12:87–96.

Iwanaga, M., K. Yamamoto, N. Higa, Y. Ichinose, N. Nakasone, and M. Tanabe. 1986. Culture conditions for stimulating cholera toxin production by *Vibrio cholerae* O1 El Tor. *Microbiol. Immunol.* 30:1075–1083.

Jobling, M. G., and R. K. Holmes. 1997. Characterization of hapR, a positive regulator of the *Vibrio cholerae* HA/protease gene hap, and its identification as a functional homologue of the *Vibrio harveyi* luxR gene. *Mol. Microbiol.* 26:1023–1034.

Jodal, M., and O. Lungren. 1986. Enterotoxin-induced fluid secretion and the enteric nervous system, p. 278. *In* J. Holmgren, A. Lindberg, and R. Molby (ed.), *Development of Vaccines and Drugs against Diarrhea*. 11th Nobel Conference, Stockholm, 1985. Studentlitteratur, Lund, Sweden.

Johnson, J. A., C. A. Salles, P. Panigrahi, M. J. Albert, A. C. Wright, R. J. Johnson, and J. G. Morris, Jr. 1994. *Vibrio cholerae* O139 synonym bengal is closely related to *Vibrio cholerae* El Tor but has important differences. *Infect. Immun.* 62:2108–2110.

Jonson, G., J. Holmgren, and A. M. Svennerholm. 1992. Analysis of expression of toxin-coregulated pili in classical and El Tor *Vibrio cholerae* O1 *in vitro* and *in vivo*. *Infect. Immun.* 60:4278–4284.

Jonson, G., M. Lebens, and J. Holmgren. 1994. Cloning and sequencing of *Vibrio cholerae* mannose-sensitive haemagglutinin pilin gene: localization of mshA within a cluster of type 4 pilin genes. *Mol. Microbiol.* 13:109–118.

Kaper, J. B., and M. M. Levine. 1981. Cloned cholera enterotoxin genes in study and prevention of cholera. *Lancet* ii:1162–1163.

Kaper, J. B., H. Lockman, M. M. Baldini, and M. M. Levine. 1984. Recombinant nontoxinogenic *Vibrio cholerae* strains as attenuated cholera vaccine candidates. *Nature* 308:655–658.

Kaper, J. B., J. G. Morris, Jr., and M. M. Levine. 1995. Cholera. *Clin. Microbiol. Rev.* 8:48–86.

Karaolis, D. K., J. A. Johnson, C. C. Bailey, E. C. Boedeker, J. B. Kaper, and P. R. Reeves. 1998. A *Vibrio cholerae* pathogenicity island associated with epidemic and pandemic strains. *Proc. Natl. Acad. Sci. USA* 95:3134–3139.

Karaolis, D. K., S. Somara, D. R. Maneval, Jr., J. A. Johnson, and J. B. Kaper. 1999. A bacteriophage encoding a pathogenicity island, a type-IV pilus and a phage receptor in cholera bacteria. *Nature* 399:375–379.

Kenne, L., B. Lindberg, P. Unger, B. Gustafsson, and T. Holme. 1982. Structural studies of the *Vibrio cholerae* O-antigen. *Carbohydr. Res.* 100:341–349.

Kenner, J. R., T. S. Coster, D. N. Taylor, A. F. Trofa, M. Barrera-Oro, T. Hyman, J. M. Adams, D. T. Beattie, K. P. Killeen, D. R. Spriggs, et al. 1995. Peru-15, an improved live attenuated oral vaccine candidate for *Vibrio cholerae* O1. *J. Infect. Dis.* 172:1126–1129.

Keyhani, N. O., and S. Roseman. 1999. Physiological aspects of chitin catabolism in marine bacteria. *Biochim. Biophys. Acta* 1473:108–122.

Kierek, K., and P. I. Watnick. 2003a. Environmental determinants of *Vibrio cholerae* biofilm development. *Appl. Environ. Microbiol.* 69:5079–5088.

Kierek, K., and P. I. Watnick. 2003b. The *Vibrio cholerae* O139 O-antigen polysaccharide is essential for Ca^{2+}-dependent biofilm development in sea water. *Proc. Natl. Acad. Sci. USA* 100:14357–14362.

Kimsey, H. H., and M. K. Waldor. 1998a. CTXf immunity: application in the development of cholera vaccines. *Proc. Natl. Acad. Sci. USA* 95:7035–7039.

Kimsey, H. H., and M. K. Waldor. 1998b. *Vibrio cholerae* hemagglutinin/protease inactivates CTXphi. *Infect. Immun.* 66:4025–4029.

Kimsey, H. H., and M. K. Waldor. 2004. The CTXphi repressor RstR binds DNA cooperatively to form tetrameric repressor-operator complexes. *J. Biol. Chem.* 279:2640–2647.

King, C. A., and W. E. van Heyningen. 1973. Deactivation of cholera toxin by a sialidase-resistant monosialosylganglioside. *J. Infect. Dis.* 127:639–647.

Kirn, T. J., N. Bose, and R. K. Taylor. 2003. Secretion of a soluble colonization factor by the TCP type 4 pilus biogenesis pathway in *Vibrio cholerae*. *Mol. Microbiol.* 49:81–92.

Kirn, T. J., M. J. Lafferty, C. M. Sandoe, and R. K. Taylor. 2000. Delineation of pilin domains required for bacterial association into microcolonies and intestinal colonization by *Vibrio cholerae*. *Mol. Microbiol.* 35:896–910.

Klose, K. E. 2000. The suckling mouse model of cholera. *Trends Microbiol.* 8:189–191.

Klose, K. E., and J. J. Mekalanos. 1998. Distinct roles of an alternative sigma factor during both free-swimming and colonizing phases of the *Vibrio cholerae* pathogenic cycle. *Mol. Microbiol.* 28:501–520.

Klose, K. E., V. Novick, and J. J. Mekalanos. 1998. Identification of multiple σ^{54}-dependent transcriptional activators in *Vibrio cholerae*. *J. Bacteriol.* 180:5256–5259.

Knirel, Y. A., L. Paredes, P. E. Jansson, A. Weintraub, G. Widmalm, and M. J. Albert. 1995. Structure of the capsular polysaccharide of *Vibrio cholerae* O139 synonym Bengal containing D-galactose 4,6-cyclophosphate. *Eur. J. Biochem.* 232:391–396.

Knirel, Y. A., G. Widmalm, S. N. Senchenkova, P. E. Jansson, and A. Weintraub. 1997. Structural studies on the short-chain lipopolysaccharide of *Vibrio cholerae* O139 Bengal. *Eur. J. Biochem.* 247:402–410.

Koch, R. 1894. An address on cholera and its bacillus. *Br. Med. J.* 2:453–459.

Kotloff, K. L., S. S. Wasserman, S. O'Donnell, G. A. Losonsky, S. J. Cryz, and M. M. Levine. 1992. Safety and immunogenicity in North Americans of a single dose of live oral cholera vaccine CVD 103-HgR: results of a randomized, placebo-controlled, double-blind crossover trial. *Infect. Immun.* 60:4430–4432.

Kovach, M. E., M. D. Shaffer, and K. M. Peterson. 1996. A putative integrase gene defines the distal end of a large cluster of ToxR-regulated colonization genes in *Vibrio cholerae*. *Microbiol.* 142:2165–2174.

Kovacikova, G., W. Lin, and K. Skorupski. 2004. *Vibrio cholerae* AphA uses a novel mechanism for virulence gene activation that involves interaction with the LysR-type regulator AphB at the tcpPH promoter. *Mol. Microbiol.* 53:129–142.

Kovacikova, G., and K. Skorupski. 2000. Differential activation of the tcpPH promoter by AphB determines biotype specificity of virulence gene expression in *Vibrio cholerae*. *J. Bacteriol.* 182:3228–3238.

Kovacikova, G., and K. Skorupski. 2001. Overlapping binding sites for the virulence gene regulators AphA, AphB and cAMP-CRP at the *Vibrio cholerae* tcpPH promoter. *Mol. Microbiol.* 41:393–407.

Kovacikova, G., and K. Skorupski. 2002. Regulation of virulence gene expression in *Vibrio cholerae* by quorum sensing: HapR functions at the aphA promoter. *Mol. Microbiol.* 46:1135–1147.

Kovacikova, G., and K. Skorupski. 1999. A *Vibrio cholerae* LysR homolog, AphB, cooperates with AphA at the tcpPH promoter to activate expression of the ToxR virulence cascade. *J. Bacteriol.* 181:4250–4256.

Krukonis, E. S., and V. J. DiRita. 2003. DNA binding and ToxR responsiveness by the wing domain of TcpP, an activator of virulence gene expression in *Vibrio cholerae*. *Mol. Cell.* 12:157–165.

Krukonis, E. S., R. R. Yu, and V. J. DiRita. 2000. The *Vibrio cholerae* ToxR/TcpP/ToxT virulence cascade: distinct roles for two membrane-localized transcriptional activators on a single promoter. *Mol. Microbiol.* 38:67–84.

Lang, H., and E. T. Palva. 1993. The ompS gene of *Vibrio cholerae* encodes a growth-phase-dependent maltoporin. *Mol. Microbiol.* 10:891–901.

Lauriano, C. M., C. Ghosh, N. E. Correa, and K. E. Klose. 2004. The sodium-driven flagellar motor controls exopolysaccharide expression in *Vibrio cholerae*. *J. Bacteriol.* 186:4864–4874.

Ledon, T., E. Valle, T. Valmaseda, B. Cedre, J. Campos, B. L. Rodriguez, K. Marrero, H. Garcia, L. Garcia, and R. Fando. 2003. Construction and characterisation of O139 cholera vaccine candidates. *Vaccine* 21:1282–1291.

Lee, C. A. 1999. *Vibrio cholerae* TCP: a trifunctional virulence factor? *Trends Microbiol.* 7:391–392.

Lee, S. H., M. J. Angelichio, J. J. Mekalanos, and A. Camilli. 1998. Nucleotide sequence and spatiotemporal expression of the *Vibrio cholerae* vieSAB genes during infection. *J. Bacteriol.* 180:2298–2305.

Lee, S. H., S. M. Butler, and A. Camilli. 2001. Selection for in vivo regulators of bacterial virulence. *Proc. Natl. Acad. Sci. USA* 98:6889–6894.

Lee, S. H., D. L. Hava, M. K. Waldor, and A. Camilli. 1999. Regulation and temporal expression patterns of *Vibrio cholerae* virulence genes during infection. *Cell* 99:625–634.

Lencer, W. I., C. Constable, S. Moe, M. G. Jobling, H. M. Webb, S. Ruston, J. L. Madara, T. R. Hirst, and R. K. Holmes. 1995. Targeting of cholera toxin and *Escherichia coli* heat labile toxin in polarized epithelia: role of COOH-terminal KDEL. *J. Cell Biol.* 131:951–962.

Lenz, D. H., K. C. Mok, B. N. Lilley, R. V. Kulkarni, N. S. Wingreen, and B. L. Bassler. 2004. The small RNA chaperone Hfq and multiple small RNAs control quorum sensing in *Vibrio harveyi* and *Vibrio cholerae*. *Cell* 118: 69–82.

Levine, M. M., and J. B. Kaper. 1993. Live oral vaccines against cholera: an update. *Vaccine* 11:207–212.

Levine, M. M., J. B. Kaper, R. E. Black, and M. L. Clements. 1983. New knowledge on pathogenesis of bacterial enteric infections as applied to vaccine development. *Microbiol. Rev.* 47: 510–550.

Levine, M. M., J. B. Kaper, D. Herrington, J. Ketley, G. Losonsky, C. O. Tacket, B. Tall, and S. Cryz. 1988a. Safety, immunogenicity, and efficacy of recombinant live oral cholera vaccines, CVD 103 and CVD 103-HgR. *Lancet* ii:467–470.

Levine, M. M., J. B. Kaper, D. Herrington, G. Losonsky, J. G. Morris, M. L. Clements, R. E. Black, B. Tall, and R. Hall. 1988b. Volunteer studies of deletion mutants of *Vibrio cholerae* O1 prepared by recombinant techniques. *Infect. Immun.* 56: 161–167.

Lilley, B. N., and B. L. Bassler. 2000. Regulation of quorum sensing in *Vibrio harveyi* by LuxO and sigma-54. *Mol. Microbiol.* 36:940–954.

Lin, W., K. J. Fullner, R. Clayton, J. A. Sexton, M. B. Rogers, K. E. Calia, S. B. Calderwood, C. Fraser, and J. J. Mekalanos. 1999. Identification of a *Vibrio cholerae* RTX toxin gene cluster that is tightly linked to the cholera toxin prophage. *Proc. Natl. Acad. Sci. USA* 96:1071–1076.

Lin, Z., K. Kumagai, K. Baba, J. J. Mekalanos, and M. Nishibuchi. 1993. *Vibrio parahaemolyticus* has a homolog of the *Vibrio cholerae* toxRS operon that mediates environmentally induced regulation of the thermostable direct hemolysin gene. *J. Bacteriol.* 175:3844–3855.

London, E., and D. A. Brown. 2000. Insolubility of lipids in triton X-100: physical origin and relationship to sphingolipid/cholesterol membrane domains (rafts). *Biochim. Biophys. Acta* 1508: 182–195.

Mahan, M. J., J. M. Slauch, and J. J. Mekalanos. 1993. Selection of bacterial virulence genes that are specifically induced in host tissues. *Science* 259:686–688.

Mahan, M. J., J. W. Tobias, J. M. Slauch, P. C. Hanna, R. J. Collier, and J. J. Mekalanos. 1995. Antibiotic-based selection for bacterial genes that are specifically induced during infection of a host. *Proc. Natl. Acad. Sci. USA* 92:669–673.

Majoul, I., K. Sohn, F. T. Wieland, R. Pepperkok, M. Pizza, J. Hillemann, and H. D. Soling. 1998. KDEL receptor (Erd2p)-mediated retrograde transport of the cholera toxin A subunit from the Golgi involves COPI, p23, and the COOH terminus of Erd2p. *J. Cell. Biol.* 143:601–612.

Majoul, I. V., P. I. Bastiaens, and H. D. Soling. 1996. Transport of an external Lys-Asp-Glu-Leu (KDEL) protein from the plasma membrane to the endoplasmic reticulum: studies with cholera toxin in Vero cells. *J. Cell. Biol.* 133:777–789.

Marsh, J. W., and R. K. Taylor. 1998. Identification of the *Vibrio cholerae* type 4 prepilin peptidase required for cholera toxin secretion and pilus formation. *Mol. Microbiol.* 29:1481–1492.

Martinez-Hackert, E., and A. M. Stock. 1997. Structural relationships in the OmpR family of winged-helix transcription factors. *J. Mol. Biol.* 269:301–312.

Mazel, D., B. Dychinco, V. A. Webb, and J. Davies. 1998. A distinctive class of integron in the *Vibrio cholerae* genome. *Science* 280:605–608.

Meibom, K. L., X. B. Li, A. T. Nielsen, C. Y. Wu, S. Roseman, and G. K. Schoolnik. 2004. The *Vibrio cholerae* chitin utilization program. *Proc. Natl. Acad. Sci. USA* 101:2524–2529.

Mekalanos, J. J. 1983. Duplication and amplification of toxin genes in *Vibrio cholerae*. *Cell* 35:253–263.

Mekalanos, J. J., R. J. Collier, and W. R. Romig. 1979. Enzymic activity of cholera toxin. I. New method of assay and the mechanism of ADP-ribosyl transfer. *J. Biol.* Chem. 254:5849–5854.

Mekalanos, J. J., D. J. Swartz, G. D. Pearson, N. Harford, F. Groyne, and M. de Wilde. 1983. Cholera toxin genes: nucleotide sequence, deletion analysis and vaccine development. *Nature* 306:551–557.

Meng, S. Y., and G. N. Bennett. 1992. Regulation of the *Escherichia coli* cad operon: location of a site required for acid induction. *J. Bacteriol.* 174:2670–2678.

Merrell, D. S., C. Bailey, J. B. Kaper, and A. Camilli. 2001. The ToxR-mediated organic acid tolerance response of *Vibrio cholerae* requires OmpU. *J. Bacteriol.* 183:2746–2754.

Merrell, D. S., S. M. Butler, F. Qadri, N. A. Dolganov, A. Alam, M. B. Cohen, S. B. Calderwood, G. K. Schoolnik, and A. Camilli. 2002a. Host-induced epidemic spread of the cholera bacterium. *Nature* 417:642–645.

Merrell, D. S., D. L. Hava, and A. Camilli. 2002b. Identification of novel factors involved in colonization and acid tolerance of *Vibrio cholerae*. *Mol. Microbiol.* 43:1471–1491.

Merrell, D. S., and A. Camilli. 1999. The cadA gene of *Vibrio cholerae* is induced during infection and plays a role in acid tolerance. *Mol. Microbiol.* 34:836–849.

Miller, M. B., K. Skorupski, D. H. Lenz, R. K. Taylor, and B. L. Bassler. 2002. Parallel quorum sensing systems converge to regulate virulence in *Vibrio cholerae*. *Cell* 110:303–314.

Miller, V. L., and J. J. Mekalanos. 1985. Genetic analysis of the cholera toxin-positive regulatory gene toxR. *J. Bacteriol.* 163: 580–585.

Miller, V. L., and J. J. Mekalanos. 1988. A novel suicide vector and its use in construction of insertion mutations: osmoregulation of outer membrane proteins and virulence determinants in *Vibrio cholerae* requires *toxR*. *J. Bacteriol.* 170:2575–2583.

Miller, V. L., and J. J. Mekalanos. 1984. Synthesis of cholera toxin is positively regulated at the transcriptional level by *toxR*. *Proc. Natl. Acad. Sci. USA* 81:3471–3475.

Miller, V. L., R. K. Taylor, and J. J. Mekalanos. 1987. Cholera toxin transcriptional activator ToxR is a transmembrane DNA binding protein. *Cell* 48:271–279.

Mitra, R., P. Figueroa, A. K. Mukhopadhyay, T. Shimada, Y. Takeda, D. E. Berg, and G. B. Nair. 2000. Cell vacuolation, a manifestation of the El Tor hemolysin of *Vibrio cholerae*. *Infect. Immun.* 68:1928–1933.

Murley, Y. M., J. Behari, R. Griffin, and S. B. Calderwood. 2000. Classical and El Tor biotypes of *Vibrio cholerae* differ in timing of transcription of tcpPH during growth in inducing conditions. *Infect. Immun.* 68:3010–3014.

Murley, Y. M., P. A. Carroll, K. Skorupski, R. K. Taylor, and S. B. Calderwood. 1999. Differential transcription of the *tcpPH* operon confers biotype-specific control of the *Vibrio cholerae* ToxR virulence regulon. *Infect. Immun.* 67:5117–5123.

Nesper, J., A. Kraiss, S. Schild, J. Blass, K. E. Klose, J. Bockemuhl, and J. Reidl. 2002a. Comparative and genetic analyses of the putative *Vibrio cholerae* lipopolysaccharide core oligosaccharide biosynthesis (wav) gene cluster. *Infect. Immun.* 70:2419–2433.

Nesper, J., S. Schild, C. M. Lauriano, A. Kraiss, K. E. Klose, and J. Reidl. 2002b. Role of *Vibrio cholerae* O139 surface polysaccharides in intestinal colonization. *Infect. Immun.* 70:5990–5996.

Nesper, J., C. M. Lauriano, K. E. Klose, D. Kapfhammer, A. Kraiss, and J. Reidl. 2001. Characterization of *Vibrio cholerae* O1 El tor galU and galE mutants: influence on lipopolysaccharide structure, colonization, and biofilm formation. *Infect. Immun.* 69:435–445.

Nichols, B. J., A. K. Kenworthy, R. S. Polishchuk, R. Lodge, T. H. Roberts, K. Hirschberg, R. D. Phair, and J. Lippincott-Schwartz.

2001. Rapid cycling of lipid raft markers between the cell surface and Golgi complex. *J. Cell Biol.* **153**:529–541.

Nikaido, H. 1996. Outer membrane, p. 29–47. *In* F. C. Neidhardt, R. Curtiss III, J. L. Ingraham, E. C. C. Lin, K. B. Low, B. Magasanik, W. S. Reznikoff, M. Riley, M. Schaechter, and H. E. Umbarger (ed.), Escherichia coli *and* Salmonella: *Cellular and Molecular Biology*, 2nd ed. ASM Press, Washington, D.C.

Nikaido, H. 1988. Structure and functions of the cell envelope of gram-negative bacteria. *Rev. Infect. Dis.* **10** (Suppl.) **2**:S279–S281.

Nikaido, H., and M. Vaara. 1985. Molecular basis of bacterial outer membrane permeability. *Microbiol. Rev.* **49**:1–32.

Nusrin, S., G. Y. Khan, N. A. Bhuiyan, M. Ansaruzzaman, M. A. Hossain, A. Safa, R. Khan, S. M. Faruque, D. A. Sack, T. Hamabata, Y. Takeda, and G. B. Nair. 2004. Diverse CTX phages among toxigenic *Vibrio cholerae* O1 and O139 strains isolated between 1994 and 2002 in an area where cholera is endemic in Bangladesh. *J. Clin. Microbiol.* **42**:5854–5856.

Nye, M. B., J. D. Pfau, K. Skorupski, and R. K. Taylor. 2000. *Vibrio cholerae* H-NS silences virulence gene expression at multiple steps in the ToxR regulatory cascade. *J. Bacteriol.* **182**:4295–4303.

Ogg, J. E., T. L. Timme, and M. M. Alemohammad. 1981. General transduction in *Vibrio cholerae*. *Infect. Immun.* **31**:737–741.

Ogierman, M. A., S. Zabihi, L. Mourtzios, and P. A. Manning. 1993. Genetic organization and sequence of the promoter-distal region of the tcp gene cluster of *Vibrio cholerae*. *Gene* **126**:51–60.

O'Shea, Y. A., and E. F. Boyd. 2002. Mobilization of the *Vibrio* pathogenicity island between *Vibrio cholerae* isolates mediated by CP-T1 generalized transduction. *FEMS Microbiol. Lett.* **214**:153–157.

Osorio, C. G., J. A. Crawford, J. Michalski, H. Martinez-Wilson, J. B. Kaper, and A. Camilli. 2005. Second-generation recombination-based in vivo expression technology for large-scale screening for *Vibrio cholerae* genes induced during infection of the mouse small intestine. *Infect. Immun.* **73**:972–980.

Osorio, C. G., H. Martinez-Wilson, and A. Camilli. 2004. The ompU paralogue vca1008 is required for virulence of *Vibrio cholerae*. *J. Bacteriol.* **186**:5167–5171.

Osorio, C. R., and K. E. Klose. 2000. A region of the transmembrane regulatory protein ToxR that tethers the transcriptional activation domain to the cytoplasmic membrane displays wide divergence among *Vibrio* species. *J. Bacteriol.* **182**:526–528.

Park, Y. K., B. Bearson, S. H. Bang, I. S. Bang, and J. W. Foster. 1996. Internal pH crisis, lysine decarboxylase and the acid tolerance response of *Salmonella typhimurium*. *Mol. Microbiol.* **20**:605–611.

Pearson, G. D., V. J. DiRita, M. B. Goldberg, S. A. Boyko, S. B. Calderwood, and J. J. Mekalanos. 1990. New attenuated derivatives of *Vibrio cholerae*. *Res. Microbiol.* **141**:893–899.

Peterson, K. M., and J. J. Mekalanos. 1988. Characterization of the *Vibrio cholerae* ToxR regulon: identification of novel genes involved in intestinal colonization. *Infect. Immun.* **56**:2822–2829.

Pollitzer, R. 1959. *Cholera*. World Health Organization, Geneva, Switzerland.

Prouty, A. M., I. E. Brodsky, J. Manos, R. Belas, S. Falkow, and J. S. Gunn. 2004. Transcriptional regulation of *Salmonella enterica* serovar Typhimurium genes by bile. *FEMS Immunol. Med. Microbiol.* **41**:177–185.

Prouty, A. M., and J. S. Gunn. 2000. *Salmonella enterica* serovar typhimurium invasion is repressed in the presence of bile. *Infect. Immun.* **68**:6763–6769.

Prouty, M. G., N. E. Correa, and K. E. Klose. 2001. The novel sigma54- and sigma28-dependent flagellar gene transcription hierarchy of *Vibrio cholerae*. *Mol. Microbiol.* **39**:1595–1609.

Provenzano, D., and K. E. Klose. 2000. Altered expression of the ToxR-regulated porins OmpU and OmpT diminishes *Vibrio cholerae* bile resistance, virulence factor expression, and intestinal colonization. *Proc. Natl. Acad. Sci. USA* **97**:10220–10224.

Provenzano, D., C. M. Lauriano, and K. E. Klose. 2001. Characterization of the role of the ToxR-modulated outer membrane porins OmpU and OmpT in *Vibrio cholerae* virulence. *J. Bacteriol.* **183**:3652–3662.

Provenzano, D., D. A. Schuhmacher, J. L. Barker, and K. E. Klose. 2000. The virulence regulatory protein ToxR mediates enhanced bile resistance in *Vibrio cholerae* and other pathogenic *Vibrio* species. *Infect. Immun.* **68**:1491–1497.

Rader, A. E., and J. R. Murphy. 1988. Nucleotide sequences and comparison of the hemolysin determinants of *Vibrio cholerae* El Tor RV79(Hly+) and RV79(Hly−) and classical 569B(Hly−). *Infect. Immun.* **56**:1414–1419.

Ramamurthy, T., S. Garg, R. Sharma, S. K. Bhattacharya, G. B. Nair, T. Shimada, T. Takeda, T. Karasawa, H. Kurazano, A. Pal, et al. 1993. Emergence of novel strain of *Vibrio cholerae* with epidemic potential in southern and eastern India. *Lancet* **341**:703–704.

Rashid, M. H., C. Rajanna, A. Ali, and D. K. Karaolis. 2003. Identification of genes involved in the switch between the smooth and rugose phenotypes of *Vibrio cholerae*. *FEMS Microbiol. Lett.* **227**:113–119.

Recchia, G. D., and R. M. Hall. 1995. Gene cassettes: a new class of mobile element. *Microbiology* **141**:3015–3027.

Recchia, G. D., and R. M. Hall. 1997. Origins of the mobile gene cassettes found in integrons. *Trends Microbiol,* **5**:389–394.

Redmond, J. W. 1979. The structure of the O-antigenic side chain of the lipopolysaccharide of *Vibrio cholerae* 569B (Inaba). *Biochim. Biophys. Acta* **584**:346–352.

Reich, K. A., and G. K. Schoolnik. 1994. The light organ symbiont *Vibrio fischeri* possesses a homolog of the *Vibrio cholerae* transmembrane transcriptional activator ToxR. *J. Bacteriol.* **176**:3085–3088.

Reidl, J., and K. E. Klose. 2002. *Vibrio cholerae* and cholera: out of the water and into the host. *FEMS Microbiol. Rev.* **26**:125–139.

Richards, A. A., E. Stang, R. Pepperkok, and R. G. Parton. 2002. Inhibitors of COP-mediated transport and cholera toxin action inhibit simian virus 40 infection. *Mol. Biol. Cell* **13**:1750–1764.

Richardson, K. 1991. Roles of motility and flagellar structure in pathogenicity of *Vibrio cholerae*: analysis of motility mutants in three animal models. *Infect. Immun.* **59**:2727–2736.

Richardson, K., J. Michalski, and J. B. Kaper. 1986. Hemolysin production and cloning of two hemolysin determinants from classical *Vibrio cholerae*. *Infect. Immun.* **54**:415–420.

Richardson, S. H., J. C. Giles, and K. S. Kruger. 1984. Sealed adult mice: new model for enterotoxin evaluation. *Infect. Immun.* **43**:482–486.

Robien, M. A., B. E. Krumm, M. Sandkvist, and W. G. Hol. 2003. Crystal structure of the extracellular protein secretion NTPase EpsE of *Vibrio cholerae*. *J. Mol. Biol.* **333**:657–674.

Rodriguez, B. L., A. Rojas, J. Campos, T. Ledon, E. Valle, W. Toledo, and R. Fando. 2001. Differential interleukin-8 response of intestinal epithelial cell line to reactogenic and nonreactogenic candidate vaccine strains of *Vibrio cholerae*. *Infect. Immun.* **69**:613–616.

Rosenberg, C. E. 1962. *The Cholera Years, the United States in 1832, 1849, and 1866.* University of Chicago Press, Chicago, Ill.

Rosenberg, E. Y., D. Bertenthal, M. L. Nilles, K. P. Bertrand, and H. Nikaido. 2003. Bile salts and fatty acids induce the expression of Escherichia coli AcrAB multidrug efflux pump through their interaction with Rob regulatory protein. *Mol. Microbiol.* **48**:1609–1619.

Roszak, D. B., D. J. Grimes, and R. R. Colwell. 1984. Viable but nonrecoverable stage of *Salmonella enteritidis* in aquatic systems. *Can. J. Microbiol.* 30:334–338.

Rowe-Magnus, D. A., A. M. Guerout, P. Ploncard, B. Dychinco, J. Davies, and D. Mazel. 2001. The evolutionary history of chromosomal super-integrons provides an ancestry for multiresistant integrons. *Proc. Natl. Acad. Sci. USA* 98:652–657.

Roy, C., and S. Mukerjee. 1962. Variability in the haemolytic property of El Tor vibrios. *Annu. Biochem. Exp. Med.* 22:295–296.

Rubin, E. J., W. Lin, J. J. Mekalanos, and M. K. Waldor. 1998. Replication and integration of a *Vibrio cholerae* cryptic plasmid linked to the CTX prophage. *Mol. Microbiol.* 28:1247–1254.

Ryan, E. T., and S. B. Calderwood. 2000. Cholera vaccines. *Clin. Infect. Dis.* 31:561–565.

Sack, D. A., R. B. Sack, G. B. Nair, and A. K. Siddique. 2004. Cholera. *Lancet* 363:223–233.

Sack, D. A., C. O. Tacket, M. B. Cohen, R. B. Sack, G. A. Losonsky, J. Shimko, J. P. Nataro, R. Edelman, M. M. Levine, R. A. Giannella, G. Schiff, and D. Lang. 1998. Validation of a volunteer model of cholera with frozen bacteria as the challenge. *Infect. Immun.* 66:1968–1972.

Sack, R. B. 1992. Colonization and pathology, p. 189–197. *In* D. Barua and W. B. Greennough III (ed.), *Cholera*. Plenum Press, New York, N.Y.

Sack, R. B., and C. C. Carpenter. 1969. Experimental canine cholera. I. Development of the model. *J. Infect. Dis.* 119:138–149.

Saha, N., and K. K. Banerjee. 1997. Carbohydrate-mediated regulation of interaction of *Vibrio cholerae* hemolysin with erythrocyte and phospholipid vesicle. *J. Biol. Chem.* 272:162–167.

Sandkvist, M., L. O. Michel, L. P. Hough, V. M. Morales, M. Bagdasarian, M. Koomey, and V. J. DiRita. 1997. General secretion pathway (eps) genes required for toxin secretion and outer membrane biogenesis in *Vibrio cholerae*. *J. Bacteriol.* 179:6994–7003.

Sandvig, K., and B. van Deurs. 2002. Membrane traffic exploited by protein toxins. *Annu. Rev. Cell Dev. Biol.* 18:1–24.

Sarkar, A., R. K. Nandy, G. B. Nair, and A. C. Ghose. 2002. Vibrio pathogenicity island and cholera toxin genetic element-associated virulence genes and their expression in non-O1 non-O139 strains of *Vibrio cholerae*. *Infect. Immun.* 70:4735–4742.

Schild, S., A. K. Lamprecht, C. Fourestier, C. M. Lauriano, K. E. Klose, and J. Reidl. 2005. Characterizing lipopolysaccharide and core lipid A mutant O1 and O139 *Vibrio cholerae* strains for adherence properties on mucus-producing cell line HT29-Rev MTX and virulence in mice. *Int. J. Med. Microbiol.* 295:243–251.

Schuhmacher, D. A., and K. E. Klose. 1999. Environmental signals modulate ToxT-dependent virulence factor expression in *Vibrio cholerae*. *J. Bacteriol.* 181:1508–1514.

Scott, M. E., Z. Y. Dossani, and M. Sandkvist. 2001. Directed polar secretion of protease from single cells of *Vibrio cholerae* via the type II secretion pathway. *Proc. Natl. Acad. Sci. USA* 98:13978–113983.

Sheahan, K. L., C. L. Cordero, and K. J. Satchell. 2004. Identification of a domain within the multifunctional *Vibrio cholerae* RTX toxin that covalently cross-links actin. *Proc. Natl. Acad. Sci. USA* 101:9798–9803.

Shimamura, T., S. Watanabe, and S. Sasaki. 1985. Enhancement of enterotoxin production by carbon dioxide in *Vibrio cholerae*. *Infect. Immun.* 49:455–456.

Shogomori, H., and A. H. Futerman. 2001. Cholera toxin is found in detergent-insoluble rafts/domains at the cell surface of hippocampal neurons but is internalized via a raft-independent mechanism. *J. Biol. Chem.* 276:9182–9188.

Silva, T. M., M. A. Schleupner, C. O. Tacket, T. S. Steiner, J. B. Kaper, R. Edelman, and R. Guerrant. 1996. New evidence for an inflammatory component in diarrhea caused by selected new, live attenuated cholera vaccines and by El Tor and Q139 *Vibrio cholerae*. *Infect. Immun.* 64:2362–2364.

Simm, R., M. Morr, A. Kader, M. Nimtz, and U. Romling. 2004. GGDEF and EAL domains inversely regulate cyclic di-GMP levels and transition from sessility to motility. *Mol. Microbiol.* 53:1123–1134.

Simonet, V. C., A. Basle, K. E. Klose, and A. H. Delcour. 2003. The *Vibrio cholerae* porins OmpU and OmpT have distinct channel properties. *J. Biol. Chem.* 278:17539–17545.

Simons, K., and E. Ikonen. 1997. Functional rafts in cell membranes. *Nature* 387:569–572.

Singleton, F. L., R. W. Attwell, M. S. Jangi, and R. R. Colwell. 1982. Influence of salinity and organic nutrient concentration on survival and growth of *Vibrio cholerae* in aquatic microcosms. *Appl. Environ. Microbiol.* 43:1080–1085.

Skorupski, K., and R. K. Taylor. 1997. Cyclic AMP and its receptor protein negatively regulate the coordinate expression of cholera toxin and toxin-coregulated pilus in *Vibrio cholerae*. *Proc. Natl. Acad. Sci. USA* 94:265–270.

Skorupski, K., and R. K. Taylor. 1999. A new level in the *Vibrio cholerae* ToxR virulence cascade: AphA is required for transcriptional activation of the *tcpPH* operon. *Mol. Microbiol.* 31:763–771.

Snow, J., W. H. Frost, and B. W. Richardson. 1936. *Snow on Cholera*. Commonwealth Fund, New York, N.Y.

Sperandio, V., J. A. Giron, W. D. Silveira, and J. B. Kaper. 1995. The OmpU outer membrane protein, a potential adherence factor of *Vibrio cholerae*. *Infect. Immun.* 63:4433–4438.

Spira, W. M., R. B. Sack, and J. L. Froehlich. 1981. Simple adult rabbit model for *Vibrio cholerae* and enterotoxigenic *Escherichia coli* diarrhea. *Infect. Immun.* 32:739–747.

Stanley, N. R., and B. A. Lazazzera. 2004. Environmental signals and regulatory pathways that influence biofilm formation. *Mol. Microbiol.* 52:917–924.

Stevenson, L. H. 1978. A case for bacterial dormancy in aquatic systems. *Microb. Ecol.* 4:127–133.

Stokes, H. W., D. B. O'Gorman, G. D. Recchia, M. Parsekhian, and R. M. Hall. 1997. Structure and function of 59-base element recombination sites associated with mobile gene cassettes. *Mol. Microbiol.* 26:731–745.

Stroeher, U. H., K. E. Jedani, and P. A. Manning. 1998. Genetic organization of the regions associated with surface polysaccharide synthesis in *Vibrio cholerae* O1, O139 and *Vibrio anguillarum* O1 and O2: a review. *Gene* 223:269–282.

Su-Arehawaratana, P., P. Singharaj, D. N. Taylor, C. Hoge, A. Trofa, K. Kuvanont, S. Migasena, P. Pitisuttitham, Y. L. Lim, G. Losonsky, et al. 1992. Safety and immunogenicity of different immunization regimens of CVD 103-HgR live oral cholera vaccine in soldiers and civilians in Thailand. *J. Infect. Dis.* 165:1042–1048.

Suharyono, C. Simanjuntak, N. Witham, N. Punjabi, D. G. Heppner, G. Losonsky, H. Totosudirjo, A. R. Rifai, J. Clemens, Y. L. Lim, et al. 1992. Safety and immunogenicity of single-dose live oral cholera vaccine CVD 103-HgR in 5-9-year-old Indonesian children. *Lancet* 340:689–694.

Sundaram, U., R. G. Knickelbein, and J. W. Dobbins. 1991a. Mechanism of intestinal secretion: effect of serotonin on rabbit ileal crypt and villus cells. *J. Clin. Invest.* 87:743–746.

Sundaram, U., R. G. Knickelbein, and J. W. Dobbins. 1991b. Mechanism of intestinal secretion: effect of cyclic AMP on rabbit ileal crypt and villus cells. *Proc. Natl. Acad. Sci. USA* 88:6249–6253.

Svennerholm, A. M., and J. Holmgren. 1986. Oral combined B subunit-whole cell cholera vaccine, p. 33–43. *In* J. Holmgren, A.

Lindberg, and R. Molby (ed.), *Development of Vaccines and Drugs against Diarrhea*. 11th Nobel Conference, Stockholm, 1985. Studentlitteratur, Lund, Sweden.

Tacket, C. O., M. B. Cohen, S. S. Wasserman, G. Losonsky, S. Livio, K. Kotloff, R. Edelman, J. B. Kaper, S. J. Cryz, R. A. Giannella, G. Schiff, and M. M. Levine. 1999. Randomized, double-blind, placebo-controlled, multicentered trial of the efficacy of a single dose of live oral cholera vaccine CVD 103-HgR in preventing cholera following challenge with *Vibrio cholerae* O1 El Tor inaba three months after vaccination. *Infect. Immun.* **67:**6341–6345.

Tacket, C. O., G. Losonsky, J. P. Nataro, S. J. Cryz, R. Edelman, A. Fasano, J. Michalski, J. B. Kaper, and M. M. Levine. 1993. Safety and immunogenicity of live oral cholera vaccine candidate CVD 110, a delta ctxA delta zot delta ace derivative of El Tor Ogawa *Vibrio cholerae*. *J. Infect. Dis.* **168:**1536–1540.

Tacket, C. O., G. Losonsky, J. P. Nataro, S. J. Cryz, R. Edelman, J. B. Kaper, and M. M. Levine. 1992. Onset and duration of protective immunity in challenged volunteers after vaccination with live oral cholera vaccine CVD 103-HgR. *J. Infect. Dis.* **166:**837–841.

Tacket, C. O., R. K. Taylor, G. Losonsky, Y. Lim, J. P. Nataro, J. B. Kaper, and M. M. Levine. 1998. Investigation of the roles of toxin-coregulated pili and mannose-sensitive hemagglutinin pili in the pathogenesis of *Vibrio cholerae* O139 infection. *Infect. Immun.* **66:**692–695.

Taylor, D. N., K. P. Killeen, D. C. Hack, J. R. Kenner, T. S. Coster, D. T. Beattie, J. Ezzell, T. Hyman, A. Trofa, M. H. Sjogren, et al. 1994. Development of a live, oral, attenuated vaccine against El Tor cholera. *J. Infect. Dis.* **170:**1518–1523.

Taylor, R., C. Shaw, K. Peterson, P. Spears, and J. Mekalanos. 1988. Safe, live *Vibrio cholerae* vaccines? *Vaccine* **6:**151–154.

Taylor, R. K., V. L. Miller, D. B. Furlong, and J. J. Mekalanos. 1987. Use of *phoA* gene fusions to identify a pilus colonization factor coordinately regulated with cholera toxin. *Proc. Natl. Acad. Sci. USA* **84:**2833–2837.

Thelin, K. H., and R. K. Taylor. 1996. Toxin-coregulated pilus, but not mannose-sensitive hemagglutinin, is required for colonization by *Vibrio cholerae* O1 El Tor biotype and O139 strains. *Infect. Immun.* **64:**2853–2856.

Tischler, A. D., and A. Camilli. 2004. Cyclic diguanylate (c-di-GMP) regulates *Vibrio cholerae* biofilm formation. *Mol. Microbiol.* **53:**857–869.

Tischler, A. D., S. H. Lee, and A. Camilli. 2002. The *Vibrio cholerae* vieSAB locus encodes a pathway contributing to cholera toxin production. *J. Bacteriol.* **184:**4104–4113.

Vance, R. E., J. Zhu, and J. J. Mekalanos. 2003. A constitutively active variant of the quorum-sensing regulator LuxO affects protease production and biofilm formation in *Vibrio cholerae*. *Infect. Immun.* **71:**2571–2576.

Voss, E., and S. R. Attridge. 1993. *In vitro* production of toxin-coregulated pili by *Vibrio cholerae* El Tor. *Microb. Pathog.* **15:**255–268.

Wai, S. N., Y. Mizunoe, A. Takade, S. I. Kawabata, and S. I. Yoshida. 1998. *Vibrio cholerae* O1 strain TSI-4 produces the exopolysaccharide materials that determine colony morphology, stress resistance, and biofilm formation. *Appl. Environ. Microbiol.* **64:**3648–3655.

Waldor, M. K., R. Colwell, and J. J. Mekalanos. 1994. The *Vibrio cholerae* O139 serogroup antigen includes an O-antigen capsule and lipopolysaccharide virulence determinants. *Proc. Natl. Acad. Sci. USA* **91:**11388–11392.

Waldor, M. K., and J. J. Mekalanos. 1994. Emergence of a new cholera pandemic: molecular analysis of virulence determinants

in *Vibrio cholerae* O139 and development of a live vaccine prototype. *J. Infect. Dis.* **170:**278–283.

Waldor, M. K., and J. J. Mekalanos. 1996. Lysogenic conversion by a filamentous phage encoding cholera toxin. *Science* **272:**1910–1914.

Waldor, M. K., E. J. Rubin, G. D. Pearson, H. Kimsey, and J. J. Mekalanos. 1997. Regulation, replication, and integration functions of the *Vibrio cholerae* CTXf are encoded by region RS2. *Mol. Microbiol.* **24:**917–926.

Waldor, M. K., H. Tschape, and J. J. Mekalanos. 1996. A new type of conjugative transposon encodes resistance to sulfamethoxazole, trimethoprim, and streptomycin in *Vibrio cholerae* O139. *J. Bacteriol.* **178:**4157–4165.

Watnick, P. I., K. J. Fullner, and R. Kolter. 1999. A role for the mannose-sensitive hemagglutinin in biofilm formation by *Vibrio cholerae* El Tor. *J. Bacteriol.* **181:**3606–3609.

Watnick, P. I., and R. Kolter. 1999. Steps in the development of a *Vibrio cholerae* biofilm. *Mol. Microbiol.* **34:**586–595.

Watnick, P. I., C. M. Lauriano, K. E. Klose, L. Croal, and R. Kolter. 2001. The absence of a flagellum leads to altered colony morphology, biofilm development and virulence in Vibrio cholerae O139. *Mol. Microbiol.* **39:**223–235.

Welch, T. J., and D. H. Bartlett. 1998. Identification of a regulatory protein required for pressure-responsive gene expression in the deep-sea bacterium *Photobacterium* species strain SS9. *Mol. Microbiol.* **27:**977–985.

Wibbenmeyer, J. A., D. Provenzano, C. F. Landry, K. E. Klose, and A. H. Delcour. 2002. *Vibrio cholerae* OmpU and OmpT porins are differentially affected by bile. *Infect. Immun.* **70:**121–126.

Wolf, A. A., Y. Fujinaga, and W. I. Lencer. 2002. Uncoupling of the cholera toxin-G(M1) ganglioside receptor complex from endocytosis, retrograde Golgi trafficking, and downstream signal transduction by depletion of membrane cholesterol. *J. Biol. Chem.* **277:**16249–16256.

Wu, Z., D. Milton, P. Nybom, A. Sjo, and K. E. Magnusson. 1996. *Vibrio cholerae* hemagglutinin/protease (HA/protease) causes morphological changes in cultured epithelial cells and perturbs their paracellular barrier function. *Microb. Pathog.* **21:**111–123.

Wu, Z., P. Nybom, and K. E. Magnusson. 2000. Distinct effects of *Vibrio cholerae* haemagglutinin/protease on the structure and localization of the tight junction-associated proteins occludin and ZO-1. *Cell. Microbiol.* **2:**11–17.

Xu, H. S., N. Roberts, F. L. Singleton, R. W. Attwell, D. J. Grimes, and R. R. Colwell. 1982. Survival and viability of nonculturable *Escherichia coli* and *Vibrio cholerae* in estuaine and marine environments. *Microb. Ecol.* **8:**313–323.

Xu, Q., M. Dziejman, and J. J. Mekalanos. 2003. Determination of the transcriptome of *Vibrio cholerae* during intraintestinal growth and midexponential phase *in vitro*. *Proc. Natl. Acad. Sci. USA* **100:**1286–1291.

Yamamoto, K., M. Al-Omani, T. Honda, Y. Takeda, and T. Miwatani. 1984. Non-O1 *Vibrio cholerae* hemolysin: purification, partial characterization, and immunological relatedness to El Tor hemolysin. *Infect. Immun.* **45:**192–196.

Yildiz, F. H., N. A. Dolganov, and G. K. Schoolnik. 2001. VpsR, a member of the response regulators of the two-component regulatory systems, is required for expression of vps biosynthesis genes and EPS(ETr)-associated phenotypes in *Vibrio cholerae* O1 El Tor. *J. Bacteriol.* **183:**1716–1726.

Yildiz, F. H., X. S. Liu, A. Heydorn, and G. K. Schoolnik. 2004. Molecular analysis of rugosity in a *Vibrio cholerae* O1 El Tor phase variant. *Mol. Microbiol.* **53:**497–515.

Yildiz, F. H., and G. K. Schoolnik. 1999. *Vibrio cholerae* O1 El Tor: identification of a gene cluster required for the rugose colony type, exopolysaccharide production, chlorine resistance, and biofilm formation. *Proc. Natl. Acad. Sci. USA* **96**:4028–4033.

Yu, R. R., and V. J. DiRita. 1999. Analysis of an autoregulatory loop controlling ToxT, cholera toxin, and toxin-coregulated pilus production in *Vibrio cholerae*. *J. Bacteriol.* **181**:2584–2592.

Yu, R. R., and V. J. DiRita. 2002. Regulation of gene expression in *Vibrio cholerae* by ToxT involves both antirepression and RNA polymerase stimulation. *Mol. Microbiol.* **43**:119–134.

Zhu, J., M. B. Miller, R. E. Vance, M. Dziejman, B. L. Bassler, and J. J. Mekalanos. 2002. Quorum-sensing regulators control virulence gene expression in *Vibrio cholerae*. *Proc. Natl. Acad. Sci. USA* **99**:3129–3134.

Zitzer, A., O. Zitzer, S. Bhakdi, and M. Palmer. 1999. Oligomerization of *Vibrio cholerae* cytolysin yields a pentameric pore and has a dual specificity for cholesterol and sphingolipids in the target membrane. *J. Biol. Chem.* **274**:1375–1380.

The Biology of Vibrios
Edited by F. L. Thompson et al.
© 2006 ASM Press, Washington, D.C.

Chapter 24

Vibrio parahaemolyticus

Tetsuya Iida, Kwan-Sam Park, and Takeshi Honda

Vibrio parahaemolyticus, a gram-negative marine bacterium, causes seafood-borne gastroenteritis in humans (Blake et al., 1980; Janda et al., 1988; Honda and Iida, 1993; McCarter, 1999; Thompson et al., 2004). Although there have been numerous studies on this bacterium, the exact mode of its pathogenic action has not been well elucidated. Recent genome analysis of *V. parahaemolyticus* (Makino et al., 2003) has shed light on unknown aspects of its pathogenic mechanism. In this chapter, we review recent progress in the study of pathogenesis by *V. parahaemolyticus*.

V. PARAHAEMOLYTICUS AND ITS MODE OF INFECTION

In 1950, Fujino and coworkers were the first to isolate *V. parahaemolyticus* as a causative agent of food-borne gastroenteritis, following a large outbreak caused by the consumption of shirasu (a semidried fish product) in Osaka, Japan (Fujino et al., 1953). In this incident, of 272 patients who suffered acute gastroenteritis, 20 died. Since then, this organism has been isolated quite often from outbreaks and sporadic cases of gastroenteritis in many countries (Joseph et al., 1982; Honda and Iida, 1993), and *V. parahaemolyticus* is now known to be a major cause of gastroenteritis throughout the world (Joseph et al., 1982; Janda et al., 1988; Honda and Iida, 1993). Recently, strains of *V. parahaemolyticus* from a few specific serotypes—probably derived from a common clonal ancestor—have caused a pandemic of gastroenteritis (Okuda et al., 1997; Arakawa et al., 1999; Bag et al., 1999; Vuddhakul et al., 2000; Wong et al., 2000, 2005; Chang et al., 2000; Chiou et al., 2000; Chowdhury et al., 2000a,b, 2004a,b; Daniels et al., 2000; Nasu et al., 2000; Iida et al., 2001; Bhuiyan et al., 2002; Khan et al., 2002; Osawa et al., 2002;

Yeung et al., 2002; DePaola et al., 2003; Hara-Kudo et al., 2003; Laohaprertthisan et al., 2003; Myers et al., 2003; Okura et al., 2003, 2004; Islam et al., 2004; Williams et al., 2004; Gonzalez-Escalona et al., 2005). Such a widespread occurrence of a few serovars of *V. parahaemolyticus* isolates had not previously been reported, and the isolates have been designated "pandemic clones" (Matsumoto et al., 2000).

The main symptoms in *V. parahaemolyticus* infection are diarrhea and abdominal pain. Patients may also show fever, vomiting, nausea, and general fatigue. The diarrhea is watery, mucoid, bloody, or mucoid and bloody. *V. parahaemolyticus* infection is usually self-limiting, and clinical symptoms may last for 2 to 10 days. A few individuals may show dehydration, collapse, and cyanosis. Possible cardiovascular disturbances, such as abnormality on electrocardiograms, are also reported. The major source of *V. parahaemolyticus* infection is contaminated seafood or seafood products (Joseph et al., 1982; Honda and Iida, 1993).

THERMOSTABLE DIRECT HEMOLYSIN (TDH) AND TDH-RELATED HEMOLYSIN (TRH)

In 1968, Wagatsuma devised a special blood agar medium for detecting the hemolytic characteristics of *V. parahaemolyticus*, and strains of this organism were classified into two types—hemolytic and non-hemolytic—on Wagatsuma agar (Joseph et al., 1982; Miyamoto et al., 1969). Strains isolated from diarrheal feces of patients were usually hemolytic, whereas those isolated from food samples were mostly non-hemolytic. In studies on a large number of isolates, Sakazaki et al. (1968) reported that the ability of this organism to cause hemolysis on Wagatsuma agar—termed the Kanagawa phenomenon (KP)—was closely associated with gastrointestinal illness.

T. Iida, K.-S. Park, and T. Honda • Department of Bacterial Infections, Research Institute for Microbial Diseases, Osaka University, 3-1 Yamadaoka, Suita, Osaka 565-0871, Japan.

Many studies have been conducted on the factor(s) responsible for the KP, and this led to the discovery of thermostable direct hemolysin (TDH) (Honda and Iida, 1993). By the late 1970s, TDH had been purified, showing that it is a protein with a homodimer structure. Purified TDH shows various biological activities (Honda et al., 1992; Honda and Iida, 1993; Huntley et al., 1993; Tang et al., 1994, 1997a,b; Iida et al., 1995; Yoh et al., 1995, 1996; Fabbri et al., 1999; Raimondi et al., 2000; Naim et al., 2001a,b; Lang et al., 2004; Hardy et al., 2004). For example, TDH lyses red blood cells from various animal sources. The toxin is also cytotoxic to various cultured cells (Honda and Iida, 1993; Tang et al., 1997b). In a rabbit ileal loop test, challenge with purified TDH at 100 to 250 µg/loop induced fluid accumulation (Miyamoto et al., 1980; Honda and Iida, 1993), which suggests that TDH is enterotoxic. In vitro experiments also provided evidence that TDH is able to cause diarrhea (Raimondi et al., 1995; Takahashi et al., 2000, 2001). In 1988, Honda et al. found that the clinical isolates of KP-negative *V. parahaemolyticus* originating in travelers from the Maldives produced a newly identified toxin, and named it TRH (for TDH-related hemolysin [Honda et al., 1988]). The amino acid sequences of TRH are ~67% homologous with TDH, and TRH shows similar biological activities (Honda and Iida, 1993). The genes that encode TDH and TRH have been cloned, and their distribution among clinical and environmental *V. parahaemolyticus* isolates has been examined (Honda and Iida, 1993; Suthienkul et al., 1995; Nishibuchi and Kaper, 1995; Osawa et al., 1996; DePaola et al., 2000). In general, most clinical *V. parahaemolyticus* strains possess the *tdh* gene, whereas few environmental strains do so. Although *trh*-positive strains are occasionally isolated from the environment (such as seawater and seafood), they are also almost exclusively pathogenic strains. Most clinical isolates in the Pacific Northwest of the United States seem to possess both *tdh* and *trh* (Kaufman et al., 2002; DePaola et al., 2003). Mutant strains of *V. parahaemolyticus* with disruptions in *tdh* or *trh* have been constructed (Nishibuchi et al., 1992; Xu et al., 1994; Park et al., 2004a). Those mutant strains present with less enterotoxicity than the wild-type strains. Thus, TDH and TRH have been considered the major virulence factors of *V. parahaemolyticus*, and studies on the pathogenic effects of the organism have mainly focused on these hemolysins in the past 2 decades.

GENOME ANALYSIS OF *V. PARAHAEMOLYTICUS*

We recently reported the whole genome sequence of the clinical *V. parahaemolyticus* strain RIMD2210633 (Makino et al., 2003). RIMD2210633 is a KP-positive strain belonging to the pandemic clones. Sequencing confirmed that the genome is composed of two circular chromosomes (see chapter 5) (Yamaichi et al., 1999; Tagomori et al., 2002; Okada et al., 2005). The larger (chromosome 1) is 3,288,558 bp in size, and the smaller (chromosome 2) is 1,877,212 bp. Most of the essential genes required for growth and viability, including genes for all the ribosomal proteins and at least one copy of all the tRNA genes, are on chromosome 1. Several genes involved in essential metabolic pathways are located only on chromosome 2, which suggests that this chromosome is also essential for growth and viability. Chromosome 2 seems to contain more genes related to transcriptional regulation and transport of various substrates than chromosome 1. Genes classified in these categories have roles in response to environmental changes. When we compared the *V. parahaemolyticus* genome with that of *V. cholerae* (Heidelberg et al., 2000), we found that, although chromosome 1 does not differ much in size between the two genomes (3.3 vs. 3.0 Mb), chromosome 2 is much larger in *V. parahaemolyticus* than in *V. cholerae* (1.9 vs. 1.1 Mb). This size difference could have arisen through acquisition of more genes or more DNA by gene duplication or horizontal transfer in *V. parahaemolyticus* during evolution, or through more frequent gene decay or deletion in *V. cholerae*. There are a few mobile genetic elements, such as transposons, phages, and DNA regions with a G+C content different from the average of the whole genome (which is indicative of recent lateral transfer) in *V. parahaemolyticus* chromosome 2. This feature suggests that recent horizontal transfer of foreign DNA was not the major factor involved in the difference in size between the *V. parahaemolyticus* and *V. cholerae* chromosomes. Comparison of the relative position of the conserved genes between the two vibrios suggested that extensive genome rearrangements occurred within and between the large and small chromosomes during evolution (Makino et al., 2003).

The most notable finding in relation to the pathogenicity of *V. parahaemolyticus* was the presence of two sets of genes for the type III secretion system (TTSS) in the genome (Fig. 1) (Makino et al., 2003). This was the first report of the presence of the TTSS genes in the genome of bacteria belonging to the *Vibrio* species. TTSS is an apparatus of gram-negative pathogenic bacteria used to secrete and translocate virulence factor proteins into the cytosol of target eukaryotic cells (Color Plate 17) (Hueck, 1998). Bacterial pathogens that cause disease by intimate interactions with eukaryotic cells, such as *Salmonella*, *Shigella*, *Yersinia*, and plant pathogens, have this secretion sys-

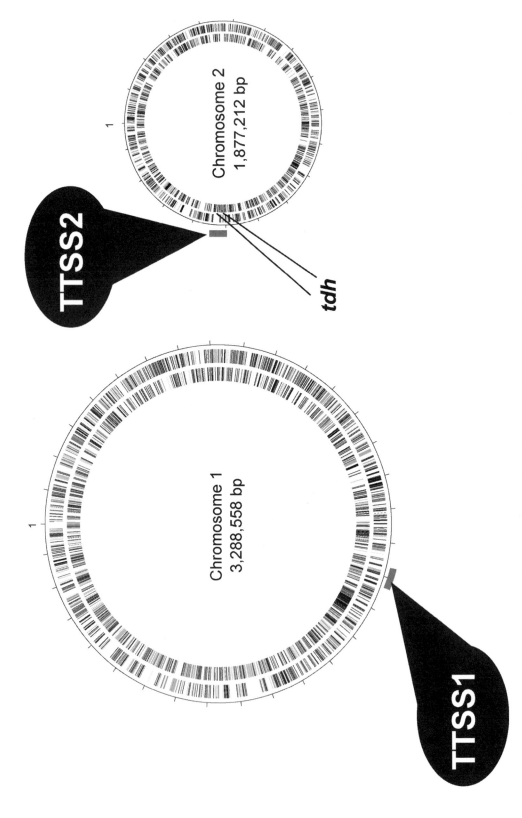

Figure 1. Location of the TTSS genes in the genome of *V. parahaemolyticus* strain RIMD2210663. Two sets of the genes for the TTSS were found by genome sequencing (Makino et al., 2003). One set (TTSS1) was located on chromosome 1, and the other (TTSS2) was identified within the PAI region on chromosome 2.

tem. In the genome of strain RIMD2210633, a set of genes for the TTSS (TTSS1) was located on chromosome 1 at a position 1.77 to 1.81 Mb from the origin of replication. In this region, nearly 30 open reading frames were identified as homologous to the TTSS-related genes of other gram-negative bacteria (Table 1) (Makino et al., 2003). The organization of the TTSS1 genes resembled those of *Yersinia* spp., but there were some differences (Park et al., 2004b). First, 12 hypothetical genes existed in *V. parahaemolyticus* TTSS1. These genes were not found in *Yersinia* spp. or other species. Second, 13 genes were encoded in the direction opposite to those of *Yersinia* species. The average G+C content in the TTSS1 region was similar to that of the rest of the genome.

Another set of genes for the TTSS was identified within the pathogenicity island (PAI) region of RIMD2210633 (Makino et al., 2003). The PAI (Groisman and Ochman, 1996) sits at a position 1.38 to 1.47 Mb from the origin of replication on chromosome 2. The G+C content of the region is 39.8%, obviously lower than that of the rest of the genome (45.4%), and suggesting that this region was obtained by recent horizontal transfer. In this region, two copies of genes for TDH are present. Within the PAI,

we could identify at least nine genes that are homologous to the TTSS apparatus genes of other bacteria (Table 1). The TTSS gene cluster on chromosome 2 was designated TTSS2. The nine TTSS apparatus genes were transcribed in the same direction. Within the TTSS2 region, there are several virulence and TTSS effector gene candidates, such as homologues of the *Yersinia* YopJ, *Escherichia coli* cytotoxic necrotizing factor, *Pseudomonas* exoenzyme T, and *Shigella* OspB. Examination of various *V. parahaemolyticus* strains from clinical and environmental sources revealed that TTSS2 was present only in the KP-positive strains, while TTSS1 was found in all the *V. parahaemolyticus* strains tested.

NEWLY IDENTIFIED TYPE III SECRETION SYSTEMS

Involvement in Bacterial Phenotypes in Relation to the Biological Activity

To determine whether the two sets of the TTSS genes identified in the genome of *V. parahaemolyticus* are functional, we constructed a series of mutant

Table 1. Comparison of the TTSS genes in *V. parahaemolyticus* and other bacteria

Yersinia spp.	*P. aeruginosa*	*Shigella flexneri*	*Salmonella enterica* serovar Typhimurium SPI-I	*Salmonella enterica* serovar Typhimurium SPI-2	EPEC	*V. parahaemolyticus* TTSS1	*V. parahaemolyticus* TTSS2
yscA	exsD	—[a]	—	—	—	vscA1	—
yscB	pscB	—	—	—	—	vscB2	—
yscC	pscC	mxiD	invG	spiA	escC	vscC1	vscC2
yscD	pscD	—	—	spiB	escD	vscD1	—
yscE	pscE	—	—	—	—	—	—
yscF	pscF	mxiH	prgI	ssaH	escF	vscF1	—
yscG	pscG	—	—	—	—	vscG1	—
yscH	pscH	—	—	—	—	vscH1	—
yscI	pscI	—	—	—	—	vscI1	—
yscJ	pscJ	mxiJ	prgK	ssaJ	escJ	vscJ1	vscJ2
yscK	pscK	—	—	—	—	vscK1	—
yscL	pscL	—	—	ssaK	—	vscL1	—
yscN	pscN	spaL/47	spaL/invC	ssaN	escN	vscN1	vscN2
yscO	pscO	spa13	spaM/invI	ssaO	—	vscO1	—
yscP	—	spa32	spaN/invJ	—	—	vscP1	—
yscQ	—	spaO/33	spaO/invK	ssaQ	—	vscQ1	vscQ2
yscR	—	spaP/24	spaP/invL	ssaR	escR	vscR1	vscR2
yscS	—	spaQ/9	spaQ/invM	ssaS	escS	vscS1	vscS2
yscT	—	spaR/29	spaR/invN	ssaT	escT	vscT1	vscT2
yscU	—	spaS/40	spaS	ssaU	escU	vscU1	vscU2
yscX	pcr3	—	—	—	—	vscX1	—
yscY	pcr4	—	—	—	—	vscY1	—
lcrD	pcrD	mxiA	invA	ssaV	escV	vcrD1	vcrD2
virG	exsB	—	—	—	—	virG1	—
yopN	popN	—	invE	—	—	vopN1	—

[a]—, there is no homologous gene in the organism.

strains from strain POR-1 (Park et al., 2004b). POR-1 was derived from RIMD2210633, having both the two *tdh* genes (*tdhA* and *tdhS*) deleted (Park et al., 2004a). We used POR-1 as the parent strain for further mutant construction to exclude the influence of TDH production on bacterial phenotypes. To examine the role of TTSS1 and TTSS2, we disrupted specific genes using in-frame deletions by homologous recombination with the positive selection suicide vector pYAK1 (Kodama et al., 2002). The genes thus disrupted were *vcrD1* and *vcrD2*, each encoding an inner membrane protein; *vscC1* and *vscC2*, each encoding an outer membrane protein; and *vscN1* and *vscN2*, each encoding a cytoplasmic protein, of the TTSS apparatus. Deletion of the target genes was confirmed by PCR and Southern hybridization analysis. The growth curves of the mutant strains were indistinguishable from those of the parent strain under in vitro culture conditions, such as in Luria Bertani medium.

V. parahaemolyticus can kill various eukaryotic cells in a short amount of time (~5 h after challenge). Therefore, we first examined whether disruption of the genes for TTSS1 or TTSS2 affects its cytotoxic effects (Park et al., 2004b). Cytotoxicity of the constructed mutant strains was analyzed 5 h after infection. That of the *vcrD2* deletion mutant was identical to that of the parent strain, whereas the *vcrD1* deletion mutant had little effect on the HeLa cells. The decrease in cytotoxicity was fully restored by an in *trans* complementation of *vcrD1*. Cells challenged by the parent or *vcrD2* mutant strains displayed rounding-up, with shrunken cytoplasm and nuclei, as illustrated by intense areas of Hoechst DNA staining. In contrast, the *vcrD1* deletion mutant caused little change in HeLa cell morphology compared with uninfected cells. The cytotoxic effects of other *vscC1* or *vscN1* mutant strains were similar to those of *vcrD1*, whereas *vscC2* or *vscN2* mutant strains were similar to the *vcrD2* mutant strain. These data suggest that TTSS1, but not TTSS2, is involved in the cytotoxic activity of the parent strain.

Another important phenotype of *V. parahaemolyticus*, in relation to the pathogenicity of the organism, is enterotoxicity (or diarrheagenicity [Honda and Iida, 1993]). To investigate the contribution of TTSS1 and TTSS2 to the enterotoxicity of *V. parahaemolyticus*, we compared the fluid-accumulating activity of the parent and mutant strains in a rabbit ileal loop test (Twedt et al., 1980). Although the amount of fluid accumulation by the *vcrD1* mutant strain was indistinguishable from that of the parent strain POR-1, the *vcrD2* deletion mutant showed little effect. Small intestinal pieces dissected from rabbit tissue infected with the parent and *vcrD1* deletion mutant showed a complete loss of villous architecture with

denudation of surface epithelium in places (Park et al., 2004b). There was massive hemorrhage with neutrophilic infiltration in the lamina propria extending to the submucosae, with congested and dilated crypts. In contrast, sections from tissues treated with the *vcrD2* deletion mutant did not manifest significant histological changes and maintained the normal villous contours. The mutant *vscC1* and *vscN1* strains were phenotypically similar to *vcrD1*, while the *vscC2* or *vscN2* mutant strains were phenotypically similar to the *vcrD2* mutant strain. These results strongly suggest that TTSS2 is involved in the enterotoxicity of *V. parahaemolyticus*, whereas TTSS1 is not.

Protein Secretion

Study with the constructed mutants demonstrated that both TTSS1 and TTSS2 are functional in the *V. parahaemolyticus* strain RIMD2210633. This prompted us to analyze whether these two TTSSs really serve to secrete bacterial proteins, and if so, whether they recognize common bacterial proteins or distinct ones for secretion. So far, we have identified several proteins of strain RIMD2210633, which are secreted in a TTSS-dependent manner (Park et al., 2004b; Iida et al., unpublished data). Among them, we analyzed the secretion of VopP (encoded by VPA 1346), a virulence-related protein encoded in the proximity of the TTSS2 gene region with homology to *Yersinia* YopP and VopD (encoded by VP1656), a protein encoded in the TTSS1 gene region with homology to *Yersinia* YopD (Park et al., 2004b). VopD was detected in the culture supernatants of the parent and TTSS2-related mutant strains such as *vcrD2*, *vscC2*, and *vscN2*. However, VopD was not detected in the culture supernatants of the TTSS1-related mutant strains such as *vcrD1*, *vscC1*, and *vscN1*. The VopP protein was detected in the culture supernatants of the TTSS1-related mutant strains, but they were not detected in the TTSS2-related mutants. These data strongly suggested that VopD is secreted in a TTSS1-dependent manner, while VopP is secreted via TTSS2. In other words, TTSS1 and TTSS2 recognize distinct proteins for secretion.

Regulation of TTSS Gene Expression

Although the genes for the two TTSSs of *V. parahaemolyticus* seem to be expressed in distinct conditions (Park et al., 2004b; T. Iida et al., unpublished data), very little is known about the regulation of the genes for the two TTSSs. Recently, Henke and Bassler (2004) reported that the TTSS1 of *V. parahaemolyticus* is negatively regulated by quorum sensing. This contrasts with other pathogenic bacteria, such as enterohemorrhagic and enteropathogenic *E. coli*, the TTSSs of which are positively regulated by quorum sensing (Sperandio et al., 1999, 2003). The fact that *V. parahaemolyticus* secretes putative effectors via TTSS1 at low cell density when they are presumably not in a community, and terminates secretion when they are at high cell density, could be important for understanding the function of TTSS1 in natural environments and its role in pathogenesis in humans, if any.

Role in Bacterial Pathogenicity

Our studies demonstrate that both TTSS1 and TTSS2 are functional in *V. parahaemolyticus* strain RIMD2210633. TTSS1 is involved in the cytotoxic effects of the organism in HeLa cells, whereas TTSS2 has a role in enterotoxicity in a rabbit model. Considering these data, we interpret the contribution of the two TTSSs to the pathogenicity of KP-positive *V. parahaemolyticus* as follows. These data raise the possibility that the two TTSSs may be involved in the pathogenic effects of the organism in humans. It is noteworthy that TTSS2 was discovered exclusively in human-pathogenic (KP-positive) *V. parahaemolyticus* strains, but not in environmental strains (Makino et al., 2003), further emphasizing the relevance of TTSS2 to pathogenicity in humans. In particular, the coincidence of the TTSS2 genes with the genes for TDH, a well-known virulence marker of pathogenic *V. parahaemolyticus* (Honda and Iida, 1993), among various *V. parahaemolyticus* strains (Makino et al., 2003) reinforces the relevance of TTSS2 for pathogenicity. On the other hand, TTSS1 is present in all the *V. parahaemolyticus* strains, irrespective of clinical or environmental isolates. In other words, even *V. parahaemolyticus* strains that are nonpathogenic to humans possess TTSS1. This raises the question of whether TTSS1 is actually involved in pathogenicity in humans. We do not have an answer to this question at present, but it is possible that TTSS1 alone is not sufficient to cause disease in humans and that it acts synergistically in pathogenesis when TTSS2 is present. This issue clearly requires further study.

The *V. cholerae* genome (Heidelberg et al., 2000) does not include TTSS; thus, *V. parahaemolyticus* seems to have a mechanism of infection distinct from that of *V. cholerae*. TTSS is a pivotal virulence factor of *Shigella* and *Salmonella* bacteria, both of which cause inflammatory diarrhea in humans, whereas *V. cholerae* causes secretory diarrhea by producing cholera toxin (Choi et al., 1996). Unlike *V. cholerae*, *V. parahaemolyticus* can cause gastroenteritis associated with inflammatory diarrhea (Daniels et al., 2000; Blake et al., 1980). These symptoms can be explained by the presence of TTSS in the genome of *V. parahaemolyticus*.

TTSS in Other Vibrios

TTSS is present not only in *V. parahaemolyticus*, but also in some other vibrios. Henke and Bassler reported a functional TTSS in *Vibrio harveyi*, a marine pathogen (Henke and Bassler, 2004). We analyzed the distribution of the TTSS genes in *Vibrio* species and found homologues to be present in *V. harveyi*, *V. alginolyticus*, and *V. tubiashii* (Park et al., 2004b). In all cases, the genes are homologues of the TTSS1 genes but not of the TTSS2 genes. These vibrios are known to live in marine or estuarine environments, often associating with phytoplankton, zooplankton, or animals, including fish and shellfish. The TTSS is a molecular apparatus that can inject bacterial proteins into eukaryotic cells. Therefore, it would be interesting to know the natural targets for protein injection by the TTSS of each vibrio. This would give a clue to a better understanding of the life cycle of these *Vibrio* species in natural environments.

NEW PARADIGM IN THE PATHOGENICITY OF *V. PARAHAEMOLYTICUS*

Genome sequencing of *V. parahaemolyticus* and subsequent studies have shed light on hitherto unknown aspects of the pathogenicity of this organism. Both of the two sets of TTSS in KP-positive *V. parahaemolyticus* are functional and are involved in distinct phenotypes. Identifying their effector proteins responsible for those phenotypes and, more precisely, analyzing the conditions for expression of and secretion by the two TTSSs should lead to better understanding of the role of the TTSSs in *V. parahaemolyticus*. Studies on the pathogenic mechanism of *V. parahaemolyticus* have so far mainly focused on TDH and TRH. The discovery of the genes for TTSSs in the *V. parahaemolyticus* genome is opening a door to a new paradigm on the pathogenicity of the organism. Further analysis of these newly identified TTSSs could lead to the development of novel preventive and therapeutic means to fight infection by this pathogen.

REFERENCES

Arakawa, E., T. Murase, T. Shimada, T. Okitsu, S. Yamai, and H. Watanabe. 1999. Emergence and prevalence of a novel *Vibrio parahaemolyticus* O3:K6 clone in Japan. *Jpn. J. Infect. Dis.* 52:246–247.

Bag, P. K., S. Nandi, R. K. Bhadra, T. Ramamurthy, S. K. Bhattacharya, M. Nishibuchi, T. Hamabata, S. Yamasaki, Y. Takeda, and G. B. Nair. 1999. Clonal diversity among recently emerged strains of *Vibrio parahaemolyticus* O3:K6 associated with pandemic spread. *J. Clin. Microbiol.* 37:2354–2357.

Bhuiyan, N. A., M. Ansaruzzaman, M. Kamruzzaman, K. Alam, N. R. Chowdhury, M. Nishibuchi, S. M. Faruque, D. A. Sack,

Y. Takeda, and G. B. Nair. 2002. Prevalence of the pandemic genotype of *Vibrio parahaemolyticus* in Dhaka, Bangladesh, and significance of its distribution across different serotypes. *J. Clin. Microbiol.* 40:284–286.

Blake, P. A., R. E. Weaver, and D. G. Hollis. 1980. Diseases of humans (other than cholera) caused by vibrios. *Annu. Rev. Microbiol.* 34:341–367.

Chang, B., S. Yoshida, H. Miyamoto, M. Ogawa, K. Horikawa, K. Ogata, M. Nishibuchi, and H. Taniguchi. 2000. A unique and common restriction fragment pattern of the nucleotide sequences homologous to the genome of vf33, a filamentous bacteriophage, in pandemic strains of *Vibrio parahaemolyticus* O3:K6 O4:K68, and O1:K untypeable. *FEMS Microbiol. Lett.* 192:231–236.

Chiou, C. S., S. Y. Hsu, S. I. Chiu, T. K. Wang, and C. S. Chao. 2000. *Vibrio parahaemolyticus* serovar O3:K6 as cause of unusually high incidence of food-borne disease outbreaks in Taiwan from 1996 to 1999. *J. Clin. Microbiol.* 38:4621–4625.

Choi, S. W., C. H. Park, T. M. Silva, E. I. Zaenker, and R. L. Guerrant. 1996. To culture or not to culture: fecal lactoferrin screening for inflammatory bacterial diarrhea. *J. Clin. Microbiol.* 34:928–932.

Chowdhury, N. R., S. Chakraborty, B. Eampokalap, W. Chaicumpa, M. Chongsa-Nguan, P. Moolasart, R. Mitra, T. Ramamurthy, S. K. Bhattacharya, M. Nishibuchi, Y. Takeda, and G. B. Nair. 2000a. Clonal dissemination of *Vibrio parahaemolyticus* displaying similar DNA fingerprint but belonging to two different serovars (O3:K6 and O4:K68) in Thailand and India. *Epidemiol. Infect.* 125:17–25.

Chowdhury, N. R., S. Chakraborty, T. Ramamurthy, M. Nishibuchi, S. Yamasaki, Y. Takeda, and G. B. Nair. 2000b. Molecular evidence of clonal *Vibrio parahaemolyticus* pandemic strains. *Emerg. Infect. Dis.* 6:631–636.

Chowdhury, A., M. Ishibashi, V. D. Thiem, D. T. Tuyet, T. V. Tung, B. T. Chien, L. von Seidlein, D. G. Canh, J. Clemens, D. D. Trach, and M. Nishibuchi. 2004a. Emergence and serovar transition of *Vibrio parahaemolyticus* pandemic strains isolated during a diarrhea outbreak in Vietnam between 1997 and 1999. *Microbiol. Immunol.* 48:319–327.

Chowdhury, N. R., O. C. Stine, J. G. Morris, and G. B. Nair. 2004b. Assessment of evolution of pandemic *Vibrio parahaemolyticus* by multilocus sequence typing. *J. Clin. Microbiol.* 42:1280–1282.

Daniels, N. A., L. MacKinnon, R. Bishop, S. Altekruse, B. Ray, R. M. Hammond, S. Thompson, S. Wilson, N. H. Bean, P. M. Griffin, and L. Slutsker. 2000. *Vibrio parahaemolyticus* infections in the United States, 1973–1998. *J. Infect. Dis.* 181:1661–1666.

DePaola, A., C. A. Kaysner, J. Bowers, and D. W. Cook. 2000. Environmental investigations of *Vibrio parahaemolyticus* in oysters after outbreaks in Washington, Texas, and New York (1997 and 1998). *Appl. Environ. Microbiol.* 66:4649–4654.

DePaola, A., J. Ulaszek, C. A. Kaysner, B. J. Tenge, J. L. Nordstrom, J. Wells, N. Puhr, and S. M. Gendel. 2003. Molecular, serological, and virulence characteristics of *Vibrio parahaemolyticus* isolated from environmental, food, and clinical sources in North America and Asia. *Appl. Environ. Microbiol.* 69:3999–4005.

Fabbri, A., L. Falzano, C. Frank, G. Donelli, P. Matarrese, F. Raimondi, A. Fasano, and C. Fiorentini. 1999. *Vibrio parahaemolyticus* thermostable direct hemolysin modulates cytoskeletal organization and calcium homeostasis in intestinal cultured cells. *Infect. Immun.* 67:1139–1148.

Fujino, T., Y. Okuno, D. Nakada, A. Aoyama, K. Fukai, T. Mukai, and T. Ueho. 1953. On the bacteriological examination of shirasu-food poisoning. *Med. J. Osaka Univ.* 4:299–304.

Gonzalez-Escalona, N., V. Cachicas, C. Acevedo, M. L. Rioseco, J. A. Vergara, F. Cabello, J. Romero, and R. T. Espejo. 2005. *Vibrio parahaemolyticus* diarrhea, Chile, 1998 and 2004. *Emerg. Infect. Dis.* 11:129–131.

Groisman, E. A., and H. Ochman. 1996. Pathogenicity islands: bacterial evolution in quantum leaps. *Cell* 87:791–794.

Hara-Kudo, Y., K. Sugiyama, M. Nishibuchi, A. Chowdhury, J. Yatsuyanagi, Y. Ohtomo, A. Saito, H. Nagano, T. Nishina, H. Nakagawa, H. Konuma, M. Miyahara, and S. Kumagai. 2003. Prevalence of pandemic thermostable direct hemolysin-producing *Vibrio parahaemolyticus* O3:K6 in seafood and the coastal environment in Japan. *Appl. Environ. Microbiol.* 69:3883–3891.

Hardy, S. P., M. Nakano, and T. Iida. 2004. Single channel evidence for innate pore-formation by *Vibrio parahaemolyticus* thermostable direct haemolysin (TDH) in phospholipid bilayers. *FEMS Microbiol. Lett.* 240:81–85.

Heidelberg, J. F., J. A. Eisen, W. C. Nelson, R. A. Clayton, M. L. Gwinn, R. J. Dodson, D. H. Haft, E. K. Hickey, J. D. Peterson, L. Umayam, S. R. Gill, K. E. Nelson, T. D. Read, H. Tettelin, D. Richardson, M. D. Ermolaeva, J. Vamathevan, S. Bass, H. Qin, I. Dragoi, P. Sellers, L. McDonald, T. Utterback, R. D. Fleishmann, W. C. Nierman, O. White, S. L. Salzberg, H. O. Smith, R. R. Colwell, J. J. Mekalanos, J. C. Venter, and C. M. Fraser. 2000. DNA sequence of both chromosomes of the cholera pathogen *Vibrio cholerae*. *Nature* 406:477–483.

Henke, J. M., and B. L. Bassler. 2004. Quorum sensing regulates type III secretion in *Vibrio harveyi* and *Vibrio parahaemolyticus*. *J. Bacteriol.* 186:3794–3805.

Honda, T., and T. Iida. 1993. The pathogenicity of *Vibrio parahaemolyticus* and the role of the thermostable direct haemolysin and related haemolysins. *Rev. Med. Microbiol.* 4:106–113.

Honda, T., Y. Ni, and T. Miwatani. 1988. Purification and characterization of a hemolysin produced by a clinical isolate of Kanagawa phenomenon-negative *Vibrio parahaemolyticus* and related to the thermostable direct hemolysin. *Infect. Immun.* 56:961–965.

Honda, T., Y. Ni, T. Miwatani, T. Adachi, and J. Kim. 1992. The thermostable direct hemolysin of *Vibrio parahaemolyticus* is a pore-forming toxin. *Can. J. Microbiol.* 38:1175–1180.

Hueck, C. J. 1998. Type III protein secretion systems in bacterial pathogens of animals and plants. *Microbiol. Mol. Biol. Rev.* 62:379–433.

Huntley, J. S., A. C. Hall, V. Sathyamoorthy, and R. H. Hall. 1993. Cation flux studies of the lesion induced in human erythrocyte membranes by the thermostable direct hemolysin of *Vibrio parahaemolyticus*. *Infect. Immun.* 61:4326–4332.

Iida, T., G.-Q. Tang, S. Suttikulpitug, K. Yamamoto, T. Miwatani, and T. Honda. 1995. Isolation of mutant toxins of *Vibrio parahaemolyticus* hemolysin by *in vitro* mutagenesis. *Toxicon* 33:209–216.

Iida, T., A. Hattori, K. Tagomori, H. Nasu, R. Naim, and T. Honda. 2001. Filamentous phage associated with recent pandemic strains of *Vibrio parahaemolyticus*. *Emerg. Infect. Dis.* 7:477–478.

Islam, M. S., R. Tasmin, S. I. Khan, H. B. Bakht, Z. H. Mahmood, M. Z. Rahman, N. A. Bhuiyan, M. Nishibuchi, G. B. Nair, R. B. Sack, A. Huq, R. R. Colwell, and D. A. Sack. 2004. Pandemic strains of O3:K6 *Vibrio parahaemolyticus* in the aquatic environment of Bangladesh. *Can. J. Microbiol.* 50:827–834.

Janda, J. M., C. Powers, R. G. Bryant, and S. L. Abbott. 1988. Current perspectives on the epidemiology and pathogenesis of clinically significant *Vibrio* spp. *Clin. Microbiol. Rev.* 1:245–267.

Joseph, S. W., R. R. Colwell, and J. B. Kaper. 1982. *Vibrio parahaemolyticus* and related halophilic vibrios. *Crit. Rev. Microbiol.* 10:77–124.

Kaufman, G. E., M. L. Myers, C. L. Pass, A. K. Bej, and C. A. Kaysner. 2002. Molecular analysis of *Vibrio parahaemolyticus* isolated from human patients and shellfish during US Pacific north-west outbreaks. *Lett. Appl. Microbiol.* **34**:155–161.

Khan, A. A., S. McCarthy, R. F. Wang, and C. E. Cerniglia. 2002. Characterization of United States outbreak isolates of *Vibrio parahaemolyticus* using enterobacterial repetitive intergenic consensus (ERIC) PCR and development of a rapid PCR method for detection of O3:K6 isolates. *FEMS Microbiol. Lett.* **206**: 209–214.

Kodama, T., Y. Akeda, G. Kono, A. Takahashi, K. Imura, T. Iida, and T. Honda. 2002. The EspB protein of enterohaemorrhagic *Escherichia coli* interacts directly with α-catenin. *Cell. Microbiol.* **4**:213–222.

Lang, P. A., S. Kaiser, S. Myssina, C. Birka, C. Weinstock, H. Northoff, T. Wieder, F. Lang, and S. M. Huber. 2004. Effect of *Vibrio parahaemolyticus* haemolysin on human erythrocytes. *Cell. Microbiol.* **6**:391–400.

Laohaprertthisan, V., A. Chowdhury, U. Kongmuang, S. Kalnauwakul, M. Ishibashi, C. Matsumoto, and M. Nishibuchi. 2003. Prevalence and serodiversity of the pandemic clone among the clinical strains of *Vibrio parahaemolyticus* isolated in southern Thailand. *Epidemiol. Infect.* **130**:395–406.

Makino, K., K. Oshima, K. Kurokawa, K. Yokoyama, T. Uda, K. Tagomori, Y. Iijima, M. Najima, M. Nakano, A. Yamashita, Y. Kubota, S. Kimura, T. Yasunaga, T. Honda, H. Shinagawa, M. Hattori., and T. Iida. 2003. Genome sequence of *Vibrio parahaemolyticus*: a pathogenic mechanism distinct from that of *V. cholerae*. *Lancet* **361**:743–749.

Matsumoto, C., J. Okuda, M. Ishibashi, M. Iwanaga, P. Garg, T. Rammamurthy, H.-C. Wong, A. DePaola, Y.-B. Kim, M. J. Albert, and M. Nishibuchi. 2000. Pandemic spread of an O3:K6 clone of *Vibrio parahaemolyticus* and emergence of related strains evidenced by arbitrarily primed PCR and toxRS sequence analyses. *J. Clin. Microbiol.* **38**:578–585.

McCarter, L. 1999. The multiple identities of *Vibrio parahaemolyticus*. *J. Mol. Microbiol. Biotechnol.* **1**:51–57.

Miyamoto, Y., T. Kato, S. Obara, S. Akiyama, K. Takizawa, and S. Yamai. 1969. In vitro hemolytic characteristic of *Vibrio parahaemolyticus*: its close correlation with human pathogenicity. *J. Bacteriol.* **100**:1147–1149.

Miyamoto, Y., Y. Obara, T. Nikkawa, S. Yamai, T. Kato, Y. Yamada, and M. Ohashi. 1980. Simplified purification and biophysicochemical characteristics of Kanagawa phenomenon-associated hemolysin of *Vibrio parahaemolyticus*. *Infect. Immun.* **28**:567–576.

Myers, M. L., G. Panicker, and A. K. Bej. 2003. PCR detection of a newly emerged pandemic *Vibrio parahaemolyticus* O3:K6 pathogen in pure cultures and seeded waters from the Gulf of Mexico. *Appl. Environ. Microbiol.* **69**:2194–2200.

Naim, R., T. Iida, A. Takahashi, and T. Honda. 2001a. Monodansylcadaverine inhibits cytotoxicity of *Vibrio parahaemolyticus* thermostable direct hemolysin on cultured rat embryonic fibroblast cells. *FEMS Microbiol. Lett.* **196**:99–105.

Naim, R., I. Yanagihara, T. Iida, and T. Honda. 2001b. *Vibrio parahaemolyticus* thermostable direct hemolysin can induce an apoptotic cell death in Rat-1 cells from inside and outside of the cells. *FEMS Microbiol. Lett.* **195**:237–244.

Nasu, H., T. Iida, T. Sugahara, Y. Yamaichi, K.-S. Park, K. Yokoyama, K. Makino, H. Shinagawa, and T. Honda. 2000. A filamentous phage associated with recent pandemic *Vibrio parahaemolyticus* O3:K6 strains. *J. Clin. Microbiol.* **38**:2156–2161.

Nishibuchi, M., A. Fasano, R. G. Russell, and J. B. Kaper. 1992. Enterotoxicity of *Vibrio parahaemolyticus* with and without genes encoding thermostable direct hemolysin. *Infect. Immun.* **60**:3539–3545.

Nishibuchi, M., and J. B. Kaper. 1995. Thermostable direct hemolysin gene of *Vibrio parahaemolyticus*: a virulence gene acquired by a marine bacterium. *Infect. Immun.* **63**:2093–2099.

Okada, K., T. Iida, K. Kita-Tsukamoto, and T. Honda. 2005. Vibrios commonly possess two chromosomes. *J. Bacteriol.* **187**:752–757.

Okuda, J., M. Ishibashi, E. Hayakawa, T. Nishino, Y. Takeda, A. K. Mukhopadhyay, S. Garg, S. K. Bhattacharya, G. B. Nair, and M. Nishibuchi. 1997. Emergence of a unique O3:K6 clone of *Vibrio parahaemolyticus* in Calcutta, India, and isolation of strains from the same clonal group from Southeast Asian travelers arriving in Japan. *J. Clin. Microbiol.* **35**:3150–3155.

Okura, M., R. Osawa, A. Iguchi, E. Arakawa, J. Terajima, and H. Watanabe. 2003. Genotypic analyses of *Vibrio parahaemolyticus* and development of a pandemic group-specific multiplex PCR assay. *J. Clin. Microbiol.* **41**:4676–4682.

Okura, M., R. Osawa, A. Iguchi, M. Takagi, E. Arakawa, J. Terajima, and H. Watanabe. 2004. PCR-based identification of pandemic group *Vibrio parahaemolyticus* with a novel group-specific primer pair. *Microbiol. Immunol.* **48**:787–790.

Osawa, R., T. Okitsu, H. Morozumi, and S. Yamai. 1996. Occurrence of urease-positive *Vibrio parahaemolyticus* in Kanagawa, Japan, with specific reference to presence of thermostable direct hemolysin (TDH) and the TDH-related-hemolysin genes. *Appl. Environ. Microbiol.* **62**:725–727.

Osawa, R., A. Iguchi, E. Arakawa, and H. Watanabe. 2002. Genotyping of pandemic *Vibrio parahaemolyticus* O3:K6 still open to question. *J. Clin. Microbiol.* **40**:2708–2709.

Park, K.-S., T. Ono, M. Rokuda, M.-H. Jang, T. Iida, and T. Honda. 2004a. Cytotoxicity and enterotoxicity of the thermostable direct hemolysin-deletion mutants of *Vibrio parahaemolyticus*. *Microbiol. Immunol.* **48**:313–318.

Park, K.-S., T. Ono, M. Rokuda, M.-H. Jang, K. Okada, T. Iida, and T. Honda. 2004b. Functional characterization of two type III secretion systems of *Vibrio parahaemolyticus*. *Infect. Immun.* **72**:6659–6665.

Raimondi, F., J. P. Y. Kao, J. B. Kaper, S. Guandalini, and A. Fasano. 1995. Calcium-dependent intestinal chloride secretion by *Vibrio parahaemolyticus* thermostable direct hemolysin in a rabbit model. *Gastroenterology* **109**:381–386.

Raimondi, F., J. P. Kao, C. Fiorentini, A. Fabbri, G. Donelli, N. Gasparini, A. Rubino, and A. Fasano. 2000. Enterotoxicity and cytotoxicity of *Vibrio parahaemolyticus* thermostable direct hemolysin in in vitro systems. *Infect. Immun.* **68**:3180–3185.

Sakazaki, T., K. Tamura, T. Kato, Y. Obara, S. Yamai, and K. Hobo. 1968. Studies on the enteropathogenic, facultatively halophilic bacteria, V. *parahaemolyticus*. III. Enteropathogenicity. *Jpn. J. Med. Sci. Biol.* **21**:325–331.

Sperandio, V., J. L. Mellies, W. Nguyen, S. Shin, and J. B. Kaper. 1999. Quorum sensing controls expression of the type III secretion gene transcription and protein secretion in enterohemorrhagic and enteropathogenic *Escherichia coli*. *Proc. Natl. Acad. Sci. USA* **96**:15196–15201.

Sperandio, V., A. G. Torres, B. Jarvis, J. P. Nataro, and J. B. Kaper. 2003. Bacteria-host communication: the language of hormones. *Proc. Natl. Acad. Sci. USA* **96**:1639–1644.

Suthienkul, O., M. Ishibashi, T. Iida, N. Nettip, S. Supavej, B. Eampokalap, M. Makino, and T. Honda. 1995. Urease production correlates with possession of the *trh* gene in *Vibrio parahaemolyticus* strains isolated in Thailand. *J. Infect. Dis.* **172**: 1405–1408.

Tagomori, K., T. Iida, and T. Honda. 2002. Comparison of genome structures of vibrios, bacteria possessing two chromosomes. *J. Bacteriol.* **184**:4351–4358.

Takahashi, A., T. Iida, R. Naim, Y. Nakaya, and T. Honda. 2001. Chloride secretion induced by thermostable direct hemolysin of

Vibrio parahaemolyticus depends on colonic cell maturation. *J. Med. Microbiol.* **50:**870–878.

Takahashi, A., Y. Sato, Y. Shiomi, V. V. Cantarelli, T. Iida, M. Lee, and T. Honda. 2000. Mechanism of chloride secretion induced by thermostable direct hemolysin of *Vibrio parahaemolyticus* in human colonic tissue and a human intestinal epithelial cell line. *J. Med. Microbiol.* **49:**801–810.

Tang, G., T. Iida, K. Yamamoto, and T. Honda. 1997a. Analysis of functional domains of *Vibrio parahaemolyticus* thermostable direct hemolysin using monoclonal antibodies. *FEMS Microbiol. Lett.* **150:**289–296.

Tang, G.-Q., T. Iida, H. Inoue, M. Yutsudo, K. Yamamoto, and T. Honda. 1997b. A mutant cell line resistant to *Vibrio parahaemolyticus* thermostable direct hemolysin (TDH): its potential in identification of putative receptor for TDH. *Biochim. Biophys. Acta* **1360:**277–282.

Tang, G.-Q., T. Iida, K. Yamamoto, and T. Honda. 1994. A mutant toxin of *Vibrio parahaemolyticus* thermostable direct hemolysin which has lost hemolytic activity but retains ability to bind to erythrocytes. *Infect. Immun.* **62:**3299–3304.

Thompson, F. L., T. Iida, and J. Swings. 2004. Biodiversity of vibrios. *Microbiol. Mol. Biol. Rev.* **68:**403–431.

Twedt, R. M., J. T. Peeler, and P. L. Spaulding. 1980. Effective ileal loop dose of Kanagawa-positive *Vibrio parahaemolyticus*. *Appl. Environ. Microbiol.* **40:**1012–1016.

Vuddhakul, V., A. Chowdhury, V. Laohaprertthisan, P. Pungrasamee, N. Patararungrong, P. Thianmontri, M. Ishibashi, C. Matsumoto, and M. Nishibuchi. 2000. Isolation of a pandemic O3:K6 clone of a *Vibrio parahaemolyticus* strain from environmental and clinical sources in Thailand. *Appl. Environ. Microbiol.* **66:**2685–2689.

Williams, T. L., S. M. Musser, J. L. Nordstrom, A. DePaola, and S. R. Monday. 2004. Identification of a protein biomarker unique to the pandemic O3:K6 clone of *Vibrio parahaemolyticus*. *J. Clin. Microbiol.* **42:**1657–1665.

Wong, H. C., S. H. Liu, T. K. Wang, C. L. Lee, C. S. Chiou, D. P. Liu, M. Nishibuchi, and B. K. Lee. 2000. Characteristics of *Vibrio parahaemolyticus* O3:K6 from Asia. *Appl. Environ. Microbiol.* **66:**3981–3986.

Wong, H. C., C. H. Chen, Y. J. Chung, S. H. Liu, T. K. Wang, C. L. Lee, C. S. Chiou, M. Nishibuchi, and B. K. Lee. 2005. Characterization of new O3:K6 strains and phylogenetically related strains of *Vibrio parahaemolyticus* isolated in Taiwan and other countries. *J. Appl. Microbiol.* **98:**572–580.

Xu, M., K. Yamamoto, and T. Honda. 1994. Construction and characterization of an isogenic mutant of *Vibrio parahaemolyticus* having a deletion in the thermostable direct hemolysin-related hemolysin gene (*trh*). *J. Bacteriol.* **176:**4757–4760.

Yamaichi, Y., T. Iida, K.-S. Park, K. Yamamoto, and T. Honda. 1999. Physical and genetic map of the genome of *Vibrio parahaemolyticus*: presence of two chromosomes in *Vibrio* species. *Mol. Microbiol.* **31:**1513–1521.

Yeung, P. S., M. C. Hayes, A. DePaola, C. A. Kaysner, L. Kornstein, and K. J. Boor. 2002. Comparative phenotypic, molecular, and virulence characterization of *Vibrio parahaemolyticus* O3:K6 isolates. *Appl. Environ. Microbiol.* **68:**2901–2909.

Yoh, M., N. Morinaga, M. Noda, and T. Honda. 1995. The binding of *Vibrio parahaemolyticus* ^{125}I-labeled thermostable direct hemolysin to erythrocytes. *Toxicon* **33:**651–657.

Yoh, M., G.-Q. Tang, T. Iida, N. Morinaga, M. Noda, and T. Honda. 1996. Phosphorylation of a 25 kDa protein is induced by thermostable direct hemolysin of *Vibrio parahaemolyticus*. *Int. J. Biochem. Cell. Biol.* **28:**1365–1369.

The Biology of Vibrios
Edited by F. L. Thompson et al.
© 2006 ASM Press, Washington, D.C.

Chapter 25

Vibrio vulnificus†

JAMES D. OLIVER

INTRODUCTION

Vibrio vulnificus constitutes part of the normal microflora of estuarine and coastal waters, occurring in especially high numbers in bivalve mollusks, such as oysters and clams. The organism is also a significant human pathogen, and its presence in bivalves presents a significant public health risk to persons who consume raw or undercooked seafood. Furthermore, the organism produces potentially fatal wound infections in persons who acquire injuries associated with coastal or estuarine waters. In this chapter I discuss the taxonomy, infections, pathogenesis, genetic heterogeneity, distribution in estuarine environments and the environmental parameters that contribute to the ecology of this organism, and methods to eliminate this pathogen from foods.

Of the several human pathogens now realized to occur naturally in seawater, the most significant, in regard to virulence, is *V. vulnificus*. Some 50 food-borne cases serious enough to require hospitalization occur each year in the United States, and it is the number one cause of seafood-borne death in this country. Indeed, in Florida, it is the number one cause of death from any food (Hlady et al., 1993). Overall, *V. vulnificus* is responsible for 95% of all seafood-related deaths in the United States. *V. vulnificus* infection carries the highest case-fatality rate of any food-borne disease (Mead et al., 1999), and fatality rates of 50 to 60% following ingestion of raw or undercooked seafood (primarily oysters) are the norm. In Korea, death rates as high as 79% have been reported (Park et al., 1991). *V. vulnificus* is also noteworthy in its ability to cause infection of preexisting wounds or wounds acquired while a person is involved in a seawater-associated activity. *V. vulnificus* infection, and/or its presence in temperate coastal and estuarine waters, has been reported in Germany, Holland, Belgium, Italy, and Spain (Baffone et al., 2001; Mertens et al., 1979; Torres et al., 2002; Veenstra et al., 1993, 1994), Scandinavia (Bock et al., 1994; Dalsgaard et al., 1996; Høi et al., 1998b; Melhus et al., 1995), South America (Matté et al., 1994), the Far East (Chan et al., 1986; Fukushima and Seki, 2004; Hsueh et al., 2004; Osaka et al., 2004; Parvathi et al., 2004; Venkateswaran et al., 1989; Yano et al., 2004), the South Pacific (Upton and Taylor, 2002), and the United States: northwestern (Kelly and Stoh, 1988), western (Kaysner et al., 1987), northeastern (Jones and Summer-Brason, 1998; O'Neill et al., 1990), eastern (Heidelberg et al., 2002a,b; Pfeffer et al., 2003), and Gulf Coast (Roberts et al., 1982).

TAXONOMY

Three biotypes of *V. vulnificus* are recognized. Whereas each is known to be a human pathogen, biotype 1 is almost exclusively associated with human disease, and this is the biotype of greatest public health concern.

Biotype 1

The Centers for Disease Control and Prevention in the United States first identified *V. vulnificus* as a new species in 1976. Prior to that, it had been referred to as the "lactose-positive" (or "L+") vibrio (Oliver, 1989). *V. vulnificus* is an obligate halophile, requiring a minimum of 0.5% (wt/vol) NaCl for growth. Differential phenotypic traits include its fermentation of lactose (without production of gas) and of both salicin and cellobiose, which, except for *Vibrio cincinnatiensis*, are traits not frequently found in *Vibrio* spp.

James D. Oliver • Department of Biology, University of North Carolina at Charlotte, Charlotte, NC 28223.

† This chapter is dedicated to the memory of Ron Siebeling, a long-time friend and valued colleague in the study of vibrios.

Phenotypic variation is so great for this species, however, that rapid identification systems (e.g., API-20E) are of little value in identifying this species (Oliver, 2005a), and the use of molecular methods for its identification is now routine. One such variation is fermentation of lactose, with ca. 15% of strains being lactose negative (despite its originally being referred to as the "lactose-positive" vibrio), whereas another 15% are sucrose positive. The latter strains present a special problem on thiosulfate citrate bile salts sucrose agar (TCBS), the most commonly employed isolation medium for *Vibrio* species, as sucrose-positive colonies appear much like those of *Vibrio cholerae* and numerous other *Vibrio* spp. (Oliver, 2003). A bioluminescent strain, originally from a fatal wound infection, has also been described (Oliver et al., 1986b).

The phenotypic variation observed in *V. vulnificus* may be the result of a fascinating aspect of this bacterium's genome. Studies conducted by many laboratories, employing a variety of methods, have all reported that no two biotype 1 strains of *V. vulnificus* strains have the identical genotypic sequence. This aspect of the biology of this pathogen is described below under "Genetic Heterogeneity."

Biotype 2

First reported in 1982 (Tison et al., 1982), biotype 2 is associated primarily with infection of cultured eels. Biotype 2 strains can be differentiated from biotype 1 strains by their negative indole and ornithine decarboxylase reactions and lack of mannitol fermentation or growth at 42°C (Biosca et al., 1993; Tison et al., 1982). Because strains of this type are homogeneous in regard to their lipopolysaccharide (LPS), which differs from that of biotype 1 strains, they have also been referred to as serogroup E (Biosca et al., 1996). Studies from Amaro's laboratory indicate, however, that other than these cell envelope differences, few differences exist between biotype 1 and 2 strains (Biosca et al., 1996, 1997).

Biotype 3

In 1999, a third biotype of *V. vulnificus* was reported by Bisharat et al. (1999). All cultures were isolated in Israel from persons who had received puncture wounds following contact with *Tilapia*. These strains could be differentiated from biotype 1 and 2 strains by five phenotypic traits: they are negative for citrate and o-nitrophenyl-β-D-galactopyranoside and are unable to ferment salicin, cellobiose, or lactose. Cameron et al. (D. N. Cameron, D. Wykstra, N. Bisharat, V. Agmon, and the Israel *Vibrio* Working Group, Abstr. Int. Conf. Emerg. Infect. Dis., 1998)

reported that all of the isolates they studied (from 50 cases of severe wound infections) were positive for the *V. vulnificus*-specific hemolysin-cytolysin gene (*vvh*), as occurs in biotype 1 strains. Unlike biotype 1 strains, however, restriction endonuclease digestion of the hemolysin gene with several enzymes yielded identical restriction fragment length polymorphism patterns.

HUMAN DISEASES CAUSED BY BIOTYPE 1 *V. VULNIFICUS*

Biotype 1 strains of *V. vulnificus* are recognized as causing three quite different human infections, as follows.

Gastroenteritis

The least significant of the three human syndromes produced by *V. vulnificus* is a relatively mild (and thus rarely reported) gastroenteritis. The first study of this syndrome was published by Johnston et al. (1986), who reported diarrhea and abdominal cramps to be the primary complaints. While symptoms subsided without antibiotic treatment, prolonged diarrhea (a month in one case) was notable. *V. vulnificus* gastroenteritis appears to be associated with consumption of raw oysters (Desenclos et al., 1991; Levine et al., 1993; Lowry et al., 1989), and a history of alcohol abuse and/or the routine and continued use of antacids may also be a factor in disease development (Johnston et al., 1986). No fatalities from *V. vulnificus* gastroenteritis have been reported.

Primary Septicemia

The most significant form of *V. vulnificus* disease is a primary septicemia, which almost always follows ingestion of raw or undercooked oysters (Hlady, 1997). Among the 113 cases that occurred in the United States between 2000 and 2003, 96% involved raw oyster consumption. The remaining 4% (5 cases) involved steamed, as opposed to raw, oysters. Similar findings were reported by Shapiro et al. (1998) in a study involving over 400 infections; 96% of those infected developed primary septicemia following oyster consumption. Cases following ingestion of raw or undercooked clams also occur occasionally, and one case of septicemia was reported after consumption of raw crab.

The incidence of oyster-associated infection with *V. vulnificus* exhibits a distinct seasonality, with the great majority of cases occurring during the warm water months of May through October (Fig. 1). Infections

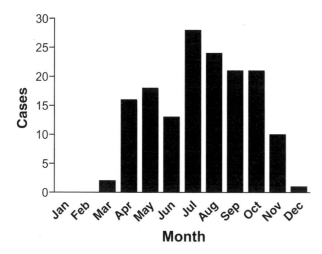

Figure 1. Seasonal distribution of *V. vulnificus* infection (primary septicemia) following ingestion of raw or undercooked oysters in the United States between 2000 and 2004 (J. D. Oliver, unpublished data).

parallel quite closely the numbers of *V. vulnificus* cells present in seawater and shellfish during these months (see "Distribution in the Marine Environment").

Development of *V. vulnificus* disease following consumption of raw molluscan shellfish occurs almost exclusively in persons with certain underlying and chronic diseases; indeed, it is quite rare to encounter primary septicemia with *V. vulnificus* in persons without such disorders (Hlady, 1997). Of the infections occurring in the United States between 2000 and 2003, 94% were reported to have one or more underlying diseases. The most common of these (ca. ≥80% of all cases) are alcoholism or alcohol abuse, or infections such as hepatitis, which generally lead to liver damage, including cirrhosis. It is believed that the actual contributing factor in such cases is the resultant elevation of serum iron, which has been shown experimentally to correlate highly with the ability of *V. vulnificus* to grow in human serum (Morris et al., 1987b; Simpson and Oliver, 1987; Wright et al., 1981). Other known risk factors, such as hemochromatosis and thalassemia major, are also known contributors to elevated serum iron levels. Diabetes, low stomach acidity (typically a result of extended use of antacids), and immunocompromising diseases such as cancer and human immunodeficiency virus infection, and high-dose corticosteroid treatment have also been reported as contributing factors. Furthermore, the great majority (>88.5% of the U.S. cases between 2000 and 2003) of all cases of primary septicemia occur in males over the age of 50. Moreover, in a study of 70 patients in Korea, 96% were in men and 90% were persons over 40 years of age (Park et al., 1991).

Symptoms of primary septicemia typically develop within 36 h, although a survey of over 100 cases included in 20 published studies revealed cases with symptoms developing as rapidly as 7 h after ingestion of oysters or as late as 10 days (unpublished data). Table 1 is a compilation from reviews by Oliver and Kaper (2001) and Strom and Paranjpye (2000) and includes >300 cases described by Oliver (1989), Blake et al. (1979), and Tacket et al. (1984), and data for the 274 *V. vulnificus* cases reported by the Food and Drug Administration (FDA) for the period 1989 to 2000, which reveal symptoms typical of *V. vulnificus* primary septicemia. The most commonly reported symptoms are fever (94%), nausea (60%), and hypotension (systolic blood pressure <85 mm Hg; 43%). A quite common and highly differential symptom associated with primary septicemia is the development of secondary lesions. These begin as fluid-filled blisters, typically on the legs and feet, and result in tissue and muscle destruction (necrotizing fasciitis), which requires surgical debridement or limb amputation. Hospital stays are often significant; in a study of 333 patients with *Vibrio* infections associated with raw oyster consumption in Florida, Hlady and Klontz (1996) reported that 94% were hospitalized, with stays of up to 43 days (mean of >8 days).

During cases of primary septicemia, *V. vulnificus* is typically isolated from blood (93% of the cases summarized in Table 1). Usually, the infections develop extremely quickly, and death occurring within hours of hospitalization is not uncommon; most deaths occur within several days of hospital admission. While the bacterium is highly sensitive to virtually all antibiotics, the time prior to antibiotic treatment appears highly correlated with eventual outcome,

Table 1. Symptoms and traits of persons contracting *V. vulnificus* infections following ingestion of oysters[a]

Trait or characteristic	Result (%)
Male	86.5
Age >40 yr	95
Recent history of raw seafood consumption	>85
History of chronic underlying disease	80
History of liver disease	86–94
Fever	94
Nausea	60
Vomiting	35
Secondary lesions of extremities	69
Hypotension (systolic pressure <85 mm)	43
Isolation of *V. vulnificus* from blood	93

[a] Data compiled from cases described by Oliver (1989), Blake et al. (1979), Tacket et al. (1984), from data on 274 *V. vulnificus* cases reported by the U.S. FDA during the period 1989 to 2000, and from reviews by Oliver and Kaper (2001) and Strom and Paranjpye (2000).

with those not receiving treatment until 72 h after onset of illness generally not surviving (Klontz et al., 1988). Those presenting with hypotension are especially at risk: up to 95% of such patients typically die of the *V. vulnificus* infection (Horré et al., 1996).

Because of the association of infection and raw oyster consumption, several states have passed laws requiring warnings, typically displayed wherever raw oysters are sold or served. An example is that required since 1998 in the State of North Carolina, which states:

> Eating raw oysters, clams or mussels may cause severe illness. People with the following conditions are at an especially high risk: liver disease, alcoholism, diabetes, cancer, stomach or blood disorders, or weakened immune system. Ask your doctor if you are unsure of your risk. If you eat shellfish and become sick, see a doctor immediately.

Wound Infections

V. vulnificus is somewhat unusual in having a second route of infection. In addition to ingestion, the bacterium is able to infect preexisting wounds or those incurred during seawater-associated activities (wound infections caused by *V. vulnificus* and other marine bacteria have recently been reviewed by Oliver, 2005b). Most common are wounds acquired while cleaning fish or shellfish (e.g., fish fin pricks of the skin or cuts sustained while shucking oysters), or immersion of a wound into the seawater where *V. vulnificus* occurs (Oliver and Kaper, 2001). The wound does not need to be significant; several cases of persons who suffered ant bites on their legs or hands before or after contact with coastal waters developed wound infections which were either fatal or resulted in limb amputation.

In an examination of >100 cases from 16 published studies (Oliver, 2005b), incubation periods were found to range from 3 h (an extreme, although 8 cases reported symptoms within 12 h after infection) to 12 days (also an extreme, although 3 cases reported symptoms within 10 to 12 days). In the great majority of these cases, symptoms began in ca. 24 h, although deaths only hours after hospital admission have been reported (Torres et al., 2002).

Symptoms of *V. vulnificus*-induced wound infections include pain, erythema, and edema at the wound site from which *V. vulnificus* is isolated. The cellulitis may proceed to deeper tissue and may also cause extensive damage (necrotizing fasciitis) to skeletal muscle. Although the secondary lesions characteristic of the primary septicemia form of infection do not occur, wound infections can spread rapidly, become necrotic, and often require amputation of the infected limbs, or at least surgical debridement of the affected tissue. In certain persons, the infection may produce a life-threatening secondary septicemia.

Unlike primary septicemia cases, most persons developing wound infections do not have the underlying chronic diseases associated with ingestion-associated disease. However, cases where wound infections progress to septicemia appear to be those in which the victims suffer chronic underlying diseases. These typically involve the liver, again indicating the important role of serum iron in *V. vulnificus* infections, regardless of mode of entry (Howard and Lieb, 1988; Penman and Lanier, 1993). Infections by this route carry fatality rates of 20 to 25% (Oliver, 1989), a result of those with infections that develop into septicemia. Soft tissue infections caused by *V. vulnificus*, as well as other vibrios, have been reviewed elsewhere (Howard and Lieb, 1988; Kaye, 1990; Penman and Lanier, 1993).

PATHOGENESIS AND VIRULENCE FACTORS

Since the first study on experimental pathogenesis in *V. vulnificus* by Poole and Oliver (1978), a considerable amount has been learned regarding the virulence factors important for *V. vulnificus* infection. Far more, however, is not understood. The pathogenesis of *V. vulnificus* has been reviewed by Linkous and Oliver (1999), Oliver and Kaper (2001), and Strom and Paranjpye (2000). The reader is directed to those reviews, with a brief description of the major virulence factors being provided here.

Capsule

To date, the best-documented virulence factor for *V. vulnificus* is the production of capsular polysaccharide (CPS). Presence of a capsule results in colonies that are opaque and that can easily be differentiated visually from nonencapsulated strains that produce translucent colonies. More than 10 capsular serotypes have been reported (Hayat et al., 1993; Simonson and Siebeling, 1993), although it is not clear if any are more important than the others for initiating human infection (Wright et al., 1999; Oliver and Kaper, 2001) or if many others remain to be described. It is clear that the capsule has antiphagocytic properties, that only the encapsulated form is virulent (Simpson et al., 1987), and that the degree of CPS expression is correlated with virulence in mice (Wright et al., 1999). In animal studies, injection of the translucent morphotype into mice results in 50% lethal doses (LD_{50}) of generally >10^6 (and often >10^8), whereas the opaque morphotype yields LD_{50} of typically $<10^2$; values as low as 1 CFU have been reported (Wright et al., 1981). Whether the capsule aids in preventing phagocytosis by oyster hemocytes is not clear (Genthner et al., 1999; Harris-Young et al., 1995).

Interestingly, cells switch from the encapsulated form to the nonencapsulated form under laboratory conditions at a relatively high frequency, typically between 0.01 and 1%, depending on the strain and growth conditions. Wright et al. (1999, 2001) reported reversion from the nonencapsulated to the encapsulated morphotype, and they consider capsule switching to be a form of phase variation. In our experience, the translucent morphotype is a highly stable trait, and we have not observed the reversion from translucent to opaque in any strain. We have recently detected the existence of a third colony type, which appears to be intermediate between the opaque and translucent morphotypes. We believe such colonies are composed of cells with intermediate expression of CPS.

Capsule production is also related to biofilm formation. Interestingly, Joseph and Wright (2004) report that expression of CPS inhibits biofilm formation in this species. Whereas an opaque strain exhibited minimal biofilm under any conditions examined, the authors speculated that the increased capacity for biofilm formation by nonencapsulated strains might predict the presence of such strains in the environment. However, as the authors point out, environmental isolates are almost always observed to be encapsulated on primary culture.

LPS

While the presence of capsule is absolutely essential for the ability of *V. vulnificus* to evade host phagocytosis, and thus initiate infection, the factor that most likely causes the shock and death associated with *V. vulnificus* infections appears to be the cell's LPS, or endotoxin. The major symptoms associated with *V. vulnificus* infections, including fever, tissue edema, hemorrhage, and especially hypotension, are classic symptoms associated with endotoxic shock. Studies from this laboratory found that injections of purified LPS extracted from *V. vulnificus* resulted in a rapid decline in heart rate and blood pressure in rats, with death resulting within 30 to 60 min (McPherson et al., 1991). A subsequent study (Elmore et al., 1992) employed an inhibitor of the LPS-induced enzyme (nitric oxide synthase) responsible for these symptoms and found complete inhibition of these toxic effects.

A rather fascinating finding regarding the role of LPS in *V. vulnificus* disease helps explain why 80% of

primary septicemia cases occur in males. In studies in which the hormone levels of rats injected with *V. vulnificus* LPS were manipulated, we observed that the female hormone, estrogen, protects females against the toxic effects of this cell wall component (Merkel et al., 2001). A more complete discussion of the mechanisms of action of the *V. vulnificus* LPS and its role in pathogenesis of its infections is contained in Linkous and Oliver (1999).

Putative Virulence Factors

A large variety of factors, including hydrophobicity, adhesion to human cell lines, and production of a variety of exoproteins, have been examined and proposed as having a role in *V. vulnificus* pathogenesis (Baffone et al., 2001; Maruo et al., 1998; Miyoshi et al., 1993; Oliver et al., 1986a; Stelma et al., 1992), although few have actually been demonstrated to be essential to these infections (DePaola et al., 2003; Miyoshi et al., 1993). As notable examples, the *V. vulnificus* cytolysin/hemolysin (product of the *vvhA* gene) and a metalloprotease (product of the *vvpE* gene) were both considered to be important virulence factors for this organism, and some studies have suggested significant cytotoxic activity of these proteins (e.g., Rho et al., 2002). However, several studies using mutations in these genes (e.g., Fan et al., 2001; Jeong et al., 2000; Wright and Morris, 1991) have indicated no apparent role for these exoproteins. It should be emphasized, however, that almost all investigators have concentrated on the role of such factors in the primary septicemia produced by *V. vulnificus*; it is very possible that these may be critical for the wound infections produced by this pathogen.

Rhee et al. (Abstr. 101st Gen. Meet. Am. Soc. Microbiol., abstr. B-285, 2002) described a new toxin, Vv-RTX, that causes depolymerization and aggregation of actin fibers, and thus cytoskeletal rearrangements, in host epithelial cells. Further, toxin production appeared to be induced by host cell contact.

A report by Kim and Rhee (2003) provides evidence that flagella may be a virulence factor, with a mutation of the *flgC* gene (encoding a flagella basal body rod protein) resulting in decreased lethality to mice. A subsequent study by Y. R. Kim et al. (2003) employed a *flgE* mutant and found it to exhibit decreased attachment to human intestinal epithelial cells and mouse virulence (5×10^6 for the mutant compared to 1×10^3 for the wild type in the iron-overload mouse model). Similarly, Lee et al. (2004) studied a *flgE*-deleted strain and found the nonmotile mutant to exhibit reduced adherence to human epithelial cells, biofilm formation, and virulence for mice.

Ion transport has also been recently linked to virulence, as Chen et al. (2004) have reported that a mutant in K$^+$ transport (*trkA*) was more serum susceptible than was the parent strain and exhibited markedly reduced lethality in an iron-overload mouse model.

An intriguing new finding indicates that *V. vulnificus*, like an increasing number of bacteria, is capable of inducing apoptosis of macrophages both in vitro and in vivo (Kashimoto et al., 2003). The authors suggested that this may be a secondary mechanism, along with capsule production, to evade the host defense (phagocytic) system. Interestingly, environmental isolates were largely incapable of inducing apoptosis. More studies are required to understand the clinical significance of this finding.

Plasmids are not commonly found in biotype 1 strains (Dalsgaard et al., 1996; Davidson and Oliver, 1986; Høi et al., 1998c) and do not appear to correlate with virulence (Dalsgaard et al., 1996; DePaola et al., 2003), although plasmids generally occur in biotype 2 strains (Biosca et al., 1997; Høi et al., 1998c).

Regulation of Pathogenesis

Recently, McDougald et al. (2001, 2003) reported that the AI-2 communication molecule plays an important role in stress responses in *V. vulnificus*, including starvation and stationary-phase phenotypes. The *luxR* (*smcR*) mutant they studied also had increased exoenzyme production, biofilm formation, and motility. In contrast, we found a significant reduction in both motility and ability to form biofilms on an abiotic surface by a *luxS* mutant produced in our laboratory (D. M. Beam and J. D. Oliver, Abstr. 104th Gen. Meet. Am. Soc. Microbiol., abstr. K-057, 2004). The reasons for these discrepancies are not yet known. Subsequently, S. Y. Kim et al. (2003) showed that a *luxS* mutant, unable to synthesize the AI-2 signal, exhibited a significant delay in protease production but an increase in hemolysin production. This mutation also resulted in decreased lethality in mice. Similarly, Shao and Hor (2001) reported that a LuxR homologue of *Vibrio harveyi* positively regulates the metalloprotease but negatively regulates the cytolysin produced by *V. vulnificus*. Such studies suggest that the AI-2 quorum-sensing system may be important for the virulence of *V. vulnificus*.

A role for the stress-responsive alternate sigma factor, RpoS, is suggested in recent studies. Kim et al. (Abstr. 101st Gen. Meet. Am. Soc. Microbiol., abstr. B-74, 2001) reported that an *rpoS* mutant of *V. vulnificus* showed impaired adherence and cytotoxicity to HeLa cells, and that mouse lethality was significantly decreased. My colleagues and I (Hülsmann et al., 2003) found that a similar mutant was excep-

tionally sensitive to H$_2$O$_2$, had greatly reduced motility, did not produce the putative virulence factors albuminase and elastase, and exhibited reduced levels of collagenase. Y. R. Kim et al. (2003) recently used in vivo-induced antigen technology and reported that another transcriptional activator, HlyU, appears to be one of the master regulators of in vivo virulence expression in *V. vulnificus*. In 2005, Kim et al. (2005) reported that the production of cyclic AMP played an essential role in regulating *V. vulnificus* virulence, and that a *cya* mutant unable to produce adenylate cyclase showed decreased motility, cytotoxicity, and production of hemolysin, protease, and CPS.

Biotype 2 and 3 Strains

The first outbreak of biotype 2-caused eel vibriosis in Europe was reported in 1991 (Biosca et al., 1991). This occurred in a densely stocked eel hatchery near Valencia, Spain. The hatchery employed tanks containing water with a salinity of 17‰ and a temperature of 22°C. The infected eels were successfully treated with antibiotics, but biotype 2 strains continue to be of special economic significance due to their virulence for eels, an important aquaculture product in several European countries. Interestingly, the isolation of biotype 2 strains from seawater, even in confined spaces during an outbreak of hemorrhagic septicemia of eel, has proven difficult. Recently, Sanjuán and Amaro (2004) found that alkaline peptone broth, the enrichment medium routinely employed for the isolation of vibrios, favors the growth of competitors, and they described a new medium containing eel serum for the successful isolation of this biotype from water.

Biotype 2 strains appear to be similar to biotype 1 strains in regard to their putative virulence factors, including production of exoproteins, uptake of various iron sources via phenolate and hydroxamate siderophores, and both LPS and capsule expression (Amaro et al., 1994; Biosca and Amaro, 1996). However, the LPS molecules of biotype 2 strains are homologous (Biosca et al., 1999), in contrast to those in biotype 1, which are heterologous (Bahrani and Oliver, 1990, 1991). Interestingly, it is the O-antigenic side chain of the LPS molecule that determines the selective virulence of biotype 2 strains for eels (Amaro et al., 1997). The bactericidal action of eel serum via complement kills biotype 1 strains and biotype 2 strains lacking the O-polysaccharide side chain, whereas wild-type biotype 2 strains resist this effect. Another major difference between these biotypes is that biotype 2 strains are virulent for eels regardless of the presence or absence of capsule (Amaro and Biosca, 1996). Based on studies reported by Biosca et al. (1999), it is clear that, as in the case of biotype 1 strains, the LPS of biotype 2 strains is highly lethal for eels, with 200-µg injections dropping blood pressure and heart rate to fatal levels in ca. 60 min. Whereas biotype 2 is not generally considered to be a human pathogen, occasional cases of human infection have been reported (Amaro and Biosca, 1996; Hlady and Klontz, 1996).

To date, biotype 3 strains of *V. vulnificus* have only been reported to cause human wound infections. Over a 2-year period (1996 to 1997), Bisharat et al. (1999) reported 62 such cases. Of these, 92% developed cellulitis, with 6% proceeding to necrotizing fasciitis, and one case involved osteomyelitis, symptoms very similar to those of biotype 1 wound infections (Colodner et al., 2002). Most cases were reported during the summer season (between August and October), with a median incubation time of 12 h (range 3 to 48) and a median hospital stay of 8 days (although up to 41 days was reported). Except for a single case where carp was implicated, all infections were in persons who had handled, and received puncture wounds from, pond-harvested *Tilapia* (St. Peter's fish). To date, cases have only been reported in Israel, where St. Peter's fish is a common aquaculture product. Interestingly, attempts to isolate biotype 3 strains from fishpond water and fish intestines resulted in the isolation of type 1 strains only. As with biotype 1 cases, underlying disorders were commonly observed (40% of the patients), and cirrhosis, diabetes, and malignant disease were the most common. Bisharat and Raz (1996) reported mortality rates of 20% during these outbreaks, and surgical debridement was required in 66% of the patients, a significantly greater rate than that recorded for biotype 1 wound infections. At the time of this writing, no studies on the virulence traits of biotype 3 strains have been reported.

HOST FACTORS

Whereas it is clear that cells of *V. vulnificus* must possess the antiphagocytic capsule, and that the LPS is responsible for much of the host damage observed during infection, the physiological state of the host is also a critical factor in determining the outcome of ingestion and wound infections. As described above, virtually all studies, whether clinical or laboratory, have revealed the importance of certain underlying diseases in the host as being predisposing factors to *V. vulnificus* infection. Many of these, e.g., those related to alcohol abuse and alcoholic cirrhosis, as well as hepatitis, a viral disease of the liver, result in serum iron overload. Others, such as certain cancers, result

in immunodeficiency, while diabetes would be expected to lead to vasculature and circulation problems. The long-term use of stomach antacids likely provides a more hospitable environment for vibrios, which are typically highly acid sensitive.

A major problem regarding the host aspect of infection is the lack of information provided to at-risk consumers. In a Florida survey, Ross et al. (1994) provided evidence for a lack of awareness among at-risk shellfish consumers, with only 26% of 167 participants having recalled being advised to avoid eating raw oysters. Of this same population, only 6.7% were provided this information by physicians. Similarly, the Centers for Disease Control and Prevention conducted a survey of liver transplant units and found that only 23% of transplant coordinators were themselves aware of the hazard to their patients, and only 16% provided their patients with information about the hazards of eating raw seafood (Tuttle et al., 1994). In Japan, which is the largest importer and biggest consumer of seafood in the world, and where >2 million people are estimated to suffer from chronic liver disease, a recent study found that only 15.7% of physicians had a basic knowledge of *V. vulnificus* infection (Osaka et al., 2004).

PROBLEMS UNDERSTANDING THE PATHOGENESIS OF INFECTIONS

Laboratory studies have indicated that, based on animal lethality, virtually all *V. vulnificus* strains are virulent. If this were true, then the number of both wound and primary septicemia cases that occur worldwide would be expected to be considerably higher than they are. For example, only 30 or fewer fatal cases typically occur in the United States each year, despite FDA estimates that 20 million Americans consume 75 million to 80 million servings of raw oysters annually and that 12 million to 30 million persons have one or more of the known risk factors for *V. vulnificus* infection. From this, it seems exceedingly likely that (i) other, as yet unknown, human factors are required to predispose a person to the serious forms of these infections, and/or (ii) laboratory studies indicating that all *V. vulnificus* strains are virulent are incorrect. It is likely that the mouse models employed to study virulence are, in fact, not valid. Further, it is likely that there are critical and essential virulence factors that many, if not most, *V. vulnificus* strains lack, making them unable to cause human infection. Indeed, all "clinical" strains must eventually derive from the environmental strains, and what virulence factors a strain must possess to allow it to become infectious are not yet fully understood.

New insight into the genetic variation that occurs naturally among *V. vulnificus* strains (see below) offers hope for providing a better understanding of the pathogenesis of this species and for answering the dilemma described above.

GENETIC HETEROGENEITY

That significant strain differences exist at the genomic sequence level is clearly the case, and such differences are likely related to the question of whether all *V. vulnificus* strains are capable of causing infection. Buchrieser et al. (1995) examined 118 strains of *V. vulnificus* isolated from three oysters and found no two strains to have the same clamped homogeneous electric field gel electrophoresis profile. Similar results have been reported using ribotyping (Dalsgaard et al., 1996; Høi et al., 1997), randomly amplified polymorphic DNA (RAPD)-PCR (Aznar et al., 1993; Warner and Oliver, 1998), pulsed-field gel electrophoresis (Tamplin et al., 1996; Hsueh et al., 2004; Wong et al., 2004), and amplified fragment length polymorphism (Arias et al., 1997). Figure 2, for example, shows a RAPD-PCR gel from our laboratory that clearly demonstrates the great genetic difference exhibited by clinical and environmental isolates of *V. vulnificus*. Despite this, Jackson et al. (1997) reported isolation of only a single *V. vulnificus* strain (based on pulsed-field gel electrophoresis) from infected human tissue. Taken together, such studies suggest that great genetic variation occurs among strains of *V. vulnificus* and that all strains of *V. vulnificus* may not be equally virulent. Indeed, the data suggest that only a single strain of the great number and variety consumed within a raw oyster may ultimately be associated with a human infection.

At least part of the reason for this heterogeneity may reside in the fact that, like *V. cholerae*, *V. parahaemolyticus*, *V. anguillarum*, and several other *Vibrio* spp., *V. vulnificus* has two chromosomes (Thompson et al., 2004). The reason for this is not evident, but it must provide some selective advantage, possibly for survival in natural environments. Chromosome 1 contains most of the housekeeping genes, while chromosome 2 contains primarily genes needed for environmental adaptation. Studies indicate that extensive rearrangements have occurred within, and between, the two chromosomes (Thompson et al., 2004). Further, as evidence for horizontal gene transfer among environmental pathogens continues to grow (Wilson and Salyers, 2003), it seems likely that some of the genetic variation seen in *V. vulnificus* may be due to such gene exchange.

The first evidence for the existence of more than one genotype for *V. vulnificus* appears to be that reported by Aznar et al. (1994). They used oligonucleotide probes targeted to the variable region of 16S rRNA to separate *V. vulnificus* strains into two groups (type A and type B). This was followed by an investigation by Kim and Jeong (2001), who found that *V. vulnificus* could be separated into two groups based on 16S rRNA analysis, with the two groups corresponding to those described by Aznar et al. (1994). In an analysis of 40 *V. vulnificus* isolates taken from oysters, seawater, and mud, Kim and Jeong (2001) found that 35% were of the A type and 65% were of the B type. Because three of the 40 isolates were biotype 2 strains, these authors suggested that assignment of *V. vulnificus* to the two biotypes (1 and 2), as defined by biochemical properties, does not reflect the genetic heterogeneity of these strains. More recently, Lin and Schwartz (2003) studied the 16S rDNA (rRNA) gene sequences, as originally described by Aznar et al. (1994), of over 200 strains isolated from Galveston Bay, Tex., over a 1-year period. They found that the two genotypes (A and B) were isolated during different months of the year (see further discussion below under "The viable but nonculturable state"), suggesting that the two genotypes correlate with seasonal culturability.

The existence of several genotypes of *V. vulnificus* is further supported by Gutacker et al. (2003),

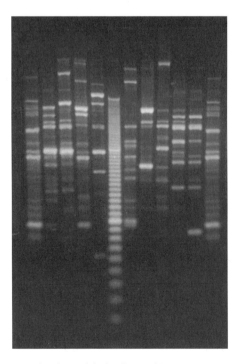

Figure 2. Randomly amplified polymorphic DNA (RAPD)-PCR of 11 individual *V. vulnificus* strains isolated from oysters (Y. Yano and J. D. Oliver, unpublished data).

who studied the genetic relationships among 62 strains, including all three biotypes, from different geographic and host origins. Using multilocus enzyme electrophoresis, RAPD-PCR, and sequence analysis of the *recA* and *glnA* genes, the authors concluded that the biotype 2 eel pathogens were a distinct genetic subgroup, and that the newly described biotype 3 (Israeli) isolates were distinct from all other *V. vulnificus* isolates. Biotype 1 strains were again found to be heterogeneous in their genetic makeup.

While these studies suggest the existence of more than one genotype of *V. vulnificus*, none have indicated any relationship to virulence. Recently, my colleagues and I (Rosche et al., 2005) reported that, using a simple and rapid PCR procedure, 55 strains of *V. vulnificus* (biotype 1) exhibited one of two major genotypes that showed a strong correlation with the source of their isolation. The great majority (90%) of the "C" genotype strains were from clinical samples, while 93% of the strains isolated from a variety of environmental sources were classified as "E-type." Nilsson et al. (2003) showed that the two groups originally identified by Aznar et al. (1994) were associated with clinical (group B) or environmental (group A) isolation. Eight of the strains we examined (Rosche et al., 2005) were also among the strains studied by Nilsson et al. (2003). Four strains that we classified as C-type were in the B group, while four strains classified as E-type were in the A group. This suggests that both classification methods concur in defining two major genotypes of *V. vulnificus*, with the "group/type B" strains of Aznar et al. (1994) and Nilsson et al. (2003) corresponding to our "C-type" strains. Whereas there is no indication that the amplicon we characterized in the C-strains is in any way involved in virulence, it appears that our PCR method, which is rapid and does not require rDNA extraction and amplification, can identify strains with a strong probability of ability to initiate human infection.

The latest entry on genomic heterogeneity is a study by Bisharat et al. (2005). In examining 62 biotype 3 strains from Israel, as well as 82 biotype 1 and 15 biotype 2 strains, they showed that biotype 3 strains evolved through a hybridization of the genomes of biotype 1 and 2 strains. This conclusion was reached after they examined the DNA sequence of 10 individual housekeeping genes (five from each of the two chromosomes) in each of the 159 strains. Whereas the biotype 1 strains were resolved into 66 sequence types, and biotype 2 strains were resolved into 4 types, all 62 biotype 3 strains were genetically identical and belonged to the same sequence type. Cluster analysis revealed that biotype 3 strains had almost equal contributions from both biotype 1 and 2 strains. This suggests, as the authors state, that this "hybrid clone"

"may have evolved by a relatively recent genome hybridization event." At the time of this writing, biotype 3 strains appear to be geographically restricted to Israel and the Mediterranean. It will be interesting to follow this type as it spreads, as it almost certainly will, throughout the world.

The study by Bisharat et al. (2005) also indicated that biotypes 1 and 2 are not entirely distinct but are present in two genetic subpopulations. Biotype 1 was present in both populations "A" and "B," while biotype 2 strains were present only in the "A" population. Biotype 3 occupied an intermediate position, as would be suggested by the hybridization event described above. In their study, an "overrepresentation" of environmental isolates occurred in population "A," while the bulk of human disease isolates resided in population "B." Thus, "A" likely corresponds to our "E" (environmental) group, while their "B" population appears to correspond to our "C" (clinical) group (Rosche et al., 2005).

The studies reported by these various groups strongly indicate that, not only are there several genotypes of *V. vulnificus*, but they exhibit differences in seasonality of isolation and of their ability to initiate human infection. Biotype 3 clearly represents an entirely new type, which only recently evolved, and the possibility that biotype 1 strains represent two species cannot be ignored.

DISTRIBUTION IN THE MARINE ENVIRONMENT

While TCBS is the most commonly employed medium for the isolation and initial differentiation of marine vibrios, most studies on the distribution of *V. vulnificus* in marine environments now employ colistin-polymyxin B-cellobiose agar or one of its modifications. This medium, originally described in 1987 (Massad and Oliver), and its modifications have been found in numerous investigations to be the most suitable medium for the isolation of this species from natural environments (e.g., Cerdà-Cuéllar et al., 2001; Høi et al., 1998a,c; Kaysner et al., 1989; Oliver et al., 1992). In one study on *V. vulnificus* in oysters, for example, over 80% of some 1,000 colonies of appropriate morphology on this medium were subsequently identified by molecular methods as isolates of this species (Sun and Oliver, 1995). Similar results were reported by Sloan et al. (1992). This medium and slightly modified variations, as well as others that have been described for the isolation of vibrios and *V. vulnificus*, have recently been compared in some detail (Harwood et al., 2004; Oliver, 2003).

Because of the significant phenotypic variation exhibited by *V. vulnificus*, it is most commonly identified through the use of molecular methods. Harwood et al. (2004) have recently summarized several of these, including the use of DNA probes, colony hybridization, PCR (including quantitative PCR), as well as immunological methods. The most commonly employed probe/primer for both hybridization and PCR is to the hemolysin/cytolysin gene. Initially developed by Wright et al. (1985), this sequence is specific for this species (Morris et al., 1987a) and is recommended in the FDA's online *Bacteriological Analytical Manual* for this purpose (http://www.cfsan.fda.gov/~ebam/bam-9.html).

Randa et al. (2004), for example, using quantitative PCR for the 23S rRNA gene, were able to detect as few as 1 to 2 cells of *V. vulnificus* in spiked water samples. Using nested PCR, they were able to detect a range of 1 to 10^7 cells/ml added to environmental samples.

Seawater

Numerous studies have documented the distribution of *V. vulnificus* in coastal and estuarine waters around the world (e.g., Kaysner et al., 1987; Oliver and Kaper, 2001; Randa et al., 2004; Roberts et al., 1982; Tamplin et al., 1982; Tilton and Ryan, 1987). The bacterium has been isolated from all coasts of the United States and from those of countries around the world but is most commonly isolated from estuarine/coastal waters during warm months. The levels reported are generally fairly low, usually in the range of <1 to 50 CFU/ml (Jones and Summer-Brason, 1998; Pfeffer et al., 2003; Tamplin et al., 1982) but occasionally as high as 10^4 CFU/ml (Heidelberg et al., 2002a,b; Vanoy et al., 1992). In a study of more than 6,000 seawater samples taken from Tokyo Harbor, Aono et al. (1997) reported only 0.2% of all isolates to be PCR-confirmable as *V. vulnificus*. This represented only a small portion of a vibrio population that constituted >22% of all bacteria in these samples. Similarly, my colleagues and I (Pfeffer et al., 2003) reported that *V. vulnificus* represented an average of 0.15% of the estuarine populations present at six sites in the eastern United States. Thus, the relative levels of this pathogen are generally low in water, although its occurrence is highly correlated with water temperature (see below), and high numbers are reached in certain localities in warm water months.

Shellfish

Whereas the number of *V. vulnificus* cells found in seawater is generally quite low, the levels in molluscan shellfish have been reported as high as 10^6 CFU/g or more, due to their filter-feeding mode for obtaining nutrients. These levels can vary dramatically, however. In one of our studies, some oysters taken from the same site harbored $>10^5$ CFU/g or more, whereas others had undetectable levels (Birkenhauer and Oliver, 2003). Figure 3 shows the results of a recent study from our lab demonstrating these differences. Such variations in load are typical (Cook, 1994; Cook and Ruple, 1989, 1992; DePaola et al., 1994).

Studies on the presence of *V. vulnificus* in shellfish, primarily oysters, have also indicated a strong correlation with temperature; in general, investigators are unable to isolate *V. vulnificus* from oysters taken from cold waters. This finding also correlates with epidemiological data indicating low numbers of oyster-associated human cases during winter months (November through March in the United States).

Other Marine Sources

Although relatively few studies have reported *V. vulnificus* in sources other than water and shellfish, it can be isolated from sediment, plankton, nonmolluscan shellfish, and fish. Oliver et al. (1983) reported that plankton taken along the entire East Coast of the United States harbored *V. vulnificus*, while Montanari et al. (1999) reported isolation from the smaller (0.45 to 200 μm) plankton taken from Italian coastal waters. Heidelberg et al. (2002b) studied *V. vulnificus* associated with zooplankton in the Chesapeake Bay of Maryland and found this species to constitute up to 57% of the *Vibrio-Photobacterium*

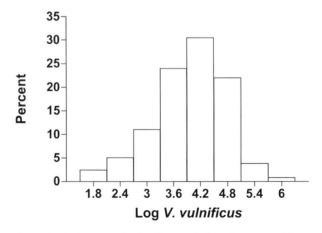

Figure 3. Variation in levels of *V. vulnificus* in oysters (*Crassostrea virginica*). Oysters (*n* = 155) were taken from a small, environmentally homogeneous site in North Carolina and were sampled for their resident load of *V. vulnificus*. Shown is the log of *V. vulnificus* loads (CFU g^{-1} wet weight) vs. the frequency (%) that each load was identified (Sokolova et al., 2005).

population attached to both large and small zooplankton species. This was higher than the proportion of *V. vulnificus* cells in the same population present in the surrounding waters. The numbers of *V. vulnificus* cells that were zooplankton associated were extremely high, from 10^5 to 10^7 cells/m^{-3}, with the largest populations occurring during the early spring and late summer. When individual zooplankton (of the >202-μm class) were studied, levels of 10^1 to 10^4/cell were found.

In a study of fish taken from Tokyo Bay, Aono et al. (1997) reported 0.3% of the 1,056 vibrios they isolated from goby to be *V. vulnificus*, based on PCR analysis. Yano et al. (2004) isolated *V. vulnificus* from 11 species of live seafood for sale in markets in China, including clams, shrimp, prawns, and fish. DePaola et al. (1994) conducted what is probably the most extensive study on the presence of *V. vulnificus* in a wide variety of fish. They studied the occurrence of this species in some 23 different fish species taken from coastal and offshore waters of the Gulf of Mexico. Of the nine offshore species, only two yielded *V. vulnificus*, consistent with the generally reported inability to isolate this bacterium from open-ocean waters. However, these authors did isolate *V. vulnificus*, often in very high numbers, from all coastal fish sampled. They were also able to isolate *V. vulnificus* from fish taken during the winter months when levels in surrounding waters and sediments were undetectable or extremely low. They thus theorized that fish might represent a reservoir during cold water months. Levels as high as 10^6/g were found in the intestinal contents of some bottom-feeding fish, especially those that consume mollusks and crustaceans. The finding is in support of an earlier study by Vanoy et al. (1992), who were able to isolate *V. vulnificus* from sediments during cold water months when this species was not detected in water, suspended particulate matter, or oysters. They thus suggested that *V. vulnificus* may also overwinter in sediments.

Despite these findings, the occurrence of *V. vulnificus* in sediments has been reported by relatively few investigators. Vanoy et al. (1992) reported *V. vulnificus* to range from <0.1 to >10^5/g of sediments taken from Galveston Bay in Texas. Similarly, a study by Williams and LaRock (1985) reported concentrations from <1 to >10^4/g, with 50% of the samples yielding 100 or fewer *V. vulnificus* cells/g. Høi et al. (1998b) were able to isolate *V. vulnificus* from ca. 57% of the sediment samples they examined, but Pfeffer et al. (2003) could only confirm a single isolate of this species in sediments over a 6-month (cold water) period. Aono et al. (1997) took >2,600 sediment samples from Tokyo Harbor

and found that, while vibrios composed 20% of the resident microflora, 0.2% could be identified as *V. vulnificus*. Thus, the levels of this species in estuarine sediments appear to vary considerably, similar to what is observed in water samples.

Factors Affecting Distribution in the Environment

Temperature

Virtually all studies have found water temperature to be a key factor in the ability to isolate *V. vulnificus* and occurrence of *V. vulnificus* infection (Høi et al., 1998b; Hsueh et al., 2004; Jones and Summer-Brason, 1998; Kelly, 1982; Randa et al., 2004; Tamplin et al., 1982; Tilton and Ryan, 1987; Vanoy et al., 1992; Veenstra et al., 1994); most investigators report isolation of this species only when water temperatures are >15°C. Investigators sampling regions that typically have low water temperatures are often unable to isolate *V. vulnificus* (Kaysner et al., 1990). Figure 4 shows the strong correlation between water temperature and *V. vulnificus* levels we found during a multiyear study of six estuarine areas in the eastern United States (K. Dyer-Blackwell and J. Oliver, unpublished). As is evident, the levels of this pathogen decrease to undetectable levels when water temperatures decrease below 15°C. Indeed, following a multiple regression analysis of the data, we have reported water temperature to account for most (47%) of the variability in concentration of *V. vulnificus* in these estuarine waters (Pfeffer et al., 2003). This is similar to the level reported by Motes et al. (1998) for the isolation of *V. vulnificus* from oysters (65%) and that reported by Randa et al. (2004) in sampling bay water (60%).

This correlation between ability to isolate *V. vulnificus* and water temperature is also paralleled by the incidence of human infection (81% of the cases described in Fig. 1 occurred between May and October). Indeed, numerous studies have reported a strong correlation between water temperature and levels of *V. vulnificus* in oysters. As is typical of such studies, *V. vulnificus* is undetectable in winter months, with loads increasing during the spring months to peak levels in the middle of the summer, when water temperatures are highest. These data are similar to a study by Motes et al. (1998) that reported a mean of 2,300 *V. vulnificus* cells per gram from oysters harvested between May and October from five Gulf Coast and Atlantic Coast sites. A gradual reduction to <10/g was observed during November and December, where levels remained until mid-March.

V. vulnificus water isolates v/s water temperature

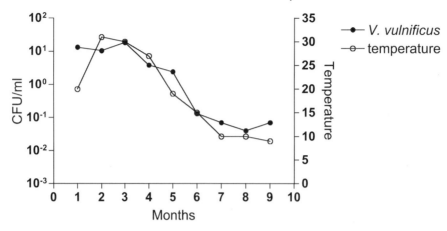

Figure 4. Correlation between water temperature (°C) and numbers (CFU/ml) of *V. vulnificus* cells isolated from the water over a 12-month period. Sampling sites were estuarine sites in eastern North Carolina (K. Dyer-Blackwell and J. Oliver, unpublished data).

Salinity

As noted above, *V. vulnificus* is an obligate halophile and is found only in waters with salinity of at least 5‰. The bacterium does not appear to survive well at elevated salinities, and it has not been isolated from open ocean waters. Indeed, Motes and DePaola (1996) found that oysters that were relayed to offshore waters (30 to 34‰) underwent reductions in *V. vulnificus* loads from 10^3 to 10^4 CFU/g to <10/g within 17 days. Most studies report *V. vulnificus* from water with salinity values between 8 and 23‰ (e.g., DePaola et al., 1994; Fukushima and Seki, 2004; Høi et al., 1998a; Tamplin et al., 1982). As an example, in a study of 21 sites in the Gulf of Mexico, *V. vulnificus* was recovered from all waters with relatively low (7 to 16‰) salinities by Kelly (1982). Oliver et al. (1983) and Parvathi et al. (2004) also reported a negative correlation with salinity. Several recent studies have provided a detailed analysis of the salinity requirements of *V. vulnificus*. Parvathi et al. (2004) and Randa et al. (2004) reported that, irrespective of temperature, the greatest concentrations of this pathogen are seen at 5 to 10‰, which appears to be the organism's optimal salinity. Randa et al. (2004) also found that, at 20 to 25‰, the abundance of *V. vulnificus* correlates positively with salinity. From the data they compiled, it appears that *V. vulnificus* can be isolated from waters with salinities of 5 to 10‰ at any temperature >10°C, but in waters of >10‰ the temperature must be >20°C.

Motes et al. (1998) found salinity was a controlling factor for the levels of *V. vulnificus* in oysters taken from Gulf and Atlantic Coast waters, with the highest numbers (>10^3/g) found in waters of intermediate salinities (5 to 25‰).

The viable but nonculturable state

The cause of the loss of culturability of *V. vulnificus* cells when water temperatures become cold is now generally believed to represent entry by this bacterium into the viable but nonculturable (VBNC) state. Now described for nearly 100 species of bacteria in more than 30 genera, this physiological state has been the subject of several reviews (Oliver, 1993, 2000a,b; Rice et al., 2000), as well as in this book (chapter 10). In this state, cells lose the ability to be cultured on routine media but retain viability; many species (including *V. vulnificus*) have been shown to be capable of resuscitating to the actively metabolizing state. Entry into the VBNC state has been suggested to be broadly protective, as cells in this state show increased resistance to a variety of stresses (Rice et al., 2000). Entry into the VBNC state by *V. vulnificus* is induced by a simple temperature downshift (to below ca. 13°C), with resuscitation accomplished by a corresponding temperature upshift. In situ studies employing membrane diffusion chambers (Oliver et al., 1995) indicated that entry of *V. vulnificus* into this state of dormancy in estuarine waters requires ca. 8 days to decline from 10^7 to <10^0 CFU/ml at 10 to 15°C, with complete resuscitation (in 16 to 19°C waters) requiring ca. 4 days. These findings are supported by the ecological study reported by Pfeffer et al. (2003) on the role of temperature in isolation of *V. vulnificus* from estuarine waters. It has also been suggested, however, that while *V. vulnificus* levels decrease in the water column during cold months, cells can be isolated from sediments and/or bottom-dwelling fish (DePaola et al., 1994; Randa et al., 2004), or when large quantities of seawater are examined (Randa et al., 2004).

In vivo studies reported by our laboratory (Oliver and Bockian, 1995) provided evidence that resuscitated cells of *V. vulnificus* retain their ability to produce fatal infections in laboratory animals. A recent study from this lab (Kong et al., 2004) indicates that entry into and resuscitation from the VBNC state by *V. vulnificus* is controlled to a great extent by the reduced activity of catalase at low temperature, an activity which is required for the cells to detoxify the H_2O_2 naturally present in routine laboratory media. Such temperature sensitivity could account, at least in part, for our inability to culture this pathogen when water temperatures are low and its apparent seasonality. Conversely, culturability would be regained when the cells are exposed to a temperature upshift, catalase activity is restored, and the cells are again able to grow on routine laboratory media.

A recent report by Lin and Schwartz (2003) presented data that offer further information regarding nonculturability of *V. vulnificus*. As a result of an analysis of the 16S rRNA (rDNA) gene sequences of strains isolated from Galveston Bay, Tex., over a 1-year period, they found that a higher proportion of the type A rDNA strains were isolated in June and July, with primarily type B strains being isolated in September. Further, after culturability of *V. vulnificus* was lost during the winter months (December through February), only strains of the A type could be isolated during the subsequent months (March through May). Thus, the occurrence of strains of the two rDNA types varied with water temperature, and the authors suggested that the 16S rDNA type may determine the culturability of *V. vulnificus* in the environment. What role the rDNA type may play in the VBNC state of this pathogen has yet to be investigated.

Other physical, chemical, and microbiological factors

A recent study on the ecology of *V. vulnificus* at six estuarine sites along the eastern coast of the United States (Pfeffer et al., 2003) included measurements of nine physicochemical parameters, as well as levels of total estuarine bacteria, *Vibrio* spp., *Escherichia coli*, and coliforms in the waters of these sites. Multiple regression analysis indicated that vibrio levels were controlled primarily by water temperature, with levels of dissolved oxygen and total vibrios playing lesser roles. An early study by Oliver et al. (1983) also found a correlation with total vibrio

levels, suggesting that *V. vulnificus* responds to most environmental parameters similarly to other *Vibrio* spp. Although dissolved oxygen had not been reported to correlate with *V. vulnificus* in previous studies (e.g., Roberts et al., 1982), a negative correlation was found by Pfeffer et al. (2003), and such a correlation seems reasonable, given that water temperature and dissolved oxygen levels are negatively correlated and that water temperature plays such a significant role in the distribution of *V. vulnificus* in estuarine waters.

Although Høi et al. (1998a) reported a correlation with total coliforms in the Danish waters they studied, most studies have not found a correlation between *V. vulnificus* and either fecal or total coliforms (e.g., Jones and Summer-Brason, 1998; Oliver et al., 1983; Pfeffer et al., 2003; Tamplin et al., 1982).

In a study of coastal waters of the Atlantic coast, Oliver et al. (1983) reported correlations between *V. vulnificus* and water pH. Heidelberg et al. (2002a), in their multiple regression analysis of vibrios in Ches-apeake Bay, found a negative correlation between *V. vulnificus* and both turbidity and tidal height. In agreement with that study, Jones and Summer-Brason (1998) used multiple regression analysis to determine that suspended solids were one of the major factors affecting the presence of *V. vulnificus* in northeastern U.S. waters. This accounted for ~27% of the variance in the presence of this species, and few or no isolates could be obtained when turbidities were high. These investigators also reported dissolved organic carbon to be a major factor determining variability in the distribution of *V. vulnificus* in the northern New England waters they studied, accounting for 44% of the variance. Randa et al. (2004) reported that *V. vulnificus* numbers in Barnegat Bay, N.J., correlated with chlorophyll *a* concentration.

Based on the above studies, it is clear that while temperature and salinity are major factors controlling the ecology of this pathogen, dissolved organic carbon, turbidity, and likely several other factors may also be important in the distribution of *V. vulnificus* in estuarine waters.

METHODS TO ELIMINATE *V. VULNIFICUS* IN FOOD

In all instances, the foods implicated in the transmission of *V. vulnificus* are raw or undercooked oysters and, to a much lesser extent, similarly prepared clams.

Cook and coworkers (Cook, 1994, 1997; Cook and Ruple, 1989) reported that *V. vulnificus* cells in oysters are unable to proliferate when kept at 13°C or below and lose culturability in a time-dependent manner when kept at 4°C or lower (Cook and Ruple, 1992). In contrast, and of great practical importance to the food industry, *V. vulnificus* cells are capable of significant growth when oysters are held at 18°C or higher (Cook, 1994, 1997).

The only method that eliminates *V. vulnificus* cells and thus provides absolute safety against infection resulting from eating oysters is complete cooking of the oysters. The FDA recommends steaming shellstock oysters for 4 to 9 min, frying shucked oysters for 10 min at 375°C, or baking at 450°C for 10 min. Such treatments, of course, result in death of the oysters, and to date there exist no methods that are capable of eliminating this pathogen from oysters while leaving the oyster alive. A number of milder "postharvest treatments" have been developed that, while still causing death of the oyster, claim to eliminate *V. vulnificus*. These include "low temperature pasteurization," "individually quick freezing," and "high hydrostatic pressure" (He et al., 2001). A promising but not yet employed treatment is irradiation. A recent study by Andrews et al. (2003) showed that *V. vulnificus* could be eliminated from oysters using 0.75 to 1.0 kGy irradiation exposure, with no apparent reduction in sensory attributes of tissue.

REFERENCES

Amaro, C., and E. G. Biosca. 1996. *Vibrio vulnificus* biotype 2, pathogenic for eels, is also an opportunistic pathogen for humans. *Appl. Environ. Microbiol.* **62:**1454–1457.

Amaro, C., E. G. Biosca, B. Fouz, A. E. Toranzo, and E. Garay. 1994. Role of iron, capsule, and toxins in the pathogenicity of *Vibrio vulnificus* biotype 2 for mice. *Infect. Immun.* **62:**759–763.

Amaro, C., B. Fouz, E. G. Biosca, E. Marco-Noales, and R. Collado. 1997. The lipopolysaccharide O side chain of *Vibrio vulnificus* serogroup E is a virulence determinant for eels. *Infect. Immun.* **65:**2475–2479.

Andrews, L., M. Jahncke, and K. Mallikarjunan. 2003. Low dose gamma irradiation to reduce pathogenic vibrios in live oysters (*Crassostrea virginica*). *J. Aquat. Food Prod. Technol.* **121:**71–82.

Aono, E., H. I. Sugita, J. Kawasaki, H. Sakakibara, T. Takahashi, K. Endo, and Y. Deguchi. 1997. Evaluation of polymerase chain reaction method for identification of *Vibrio vulnificus* isolated from marine environments. *J. Food Prot.* **60:**81–83.

Arias, C. R., L. Verdonck, J. Swings, E. Garay, and R. Aznar. 1997. Intraspecific differentiation of *Vibrio vulnificus* biotypes by amplified fragment length polymorphism and ribotyping. *Appl. Environ. Microbiol.* **63:**2600–2606.

Aznar, R., W. Ludwig, and H.-H. Schleifer. 1993. Ribotyping and randomly amplified polymorphic DNA analysis of *Vibrio vulnificus* biotypes. *Syst. Appl. Microbiol.* **16:**303–309.

Aznar, R., W. Ludwig, R. I. Amann, and K. H. Schafer. 1994. Sequence determination of rRNA genes of pathogenic *Vibrio* species and whole-cell identification of *Vibrio vulnificus* with targeted oligonucleotide probes. *Int. J. Syst. Bacteriol.* **44:**330–337.

Baffone, W., B. Citterio, E. Vittoria, A. Casaroli, A. Pianetti, R. Campana, and F. Bruscolini. 2001. Determination of several potential virulence factors in *Vibrio* spp. isolated from sea water. *Food Microbiol.* **18:**479–488.

Bahrani, F., and J. D. Oliver. 1990. Studies on the lipopolysaccharide of a virulent and an avirulent strain of *Vibrio vulnificus*. *Biochem. Cell Biol.* **68:**547–551.

Bahrani, F., and J. D. Oliver. 1991. Electrophoretic analysis of lipopolysaccharide isolated from opaque and translucent colony variants of *Vibrio vulnificus* using various extraction methods. *Microbios* **66:**83–93.

Biosca, E. G., and C. Amaro. 1996. Toxic and enzymatic activities of *Vibrio vulnificus* biotype 2 with respect to host specificity. *Appl. Environ. Microbiol.* **62:**2331–2337.

Biosca, E. G., J. D. Oliver, and C. Amaro. 1996. Phenotypic characterization of *Vibrio vulnificus* biotype 2, a lipopolysaccharide-based homogeneous O serogroup within *Vibrio vulnificus*. *Appl. Environ. Microbiol.* **62:**918–927.

Biosca, E. G., C. Esteve, E. Garay, and C. Amaro. 1993. Evaluation of the API-20E system for identification and discrimination of *Vibrio vulnificus* biotypes 1 and 2. *J. Fish Dis.* **16:**79–82.

Biosca, E. G., C. Amaro, J. L. Larsen, and K. Pedersen. 1997. Phenotypic and genotypic characterization of *Vibrio vulnificus*: proposal for the substitution of the subspecific taxon biotype for serovar. *Appl. Environ. Microbiol.* **63:**1460–1466.

Biosca, E. G., R. M. Collado, J. D. Oliver, and C. Amaro. 1999. Comparative study of biological properties and electrophoretic characteristics of lipopolysaccharide from eel-virulent and eel-avirulent *Vibrio vulnificus* strains. *Appl. Environ. Microbiol.* **65:**856–858.

Biosca, E. G., C. Amarao, C. Esteve, E. Alcaide, and E. Garay. 1991. First record of *Vibrio vulnificus* biotype 2 from diseased European eel, *Anguilla anguilla* L. *J. Fish Dis.* **14:**103–109.

Birkenhauer, J. M., and J. D. Oliver. 2003. Use of diacetyl to reduce the load of *Vibrio vulnificus* in the eastern oyster, *Crassostrea virginica*. *J. Food Prot.* **66:**38–43.

Bisharat, N., and R. Raz. 1996. *Vibrio* infection in Israel due to changes in fish marketing. *Lancet* **348:**1585–1586.

Bisharat, N., D. I. Cohen, R. M. Harding, D. Falush, D. W. Crook, T. Peto, and M. C. Maiden. 2005. Hybrid *Vibrio vulnificus*. *Emerg. Infect. Dis.* **11:**30–35.

Bisharat, N., V. Agmon, R. Finkelstein, R. Raz, G. Ben-Dror, L. Lerner, S. Soboh, R. Colodner, D. N. Cameron, D. L. Wykstra, D. L. Swerdlow, J. J. Farmer III, and the Israel *Vibrio* Study Group. 1999. Clinical, epidemiological, and microbiological features of *Vibrio vulnificus* biogroup 3 causing outbreaks of wound infection and bacteraemia in Israel. *Lancet* **354:**1421–1424.

Blake, P. A., M. H. Merson, R. E. Weaver, D. G. Hollis, and P. C. Heublein. 1979. Disease caused by a marine *Vibrio*. *New Engl. J. Med.* **300:**1–5.

Bock, T., N. Christensen, N. H. R. Eriksen, S. Winter, H. Rygaard, and F. Jørgensen. 1994. The first fatal case of *Vibrio vulnificus* infection in Denmark. *Acta Pathol. Microbiol. Immunol. Scand.* **102:**874–876.

Buchrieser, C., V. V. Gangar, R. L. Murphree, M. L. Tamplin, and C. W. Kaspar. 1995. Multiple *Vibrio vulnificus* strains in oysters as demonstrated by clamped homogeneous electric field gel electrophoresis. *Appl. Environ. Microbiol.* **61:**1163–1168.

Cameron, D. N. 1998. *Proc. Int. Conf. Emerg. Infect. Dis.* Atlanta, Ga.

Cerdà-Cuéllar, M., L. Permin, J. L. Larsen, and A. R. Blanch. 2001. Comparison of selective media for the detection of *Vibrio vulnificus* in environmental samples. *J. Appl. Microbiol.* **91:**322–327.

Chan, K. Y., M. L. Woo, K. W. Lo, and G. L. French. 1986. Occurrence and distribution of halophilic vibrios in subtropical coastal waters of Hong Kong. *Appl. Environ. Microbiol.* **52:** 1407–1411.

Chen, Y.-C., Y.-C. Chuang, C.-C. Chang, C.-L. Jeang, and M.-C. Chang. 2004. A K+ uptake protein, TrkA, is required for serum, protamine, and polymyxin B resistance in *Vibrio vulnificus. Infect. Immun.* **72:**629–636.

Colodner, R., B. Chazan, J. Kopelowitz, Y. Keness, and R. Raz. 2002. Unusual portal of entry of *Vibrio vulnificus*: evidence of its prolonged survival on the skin. *Clin. Infect. Dis.* **34:**714–715.

Cook, D. W. 1994. Effect of time and temperature on multiplication of *Vibrio vulnificus* in postharvest Gulf Coast shellstock oysters. *Appl. Environ. Microbiol.* **60:**3483–3484.

Cook, D. W. 1997. Refrigeration of oyster shellstock: conditions which minimize the outgrowth of *Vibrio vulnificus. J. Food Prot.* **60:**349–352.

Cook, D. W., and A. D. Ruple. 1989. Indicator bacteria and *Vibrionaceae* multiplication in post-harvest shellstock oysters. *J. Food Prot.* **52:**343–349.

Cook, D. W., and A. D. Ruple. 1992. Cold storage and mild heat treatment as processing aids to reduce the numbers of *Vibrio vulnificus* in raw oysters. *J. Food Prot.* **55:**985–989.

Dalsgaard, A., N. Frimodt-Møllere, B. Bruun, L. Høi, and J. L. Larsen. 1996. Clinical manifestations and molecular epidemiology of *Vibrio vulnificus* infections in Denmark. *Eur. J. Clin. Microbiol. Infect. Dis.* **15:**227–232.

Davidson, L., and J. D. Oliver. 1986. Plasmid carriage in *Vibrio vulnificus* and other lactose-fermenting marine vibrios. *Appl. Environ. Microbiol.* **51:**211–213.

DePaola, A., G. M. Capers, and D. Alexander. 1994. Densities of *Vibrio vulnificus* in the intestines of fish from the U.S. Gulf Coast. *Appl. Environ. Microbiol.* **60:**984–988.

DePaola, A., J. L. Nordstrom, A. Dalsgaard, A. Forslund, J. D. Oliver, T. Bates, K. L. Bourdage, and P. A. Gulig. 2003. Analysis of *Vibrio vulnificus* from market oysters and septicemia cases for virulence markers. *Appl. Environ. Microbiol.* **69:**4006–4011.

Desenclos, J.-C., K. C. Klontz, L. E. Wolfe, and S. Hoecheeri. 1991. The risk of *Vibrio* illness in the Florida raw oyster eating population, 1981–1988. *Am. J. Epidemiol.* **134:**290–297.

Elmore, S. P., J. A. Watts, L. M. Simpson, and J. D. Oliver. 1992. Reversal of hypotension induced by *Vibrio vulnificus* lipopolysaccharide in the rat by inhibition of nitric oxide synthase. *Microb. Pathog.* **13:**39–397.

Fan, J.-J., C. H. Shao, Y.-C. Ho, C.-K. Yu, and L.-I. Hor. 2001. Isolation and characterization of a *Vibrio vulnificus* mutant deficient in both extracellular metalloprotease and cytolysin. *Infect. Immun.* **69:**5943–5948.

Fukushima, H., and R. Seki. 2004. Ecology of *Vibrio vulnificus* and *V. parahaemolyticus* in brackish environments of the Sada River in Shimane Prefecture, Japan. *FEMS Microbiol. Ecol.* **48:**221–229.

Genthner, F. J., A. K. Volety, L. M. Oliver, and W. S. Fisher. 1999. Factors influencing in vitro killing of bacteria by hemocytes of the Eastern oyster (*Crassostrea virginica*). *Appl. Environ. Microbiol.* **65:**3015–3020.

Gutacker, M., N. Conza, C. Benagli, A. Pedroli, M. V. Bernasconi, L. Permin, R. Aznar, and J.-C. Piffaretti. 2003. Population genetics of *Vibrio vulnificus*: identification of two divisions and a distinct eel-pathogenic clone. *Appl. Environ. Microbiol.* **69:** 3203–3212.

Harris-Young, L., M. L. Tamplin, J. W. Mason, H. C. Aldrich, and J. K. Jackson. 1995. Viability of *Vibrio vulnificus* in association with hemocytes of the American oyster (*Crassostrea virginica*). *Appl. Environ. Microbiol.* **61:**52–57.

Harwood, V. J., J. P. Gandhi, and A. C. Wright. 2004. Methods for isolation and confirmation of *Vibrio vulnificus* from oysters and environmental sources: a review. *J. Microbiol. Methods* **59:**301–316.

Hayat, U., G. P. Reddy, C. A. Bush, J. A. Johnson, A. C. Wright, and J. G. Morris, Jr. 1993. Capsular types of *Vibrio vulnificus*: an analysis of strains from clinical and environmental sources. *J. Infect. Dis.* **168:**758–762.

He, H., R. M. Adams, D. F. Farkas, and M. T. Morrissey. 2001. The use of high hydrostatic pressure to shuck oysters and extend shelf-life. *J. Shellfish Res.* **20:**1299–1300.

Heidelberg, J. F., K. B. Heidelberg, and R. R. Colwell. 2002a. Seasonality of Chesapeake Bay bacterioplankton species. *Appl. Environ. Microbiol.* **68:**5488–5497.

Heidelberg, J. F., K. B. Heidelberg, and R. R. Colwell. 2002b. Bacteria of the γ-subclass *Proteobacteria* associated with zooplankton in Chesapeake Bay. *Appl. Environ. Microbiol.* **68:**5498–5507.

Hlady, W. G. 1997. *Vibrio* infections associated with raw oyster consumption in Florida, 1981-1994. *J. Food Prot.* **60:**353–357.

Hlady, W. G., and K. C. Klontz. 1996. The epidemiology of *Vibrio* infections in Florida, 1981–1993. *J. Infect. Dis.* **173:**1176–1183.

Hlady, W. G., R. C. Mullen, and R. S. Hopkins. 1993. *Vibrio vulnificus* from raw oysters. Leading cause of reported deaths from foodborne illness in Florida. *J. Flor. Med. Assoc.* **80:**536–538.

Høi, L., J. L. Larsen, I. Dalsgaard, and A. Dalsgaard. 1998a. Occurrence of *Vibrio vulnificus* in Danish marine environments. *Appl. Environ. Microbiol.* **64:**7–13.

Høi, L., I. Dalsgaard, A. DePaola, R. J. Siebeling, and A. Dalsgaard. 1998b. Heterogeneity among isolates of *Vibrio vulnificus* recovered from eels (*Anguilla anguilla*) in Denmark. *Appl. Environ. Microbiol.* **64:**4676–4682.

Høi, L., I. Dalsgaard, and A. Dalsgaard. 1998c. Improved isolation of *Vibrio vulnificus* from seawater and sediment with cellobiose-colistin agar. *Appl. Environ. Microbiol.* **64:**1721–1724.

Høi, L., A. Dalsgaard, J. L. Larsen, J. M. Warner, and J. D. Oliver. 1997. Comparison of ribotyping and randomly amplified polymorphic DNA PCR for characterization of *Vibrio vulnificus. Appl. Environ. Microbiol.* **63:**1674–1678.

Horré, R., G. Marklein, and K. P. Schall. 1996. *Vibrio vulnificus*, an emerging human pathogen. *Zentbl. Bakteriol.* **284:**273–284.

Howard, R. J., and S. Lieb. 1988. Soft-tissue infections caused by halophilic marine vibrios. *Arch. Surg.* **123:**245–249.

Hsueh, P.-R., C.-Y. Lin, H.-J. Tang, H.-C. Lee, J.-W. Liu, Y.-C. Liu, and Y.-C. Chuang. 2004. *Vibrio vulnificus* in Taiwan. *Emerg. Infect. Dis.* **10:**1363–1368.

Hülsmann, A., T. M. Rosche, I.-S. Kong, H. M. Hassan, D. M. Beam, and J. D. Oliver. 2003. RpoS-dependent stress response and exoenzyme production in *Vibrio vulnificus. Appl. Environ. Microbiol.* **69:**6114–6120.

Jackson, J. K., R. L. Murphree, and M. L. Tamplin. 1997. Evidence that mortality from *Vibrio vulnificus* infection results from single strains among heterogeneous populations in shellfish. *J. Clin. Microbiol.* **35:**2098–2101.

Johnston, J. M., S. F. Becker, and L. M. McFarland. 1986. Gastroenteritis in patients with stool isolations of *Vibrio vulnificus. Am. J. Med.* **80:**336–338.

Jones, S. H., and B. Summer-Brason. 1998. Detection of pathogenic *Vibrio* spp. in a northern New England estuary, USA. *J. Shellfish Res.* **17:**1665–1669.

Joseph, L. A., and A. C. Wright. 2004. Expression of *Vibrio vulnificus* capsular polysaccharide inhibits biofilm formation. *J. Bacteriol.* **186:**889–893.

Kaysner, C. A., C. Abeyta, Jr., M. M. Wekell, A. DePaola, Jr., R. F. Stott, and J. M. Leitch. 1987. Virulent strains of *Vibrio vulnificus* isolated from estuaries of the United States west coast. *Appl. Environ. Microbiol.* **53:**1349–1351.

Kaysner, C. A., M. L. Tamplin, M. M. Wekell, R. F. Stott, and K. G. Colburn. 1989. Survival of *Vibrio vulnificus* in shellstock and shucked oysters (*Crassostrea gigas* and *Crassostrea virginica*) and effects of isolation medium on recovery. *Appl. Environ. Microbiol.* 55:3072–3079.

Kaysner, C. A., C. Abeyta, Jr., R. F. Stott, M. H. Krane, and M. M. Wekell. 1990. Enumeration of *Vibrio* species, including *V. cholerae*, from samples of an oyster growing area, Grays Harbor, Washington. *J. Food Prot.* 53:300–302.

Kelly, M. T. 1982. Effect of temperature and salinity on *Vibrio* (*Beneckea*) *vulnificus* occurrence in a Gulf Coast environment. *Appl. Environ. Microbiol.* 44:820–824.

Kelly, M. T., and E. M. D. Stroh. 1988. Occurrence of *Vibrionaceae* in natural and cultivated oyster populations in the Pacific northwest. *Diagn. Microbiol. Infect. Dis.* 9:1–5.

Kim, M. S., and H. D. Jeong. 2001. Development of 16S rRNA targeted PCR methods for the detection and differentiation of *Vibrio vulnificus* in marine environments. *Aquaculture* 193:199–211.

Kim, R. Y., and J. H. Rhee. 2003. Flagellar basal body *flg* operon as a virulence determinant of *Vibrio vulnificus*. *Biochem. Biophys. Res. Commun.* 304:405–410.

Kim, S. Y., S. E. Lee, Y. R. Kim, C. M. Kim, P. Y. Ryu, H. E. Choy, S. S. Chung, and J. H. Rhee. 2003. Regulation of *Vibrio vulnificus* virulence by the LuxS quorum-sensing system. *Mol. Microbiol.* 48:1647–1664.

Kim, Y. R., S. E. Lee, C. M. Kim, S. Y. Kim, E. K. Shin, D. H. Shin, S. S. Chung, H. E. Choy, A. Progulske-Fox, J. D. Hillman, M. Handfield, and J. H. Rhee. 2003. Characterization and pathogenic significance of *Vibrio vulnificus* antigens preferentially expressed in septicemic patients. *Infect. Immun.* 71:5461–5471.

Kim, Y. R., S. Y. Kim, C. M. Kim, S. E. Lee, and J. H. Rhee. 2005. Essential role of an adenylate cyclase in regulating *Vibrio vulnificus* virulence. *FEMS Microbiol. Lett.* 243:497–503.

Klontz, K. C., S. Lieb, M. Schreiber, H. T. Janowski, L. M. Baldy, and R. A. Gunn. 1988. Syndromes of *Vibrio vulnificus* infections. Clinical and epidemiological features in Florida cases, 1981-1987. *Ann. Intern. Med.* 109:318–323.

Kong, I.-S., T. C. Bates, A. Hülsmann, H. Hassan, and J. D. Oliver. 2004. Role of catalase and *oxyR* in the viable but nonculturable state of *Vibrio vulnificus*. *FEMS Microbiol. Ecol.* 50:133–142.

Lee, J.-H., J. B. Rho, K.-J. Park, C. B. Kim, Y.-S. Han, S. H. Choi, K.-H. Lee, and S.-J. Park. 2004. Role of flagellum and motility in pathogenesis of *Vibrio vulnificus*. *Infect. Immun.* 72:4905–4910.

Levine, W. C., P. M. Griffin, and the Gulf Coast *Vibrio* Working Group. 1993. *Vibrio* infections on the Gulf Coast: results of first year of regional surveillance. *J. Infect. Dis.* 167:479–483.

Lin, M., and J. R. Schwartz. 2003. Seasonal shifts in population structure of *Vibrio vulnificus* in an estuarine environment as revealed by partial 16S ribosomal DNA sequencing. *FEMS Microbiol. Ecol.* 45:23–27.

Linkous, D. A., and J. D. Oliver. 1999. Pathogenesis of *Vibrio vulnificus*. *FEMS Microbiol. Lett.* 174:207–214.

Lowry, P. W., L. M. McFarland, B. H. Peltier, N. C. Roberts, H. B. Bradford, J. L. Herndon, D. F. Stroup, J. B. Mathison, P. A. Blake, and R. A. Gunn. 1989. *Vibrio* gastroenteritis in Louisiana: a prospective study among attendees of a scientific congress in New Orleans. *J. Infect. Dis.* 160:978–984.

Maruo, K., T. Akaike, T. Ono, and H. Maeda. 1998. Involvement of bradykinin generation in intravascular dissemination of *Vibrio vulnificus* and prevention of invasion by a bradykinin antagonist. *Infect. Immun.* 66:866–869.

Massad, G., and J. D. Oliver. 1987. New selective and differential plating medium for *Vibrio vulnificus* and *Vibrio cholerae*. *Appl. Environ. Microbiol.* 53:2262–2264.

Matté, G. R., M. H. Matté, I. G. Rivera, and M. T. Martins. 1994. Distribution of potentially pathogenic vibrios in oysters from a tropical region. *J. Food Prot.* 57:870–873.

McDougald, D., S. A. Rice, and S. Kjelleberg. 2001. SmcR-dependent regulation of adaptive phenotypes in *Vibrio vulnificus*. *J. Bacteriol.* 183:758–762.

McDougald, D., S. Srinivasan, S. A. Rice, and S. Kjelleberg. 2003. Signal-mediated cross-talk regulates stress adaptation in *Vibrio* species. *Microbiology* 149:1923–1933.

McPherson, V. L., J. A. Watts, L. M. Simpson, and J. D. Oliver. 1991. Physiological effects of the lipopolysaccharide of *Vibrio vulnificus* on mice and rats. *Microbios* 67:272–273.

Mead, P. S., L. Slusker, V. Dietz, L. F. McCaig, J. S. Bresee, C. Shapiro, P. M. Griffin, and R. B. V. Tauxe. 1999. Food-related illness and death in the United States. *Emerg. Infect. Dis.* 5:607–625.

Melhus, Å., T. Holmdahl, and I. Tjernberg. 1995. First documented case of bacteremia with *Vibrio vulnificus* in Sweden. *Scand. J. Infect. Dis.* 27:81–82.

Merkel, S. M., S. Alexander, E. Zufall, J. D. Oliver, and Y. M. Huet-Hudson. 2001. Essential role for estrogen in protection against *Vibrio vulnificus* induced endotoxic shock. *Infect. Immun.* 69:6119–6122.

Mertens, A., J. Nagler, W. Hansen, and E. Gepts-Friedenreich. 1979. Halophilic, lactose-positive *Vibrio* in a case of fatal septicemia. *J. Clin. Microbiol.* 9:233–235.

Miyoshi, S., E. G. Oh, K. Hirata, and S. Shinoda. 1993. Exocellular toxic factors produced by *Vibrio vulnificus*. *J. Toxicol. Toxin Rev.* 12:253–288.

Montanari, M. P., C. Pruzzo, L. Pane, and R. R. Colwell. 1999. Vibrios associated with plankton in a coastal zone of the Adriatic Sea (Italy). *FEMS Microbiol. Ecol.* 29:241–247.

Morris, J. G., Jr., A. C. Wright, D. M. Roberts, P. K. Wood, L. M. Simpson, and J. D. Oliver. 1987a. Identification of environmental *Vibrio vulnificus* isolates with a DNA probe for the cytotoxin-hemolysin gene. *Appl. Environ. Microbiol.* 53:193–195.

Morris, J. G., Jr., A. C. Wright, L. M. Simpson, P. K. Wood, D. E. Johnson, and J. D. Oliver. 1987b. Virulence of *Vibrio vulnificus*: association with utilization of transferrin-bound iron, and lack of correlation with levels of cytotoxin or protease production. *FEMS Microbiol. Lett.* 40:55–59.

Motes, M. L., and A. DePaola. 1996. Offshore suspension relaying to reduce levels of *Vibrio vulnificus* in oysters (*Crassostrea virginica*). *Appl. Environ. Microbiol.* 62:3875–3877.

Motes, M. L., A. DePaola, D. W. Cook, J. E. Veazey, J. C. Hunsucker, W. E. Garthright, R. J. Blodgett, and S. Chirtel. 1998. Influence of water temperature and salinity on *Vibrio vulnificus* in Northern Gulf and Atlantic Coast oysters (*Crassostrea virginica*). *Appl. Environ. Microbiol.* 64:1459–1465.

Nilsson, W. B., R. N. Paranjype, A. DePaola, and M. S. Strom. 2003. Sequence polymorphism of the 16S rRNA gene of *Vibrio vulnificus* is a possible indicator of strain virulence. *J. Clin. Microbiol.* 41:442–446.

Oliver, J. D. 1989. *Vibrio vulnificus*, p. 569–599. *In* M. P. Doyle (ed.), *Foodborne Bacterial Pathogens*. Marcel Dekker, Inc., New York, N.Y.

Oliver, J. D. 1993. Formation of viable but nonculturable cells, p. 239–272. *In* S. Kjelleberg (ed.), *Starvation in Bacteria*. Plenum Press, New York, N.Y.

Oliver, J. D. 2000a. Problems in detecting dormant (VBNC) cells and the role of DNA elements in this response, p. 1–15. *In* J. K. Jansson, J. D. van Elsas, and M. J. Bailey (ed.), *Tracking Genetically-Engineered Microorganisms*. Landes Biosciences, Georgetown, Tex.

Oliver, J. D. 2000b. The public health significance of viable but nonculturable bacteria, p. 277–300. *In* R. R. Colwell and D. J.

Grimes (ed.), *Nonculturable Microorganisms in the Environment*. ASM Press, Washington, D.C.

Oliver, J. D. 2003. Culture media for the isolation and enumeration of pathogenic *Vibrio* species in foods and environmental samples, p. 249–269. *In* J. E. L. Corry, G. D. W. Curtis, and R. M. Baird (ed.), *Culture Media for Food Microbiology*, 2nd ed. Elsevier, Amsterdam, The Netherlands.

Oliver, J. D. 2005a. *Vibrio vulnificus*. *In* S. Belkin and R. Colwell (ed.), *Oceans and Health: Pathogens in the Marine Environment*. Springer, New York, N.Y.

Oliver, J. D. 2005b. Wound infections caused by *Vibrio vulnificus* and other marine bacteria. *Epidemiol. Infect.* **133**:383–391.

Oliver, J. D., and R. Bockian. 1995. In vivo resuscitation, and virulence towards mice, of viable but nonculturable cells of *Vibrio vulnificus*. *Appl. Environ. Microbiol.* **61**:2620–2623.

Oliver, J. D., and J. B. Kaper. 2001. *Vibrio* species, p. 263–300. *In* M. P. Doyle, L. R. Beuchat, and T. J. Montville (ed.), *Food Microbiology: Fundamentals and Frontiers*. ASM Press, Washington, D.C.

Oliver, J. D., R. A. Warner, and R. R. Cleland. 1983. Distribution of *Vibrio vulnificus* and other lactose-fermenting vibrios in the marine environment. *Appl. Environ. Microbiol.* **45**:985–998.

Oliver, J. D., J. E. Wear, M. B. Thomas, M. Warner, and K. Linder. 1986a. Production of extracellular enzymes and cytotoxicity by *Vibrio vulnificus*. *Diagn. Microbiol. Infect. Dis.* **5**:99–111.

Oliver, J. D., D. M. Roberts, V. K. White, M. A. Dry, and L. M. Simpson. 1986b. Bioluminescence in a strain of the human bacterial pathogen, *Vibrio vulnificus*. *Appl. Environ. Microbiol.* **52**: 1209–1211.

Oliver, J. D., F. Hite, D. McDougald, N. L. Andon, and L. M. Simpson. 1995. Entry into, and resuscitation from, the viable but nonculturable state by *Vibrio vulnificus* in an estuarine environment. *Appl. Environ. Microbiol.* **61**:2624–2630.

Oliver, J. D., K. Guthrie, J. Preyer, A. Wright, L. M. Simpson, R. Siebeling, and J. G. Morris, Jr. 1992. Use of colistin-polymyxin B-cellobiose agar in the isolation of *Vibrio vulnificus* from the environment. *Appl. Environ. Microbiol.* **58**:737–739.

O'Neill, K. R., S. H. Jones, and D. J. Grimes. 1990. Incidence of *Vibrio vulnificus* in northern New England water and shellfish. *FEMS Microbiol. Lett.* **72**:163–168.

Osaka, K., M. Komatsuzaki, H. Takahashi, S. Sakano, and N. Okabe. 2004. *Vibrio vulnificus* septicaemia in Japan: an estimated number of infections and physicians' knowledge of the syndrome. *Epidemiol. Infect.* **132**:993–996.

Park, S. D., H. S. Shon, and N. J. Joh. 1991. *Vibrio vulnificus* septicemia in Korea: clinical and epidemiological findings in seventy patients. *J. Am. Acad. Dermatol.* **24**:397–403.

Parvathi, A., H. S. Kumar, I. Karunasagar, and I. Karunasagar. 2004. Detection and enumeration of *Vibrio vulnificus* in oysters from two estuaries along the southwest coast of India, using molecular methods. *Appl. Environ. Microbiol.* **70**:6909–6913.

Penman, A. B., and D. C. Lanier, Jr. 1993. *Vibrio vulnificus* wound infections from the Mississippi Gulf Coast waters: June to August 1993. *South. Med. J.* **88**:531–533.

Pfeffer, C. S., M. F. Hite, and J. D. Oliver. 2003. Ecology of *Vibrio vulnificus* in estuarine waters of eastern North Carolina. *Appl. Environ. Microbiol.* **69**:3526–3531.

Poole, M. D., and J. D. Oliver. 1978. Experimental pathogenicity and mortality in ligated ileal loop studies of the newly reported halophilic lactose-positive *Vibrio* species. *Infect. Immun.* **20**:126–129.

Randa, M. A., M. F. Potz, and E. Lim. 2004. Effects of temperature and salinity on *Vibrio vulnificus* population dynamics as assessed by quantitative PCR. *Appl. Environ. Microbiol.* **70**:5469–5476.

Rho, H.-W., M.-J. Choi, J.-N. Lee, J.-W. Park, J. S. Kim, B.-H. Park, H.-S. Sohn, and H.-R. Kim. 2002. Cytotoxic mechanism of *Vibrio vulnificus* cytolysin in CPAE cells. *Life Sci.* **70**:1923–1934.

Rice, S. A., D. McDougald, and S. Kjelleberg. 2000. *Vibrio vulnificus*: a physiological and genetic approach to the viable but nonculturable response. *J. Infect. Chemother.* **6**:115–120.

Roberts, N. C., R. J. Siebeling, J. B. Kaper, and H. B. Bradford, Jr. 1982. Vibrios in the Louisiana Gulf Coast environment. *Microb. Ecol.* **8**:299–312.

Rosche, T. M., Y. Yano, and J. D. Oliver. 2005. A rapid and simple PCR analysis indicates there are two subgroups of *Vibrio vulnificus* which correlate with clinical or environmental isolation. *Microbiol. Immunol.* **49**:381–389.

Ross, E. E., L. Guyer, J. Varnes, and G. Rodrick. 1994. *Vibrio vulnificus* and molluscan shellfish: the necessity of education for high risk individuals. *J. Am. Diet. Assoc.* **94**:312–314.

Sanjuán, E., and C. Amaro. 2004. Protocol for specific isolation of virulent strains of *Vibrio vulnificus* serovar E (biotype 2) from environmental samples. *Appl. Environ. Microbiol.* **70**:7024–7032.

Shao, C.-P., and L. I. Hor. 2001. Regulation of metalloprotease gene expression in *Vibrio vulnificus* by a *Vibrio harveyi* LuxR homologue. *J. Bacteriol.* **183**:1369–1375.

Shapiro, R. L., S. Altekruse, L. Hutwagner, R. Bishop, R. Hammond, S. Wilson, B. Ray, S. Thompson, R. V. Tauxe, and P. M. Griffin. 1998. The role of Gulf Coast oysters harvested in warmer months in *Vibrio vulnificus* infections in the United States, 1988–1996. *J. Infect. Dis.* **178**:752–759.

Simonson, J. G., and R. J. Siebeling. 1993. Immunogenicity of *Vibrio vulnificus* capsular polysaccharides and polysaccharide-protein conjugates. *Infect. Immun.* **61**:2053–2058.

Simpson, L. M., and J. D. Oliver. 1987. Ability of *Vibrio vulnificus* to obtain iron from transferrin and other iron-binding proteins. *Curr. Microbiol.* **15**:155–157.

Simpson, L. M., V. K. White, S. F. Zane, and J. D. Oliver. 1987. Correlation between virulence and colony morphology in *Vibrio vulnificus*. *Infect. Immun.* **55**:269–272.

Sloan, E. M., C. J. Hagen, G. A. Lancette, J. T. Peeler, and J. N. Sofos. 1992. Comparison of five selective enrichment broths and two selective agars for recovery of *Vibrio vulnificus* from oysters. *J. Food Prot.* **55**:356–359.

Sokolova, I. M., L. Leamy, M. Harrison, and J. D. Oliver. 2005. Intrapopulational variation in *Vibrio vulnificus* levels in *Crassostrea virginica* (Gmelin 1971) is associated with the host size but not with disease status or developmental stability. *J. Shellfish Res.* **24**:503–508.

Stelma, G. N., Jr., A. L. Reyes, J. T. Peeler, C. H. Johnson, and P. L. Spaulding. 1992. Virulence characteristics of clinical and environmental isolates of *Vibrio vulnificus*. *Appl. Environ. Microbiol.* **58**:2776–2782.

Strom, M. S., and R. N. Paranjpye. 2000. Epidemiology and pathogenesis of *Vibrio vulnificus*. *Microbes Infect.* **2**:177–188.

Sun, Y., and J. D. Oliver. 1995. The value of CPC agar for the isolation of *Vibrio vulnificus* from oysters. *J. Food Prot.* **58**:439–440.

Tacket, C. O., F. Brenner, and P. A. Blake. 1984. Clinical features and an epidemiological study of *Vibrio vulnificus* infections. *J. Infect. Dis.* **149**:558–561.

Tamplin, M., G. E. Rodrick, N. J. Blake, and T. Cuba. 1982. Isolation and characterization of *Vibrio vulnificus* from two Florida estuaries. *Appl. Environ. Microbiol.* **44**:1466–1470.

Tamplin, M. L., J. K. Jackson, C. Buchrieser, R. L. Murphree, K. M. Portier, V. Gangar, L. G. Miller, and C. W. Kaspar. 1996. Pulsed-field gel electrophoresis and ribotype profiles of clinical and environmental *Vibrio vulnificus* isolates. *Appl. Environ. Microbiol.* **62**:3572–3580.

Thompson, F. L., T. Iida, and J. Swings. 2004. Biodiversity of vibrios. *Microbiol. Mol. Biol. Rev.* **68**:403–431.

Tilton, R. C., and R. W. Ryan. 1987. Clinical and ecological characterisics of *Vibrio vulnificus* in the northeastern United States. *Diagn. Microbiol. Infect. Dis.* **6**:109–117.

Tison, D. L., M. Nishibuchi, J. D. Greenwood, and R. J. Seidler. 1982. *Vibrio vulnificus* biogroup 2: new biogroup pathogenic for eels. *Appl. Environ. Microbiol.* **44:**640–646.

Torres, L., S. Escobar, A. I. López, M. L. Marco, and V. Pobo. 2002. Wound infection due *Vibrio vulnificus* in Spain. *Eur. J. Clin. Microbiol. Infect. Dis.* **21:**537–538.

Tuttle, J., S. Kellerman, and R. V. Tauxe. 1994. The risks of raw shellfish: what every transplant patient should know. *J. Transplant. Coord.* **4:**60–63.

Upton, A., and S. Taylor. 2002. *Vibrio vulnificus* necrotizing fasciitis and septicaemia. *N. Z. Med. J.* **115:**108–109.

Vanoy, R. W., M. L. Tamplin, and J. R. Schwarz. 1992. Ecology of *Vibrio vulnificus* in Galveston Bay oysters, suspended particulate matter, sediment and seawater: detection by monoclonal antibody − immunoassay − most probable number procedures. *J. Ind. Microbiol.* **9:**219-223.

Veenstra, J., P. J. G. M. Rietra, J. M. Coster, E. Slaats, and S. Dirks-Go. 1994. Seasonal variations in the occurrence of *Vibrio vulnificus* along the Dutch coast. *Epidemiol. Infect.* **112:**285–290.

Veenstra, J., P. J. G. M. Rietra, J. M. Coster, C. P. Stoutenbeek, E. A. Ter Laak, O. L. M. Haenen, H. H. W. De Gier, and S. Dirks-Go. 1993. Human *Vibrio vulnificus* infections and environmental isolates in the Netherlands. *Aquacult. Fish. Manag.* **24:**119–122.

Venkateswaran, K., H. Nakano, T. Okabe, K. Takayama, O. Matsuda, and H. Hashimoto. 1989. Occurrence and distribution of *Vibrio* spp., *Listonella* spp., and *Clostridium botulinum* in the Seto Inland Sea of Japan. *Appl. Environ. Microbiol.* **55:**559–567.

Warner, J. M., and J. D. Oliver. 1998. Randomly amplified polymorphic DNA (RAPD) analysis of clinical and environmental isolates of *Vibrio vulnificus* and other *Vibrio* species. *Appl. Environ. Microbiol.* **65:**1141–1144.

Williams, L. A., and P. LaRock. 1985. Temporal occurrence of *Vibrio* species and *Aeromonas hydrophila* in estuarine sediments. *Appl. Environ. Microbiol.* **50:**1490–1495.

Wilson, B. A., and A. A. Salyers. 2003. Is the evolution of bacterial pathogens an out-of-body experience? *Trends Microbiol.* **11:**347–350.

Wong, H.-C., S. Y. Chen, M.-Y. Chen, J. D. Oliver, L.-I. Hor, and W.-C. Tsai. 2004. Pulsed-field electrophoresis analysis of *Vibrio vulnificus* strains isolated from Taiwan and the United States. *Appl. Environ. Microbiol.* **70:**5153–5158.

Wright, A. C., and J. G. Morris, Jr. 1991. The extracellular cytolysin of *Vibrio vulnificus*: inactivation and relationship to virulence in mice. *Infect. Immun.* **59:**192–197.

Wright, A. C., L. M. Simpson, and J. D. Oliver. 1981. Role of iron in the pathogenesis of *Vibrio vulnificus* infections. *Infect. Immun.* **34:**503–507.

Wright, A. C., J. G. Morris, Jr., D. R. Maneval, Jr., K. Richardson, and J. B. Kaper. 1985. Cloning of the cytotoxin-hemolysin gene of *Vibrio vulnificus*. *Infect. Immun.* **50:**922–924.

Wright, A. C., J. L. Powell, M. K. Tanner, L. A. Ensor, A. B. Karpas, J. G. Morris, Jr., and M. B. Sztein. 1999. Differential expression of *Vibrio vulnificus* capsular polysaccharide. *Infect. Immun.* **67:**2250–2257.

Yano, Y., M. Yokoyama, M. Satomi, H. Oikawa, and S.-S. Chen. 2004. Occurrence of *Vibrio vulnificus* in fish and shellfish available from markets in China. *J. Food Prot.* **67:**1617–1623.

The Biology of Vibrios
Edited by F. L. Thompson et al.
© 2006 ASM Press, Washington, D.C.

Chapter 26

Miscellaneous Human Pathogens

MITSUAKI NISHIBUCHI

INTRODUCTION

Besides three clinically important species in the genus *Vibrio*—*V. cholerae*, *V. parahaemolyticus*, and *V. vulnificus*—nine other *Vibrio* species have been isolated from human infections (Table 1). They include *V. mimicus*, *Grimontia hollisae*, *V. fluvialis*, and *V. furnissii*, which are solely or principally isolated from gastroenteritis cases, and *V. alginolyticus* and *Photobacterium damselae*, which are chiefly associated with wound infections. In addition, *V. metschnikovii*, *V. cincinnatiensis*, and *V. harveyi* are chiefly isolated from extraintestinal infections (i.e., septicemias and wounds), but these isolations are less frequent. All of these species, except for *V. alginolyticus* and *V. metschnikovii*, were first described after 1981. Whether all or some strains from the nine species are human pathogens is not clear. Although possible virulence factors are known, the exact pathogenic mechanism, if any, is not understood. The nine species are currently considered to be potential human pathogens and are reportable in public health surveys from the marine environment and seafood. Physicians are alerted to possible infections with these species, particularly among patients with underlying diseases. Three species are considered to be pathogens of aquatic animals important to aquaculture. Phenotypic characteristics useful for differentiation are shown in Table 2. Each species is described further below.

Grimontia (Vibrio) hollisae

The name *Vibrio hollisae* was given to 16 strains (designated as EF-13) isolated from the stools of patients with diarrhea in the United States in 1982 (Hickman et al., 1982). Subsequently, human infections due to this species were reported infrequently in the United States. Thirty cases of *V. hollisae* infec-

tion were reviewed in 1994 (Abbott and Janda, 1994). Most illnesses (87%) were gastroenteritis among otherwise healthy individuals who consumed raw oysters, clams, or shrimp, and the rest were septicemia or wound infections. Reports of *V. hollisae* infection from France (Gras-Rouzet et al., 1996) and Indonesia (Lesmana et al., 2002) were subsequently published, but the number of infections is still small. Based on analysis of the 16S rRNA gene sequence, Thompson et al. (2003) proposed the transfer of the species from *Vibrio* to a new genus, *Grimontia*. Isolation of this species from environmental samples is limited. This organism can be isolated from fish intestine, cultivated oyster beds, and mussels (Nishibuchi et al., 1988; Kelly and Stroh, 1988; Cavallo and Stabili, 2002).

The rare isolation of this species is partly due to its unique growth and biochemical characteristics. *G. hollisae* does not grow, or grows very poorly, on selective isolation media for enteric pathogens, including thiosulfate citrate bile salt sucrose (TCBS) agar and MacConkey agar (Hickman et al., 1982). Therefore, *G. hollisae* is rarely isolated in routine screening of clinical specimens. Identification by standard biochemical tests is not easy, because it is biochemically inert where addition of 1% (wt/vol) NaCl to test media is recommended (Hickman et al., 1982; Janda et al., 1988). Therefore, Vuddhakul et al. (2000) developed PCR assays to detect unique bases in the *toxR* and *gyrB* genes to accurately identify isolates. A differential medium designed to isolate *G. hollisae* (Nishibuchi et al., 1988) and PCR identification methods would help confirm the presence of *G. hollisae* in the environment and diagnose infection. The unique biochemical characteristics suggest that *G. hollisae* is distinct from other vibrios.

Two factors in *G. hollisae* that may play roles in diarrhea were studied. *G. hollisae* produces a hemo-

Mitsuaki Nishibuchi • Center for Southeast Asian Studies, Kyoto University, 46 Shimoadachi-cho, Yoshida, Sakyo-ku, Kyoto 606-8501, Japan.

Table 1. Reports of infections by potentially pathogenic *Vibrio* species[a]

Species	Infection of humans				Infection of aquatic animals
	Gastroenteritis/diarrhea	Wound	Septicemia	Ear/eyes	
V. mimicus	++	+	+	+	
V. (Grimontia) hollisae	++	+	+		
V. fluvialis	++	+	+	+	
V. furnissii	+				
V. alginolyticus	+	++	+	++	+
P. damselae		++	+		++
V. metschnikovii	+	+	+		
V. cincinnatiensis	+		+		
V. harveyi (V. carchariae)		+			++

[a] ++, Infection reported and relatively important; +, infection reported. Frequency of the infection is described in more detail in the text.

lysin similar to the TDH (thermostable direct hemolysin) of *V. parahaemolyticus*. However, the hemolysin of *G. hollisae* is heat-labile, but its biological characteristics are similar to those of *V. parahaemolyticus* TDH (Yoh et al., 1986, 1988). The *tdh* gene coding for this hemolysin is 93% identical to the *tdh* gene of *V. parahaemolyticus* (Yamasaki et al., 1991). The *G. hollisae tdh* gene was detected in all strains tested, including a rare environmental isolate (Nishibuchi et al., 1985, 1996). A *G. hollisae* strain produces the heat-labile factor that elongates CHO cells and induces intestinal fluid accumulation in suckling mice, but it is distinct from cholera toxin (Kothary et al., 1995). The ability to adhere to and invade cultured cells was studied as a possible virulence factor involved in systemic infections (Miliotis et al., 1995). In addition, Okujo and Yamamoto (1994) studied a siderophore produced by *G. hollisae*.

Photobacterium damselae

P. damselae is responsible for rare infections in humans and aquatic animals. It was first described as a pathogen causing skin ulcers on damselfish (Love et al., 1981). *P. damselae* strains, isolated from six wound infections, were received by the Centers for Disease Control and Prevention between 1971 and 1981 (Morris et al., 1982). Most of the patients had a history of exposure to seawater or brackish water. Overall, the wounds were not severe, although surgical debridement was performed in most cases. Subsequently, *P. damselae* was isolated from more severe infections, e.g., a fatal wound infection (Clarridge and Zighelboim-Daum, 1985), cellulitis later requiring fasciotomy and amputation (Coffey et al., 1986), and fatal necrotizing fasciitis (Yuen et al., 1993; Goodell et al., 2004; Yamane et al., 2004). It was reported

that some of the patients had underlying diseases. For example, *P. damselae* was isolated from a fatal primary septicemia possibly associated with consumption of raw eels, and the patient had severe underlying disease (Shin et al., 1996). However, the frequency of wound infection by *P. damselae* is low compared with that of other pathogenic *Vibrio* species. As an example, of 29 cases of wound infections by *Vibrio* spp. reported from the U.S. Gulf Coast states in 1989, only one case was attributed to *P. damselae* whereas five cases were caused by *V. alginolyticus* (Levine et al., 1993). Furthermore, 1 and 11 cases were attributable to infection by *P. damselae* and *V. alginolyticus*, respectively, among 56 wound infections by *Vibrio* spp. in Florida between 1981 and 1988 (Desenclos et al., 1991).

Besides being a pathogen for humans, *P. damselae* is a potential pathogen for various aquatic animals, including damselfish, yellowtail, sea bream, barramundi, brown shark, turbot, rainbow trout, eels, bottlenose dolphins, and turtles (Love et al., 1981; Ketterer and Eaves, 1992; Pedersen et al., 1997).

The organism is distributed in the marine environment throughout the world, being isolated from a water sample from the Gulf Coast (Love et al., 1981) and from oysters from the Pacific Northwest (Kelly and Stroh, 1988). Environmental surveys in Senegal (Schandevyl et al., 1984), Spain (Alcaide, 2003), Turkey (Yalcinkaya et al., 2003), and Malaysia (Elhadi et al., 2004) revealed the presence of *P. damselae* as well as other potentially pathogenic *Vibrio* species.

Wide distribution in the environment and rare cases of infection in humans and animals suggest that host susceptibility is an important factor, and all strains may not be equally virulent. Damselysin, an extracellular cytolytic toxin of *P. damselae*, was studied as a potentially important virulence factor. This

Table 2. Phenotypic characteristics useful for differentiation of human pathogenic *Vibrio* species[a]

Result for:

Characteristic	*V. cholerae*	*V. parahae-molyticus*	*V. vulnificus*	*V. mimicus*	*V. (Grimontia) hollisae*	*V. fluvialis*	*V. furnissii*	*V. alginolyticus*	*P. damselae*	*V. metschnikovii*	*V. cincinnatiensis*	*V. harveyi (carchariae)*
Oxidase	+	+	+	+	+	+	+	+	+	−	+	+
Gas from glucose	−	−	−	−	−	d	+	−	−	+	−	−
O129 sensitivity	+	[−]	+	+	d	+	−	[−]	+	d	[−]	+
Arginine dihydrolase	−	−	−	−	−	+	+	−	+	d	−	−
Lysine decarboxylase	+	+	+	+	−	−	−	+	d	+	+	+
Ornithine decarboxylase	+	+	d	+	−	−	−	d	−	−	−	d
Sucrose fermentation	+	−	[−]	−	−	+	+	+	−	+	+	−
Lactose fermentation	−	−	[+]	[−]	−	−	−	−	−	d	−	−
myo-inositol fermentation	−	−	−	−	−	−	−	−	−	d	+	−
Cellobiose fermentation	−	−	+	−	−	d	[−]	−	−	−	+	d
Salicin fermentation	−	−	+	−	−	−	−	−	−	−	+	d
Voges-Proskauer reaction	d	−	−	+	−	−	−	+	+	+	−	d
Nitrate reduction	+	+	+	+	+	+	+	+	+	−	+	+
Growth with 0% NaCl	+	−	−	+	−	−	−	−	−	−	−	−
Growth with 8% NaCl	−	[+]	−	−	−	d	[+]	+	−	d	+	−

[a] Data from Holt et al. (1994). −, 0–10% positive; [−], 11–25% positive; d, 26–75% positive; [+], 76–89% positive; +, 90–100% positive.

toxin has hemolytic activity, is lethal to mice and cytotoxic to CHO cells, and produces local edema by subcutaneous injection. Production of this toxin and experimental pathogenicity to mice are correlated (Kreger, 1984; Kothary and Kreger, 1985). Damselysin was shown to be phospholipase D (Kreger et al., 1987). The *dly* gene coding for damselysin is present in hemolytic strains but absent in nonhemolytic strains of *P. damselae* and other *Vibrio* species (Cutter and Kreger, 1990). Other *P. damselae* factors studied include a siderophore-mediated iron-uptake system (Fouz et al., 1997), a neuraminidase (Sugita et al., 2000), and an outer membrane porin protein (Gribun et al., 2004). Involvement of these possible virulence factors in infection and pathological processes remains to be determined. The type strain of *P. damselae* produces histamine (Kimura et al., 2000), and it may be of concern in seafood consumption.

P. damselae grows well on TCBS agar (Farmer and Hickman-Brenner, 1992). The unusual biochemical characteristics of *P. damselae* help in identification: *P. damselae*, *V. fluvialis*, and *V. furnissii* are arginine dihydrolase positive and ornithine decarboxylase negative; *V. furnissii* and some strains of *P. damselae* produce gas from glucose (Farmer and Hickman-Brenner, 1992). *V. damselae* was transferred to the genus *Listonella* based on analysis of the 5S rRNA sequence (MacDonell and Colwell, 1985) and then to the genus *Photobacterium* on the basis of phenotypic characteristics (Smith et al., 1991). Currently, *P. damselae* comprises two subspecies, *P. damselae* subsp. *damselae* and *P. damselae* subsp. *piscicida*. *P. damselae* subsp. *damselae* is the former *V. damselae*, and *P. damselae* subsp. *piscicida* is the former *Pasteurella piscicida*, a fish pathogen. A classification to include *Pasteurella piscicida* in *P. damselae* was based on the results of DNA-DNA hybridization and 16S rRNA analysis (Gauthier et al., 1995). Thyssen et al. (1998) proposed that the two subspecies be clearly distinguished morphologically and biochemically. A PCR targeting the capsular polysaccharide gene identifies *P. damselae* but does not differentiate the two subspecies (Rajan et al., 2003); this may be achieved using a multiplex PCR targeting the 16S rRNA sequence and the *ureC* gene (Osorio et al., 2000).

Vibrio alginolyticus

V. alginolyticus was formerly classified as a biotype of *V. parahaemolyticus*; however, it was proposed as a species distinct from *V. parahaemolyticus*, based on sucrose fermentation and other phenotypic characteristics (Sakazaki, 1968). Subsequent genetic studies confirmed that *V. alginolyticus* is genetically distinct from other *Vibrio* species but is closely related to *V. parahaemolyticus* (Staley and Colwell, 1973a,b). *V. alginolyticus* is prevalent in the marine environment and is generally recognized as a saprophytic organism, producing yellow colonies on TCBS (Farmer and Hickman-Brenner, 1992). *V. alginolyticus* has been found to be the predominant *Vibrio* species in many environmental surveys in various parts of the world (Chan et al., 1989; Matte et al., 1994; Barbieri et al., 1999; Cavallo and Stabili, 2002). However, *V. alginolyticus* may be potentially pathogenic; it was first recognized as a human pathogen by Zen-Yoji et al. (1973). Thus, it was isolated from localized tissue infections and first presumed to be *V. parahaemolyticus* but later found to be *V. alginolyticus*. This species has been isolated from wound and ear infections but rarely from blood or eye infections. *V. alginolyticus* is usually found in mixed culture and is infrequently isolated in pure culture from wound and blood (Schmidt et al., 1979; Blake et al., 1980). The organism is self-limiting to 1 or 2 days in most wound infections, where the infections are usually not life-threatening (Blake et al., 1980). *V. alginolyticus* is sensitive to most antibiotics but resistant to penicillins and ampicillin (Schmidt et al., 1979; Joseph et al., 1982; Zanetti et al., 2001). It is difficult, therefore, to evaluate whether antibiotic treatment is effective in the self-limiting wound infections (Blake et al., 1980). Severe necrotizing fasciitis attributed to *V. alginolyticus* has occurred in rare cases when the patients had cirrhosis (Ho et al., 1998) or a history of asthma (Gomez et al., 2003) or were otherwise healthy (Howard et al., 1985). Of the 121 *Vibrio* infections reported from the Gulf Coast states of the United States in 1989, 29 were wound infections attributed to *V. vulnificus* (8 cases), *V. parahaemolyticus* (6 cases), *V. alginolyticus* (5 cases), and others (10 cases) (Levine et al., 1993). Thirty cases of extraintestinal infections by *V. alginolyticus* (17 cases) and *V. parahaemolyticus* (13 cases) occurred in Denmark between 1987 and 1992. Interestingly, all cases attributed to *V. alginolyticus* caused ear infections (Hornstrup and Gahrn-Hansen, 1993). Exposure to seawater is regarded as an important risk factor for the extraintestinal infections by *V. alginolyticus* (Blake et al., 1980; Joseph et al., 1982; Levine et al., 1993; Hornstrup and Gahrn-Hansen, 1993). Reports of the isolation of *V. alginolyticus* from cases of diarrhea or gastroenteritis are rare (Hiratsuka et al., 1980; Aggarwal et al., 1986; Chapman, 1987; Reina et al., 1995; Caccamese and Rastegar, 1999; Uh et al., 2001), suggesting that the enteropathogenic potential of *V. alginolyticus* is, in fact, very low.

Isolation from diseased marine animals and/or the results of experimental infections indicate that *V. alginolyticus* is also potentially pathogenic to various marine animals, including fish (Lee, 1995; Balebona et al., 1998), prawns (Lee et al., 1996), shrimp (Wang and Chen, 2005), bivalve mollusk larvae (Luna-Gonzalez et al., 2002; Gómez-Leon et al., 2005), cuttlefish (Sangster and Smolowitz, 2003), abalone (Liu et al., 2001), and dolphin (Schroeder et al., 1985).

It is not clear whether all or only some strains of *V. alginolyticus* are potentially pathogenic to humans and marine animals. Isolates have been examined for possible virulence factors, with the results highlighting lethality to mice, cytotoxicity, adhesiveness, the ability to induce intestinal fluid accumulation (in rats), and the production of various extracellular enzymes, including caseinase, lipase, amylase, gelatinase, lecithinase, or hemolysin (Molitoris et al., 1985; Hörmansdorfer et al., 2000; Baffone et al., 2003). In particular, the proteolytic extracellular enzymes collagenase and alkaline protease have been well characterized (Reid et al., 1980; Long et al., 1981; Takeuchi et al., 1992; Deane et al., 1989). Using PCR, Sechi et al. (2000) showed that some strains carry the homologues of the virulence pathogenicity island (*vpi*), the *ace* gene, or the *zot* gene of *V. cholerae*. However, it is unclear whether any of these virulence factors are actually associated with pathogenesis. Also, cultures producing histamine (Yoshinaga and Frank, 1982) or tetrodotoxin (Yu et al., 2004) have been recognized. These factors are of public health concern when associated with the consumption of seafood.

Kourany (1983) designed an isolation medium that differentiates *V. alginolyticus* and *V. parahaemolyticus*. Four biochemical tests—fermentation of lactose and sucrose, the Voges-Proskauer test, and tolerance to NaCl—are commonly used to differentiate *V. vulnificus*, *V. parahaemolyticus*, and *V. alginolyticus* (Blake et al., 1979). However, strains showing atypical biochemical characteristics can complicate the differentiation (Schmidt et al., 1979). PCR tests to identify these species are available and involve targeting the 16S rRNA gene (Liu et al., 2004) and a collagenase gene (Di Pinto et al., 2005).

It is relevant to mention that *V. alginolyticus* serves as a representative marine bacterium in many physiological studies. The electrochemical Na^+ gradient (Nakamura et al., 1995; Enomoto et al., 1998; Hayashi et al., 2001), the roles of Na^+ in the active uptake of nutrients (Tokuda et al., 1982), and motility by the polar flagellum (McCarter, 2001; Yorimitsu and Homma, 2001) have all been studied.

Vibrio cincinnatiensis

V. cincinnatiensis was established as a new species in 1986 when it was isolated from a 70-year-old person with septicemia and meningitis (Brayton

et al., 1986). The patient was immunocompetent and had no history of exposure to seafood or seawater (Bode et al., 1986). Subsequently, Wuthe et al. (1993) isolated *V. cincinnatiensis* from the diarrheal stool of a 67-year-old immunocompromised patient and the stomachs of two aborted bovine fetuses from different herds. The sources of these infections were unknown. Overall, *V. cincinnatiensis* has rarely been isolated from clinical cases, and its pathogenic significance is still questionable.

V. cincinnatiensis was initially considered to grow very poorly on TCBS agar medium (Farmer and Hickman-Brenner, 1992). Wuthe et al. (1993) isolated the organism on cefsulodin-irgasan-novobiocin agar, which is used for the isolation of *Yersinia enterocolitica*. However, these workers reported that their cultures grew well on TCBS agar. *V. cincinnatiensis* has been isolated from environmental samples on TCBS agar with or without enrichment culture in alkaline peptone water. Three of 361 *Vibrio* cultures isolated from river water in Japan were *V. cincinnatiensis* (Venkateswaran et al., 1989), and strains were isolated from ~3% of mussel samples examined in Italy (Ripabelli et al., 1999; Cavallo and Stabili, 2002). A quantitative hybridization analysis using a species-specific oligonucleotide probe demonstrated that water samples taken from Chesapeake Bay during April to December 1996 contained 0.5×10^5 to 8×10^5 cells of *V. cincinnatiensis* per liter; 0.1 to 3% of *Vibrio* and *Photobacterium* populations associated with zooplankton consisted of *V. cincinnatiensis* in the same period (Heidelberg et al., 2002a,b).

V. cincinnatiensis is a halophilic vibrio requiring NaCl for growth (Brayton et al., 1986). The ability to ferment *myo*-inositol is unique among human pathogenic species (Brayton et al., 1986). This characteristic and the ability to ferment cellobiose and salicin are useful for the identification of the organism (Wuthe et al., 1993).

The cultures isolated from river water in Japan demonstrated hemolytic and cytotoxic activities (Venkateswaran et al., 1989). However, the pathological significance of the hemolytic/cytotoxic factor(s) is unknown. Wuthe et al. (1993) used standard assays to examine their clinical strains for the production of known virulence factors, and they determined that there was a lack of any heat-stable enterotoxin (a suckling mouse assay), cholera toxin (an immunological assay), or Shiga-like toxins (oligonucleotide probes), or the ability to invade the conjunctiva of the guinea pig.

Vibrio fluvialis

A bacterial strain isolated from a patient with diarrhea from Bahrain in 1975 and phenotypically similar strains isolated from aquatic environments were referred to as "group F" vibrios (Furniss et al., 1977). Similar strains were isolated from >500 patients with diarrhea in Bangladesh between 1976 and 1977 and were designated as "group EF-6" (Huq et al., 1980). These strains grow on TCBS agar and MacConkey agar (Furniss et al., 1977). They are similar to *Aeromonas* because they show intermediate sensitivity to the vibriostatic agent (O129) and are arginine dihydrolase positive and ornithine decarboxylase negative; other characteristics, including NaCl requirement for growth, support their close relationship to *Vibrio* (Furniss et al., 1977; Huq et al., 1980). These strains were eventually classified as a new *Vibrio* species, *V. fluvialis*, in 1981 (Seidler et al., 1980; Lee et al., 1981). *V. fluvialis* may be differentiated into various O antigen groups (Shimada and Sakazaki, 1983; Iguchi et al., 1993; Shimada et al., 1999) and numerous phage types (Suthienkul, 1993), but they share the same H antigen (Shimada and Sakazaki, 1983). The lateral flagella antigen of *V. fluvialis* is common to this species and a closely related group, *V. furnissii*, but not to other *Vibrio* species (Shinoda et al., 1984, 1992). Some O antigens of *V. fluvialis* are identical to those of some strains of *V. cholerae* (Shimada and Sakazaki, 1983).

V. fluvialis is widely distributed in the aquatic environment, where it has been isolated from water, sediment, aquatic animals, and seafoods from various parts of the world (Lee et al., 1981; West et al., 1986; Maugeri et al., 2000; Dumontet et al., 2000; Mazari-Hiriart et al., 2001; Yalcinkaya et al., 2003; Elhadi et al., 2004).

The organism has been implicated in enteric infections. In particular, it was isolated from a large outbreak of diarrhea cases in Bangladesh (Huq et al., 1980) and sporadic cases of diarrhea or gastroenteritis in the United States and other countries (Furniss et al., 1977; Tacket et al., 1982; Chatterjee et al., 1989; Magalhães et al., 1990; Levine et al., 1993; Klontz et al., 1994; Lesmana et al., 2002). Some studies implicated *V. fluvialis* in cases of gastroenteritis after the consumption of contaminated raw oysters and other seafood (Desenclos et al., 1991; Levine et al., 1993). Also, the organism was isolated from some rare extraintestinal infections: wound infections (Varghese et al., 1996), septicemias (Albert et al., 1991), eye infections (Penland et al., 2000), and peritonitis (Ratnaraja et al., 2005).

To determine if *V. fluvialis* produces an enterotoxic factor(s), the ability to induce intestinal fluid accumulation in suckling mice and rabbit ileal loops and to affect CHO cells was assessed. Lockwood et al. (1982) showed that *V. fluvialis* produces four factors affecting CHO cells, including a cytolytic (hemolytic) factor. Nishibuchi and Seidler (1983) reported that *V. fluvialis*, as well as *V. cholerae* non-O1 and *V. mimicus*, produces a heat-labile factor that is active in the suckling mouse assay. Moreover, Kothary et al. (2003) purified a hemolysin showing activity in the suckling mouse assay. This hemolysin was similar to those of *V. cholerae* and *V. mimicus* and is most probably the enterotoxic factor of *V. fluvialis*. However, its role in human diarrhea must be interpreted with some caution, as the role of the *V. cholerae* hemolysin in diarrhea is questioned (Kaper et al., 1995). Miyoshi et al. (2002) showed that the metalloprotease of *V. fluvialis* is similar to that of *V. vulnificus* and has hemagglutinating, permeability-enhancing, and hemorrhagic activities. The siderophores (Yamamoto et al., 1993) and a hemin-binding outer membrane protein (Ahn et al., 2005) are associated with iron acquisition by *V. fluvialis*. However, the importance of these possible virulence factors in intestinal and extraintestinal infections remains to be elucidated.

Vibrio furnissii

V. furnissii was formally classified as a biogroup of *V. fluvialis* (Lee et al., 1981). This biogroup produces gas from glucose and was thus termed aerogenic, whereas the other is anaerogenic. This gas production from glucose is unusual for representatives of the genus *Vibrio*. Both biogroups are arginine dihydrolase positive and ornithine decarboxylase negative, and thus are similar to *Aeromonas*. Subsequently, the two biogroups were shown to be genetically distant: the aerogenic biogroup was established as a new species of the genus *Vibrio*, i.e., *V. furnissii*, and the anaerogenic cultures were retained as *V. fluvialis* (Brenner et al., 1983). *V. furnissii* and *V. fluvialis* share the lateral flagella antigen that is distinct from those of other *Vibrio* species (Shinoda et al., 1992). As with *V. fluvialis*, *V. furnissii* strains are differentiated into various O serogroups, which have use for epidemiological investigations (Shimada et al., 1999; Dalsgaard et al., 1997).

V. furnissii has been isolated from the environment and cases of gastroenteritis from different parts of the world, but the isolation frequency is less than that for *V. fluvialis* (Lee et al., 1981; Brenner et al., 1983; Hickman-Brenner et al., 1984; Schandevyl et al., 1984; West et al., 1986; Magalhães et al., 1990; Wong et al., 1992; Matte et al., 1994; Dalsgaard et al., 1997).

Extracellular proteins (ECPs) and cell-associated factors have been studied in experimental models to determine if *V. furnissii* produces a possible diarrheagenic factor(s), with results indicating that culture filtrates of some strains cause diarrhea and mortality in suckling mice (Nishibuchi et al., 1983; Chikahira and Hamada, 1988). Cultures of some strains induce fluid accumulation in rabbit ileal loops or elongate CHO cells (Chikahira and Hamada, 1988). Magalhães et al. (1993) proposed that a cytolytic (hemolytic) factor was a virulence factor of *V. furnissii*. It has been established that *V. furnissii* produces a protease the antigenicity of which is shared by the proteases of several *Vibrio* species, including *V. fluvialis* (Alam et al., 1995). Furthermore, Kim et al. (2003) demonstrated that the O antigen plays some role in proliferation of *V. furnissii* in the mouse intestine by mutating the gene coding for phosphomannomutase. However, it is uncertain whether all or only some of the strains of *V. furnissii* exert a pathogenic role in gastroenteritis.

V. furnissii was selected to study the complex chitin utilization mechanism of marine bacteria. Thus, the chemotaxis and adhesion to—and catabolism of—chitin was studied (Bassler et al., 1991; Yu et al., 1991; Keyhani and Roseman, 1996a,b; Chitlaru and Roseman, 1996; Bouma and Roseman, 1996a,b; Keyhani et al., 2000; Park et al., 2000).

Vibrio harveyi

Grimes et al. (1984) isolated a halophilic *Vibrio* from a shark that died in captivity. By DNA-DNA hybridization, they showed that the organism constituted a new *Vibrio* species, which they named *V. carchariae*. Subsequently, there has been one report of a human infection by this organism, when an 11-year-old girl was attacked by a shark off the South Carolina coast and *V. carchariae* was isolated in pure culture from the wound 5 days later (Pavia et al., 1989). Consequently, *V. carchariae* became added to the list of potentially pathogenic vibrios. The culture isolated from this patient grew on chocolate, MacConkey, and TCBS agar and was not hemolytic (Pavia et al., 1989).

V. carchariae is recognized as a member of the autochthonous flora in the mouth of sharks (Grimes et al., 1985), and it has been attributed to be the cause of gastroenteritis or infectious necrotizing enteritis in various cultured marine fish. A serine protease has been suggested to be an important virulence factor in these infections (Lee et al., 2002). Pathogenicity has also been demonstrated to abalone (Nicolas et al., 2002).

Polyphasic taxonomic studies showed that *V. carchariae* is a junior synonym of *V. harveyi* (Pedersen et al., 1998; Gauger and Gomez-Chiarri, 2002).

As a result, the name *V. carchariae* was replaced by *V. harveyi* in the list of potentially pathogenic vibrios. Thus, Yalcinkaya et al. (2003) reported the isolation of *V. harveyi* and other vibrios from blue crab during a survey for human-pathogenic vibrios in Turkey.

V. harveyi is a well-known luminous, halophilic bacterium widely distributed in coastal and ocean water, fish, and squid (Farmer and Hickman-Brenner. 1992). The *lux* genes of *V. harveyi* have been well studied, and its luciferase serves as a useful marker for bioscience research (Greer and Szalay, 2002). Moreover, *V. harveyi* is used in bioassays for the detection of toxic chemicals and mutagenic substances (Czyż et al., 2002; Peinado et al., 2002). Also, *V. harveyi* is recognized as an important pathogen in aquaculture, being pathogenic to penaeid shrimp (Karunasagar et al., 1994), rock lobsters (Diggles et al., 2000), various finfish (Zhang and Austin, 2000; Zorrilla et al., 2003; Pujalte et al., 2003), and pearl oysters (Pass et al., 1987). A comparison of different isolates suggests that ECPs, in particular, hemolysins (Zhang et al., 2001), metalloprotease (Teo et al., 2003a), and lipase (Teo et al., 2003b), are associated with pathogenicity in marine animals (Liu et al., 1996). A myovirus-like bacteriophage harbored by *V. harveyi* strains stimulates hemolysin production and excretion of proteins from the cells and therefore contributes to the expression of virulence (Munro et al., 2003; Austin et al., 2003).

Although PCRs to identify *V. harveyi* have been reported (Oakey et al., 2003; Conejero and Hedreyda, 2003, 2004; Hernandez and Olmos, 2004), the distinction of *V. harveyi* from related species, such as *V. campbellii*, is sometimes a difficult task (Thompson et al., 2002; Gomez-Gil et al., 2004). Nevertheless, the *V. carchariae* infection by the shark bite is the only human infection reported for *V. harveyi* and its related species. This suggests that the potential pathogenicity of *V. harveyi* for humans is extremely low. Even if *V. harveyi* is introduced into the human body through injury, the mechanism of virulence may be different from that of aquatic animals.

Vibrio metschnikovii

V. metschnikovii has long been known as an inhabitant of the marine environment (Farmer and Hickman-Brenner, 1992). Lee et al. (1978) defined the characteristics of *V. metschnikovii* using a numerical taxonomic approach when they systematically compared oxidase-negative strains that grow on TCBS agar. Thus, the strains were isolated from fowl, lobsters, cockles, prawns, river water, and sewage. Cultures isolated from crab and shrimp were also identified as *V. metschnikovii* by DNA-DNA hybridization (Farmer et al., 1988). To date, *V. metschnikovii* has been isolated from various seafoods in environmental surveys carried out in Turkey, Italy, and Malaysia (Scoglio et al., 2001; Yalcinkaya et al., 2003; Elhadi et al., 2004). In addition, *V. metschnikovii*, *V. mimicus*, and *Aeromonas* spp. have been isolated from a drinking water reservoir in Russia (Ivanova et al., 2001), indicating that *V. metschnikovii* can share the habitat with freshwater bacteria.

V. metschnikovii was added to the list of potentially human pathogenic *Vibrio* spp. after Jean-Jacques et al. (1981) reported its isolation from the blood of an 82-year-old patient with cholecystitis. Human infections were subsequently reported from various cases, albeit at low frequency. For example, three cases of septicemia involving *V. metschnikovii* were reported (Hansen et al., 1993; Hardardottir et al., 1994). These patients were >70 years old, and one who had underlying disease died. Furthermore, *V. metschnikovii* was isolated from a wound infection in a 64-year-old patient after surgery (Linde et al., 2004). The organism was also isolated from diarrheal fecal specimens in surveys for cholera or diarrhea: five infantile samples out of the samples examined in two hospitals in Peru during 2 years (Dalsgaard et al., 1996), six samples among 4,000 samples in Brazil (Magalhães et al., 1996), and three samples among 4,820 samples from adults in Indonesia (Lesmana et al., 2002). It was difficult to infer the route of infection in these cases. Linde et al. (2004) speculated that *V. metschnikovii* may be responsible for zoonoses, because it was isolated from animals in Germany where humans may be infected through the consumption of contaminated food.

ECPs may play some role in the disease process. Thus, Miyake et al. (1989) reported a unique cytolysin that was hemolytic and cytotoxic and was able to induce intestinal fluid accumulation in infant mice and to stimulate vascular permeability in rabbit skin. Alam et al. (1995) suggested that *V. metschnikovii* may produce more than one protease. Indeed, Kwon et al. (1994, 1995) reported six alkaline serine proteases and a metalloprotease. Moreover, *V. metschnikovii* was found to contain large plasmids, but their functions were not investigated (Dalsgaard et al., 1996; Linde et al., 2004). Interestingly, the *V. metschnikovii* strain isolated from mussels in Italy was not hemolytic, did not produce mucinase and elastase, but produced caseinase and lecithinase (Scoglio et al., 2001). Overall, the results suggest strain variation in the virulence of *V. metschnikovii*.

V. metschnikovii usually grows as yellow colonies on TCBS agar (Lee et al., 1978; Jean-Jacques et al., 1981). It also grows on blood agar containing 0.5% (wt/vol) NaCl due to a requirement for low concentrations of NaCl (Jean-Jacques et al., 1981). These growth characteristics and awareness of this species as a potential pathogen have helped in the isolation of *V. metschnikovii* during recent surveys for cholera and from wound infections and septicemias. The organism can be differentiated from other human pathogenic vibrios by negative oxidase and nitrate reduction tests (Lee et al., 1978; Farmer et al., 1988).

Vibrio mimicus

V. mimicus is similar to *V. cholerae*, particularly the non-O1/O139 serotypes (Davis et al., 1981). They share most biochemical pathways and some O antigens (Davis et al., 1981; Shinoda et al., 2004). *V. mimicus* strains grow as green to blue colonies on TCBS agar and are differentiated from *V. cholerae* as "sucrose-negative" strains (Davis et al., 1981). *V. mimicus* and *V. cholerae* are separated at a genetic level (Vieira et al., 2001), but the ecology and pathology of the two species are analogous. *V. mimicus* is widely distributed in freshwater and estuarine environments (Davis et al., 1981; Chowdhury et al., 1989). This species is isolated from patients with diarrhea but not as frequently as *V. cholerae* non-O1/O139, but is seldom isolated from extraintestinal infections (i.e., ear infections, wounds, and septicemia) (Davis et al., 1981; Desenclos et al., 1991; Hlady and Klontz, 1996; Morris and Black, 1985). Diarrhea cases are usually sporadic, although outbreaks of seafoodborne diarrhea are reported (Oliver and Kaper, 1997; Campos et al., 1996).

Virulence factors that are important in other enteric pathogens are detected in some clinical strains of *V. mimicus* and were studied at the genetic level. The genes are homologous to the *ctx* (cholera toxin), *zot* (zonula occludens toxin), and *ace* (accessory cholera enterotoxin) genes in the CTX genetic element and the toxin-coregulated pilus operon of *V. cholerae* (Chowdhury et al., 1994; Shi et al., 1998; Boyd et al., 2000; Bi et al., 2001); similar to the *tdh* (thermostable direct hemolysin) gene of *V. parahaemolyticus* (Terai et al., 1990); and similar to the *st* (heat-stable enterotoxin) gene of enterotoxigenic *Escherichia coli* (Shi et al., 1998). The homologues are also detected in some strains of *V. cholerae* non-O1/O139 (Nair et al., 1988; Ogawa et al., 1990; Baba et al., 1991; Ghosh et al., 1997; Jiang et al., 2003). As observed for some strains of *V. cholerae* non-O1/O139 (Ghosh et al., 1997), atypical strains carrying the *ctx* gene but lacking the *tcpA* gene are reported in some clinical strains of *V. mimicus* (Bi et al., 2001). Shinoda et al. (2004) examined the *V. mimicus* strains isolated from four countries and found the *ctx*, *tdh*, and *st* genes in 5,

33, and 19% of 42 clinical strains and 0, 0, and 16% of 58 environmental isolates, respectively. These virulence genes are associated with bacteriophage or other mobile genetic elements. A cholera toxin-non-producing strain of *V. mimicus* produced this toxin after experimental lysogenization with the CTXφ phage carrying the *ctx* gene (Faruque et al., 1999). These results suggest that some strains of *V. mimicus* and *V. cholerae* non-O1/O139 in the environment may have acquired the virulence genes by horizontal transfer, and the acquired genes may play major roles in diarrhea caused by these microorganisms.

The *V. mimicus* hemolysin (Kim et al., 1997), phospholipase (Kang et al., 1998), metalloprotease (Lee et al., 1998), siderophore (Okujo and Yamamoto, 1994), and hemagglutinins (Alam et al., 1996) were studied at the molecular genetic level. Whether these factors play major or auxiliary roles in intestinal and extraintestinal infections is not certain.

CONCLUSIONS

The species described in this chapter seem to make up the autochthonous component of the aquatic environment, being distributed in brackish, marine, or fresh water. An exception is *G. hollisae*, which does not grow well, if at all, on routine isolation media. Further studies using a more suitable isolation procedure should clarify whether *G. hollisae* is also widely distributed in the natural environment.

Despite the wide distribution of these species in the aquatic environment, only small numbers of strains have been isolated from human infections. There is insufficient information to determine the actual pathogenicity of these strains. Certainly, some cultures of *V. mimicus* and most, if not all, isolates of *G. hollisae* seem to have acquired the virulence determinants of the important enteropathogens by horizontal transfer mechanisms, e.g., the homologues of the *ctx* gene of *V. cholerae*, the *tdh* gene of *V. parahaemolyticus*, or the *st* gene of *E. coli*. Thus, virulence is most likely strain dependent. The virulence of other cultures may depend on the production of other virulence factors, the nature of which differs from strain to strain.

Another important factor is the susceptibility of the host to infection. Almost all of the cultures may well be opportunistic pathogens in susceptible hosts with underlying diseases. Determination of a possible virulence mechanism(s) in clinical strains and epidemiological studies comparing clinical and environmental cultures are needed to better understand the virulence. Molecular genetic techniques would be helpful in detecting defined virulence genes.

REFERENCES

Abbott, S. L., and J. M. Janda. 1994. Severe gastroenteritis associated with *Vibrio hollisae* infection: report of two cases and review. *Clin. Infect. Dis.* **18**:310–312.

Aggarwal, P., M. Singh, and S. Kumari. 1986. Isolation of *Vibrio alginolyticus* from two patients of acute gastroenteritis. *J. Diarrhoeal Dis. Res.* **4**:30.

Ahn, S. H., J. H. Han, J. H. Lee, K. J. Park, and I. S. Kong. 2005. Identification of an iron-regulated hemin-binding outer membrane protein, HupO, in *Vibrio fluvialis*: effects on hemolytic activity and the oxidative stress response. *Infect. Immun.* **73**:722–729.

Alam, M., S. Miyoshi, and S. Shinoda. 1995. Production of antigenically related exocellular elastolytic proteases mediating hemagglutination by vibrios. *Microbiol. Immunol.* **39**:67–70.

Alam, M., S. I. Miyoshi, K. I. Tomochika, and S. Shinoda. 1996. Purification and characterization of novel hemagglutinins from *Vibrio mimicus*: a 39-kilodalton major outer membrane protein and lipopolysaccharide. *Infect. Immun.* **64**:4035–4041.

Albert, M. J., M. A. Hossain, K. Alam, I. Kabir, P. K. Neogi, and S. Tzipori. 1991. A fatal case associated with shigellosis and *Vibrio fluvialis* bacteremia. *Diagn. Microbiol. Infect. Dis.* **14**:509–510.

Alcaide, E. 2003. Numerical taxonomy of *Vibrionaceae* isolated from cultured amberjack (*Seriola dumerili*) and surrounding water. *Curr. Microbiol.* **46**:184–189.

Austin, B., A. C. Pride, and G. A. Rhodie. 2003. Association of a bacteriophage with virulence in *Vibrio harveyi*. *J. Fish. Dis.* **26**:55–58.

Baba, K., H. Shirai, A. Terai, K. Kumagai, Y. Takeda, and M. Nishibuchi. 1991. Similarity of the *tdh* gene-bearing plasmids of *Vibrio cholerae* non-O1 and *Vibrio parahaemolyticus*. *Microb. Pathog.* **10**:61–70.

Baffone, W., B. Citterio, E. Vittoria, A. Casaroli, R. Campana, L. Falzano, and G. Donelli. 2003. Retention of virulence in viable but non-culturable halophilic *Vibrio* spp. *Int. J. Food Microbiol.* **89**:31–39.

Balebona, M. C., M. J. Andreu, M. A. Bordas, I. Zorrilla, M. A. Morinigo, and J. J. Borrego. 1998. Pathogenicity of *Vibrio alginolyticus* for cultured gilt-head sea bream (*Sparus aurata* L.). *Appl. Environ. Microbiol.* **64**:4269–4275.

Barbieri, E., L. Falzano, C. Fiorentini, A. Pianetti, W. Baffone, A. Fabbri, P. Matarrese, A. Casiere, M. Katouli, I. Kühn, R. Möllby, F. Bruscolini, and G. Donelli. 1999. Occurrence, diversity, and pathogenicity of halophilic *Vibrio* spp. and non-O1 *Vibrio cholerae* from estuarine waters along the Italian Adriatic coast. *Appl. Environ. Microbiol.* **65**:2748–2753.

Bassler, B. L., P. J. Gibbons, C. Yu, and S. Roseman. 1991. Chitin utilization by marine bacteria. Chemotaxis to chitin oligosaccharides by *Vibrio furnissii*. *J. Biol. Chem.* **266**:24268–24275.

Bi, K., S. I. Miyoshi, K. I. Tomochika, and S. Shinoda. 2001. Detection of virulence associated genes in clinical strains of *Vibrio mimicus*. *Microbiol. Immunol.* **45**:613–616.

Blake, P. A., M. H. Merson, R. E. Weaver, D. G. Hollis, and P. C. Heublein. 1979. Disease caused by a marine vibrio: clinical characteristics and epidemiology. *N. Engl. J. Med.* **300**:1–5.

Blake, P. A., R. E. Weaver, and D. G. Hollis. 1980. Diseases of humans (other than cholera) caused by vibrios. *Annu. Rev. Microbiol.* **34**:341–367.

Bode, R. B., P. R. Brayton, R. R. Colwell, F. M. Russo, and W. E. Bullock. 1986. A new *Vibrio* species, *Vibrio cincinnatiensis*, causing meningitis: successful treatment in an adult. *Ann. Intern. Med.* **104**:55–56.

Bouma, C. L., and S. Roseman. 1996a. Sugar transport by the marine chitinolytic bacterium *Vibrio furnissii*. Molecular cloning and analysis of the mannose/glucose permease. *J Biol. Chem.* 271:33468–33475.

Bouma, C. L., and S. Roseman. 1996b. Sugar transport by the marine chitinolytic bacterium *Vibrio furnissii*. Molecular cloning and analysis of the glucose and N-acetylglucosamine permeases. *J. Biol. Chem.* 271:33457–33467.

Boyd, E. F., K. E. Moyer, L. Shi, and M. K. Waldor. 2000. Infectious CTXφ and the vibrio pathogenicity island prophage in *Vibrio mimicus*: evidence for recent horizontal transfer between *V. mimicus* and *V. cholerae*. *Infect. Immun.* 68:1507–1513.

Brayton, P. R., R. B. Bode, R. R. Colwell, M. T. MacDonell, H. L. Hall, D. J. Grimes, P. A. West, and T. N. Bryant. 1986. *Vibrio cincinnatiensis* sp. nov., a new human pathogen. *J. Clin. Microbiol.* 23:104–108.

Brenner, D. J., F. W. Hickman-Brenner, J. V. Lee, A. G. Steigerwalt, G. R. Fanning, D. G. Hollis, J. J. Farmer III, R. E. Weaver, S. W. Joseph, and R. J. Seidler. 1983. *Vibrio furnissii* (formerly aerogenic biogroup of *Vibrio fluvialis*), a new species isolated from human feces and the environment. *J. Clin. Microbiol.* 18:816–824.

Caccamese, S. M., and D. A. Rastegar. 1999. Chronic diarrhea associated with *Vibrio alginolyticus* in an immunocompromised patient. *Clin. Infect. Dis.* 29:946–947.

Campos, E., H. Bolanos, M. T. Acuna, G. Diaz, M. C. Matamoros, H. Raventos, L. M. Sanchez, O. Sanchez, and C. Barquero. 1996. *Vibrio mimicus* diarrhea following ingestion of raw turtle eggs. *Appl. Environ. Microbiol.* 62:1141–1144.

Cavallo, R. A., and L. Stabili. 2002. Presence of vibrios in seawater and *Mytilus galloprovincialis* (Lam.) from the Mar Piccolo of Taranto (Ionian Sea). *Water Res* 36:3719–3726.

Chan, K. Y., M. L. Woo, L. Y. Lam, and G. L. French. 1989. *Vibrio parahaemolyticus* and other halophilic vibrios associated with seafood in Hong Kong. *J. Appl. Bacteriol.* 66:57–64.

Chapman, P. A. 1987. *Vibrio alginolyticus* and diarrhoeal disease. *J. Diarrhoeal Dis. Res.* 5:40.

Chatterjee, B. D., G. Thawani, and S. N. Sanyal. 1989. Etiology of acute childhood diarrhoea in Calcutta. *Trop. Gastroenterol.* 10:158–166.

Chikahira, M., and K. Hamada. 1988. Enterotoxigenic substance and other toxins produced by *Vibrio fluvialis* and *Vibrio furnissii*. *Nippon Juigaku Zasshi* 50:865–873.

Chitlaru, E., and S. Roseman. 1996. Molecular cloning and characterization of a novel beta-N-acetyl-D-glucosaminidase from *Vibrio furnissii*. *J. Biol. Chem.* 271:33433–33439.

Chowdhury, M. A., R. T. Hill, and R. R. Colwell. 1994. A gene for the enterotoxin zonula occludens toxin is present in *Vibrio mimicus* and *Vibrio cholerae* O139. *FEMS Microbiol. Lett.* 119:377–380.

Chowdhury, M. A., H. Yamanaka, S. Miyoshi, K. M. Aziz, and S. Shinoda. 1989. Ecology of *Vibrio mimicus* in aquatic environments. *Appl. Environ. Microbiol.* 55:2073–2078.

Clarridge, J. E., and S. Zighelboim-Daum. 1985. Isolation and characterization of two hemolytic phenotypes of *Vibrio damsela* associated with a fatal wound infection. *J. Clin. Microbiol.* 21:302–306.

Coffey, J. A., Jr., R. L. Harris, M. L. Rutledge, M. W. Bradshaw, and T. W. Williams, Jr. 1986. *Vibrio damsela*: another potentially virulent marine vibrio. *J. Infect. Dis.* 153:800–802.

Conejero, M. J. U., and C. T. Hedreyda. 2003. Isolation of partial *toxR* gene of *Vibrio harveyi* and design of *toxR*-targeted PCR primers for species detection. *J. Appl. Microbiol.* 95:602–611.

Conejero, M. J. U., and C. T. Hedreyda. 2004. PCR detection of hemolysin (*vhh*) gene in *Vibrio harveyi*. *J. Gen. Appl. Microbiol.* 50:137–142.

Cutter, D. L., and A. S. Kreger. 1990. Cloning and expression of the damselysin gene from *Vibrio damsela*. *Infect. Immun.* 58:266–268.

Czyż, A., H. Szpilewska, R. Dutkiewicz, W. Kowalska, A. Biniewska-Godlewska, and G. Węgrzyn. 2002. Comparison of the Ames test and a newly developed assay for detection of mutagenic pollution of marine environments. *Mutat. Res.* 519:67–74.

Dalsgaard, A., A. Alarcon, C. F. Lanata, T. Jensen, H. J. Hansen, F. Delgado, A. I. Gil, M. E. Penny, and D. Taylor. 1996. Clinical manifestations and molecular epidemiology of five cases of diarrhoea in children associated with *Vibrio metschnikovii* in Arequipa, Peru. *J. Med. Microbiol.* 45:494–500.

Dalsgaard, A., P. Glerup, L. L. Hoybye, A. M. Paarup, R. Meza, M. Bernal, T. Shimada, and D. N. Taylor. 1997. *Vibrio furnissii* isolated from humans in Peru: a possible human pathogen? *Epidemiol. Infect.* 119:143–149.

Davis, B. R., G. R. Fanning, J. M. Madden, A. G. Steigerwalt, H. B. Bradford, Jr., H. L. Smith, Jr., and D. J. Brenner. 1981. Characterization of biochemically atypical *Vibrio cholerae* strains and designation of a new pathogenic species, *Vibrio mimicus*. *J. Clin. Microbiol.* 14:631–639.

Deane, S. M., F. T. Robb, S. M. Robb, and D. R. Woods. 1989. Nucleotide sequence of the *Vibrio alginolyticus* calcium-dependent, detergent-resistant alkaline serine exoprotease A. *Gene* 76:281–288.

Desenclos, J. C. A., K. C. Klontz, L. E. Wolfe, and S. Hoecherl. 1991. The risk of *Vibrio* illness in the Florida raw oyster eating population, 1981–1988. *Am. J. Epidemiol.* 134:290–297.

Diggles, B. K., G. A. Moss, J. Carson, and C. D. Anderson. 2000. Luminous vibriosis in rock lobster *Jasus verreauxi* (Decapoda: Palinuridae) phyllosoma larvae associated with infection by *Vibrio harveyi*. *Dis. Aquat. Organ.* 43:127–137.

Di Pinto, A., G. Ciccarese, G. Tantillo, D. Catalano, and V. T. Forte. 2005. A collagenase-targeted multiplex PCR assay for identification of *Vibrio alginolyticus*, *Vibrio cholerae*, and *Vibrio parahaemolyticus*. *J. Food Prot.* 68:150–153.

Dumontet, S., K. Krovacek, S. B. Svenson, V. Pasquale, S. B. Baloda, and G. Figliuolo. 2000. Prevalence and diversity of *Aeromonas* and *Vibrio* spp. in coastal waters of Southern Italy. *Comp. Immunol. Microbiol. Infect. Dis.* 23:53–72.

Elhadi, N., S. Radu, C. J. Chen, and M. Nishibuchi. 2004. Prevalence of potentially pathogenic *Vibrio* species in the seafood marketed in Malaysia. *J. Food Prot.* 67:1469–1475.

Enomoto, H., T. Unemoto, M. Nishibuchi, E. Padan, and T. Nakamura. 1998. Topological study of *Vibrio alginolyticus* NhaB Na⁺/H⁺ antiporter using gene fusions in *Escherichia coli* cells. *Biochim. Biophys. Acta* 1370:77–86.

Farmer, J. J., III, and F. W. Hickman-Brenner. 1992. The genera *Vibrio* and *Photobacterium*, p. 2952–3011. *In* A. Balows, H. G. Trüper, M. Dworkin, W. Harder, and K. Schleifer (ed.), *The Prokaryotes: a Handbook on the Biology of Bacteria: Ecophysiology, Isolation, Identification, Applications*, 2nd ed. Springer-Verlag, New York, N.Y.

Farmer, J. J., III, F. W. Hickman-Brenner, G. R. Fanning, C. M. Gordon, and D. J. Brenner. 1988. Characterization of *Vibrio metschnikovii* and *Vibrio gazogenes* by DNA-DNA hybridization and phenotype. *J. Clin. Microbiol.* 26:1993–2000.

Faruque, S. M., M. M. Rahman, Asadulghani, K. M. N. Islam, and J. J. Mekalanos. 1999. Lysogenic conversion of environmental *Vibrio mimicus* strains by CTXφ. *Infect. Immun.* 67:5723–5729.

Fouz, B., E. G. Biosca, and C. Amaro. 1997. High affinity iron-uptake systems in *Vibrio damsela*: role in the acquisition of iron from transferrin. *J. Appl. Microbiol.* 82:157–167.

Furniss, A. L., J. V. Lee, and T. J. Donovan. 1977. Group F, a new vibrio? *Lancet* ii:565–566.

Gauger, E. J., and M. Gomez-Chiarri. 2002. 16S ribosomal DNA sequencing confirms the synonymy of *Vibrio harveyi* and *V. carchariae*. *Dis. Aquat. Organ.* 52:39–46.

Gauthier, G., B. Lafay, R. Ruimy, V. Breittmayer, J. L. Nicolas, M. Gauthier, and R. Christen. 1995. Small-subunit rRNA sequences and whole DNA relatedness concur for the reassignment of *Pasteurella piscicida* (Snieszko et al.) Janssen and Surgalla to the genus *Photobacterium* as *Photobacterium damsela* subsp. *piscicida* comb. nov. *Int. J. Syst. Bacteriol.* **45:**139–144.

Ghosh, C., R. K. Nandy, S. K. Dasgupta, G. B. Nair, R. H. Hall, and A. C. Ghose. 1997. A search for cholera toxin (CT), toxin coregulated pilus (TCP), the regulatory element ToxR and other virulence factors in non-O1/non-O139 *Vibrio cholerae*. *Microb. Pathog.* **22:**199–208.

Gomez, J. M., R. Fajardo, J. F. Patiño, and C. A. Arias. 2003. Necrotizing fasciitis due to *Vibrio alginolyticus* in an immunocompetent patient. *J. Clin. Microbiol.* **41:**3427–3429.

Gomez-Gil, B., S. Soto-Rodriguez, A. Garcia-Gasca, A. Roque, R. Vazquez-Juarez, F. L. Thompson, and J. Swings. 2004. Molecular identification of *Vibrio harveyi*-related isolates associated with diseased aquatic organisms. *Microbiology* **150:**1769–1777.

Gómez-León, J., L. Villamil, M. L. Lemos, B. Novoa, and A. Figueras. 2005. Isolation of *Vibrio alginolyticus* and *Vibrio splendidus* from aquacultured carpet shell clam (*Ruditapes decussatus*) larvae associated with mass mortalities. *Appl. Environ. Microbiol.* **71:**98–104.

Goodell, K. H., M. R. Jordan, R. Graham, C. Cassidy, and S. A. Nasraway. 2004. Rapidly advancing necrotizing fasciitis caused by *Photobacterium* (*Vibrio*) *damsela*: a hyperaggressive variant. *Crit. Care Med.* **32:**278–281.

Gras-Rouzet, S., P. Y. Donnio, F. Juguet, P. Plessis, J. Minet, and J. L. Avril. 1996. First European case of gastroenteritis and bacteremia due to *Vibrio hollisae*. *Eur. J. Clin. Microbiol. Infect. Dis.* **15:**864–866.

Greer, L. F., III, and A. A. Szalay. 2002. Imaging of light emission from the expression of luciferases in living cells and organisms: a review. *Luminescence* **17:**43–74.

Gribun, A., D. J. Katcoff, G. Hershkovits, I. Pechatnikov, and Y. Nitzan. 2004. Cloning and characterization of the gene encoding for OMP-PD porin: the major *Photobacterium damsela* outer membrane protein. *Curr. Microbiol.* **48:**167–174.

Grimes, D. J., P. Brayton, R. R. Colwell, and S. H. Gruber. 1985. Vibrios as autochthonous flora of neritic sharks. *Syst. Appl. Microbiol.* **6:**221–226.

Grimes, D. J., J. Stemmler, H. Hada, E. B. May, D. Maneval, F. M. Hetrick, R. T. Jones, M. Stockopf, and R. R. Colwell. 1984. *Vibrio* species associated with mortality of sharks held in captivity. *Microb. Ecol.* **10:**271–282.

Hansen, W., J. Freney, H. Benyagoub, M. N. Letouzey, J. Gigi, and G. Wauters. 1993. Severe human infections caused by *Vibrio metschnikovii*. *J. Clin. Microbiol.* **31:**2529–2530.

Hardardottir, H., K. Vikenes, A. Digranes, J. Lassen, and A. Halstensen. 1994. Mixed bacteremia with *Vibrio metschnikovii* in an 83-year-old female patient. *Scand. J. Infect. Dis.* **26:**493–494.

Hayashi, M., Y. Nakayama, and T. Unemoto. 2001. Recent progress in the Na$^+$-translocating NADH-quinone reductase from the marine *Vibrio alginolyticus*. *Biochim. Biophys. Acta* **1505:**37–44.

Heidelberg, J. F., K. B. Heidelberg, and R. R. Colwell. 2002a. Seasonality of Chesapeake Bay bacterioplankton species. *Appl. Environ. Microbiol.* **68:**5488–5497.

Heidelberg, J. F., K. B. Heidelberg, and R. R. Colwell. 2002b. Bacteria of the γ-subclass *Proteobacteria* associated with zooplankton in Chesapeake Bay. *Appl. Environ. Microbiol.* **68:**5498–5507.

Hernández, G., and J. Olmos. 2004. Molecular identification of pathogenic and nonpathogenic strains of *Vibrio harveyi* using PCR and RAPD. *Appl. Microbiol. Biotechnol.* **63:**722–727.

Hickman, F. W., J. J. Farmer III, D. G. Hollis, G. R. Fanning, A. G. Steigerwalt, R. E. Weaver, and D. J. Brenner. 1982. Identification of *Vibrio hollisae* sp. nov. from patients with diarrhea. *J. Clin. Microbiol.* **15:**395–401.

Hickman-Brenner, F. W., D. J. Brenner, A. G. Steigerwalt, M. Schreiber, S. D. Holmberg, L. M. Baldy, C. S. Lewis, N. M. Pickens, and J. J. Farmer III. 1984. *Vibrio fluvialis* and *Vibrio furnissii* isolated from a stool sample of one patient. *J. Clin. Microbiol.* **20:**125–127.

Hiratsuka, M., Y. Saito, and N. Yamane. 1980. The isolation of *Vibrio alginolyticus* from a patient with acute entero-colitis. *Tohoku J. Exp. Med.* **132:**469–472.

Hlady, W. G., and K. C. Klontz. 1996. The epidemiology of *Vibrio* infection in Florida, 1981–1993. *J. Infect. Dis.* **173:**1176–1183.

Ho, P. L., W. M. Tang, K. S. Lo, and K. Y. Yuen. 1998. Necrotizing fasciitis due to *Vibrio alginolyticus* following an injury inflicted by a stingray. *Scand. J. Infect. Dis.* **30:**192–193.

Holt, J. G., N. R. Krieg, P. H. A. Sneath, J. T. Staley, and S. T. Williams (ed.). 1994. *Bergey's Manual of Determinative Bacteriology*, 9th ed. Lippincott Williams & Wilkins, Philadelphia, Pa.

Hörmansdorfer, S., H. Wentges, K. Neugebaur-Büchler, and J. Bauer. 2000. Isolation of *Vibrio alginolyticus* from seawater aquaria. *Int. J. Hyg. Environ. Health* **203:**169–175.

Hornstrup, M. K., and B. Gahrn-Hansen. 1993. Extraintestinal infections caused by *Vibrio parahaemolyticus* and *Vibrio alginolyticus* in a Danish county, 1987–1992. *Scand. J. Infect. Dis.* **25:**735–740.

Howard, R. J., M. E. Pessa, B. H. Brennaman, and R. Ramphal. 1985. Necrotizing soft-tissue infections caused by marine vibrios. *Surgery* **98:**126–130.

Huq, M. I., A. K. Alam, D. J. Brenner, and G. K. Morris. 1980. Isolation of *Vibrio*-like group, EF-6, from patients with diarrhea. *J. Clin. Microbiol.* **11:**621–624.

Iguchi, T., S. Kondo, and K. Hisatsune. 1993. A chemotaxonomic study of *Vibrio fluvialis* based on the sugar composition of the polysaccharide portion of the lipopolysaccharides. *Microbiol. Immunol.* **37:**153–157.

Ivanova, E. P., N. V. Zhukova, N. M. Gorshkova, and E. L. Chaikina. 2001. Characterization of *Aeromonas* and *Vibrio* species isolated from a drinking water reservoir. *J. Appl. Microbiol.* **90:**919–927.

Janda, J. M., C. Powers, R. Bryant, and S. L. Abbott. 1988. Current perspectives on the epidemiology and pathogenesis of clinically significant *Vibrio* spp. *Clin. Microbiol. Rev.* **1:**245–267.

Jean-Jacques, W., K. R. Rajashekaraiah, J. J. Farmer III, F. W. Hickman, J. G. Morris, and C. A. Kallick. 1981. *Vibrio metschnikovii* bacteremia in a patient with cholecystitis. *J. Clin. Microbiol.* **14:**711–712.

Joseph, S. W., R. R. Colwell, and J. B. Kaper. 1982. *Vibrio parahaemolyticus* and related halophilic vibrios. *Crit. Rev Microbiol.* **10:**77–123.

Jiang, S., W. Chu, and W. Fu. 2003. Prevalence of cholera toxin genes (*ctxA* and *zot*) among non-O1/O139 *Vibrio cholerae* strains from Newport Bay, California. *Appl. Environ. Microbiol.* **69:**7541–7544.

Kang, J. H., J. H. Lee, J. H. Park, S. H. Huh, and I. S. Kong. 1998. Cloning and identification of a phospholipase gene from *Vibrio mimicus*. *Biochim. Biophys. Acta* **1394:**85–89.

Kaper, J. B., J. G. Morris, Jr., and M. M. Levine. 1995. Cholera. *Clin. Microbiol. Rev.* **8:**48–86.

Karunasagar, I., R. Pai, G. R. Malathi, and I. Karunasagar. 1994. Mass mortality of *Penaeus monodon* larvae due to antibiotic-resistant *Vibrio harveyi* infection. *Aquaculture* **128:**203–209.

Kelly, M. T., and E. M. D. Stroh. 1988. Occurrence of *Vibrionaceae* in natural and cultivated oyster populations in the Pacific Northwest. *Diagn. Microbiol. Infect. Dis.* **9:**1–5.

Ketterer, P. J., and L. E. Eaves. 1992. Deaths in captive eels (*Anguilla reinhardtii*) due to *Photobacterium* (*Vibrio*) *damsela*. *Aust. Vet. J.* **69:**203–204.

Keyhani, N. O., and S. Roseman. 1996a. The chitin catabolic cascade in the marine bacterium *Vibrio furnissii*. Molecular cloning, isolation, and characterization of a periplasmic chitodextrinase. *J. Biol. Chem.* **271:**33414–33424.

Keyhani, N. O., and S. Roseman. 1996b. The chitin catabolic cascade in the marine bacterium *Vibrio furnissii*. Molecular cloning, isolation, and characterization of a periplasmic beta-N-acetyl-glucosaminidase. *J. Biol. Chem.* **271:**33425–33432.

Keyhani, N. O., X. B. Li, and S. Roseman. 2000. Chitin catabolism in the marine bacterium *Vibrio furnissii*. Identification and molecular cloning of a chitoporin. *J. Biol. Chem.* **275:**33068–33076.

Kim, G. T., J. Y. Lee, S. H. Huh, J. H. Yu, and I. S. Kong. 1997. Nucleotide sequence of the *vmhA* gene encoding hemolysin from *Vibrio mimicus*. *Biochim. Biophys. Acta* **1360:**102–104.

Kim, S. H., S. H. Ahn, J. H. Lee, E. M. Lee, N. H. Kim, K. J. Park, and I. S. Kong. 2003. Genetic analysis of phosphomannomutase/phosphoglucomutase from *Vibrio furnissii* and characterization of its role in virulence. *Arch. Microbiol.* **180:**240–250.

Kimura, B., S. Hokimoto, H. Takahashi, and T. Fujii. 2000. *Photobacterium histaminum* Okuzumi et al. 1994 is a later subjective synonym of *Photobacterium damselae* subsp. *damselae* (Love et al. 1981) Smith et al. 1991. *Int. J. Syst. Evol. Microbiol.* **50:**1339–1342.

Klontz, K. C., D. E. Cover, F. N. Hyman, and R. C. Mullen. 1994. Fatal gastroenteritis due to *Vibrio fluvialis* and nonfatal bacteremia due to *Vibrio mimicus*: unusual vibrio infections in two patients. *Clin. Infect. Dis.* **19:**541–542.

Kothary, M. H., E. F. Claverie, M. D. Miliotis, J. M. Madden, and S. H. Richardson. 1995. Purification and characterization of a Chinese hamster ovary cell elongation factor of *Vibrio hollisae*. *Infect. Immun.* **63:**2418–2423.

Kothary, M. H., and A. S. Kreger. 1985. Purification and characterization of an extracellular cytolysin produced by *Vibrio damsela*. *Infect. Immun.* **49:**25–31.

Kothary, M. H., H. Lowman, B. A. McCardell, and B. D. Tall. 2003. Purification and characterization of enterotoxigenic El Tor-like hemolysin produced by *Vibrio fluvialis*. *Infect. Immun.* **71:**3213–3220.

Kourany, M. 1983. Medium for isolation and differentiation of *Vibrio parahaemolyticus* and *Vibrio alginolyticus*. *Appl. Environ. Microbiol.* **45:**310–312.

Kreger, A. S. 1984. Cytolytic activity and virulence of *Vibrio damsela*. *Infect. Immun.* **44:**326–331.

Kreger, A. S., A. W. Bernheimer, L. A. Etkin, and L. W. Daniel. 1987. Phospholipase D activity of *Vibrio damsela* cytolysin and its interaction with sheep erythrocytes. *Infect. Immun.* **55:**3209–3212.

Kwon, Y. T., J. O. Kim, S. Y. Moon, H. H. Lee, and H. M. Rho. 1994. Extracellular alkaline serine proteases from alkalophilic *Vibrio metschnikovii* strain RH530. *Biotechnol. Lett.* **16:**413–418.

Kwon, Y. T., J. O. Kim, S. Y. Moon, Y. D. Yoo, and H. M. Rho. 1995. Cloning and characterization of the gene encoding an extracellular alkaline protease from *Vibrio metschnikovii* strain RH530. *Gene* **152:**59–63.

Lee, J. H., G. T. Kim, J. Y. Lee, H. K. Jun, J. H. Yu, and I. S. Kong. 1998. Isolation and sequence analysis of metalloprotease gene from *Vibrio mimicus*. *Biochim. Biophys. Acta* **1384:**1–6.

Lee, J. V., T. J. Donovan, and A. L. Furniss. 1978. Characterization, taxonomy, and emended description of *Vibrio metschnikovii*. *Int. J. Syst. Bacteriol.* **28:**99–111.

Lee, J. V., P. Shread, A. L. Furniss, and T. N. Bryant. 1981. Taxonomy and description of *Vibrio fluvialis* sp. nov. (synonym group F vibrios, group EF6). *J. Appl. Bacteriol.* **50:**73–94.

Lee, K. K. 1995. Pathogenesis studies on *Vibrio alginolyticus* in the grouper, *Epinephelus malabaricus*, Bloch et Schneider. *Microb. Pathog.* **19:**39–48.

Lee, K. K., P. C. Liu, and W. H. Chuang. 2002. Pathogenesis of gastroenteritis caused by *Vibrio carchariae* in cultured marine fish. *Mar. Biotechnol.* **4:**267–277.

Lee, K. K., S. R. Yu, F. R. Chen, T. I. Yang, and P. C. Liu. 1996. Virulence of *Vibrio alginolyticus* isolated from diseased tiger prawn, *Penaeus monodon*. *Curr. Microbiol.* **32:**229–231.

Lesmana, M., D. S. Subekti, P. Tjaniadi, C. H. Simanjuntak, N. H. Punjabi, J. R. Campbell, and B. A. Oyofo. 2002. Spectrum of *vibrio* species associated with acute diarrhea in North Jakarta, Indonesia. *Diagn. Microbiol. Infect. Dis.* **43:**91–97.

Levine, W. C., P. M. Griffin, and the Gulf Coast Vibrio Working Group. 1993. *Vibrio* infections on the Gulf Coast: results of first year of regional surveillance. *J. Infect. Dis.* **167:**479–483.

Linde, H. J., R. Kobuch, S. Jayasinghe, U. Reischl, N. Lehn, S. Kaulfuss, and L. Beutin. 2004. *Vibrio metschnikovii*, a rare cause of wound infection. *J. Clin. Microbiol.* **42:**4909–4911.

Liu, C. H., W. Cheng, J. P. Hsu, and J. C. Chen. 2004. *Vibrio alginolyticus* infection in the white shrimp *Litopenaeus vannamei* confirmed by polymerase chain reaction and 16S rDNA sequencing. *Dis. Aquat. Organ.* **61:**169–174.

Liu, P. C., Y. C. Chen, and K. K. Lee. 2001. Pathogenicity of *Vibrio alginolyticus* isolated from diseased small abalone *Haliotis diversicolor* supertexta. *Microbios* **104:**71–77.

Liu, P. C., K. K. Lee, and S. N. Chen. 1996. Pathogenicity of different isolates of *Vibrio harveyi* in tiger prawn, *Penaeus monodon*. *Lett. Appl. Microbiol.* **22:**413–416.

Lockwood, D. E., A. S. Kreger, and S. H. Richardson. 1982. Detection of toxins produced by *Vibrio fluvialis*. *Infect. Immun.* **35:**702–708.

Long, S., M. A. Mothibeli, F. T. Robb, and D. R. Woods. 1981. Regulation of extracellular alkaline protease activity by histidine in a collagenolytic *Vibrio alginolyticus* strain. *J. Gen. Microbiol.* **127:**193–199.

Love, M., D. Teebken-Fisher, J. E. Hose, J. J. Farmer III, F. W. Hickman, and G. R. Fanning. 1981. *Vibrio damsela*, a marine bacterium, causes skin ulcers on the damselfish *Chromis punctipinnis*. *Science* **214:**1139.

Luna-Gonzalez, A., A. N. Maeda-Martinez, J. C. Sainz, and F. Ascencio-Valle. 2002. Comparative susceptibility of veliger larvae of four bivalve mollusks to a *Vibrio alginolyticus* strain. *Dis. Aquat. Organ.* **49:**221–226.

MacDonell, M. T., and R. R. Colwell. 1985. Phylogeny of the Vibrionaceae, and recommendation for two new genera, *Listonella* and *Shewanella*. *Syst. Appl. Microbiol.* **6:**171–182.

Magalhães, V., A. Branco, R. de Andrade Lima, and M. Magalhães. 1996. *Vibrio metschnikovii* among diarrheal patients during cholera epidemic in Recife Brazil. *Rev. Inst. Med. Trop. São Paulo* **38:**1–3.

Magalhães, V., A. Castello Filho, M. Magalhães, and T. T. Gomes. 1993. Laboratory evaluation on pathogenic potentialities of *Vibrio furnissii*. *Mem. Inst. Oswaldo Cruz.* **88:**593–597.

Magalhães, M., G. P. de Silva, V. Magalhães, M. G. Antas, M. A. Andrade, and S. Tateno. 1990. *Vibrio fluvialis* and *Vibrio furnissii* associated with infantile diarrhea. *Rev. Microbiol.*, Sao Paulo **21:**295–298.

Matte, G. R., M. H. Matte, M. I. Sato, P. S. Sanchez, I. G. Rivera, and M. T. Martins. 1994. Potentially pathogenic vibrios associated with mussels from a tropical region on the Atlantic coast of Brazil. *J. Appl. Bacteriol.* **77:**281–287.

Maugeri, T. L., D. Caccamo, and C. Gugliandolo. 2000. Potentially pathogenic vibrios in brackish waters and mussels. *J. Appl. Microbiol.* **89:**261–266.

Mazari-Hiriart, M., Y. Lopez-Vidal, G. Castillo-Rojas, S. Ponce de Leon, and A. Cravioto. 2001. *Helicobacter pylori* and other enteric bacteria in freshwater environments in Mexico City. *Arch. Med. Res.* **32:**458–467.

McCarter, L. L. 2001. Polar flagellar motility of the *Vibrionaceae*. *Microbiol. Mol. Biol. Rev.* **65:**445–462.

Miliotis, M. D., B. D. Tall, and R. T. Gray. 1995. Adherence to and invasion of tissue culture cells by *Vibrio hollisae*. *Infect. Immun.* **63:**4959–4963.

Miyake, M., T. Honda, and T. Miwatani. 1989. Effects of divalent cations and saccharides on *Vibrio metschnikovii* cytolysin-induced hemolysis of rabbit erythrocytes. *Infect. Immun.* **57:**158–163.

Miyoshi, S., Y. Sonoda, H. Wakiyama, M. M. Rahman, K. Tomochika, S. Shinoda, S. Yamamoto, and K. Tobe. 2002. An exocellular thermolysin-like metalloprotease produced by *Vibrio fluvialis*: purification, characterization, and gene cloning. *Microb. Pathog.* **33:**127–134.

Molitoris, E., S. W. Joseph, M. I. Krichevsky, W. Sindhuhardja, and R. R. Colwell. 1985. Characterization and distribution of *Vibrio alginolyticus* and *Vibrio parahaemolyticus* isolated in Indonesia. *Appl. Environ. Microbiol.* **50:**1388–1394.

Morris, J. G., Jr., and R. E. Black. 1985. Cholera and other vibrioses in the United States. *N. Engl. J. Med.* **312:**343–350.

Morris, J. G., Jr., H. G. Miller, R. Wilson, C. O. Tacket, D. G. Hollis, F. W. Hickman, R. E. Weaver, and P. A. Blake. 1982. Illness caused by *Vibrio damsela* and *Vibrio hollisae*. *Lancet* **i:**1294–1297.

Munro, J., J. Oakey, E. Bromage, and L. Owens. 2003. Experimental bacteriophage-mediated virulence in strains of *Vibrio harveyi*. *Dis. Aquat. Organ.* **54:**187–194.

Nair, G. B., Y. Oku, Y. Takeda, A. Ghosh, R. K. Ghosh, S. Chattopadhyyay, S. C. Pal, J. B. Kaper, and T. Takeda. 1988. Toxin profiles of *Vibrio cholerae* non-O1 from environmental sources in Calcutta, India. *Appl. Microbiol.* **54:**3180–3182.

Nakamura, T., Y. Komano, and T. Unemoto. 1995. Three aspartic residues in membrane-spanning regions of Na^+/H^+ antiporter from *Vibrio alginolyticus* play a role in the activity of the carrier. *Biochim. Biophys. Acta* **1230:**170–176.

Nicolas, J. L., O. Basuyaux, J. Mazurie, and A. Thebault. 2002. *Vibrio carchariae*, a pathogen of the abalone *Haliotis tuberculata*. *Dis. Aquat. Organ.* **50:**35–43.

Nishibuchi, M., S. Doke, S. Toizumi, T. Umeda, M. Yoh, and T. Miwatani. 1988. Isolation from a coastal fish of *Vibrio hollisae* capable of producing a hemolysin similar to the thermostable direct hemolysin of *Vibrio parahaemolyticus*. *Appl. Environ. Microbiol.* **54:**2144–2146.

Nishibuchi, M., M. Ishibashi, Y. Takeda, and J. B. Kaper. 1985. Detection of the thermostable direct hemolysin gene and related DNA sequence in *Vibrio parahaemolyticus* and other *Vibrio* species by the DNA colony hybridization test. *Infect. Immun.* **49:**481–486.

Nishibuchi, M., J. M. Janda, and T. Ezaki. 1996. The thermostable direct hemolysin gene (*tdh*) of *Vibrio hollisae* is dissimilar in prevalence to and phylogenetically distant from the *tdh* genes of the vibrios: implications in the horizontal transfer of the *tdh* gene. *Microbiol. Immunol.* **40:**59–65.

Nishibuchi, M., and R. J. Seidler. 1983. Medium-dependent production of extracellular enterotoxins by non-O-1 *Vibrio cholerae*, *Vibrio mimicus*, and *Vibrio fluvialis*. *Appl. Environ. Microbiol.* **45:**228–231.

Nishibuchi, M., R. J. Seidler, D. M. Rollins, and S. M. Joseph. 1983. *Vibrio* factors cause rapid fluid accumulation in suckling mice. *Infect. Immun.* **40:**1083–1091.

Oakey, H. J., N. Levy, D. G. Bourne, B. Cullen, and A. Thomas. 2003. The use of PCR to aid in the rapid identification of *Vibrio harveyi* isolates. *J. Appl. Microbiol.* **95:**1293–1303.

Ogawa, A., J.-I. Kato, H. Watanabe, G. B. Nair, and T. Takeda. 1990. Cloning and nucleotide sequence of a heat-stable enterotoxin gene from *Vibrio cholerae* non-O1 isolated from a patient with traveler's diarrhea. *Infect. Immun.* **58:**3325–3329.

Okujo, N., and S. Yamamoto. 1994. Identification of the siderophores from *Vibrio hollisae* and *Vibrio mimicus* as aerobactin. *FEMS Microbiol. Lett.* **118:**187–192.

Oliver, J. D., and J. B. Kaper. 1997. *Vibrio* species, p. 228–264. *In* M. P. Doyle, L. R. Beuchat, and T. J. Montville (ed.), *Food Microbiology: Fundamentals and Frontiers*. ASM Press, Washington, D.C.

Osorio, C. R., A. E. Toranzo, J. L. Romalde, and J. L. Barja. 2000. Multiplex PCR assay for *ureC* and 16S rRNA genes clearly discriminates between both subspecies of *Photobacterium damselae*. *Dis. Aquat. Organ.* **40:**177–183.

Park, J. K., N. O. Keyhani, and S. Roseman. 2000. Chitin catabolism in the marine bacterium *Vibrio furnissii*. Identification, molecular cloning, and characterization of a *N, N*'-diacetylchitobiose phosphorylase. *J. Biol. Chem.* **275:**33077–33083.

Pass, D. A., R. Dybadahl, and M. M. Mannion. 1987. Investigations into the causes of mortality of the pearl oyster, *Pintada maxima* (Jamson), in Western Australia. *Aquaculture* **65:**149–169.

Pavia, A. T., J. A. Bryan, K. L. Maher, T. R. Hester, Jr., and J. J. Farmer III. 1989. *Vibrio carchariae* infection after a shark bite. *Ann. Intern. Med.* **111:**85–86.

Pedersen, K., L. Verdonck, B. Austin, D. A. Austin, A. R. Blanch, P. A. D. Grimont, J. Jofre, S. Koblavi, J. L. Larsen, T. Tianen, M. Vigneulle, and J. Swings. 1998. Taxonomic evidence that *Vibrio carchariae* (Grimes et al. 1985) is a junior synonym of *Vibrio harveyi* (Johnson and Shunk 1936) Baumam et al. 1981. *Int. J. Syst. Bacteriol.* **48:**749–758.

Pedersen, K., I. Dalsgaard, and J. L. Larsen. 1997. *Vibrio damsela* associated with diseased fish in Denmark. *Appl. Environ. Microbiol.* **63:**3711–3715.

Peinado, M. T., A. Mariscal, M. Carnero-Varo, and J. Fernández-Crehuet. 2002. Correlation of two bioluminescence and one fluorogenic bioassay for the detection of toxic chemicals. *Ecotoxicol. Environ. Saf.* **53:**170–177.

Penland, R. L., M. Boniuk, and K. R. Wilhelmus. 2000. Vibrio ocular infections on the U.S. Gulf Coast. *Cornea* **19:**26–29.

Pujalte, M. J., A. Sitjà-Bobadilla, M. C. Macián, C. Belloch, P. Álvarez-Pellitero, J. Pérez-Sánchez, F. Uruburu, and E. Garay. 2003. Virulence and molecular typing of *Vibrio harveyi* strains isolated from cultured dentex, gilthead sea bream and European sea bass. *Syst. Appl. Microbiol.* **26:**284–292.

Rajan, P. R., J. H. Lin, M. S. Ho, and H. L. Yang. 2003. Simple and rapid detection of *Photobacterium damselae* ssp. *piscicida* by a PCR technique and plating method. *J. Appl. Microbiol.* **95:**1375–1380.

Ratnaraja, N., T. Blackmore, J. Byrne, and S. Shi. 2005. *Vibrio fluvialis* peritonitis in a patient receiving continuous ambulatory peritoneal dialysis. *J. Clin. Microbiol.* **43:**514–515.

Reid, G. C., D. R. Woods, and F. T. Robb. 1980. Peptone induction and rifampin-insensitive collagenase production by *Vibrio alginolyticus*. *J. Bacteriol.* **142:**447–454.

Reina, J., V. Fernandez-Baca, and A. Lopez. 1995. Acute gastroenteritis caused by *Vibrio alginolyticus* in an immunocompetent patient. *Clin. Infect. Dis.* **21:**1044–1045.

Ripabelli, G., M. L. Sammarco, G. M. Grasso, I. Fanelli, A. Caprioli, and I. Luzzi. 1999. Occurrence of *Vibrio* and other pathogenic bacteria in *Mytilus galloprovincialis* (mussels) harvested from Adriatic Sea, Italy. *Int. J. Food Microbiol.* **49:**43–48.

Sakazaki, R. 1968. Proposal of *Vibrio alginolyticus* for the biotype 2 of *Vibrio parahaemolyticus*. *Jpn. J. Med. Sci. Biol.* **21:**359–362.

Sangster, C. R., and R. M. Smolowitz. 2003. Description of *Vibrio alginolyticus* infection in cultured *Sepia officinalis*, *Sepia apama*, and *Sepia pharaonis*. *Biol. Bull.* **205:**233–234.

Schandevyl, P., E. Van Dyck, and P. Piot. 1984. Halophilic *Vibrio* species from seafish in Senegal. *Appl. Environ. Microbiol.* **48:**236–238.

Schmidt, U., H. Chmel, and C. Cobbs. 1979. *Vibrio alginolyticus* infections in humans. *J. Clin. Microbiol.* **10**:666–668.

Schroeder J. P., J. G. Wallace, M. B. Cates, S. B. Greco, and P. W. Moore. 1985. An infection by *Vibrio alginolyticus* in an Atlantic bottlenose dolphin housed in an open ocean pen. *J. Wildl. Dis.* **21**:437–438.

Scoglio, M. E., A. Di Pietro, I. Picerno, S. Delia, A. Mauro, and P. Laganà. 2001. Virulence factors in vibrios and aeromonads isolated from seafood. *New Microbiol.* **24**:273–280.

Sechi, L. A., I. Duprè, A. Deriu, G. Fadda, and S. Zanetti. 2000. Distribution of *Vibrio cholerae* virulence genes among different *Vibrio* species isolated in Sardinia, Italy. *J. Appl. Microbiol.* **88**:475–481.

Seidler, R. J., D. A. Allen, R. R. Colwell, S. W. Joseph, and O. P. Daily. 1980. Biochemical characteristics and virulence of environmental group F bacteria isolated in the United States. *Appl. Environ. Microbiol.* **40**:715–720.

Shi, L., S. Miyoshi, M. Hiura, K. Tomochika, T. Shimada, and S. Shinoda. 1998. Detection of genes encoding cholera toxin (CT), zonula occludens toxin (ZOT), accessory cholera enterotoxin (ACE) and heat-stable enterotoxin (ST) in *Vibrio mimicus* clinical strains. *Microbiol. Immunol.* **42**:823–828.

Shimada, T., E. Arakawa, T. Okitsu, S. Yamai, S. Matsushita, Y. Kudoh, and N. Okamura. 1999. Additional O antigens of *Vibrio fluvialis* and *Vibrio furnissii. Jpn. J. Infect. Dis.* **52**:124–126.

Shimada, T., and R. Sakazaki. 1983. Serological studies on *Vibrio fluvialis. Jpn. J. Med. Sci. Biol.* **36**:315–323.

Shin, J. H., M. G. Shin, S. P. Suh, D. W. Ryang, J. S. Rew, and F. S. Nolte. 1996. Primary *Vibrio damsela* septicemia. *Clin. Infect. Dis.* **22**:856–857.

Shinoda, S., N. Nakahara, and H. Kane. 1984. Lateral flagellum of *Vibrio fluvialis*: a species-specific antigen. *Can. J. Microbiol.* **30**:1525–1529.

Shinoda, S., T. Nakagawa, L. Shi, K. Bi, Y. Kanoh, K. Tomochika, S. Miyoshi, and T. Shimada. 2004. Distribution of virulence-associated genes in *Vibrio mimicus* isolates from clinical and environmental origins. *Microbiol. Immunol.* **48**:547–551.

Shinoda, S., I. Yakiyama, S. Yasui, Y. M. Kim, B. Ono, and S. Nakagami. 1992. Lateral flagella of vibrios: serological classification and genetical similarity. *Microbiol. Immunol.* **36**:303–309.

Smith, S. K., D. C. Sutton, J. A. Fuerst, and J. L. Reichelt. 1991. Evaluation of the genus *Listonella* and reassignment of *Listonella damsela* (Love et al.) MacDonell and Colwell to the genus *Photobacterium* as *Photobacterium damsela* comb. nov. with an emended description. *Int. J. Syst. Bacteriol.* **41**:529–534.

Staley, T. E., and R. R. Colwell. 1973a. Polynucleotide sequence relationships among Japanese and American strains of *Vibrio parahaemolyticus. J. Bacteriol.* **114**:916–927.

Staley, T. E., and R. R. Colwell. 1973b. Deoxyribonucleic acid reassociation among members of the genus *Vibrio. Int. J. Syst. Bacteriol.* **23**:316–332.

Sugita, H., Y. Shinagawa, and R. Okano. 2000. Neuraminidase-producing ability of intestinal bacteria isolated from coastal fish. *Lett. Appl. Microbiol.* **31**:10–13.

Suthienkul, O. 1993. Bacteriophage typing of *Vibrio fluvialis. Southeast Asian J. Trop. Med. Public Health* **24**:449–454.

Tacket, C. O., F. Hickman, G. V. Pierce, and L. F. Mendoza. 1982. Diarrhea associated with *Vibrio fluvialis* in the United States. *J. Clin. Microbiol.* **16**:991–992.

Takeuchi, H., Y. Shibano, K. Morihara, J. Fukushima, S. Inami, B. Keil, A. M. Gilles, S. Kawamoto, and K. Okuda. 1992. Structural gene and complete amino acid sequence of *Vibrio alginolyticus* collagenase. *Biochem. J.* **281**:703–708.

Teo, J. W. P., L.-H. Zhang, and C. L. Poh. 2003a. Cloning and characterization of a metalloprotease from *Vibrio harveyi* strain AP6. *Gene* **303**:147–156.

Teo, J. W. P., L.-H. Zhang, and C. L. Poh. 2003b. Cloning and characterization of a novel lipase from *Vibrio harveyi* strain AP6. *Gene* **312**:181–188.

Terai, A., H. Shirai, O. Yoshida, Y. Takeda, and M. Nishibuchi. 1990. Nucleotide sequence of the thermostable direct hemolysin gene (*tdh* gene) of *Vibrio mimicus* and its evolutionary relationship with the *tdh* genes of *Vibrio parahaemolyticus. FEMS Microbiol. Lett.* **71**:319–324.

Thompson, F. L., B. Hoste, K. Vandemeulebroecke, K. Engelbeen, R. Denys, and J. Swings. 2002. *Vibrio trachuri* Iwamoto et al. 1995 is a junior synonym of *Vibrio harveyi* (Johnson and Shunk 1936) Baumann et al. 1981. *Int. J. Syst. Evol. Microbiol.* **52**:973–976.

Thompson, F. L., B. Hoste, K. Vandemeulebroecke, and J. Swings. 2003. Reclassification of *Vibrio hollisae* as *Grimontia hollisae* gen. nov., comb. nov. *Int. J. Syst. Evol. Microbiol.* **53**:1615–1617.

Thyssen, A., L. Grisez, R. van Houdt, and F. Ollevier. 1998. Phenotypic characterization of the marine pathogen *Photobacterium damselae* subsp. *piscicida. Int. J. Syst. Bacteriol.* **48**:1145–1151.

Tokuda, H., M. Sugasawa, and T. Unemoto. 1982. Roles of Na$^+$ and K$^+$ in alpha-aminoisobutyric acid transport by the marine bacterium *Vibrio alginolyticus. J. Biol. Chem.* **257**:788–794.

Uh, Y., J. S. Park, G. Y. Hwang, I. H. Jang, K. J. Yoon, H. C. Park, and S. O. Hwang. 2001. *Vibrio alginolyticus* acute gastroenteritis: report of two cases. *Clin. Microbiol. Infect.* **7**:104–106.

Varghese, M. R., R. W. Farr, M. K. Wax, B. J. Chafin, and R. M. Owens. 1996. *Vibrio fluvialis* wound infection associated with medicinal leech therapy. *Clin. Infect. Dis.* **22**:709–710.

Venkateswaran, K., C. Kiiyukia, M. Takaki, H. Nakano, H. Matsuda, H. Kawakami, and H. Hashimoto. 1989. Characterization of toxigenic vibrios isolated from the freshwater environment of Hiroshima, Japan. *Appl. Environ. Microbiol.* **55**:2613–2618.

Vieira, V. V., L. F. Teixeira, A. C. Vicente, H. Momen, and C. A. Salles. 2001. Differentiation of environmental and clinical isolates of *Vibrio mimicus* from *Vibrio cholerae* by multilocus enzyme electrophoresis. *Appl. Environ. Microbiol.* **67**:2360–2364.

Vuddhakul, V., T. Nakai, C. Matsumoto, T. Oh, T. Nishino, C.-H. Chen, M. Nishibuchi, and J. Okuda. 2000. Analysis of *gyrB* and *toxR* gene sequences of *Vibrio hollisae* and development of *gyrB*- and *toxR*-targeted PCR methods for isolation of *Vibrio hollisae* from the environment and its identification. *Appl. Environ. Microbiol.* **66**:3506–3514.

Wang, L. U., and J. C. Chen. 2005. The immune response of white shrimp *Litopenaeus vannamei* and its susceptibility to *Vibrio alginolyticus* at different salinity levels. *Fish Shellfish Immunol.* **18**:269–278.

West, P. A., P. R. Brayton, T. N. Bryant, and R. R. Colwell. 1986. Numerical taxonomy of vibrios isolated from aquatic environments. *Int. J. Syst. Bacteriol.* **36**:531–543.

Wong, H. C., S. H. Ting, and W. R. Shieh. 1992. Incidence of toxigenic vibrios in foods available in Taiwan. *J. Appl. Bacteriol.* **73**:197–202.

Wuthe, H. H., S. Aleksic, and W. Hein. 1993. Contribution to some phenotypical characteristics of *Vibrio cincinnatiensis*. Studies in one strain of a diarrhoeic human patient and in two isolates from aborted bovine fetuses. *Zentbl. Bakteriol.* **279**:458–465.

Yalcinkaya, F., C. Ergin, C. Agalar, S. Kaya, and M. Y. Aksoylar. 2003. The presence and antimicrobial susceptibilities of human-pathogen Vibrio spp. isolated from blue crab (*Callinectes sapidus*) in Belek tourism coast, Turkey. *Int. J. Environ. Health Res.* **13**:95–98.

Yamamoto, S., N. Okujo, Y. Fujita, M. Saito, T. Yoshida, and S. Shinoda. 1993. Structures of two polyamine-containing catecholate siderophores from *Vibrio fluvialis. J. Biochem. (Tokyo)* **113**:538–544.

Yamane, K., J. Asato, N. Kawade, H. Takahashi, B. Kimura, and Y. Arakawa. 2004. Two cases of fatal necrotizing fasciitis caused by *Photobacterium damsela* in Japan. *J. Clin. Microbiol.* **42:**1370–1372.

Yamasaki, S., H. Shirai, Y. Takeda, and M. Nishibuchi. 1991. Analysis of the gene of *Vibrio hollisae* encoding the hemolysin similar to the thermostable direct hemolysin of *Vibrio parahaemolyticus*. *FEMS Microbiol. Lett.* **80:**259–264.

Yoh, M., T. Honda, and T. Miwatani. 1986. Purification and partial characterization of a *Vibrio hollisae* hemolysin that relates to the thermostable direct hemolysin of *Vibrio parahaemolyticus*. *Can. J. Microbiol.* **32:**632–636.

Yoh, M., T. Honda, and T. Miwatani. 1988. Comparison of hemolysins of *Vibrio cholerae* non-O1 and *Vibrio hollisae* with thermostable direct hemolysin of *Vibrio parahaemolyticus*. *Can. J. Microbiol.* **34:**1321–1324.

Yorimitsu, T., and M. Homma. 2001. Na$^+$-driven flagellar motor of *Vibrio*. *Biochim. Biophys. Acta* **1505:**82–93.

Yoshinaga, D. H., and H. A. Frank. 1982. Histamine-producing bacteria in decomposing skipjack tuna (*Katsuwonus pelamis*). *Appl. Environ. Microbiol.* **44:**447–452.

Yu, C., A. M. Lee, B. L. Bassler, and S. Roseman. 1991. Chitin utilization by marine bacteria. A physiological function for bacterial adhesion to immobilized carbohydrates. *J. Biol. Chem.* **266:**24260–24267.

Yu, C. F., P. H. Yu, P. L. Chan, Q. Yan, and P. K. Wong. 2004. Two novel species of tetrodotoxin-producing bacteria isolated from toxic marine puffer fishes. *Toxicon* **44:**641–647.

Yuen, K. Y., L. Ma, S. S. Wong, and W. F. Ng. 1993. Fatal necrotizing fasciitis due to *Vibrio damsela*. *Scand. J. Infect. Dis.* **25:**659–661.

Zanetti, S., T. Spanu, A. Deriu, L. Romano, L. A. Sechi, and G. Fadda. 2001. In vitro susceptibility of *Vibrio* spp. isolated from the environment. *Int. J. Antimicrob. Agents* **17:**407–409.

Zen-Yoji, H., R. A. Le Clair, K. Ota, and T. S. Montague. 1973. Comparison of *Vibrio parahaemolyticus* cultures isolated in the United States with those isolated in Japan. *J. Infect. Dis.* **127:**237–241.

Zhang, X.-H., and B. Austin. 2000. Pathogenicity of *Vibrio harveyi* to salmonids. *J. Fish Dis.* **23:**93–102.

Zhang, X. H., P. G. Meaden, and B. Austin. 2001. Duplication of hemolysin genes in a virulent isolate of *Vibrio harveyi*. *Appl. Environ. Microbiol.* **67:**3161–3167.

Zorrilla, I., S. Arijo, P. Chabrillon, P. Diaz, E. Martinez-Manzanares, M. C. Balebona, and M. A. Morinigo. 2003. *Vibrio* species isolated from diseased farmed sole, *Solea senegalensis* (Kaup), and evaluation of the potential virulence role of their extracellular products. *J. Fish Dis.* **26:**103–108.

IX. EPIDEMIOLOGY

The Biology of Vibrios
Edited by F. L. Thompson et al.
© 2006 ASM Press, Washington, D.C.

Chapter 27

Epidemiology

SHAH M. FARUQUE AND G. BALAKRISH NAIR

INTRODUCTION

The epidemiological importance of pathogenic *Vibrio* species was not recognized until the fifth pandemic of cholera, when Robert Koch isolated a motile rod-shaped bacterium, *Vibrio cholerae* (then referred to as "comma bacilli") from the stools of cholera patients in Egypt in 1883 and later in India in 1884 (Koch, 1894). However, the classic epidemiological study of John Snow in 1854 in London showed the association of the disease with contaminated drinking water even before any bacteria were known to exist (Snow, 1855). *Vibrio* species are now known to be ubiquitous in marine, estuarine, and freshwater environments and to encompass a diverse group of bacteria. Currently, the genus *Vibrio* consists of 65 species, of which >12 species are known to be associated with human disease (Blake et al., 1979; Tacket et al., 1984; Perez-Tirse et al., 1993; Abbot and Janda, 1994; Mitra et al., 1996; Janda et al., 1998). Of these species, *V. cholerae*, *V. parahaemolyticus*, *V. fluvialis*, *V. furnissii*, *V. mimicus*, and *Grimontia hollisae* are primarily associated with diarrheal diseases, whereas *V. alginolyticus* and *Photobacterium damselae* generally cause wound infections, and *V. vulnificus* is an important cause of septicemia in alcoholics and immunosuppressed hosts. The significance of isolation of three other *Vibrio* species, including *V. cincinnatiensis*, *V. harveyi*, and *V. metschnikovii*, from humans remains to be determined. Some vibrios have multiple lifestyles and habitats, which include either a free-swimming state or a sessile existence attached to plankton, shellfish, or other surfaces in the aquatic environment, and a pathogenic state in which these *Vibrio* species infect humans, causing intestinal or extraintestinal diseases (Colwell and Spira, 1992; Colwell and Huq, 1994). Some of the pathogenic strains produce virulence factors, such as toxins, or colonization factors that have been associated with the pathogenic symptoms caused by these organisms (Kaper et al., 1995: Faruque et al., 1998).

Vibrio cholerae

Of the various *Vibrio* species associated with human disease, *V. cholerae* strains belonging to the O1 and O139 serogroups cause the severest form of the clinical-epidemiological syndrome known as cholera, which usually occurs as explosive outbreaks. The clinical disease is characterized by the passage of voluminous watery stools that rapidly causes dehydration, hypovolemic shock, and acidosis and can lead to death if prompt and appropriate treatment is not initiated. Historically, seven distinct pandemics of cholera have been documented since the beginning of the first pandemic in 1817 (Kaper et al., 1995). Because of the large numbers of cases and deaths during these pandemics, the disease was viewed as a major public health disaster requiring governmental intervention. The New York cholera epidemic led to the first board of health in the United States in 1866, and cholera became the first reportable disease (Duffy, 1971).

Pathogenic *V. cholerae* enters its host through an oral route of infection, colonizes the small intestine, and produces a potent enterotoxin known as cholera toxin (CT), which is mainly responsible for the manifestation of the disease (Faruque et al., 1995; Kaper et al., 1995). In toxigenic *V. cholerae*, CT is encoded by a lysogenic phage referred to as CTXφ (Waldor and Mekalanos, 1996). In addition to CT, the ability of pathogenic *V. cholerae* to cause disease depends primarily on the expression of a pilus colonization factor known as toxin coregulated pilus (TCP), so named because expression of TCP is under the same genetic control as that of CT (Faruque et al., 1998,

Shah M. Faruque • Molecular Genetics Laboratory, International Centre for Diarrhoeal Disease Research, Bangladesh, Mohakhali, Dhaka-1212, Bangladesh. **G. Balakrish Nair** • Laboratory Sciences Division, International Centre for Diarrhoeal Disease Research, Bangladesh, Mohakhali, Dhaka-1212, Bangladesh.

2004a). Both these pathogenic factors and their genetic determinants have been characterized to a considerable extent. However, many aspects of the organism, particularly the genetic mechanisms that allow the bacterium to adapt both to the aquatic environment and to the human intestine, are not adequately understood.

V. cholerae strains belonging to serogroups other than O1 and O139, collectively referred to as the non-O1/non-O139 vibrios, have also been implicated as etiologic agents of moderate to severe human gastroenteritis (Morris, 1990; Bagchi et al., 1993; Ramamurthy et al., 1993a; Dalsgaard et al., 1996; Rudra et al., 1996). These bacteria most often cause watery diarrhea, but they may rarely cause systemic infections (bacteremia or meningitis) and wound infections. Some non-O1/non-O139 strains produce CT or heat-stable enterotoxin (NAG-ST). A wide range of other factors, including cytotoxins, hemolysins, colonization factors, and a CT-like toxin, have also been described as virulence factors for non-O1/non-O139 *V. cholerae* (Davis et al., 1981; Bagchi et al., 1993; Ramamurthy et al., 1993a). Recently, there has been an increasing incidence of diarrhea associated with these organisms. For example, an unusual upsurge of diarrhea associated with *V. cholerae* non-O1/non-O139 occurred in Calcutta, India (Sharma et al., 1998). Although *V. cholerae* non-O1/non-O139 usually causes sporadic cases, several localized outbreaks of diarrhea caused by non-O1/non-O139 have been described. These include an outbreak caused by *V. cholerae* O10 and O12 in February 1994 in Lima, Peru (Dalsgaard et al., 1995), another caused by *V. cholerae* O10 in East Delhi, India (Rudra et al., 1996), and an epidemic caused by non-O1 *V. cholerae* that produced heat-stable enterotoxin among Khmers in a camp in Thailand (Bagchi et al., 1993). Other nonepidemic serogroups of *V. cholerae*, including toxigenic O141 strains (Dalsgaard et al., 2001), have been associated primarily with sporadic cases.

Other Pathogenic *Vibrio* Species

Vibrio mimicus was first proposed to include biochemically atypical *V. cholerae* non-O1 isolates, being readily differentiated on the basis of negative reactions in sucrose, the Voges-Proskauer reaction, corn oil, and Jordan tartarate reactions (Davis et al., 1981). Some strains of *V. mimicus* produce a heat-labile enterotoxin similar to CT as well as other toxins and toxic substances that might contribute to its pathogenesis. CT, however, is the main factor responsible for the severity of the diarrhea (Chowdhury et al., 1987, 2000). Recent studies have shown that some nontoxigenic *V. mimicus* strains are susceptible to infection by the cholera toxin-converting bacteriophage CTXφ and thus become toxigenic (Faruque et al., 1999a). However, *V. mimicus* differs epidemiologically from *V. cholerae* in that the former does not cause epidemics. Consumption of *V. mimicus*-contaminated shellfish has been linked to the development of gastroenteritis. Moreover, turtle eggs serve as potential sources of *V. mimicus* diarrhea in tropical countries where turtle eggs are used for human consumption (Campos et al., 1996).

Vibrio parahaemolyticus is a marine seafood-borne pathogen that causes gastroenteritis in humans (Miwatani and Takeda, 1975). Unlike *V. cholerae*, where only two serogroups (O1 and O139) are involved in epidemic and pandemic disease, *V. parahaemolyticus* gastroenteritis is a multiserogroup affliction, and as many as 76 different combinations of O and K serotypes are currently recognized and known to be associated with gastroenteritis. Most patients with *V. parahaemolyticus* infection have acute watery diarrhea, but occasionally systemic infections (e.g., sepsis and wound infections) may occur (Pal et al., 1984), especially in patients at increased risk because of immunodeficiency or liver cirrhosis.

In contrast to other *Vibrio* species, *Vibrio vulnificus* has been associated primarily with systemic infections and rarely diarrhea (Tacket et al., 1984; Morris, 1988; Linkous and Oliver, 1999; Chiang and Chuang, 2003). The organism may also infect wounds, leading to severe necrotic wound infections and sepsis, also associated with high case-fatality rates. The wound infections frequently have bullous lesions, which provide a clue to the etiology, and the organism can often be identified in the bullous fluid.

A few other halophilic vibrios, including *Vibrio fluvialis*, *Grimontia hollisae*, and *Vibrio furnissii*, may also cause moderate to severe diarrhea, whereas *G. hollisae* may rarely cause sepsis in susceptible hosts (Klontz and Desenclos, 1990; Abbot and Janda, 1994). *V. alginolyticus* and *P. damselae* are also rare causes of sepsis or wound infections, and are associated with exposure to seawater or seafood (Perez-Tirse et al., 1993). The other vibrios of possible clinical importance include *V. metschnikovii*, *V. cincinnatiensis*, and *V. harveyi*, but infections due to these species are extremely rare. Although there are at least 12 known *Vibrio* species pathogenic to humans, *V. cholerae* strains belonging to the O1 and O139 serogroups cause cholera, the most severe form of diarrheal disease, and the most important in terms of pandemic spread. In this chapter we will summarize available information on the epidemiology of major clinically significant *Vibrio* species, including *V. cholerae*, *V.*

parahaemolyticus, and *V. vulnificus* with an especial emphasis on cholera.

EPIDEMIOLOGY OF CHOLERA

Cholera is primarily a waterborne disease, and populations interacting with contaminated surface water may be affected at a high rate. The disease is endemic in Southern Asia and parts of Africa and Latin America, where outbreaks occur widely and are particularly associated with poverty and poor sanitation (Faruque et al., 1998; Kaper et al., 1995). The fecal-oral transmission of cholera usually occurs due to ingestion of fecally contaminated water by susceptible individuals. Besides drinking water, food has also been recognized to be an important vehicle of transmission of cholera.

After ingestion of contaminated water or food containing *V. cholerae*, the organism passes through the acid barrier of the stomach, colonizes the small intestine, and produces CT, which is mainly responsible for the manifestation of the disease (Kaper et al., 1995). CT acts as a classical A–B type toxin, leading to ADP-ribosylation of a small G protein and constitutive activation of adenylate cyclase, thus giving rise to increased levels of cyclic AMP within the host cell. This results in the rapid efflux of chloride ions and water from host intestinal cells. The subsequent loss of water and electrolytes leads to the severe diarrhea and vomiting characteristic of cholera. Massive outpouring of fluid and electrolytes leads to severe dehydration, electrolyte abnormalities, and metabolic acidosis (Kaper et al., 1995). In severe disease, death may occur in as high as 50 to 70% of cases if they are not adequately hydrated. Hallmarks of the epidemiology of cholera include a high degree of clustering of cases by location and season, with highest rates of infection in children 1 to 5 years of age in endemic areas, and protection against the disease being afforded by improved sanitation/hygiene and preexisting immunity. *V. cholerae* strains associated with epidemics also undergo frequent genetic and phenotypic changes. These include antibiotic resistance patterns that frequently change and changes in epidemiological markers showing clonal diversity among epidemic strains (Faruque et al., 1998). Cholera can spread rapidly and in explosive epidemics from one region to another, affecting large numbers of people. The epidemic strains can also spread across countries and continents over time, giving rise to cholera pandemics. At least seven distinct pandemics of cholera have occurred since the onset of the first pandemic in 1817 (Pollitzer, 1959). Except for the seventh pandemic, which originated on the island of Sulawesi in Indonesia (Kamal, 1974), the other six pandemics arose from the Ganges delta region of the Indian subcontinent and spread to reach other continents, affecting many countries and extending over many years. Details and duration of these early pandemics have been reviewed by Kaper et al. (1995). The ongoing pandemic of cholera is the seventh, which is the most extensive in terms of duration and geographical spread, and hence warrants more detailed discussion.

The Seventh Pandemic of Cholera

The causative agent of the current, seventh pandemic is *V. cholerae* O1 of the El Tor biotype, as opposed to the classical biotype, which caused the sixth and presumably the earlier pandemics. The pandemic started in Sulawesi, Indonesia, and spread to the entire Southeast Asian archipelago by the end of 1962 (Kamal, 1974). During 1963 to 1969, the pandemic spread to the Asian mainland. By 1970, the pandemic reached the Arabian Peninsula and spread to sub-Saharan West Africa, causing explosive outbreaks resulting in more than 400,000 cases of cholera with a high case-fatality rate, due mainly to a lack of background immunity in the population and inadequacies in the health care infrastructure (Cohen et al., 1971; Goodgame and Greenough, 1975). The seventh pandemic reached South America in 1991, in the form of an explosive epidemic that began in Peru and spread to neighboring Ecuador, and then Colombia and Chile (Levine, 1991; Pan American Health Organization, 1991; Ries et al., 1992). Cholera cases began to occur along the Pacific coast of South America and progressively entered more countries in South and Central America. It was estimated that during 1991 to 1992, there were 750,000 cases of cholera in the Americas with 6,500 deaths (Pan American Health Organization, 1991).

Recently, one of the worst cholera outbreaks occurred in Goma, Eastern Zaire, in July 1994 (Siddique et al., 1995). Conflicts between tribes in neighboring Rwanda had displaced nearly a million people who were sheltered in refugee camps in Zaire. Outbreak of cholera in the poverty-stricken refugee camps led to the death of ~12,000 Rwandan refugees (Siddique et al., 1995). The seventh pandemic is ongoing, with seasonal outbreaks of cholera in many developing countries of Asia, Africa, and Latin America, and imported cases occur in many developed countries (Color Plate 18). While the cholera pandemic due to *V. cholerae* O1 of the El Tor biotype continues, a new strain of toxigenic *V. cholerae*, later designated *V. cholerae* O139, emerged in 1992 as the first non-O1 strain to cause epidemics of cholera (Ramamurthy et al., 1993b; Cholera Working Group, 1993).

V. cholerae O139

In late 1992, cholera epidemics were reported in Madras and elsewhere in India and in southern Bangladesh. Remarkably, although the clinical syndrome was typical of cholera, the causative agent was a *V. cholerae* non-O1 strain, which was later serogrouped as O139 (Cholera Working Group, 1993). The epidemic continued through 1993, and *V. cholerae* O139 spread throughout Bangladesh and India and neighboring countries. Outbreaks or cases due to *V. cholerae* O139 were reported in Pakistan, Nepal, China, Thailand, Kazakhastan, Afghanistan, and Malaysia (Centers for Disease Control, 1993; Public Health Laboratory Service, 1993; Chongsanguan et al., 1993; Swerdlow and Ries, 1993). Imported cases were also reported from the United Kingdom and the United States (Centers for Disease Control, 1993; Public Health Laboratory Service, 1993). *V. cholerae* O139, which caused explosive epidemics throughout Bangladesh, India, and neighboring countries, was presumed to be the causative agent of an eighth pandemic of cholera (Swerdlow and Ries, 1993). In the beginning, the new strain totally displaced the existing *V. cholerae* O1 strains. However, during 1994 and until the middle of 1995, in most northern and central areas of Bangladesh, including the capital city of Dhaka, the O139 vibrios were replaced by a new clone of *V. cholerae* O1 of the El Tor biotype, whereas in the southern coastal regions the O139 vibrios continued to exist (Faruque et al., 1997; Siddique et al., 1996). During the second half of 1995 and in 1996, nearly 4 years after the initial detection of O139 vibrios, cases due to both *V. cholerae* O1 and O139 strains were detected in various regions of Bangladesh. A resurgence of *V. cholerae* O139 infection for the first time since its initial predominance in 1993 was also reported in Calcutta, India (Mitra et al., 1996). Recent surveillance has shown that *V. cholerae* O139 continues to cause cholera outbreaks in India and Bangladesh and coexists with the El Tor vibrios. Between March and May 2002, there was a marked increase in cholera cases associated with *V. cholerae* O139 in Bangladesh, when an estimated 35,000 cases occurred in and around Dhaka (Faruque et al., 2003).

Seasonal Epidemics of Cholera

Once endemicity is established in an area, cholera tends to settle into a seasonal pattern. Surveillance of cholera in Bangladesh has provided important information regarding the epidemiology and seasonality of the disease (Faruque et al., 1998). Bangladesh is situated in the Ganges delta, and be-cause of its low-lying deltaic environment, the water volumes in ponds and rivers in Bangladesh change between the dry and wet seasons. Besides raising the river levels, the monsoons also wash the sewage from the villages into the rivers. In rural areas, the people are in direct contact with the surface water for drinking, bathing, cooking, and irrigation. In Bangladesh, epidemic outbreaks usually occur twice a year, with the highest number of cases just after the monsoon during September to December, and a somewhat smaller peak of cases in the spring between March and May (Glass et al., 1982, Baqui et al., 1991; Siddique et al., 1992). Seasonal patterns also differ in different geographical areas. For example, in Calcutta, India, the highest number of cases occurs in April, May, and June. Cholera in South America also developed a periodicity, with more cases in the summer months of January and February. This epidemiological observation suggests a role of environmental and climatic factors in the occurrence of epidemics. However, since infection due to *V. cholerae* occurs exclusively through the oral route, contaminated food and water are the direct sources of human infection, while environmental factors are likely to influence the seasonal prevalence of *V. cholerae* in the aquatic environment, as well as their survival and epidemic spread.

Susceptible Population

Although cholera occurs in individuals of all ages, children aged 2 to 9 years have the highest incidence of the disease in endemic areas, such as Bangladesh (Glass et al., 1982). The decreased rates in children under the age of 1 year may relate to decreased exposure and to the protective effect of breast-feeding or breast milk (Clemens et al., 1990). These assumptions have been supported by the observation that, when introduced into populations lacking prior exposure to the disease, cholera tends to occur with equal frequency in all age groups (Holmberg et al., 1984; Glass and Black, 1992), and this was clearly observed in the South American epidemics (Swerdlow et al., 1992). The newly emerged *V. cholerae* O139 strains in Bangladesh and India caused the majority of cases in adults, suggesting that preexisting immunity against *V. cholerae* O1 did not protect against O139 cholera (Cholera Working Group, 1993). Susceptibility to cholera also depends on largely unknown host factors. In natural infection, as well as with experimental infection, individuals of blood group O have been demonstrated to be at increased risk of more severe cholera (Glass et al., 1985; Clemens et al., 1989).

Infectious Dose and Mode of Transmission

Volunteer challenge studies with *V. cholerae* have shown that approximately 10^{11} *V. cholerae* organisms are required to induce diarrhea in fasting North American volunteers, unless sodium bicarbonate is administered to neutralize gastric acid (Cash et al., 1974). This finding suggests that generally a high dose of the organism is required for pathogenesis. However, when stomach acidity is neutralized with sodium bicarbonate, administration of an inoculum of 10^6 *V. cholerae* cells induces diarrhea in 90% of volunteers (Cash et al., 1974; Levine et al., 1981). The incubation period generally varies between 25 and 33 h. Most volunteers, who received as few as 10^3 to 10^4 organisms with buffer, also developed diarrhea, although with diminished severity of the disease, and required a longer incubation period (Levine et al., 1981). Since the presence of toxigenic *V. cholerae* in environmental waters is usually far less than the required dose for a severe infection, the majority of natural infections due to *V. cholerae* are expected to be asymptomatic or cause mild disease (McCormack et al., 1969). It has been suggested, however, that a pre-enrichment of the organism through asymptomatic infection of humans may lead to subsequent high prevalence of *V. cholerae*, which leads to cholera outbreaks (Faruque et al., 2004b). It also seems likely that a pre-enrichment of the organism in contaminated foods may lead to the first case of cholera, which subsequently leads to an epidemic fostered by contaminated water and poor sanitation.

In developing countries with poverty and poor sanitation, fecal contamination of domestic and commercial food is likely to occur. Patients with acute cholera may excrete 10^7 to 10^8 *V. cholerae* cells/g of stool (Levine et al., 1981), and the total output by a patient can be in the range of 10^{11} to 10^{13} highly infectious *V. cholerae* cells. This large number of organisms can contaminate environmental waters, and people consuming such water for household use are likely to contaminate food stuffs. Even after cessation of symptoms, patients who have not been treated with antibiotics may continue to excrete vibrios for 1 to 2 weeks (Levine et al., 1988), and a small minority of patients may continue to excrete the organism for even longer periods. Asymptomatic carriers are most commonly identified among household members of persons with acute illness. It is likely that water may serve as a source of secondary contamination of food during its preparation. In areas of endemic disease, transmission of cholera through contaminated foods served by street vendors and restaurants should be considered. In Piura, Peru, drinking unboiled water, eating food from a street vendor, and

eating rice after 3 h without reheating were all independently associated with illness (Ries et al., 1992).

In developed countries, food-borne outbreaks of cholera have on many occasions occurred due to consumption of contaminated seafood. It is clear that CT-producing *V. cholerae* O1 or O139 can persist in the environment in the absence of known human disease. Periodic introduction of such environmental isolates into the human population through ingestion of uncooked or undercooked shellfish appears to be responsible for isolated foci of endemic disease along the Gulf Coast in the United States and in Australia (Blake et al., 1980; Chen et al., 1991). Environmental isolates contaminating seafood may also have been responsible for the initial cases in the South American epidemic.

Molecular Epidemiology

Epidemiological surveillance of cholera has provided important information regarding temporal changes in the properties of *V. cholerae* strains isolated from different epidemics. These include not only the prevalence of particular biotypes or serotypes in different regions over time, and their changing antibiotic resistance pattern, but also clonal variation within particular serotypes. Analysis of toxigenic El Tor strains by multilocus enzyme electrophoresis has been used to group the El Tor strains into major clonal groups. The clones seem to reflect broad geographical and epidemiological associations (Wachsmuth et al., 1994). Developments in DNA analysis techniques led to the introduction of several new typing methods that have enabled the study of the epidemiology of *V. cholerae* on a larger global scale (Chen et al., 1991; Faruque et al., 1993, 1994, 1995; Faruque and Albert, 1992; Kaper et al., 1982; Wachsmuth et al., 1991, 1993, 1994). These techniques include the analysis of restriction fragment length polymorphisms (RFLPs) in different genes. The use of gene probes to study RFLPs in the *ctxAB* genes and their flanking DNA sequences, which are part of the CTX prophage (Waldor and Mekalanos, 1996), indicated that the U.S. Gulf Coast isolates of toxigenic *V. cholerae* are clonal and that they are different from other seventh pandemic isolates (Kaper et al., 1982). RFLPs in conserved rRNA genes have also been used to differentiate *V. cholerae* strains into different ribotypes. Analysis of isolates from the Latin American epidemic in 1991 showed that they were related to the seventh pandemic isolates from other parts of the world and that the Latin American cholera epidemic was an extension of the seventh pandemic (Faruque and Albert, 1992; Wachsmuth et al., 1993).

Molecular analysis of epidemic isolates of *V. cholerae* between 1961 and 2002 in Bangladesh revealed clonal diversity among strains isolated during different epidemics and demonstrated the transient appearance and disappearance of more than six ribotypes among classical vibrios, more than five ribotypes of El Tor vibrios, and three ribotypes of *V. cholerae* O139 (Faruque et al., 1998, 1999b, 2000, 2003). Numerical analysis of ribotype patterns has revealed that *V. cholerae* strains belonging to the non-O1/non-O139 serogroups diverge widely from the O1 or O139 *V. cholerae* strains (Faruque et al., 2004b). Overall, molecular epidemiological analysis indicates that there had been a continual emergence of new clones of toxigenic *V. cholerae*, which replaced existing clones, possibly through natural selection involving unidentified environmental factors and immunity of the host population.

Evolution of Pathogenic *V. cholerae* and Emerging Variants

Since *V. cholerae*, a well-defined species on the basis of biochemical tests and DNA homology, is part of the natural flora of the aquatic environment (Colwell and Spira, 1992), it is obvious that the pathogenic clones have been evolved from aquatic forms that attained the ability to colonize the human intestine by progressive acquisition of genetic information. The genome of *V. cholerae* is not a single chromosome but is composed of two unique and separate circular megareplicons (Trucksis et al., 1998). The two circular chromosomes, designated ChrI and ChrII, consist of 2,961,146 bp and 1,072,314 bp, respectively, and together encode 3,885 open reading frames (Heidelberg et al., 2000). The larger chromosome contains most of the recognizable genes required for essential cell functions. Genes involved in pathogenesis, including those encoding toxins, surface antigens, and adhesins, are also located on the large chromosome. In contrast, the small chromosome contains a larger fraction (59%) of hypothetical genes compared to the large chromosome (42%). The small chromosome of *V. cholerae* also contains a distinctive class of integrons, which constitutes an efficient gene capture system (Mazel et al., 1998; Heidelberg et al., 2000). Integrons are gene expression elements that acquire open reading frames and convert them to functional genes. Nearly 12% of the small chromosome of *V. cholerae* constitutes a large gene cluster known as the "integron island." The contribution of integrons to the evolutionary biology of *V. cholerae* has yet to be fully explored.

Identification of chromosomal regions comprising clusters of genes associated with virulence, which are absent in nonpathogenic strains, was crucial in our perception of the evolution of pathogenic *V. cholerae* from its aquatic progenitors. The major virulence gene clusters include the TCP pathogenicity island (Kovach et al., 1996) encoding the colonization factor TCP and the CTX prophage encoding CT (Waldor and Mekalanos, 1996). A region of ~40 kb, including the TCP-ACF gene clusters flanked on both sides by a putative 20-bp *att*-like attachment sequence and carrying a putative integrase gene and a transposase gene, constitutes the TCP pathogenicity island (Kovach et al., 1996). Remarkably, the CTXφ that encodes CT uses TCP as its receptor for infecting new strains, and thus these two horizontally moving elements are linked evolutionarily.

Other clusters of genes having putative additional roles in pathogenesis include the *V. cholerae* RTX toxin gene cluster (Lin et al., 1999) and genes for a new type IV pilus (Fullner and Mekalanos, 1999). The RTX (repeat in toxin) family includes a group of related exotoxins produced by a variety of pathogenic gram-negative bacteria. Recently, comparative genomic analysis of *V. cholerae* isolates using DNA microarrays identified seventh pandemic-specific genes that are grouped into two chromosomal islands, the VSP-1 and VSP-2 (*Vibrio* seventh pandemic island-1 and -2) (Dziejman et al., 2002). It has been suggested that these genes might have been involved in the epidemiological success of the seventh pandemic clone (Faruque and Mekalanos, 2003). Recently, a 57-kb chromosomal insert encoding genes for neuraminidase (*nanH*) and amino sugar metabolism was proposed to correspond to a pathogenicity island designated VPI-2 (Jermyn and Boyd, 2002). The structure, G+C content, and codon usage within the known pathogenicity island of *V. cholerae* suggest that these islands were recently acquired by *V. cholerae*, thus further supporting the assumption that the pathogenic *V. cholerae* strains have evolved from nonpathogenic strains. Although some of the virulence gene clusters have been characterized, the potential exists for identification of yet new genes, which may influence the pathogenicity and epidemiological characteristics of *V. cholerae*.

Besides acquiring large virulence gene clusters by nonpathogenic *V. cholerae* and thus making quantum leaps toward attaining virulent forms, pathogenic strains of *V. cholerae* appear to undergo temporal genetic and phenotypic changes leading to the emergence of diverse epidemic strains. The transformation of *V. cholerae* O1 El Tor strains to O139 occurred by one or more horizontal gene transfer events resulting in deletion and replacements of the gene cluster-encoding enzymes involved in the lipopolysaccharide O-side chain synthesis (Bik et al., 1995). This serogroup transformation of an epidemic clone provides one of the best examples of how pathogenic *V. chol-*

erae might attain greater evolutionary fitness through continued gene acquisition.

O139 strains continue to undergo rapid genetic changes, and the remerged O139 strains in Calcutta in 1996 carried a new CTX prophage referred to as the Calcutta type CTX prophage (CTX^{Calc}) (Kimsey et al., 1998). Analysis of epidemic *V. cholerae* O139 strains in Bangladesh in 2002 revealed that these cultures had undergone changes by acquisition of a new CTX prophage. Analysis of the repressor gene of the CTX prophage, *rstR*, carried by these strains showed that, whereas the initial strains of 1993 carried an El Tor type CTX^{ET} prophage, the new cultures carry at least one copy of the Calcutta type CTX^{Calc} prophage in addition to the CTX^{ET} prophage (Faruque et al., 2003; Kimsey et al., 1998).

Recent studies have revealed the emergence of *V. cholerae* O1 strains with hybrid properties. These strains, which possess partially classical and partially El Tor biotype properties, were initially isolated from cholera patients in Matlab, a rural area of Bangladesh (Nair et al., 2002). Later, similar strains were isolated from outbreaks of cholera in Mozambique (Ansaruzzaman et al., 2004). In summary, temporal changes in genetic or phenotypic properties have been noted among *V. cholerae* strains isolated during different epidemics of cholera. These events suggest that frequent genetic exchange occurs among *V. cholerae* strains, leading to a continual emergence of new toxigenic clones. The observed genetic changes in pathogenic *V. cholerae* probably reflect the selective process associated with evading growing immunity in an endemic population.

Environmental Reservoirs and Ecology

Although toxigenic *V. cholerae* O1 and O139 strains are human pathogens, these bacteria belong to a group of organisms that are normal inhabitants of the aquatic environment (Colwell and Spira, 1992). The physicochemical conditions for the survival of *V. cholerae* O1 have been investigated, and the possibility of survival of the organism in an estuarine environment and other brackish waters is widely accepted. The survival may be dependent on several factors, such as occurrence of particular physicochemical conditions, association of the bacteria with aquatic plants or animals, and/or the existence of specific ecological association involving several components of the aquatic environment. It has been postulated that under unfavorable environmental conditions the vibrios are converted to a viable but nonculturable (VBNC) form that cannot be recovered by standard culture techniques, and that such forms are able to produce infection and can revert to the culturable

form (Colwell and Huq, 1994). Contrary to this proposition, laboratory-based studies on a marine vibrio strain (Novitsky and Morita, 1976, 1977, 1978) showed that the organism responds to starvation by reducing metabolic activities and inducing morphological changes, e.g., rod-shaped to a coccoid shape, and producing progeny cells significantly decreased in volume, but that it still remains culturable. However, it has been argued that in this investigation only those cells that remained culturable were studied, and that cells that were possibly VBNC were not recognized (Colwell and Huq, 1994). The public health and ecological importance of the possible survival forms, such as VBNC, depends on whether these forms are (re)convertible to live infectious bacteria. Hence, there is considerable scope to further investigate the role of the postulated VBNC forms of *V. cholerae* through carefully controlled studies. The concept of an aquatic reservoir of *V. cholerae* O1 or O139 implies not only that the vibrios survive but that they form an essential component of the ecosystem. Laboratory studies of microcosms have illustrated the ability of *V. cholerae* O1 to associate with a variety of zooplankton, phytoplankton, and algae (Islam et al., 1994). The associations prolong survival, and presumably the vibrios gain nutrients from the host.

Although water is clearly a vehicle for transmission of *V. cholerae*, the physical, chemical, and biological parameters that support the seasonal pattern of epidemics are not clear. Several models have been proposed to explain this epidemiological observation. Thus, during interepidemic periods, toxigenic *V. cholerae* may exist in an unexplained ecological association with aquatic organisms in a possible nonculturable form until the next epidemic season, when environmental factors trigger the dormant bacteria to multiply and lead to cholera outbreaks (Colwell and Spira, 1992; Colwell and Huq, 1994). For example, *V. cholerae* may colonize copepods and other zooplankton, and zooplankton blooms promoted by climatic factors may play a key role in this process (Lobitz et al., 2000). Remotely sensed satellite data for the Bay of Bengal were used to monitor the timing and spread of cholera in Bangladesh from 1992 to 1995, and it was found that sea surface temperature revealed an annual cycle similar to the cholera case data. Colwell and coworkers suggested that sea surface height may be an indicator of incursion of plankton-laden water inland, e.g., tidal rivers, suggesting that cholera epidemics are also climate-linked (Lobitz et al., 2000). However, it is not clear how *V. cholerae* strains with epidemic potential are selectively enriched, prior to an epidemic, from the vast majority of environmental strains, which do not appear to have epidemic potential. Subsequently, it was proposed that, in addition to other possible seasonal factors causing a bloom of diverse *V. cholerae* in the environment, epidemics may be preceded by a gradual enrichment of pathogenic strains through passage in human beings who consume surface water (Faruque et al., 2004b). The model involving both environmental and host factors in the initiation of seasonal epidemics is presented in Color Plate 19. Also, it has been shown that human colonization creates a hyperinfectious bacterial state that is maintained after dissemination and that may contribute to the epidemic spread of cholera (Merrell et al., 2002). A more recent study (Faruque et al., 2005) has shown that seasonal epidemics of cholera are also influenced by the environmental prevalence of lytic phages acting on epidemic *V. cholerae* strains.

In summary, studies so far suggest that causation of cholera in humans is also linked with a natural process of enrichment of toxigenic *V. cholerae*, and partly explains the benefit imparted to the pathogen during the disease in humans. However, to further understand the general epidemiological behavior of *V. cholerae*, which includes mechanisms leading to seasonal pattern of epidemics, transient appearance and disappearance of different clones, and emergence of new epidemic clones, it is important to study the interactions among the bacteria, genetic elements mediating the transfer of virulence genes, the human host, and possible environmental factors.

EPIDEMIOLOGY OF OTHER CLINICALLY SIGNIFICANT VIBRIOS

V. parahaemolyticus

Gastroenteritis due to *V. parahaemolyticus* is particularly common among people who eat raw or undercooked seafood. As opposed to *V. cholerae* infection, which is highly associated with poverty and poor sanitation, *V. parahaemolyticus* infection is also high among people with good socioeconomic status who can afford seafood (Tuyet et al., 2002). The most common clinical syndrome is gastroenteritis with watery diarrhea, abdominal cramps, nausea, vomiting, headache, and low-grade fever; rarely, sudden cardiac arrhythmia may also occur (Honda et al., 1976; Honda and Iida, 1993). Sometimes the diarrhea is bloody with reddish and watery stools. The onset of illness usually takes place 4 to 96 h after consumption of contaminated foods. Under appropriate conditions, the bacterium can grow extremely fast, with a generation time ranging from 8 to 12 min; a generation time of 27 min has been reported in crabmeat (Liston, 1974). *V. parahaemolyticus* also causes

other syndromes of clinical illness, including wound infections and septicemia. Infection can cause serious illness in persons with underlying disease, especially in immunocompromised individuals, e.g., those with leukemia, liver disease, diabetes, and those infected with human immunodeficiency virus-AIDS, where it can cause serious systemic infections (Hlady and Klontz, 1996).

V. parahaemolyticus is widely present in estuarine, marine, and coastal environments throughout the world (Vasconcelos et al., 1975; Joseph et al., 1982; Eko et al., 1994; Barbieri et al., 1999; Daniels et al., 2000) and has been reported either as a source of human disease or in the environment along the North American, African, and Mediterranean coasts. Water temperature, salinity, zooplankton blooms, tidal flushing, and dissolved oxygen may affect their spatial and temporal distribution (Kaneko and Colwell, 1978). Water temperatures have been shown to influence the growth of *V. parahaemolyticus*; the importance of water temperature in the epidemiology of infections is reflected by the fact that most outbreaks occur during the warmer months (Kaper et al., 1981).

Virulence factors

Diarrheal disease due to *V. parahaemolyticus* is toxin-mediated. At least two toxins have been identified as potential virulence factors, thermostable direct hemolysin (TDH) and TDH-related hemolysin (TRH) (Shirai et al., 1990). The TDH-positive strains are beta-hemolytic on Wagatsuma agar and are known as Kanagawa-positive strains, but those that produce TRH may be Kanagawa negative. The presence of either or both of the virulence genes, namely, *tdh* or *trh*, differentiates the pathogenic from nonpathogenic strains. A strong correlation between urease production, which is an unusual phenotype for *V. parahaemolyticus*, and the presence of *trh* has been demonstrated (Okuda et al., 1997a). Because most environmental strains, as well as other vibrios, are negative for both TDH and TRH, detection of these markers helps in differentiating pathogenic from nonpathogenic strains. The exact mechanism by which TDH or TRH causes gastroenteritis is not clearly known. *V. parahaemolyticus* also produces a capsular polysaccharide whose role in pathogenesis is unknown but whose antigenic properties have been used for serotyping. Studies in the rabbit model have shown that the bacterium is enteroinvasive, and it colonizes and causes inflammation in the small intestine (Chatterjee et al., 1984). However, the overall mechanism of pathogenesis by *V. parahaemolyticus* remains unclear.

Pandemic potential

V. parahaemolyticus used to be commonly isolated in Japan, but the species is now recognized as a pathogen in all parts of the world, including both industrialized and developing countries (Zen-Yoji et al., 1965; World Health Organization, 1999). *V. parahaemolyticus* is an important etiological agent of diarrhea in Calcutta, India, where gastroenteritis ranks second to cholera; epidemiological studies have also revealed a high incidence of human carriers of *V. parahaemolyticus* in this population (Chatterjee et al., 1970; Pal et al., 1984).

Before 1996, *V. parahaemolyticus* was sporadically isolated, with diverse serogroups being involved in disease in different geographical areas. Beginning in February 1996, a new clone of *tdh*-positive and *trh*-negative *V. parahaemolyticus* belonging to the O3:K6 serotype was responsible for a dramatic increase in the number of cases of diarrhea in Calcutta (Okuda et al., 1997b). An increase in incidence of *V. parahaemolyticus* food poisoning observed during 1997 to 1998 was also ascribed to increased incidence of O3:K6 food poisoning in Japan (World Health Organization, 1999). Evidence supporting the hypothesis that the O3:K6 clone emerged recently and had pandemic potential was recently presented (Matsumoto et al., 2000). The *V. parahaemolyticus* pandemic has now spread into more than eight countries, and the emergence of several other serotypes possessing the pandemic potential has been documented (Chowdhury et al., 2000). Three large outbreaks have been reported in the United States on the Gulf, Atlantic, and Pacific coasts, and molecular analysis of these strains indicates that they may have diverged from the pandemic O3:K6 strains by alteration of the O:K antigens (Chowdhury et al., 2000; World Health Organization, 1999). Thus, *V. parahaemolyticus* is an emerging pathogen that has acquired the potential of causing a pandemic.

V. vulnificus

V. vulnificus was first identified and described by the Centers for Disease Control and Prevention in the United States in 1976, and it is now recognized as the most common *Vibrio* causing serious morbidity and mortality in the United States, with 95% of all seafood-related deaths being due to this species (Morris, 1988; Hlady et al., 1993). *V. vulnificus* is associated primarily with systemic infections. Persons with an underlying illness, e.g., chronic cirrhosis, hemochromatosis, thalassemia hemochromatosis, elevated serum iron level, immune function abnormalities, chronic renal insufficiency, and human immunodeficiency virus-AIDS, appear to be predisposed to infection if they ingest foods con-taminated with the bacteria (Tacket et al., 1984; Johnston et al., 1985). In Taiwan, *V. vulnificus* infections are rising; one of the factors associated with this increase is the high prevalence of hepatitis B or C virus infection-related hepatic diseases, including liver cirrhosis and hepatoma (Hsueh et al., 2004). The bacteria can invade and cause severe sepsis associated with as high as 50% case-fatality rates. Diseases associated with *V. vulnificus* infection have been found to present in two patterns that include localized wound infections acquired through exposure of a wound to salt water or shellfish, and primary septicemia acquired through oral ingestion of the organisms, with raw oysters as the most common vehicle (Blake et al., 1979). Opportunistic infections in susceptible individuals typically cause death within 24 to 48 h of exposure. Wound infection can occur in the absence of predisposing conditions but progresses more frequently to septicemia and has a higher mortality rate in those who are predisposed. A characteristic of *V. vulnificus* infection is the fulminant reaction caused by the invading bacteria in connective tissues, displayed as blisters and hemorrhagic necrosis, and even in nonfatal cases, *V. vulnificus* infection evokes intensive tissue damage (Tacket et al., 1984; Linkous and Oliver, 1999; Chiang and Chuang, 2003).

Virulence factors

V. vulnificus is a highly invasive pathogen, able to reach the bloodstream and cause septicemia via translocation across the intact intestinal wall. *V. vulnificus* belongs to three biotypes based on differences in biochemical and biological properties (Linkous and Oliver, 1999). Among the three biotypes of *V. vulnificus*, strains belonging to biotype 1 (indole-negative) are most frequently isolated from clinical specimens. Strains produce a variety of factors that have been implicated in bacterial virulence and pathogenesis, including capsular polysaccharide, cytolysin, metalloprotease (protease), phospholipases, and siderophores (Kreger and Lockwood, 1981; Testa et al., 1984; Gray and Kreger, 1985; Kothary and Kreger, 1987; Miyoshi et al., 1987; Wright et al., 1990; Okujo and Yamamoto, 1994). *V. vulnificus* also produces several enzymes, including lipase, hyaluronidase, mucinase, and DNase (Hayat et al., 1993). A hemolytic toxin has also been identified (Wright et al., 1985), but mutants that do not express the toxin also remain virulent in animal models. The primary virulence factor is the polysaccharide capsule, which prevents phagocytosis and activation of complement (Tamplin et al., 1985; Yoshida et al., 1985; Shinoda et al., 1987; Wright et al., 1990). Collectively, the cytolysin and the protease are thought to be important for the pathogenesis of *V. vulnificus*. Biochemical and genetic studies suggest that extracellular proteins released by the invading bacteria mediate the pathogenesis process of penetrating cellular barriers, vascular dissemination, and local destruction of affected tissues. The ability to acquire iron from the host via siderophore production is also an essential virulence attribute (Litwin et al., 1996). Thus, multifactor interaction in bacterial virulence is likely to produce the severe infection caused by *V. vulnificus* (Park et al., 1991).

Environmental prevalence

V. vulnificus is present in tropical and temperate estuarine ecosystems throughout the world. Infection due to this organism has been reported from the United States, Europe, Korea, Taiwan, and other countries (Park et al., 1991; Chuang et al., 1992; Hlady and Klontz, 1996). One factor that influences the incidence of disease caused by *V. vulnificus* is the prevalence of this organism in the environment. *V. vulnificus* disease parallels its concentration in oyster tissues, with the greatest number of infections occurring during the summer months when seawater temperature ranges between 20 and 30°C (Howard et al., 1988; Hlady et al., 1993). The relationship between environmental factors and *V. vulnificus* densities in oysters collected monthly in 14 states in the United States showed that the levels ranged from none detected to 1.1×10^6/g (Zen-Yoji et al., 1965). The concentration of *V. vulnificus* in oysters across the northern Gulf Coast is influenced primarily by water temperature and salinities <25% (Motes et al., 1998). Variations in surface water temperature above 26°C have little effect on densities, but the densities decline rapidly as temperature declines below 26°C. The relationship between disease and high infective dose is also supported by data showing that *V. vulnificus* infections do not occur during cold months when the numbers of organisms are very low, even though greater numbers of oysters are consumed in winter months than in summer months (Howard et al., 1988; Hlady et al., 1993).

lesser extent, include *V. alginolyticus*, *V. mimicus*, *P. damselae*, *G. hollisae*, and *V. fluvialis*. Since transmission of cholera occurs by the fecal-oral route, and untreated diarrheal stools from cholera patients are the primary source of contamination, effective public health measures, including sanitary disposal and sewage treatment systems, can prevent fecal contamination of water and interrupt the transmission cycle. Besides water, foods constitute an important vehicle for the transmission of cholera. However, *V. cholerae* cells are very sensitive to heat and are rapidly killed when exposed to 100°C. In epidemic situations, particularly in areas where drinking water may be contaminated, drinking water pretreated by boiling and consumption of hot food can reduce transmission.

V. parahaemolyticus and *V. vulnificus* infections are spread through consumption of contaminated seafood. Of all food-borne infectious diseases in the United States, *V. vulnificus* has the highest case-fatality rate (Mead et al., 1999). There is a marked seasonality of *V. vulnificus* infections, with most cases occurring during the warm months of the year. Persons with a predisposing risk factor should be warned against consumption of raw or undercooked seafood, especially during the summer. Recent events underscoring the increasing importance of avoiding raw or undercooked seafoods include the sudden appearance of specific serotypes of *V. parahaemolyticus* that have lately caused a pandemic of gastroenteritis and the escalation of *V. vulnificus* infection in the United States and Taiwan.

It is now widely accepted that pathogenic strains of *Vibrio* species have evolved from nonpathogenic environmental strains. This adaptation of a normally marine or brackish water species to the human intestine possibly reflects a need to find a niche where the organisms can rapidly amplify. Recently, *V. parahaemolyticus* has acquired pandemic potential, as the second *Vibrio* species after *V. cholerae*. Preventive measures to interrupt transmission of pathogenic vibrios and thus reduce the disease burden are possibly the most effective disruption to the emergence or rapid evolution of *Vibrio* species toward enhanced virulence.

CONCLUSIONS

Vibrio species are responsible for the majority of human diseases attributed to the natural bacterial flora of the aquatic environment or seafoods. Of these *V. cholerae*, *V. parahaemolyticus*, and *V. vulnificus* are the major human pathogens. Other vibrios associated with food poisoning, although to a much

REFERENCES

Abbot, S. L., and J. M. Janda. 1994. Severe gastroenteritis associated with *Vibrio hollisae* infection: report of two cases and review. *Clin. Infect. Dis.* **18**:310–312.

Ansaruzzaman, M., N. A. Bhuiyan, G. B. Nair, D. A. Sack, M. Lucas, J. L. Deen, J. Ampuero, and C. L. Chaignat. 2004. Cholera in Mozambique, variant of *Vibrio cholerae*. *Emerg. Infect. Dis.* **10**:2057–2059.

Bagchi, K., P. Echeverria, J. D. Arthur, O. Sethabutr, O. Serichanta-
lergs, and C. W. Hoge. 1993. Epidemic of diarrhea caused by
Vibrio cholerae non-O1 that produced heat-stable toxin among
Khmers in a camp in Thailand. *J. Clin. Microbiol.* **31**:1315–1317.

Baqui, A. H., M. D. Yunus, K. Zaman, A. K. Mitra, and K. M. B.
Hossain. 1991. Surveillance of patients attending a rural diarrhoea
treatment centre in Bangladesh. *Trop. Geog. Med.* **43**:17–22.

Barbieri, E., L. Falzano, C. Fiorentini, A. Pianetti, W. Baffone,
A. Fabbri, P. Matarrese, A. Casiere, M. Katouli, I. Kuhn,
R. Mollby, F. Bruscolini, and G. Donelli. 1999. Occurrence, di-
versity, and pathogenicity of halophilic *Vibrio* spp. and non-O1
Vibrio cholerae from estuarine waters along the Italian Adriatic
coast. *Appl. Environ. Microbiol.* **65**:2448–2753.

Bik, E. M., A. E. Bunschoten, R. D. Gouw, and F. R. Mooi. 1995.
Genesis of the novel epidemic *Vibrio cholerae* O139 strain: evidence for horizontal transfer of genes involved in polysaccharide
synthesis. *EMBO J.* **14**:209–216.

Blake, P. A., D. T. Allegra, J. D. Snyder, T. J. Barrett, L. McFarland, C. T. Caraway, J. C. Feeley, J. P. Craig, J. V. Lee, N. D.
Puhr, and R. A. Feldman. 1980. Cholera—a possible endemic
focus in the United States. *N. Engl. J. Med.* **302**:305–309.

Blake, P. A., M. H. Merson, R. E. Weaver, D. G. Hollis, and P. C.
Heublein. 1979. Disease caused by a marine *Vibrio*. Clinical
characteristics and epidemiology. *N. Engl. J. Med.* **300**:1–5.

Campos, E., H. Bolanos, M. T. Acuna, G. Diaz, M. C. Matamoros,
H. Raventos, L. M. Sanchez, O. Sanchez, and C. Barquero.
1996. *Vibrio mimicus* diarrhea following ingestion of raw turtle
eggs. *Appl. Environ. Microbiol.* **62**:1141–1144.

Cash, R. A., S. I. Music, J. P. Libonati, M. J. Snyder, R. P. Wenzel,
and R. B. Hornick. 1974. Response of man to infection with
Vibrio cholerae 1: Clinical, serologic, and bacteriologic responses
to a known inoculum. *J. Infect. Dis.* **129**:45–52.

Centers for Disease Control. 1993. Imported cholera associated
with a newly described toxigenic *Vibrio cholerae* O139 strain
California, 1993. *Morb. Mortal. Wkly. Rep.* **42**:501–503.

Chatterjee, B. D., A. Mukherjee, and S. N. Sanyal. 1984. Enteroinvasive model of *Vibrio parahaemolyticus*. *Indian J. Med.
Res.* **79**:151–158.

Chatterjee, B. D., K. N. Neogy, and S. L. Gorbach. 1970. Study of
Vibrio parahaemolyticus from cases of diarrhoea in Calcutta.
Indian J. Med. Res. **58**:235–239.

Chen, F., G. M. Evins, W. L. Cook, R. Almeida, H. N. Bean, and
I. K. Wachsmuth. 1991. Genetic diversity among toxigenic and
non-toxigenic *Vibrio cholerae* O1 isolated from the Western
Hemisphere. *Epidemiol. Infect.* **107**:225–233.

Chiang, S. R., and Y. C. Chuang. 2003. *Vibrio vulnificus* infection:
clinical manifestations, pathogenesis, and antimicrobial therapy.
J. Microbiol. Immunol. Infect. **36**:81–88.

Cholera Working Group, International Centre for Diarrhoeal Diseases Research, Bangladesh. 1993. Large epidemic of cholera-
like disease in Bangladesh caused by *Vibrio cholerae* O139 synonym Bengal. *Lancet* **342**:387–390.

Chongsanguan, M., W. Chaicumpa, P. Moolasart, P. Kandhas-
ingha, T. Shimada, H. Kurazono, and Y. Takeda. 1993. *Vibrio
cholerae* O139 Bengal in Bangkok. *Lancet* **342**:430–431.

Chowdhury, M. A., K. M. Aziz, B. A. Kay, and Z. Rahim. 1987.
Toxin production by *Vibrio mimicus* strains isolated from human and environmental sources in Bangladesh. *J. Clin. Microbiol.* **25**:2200–2203.

Chowdhury, N. R., S. Chakraborty, and T. Ramamurthy. 2000.
Molecular evidence of clonal *Vibrio parahaemolyticus* pandemic
strains. *Emerg. Infect. Dis.* **6**:631–636.

Chuang, Y. C., C. Y. Yuan, C. Y. Liu, C. K. Lan, and A. H.
Huang. 1992. *Vibrio vulnificus* infection in Taiwan: report of
28 cases and review of clinical manifestations and treatment.
Clin. Infect. Dis. **15**:271–276.

Clemens, J. D., D. A. Sack, J. R. Harris, M. R. Khan, J.
Chakraborty, S. Chowdhury, M. R. Rao, F. P. L. Van Loon, B.
F. Stanton, M. Yunus, M. Ali, M. Ansaruzzaman, A. M. Svennerholm, and J. Holmgren. 1990. Breast feeding and the risk of
severe cholera in rural Bangladesh children. *Am. J. Epidemiol.*
131:400–411.

Clemens, J. D., D. A. Sack, J. R. Harris, J. Chakraborty, M. R.
Khan, S. Huda, F. Ahmed, J. Gomes, M. R. Rao, A. M. Svennerholm, and J. Holmgren. 1989. ABO blood groups and
cholera: new observations on specificity of risk and modification
of vaccine efficacy. *J. Infect. Dis.* **159**:770–773.

Cohen, J., T. Schwartz, R. Klasmer, D. Pridan, H. Ghalayini, and
A. M. Davies. 1971. Epidemiological aspects of cholera El Tor
outbreak in a non-endemic area. *Lancet* **10**:86–89.

Colwell, R. R., and A. Huq. 1994. Vibrios in the environment:
viable but non-culturable *Vibrio cholerae*, p. 117–133. *In* I. K.
Wachsmuth, P. A. Blake, and O. Olsvik (ed.), Vibrio cholerae
and Cholera: Molecular to Global Perspectives. ASM Press,
Washington, D.C.

Colwell, R. R., and W. M. Spira. 1992. The ecology of *Vibrio
cholerae*, p. 107–127. *In* D. Barua and W. B. Greenough III
(ed.), *Cholera.* Plenum Press, New York, N.Y.

Dalsgaard, A., M. J. Albert, D. N. Taylor, T. Shimada, R. Meza,
O. Serichantalergs, and P. Echeverria. 1995. Characterization
of *Vibrio cholerae* non-O1 serogroup obtained from an outbreak of diarrhea in Lima, Peru. *J. Clin. Microbiol.* **33**:2715–
2722.

Dalsgaard, A., O. Serichantalergs, A. Forslund, W. Lin, J. Mekalanos, E. Mintz, T. Shimada, and J. G. Wells. 2001. Clinical and
environmental isolates of *Vibrio cholerae* serogroup O141 carry
the CTX phage and the genes encoding the toxin-coregulated
pili. *J. Clin. Microbiol.* **39**:4086–4092.

Daniels, N. A., L. MacKinnon, R. Bishop, S. Altekruse, B. Ray, R.
M. Hammond, S. Thompson, S. Wilson, N. H. Bean, P. M. Griffin, and L. Slutsker. 2000. *Vibrio parahaemolyticus* infections in
the United States, 1973-1998. *J. Infect. Dis.* **181**:1661–1666.

Davis, B. R., G. R. Fanning, J. M. Madden, A. G. Steigerwalt,
H. B. Bradford, Jr., H. L. Smith, Jr., and D. J. Brenner. 1981.
Characterization of biochemically atypical *Vibrio cholerae* strains
and designation of a new pathogenic species, *Vibrio mimicus.*
J. Clin. Microbiol. **14**:631–639.

Duffy, J. 1971. The history of Asiatic cholera in the United States.
Bull. N. Y. Acad. Med. **47**:1152–1168.

Dziejman, M., E. Balon, D. Boyd, C. M. Fraser, J. F. Heidelberg,
and J. J. Mekalanos. 2002. Comparative genomic analysis of
Vibrio cholerae: genes that correlate with cholera endemic and
pandemic disease. *Proc. Natl. Acad. Sci. USA* **99**:1556–1561.

Eko, F. O., S. M. Udo, and O. E. Antia-Obang. 1994. Epidemiology and spectrum of *vibrio* diarrheas in the lower cross river
basin of Nigeria. *Cent. Eur. J. Public Health* **2**:37–41.

Faruque, S. M., M. M. Rahman, Asadulghani, K. M. N. Islam,
and J. J. Mekalanos. 1999a. Lysogenic conversion of environmental *Vibrio mimicus* strains by CTXΦ. *Infect. Immun.* **67**:
5723–5729.

Faruque, S. M., A. K. Siddique, M. N. Saha, Asadulghani, M. M.
Rahman, K. Zaman, M. J. Albert, D. A. Sack, and R. B. Sack.
1999b. Molecular characterization of a new ribotype of *Vibrio
cholerae* O139 Bengal associated with an outbreak of cholera in
Bangladesh. *J. Clin. Microbiol.* **37**:1313–1318.

Faruque, S. M., A. R. M. A. Alim, M. M. Rahman, A. K. Siddique,
R. B. Sack, and M. J. Albert. 1993. Clonal relationships among
classical *Vibrio cholerae* O1 strains isolated between 1961 and
1992 in Bangladesh. *J. Clin. Microbiol.* **31**:2513–2516.

Faruque, S. M., A. R. M. A. Alim, S. K. Roy, F. Khan, G. B. Nair,
R. B. Sack, and M. J. Albert. 1994. Molecular analysis of rRNA
and cholera toxin genes carried by the new epidemic strain of

toxigenic *Vibrio cholerae* O139 synonym Bengal. *J. Clin. Microbiol.* 32:1050–1053.

Faruque, S. M., and J. J. Mekalanos. 2003. Pathogenicity islands and phages in *Vibrio cholerae* evolution. *Trends Microbiol.* 11:505–510.

Faruque, S. M., and M. J. Albert. 1992. Genetic relation between *Vibrio cholerae* O1 strains in Ecuador and Bangladesh. *Lancet* 339:740–741.

Faruque, S. M., G. B. Nair, and J. J. Mekalanos. 2004a. Genetics of stress-adaptation and virulence in toxigenic *Vibrio cholerae*. *DNA Cell Biol.* 23:723–741.

Faruque, S. M., N. Chowdhury, M. Kamruzzaman, M. Dziejman, M. H. Rahman, D. A. Sack, G. B. Nair, and J. J. Mekalanos. 2004b. Genetic diversity and virulence potential of environmental *Vibrio cholerae* population in a cholera-endemic area. *Proc. Natl. Acad. Sci. USA* 101:2123–2128.

Faruque, S. M., K. M. Ahmed, A. R. M. A. Alim, F. Qadri, A. K. Siddique, and M. J. Albert. 1997. Emergence of a new clone of toxigenic *Vibrio cholerae* biotype El Tor displacing *V. cholerae* O139 Bengal in Bangladesh. *J. Clin. Microbiol.* 35:624–630.

Faruque, S. M., M. J. Albert, and J. J. Mekalanos. 1998. Epidemiology, genetics and ecology of toxigenic *Vibrio cholerae*. *Microbiol. Mol. Biol. Rev.* 62:1301–1314.

Faruque, S. M., M. N. Saha, Asadulghani, P. K. Bag, R. K. Bhadra, S. K. Bhattacharya, R. B. Sack, Y. Takeda, and G. B. Nair. 2000. Genomic diversity among *Vibrio cholerae* O139 strains isolated in Bangladesh and India between 1992 and 1998. *FEMS Microbiol. Lett.* 184:279–284.

Faruque, S. M., N. Chowdhury, M. Kamruzzaman, Q. S. Ahmad, A. S. G. Faruque, M. A. Salam, T. Ramamurthy, G. B. Nair, A. Weintraub, and D. A. Sack. 2003. Reemergence of epidemic *Vibrio cholerae* O139, Bangladesh. *Emerg. Infect. Dis.* 9:1116–1122.

Faruque, S. M., S. K. Roy, A. R. M. A. Alim, A. K. Siddique, and M. J. Albert. 1995. Molecular epidemiology of toxigenic *V. cholerae* in Bangladesh studied by numerical analysis of rRNA gene restriction patterns. *J. Clin. Microbiol.* 33:2833–2838.

Faruque, S. M., I. B. Naser, M. J. Islam, A. S. G. Faruque, A. N. Ghosh, G. B. Nair, D. A. Sack, and J. J. Mekalanos. 2005. Seasonal epidemics of cholera inversely correlate with the prevalence of environmental cholera phages. *Proc. Natl. Acad. Sci. USA* 102:1702–1707.

Fullner, K. J., and J. J. Mekalanos. 1999. Genetic characterization of a new type IV-A pilus gene cluster found in both classical and El Tor biotypes of *Vibrio cholerae*. *Infect. Immun.* 67:1393–1404.

Glass, R. I., and R. E. Black. 1992. The epidemiology of cholera. *In* D. Barua and W. B. Greenough III (ed.), *Cholera*. Plenum Press, New York, N.Y.

Glass, R. I., J. Holmgren, C. E. Haley, M. R. Khan, A. M. Svennerholm, B. J. Stoll, K. M. B. Hossain, R. E. Black, M. Yunus, and D. Barua. 1985. Predisposition for cholera of individuals with O blood group. Possible evolutionary significance. *Am. J. Epidemiol.* 121:791–796.

Glass, R. I., S. Becker, M. I. Huq, B. J. Stoll, M. U. Khan, M. H. Merson, J. V. Lee, and R. E. Black. 1982. Endemic cholera in rural Bangladesh, 1966–1980. *Am. J. Epidemiol.* 116:959–970.

Goodgame, R. W., and W. B. Greenough. 1975. Cholera in Africa: a message for the West. *Ann. Intern. Med.* 82:101–106.

Gray, L. D., and A. S. Kreger. 1985. Purification and characterization of an extracellular cytolysin produced by *Vibrio vulnificus*. *Infect. Immun.* 48:62–72.

Hayat, U., G. P. Reddy, and C. A. Bush. 1993. Capsular types of *Vibrio vulnificus*: an analysis of strains from clinical and environmental sources. *J. Infect. Dis.* 168:758–762.

Heidelberg, J. F., J. A. Eisen, W. C. Nelson, R. A. Clayton, M. L. Gwinn, R. J. Dodson, D. H. Haft, E. K. Hickey, J. D. Peterson, L. Umayam, S. R. Gill, K. E. Nelson, T. D. Read, H. Tettelin, D. Richardson, M. D. Ermolaeva, J. Vamathevan, S. Bass, H. Qin, I. Dragoi, P. Sellers, L. McDonald, T. Utterback, R. D. Fleishman, W. C. Nierman, O. White, S. L. Salzberg, H. O. Smith, R. R. Colwell, J. J. Mekalanos, J. C. Ventor, and C. M. Frasier. 2000. DNA sequence of both chromosomes of the cholera pathogen *Vibrio cholerae*. *Nature* 406:477–483.

Hlady, W. G., and K. C. Klontz. 1996. The epidemiology of *Vibrio infections* in Florida (1981–1993). *J. Infect. Dis.* 173:1176–1183.

Hlady, W. G., R. C. Mullen, and R. S. Hopkin. 1993. *Vibrio vulnificus* from raw oysters. Leading cause of reported deaths from foodborne illness in Florida. *J. Fla. Med. Assoc.* 80:536–538.

Holmberg, S. D., J. R. Harris, D. E. Kay, N. T. Hargrett, R. D. R. Parker, N. Kansou, N. U. Rao, and P. A. Blake. 1984. Foodborne transmission of cholera in Micronesian households. *Lancet* ii:325–328.

Honda, T., and T. Iida. 1993. The pathogenicity of *Vibrio parahaemolyticus* and the role of the thermostable direct haemolysin and related haemolysin. *Rev. Med. Microbiol.* 4:106–113.

Honda, T., Y. Takeda, T. Miwatani, K. Kato, and Y. Nimura. 1976. Clinical features of patients suffering from food poisoning due to *Vibrio parahaemolyticus*, especially on changes in electrocardiograms. *J. Jpn. Assoc. Infect. Dis.* 50:216–223.

Howard, R., B. Brennman, and S. Lieb. 1988. Soft tissue infections in Florida due to marine *vibrio* bacteria. *J. Fla. Med. Assoc.* 73:29–34.

Hsueh, P. R., C. Y. Lin, H. J. Tang, H. C. Lee, J. W. Liu, Y. C. Liu, and Y. C. Chuang. 2004. *Vibrio vulnificus* in Taiwan. *Emerg. Infect. Dis.* 10:1363–1368.

Islam, M. S., B. S. Drasar, and R. B. Sack. 1994. The aquatic flora and fauna as reservoirs of *Vibrio cholerae*: a review. *J. Diarrhoeal Dis. Res.* 12:87–96.

Janda, J. M., C. Powers, R. G. Bryant, and S. L. Abbott. 1988. Current perspective on the epidemiology and pathogenesis of clinically significant *Vibrio* spp. *Clin. Microbiol. Rev.* 1:245–267.

Jermyn, W. S., and E. F. Boyd. 2002. Characterization of a novel *Vibrio* pathogenicity island (VPI-2) encoding neuraminidase (nanH) among toxigenic *Vibrio cholerae* isolates. *Microbiology* 148:3681–3693.

Johnston, J. M., S. F. Becker, and L. M. McFarland. 1985. *Vibrio vulnificus*. Man and the sea. *JAMA* 253:2850–2853.

Joseph, S. W., R. R. Colwell, and J. B. Kaper. 1982. *Vibrio parahaemolyticus* and related halophilic vibrios. *Crit. Rev. Microbiol.* 10:77–124.

Kamal, A. M. 1974. The seventh pandemic of cholera, p. 1–14. *In* D. Barua and W. Burrows (ed.), *Cholera*. The W. B. Saunders Co., Philadelphia, Pa.

Kaneko, T., and R. R. Colwell. 1978. The annual cycle of *Vibrio parahaemolyticus* in Chesapeake Bay. *Microb. Ecol.* 4:135–155.

Kaper, J. B., E. F. Remmers, H. Lockman, and R. R. Colwell. 1981. Distribution of *Vibrio parahaemolyticus* in Chesapeake Bay during the summer season. *Estuaries* 4:321–327.

Kaper, J. B., H. B. Bradford, N. C. Roberts, and S. Falkow. 1982. Molecular epidemiology of *Vibrio cholerae* in the U.S. Gulf Coast. *J. Clin. Microbiol.* 16:129–134.

Kaper, J. B., J. G. Morris, and M. M. Levine. 1995. Cholera. *Clin. Microbiol. Rev.* 8:48–86.

Kimsey, H. H., G. B. Nair, A. Ghosh, and M. K. Waldor. 1998. Diverse CTXΦ and evolution of new pathogenic *Vibrio cholerae*. *Lancet* 352:457–458.

Klontz, K. C., and J. C. Desenclos. 1990. Clinical and epidemiological features of sporadic infections with *Vibrio fluvialis* in Florida, USA. *J. Diarrhoeal Dis. Res.* 8:24–26.

Koch, R. 1894. An address on cholera and its bacillus. *Br. Med. J.* 2:453–459.

Kothary, M. H., and A. S. Kreger. 1987. Purification and characterization of an elastolytic protease of *Vibrio vulnificus*. *J. Gen. Microbiol.* **133:**1783–1791.

Kovach, M. E., M. D. Shaffer, and K. M. Peterson. 1996. A putative integrase gene defines the distal end of a large cluster of ToxR-regulated colonization genes in *Vibrio cholerae*. *Microbiology* **142:**2165–2174.

Kreger, A., and D. Lockwood. 1981. Detection of extracellular toxin(s) produced by *Vibrio vulnificus*. *Infect. Immun.* **33:**583–590.

Levine, M. M. 1991. South America: the return of cholera. *Lancet* **338:**45–46.

Levine, M. M., J. B. Kaper, D. Herrington, G. Losonsky, J. G. Morris, M. L. Clements, R. E. Black, B. Tell, and R. Hall. 1988. Volunteer studies of deletion mutants of *Vibrio cholerae* O1 prepared by recombinant techniques. *Infect. Immun.* **56:**161–167.

Levine, M. M., R. E. Black, M. L. Clements, D. R. Nalin, L. Cisneros, and R. A. Finkelstein. 1981. Volunteer studies in development of vaccines against cholera and enterotoxigenic *Escherichia coli*: a review, p. 443–459. *In* T. Holme, J. Holmgren, M. H. Merson, and R. Mollby (ed.), *Acute Enteric Infections in Children: New Prospects for Treatment and Prevention*. Elsevier/North-Holland Biomedical Press, Amsterdam, The Netherlands.

Lin, W., K. J. Fullner, R. Clayton, J. A. Sexton, M. B. Rogers, K. E. Calia, S. B. Calderwood, C. Fraser, and J. J. Mekalanos. 1999. Identification of a vibrio cholerae RTX toxin gene cluster that is tightly linked to the cholera toxin prophage. *Proc. Natl. Acad. Sci. USA* **96:**1071–1076.

Linkous, D. A., and J. D. Oliver. 1999. Pathogenesis of *Vibrio vulnificus*. *FEMS Microbiol. Lett.* **174:**207–214.

Liston, J. 1974. Influence of U.S. seafood handling procedures on *Vibrio parahaemolyticus*, p. 123–128. *In* T. Fujino, G. Sakaguchi, R. Sakazaki, and Y. Takeda (ed.), *International Symposium on* Vibrio parahaemolyticus. Saikon Publishing Co., Tokyo, Japan.

Litwin, C. M., T. W. Rayback, and J. Skinner. 1996. Role of catechol siderophore synthesis in *Vibrio vulnificus* virulence. *Infect. Immun.* **64:**283–288.

Lobitz, B., L. Beck, A. Huq, B. Wood, G. Fuchs, A. S. G. Faruque, and R. Colwell. 2000. Climate and infectious disease: use of remote sensing for detection of *Vibrio cholerae* by indirect measurement. *Proc. Natl. Acad. Sci. USA* **97:**1438–1443.

Matsumoto, C., J. Okuda, and M. Ishibashi. 2000. Pandemic spread of an O3:K6 clone of *Vibrio parahaemolyticus* and emergence of related strains evidenced by arbitrarily primed PCR and *toxRS* sequence analyses. *J. Clin. Microbiol.* **38:**578–585.

Mazel, D., B. Dychinco, V. A. Webb, and J. Davies. 1998. A distinctive class of integron in the *Vibrio cholerae* genome. *Science* **280:**605–608.

McCormack, W. M., M. S. Islam, M. Fahimuddin, and W. H. Mosley. 1969. A community study of inapparent cholera infections. *Am. J. Epidemiol.* **89:**658–664.

Mead, P. S., L. Slutsker, V. Dietz, L. F. McGaig, J. S. Bresee, C. Sharpiro, P. M. Griffin, and R. V. Tauxe. 1999. Food-related illness and death in the United States. *Emerg. Infect. Dis.* **5:**607–625.

Merrell, D. S., S. M. Butler, F. Qadri, N. A. Dolganov, A. Alam, M. B. Cohen, S. B. Calderwood, G. K. Schoolnik, and A. Camilli. 2002. Host-induced epidemic spread of the cholera bacterium. *Nature* **417:**642–645.

Mitra, R., A. Basu, D. Dutta, G. B. Nair, and Y. Takeda. 1996. Resurgence of *Vibrio cholerae* O139 Bengal with altered antibiogram in Calcutta, India. *Lancet* **348:** 181.

Miwatani, T., and Y. Takeda. 1975. *Vibrio parahaemolyticus* epidemiology ecology and biology, p. 22–24. *In* T. Miwatani and Y. Takeda (ed.), Vibrio parahaemolyticus: *a Causative Bacterium of Seafood Poisoning*. Saiko, Tokyo, Japan.

Miyoshi, N., C. Shimizu, I. Miyoshi, and S. Shinoda. 1987. Purification and characterization of *Vibrio vulnificus* protease. *Microbiol. Immunol.* **31:**13–25.

Morris, J. G., Jr. 1988. *Vibrio vulnificus*: a new monster of the deep? *Ann. Intern. Med.* **109:**261–263.

Morris, J. G., Jr. 1990. Non-O group 1 *Vibrio cholerae*: a look at the epidemiology of an occasional pathogen. *Epidemiol. Rev.* **12:**179–191.

Motes, M. L., A. DePaola, D. W. Cook, J. E. Veazey, J. C. Hunsucher, W. E. Garthright, R. J. Blodgett, and S. J. Chirtel. 1998. Influence of water temperature and salinity on *Vibrio vulnificus* in northern Gulf and Atlantic Coast oysters *(Crassostrea virginica)*. *Appl. Environ. Microbiol.* **64:**1459–1465.

Nair, G. B., S. M. Faruque, N. A. Bhuiyan, M. Kamruzzaman, A. K. Siddique, and D. A. Sack. 2002. New variants of *Vibrio cholerae* O1 biotype El Tor with attributes of the classical biotype from hospitalized patients with acute diarrhea in Bangladesh. *J. Clin. Microbiol.* **40:**3296–3299.

Novitsky, J. A., and R. Y. Morita. 1976. Morphological characterization of small cells resulting from nutrient starvation of a psychrophilic marine vibrio. *Appl. Environ. Microbiol.* **32:**617–622.

Novitsky, J. A., and R. Y. Morita. 1977. Survival of a psychrophilic marine vibrio under long-term nutrient starvation. *Appl. Environ. Microbiol.* **33:**635–641.

Novitsky, J. A., and R. Y. Morita. 1978. Possible strategy for the survival of marine bacteria under starvation conditions. *Mar. Biol.* **48:**289–295.

Okuda, J., M. Ishibashi, S. L. Abbott, J. M. Janda, and M. Nishibuchi. 1997a. Analysis of the thermostable direct hemolysin *(tdh)* gene and the tdh-related hemolysin *(trh)* genes in urease-positive strains of *Vibrio parahaemolyticus* isolated on the West Coast of the United States. *J. Clin. Microbiol.* **35:**1965–1971.

Okuda, J., M. Ishibashi, and E. Hayakawa. 1997b. Emergence of a unique O3:K6 clone of *Vibrio parahaemolyticus* in Calcutta, India, and isolation of strains from the same clonal group from southeast Asian travellers arriving in Japan. *J. Clin. Microbiol.* **35:**3150–3155.

Okujo, N., and S. Yamamoto. 1994. Identification of the siderophores from *Vibrio hollisae* and *Vibrio mimicus* as aerobactin. *FEMS Microbiol. Lett.* **118:**187–192.

Pal, S. C., B. K. Sircar, G. B. Nair, and B. C. Deb. 1984. Epidemiology of bacterial diarrhoeal diseases in India with special reference to *Vibrio parahaemolyticus* infections, p. 65–73. *In* Y. Takeda and T. Miwatani (ed.), *Bacterial Diarrhoeal Disease*. KTK Scientific Publishers, Tokyo, Japan.

Pan American Health Organization. 1991. Cholera situation in the Americas. *Epidemiol. Bull.* **12:**1–4.

Park, S. D., H. S. Shon, and N. J. Joh. 1991. *Vibrio vulnificus* septicemia in Korea: clinical and epidemiologic findings in seventy patients. *J. Am. Acad. Dermatol.* **24:**397–403.

Perez-Tirse, J., J. F. Levine, and M. Mecca. 1993. *Vibrio damsela*. A cause of fulminant septicemia. *Arch. Intern. Med.* **153:**1838–1840.

Pollitzer, R. 1959. History of the disease, p. 11–50. *In* R. Pollitzer (ed.), *Cholera*. World Health Organization, Geneva, Switzerland.

Ramamurthy, T., P. K. Bag, A. Pal, S. K. Bhattacharya, M. K. Bhattacharya, T. Shimada, T. Takeda, T. Karasawa, H. Kurazono, and Y. Takeda. 1993a. Virulence patterns of *Vibrio cholerae* non-O1 strains isolated from hospitalized patients with acute diarrhea in Calcutta, India. *J. Med. Microbiol.* **39:**310–317.

Ramamurthy, T., S. Garg, R. Sharma, S. K. Bhattacharya, G. B. Nair, T. Shimada, T. Takeda, T. Karasawa, H. Kurazano, A. Pal, and Y. Takeda. 1993b. Emergence of a novel strain of *Vibrio cholerae* with epidemic potential in Southern and Eastern India. *Lancet* **341:**703–704.

Ries, A. A., D. J. Vugia, L. Beingolea, A. M. Palacios, E. Vasquez, J. G. Wells, N. G. Baca, D. L. Swerdlow, M. Pollack, N. H. Bean, L. Seminario, and R. V. Tauxe. 1992. Cholera in Piura, Peru: a modern urban epidemic. *J. Infect. Dis.* **166:**1429–1433.

Rudra, S., R. Mahajan, M. Mathur, K. Kathuria, and V. Talwar. 1996. Cluster of cases of clinical cholera due to *Vibrio cholerae* O10 in East Delhi. *Indian J. Med. Res.* **103:**71–73.

Sharma, C., M. Thungapathra, A. Ghosh, A. K. Mukhopadhyay, A. Basu, R. Mitra, I. Basu, S. K. Bhattacharya, T. Shimada, T. Ramamurthy, T. Takeda, S. Yamasaki, Y. Takeda, and G. B. Nair. 1998. Molecular analysis of non-O1, non-O139 *Vibrio cholerae* associated with an unusual upsurge in the incidence of cholera-like disease in Calcutta, India. *J. Clin. Microbiol.* **36:**756–763

Shinoda, S., M. Kobayashi, H. Yamada, S. Yoshida, M. Ogawa, and Y. Mizuguchi. 1987. Inhibitory effect of capsular antigen of *Vibrio vulnificus* on bactericidal activity of human serum. *Microbiol. Immunol.* **31:**393–401.

Shirai H., H. Ito, T. Hirayama, Y. Nakamoto, N. Nakabayashi, K. Kumagai, Y. Takeda, and M. Nishibuchi. 1990. Molecular epidemiologic evidence for association of thermostable direct hemolysin (TDH) and TDH-related hemolysin of *Vibrio parahaemolyticus* with gastroenteritis. *Infect. Immun.* **58:**3568–3573.

Siddique, A. K., A. Salam, M. S. Islam, K. Akram, R. N. Majumdar, K. Zaman, N. Fronczak, and S. Laston. 1995. Why treatment centres failed to prevent cholera deaths among Rwandan refugees in Goma, Zaire. *Lancet* **345:**359–361.

Siddique, A. K., K. Zaman, A. H. Baqui, K. Akram, P. Mutsuddy, A. Eusof, K. Haider, S. Islam, and R. B. Sack. 1992. Cholera epidemics in Bangladesh: 1985-1991. *J. Diarrhoeal Dis. Res.* **10:**79–86.

Siddique, A. K., K. Akram, K. Zaman, P. Mutsuddy, A. Eusof, and R. B. Sack. 1996. *Vibrio cholerae* O139: how great is the threat of a pandemic? *Trop. Med. Int. Health* **1:**393–398.

Snow, J. 1855. *On the Mode of Communication of Cholera*, 2nd ed. J. Churchill, London, England.

Swerdlow, D. L., and A. A. Ries. 1993. *Vibrio cholerae* non-O1—the eighth pandemic? *Lancet* **342:**382–383.

Swerdlow, D. L., E. D. Mintz, M. Rodriguez, E. Tejada, C. Ocampo, L. Espejo, K. D. Greene, W. Saldana, L. Seminario, R. V. Tauxe, J. G. Wells, N. H. Bean, A. A. Ries, M. Pollack, B. Vertiz, and P. A. Blake. 1992. Waterborne transmission of epidemic cholera in Trujillo, Peru: lessons for a continent at risk. *Lancet* **340:**28–32.

Tacket, C. O., F. Brenner, and P. A. Blake. 1984. Clinical features and an epidemiological study of *Vibrio vulnificus* infections. *J. Infect. Dis.* **149:**558–561.

Tamplin, M. L., S. Specter, G. E. Rodrick, and H. Friedman. 1985. *Vibrio vulnificus* resists phagocytosis in the absence of serum opsonins. *Infect. Immun.* **49:**715–718.

Testa, J., L. W. Daniel, and A. S. Kreger. 1984. Extracellular phospholipase A2 and lysophospholipase produced by *Vibrio vulnificus. Infect. Immun.* **45:**458–463.

Trucksis, M., J. Michalski, Y. K. Deng, and J. B. Kaper. 1998. The *Vibrio cholerae* genome contains two unique circular chromosomes. *Proc. Natl. Acad. Sci. USA* **95:**14464–14469.

Tuyet, D. T., V. D. Thiem, L. V. Seidlein, C. Ashrafuzzaman, E. Park, D. G. Canh, B. T. Chien, T. V. Gung, A. Naficy, M. R. Rao, M. Ali, Y. Lee, T. S. Hung, M. Nichibuchi, J. Clemens, and D. D. Trach. 2002. Clinical epidemiological and socioeconomic analysis of an outbreak of *Vibrio parahaemolyticus* in Khanh Hoa Province, Vietnam. *J. Infect. Dis.* **186:**1615–1620.

U.K. Public Health Laboratory Service Communicable Disease Surveillance Centre. 1993. *Vibrio cholerae* O139 and epidemic cholera. *Commun. Dis. Rep. CDR Wkly.* **3:**173. [Online.] http://www.hpa.org.uk/cdr/.

Vasconcelos, F. J., W. J. Stang, and R. H. Laidlaw. 1975. Isolation of *Vibrio parahaemolyticus* and *Vibrio alginolyticus* from estuarine areas of Southeastern Alaska. *Appl. Microbiol.* **29:**557–559.

Wachsmuth, I. K., C. A. Bopp, P. I. Fields, and C. Carrillo. 1991. Difference between toxigenic *Vibrio cholerae* O1 from South America and US gulf coast. *Lancet* **337:**1097–1098.

Wachsmuth, I. K., G. M. Evins, P. I. Fields, O. Olsvik, T. Popovic, C. A. Bopp, J. G. Wells, C. Carrillo, and P. A. Blake. 1993. The molecular epidemiology of cholera in Latin America. *J. Infect. Dis.* **167:**621–626.

Wachsmuth, I. K., O. Olsvik, G. M. Evins, and T. Popovic. 1994. Molecular epidemiology of cholera, p. 357–370. *In* K. Wachsmuth, P. A. Blake, and O. Olsvik (ed.), Vibrio cholerae *and Cholera: Molecular to Global Perspectives.* ASM Press, Washington, D.C.

Waldor, M. K., and J. J. Mekalanos. 1996. Lysogenic conversion by a filamentous bacteriophage encoding cholera toxin. *Science* **272:**1910–1914.

World Health Organization. 1999. *Vibrio parahaemolyticus*, Japan, 1996-1998. *Wkly. Epidemiol. Rec.* **74:**357–364.

Wright, A. C., L. M. Simpson, J. D. Oliver, and J. G. Morris, Jr. 1990. Phenotypic evaluation of acapsular transposon mutants of *Vibrio vulnificus. Infect. Immun.* **58:**1769–1773.

Wright, A. C., J. G. Morris, Jr., D. R. Maneval, Jr., K. Richardson, and J. B. Kaper. 1985. Cloning of the cytotoxin-hemolysin gene of *Vibrio vulnificus. Infect. Immun.* **50:**922–924.

Yoshida, S., M. Ogawa, and Y. Mizuguchi. 1985. Relation of capsular materials and colony opacity to virulence of *Vibrio vulnificus. Infect. Immun.* **47:**446–451.

Zen-Yoji, H., S. Sakai, T. Terayama, Y. Kudo, T. Ito, M. Benoki, and M. Nagasaki. 1965. Epidemiology, enteropathogenicity, and classification of *Vibrio parahaemolyticus. J. Infect. Dis.* **115:**436–444.

X. APPLICATIONS

The Biology of Vibrios
Edited by F. L. Thompson et al.
© 2006 ASM Press, Washington, D.C.

Chapter 28

Biotechnological Applications

J. Grant Burgess

INTRODUCTION

On first inspection, members of the genus *Vibrio* are more notable for their ability to cause disease than for their potential as useful or beneficial bacteria. However, their diverse physiological properties and their prevalence in a wide range of ecological habitats have allowed a number of biotechnological applications for these bacteria to be developed. This chapter provides a summary of the main applications which have been commercialized or which are being developed as biotechnological products or processes.

MEDICAL APPLICATIONS

Cholera Vaccines

Although efficient hygiene and sanitation are the most successful methods for controlling cholera, effective vaccines are still necessary as an additional control measure; as such, the World Health Organization has been overseeing the introduction and use of a number of cholera vaccines over the years. The early parenteral vaccines (not administered through the gut), which were based on inactivated *V. cholerae* O1 cells, were not very effective. They provided modest protection over a short duration and furthermore did not prevent further transmission of the disease. However, with the emergence of newer oral vaccines there have been significant improvements. Two oral vaccines are available. First is the WC/rBS vaccine, which consists of whole cells of *V. cholerae* O1 plus a recombinant subunit of the cholera toxin. This provides good protection. Second, a live attenuated vaccine is available which is a genetically manipulated *V. cholerae* strain CVD103-HgR (Levine et al., 1988). Neither of these two vaccines, however, provided protection in young children. More effective protection was later seen in children and adults in sub-Saharan

Africa, in populations that were also showing a high proportion of human immunodeficiency virus infection (Lucas et al., 2005). Vaccination is also of enormous benefit in preventing infection in travelers (Calain et al., 2004). Peru-15 is a single-dose, recombinant cholera vaccine under development by AVANT Immunotherapeutics. As of September 2003, AVANT was planning a phase III trial in a developing country and phase IIb and phase III challenge studies in travelers (Jones, 2004). AVANT announced promising results from the vaccine in adults and in late 2004 was initiating clinical trials with children in Bangladesh. Thus, as is often the case with bacteria, we can find benefits from their use in addition to suffering from their negative effects. As well as being useful for the development of cholera vaccines, there are also some interesting applications being looked at which use *Vibrio*-derived toxins in other ways.

Toxins

Additional potential uses of the cholera toxin

The cholera toxin B subunit is a strong mucosal adjuvant which enhances the immune response to mucosally administered antigens. Thus, it may find use as an adjuvant for effective administration of nasal influenza vaccines (Lebens and Holmgren, 1994; Matsuo et al., 2000; Tamura and Kurata, 2000). This approach has also found application with other mucosal immunotherapy interventions in diabetes and other autoimmune and allergic disorders. The use of cholera toxin- and cholera toxin B subunit-coadministered antigens for dendritic cell vaccination or immunotherapy purposes is also undergoing investigation (Holmgren et al., 2003). For example, the oral administration of cholera toxin–insulin conjugates protected mice against spontaneous autoimmune development of diabetes (Bergerot et al., 1997; Ploix et al., 1999). In humans, however, trials have not shown

---

J. Grant Burgess • School of Marine Science and Technology, University of Newcastle upon Tyne, Newcastle NE1 7R4, United Kingdom.

promise compared with placebos (Skyler, 2002). The use of engineered cholera toxin as an adjuvant is also being investigated for the development of nasal vaccines against mucosal infections, including sexually transmitted diseases, such as AIDS/human immunodeficiency virus (Yoshino et al., 2004), and against parasitic infections caused by *Schistosoma mansoni* and *Leishmania* spp. (Holmgren et al., 2003).

The *Vibrio* heat-stable enterotoxin is similar to the heat-stable toxins found in *Escherichia coli* that cause diarrhea by a similar mechanism. The toxin binds to glycoprotein receptors on the intestinal cells of the brush border, where they mimic the action of the peptide hormone guanylin in stimulating guanyl cyclase. This leads to an elevation of cyclic GMP and a massive release of fluid and ions, resulting in diarrhea. Apart from their use in research on intestine physiology, there are few reported applications of these toxins.

Tetrodotoxin

Tetrodotoxin (TTX) is a well-known neurotoxin found in many species of puffer fish and certain amphibian species, such as newts, and other marine creatures, such as crabs and the blue-banded octopus. Named after the *Tetraodontidae* (four-toothed) family of fish, this toxin, although present in many internal organs of these fish, is actually produced by several species of bacteria, including *Pseudoalteromonas tetraodonis* and *Vibrio* spp.

The toxin has a very specific mechanism of action insofar as it binds to nerve cell sodium channels, and, therefore, it has found widespread use and applications as a research reagent in neurobiology, along with the related saxitoxin (Evans, 1972). For example, TTX has been used to differentiate between the different sodium channel proteins involved in nerve transmission, and this has allowed the identification of a TTX-insensitive voltage-gated sodium channel (Na(v) 1.8; SNS/PN3) which is being used as a novel target to find specific channel blockers for use in nonopioid pain control (Porreca et al., 1999; LoGrasso and McKelvy, 2003). In addition, TTX is in some cases now used as a pain control therapeutic (Anonymous, 2004). Saxitoxin and TTX have the interesting property that, when mixed in small quantities with many classes of anaesthetics, the effectiveness of the anaesthetic dramatically increases (Schantz and Johnson, 1992). These toxins have over the years been used extensively to characterize sodium channels in myelinated and nonmyelinated nerves and have allowed research to progress, for example, in the study of multiple sclerosis (Schantz and Johnson, 1992).

A recent look at the biotechnology of TTX shows one company to be particularly active in the use of the compound for pain management and, in particular, for the management of pain associated with withdrawal from heroin and other opioid drugs and for subjects who are receiving methadone. International WEX Technologies, Inc., has a lead product, Tectin, for the reduction of cancer pain and possibly other broad pain treatments in the future (http://www.wextech.ca). Phase II clinical trial results demonstrated that Tectin could relieve pain in refractory cancer pain patients. A recently initiated phase IIb/III trial is also under way. Tectin is administered via intramuscular injection. The study results are expected in 2006. The source of the TTX used in these applications appears to be pufferfish. It is clear that a possible microbial fermentation method using a *Vibrio* TTX producer may also be used (Lee et al., 2000). TTX-producing marine bacteria, including *Vibrio alginolyticus*, continue to be isolated (Simidu et al., 1987, 1990; Kogure et al., 1988; Yu et al., 2004), but as yet there is little information available on the precise mechanism of biosynthesis or on the genes required for production of this compound. Neither is its possible ecological role understood.

Low-Molecular-Weight Bioactive Compounds

The ability of marine bacteria to produce brominated compounds is well known, as is the effect of bromine on the production of antibiotics (Marwick et al., 1999). Several strains of *Vibrio* were isolated from the marine sponge *Dysidea* sp. that synthesized cytotoxic and antibacterial tetrabromodiphenyl ethers (Elyakov et al., 1991; Voinov et al., 1991; Marwick et al., 1999). Other bioactive compounds isolated from species of *Vibrio* include the anticyanobacterial compound beta-cyanoalanine (Yoshikawa et al., 2000). This compound was discovered in many strains of marine bacteria during a screening for compounds that prevent algal blooms. Although this compound was active against cyanobacteria, the authors reported no activity against any of the eukaryotic algae tested. In addition, *Vibrio* isolates associated with sponges have been reported to be able to produce antimicrobial compounds (Hentschel et al., 2001).

Many species of *Vibrio* produce very effective iron-chelating compounds or siderophores. These compounds have been studied for their biotechnological application in the fields of metal uptake, gallium uptake, and cobalt recovery. Siderophores (and bacteria that produce them) are also being developed as biocontrol agents for plant root pathogens (Thomashow, 1996) and as potential drug delivery agents (Roosenberg et al., 2000; Vergne et al., 2000). It will, therefore, be interesting to see if siderophores produced by *Vibrio* species, for example, anguibactin, may find similar applications in the future.

Interfering with Cell-Cell Communication

The discovery of quorum sensing in species of *Vibrio* in the 1980s is one of the most important developments in the history of microbiology (Fuqua et al., 1994, 1996; Miller and Bassler, 2001; Winzer and Williams, 2001). Although initially confined to the study of bioluminescence, microbiologists quickly realized that communication between individual cells of the same species and, later, different species allowed them to coordinate complex activities carried out by populations of cells rather than individual cells (Gray, 1997; O'Toole et al., 2000; Bonner, 2003). Thus, the evolution of quorum sensing was a major step which allowed bacteria to secure some of the adaptive advantages of more complex multicelled organisms. Importantly, it is now clear that quorum sensing and cell-cell communication are required for the control of many types of virulence in diseases, both in plants (Von Bodman et al., 2003) and animals (Hardman et al., 1998; Pearson et al., 2000; Williams et al., 2000), including cholera (Miller et al., 2002; Zhu et al., 2002) and marine organisms (Rasch et al., 2004). Examples include the need for quorum-sensing regulation to allow infection to progress in cholera and other diseases. Now that these mechanisms are being elucidated, it is also clear that one method by which infection may be slowed or inhibited is to interrupt cell-cell communication signals. The use of cell signaling as a target for new therapeutics has been recognized by many groups (Finch et al., 1998; Raffa et al., 2005). Indeed, this is the strategy that several companies are using to find new drugs—not traditional antibiotics but quorum "quenchers" that can be used to fight disease. Examples include Quorex Pharmaceuticals, which developed compounds that block quorum sensing to treat infection. A naturally occurring compound, furanone, from the marine alga *Delisea pulchra* was reported to inhibit quorum sensing in bacteria (Manefield et al., 2001) and can be used to prevent infection (Manefield et al., 2000; Rasmussen et al., 2000). These discoveries have led to the formation of a company, Biosignal Ltd. (http://www.biosignal.com.au), which is attempting to commercialize quorum-quenching molecules as therapeutic agents. These compounds may find application in the cleaning and prevention of biofilm buildup on biomedical devices and contact lenses and in the prevention of biofouling on marine surfaces.

Effective cell-cell signaling systems are required to orchestrate the complex development of biofilms on surfaces, and an understanding of these systems may help in the discovery of new bioactive molecules from marine bacteria (Mearns-Spragg et al., 1998; Burgess et al., 1999, 2003; Yan et al., 2003).

Enzymes

Vibrio spp. have been found to produce a variety of extracellular proteases. *V. alginolyticus*, for example, produces six proteases, including an unusual detergent-resistant alkaline serine exoprotease. This marine bacterium also produces collagenase, an enzyme with a variety of industrial and commercial applications, including the dispersion of cells in tissue culture studies. In addition, a *Vibrio* sp. was found to produce a neutral protease, vimelysin (Ikeuchi et al., 2000). *Vibrio* proteases have also been developed for the breakdown of feather waste (Sangali and Brandelli, 2000). New restriction enzymes have been isolated from *Vibrio nereis* (VneI) and *Vibrio* sp. (VspI), patented in Russia, and currently sold to international distributors by the Russian company SibEnzyme (http://www.SibEnzyme.com), which is based in Novosibirsk, Siberia (Elyakov et al., 1994).

Biomonitoring Using *Vibrio* Bioluminescence as a Reporter System

Marine bacteria may be used to develop biosensors and diagnostic devices for medicine, aquaculture, and environmental monitoring. One type of biosensor uses the enzymes responsible for bioluminescence. The *lux* genes, which encode these enzymes, have been cloned from marine bacteria, such as *Vibrio fischeri*, and transferred successfully to a variety of plants and other bacteria. The *lux* genes are usually cloned into a gene sequence or operon that is functional only when stimulated by a defined environmental feature. The enzymes responsible for toluene degradation, for example, are synthesized only in the presence of toluene. When *lux* genes are inserted into a toluene operon, the engineered bacterium glows yellow-green in the presence of toluene. This genetically engineered system "reports" that biodegradation of a specific chemical, in this case toluene, is proceeding (Powell, 1995; Applegate et al., 1997, 1998). The advantage of using the *Vibrio lux* system is that it can provide real-time nondestructive monitoring, and it has therefore been applied over the last 20 years to the study of an enormous variety of systems (Roda et al., 2004). These include monitoring and biodegradation of naphthalene (Burlage et al., 1990) and online groundwater monitoring (Heitzer et al., 1994), monitoring alginate production (Applegate et al., 1998), and many others.

The most widely used bioluminescence system is the commercially available Microtox system. This method, developed by Bulich (Bulich, 1979; Bulich and Isenberg, 1981), relies on the output of light by cells of *V. fischeri*. Emission of light is reduced if the cells are exposed to toxic materials, which inhibit bacterial respiration. This test, used internationally and standardized for use in a wide variety of applications, was commercialized in the 1980s by Microbics Corporation, which later became Azur Environmental Technologies. The Microtox technology is now owned and sold by Strategic Diagnostics Inc. The technology was built on the use of freeze-drying techniques, which were used to make suspensions of *V. fischeri* cells easy to store and transport. Strain NRRL B-11177 was originally chosen for its sensitivity to a wide variety of chemicals.

Diagnostics

Until recently, there was no rapid method for testing environmental and biological samples for cholera. However, an enzyme-linked immunosorbent assay for the detection of *V. cholerae* in a variety of homogenized samples has been developed. Within 10 min, a hand-held kit produces a visible color reaction if *V. cholerae* is present. In 1992, these kits were sent to South American countries to assist health officials in combating the cholera pandemic. There are now a number of diagnostic kits available on the market for detection of cholera in environmental samples.

Probiotics

The use of strains of *Vibrio* in probiotics is also increasing. Recently, Iannacone and colleagues (unpublished data) screened marine bacteria for potential use in probiotic feed formulations for sea bass aquaculture. Probiotics are often developed to reduce the disease-causing effects of many species of *Vibrio*. In this regard, many antimicrobial peptides have been examined for their use as anti-*Vibrio* agents. These include peptides produced by the tiger shrimp (Chiou et al., 2005) and other species of bacteria (Ruiz-Ponte et al., 1998, 1999; Longeon et al., 2005). However, the beneficial probiotic properties of some strains of *Vibrio* have also often been observed and found to be effective in reducing *Vibrio*-induced infections (Austin et al., 1995). The screening of *Vibrio* for the production of bacteriocidal compounds and their development as probiotics have also been carried out by Jorquera et al. (2000) while investigating methods for disease control in scallop aquaculture.

OTHER APPLICATIONS

Bioterrorism

Although biological warfare can be traced back for many centuries, the best evidence stems from the 20th century. It is apparent that such activities have been used since the 1920s, throughout the cold war, and, more recently, in the deaths in 2001 of five U.S. citizens, which were caused by the deliberate release of anthrax spores. These events have increased attention on the possible use of biological warfare in the 21st century. In particular, the Centers for Disease Control and Prevention lists *V. cholerae, V. parahaemolyticus, and V. vulnificus* as category B bioterrorist threats (Rotz and Hughes, 2004). Thus, there is a risk that these microorganisms could be used to attack civilian populations through the water supply (Christen, 2001; Khan et al., 2001; Green et al., 2003). This has important implications for the development of effective monitoring and diagnostic methods for these bacteria, though it should be underlined that the number of deaths due to biological agents in recent years remains very small.

CONCLUSIONS

As is clear that there are a wide variety of uses for *Vibrio* spp., their proteins or products, or the harnessing of their specific metabolic properties. Clearly, as more and more species of *Vibrio* are uncovered from extreme environments and the mechanisms of their biology are elucidated, then the number and complexity of their uses and applications must also surely be set to rise.

Acknowledgment. I thank the UK NERC for research funding (www.esmb.org/JGBhome.htm).

REFERENCES

Anonymous. 2004. Tetrodotoxin is safe and effective for severe, refractory cancer pain. *J. Support. Oncol.* 2:18.

Applegate, B., C. Kelly, L. Lackey, J. McPherson, S. Kehrmeyer, F. M. Menn, P. Bienkowski, and G. Sayler. 1997. *Pseudomonas putida* B2: a tod-lux bioluminescent reporter for toluene and trichloroethylene co-metabolism. *J. Ind. Microbiol. Biotechnol.* 18:4–9.

Applegate, B. M., S. R. Kehrmeyer, and G. S. Sayler. 1998. A chromosomally based tod-luxCDABE whole-cell reporter for benzene, toluene, ethybenzene, and xylene (BTEX) sensing. *Appl. Environ. Microbiol.* 64:2730–2735.

Austin, B., L. F. Stuckey, P. A. W. Robertson, I. Effendi, and D. R. W. Griffith. 1995. A probiotic strain of *Vibrio alginolyticus* effective in reducing diseases caused by *Aeromonas salmonicida*, *Vibrio anguillarum* and *Vibrio ordalii*. *J. Fish Dis.* 18:93–96.

Bergerot, I., C. Ploix, J. Petersen, V. Moulin, C. Rask, N. Fabien, M. Lindblad, A. Mayer, C. Czerkinsky, J. Holmgren, and C.

Thivolet. 1997. A cholera toxoid-insulin conjugate as an oral vaccine against spontaneous autoimmune diabetes. *Proc. Natl. Acad. Sci. USA* **94:**4610–4614.

Bonner, J. T. 2003. Evolution of development in the cellular slime molds. *Evol. Dev.* **5:**305–313.

Bulich, A. A. 1979. Use of luminescent bacteria for determining relative toxicity in aquatic environments, p. 98–106. *In* L. L. Marking and R. A. Kimerle (ed.), *Aquatic Toxicology*. ASTM STP 667. American Society for Testing and Materials, Philadelphia, Pa.

Bulich, A. A., and D. L. Isenberg. 1981. Use of the luminescent bacterial system for the rapid assessment of aquatic toxicity. *ISA Trans.* **20:**29–33.

Burgess, J. G., K. G. Boyd, E. Armstrong, Z. Jiang, L. Yan, M. Berggren, U. May, T. Pisacane, A. Granmo, and D. R. Adams. 2003. The development of a marine natural product-based antifouling paint. *Biofouling* **19**(Suppl.):197–205.

Burgess, J. G., E. M. Jordan, M. Bregu, A. Mearns-Spragg, and K. G. Boyd. 1999. Microbial antagonism: a neglected avenue of natural products research. *J. Biotechnol.* **70:**27–32.

Burlage, R. S., G. S. Sayler, and F. Larimer. 1990. Monitoring of naphthalene catabolism by bioluminescence with nah-lux transcriptional fusions. *J. Bacteriol.* **172:**4749–4757.

Calain, P., J. P. Chaine, E. Johnson, M. L. Hawley, M. J. O'Leary, H. Oshitani, and C. L. Chaignat. 2004. Can oral cholera vaccination play a role in controlling a cholera outbreak? *Vaccine* **22:**2444–2451.

Chiou, T. T., J. L. Wu, T. T. Chen, and J. K. Lu. 2005. Molecular cloning and characterization of cDNA of penaeidin-like antimicrobial peptide from tiger shrimp *(Penaeus monodon)*. *Mar. Biotechnol.* (N.Y.) **7:**119–127.

Christen, K. 2001. Bioterrorism and waterborne pathogens: how big is the threat? *Environ. Sci. Technol.* **35:**396A–397A.

Elyakov, G. B., T. Kuznetsova, V. V. Mikhailov, I. I. Maltsev, V. G. Voinov, and S. A. Fedoryev. 1991. Brominated diphenyl ethers from a marine bacterium associated with the sponge *Dysidea* sp. *Experientia* **47:**632–633.

Elyakov, G. B., T. A. Kuznetsova, V. A. Sasunkevich, and V. V. Mikhailov. 1994. New trends of marine biotechnology development. *Pure Appl. Chem.* **66:**811–818.

Evans, M. H. 1972. Tetrodotoxin, saxitoxin, and related substances: their applications in neurobiology. *Int. Rev. Neurobiol.* **15:**83–166.

Finch, R. G., D. I. Pritchard, B. W. Bycroft, P. Williams, and G. S. Stewart. 1998. Quorum sensing: a novel target for anti-infective therapy. *J. Antimicrob. Chemother.* **42:**569–571.

Fuqua, W. C., S. C. Winans, and E. P. Greenberg. 1994. Quorum sensing in bacteria: the LuxR-LuxI family of cell density-responsive transcriptional regulators. *J. Bacteriol.* **176:**269–275.

Fuqua, W. C., S. C. Winans, and E. P. Greenberg. 1996. Census and consensus in bacterial ecosystems: the LuxR-LuxI family of quorum-sensing transcriptional regulators. *Annu. Rev. Microbiol.* **50:**727–751.

Gray, K. M. 1997. Intercellular communication and group behavior in bacteria. *Trends Microbiol.* **5:**184–188.

Green, U., J. H. Kremer, M. Zillmer, and C. Moldaenke. 2003. Detection of chemical threat agents in drinking water by an early warning real-time biomonitor. *Environ. Toxicol.* **18:**368–374.

Hardman, A. M., G. S. Stewart, and P. Williams. 1998. Quorum sensing and the cell-cell communication dependent regulation of gene expression in pathogenic and non-pathogenic bacteria. *Antonie Leeuwenhoek* **74:**199–210.

Heitzer, A., K. Malachowsky, J. E. Thonnard, P. R. Bienkowski, D. C. White, and G. S. Sayler. 1994. Optical biosensor for environmental on-line monitoring of naphthalene and salicylate bioavailability with an immobilized bioluminescent catabolic reporter bacterium. *Appl. Environ. Microbiol.* **60:**1487–1494.

Hentschel, U., M. Schmid, M. Wagner, L. Fieseler, C. Gernert, and J. Hacker. 2001. Isolation and phylogenetic analysis of bacteria with antimicrobial activities from the Mediterranean sponges *Aplysina aerophoba* and *Aplysina cavernicola*. *FEMS Microbiol. Ecol.* **35:**305–312.

Holmgren, J., C. Czerkinsky, K. Eriksson, and A. Mharandi. 2003. Mucosal immunisation and adjuvants: a brief overview of recent advances and challenges. *Vaccine* **21**(Suppl. 2):S89–S95.

Ikeuchi, H., S. Kunugi, and K. Oda. 2000. Activity and stability of a neutral protease from *Vibrio* sp. (vimelysin) in a pressure-temperature gradient. *Eur. J. Biochem.* **267:**979–983.

Jones, T. 2004. Peru-1 5 (AVANT). *Curr. Opin. Investig. Drugs* **5:**887–891.

Jorquera, M. A., C. E. Riquelme, L. A. Loyola, and L. F. Munoz. 2000. Production of bactericidal substances by a marine vibrio isolated from cultures of the scallop *Argopecten purpuratus*. *Aquacult. Int.* **7:**433–448.

Khan, A. S., D. L. Swerdlow, and D. D. Juranek. 2001. Precautions against biological and chemical terrorism directed at food and water supplies. *Public Health Rep.* **116:**3–14.

Kogure, K., M. L. Tamplin, U. Simidu, and R. R. Colwell. 1988. A tissue culture assay for tetrodotoxin, saxitoxin and related toxins. *Toxicon* **26:**191–197.

Lebens, M., and J. Holmgren. 1994. Mucosal vaccines based on the use of cholera toxin B subunit as immunogen and antigen carrier. *Dev. Biol. Stand.* **82:**215–227.

Lee, M. J., D. Y. Jeong, W. S. Kim, H. D. Kim, C. H. Kim, W. W. Park, Y. H. Park, K. S. Kim, H. M. Kim, and D. S. Kim. 2000. A tetrodotoxin-producing *Vibrio* strain, LM-1, from the puffer fish *Fugu vermicularis radiatus*. *Appl. Environ. Microbiol.* **66:**1698–1701.

Levine, M. M., J. B. Kaper, D. Herrington, J. Ketley, G. Losonsky, C. O. Tacket, B. Tall, and S. Cryz. 1988. Safety, immunogenicity, and efficacy of recombinant live oral cholera vaccines, CVD 103 and CVD 103-HgR. *Lancet* **2:**467–470.

LoGrasso, P., and J. McKelvy. 2003. Advances in pain therapeutics. *Curr. Opin. Chem. Biol.* **7:**452–456.

Longeon, A., J. Peduzzi, M. Barthelemy, S. Corre, J. L. Nicolas, and M. Guyot. 2005. Purification and partial identification of novel antimicrobial protein from marine bacterium *Pseudoalteromonas* species strain X153. *Mar. Biotechnol.* (N.Y.) **6:**633–641.

Lucas, M. E., J. L. Deen, L. von Seidlein, X. Y. Wang, J. Ampuero, M. Puri, M. Ali, M. Ansaruzzaman, J. Amos, A. Macuamule, P. Cavailler, P. J. Guerin, C. Mahoudeau, P. Kahozi-Sangwa, C. L. Chaignat, A. Barreto, F. F. Songane, and J. D. Clemens. 2005. Effectiveness of mass oral cholera vaccination in Beira, Mozambique. *N. Engl. J. Med.* **352:**757–767.

Manefield, M., L. Harris, S. A. Rice, R. de Nys, and S. Kjelleberg. 2000. Inhibition of luminescence and virulence in the black tiger prawn *(Penaeus monodon)* pathogen *Vibrio harveyi* by intercellular signal antagonists. *Appl. Environ. Microbiol.* **66:**2079–2084.

Manefield, M., M. Welch, M. Givskov, G. P. Salmond, and S. Kjelleberg. 2001. Halogenated furanones from the red alga, *Delisea pulchra*, inhibit carbapenem antibiotic synthesis and exoenzyme virulence factor production in the phytopathogen *Erwinia carotovora*. *FEMS Microbiol. Lett.* **205:**131–138.

Marwick, J. D., P. C. Wright, and J. G. Burgess. 1999. Bioprocess intensification for production of novel marine bacterial antibiotics through bioreactor operation and design. *Mar. Biotechnol.* (N. Y.) **1:**495–507.

Matsuo, K., T. Yoshikawa, H. Asanuma, T. Iwasaki, Y. Hagiwara, Z. Chen, S. E. Kadowaki, H. Tsujimoto, T. Kurata, and S. I. Tamura. 2000. Induction of innate immunity by nasal influenza vaccine administered in combination with an adjuvant (cholera toxin). *Vaccine* **18:**2713–2722.

Mearns-Spragg, A., M. Bregu, K. G. Boyd, and J. G. Burgess. 1998. Cross-species induction and enhancement of antimicrobial activity produced by epibiotic bacteria from marine algae and invertebrates, after exposure to terrestrial bacteria. *Lett. Appl. Microbiol.* **27:**142–146.

Miller, M. B., and B. L. Bassler. 2001. Quorum sensing in bacteria. *Annu. Rev. Microbiol.* **55:**165–199.

Miller, M. B., K. Skorupski, D. H. Lenz, R. K. Taylor, and B. L. Bassler. 2002. Parallel quorum sensing systems converge to regulate virulence in *Vibrio cholerae. Cell* **110:**303–314.

O'Toole, G., H. B. Kaplan, and R. Kolter. 2000. Biofilm formation as microbial development. *Annu. Rev. Microbiol.* **54:**49–79.

Pearson, J. P., M. Feldman, B. H. Iglewski, and A. Prince. 2000. *Pseudomonas aeruginosa* cell-to-cell signaling is required for virulence in a model of acute pulmonary infection. *Infect. Immun.* **68:**4331–4334.

Ploix, C., I. Bergerot, A. Durand, C. Czerkinsky, J. Holmgren, and C. Thivolet. 1999. Oral administration of cholera toxin B-insulin conjugates protects NOD mice from autoimmune diabetes by inducing CD4+ regulatory T-cells. *Diabetes* **48:**2150–2156.

Porreca, F., J. Lai, D. Bian, S. Wegert, M. H. Ossipov, R. M. Eglen, L. Kassotakis, S. Novakovic, D. K. Rabert, L. Sangameswaran, and J. C. Hunter. 1999. A comparison of the potential role of the tetrodotoxin-insensitive sodium channels, PN3/SNS and NaN/SNS2, in rat models of chronic pain. *Proc. Natl. Acad. Sci. USA* **96:**7640–7644.

Powell, L. 1995. *Biotechnology for the 21st Century: New Horizons.* National Science and Technology Council, Biotechnology Research Subcommittee, London, England.

Raffa, R. B., J. R. Iannuzzo, D. R. Levine, K. K. Saeid, R. C. Schwartz, N. T. Sucic, O. D. Terleckyj, and J. M. Young. 2005. Bacterial communication ("quorum sensing") via ligands and receptors: a novel pharmacologic target for the design of antibiotic drugs. *J. Pharmacol. Exp. Ther.* **312:**417–423.

Rasch, M., C. Buch, B. Austin, W. J. Slierendrecht, K. S. Ekmann, J. L. Larsen, C. Johansen, K. Riedel, L. Eberl, M. Givskov, and L. Gram. 2004. An inhibitor of bacterial quorum sensing reduces mortalities caused by vibriosis in rainbow trout (*Oncorhynchus mykiss,* Walbaum). *Syst. Appl. Microbiol.* **27:**350–359.

Rasmussen, T. B., M. Manefield, J. B. Andersen, L. Eberl, U. Anthoni, C. Christophersen, P. Steinberg, S. Kjelleberg, and M. Givskov. 2000. How *Delisea pulchra* furanones affect quorum sensing and swarming motility in *Serratia liquefaciens* MG1. *Microbiology* **146:**3237–3244.

Roda, A., P. Pasini, M. Mirasoli, E. Michelini, and M. Guardigli. 2004. Biotechnological applications of bioluminescence and chemiluminescence. *Trends Biotechnol.* **22:**295–303.

Roosenberg, J. M., 2nd, Y. M. Lin, Y. Lu, and M. J. Miller. 2000. Studies and syntheses of siderophores, microbial iron chelators, and analogs as potential drug delivery agents. *Curr. Med. Chem.* **7:**159–197.

Rotz, L. D., and J. M. Hughes. 2004. Advances in detecting and responding to threats from bioterrorism and emerging infectious disease. *Nat. Med.* **10**(12 Suppl.):S130–S136.

Ruiz-Ponte, C., V. Cilia, C. Lambert, and J. L. Nicolas. 1998. *Roseobacter gallaeciensis* sp. nov., a new marine bacterium isolated from rearings and collectors of the scallop *Pecten maximus. Int. J. Syst. Bacteriol.* **48:**537–542.

Ruiz-Ponte, C., J. F. Samain, J. L. Sanchez, and J. L. Nicolas. 1999. The benefit of a *Roseobacter* species on the survival of scallop larvae. *Mar. Biotechnol. (N.Y.)* **1:**52–59.

Sangali, S., and A. Brandelli. 2000. Feather keratin hydrolysis by a *Vibrio* sp. strain kr2. *J. Appl. Microbiol.* **89:**735–743.

Schantz, E. J., and E. A. Johnson. 1992. Properties and use of botulinum toxin and other microbial neurotoxins in medicine. *Microbiol Rev.* **56:**80–99.

Simidu, U., K. Kita-Tsukamoto, T. Yasumoto, and M. Yotsu. 1990. Taxonomy of four marine bacterial strains that produce tetrodotoxin. *Int. J. Syst. Bacteriol.* **40:**331–336.

Simidu, U., T. Noguchi, D. F. Hwang, Y. Shida, and K. Hashimoto. 1987. Marine bacteria which produce tetrodotoxin. *Appl. Environ. Microbiol.* **53:**1714–1715.

Skyler, J. S. 2002. Effects of insulin in relatives of patients with type 1 diabetes mellitus. *N. Engl. J. Med.* **346:**1685–1691.

Tamura, S. I., and T. Kurata. 2000. A proposal for safety standards for human use of cholera toxin (or *Escherichia coli* heat-labile enterotoxin) derivatives as an adjuvant of nasal inactivated influenza vaccine. *Jpn. J. Infect. Dis.* **53:**98–106.

Thomashow, L. S. 1996. Biological control of plant root pathogens. *Curr. Opin. Biotechnol.* **7:**343–347.

Vergne, A. F., A. J. Walz, and M. J. Miller. 2000. Iron chelators from mycobacteria (1954–1999) and potential therapeutic applications. *Nat. Prod. Rep.* **17:**99–116.

Voinov, V. G., Y. N. El'kin, T. A. Kuznetsova, I. I. Mal'tsev, V. V. Mikhailov, and V. A. Sasunkevich. 1991. Use of mass spectrometry for the detection and identification of bromine-containing diphenyl ethers. *J. Chromatogr.* **586:**360–362.

Von Bodman, S. B., W. D. Bauer, and D. L. Coplin. 2003. Quorum sensing in plant-pathogenic bacteria. *Annu. Rev. Phytopathol.* **41:**455–482.

Williams, P., M. Camara, A. Hardman, S. Swift, D. Milton, V. J. Hope, K. Winzer, B. Middleton, D. I. Pritchard, and B. W. Bycroft. 2000. Quorum sensing and the population-dependent control of virulence. *Philos. Trans. R. Soc. Lond. B Biol. Sci.* **355:**667–680.

Winzer, K., and P. Williams. 2001. Quorum sensing and the regulation of virulence gene expression in pathogenic bacteria. *Int. J. Med. Microbiol.* **291:**131–143.

Yan, L., K. G. Boyd, D. R. Adams, and J. G. Burgess. 2003. Biofilm-specific cross-species induction of antimicrobial compounds in bacilli. *Appl. Environ. Microbiol.* **69:**3719–3727.

Yoshikawa, K., K. Adachi, M. Nishijima, T. Takadera, S. Tamaki, K. Harada, K. Mochida, and H. Sano. 2000. Beta-cyanoalanine production by marine bacteria on cyanide-free medium and its specific inhibitory activity toward cyanobacteria. *Appl. Environ. Microbiol.* **66:**718–722.

Yoshino, N., F. X. Lu, K. Fujihashi, Y. Hagiwara, K. Kataoka, D. Lu, L. Hirst, M. Honda, F. W. van Ginkel, Y. Takeda, C. J. Miller, H. Kiyono, and J. R. McGhee. 2004. A novel adjuvant for mucosal immunity to HIV-1 gp120 in nonhuman primates. *J. Immunol.* **173:**6850–6857.

Yu, C. F., P. H. Yu, P. L. Chan, Q. Yan, and P. K. Wong. 2004. Two novel species of tetrodotoxin-producing bacteria isolated from toxic marine puffer fishes. *Toxicon* **4:**641–647.

Zhu, J., M. B. Miller, R. E. Vance, M. Dziejman, B. L. Bassler, and J. J. Mekalanos. 2002. Quorum-sensing regulators control virulence gene expression in *Vibrio cholerae. Proc. Natl. Acad. Sci. USA* **99:**3129–3134.

XI. CONCLUSIONS

The Biology of Vibrios
Edited by F. L. Thompson et al.
© 2006 ASM Press, Washington, D.C.

Chapter 29

Conclusions

FABIANO L. THOMPSON, BRIAN AUSTIN, AND JEAN SWINGS

There has clearly been an upturn of interest in the vibrios since Rita Colwell's edited text appeared in 1986, with scientists highlighting taxonomic, ecological, and pathogenicity aspects of the bacteria. It is apparent that the fields of (post)genomics and proteomics refined and enhanced our view of the biology of vibrios in many different ways over the last few years. Apart from the ages-old interest in human diseases, such as cholera, there has been recognition of the increasing role of vibrios in aquatic animal diseases, including those of fish, crustacea, and bivalves (e.g., Austin and Austin, 1999). More recently, *Vibrio* species have become associated with diseases of corals, including bleaching (Rosenberg and Loya, 2004), skeletal tumors (Breitbart et al., 2005), and yellow band (Cervino et al., 2004). These conditions are undoubtedly contributing to the destruction of coral reefs worldwide (Rosenberg and Loya, 2004), where global warming and interactions with their hosts may trigger the expression of virulence factors abundant in vibrio genomes and in their "parasites," i.e., vibriophages (Faruque et al., 2005a,b). And what will the future hold? It is clear that scientists have researched some fundamental aspects of vibrio biology, but many more questions remain to be resolved.

CULTURING

The limitation of agar plates to recover microorganisms from environmental samples was realized over a century ago by Winogradsky. Indeed, the great majority of the marine microbiota is elusive to growth on conventional laboratory media, probably due to the artificially high nutrient load of commercial media and incompatibility of the media with the environmental conditions. Such nutrient-rich conditions would not support oligotrophs. Of course, it is unrealistic to believe that any single medium and incubation condition would allow the recovery of the majority of marine bacteria. As an example, incubation of inoculated media in the presence of air would preclude the recovery of anaerobes. In addition, a fraction of the bacterial populations may be in the so-called VBNC (viable but nonculturable) state, thereby hampering their recovery using conventional approaches.

It is clear that there is still a widespread reliance on thiosulfate citrate bile salt sucrose (TCBS) agar for the recovery of vibrios, despite its formulation being specifically for cholera. It is essential to ask, how effective is it, really, for the recovery of the wider range of vibrios that occur in the aquatic environment? Indeed, some comparative work from Japan during the 1970s revealed that TCBS was not the best medium for the recovery of the greatest number of vibrios from the aquatic environment. In particular, some vibrios, e.g., *V. penaeicida*, do not grow on TCBS at all, whereas others, e.g., *V. cincinnatiensis*, *V. metschnikovii*, and *Grimontia hollisae*, grow poorly, if at all. Perhaps the main problem with TCBS is its lack of selectivity, insofar as different *Proteobacteria*, including strains of *Pseudomonas*, *Proteus*, and *Providencia*, may also grow on this medium.

Recent studies have shown that quantitative data obtained using TCBS parallel the molecular counts by quantitative PCR (Randa et al., 2004; J. R. Thompson et al., 2004, 2005). In addition to this new strategy to locate vibrios in the environment, real-time PCR (Panicker et al., 2004) and flow cytometry (Wallner et al., 1995) offer promise for the rapid detection and quantification of vibrios as part of environmental monitoring of coastal and ballast water, and for ecological studies. However, care should be exercised on the choice of gene targets used in these

Fabiano L. Thompson • Microbial Resources Division and Brazilian Collection of Environmental and Industrial Micro-organisms (CBMAI), CPQBA, UNICAMP, Alexandre Caselatto 999, CEP 13140000, Paulínia, Brazil. **Brian Austin** • School of Life Sciences, John Muir Building, Heriot-Watt University, Riccarton, Edinburgh EH14 4AS, Scotland, United Kingdom. **Jean Swings** • Laboratory of Microbiology and BCCM/LMG Bacteria Collection, Ghent University, K.L. Ledeganckstraat 35, B-9000 Ghent, Belgium.

techniques, insofar as different *Vibrio* species may have identical alleles for currently used genes, e.g., 16S rRNA. Recent culture-independent studies targeting 16S rRNA sequences of fish (Jensen et al., 2004) and coral microbiota (Bourne and Munn, 2005) revealed the high abundance of vibrios, including *V. harveyi*- and *V. coralliilyticus*-related species.

In certain periods of the year, vibrios may be hard to recover from the environment by traditional culturing techniques. One explanation evoked by several researchers is the establishment of a VBNC state in order to endure harsh environmental conditions. It remains to be clearly and unambiguously determined in future work what the genetic program underlying the yet obscure VBNC state is (McDougald et al., 1998). In many instances, the cells have irreversible nucleic acid degradation, becoming really debilitated nonviable cells (senescent/dying?), whereas clear-cut resuscitation has not been shown yet. The VBNC concept clearly needs further challenging. Molecular approaches may also fail to recover vibrios in the (water) environment, but it is important to keep in mind that certain vibrios may have a complex life cycle. For example, *V. shilonii* (*V. mediterranei*) has a winter reservoir, the fireworm *Hermodice caranculata* (Sussman et al., 2003). Another possibility is the persistence of vibrios in nature through the formation of biofilms in or on living or nonliving surfaces as revealed from laboratory experiments. Biofilm formation is an important feature in the ecology (and pathogenesis) of vibrios. The development of biofilms is under orchestrated and complex genetic control involving several independent loci (Lauriano et al., 2004; Reguera and Kolter, 2005; Tischler and Camilli, 2004). In addition to the expression of essential exopolysaccharides, recently gathered data suggest that toxin-coregulated pili enhance biofilm formation, but data on biofilm development need to be further challenged with true environmental studies to confirm what has been found in the laboratory.

What exactly is the VBNC state from a genetic/genomic perspective? Does it represent a normal survival state, an altered form, e.g., senescent, dying, or morphologically fragile (L-form), or does it reflect the inability of scientists to culture unusual forms or to recognize minimal growth? Why should aquatic bacteria, which are present in a comparatively nutrient-deprived environment, grow on agar plates, which contain unnatural ingredients in unusually high concentrations compared to the levels in the natural environment? Why should microorganisms produce visible growth, i.e., colonies on agar plates? Could we recognize limited growth (microcolonies)? Do we need to develop approaches that more closely resem-

ble the natural environment? These are questions which are being addressed and to some extent answered in the study of the natural form of microorganisms in the aquatic environment.

TAXONOMY AND PHYLOGENOMICS

It is clear that many more new species of vibrios are likely to be described in the near future as surveys on different environmental settings continue to be carried out (Dunlap and Ast, 2005; F. L. Thompson et al., 2005; Seo et al., 2005a,b). With the exception of the *Vibrionaceae*, which has been under massive taxonomic scrutiny in the last years, the study of the biodiversity of *Enterovibrionaceae*, *Photobacteriaceae*, and *Salinivibrionaceae* offers the opportunity to find a great untapped genetic diversity. Also, there are likely to be further refinements in the taxonomic processes that may well lead to reevaluation, including the amalgamation of some taxa, the splitting of others, and the concomitant proposals for new names. Phenotyping techniques will probably give place to molecular tools for the needed screening of massive numbers of isolates. One of the main challenges for vibrio taxonomists in the coming years will be the development of a standard, electronic online-based taxonomy using multilocus sequence analysis (MLSA) (http://lmg.ugent.be/bnserver/MLSA/Vibrionaceae/; F. L. Thompson et al., 2005). This approach will be built up on the traditional polyphasic taxonomy, and will enable end-users to readily identify their isolates through curated online databases. New phylogenetic insights will be gained as more genes are analyzed for the whole of the vibrios. The phylogenetic picture obtained by trees with concatenated sequences of the 16S rDNA and a few other housekeeping genes (e.g., recA, rpoA, and pyrH) will be refined. Moreover, it will be possible to evaluate whether certain species groups, e.g., *V. splendidus*, form a genetic continuum rather than discrete gene clusters. MLSA and whole-genome sequence data will also allow vibrio taxonomists to address questions concerning the species definition and concept, a highly debatable issue in the current prokaryotic taxonomy (Gevers et al., 2005; Konstantinidis and Tiedje, 2005a,b; Ochman et al., 2005). Can we assume that a bifurcating phylogenetic tree really expresses the history of vibrios, or should we accept a netted phylogeny for this group? Gevers et al. (2005) encouraged further large-scale MLSA analyses on well-studied genera, e.g., vibrios, in order to understand the prokaryotic species. Instead of using preset cut-offs and rules to define species, these authors suggested that clusters reflecting ecologically distinct populations may be named ecotypes (species). Konstantinidis and Tiejde (2005a) also suggested that "it may be realistic from the bacterial perspective" to consider as separate species sequence clusters that share high similarity (>99%) but have distinguishable ecology and/or distribution. Clearly, much more genomic data are needed in order to extend this type of perception for vibrios. In some cases, different *Vibrio* species have seemly overlapping ecological niches. Gathering more representative whole-genome sequences from different vibrios may eventually lead to a phylogenomic approach through which different aspects of the biology of vibrios, including taxonomy, phylogeny, pathogenicity, and metabolism, could be understood from both an evolutionary and a genomic perspective (Eisen and Fraser, 2003).

SYSTEMS MICROBIOLOGY

Several important insights related to genome composition, configuration, and plasticity have been gained in the last few years as a result of whole-genome sequence projects (e.g., Heidelberg et al., 2000). Recent studies have shown a striking similarity between a pathogen, i.e., *V. cholerae*, and a mutualist, namely, a *V. fischeri* strain, concerning virulence factors (Ruby et al., 2005). Overall, a considerable proportion of the genes in vibrio genomes have no known biological function, indicating that much has to be learned about the biology of these organisms. Even in the best-studied model, i.e., *V. cholerae*, several genes expressed during the onset of disease have no known function, but may be related to stress adaptation and pathogenesis (Larocque et al., 2005; Xu et al., 2003). A type III secretion system has been recently discovered in *V. cholerae* non-O1 and non-O139 strains through genomics (Dziejman et al., 2005). Further postgenomic studies are likely to unravel the role of such genes in the pathogenesis and metabolism of vibrios under different environmental conditions (Bina et al., 2003; Meibom et al., 2004). Nonchemotactic *V. cholerae* mutants may have increased infectiveness, suggesting that chemotaxis is detrimental for pathogenesis (Butler and Camilli, 2005).

Microarray analyses have shed light on the regulation of quorum sensing in *V. fischeri* and revealed that much remains to be understood in the host colonization process in this well-studied microbe (Lupp and Ruby, 2005). *V. fischeri* influences the developmental program of its host through several molecules, for example, a peptidoglycan monomer (PGM) that stimulates regression of specific light organ cells (Ko-ropatnick et al., 2004). The PGM molecule produced by *V. fischeri* is identical to cytotoxins of known human pathogens, such as *Bordetella pertussis* and *Neisseria gonorrhoeae*. This striking similarity reveals that mutualistic and pathogenic bacteria–animal interactions are separated by a tenuous line. Much work is still needed at the transcriptional and proteomic level in order to fully understand mutual partnerships involving vibrios.

We cannot assume that the eight currently available whole genomes represent entirely the genomic armamentarium of vibrios. For example, we do not have as yet any representative genome of *V. harveyi* or *V. anguillarum*, which are among the most important marine animal pathogens. In this respect, new and emerging pathogens, such as *V. coralliilyticus*, deserve a high level of priority for sequencing, but we could as well point out that *Salinivibrio costicola* represents one of the earliest phylogenetic branches of vibrios.

The impact of mobile genetic elements and bacteriophages needs to be further evaluated in the (co)evolution of vibrio genomes. It is clear that phages have had a tremendous impact on the evolution of pathogenesis of some vibrios, including *V. cholerae*, *V. harveyi*, and *V. parahaemolyticus*. What is the long-term quantitative and qualitative impact of these life forms on ecology (Paul et al., 2005; Faruque et al., 2005a,b), speciation, and diversification of vibrios (Weinbauer and Rassoulzadegan, 2004)? It is remarkable that *V. parahaemolyticus* strains show a considerable variation in genome content because of their prophage content (T. Iida, personal communication). Several open reading frames of the recently sequenced *V. pelagius* phage share high similarity with phages of taxonomically unrelated pathogenic bacteria, including *Streptococcus pyogenes*, *Lactobacillus* spp., and *Staphylococcus* spp. (Paul et al., 2005), clearly indicating the plasticity of vibrio genomes.

Experimental approaches to vibrio plasticity and evolution are badly needed. New algorithms for analyzing horizontal gene transfer and genome rearrangements will help to better understand the evolution of vibrios (Lin et al., 2005), but much remains to be done on the computational biology of vibrios. The usefulness of genome-based approaches to taxonomy, using amino acid identity of MLSA data, may offer insights on vibrio systematics (Konstantinidis and Tiedje, 2005b). Clearly, vibrios offer an excellent tractable model for systems microbiology. In the near future, different kinds of data, including phenotype, genotype, pathogenicity, metabolic and proteomic maps, but also environmental data related to well-studied and documented strain collections (F. L. Thompson et al., 2004) could be easily fed into online databases. This would certainly represent a step forward in our current way of doing research on the various aspects of the biology of vibrios, and would eventually lead to the establishment of an authoritative and ever-evolving online resource for the systems microbiology of the vibrios.

ECOLOGY

Although vibrios represent only a small fraction of the total marine microbiota, much light has been shed on several important aspects of microbial ecology through the study of these microbes. This includes the study of bioluminescence, quorum sensing, biofilm formation, mutualism, and the VBNC state. More recently, the use of remote sensing to predict the growth/abundance of V. cholerae has opened up new avenues for epidemiological and environmental research (Lobitz et al., 2000). Can we gain more insight into environmental functioning through the study of the biology of vibrios, or should we change direction toward the uncultured majority observed in culture-independent surveys? We argue that much can still be learned from studying vibrios in an ecological context. For example, the idea that "everything is everywhere, the environment selects," suggesting that microbial species have ubiquitous distribution, needs to be further challenged. In this respect, vibrios offer a very useful and tractable study model. Although some studies have already indicated that certain vibrios, e.g., V. coralliilyticus, V. cholerae, V. harveyi, and V. splendidus, may be ubiquitous, we do not know much about the spatial scaling of most Vibrio species (Green et al., 2004; Horner-Devine et al., 2004a,b). In one example, the coral Porites harbored the same microbiota irrespective of time and space (Rohwer et al., 2002). What is the actual α and β diversity of vibrios within different environmental settings? Recent data indicate high intraspecific genomic variability of planktonic V. splendidus, which was suggested to be neutral (J. R. Thompson et al., 2005).

Most vibrios are isolated from a variety of habitats, including open-ocean (oligotrophic) waters and the deep sea, contrasting with the idea that these organisms are adapted to only nutrient-rich settings, namely, coastal marine waters, estuaries, and aquacultural locations. The role of vibrios in the nutrient cycling of oligotrophic regions remains to be determined. Molecular surveys suggest that vibrios represent only a small fraction ($\sim 10^3$ cell/ml) of the total bacterial community and that the abundance and diversity pattern follow seasonal variations in, for example, temperature (J. R. Thompson et al., 2005). Certainly, vibrios have the metabolic machinery to re-

spond rapidly to carbon and nutrient enrichments. Indeed, these microbes have a broad metabolism including various extracellular proteases, chitinases, lipases, keratinases, oxygenases, and polycyclic aromatic hydrocarbon-degrading enzymes. Because of their versatile cell machinery, vibrios may be a most efficient link in food webs, particularly during pulses of nutrient input into the environment. Therefore, under optimum environmental conditions (i.e., high temperature and nutrient load), vibrios may disproportionately contribute to nutrient turnover. Further work linking function and (micro)diversity is badly needed in order to shed more light on the role of vibrios in nutrient cycling (Dumont and Murrell, 2005).

Marine fish, shellfish, cnidarians, and porifers harbor a wealth of vibrios on their skins and/or within their body cavities. It is possible that in several cases, e.g., in the gut of abalones, vibrios contribute to the nourishment of their hosts via production of volatile fatty acids and other end products of the fermentation of alginate. Recent work on the shrimp Litopenaeus vannamei suggests that vibrios are actively removed from the hemolymph through the lymphoid organ (Burgents et al., 2005), putting into doubt the notion that healthy shrimps may tolerate vibrios within their body fluids (Gomez-Gil et al., 1998). Further work is clearly needed to unravel the role of these vibrios in the overall health of shrimps. The immunological mechanisms by which the squid E. scolopes detects and allows tissue colonization only by V. fischeri are being revealed (Nyholm and McFall-Ngai, 2004). Lipopolysaccharide and PGM molecules normally trigger antimicrobial responses by the innate immune system of animals, but in the specific case of V. fischeri and E. scolopes, the host seems to tolerate the benign infection (Koropatnick et al., 2004). Studies of this type may well aid a better understanding of other potential mutual partnership models, such as the abalone–V. halioticoli.

HUMAN AND ANIMAL DISEASES AND THEIR CONTROL

Scientists should not lose sight of the need to relate artificial laboratory findings to those of the real world. This is particularly true in the study of pathogenicity. It is apparent that there is an unfortunate tendency to work with single bacterial cultures, and as an added complication, it is then difficult to relate the findings of one laboratory to another. Pathogenicity appears to be multifactorial, including interaction with bacteriophage and plasmids in the production of a wide range of extracellular and intracellular pathogenicity factors. Further work is needed to unravel the emergence of new virulence traits by horizontal gene transfer in the various vibrios. Scientists must be aware that organisms may produce different pathogenicity factors in the host compared to laboratory findings. Also, synergism among different pathogenic strains and species may lead to more acute disease outbreaks in the case of marine organisms. Whereas it is acknowledged that there has been a great deal of progress in the understanding of the pathogenicity of a few *Vibrio* spp. (notably *V. cholerae*), much needs to be learned about others, especially those taxa that have been recognized only recently to be pathogenic, e.g. *V. brasiliensis*, *V. coralliilyticus*, and *V. neptunius* (Austin et al., 2005). As yet uncharacterized toxins produced by these vibrios act over a wide spectrum of marine organisms.

Will any of the current approaches to vaccine development lead to a successful candidate for the control of cholera? Undoubtedly, *V. cholerae* has been one of the best-studied human pathogens (Faruque and Mekalanos, 2003). Yet no effective measures for epidemic control have been achieved to date. Indeed, ~120,000 cholera cases occur each year (World Health Organization, 2003, 2004), mainly in Africa, with the majority of incidences attributable to *V. cholerae* O1, biotype El Tor. Nevertheless, sanitation and improved hygiene measures have long been recognized as the most efficient preventive measures for reducing the disease, but clearly, several developing countries cannot afford this action. Societies continue to hope for the availability of an effective vaccine, many formulations of which have been tested with varying results over the last few decades. Recent trials using live attenuated cholera vaccines have shown promise for the control of disease under experimental conditions, yet much needs to be learned about their broad application in cholera endemic regions (Garcia et al., 2005; Qadri et al., 2005).

An ever-increasing range of pathogens, sometimes reflecting the introduction of nonnative species into new environments or contaminated seeds and broodstock, have resulted in the spread of disease caused by particular *Vibrio* strains. Also, ballast water may well represent a source of spread of pathogenic strains of vibrios worldwide. Will any of the current approaches to vaccine development lead to a successful candidate for the control of vibriosis? Time will surely tell. Clearly, there is an urgent need to move away from reliance on antibiotics, especially in nonmedical use, and adopt other approaches, such as improved hygiene/management practices, immunostimulants, dietary supplements, probiotics, phage therapy, and modulation of virulence by exploiting quorum-sensing mechanisms. The use of phages to control the number of pathogenic vibrios

(Faruque et al., 2005a,b) and to prevent disease in reared fish and shellfish (Cerveny et al., 2002) holds promise, although much work needs to be done before this strategy may replace or even supplement the use of antibiotics (Brüssow, 2005). Little is known about the impact of phage introduction on the receiving environment, and this aspect needs to be considered. Also, there may be value in the consideration of polyvalent preparation of phages, which may well kill both pathogenic and mutualistic vibrio strains in rearing systems. Certainly, probiotics have been used by aquaculturists in several countries, including Ecuador and China, and much has been advocated about their apparent miraculous properties relating to health improvement (Verschuere et al., 2000). However, it should not be forgotten that miracles are very rare phenomena in real life. Overall, the scientific basis of using probiotics is still debatable and needs further research, with possible modes of action including competitive exclusion, dietary modification, nutritional supplementation, and stimulation of the innate immune response (see Irianto and Austin, 2002). Another strategy involving quorum-sensing antagonists (quenching molecules) has been evaluated to reduce shrimp (Defoirdt et al., 2005) and fish mortalities (Rasch et al., 2004) caused by vibriosis. However, these molecules have not led to full protection of the host animals. This suggests that other, as yet unknown pathogenesis mechanisms, not regulated by quorum sensing, are present in these pathogens, although quorum-sensing systems may well have a central role in the coordination of the physiology of vibrios (including biofilm formation, starvation adaptation, and virulence factors production). Nevertheless, this approach is worthy of further development.

BIOTECHNOLOGICAL APPLICATIONS: EXPLOITING THE NATURAL BIODIVERSITY, EXPLOITING THE GENOMIC DATA MINES, AND THE ROLE OF PUBLICLY ACCUMULATED WORLDWIDE CULTURE COLLECTIONS AND DATABASES

With the impetus toward developing marine biotechnology, the future may hold promise for the development of novel anti-infectives, antitumor compounds, enzymes, and polymers/substrates from vibrios. Clearly, scientists need to have a much better understanding of the biotechnological potential of vibrios that inhabit extreme environments, including the deep sea and hydrothermal vents. A few successful surveys have shown the potential of vibrios in the production of exopolysaccharides (Guezennec, 2002). The genetic and metabolic potential of certain vibrios isolated from diverse environments, different extreme ocean ecosystems, as well as from marine organisms such as corals and porifera, has been highlighted throughout this book. However, their biotechnological exploitation remains largely unknown and thus might represent an exciting endeavor for further secondary metabolite and drug discovery through, for example, the screening of the natural biodiversity and/or metabolic engineering. Also, new whole-genome sequences and advances in bioinformatics may well underpin future developments in biotechnology and metabolic engineering (Covert et al., 2004; Price et al., 2004). Finally, it is important to highlight that internationally recognized culture collections can serve both as untapped genetic resources for bioprospecting projects and as authoritative repositories of high-quality authentic biological material uncovered in environmental surveys. Paradoxically, several of those countries with the potentially largest and yet unexplored microbial (vibrio) diversity are those that lack such repositories.

Acknowledgments. F.L.T. acknowledges a young researcher grant (2004/00814-9) from FAPESP, Brazil. J.S. acknowledges grants of the Fund for Scientific Research (FWO), Belgium. We thank Prof. K. Klose (UTSA) for valuable comments.

REFERENCES

Austin, B., and D. A. Austin. 1999. *Bacterial Fish Pathogens: Disease of Farmed and Wild Fish*, 3rd (rev.) ed. Springer-Praxis, Godalming, England.

Austin, B., D. Austin, R. Sutherland, F. L. Thompson, and J. Swings. 2005. Pathogenicity of vibrios to rainbow trout (*Oncorhynchus mykiss*, Walbaum) and *Artemia* nauplii. *Environ. Microbiol.* 7: 1488–1495.

Bina, J., J. Zhu, M. Dziejman, S. Faruque, S. Calderwood, and J. Mekalanos. 2003. ToxR regulon of *Vibrio cholerae* and its expression in vibrios shed by cholera patients. *Proc. Natl. Acad. Sci. USA* 100:2801–2806.

Bourne, D. G., and C. B. Munn. 2005. Diversity of bacteria associated with the coral *Pocillopora damicornis* from the Great Barrier Reef. *Environ. Microbiol.* 7:1162–1174.

Breitbart, M., R. Bhagooli, S. Griffin, I. Johnston, and F. Rohwer. 2005. Microbial communities associated with skeletal tumors on *Porites compressa*. *FEMS Microbiol. Lett.* 243:431–436.

Brüssow, H. 2005. Phage therapy: the *Escherichia coli* experience. *Microbiology* 151:2133–2140.

Burgents, J. E., L. E. Burnett, E. V. Stabb, and K. G. Burnett. 2005. Localization and bacteriostasis of *Vibrio* introduced into the Pacific white shrimp, *Litopenaeus vannamei*. *Dev. Comp. Immunol.* 29:681–691.

Butler, S. M., and A. Camilli. 2005. Going against the grain: chemotaxis and infection in Vibrio cholerae. *Nat. Rev. Microbiol.* 3:611–620.

Cerveny, K. E., A. DePaola, D. H. Duckworth, and P. A. Gulig. 2002. Phage therapy of local and systemic disease caused by *Vibrio vulnificus* in iron-dextran-treated mice. *Infect Immun.* 70:6251–6262.

Cervino, J. M., R. L. Hayes, S. W. Polson, S. C. Polson, T. J. Goreau, R. J. Martinez, and G. W. Smith. 2004. Relationship of *Vibrio* species infection and elevated temperatures to yellow blotch/band disease in Caribbean corals. *Appl. Environ. Microbiol.* 70:6855–6864.

Covert, M. W., E. M. Knight, J. L. Reed, M. J. Herrgard, and B. O. Palsson. 2004. Integrating high-throughput and computational data elucidates bacterial networks. *Nature* 429:92–96.

Defoirdt, T., P. Bossier, P. Sorgeloos, and W. Verstraete. 2005. The impact of mutations in the quorum sensing systems of *Aeromonas hydrophila*, *Vibrio anguillarum* and *Vibrio harveyi* on their virulence towards gnotobiotically cultured *Artemia franciscana*. *Environ. Microbiol.* 7:1239–1247.

Dumont, M. G., and J. C. Murrell. 2005. Stable isotope probing: linking microbial identity to function. *Nat. Rev. Microbiol.* 3: 499–504.

Dunlap, P. V., and J. C. Ast. 2005. Genomic and phylogenetic characterization of luminous bacteria symbiotic with the deep-sea fish *Chlorophthalmus albatrossis* (Aulopiformes: Chlorophthalmidae). *Appl. Environ. Microbiol.* 71:930–939.

Dziejman, M., D. Serruto, V. C. Tam, D. Sturtevant, P. Diraphat, S. M. Faruque, M. H. Rahman, J. F. Heidelberg, J. Decker, L. Li, K. T. Montgomery, G. Grills, R. Kucherlapati, and J. J. Mekalanos. 2005. Genomic characterization of non-O1, non-O139 Vibrio cholerae reveals genes for a type III secretion system. *Proc. Natl. Acad. Sci. USA* 102:3465–3470.

Eisen, J. A., and C. M. Fraser. 2003. Phylogenomics: intersection of evolution and genomics. *Science* 300:1706–1707.

Faruque, S. M., and J. J. Mekalanos. 2003. Pathogenicity islands and phages in *Vibrio cholerae* evolution. *Trends Microbiol.* 11: 505–510.

Faruque, S. M., M. J. Islam, Q. S. Ahmad, A. S. Faruque, D. A. Sack, G. B. Nair, and J. J. Mekalanos. 2005a. Self-limiting nature of seasonal cholera epidemics: role of host-mediated amplification of phage. *Proc. Natl. Acad. Sci. USA* 102:6119–6124.

Faruque, S. M., I. B. Naser, M. J. Islam, A. S. Faruque, A. N. Ghosh, G. B. Nair, D. A. Sack, and J. J. Mekalanos. 2005b. Seasonal epidemics of cholera inversely correlate with the prevalence of environmental cholera phages. *Proc. Natl. Acad. Sci. USA* 102:1702–1707.

Garcia, L., M. D. Jidy, H. Garcia, B. L. Rodriguez, R. Fernandez, G. Ano, B. Cedre, T. Valmaseda, E. Suzarte, M. Ramirez, Y. Pino, J. Campos, J. Menendez, R. Valera, D. Gonzalez, I. Gonzalez, O. Perez, T. Serrano, M. Lastre, F. Miralles, J. Del Campo, J. L. Maestre, J. L. Perez, A. Talavera, A. Perez, K. Marrero, T. Ledon, and R. Fando. 2005. The vaccine candidate *Vibrio cholerae* 638 is protective against cholera in healthy volunteers. *Infect. Immun.* 73:3018–3024.

Gevers, D., F. M. Cohan, J. G. Lawrence, B. G. Spratt, T. Coenye, E. J. Feil, E. Stackebrandt, Y. Van de Peer, P. Vandamme, F. L. Thompson, and J. Swings. 2005. Opinion: reevaluating prokaryotic species. *Nat. Microbiol. Rev.* 3:5107–5115.

Gomez-Gil B., L. Tron-Mayén, J. F. Turnbull, V. Inglis, and A. L. Guerra-Flores. 1998. Species of *Vibrio* spp. isolated from hepatopancreas, hemolymph and digestive tract of a population of healthy juvenile *Penaeus vannamei*. *Aquaculture* 163:1–9.

Green, J. L., A. J. Holmes, M. Westoby, I. Oliver, D. Briscoe, M. Dangerfield, M. Gillings, and A. J. Beattie. 2004. Spatial scaling of microbial eukaryote diversity. *Nature* 432:747–750.

Guezennec, J. 2002. Deep-sea hydrothermal vents: a new source of innovative bacterial exopolysaccharides of biotechnological interest? *J. Ind. Microbiol. Biotechnol.* 29:204–208.

Heidelberg, J. F., J. A. Eisen, W. C. Nelson, R. A. Clayton, M. L. Gwinn, R. J. Dodson, D. H. Haft, E. K. Hickey, J. D. Peterson,

L. Umayam, S. R. Gill, K. E. Nelson, T. D. Read, H. Tettelin, D. Richardson, M. D. Ermolaeva, J. Vamathevan, S. Bass, H. Qin, I. Dragoi, P. Sellers, L. McDonald, T. Utterback, R. D. Fleishmann, W. C. Nierman, and O. White. 2000. DNA sequence of both chromosomes of the cholera pathogen *Vibrio cholerae*. *Nature* **406**:477–483.

Horner-Devine, M. C., M. Lage, J. B. Hughes, and B. J. Bohannan. 2004a. A taxa-area relationship for bacteria. *Nature* **432**:750–753.

Horner-Devine, M. C., K. M. Carney, and B. J. Bohannan. 2004b. An ecological perspective on bacterial biodiversity. *Proc. Biol. Sci.* **271**:113–122.

Irianto, A., and B. Austin. 2002. Probiotics in aquaculture. *J. Fish Dis.* **25**:633–642.

Jensen, S., L. Ovreas, O. Bergh, and V. Torsvik. 2004. Phylogenetic analysis of bacterial communities associated with larvae of the Atlantic halibut propose succession from a uniform normal flora. *Syst. Appl. Microbiol.* **27**:728–736.

Konstantinidis, K. T., and J. M. Tiedje. 2005a. Genomic insights that advance the species definition for prokaryotes. *Proc. Natl. Acad. Sci. USA* **102**:2567–2572.

Konstantinidis, K. T., and J. M. Tiedje. 2005b. Towards a genome-based taxonomy for prokaryotes. *J. Bacteriol.* **187**:6258–6264.

Koropatnick, T. A., J. T. Engle, M. A. Apicella, E. V. Stabb, W. E. Goldman, and M. J. McFall-Ngai. 2004. Microbial factor-mediated development in a host-bacterial mutualism. *Science* **12**:1186–1188.

Larocque, R. C., J. B. Harris, M. Dziejman, X. Li, A. I. Khan, A. S. Faruque, S. M. Faruque, G. B. Nair, E. T. Ryan, F. Qadri, J. J. Mekalanos, and S. B. Calderwood. 2005. Transcriptional profiling of *Vibrio cholerae* recovered directly from patient specimens during early and late stages of human infection. *Infect. Immun.* **73**:4488–4493.

Lauriano, C. M., C. Ghosh, N. E. Correa, and K. E. Klose. 2004. The sodium-driven flagellar motor controls exopolysaccharide expression in *Vibrio cholerae*. *J Bacteriol.* **186**:4864–4874.

Lin, Y. C., C. L. Lu, H. Y. Chang, and C. Y. Tang. 2005. An efficient algorithm for sorting by block-interchanges and its application to the evolution of vibrio species. *J. Comput. Biol.* **12**:102–112.

Lobitz, B., L. Beck, A. Huq, B. Wood, G. Fuchs, A. S. Faruque, and R. R. Colwell. 2000. Climate and infectious disease: use of remote sensing for detection of *Vibrio cholerae* by indirect measurement. *Proc. Natl. Acad. Sci. USA* **97**:1438–1443.

Lupp, C., and E. G. Ruby. 2005. *Vibrio fischeri* uses two quorum-sensing systems for the regulation of early and late colonization factors. *J. Bacteriol.* **187**:3620–3629.

McDougald, D., S. A. Rice, D. Weichart, and S. Kjelleberg. 1998. Nonculturability: adaptation or debilitation? *FEMS Microbiol. Ecol.* **25**:1–9.

Meibom, K. L., X. B. Li, A. T. Nielsen, C. Y. Wu, S. Roseman, and G. K. Schoolnik. 2004. The *Vibrio cholerae* chitin utilization program. *Proc. Natl. Acad. Sci. USA* **101**:2524–2529.

Nyholm, S. V., and M. J. McFall-Ngai. 2004. The winnowing: establishing the squid-vibrio symbiosis. *Nat. Rev. Microbiol.* **2**:632–642.

Ochman, H., E. Lerat, and V. Daubin. 2005. Examining bacterial species under the specter of gene transfer and exchange. *Proc. Natl. Acad. Sci. USA* **102**:6595–6599.

Panicker, G., M. L. Myers, and A. K. Bej. 2004. Rapid detection of *Vibrio vulnificus* in shellfish and Gulf of Mexico water by real-time PCR. *Appl. Environ. Microbiol.* **70**:498–507.

Paul, J. H., S. J. Williamson, A. Long, R. N. Authement, D. John, A. M. Segall, F. L. Rowher, M. Androlewicz, and S. Patterson. 2005. Complete genome sequence of phiHSIC, a pseudotemperate marine phage of *Listonella pelagia*. *Appl. Environ. Microbiol.* **71**:3311–3320.

Price, N. D., J. L. Reed, and B. O. Palsson. 2004. Genome-scale models of microbial cells: evaluating the consequences of constraints. *Nat. Rev. Microbiol.* **2**:886–897.

Qadri, F., M. I. Chowdhury, S. M. Faruque, M. A. Salam, T. Ahmed, Y. A. Begum, A. Saha, M. S. Alam, K. Zaman, L. V. Seidlein, E. Park, K. P. Killeen, J. J. Mekalanos, J. D. Clemens, and D. A. Sack; Peru-15 Study Group. 2005. Randomized, controlled study of the safety and immunogenicity of Peru-15, a live attenuated oral vaccine candidate for cholera, in adult volunteers in Bangladesh. *J. Infect. Dis.* **192**:573–579.

Randa, M. A., M. F. Polz, and E. Lim. 2004. Effects of temperature and salinity on *Vibrio vulnificus* population dynamics as assessed by quantitative PCR. *Appl. Environ. Microbiol.* **70**:5469–5476.

Rasch, M., C. Buch, B. Austin, W. J. Slierendrecht, K. S. Ekmann, J. L. Larsen, C. Johansen, K. Riedel, L. Eberl, M. Givskov, and L. Gram. 2004. An inhibitor of bacterial quorum sensing reduces mortalities caused by vibriosis in rainbow trout (*Oncorhynchus mykiss*, Walbaum). *Syst. Appl. Microbiol.* **27**:350–359.

Reguera, G., and R. Kolter. 2005. Virulence and the environment: a novel role for *Vibrio cholerae* toxin-coregulated pili in biofilm formation on chitin. *J. Bacteriol.* **187**:3551–3555.

Rohwer, F., V. Seguritan, F. Azam, and N. Knowlton. 2002. Diversity and distribution of coral-associated bacteria. *Mar. Ecol. Prog. Ser.* **243**:1–10.

Rosenberg, E., and Y. Loya. 2004. *Coral Health and Disease.* Springer-Verlag, New York, N.Y.

Ruby, E. G., M. Urbanowski, J. Campbell, A. Dunn, M. Faini, R. Gunsalus, P. Lostroh, C. Lupp, J. McCann, D. Millikan, A. Schaefer, E. Stabb, A. Stevens, K. Visick, C. Whistler, and E. P. Greenberg. 2005. Complete genome sequence of *Vibrio fischeri*: a symbiotic bacterium with pathogenic congeners. *Proc. Natl. Acad. Sci. USA* **102**:3004–3009.

Seo, H. J., S. S. Bae, S. H. Yang, J.-H. Lee, and S.-J. Kim. 2005a. *Photobacterium aplysiae* sp. nov., a lipolytic marine bacterium isolated from eggs of the sea hare *Aplysia kurodai*. *Int. J. Syst. Evol. Microbiol.* **55**:2293–2296.

Seo, H. J., S. S. Bae, J.-H. Lee, and S.-J. Kim. 2005b. *Photobacterium frigidiphilum* sp. nov., a psychrophilic, lipolytic bacterium isolated from deep-sea sediments of Edison Seamount. *Int. J. Sys. Evol. Microbiol.* **55**:1661–1666.

Sussman, M., Y. Loya, M. Fine, and E. Rosenberg. 2003. The marine fireworm *Hermodice carunculata* is a winter reservoir and spring-summer vector for the coral-bleaching pathogen *Vibrio shiloi*. *Environ. Microbiol.* **5**:250–255.

Thompson, F. L., T. Iida, and J. Swings. 2004. Biodiversity of vibrios. *Microbiol. Mol. Biol. Rev.* **68**:403–431.

Thompson, F. L., D. Gevers, C. C. Thompson, P. Dawyndt, S. Naser, B. Hoste, C. B. Munn, and J. Swings. 2005. Phylogeny and molecular identification of vibrios on the basis of multilocus sequence analysis (MLSA). *Appl. Environ. Microbiol.* **71**:5107–5115.

Thompson, J. R., M. A. Randa, L. A. Marcelino, A. Tomita-Mitchell, E. Lim, and M. F. Polz. 2004. Diversity and dynamics of a North Atlantic coastal *Vibrio* community. *Appl. Environ. Microbiol.* **70**:4103–4110.

Thompson, J. R, S. Pacocha, C. Pharino, V. Klepac-Ceraj, D. E. Hunt, J. Benoit, R. Sarma-Rupavtarm, D. L. Distel, and M. F. Polz. 2005. Genotypic diversity within a natural coastal bacterioplankton population. *Science* **307**:1311–1313.

Tischler, A. D., and A. Camilli. 2004. Cyclic diguanylate (c-di-GMP) regulates *Vibrio cholerae* biofilm formation. *Mol. Microbiol.* **53**:857–869.

Verschuere, L., G. Rombaut, P. Sorgeloos, and W. Verstraete. 2000. Probiotic bacteria as biological control agents in aquaculture. *Microbiol. Mol. Biol. Rev.* **64**:655–671.

Wallner, G., R. Erhart, and R. Amann. 1995. Flow cytometric analysis of activated sludge with rRNA-targeted probes. *Appl. Environ. Microbiol.* **61:**1859–1866.

Weinbauer, M. G., and F. Rassoulzadegan. 2004. Are viruses driving microbial diversification and diversity? *Environ. Microbiol.* **6:**1–11.

World Health Organization. Epidemiological Surveillance of Communicable Diseases. 2003. Cholera, 2002. *Wkly. Epidemiol. Rec.* **78:**269–276.

World Health Organization. Epidemiological Surveillance of Communicable Diseases. 2004. Cholera, 2003. *Wkly. Epidemiol. Rec.* **79:**281–288.

Xu, Q., M. Dziejman, and J. J. Mekalanos. 2003. Determination of the transcriptome of *Vibrio cholerae* during intraintestinal growth and midexponential phase *in vitro*. *Proc. Natl. Acad. Sci. USA* **100:**1286–1291.

INDEX